An Introduction to Mechanics

In the years since it was first published, this classic introductory textbook has established itself as one of best-known and most highly regarded descriptions of Newtonian mechanics.

Intended for undergraduate students with foundation skills in mathematics and a deep interest in physics, it systematically lays out the principles of mechanics: vectors, Newton's laws, momentum, energy, rotational motion, angular momentum, and noninertial systems, and includes chapters on central force motion, the harmonic oscillator, and relativity.

Numerous worked examples demonstrate how the principles can be applied to a wide range of physical situations, and more than 600 figures illustrate methods for approaching physical problems. The book also contains over 200 challenging problems to help the student develop a strong understanding of the subject. Password-protected solutions are available for instructors at www.cambridge.org/9780521198219.

DANIEL KLEPPNER is Lester Wolfe Professor of Physics, Emeritus, at Massachusetts Institute of Technology. For his contributions to teaching, he has been awarded the Oersted Medal by the American Association of Physics Teachers and the Lilienfeld Prize of the American Physical Society. He has also received the Wolf Prize and the National Medal of Science.

ROBERT KOLENKOW was Associate Professor of Physics at Massachusetts Institute of Technology. Renowned for his skills as a teacher, Kolenkow was awarded the Everett Moore Baker Award for Outstanding Teaching. He has since retired.

Daniel Kleppner
Lester Wolfe Professor of Physics, Massachusetts Institute of Technology
Robert J. Kolenkow
Formerly Associate Professor of Physics, Massachusetts Institute of Technology

AN INTRODUCTION TO MECHANICS

CAMBRIDGE
UNIVERSITY PRESS

CAMBRIDGE UNIVERSITY PRESS
Cambridge, New York, Melbourne, Madrid, Cape Town, Singapore,
São Paulo, Delhi, Dubai, Tokyo

Cambridge University Press
The Edinburgh Building, Cambridge CB2 8RU, UK

Published in the United States of America by Cambridge University Press, New York

www.cambridge.org
Information on this title: www.cambridge.org/9780521198219

Previously published by McGraw-Hill Education 1973

First published by Cambridge University Press 2010

Printed in the United Kingdom at the University Press, Cambridge

A catalog record for this publication is available from the British Library

Library of Congress Cataloging in Publication Data

ISBN 978-0-521-19821-9 Hardback

Additional resources for this publication at www.cambridge.org/9780521198219

To our parents

Beatrice and Otto
Katherine and John

CONTENTS

LIST OF EXAMPLES xi
PREFACE xv
TO THE TEACHER xix

1 VECTORS AND KINEMATICS—A FEW MATHEMATICAL PRELIMINARIES

1.1 INTRODUCTION 2
1.2 VECTORS 2
Definition of a Vector, The Algebra of Vectors, 3.
1.3 COMPONENTS OF A VECTOR 8
1.4 BASE VECTORS 10
1.5 DISPLACEMENT AND THE POSITION VECTOR 11
1.6 VELOCITY AND ACCELERATION 13
Motion in One Dimension, 14; *Motion in Several Dimensions,* 14; *A Word about Dimensions and Units,* 18.
1.7 FORMAL SOLUTION OF KINEMATICAL EQUATIONS 9
1.8 MORE ABOUT THE DERIVATIVE OF A VECTOR 23
1.9 MOTION IN PLANE POLAR COORDINATES 27
Polar Coordinates, 27; *Velocity in Polar Coordinates,* 27; *Evaluating* $d\hat{r}/dt$, 31; *Acceleration in Polar Coordinates,* 36.
Note 1.1 MATHEMATICAL APPROXIMATION METHODS 39
The Binomial Series, 41; *Taylor's Series,* 42; *Differentials,* 45.
Some References to Calculus Texts, 47.
PROBLEMS 47

2 NEWTON'S LAWS—THE FOUNDATIONS OF NEWTONIAN MECHANICS

2.1 INTRODUCTION 52
2.2 NEWTON'S LAWS 53
Newton's First Law, 55; *Newton's Second Law,* 56; *Newton's Third Law,* 59.
2.3 STANDARDS AND UNITS 64
The Fundamental Standards, 64; *Systems of Units,* 67.
2.4 SOME APPLICATIONS OF NEWTON'S LAWS 68
2.5 THE EVERYDAY FORCES OF PHYSICS 79
Gravity, Weight, and the Gravitational Field, 80; *The Electrostatic Force,* 86; *Contact Forces,* 87; *Tension—The Force of a String,* 87; *Tension and Atomic Forces,* 91; *The Normal Force,* 92; *Friction,* 92; *Viscosity,* 95; *The Linear Restoring Force: Hooke's Law, the Spring, and Simple Harmonic Motion,* 97.
Note 2.1 THE GRAVITATIONAL ATTRACTION OF A SPHERICAL SHELL 101
PROBLEMS 103

3 MOMENTUM

3.1 INTRODUCTION 112
3.2 DYNAMICS OF A SYSTEM OF PARTICLES 113
Center of Mass, 116.
3.3 CONSERVATION OF MOMENTUM 122
Center of Mass Coordinates, 127.
3.4 IMPULSE AND A RESTATEMENT OF THE MOMENTUM RELATION 130
3.5 MOMENTUM AND THE FLOW OF MASS 133

3.6 MOMENTUM TRANSPORT 139
Note 3.1 CENTER OF MASS 145
PROBLEMS 147

4 WORK AND ENERGY
4.1 INTRODUCTION 152
4.2 INTEGRATING THE EQUATION OF MOTION IN ONE DIMENSION 153
4.3 THE WORK-ENERGY THEOREM IN ONE DIMENSION 156
4.4 INTEGRATING THE EQUATION OF MOTION IN SEVERAL DIMENSIONS 158
4.5 THE WORK-ENERGY THEOREM 160
4.6 APPLYING THE WORK-ENERGY THEOREM 162
4.7 POTENTIAL ENERGY 168
Illustrations of Potential Energy, 170.
4.8 WHAT POTENTIAL ENERGY TELLS US ABOUT FORCE 173
Stability, 174.
4.9 ENERGY DIAGRAMS 176
4.10 SMALL OSCILLATIONS IN A BOUND SYSTEM 178
4.11 NONCONSERVATIVE FORCES 182
4.12 THE GENERAL LAW OF CONSERVATION OF ENERGY 184
4.13 POWER 186
4.14 CONSERVATION LAWS AND PARTICLE COLLISIONS 187
Collisions and Conservation Laws, 188; *Elastic and Inelastic Collisions,* 188; *Collisions in One Dimension,* 189; *Collisions and Center of Mass Coordinates,* 190.
PROBLEMS 194

5 SOME MATHEMATICAL ASPECTS OF FORCE AND ENERGY
5.1 INTRODUCTION 202
5.2 PARTIAL DERIVATIVES 202
5.3 HOW TO FIND THE FORCE IF YOU KNOW THE POTENTIAL ENERGY 206
5.4 THE GRADIENT OPERATOR 207
5.5 THE PHYSICAL MEANING OF THE GRADIENT 210
Constant Energy Surfaces and Contour Lines, 211.
5.6 HOW TO FIND OUT IF A FORCE IS CONSERVATIVE 215
5.7 STOKES' THEOREM 225
PROBLEMS 228

6 ANGULAR MOMENTUM AND FIXED AXIS ROTATION
6.1 INTRODUCTION 232
6.2 ANGULAR MOMENTUM OF A PARTICLE 233
6.3 TORQUE 238
6.4 ANGULAR MOMENTUM AND FIXED AXIS ROTATION 248
6.5 DYNAMICS OF PURE ROTATION ABOUT AN AXIS 253
6.6 THE PHYSICAL PENDULUM 255
The Simple Pendulum, 253; *The Physical Pendulum,* 257.
6.7 MOTION INVOLVING BOTH TRANSLATION AND ROTATION 260
The Work-energy Theorem, 267.
6.8 THE BOHR ATOM 270
Note 6.1 CHASLES' THEOREM 274
Note 6.2 PENDULUM MOTION 276
PROBLEMS 279

7 RIGID BODY MOTION AND THE CONSERVATION OF ANGULAR MOMENTUM

7.1 INTRODUCTION 288
7.2 THE VECTOR NATURE OF ANGULAR VELOCITY AND ANGULAR MOMENTUM 288
7.3 THE GYROSCOPE 295
7.4 SOME APPLICATIONS OF GYROSCOPE MOTION 300
7.5 CONSERVATION OF ANGULAR MOMENTUM 305
7.6 ANGULAR MOMENTUM OF A ROTATING RIGID BODY 308
Angular Momentum and the Tensor of Inertia, 308; *Principal Axes,* 313; *Rotational Kinetic Energy,* 313; *Rotation about a Fixed Point,* 315.
7.7 ADVANCED TOPICS IN THE DYNAMICS OF RIGID BODY ROTATION 316
Introduction, 316; *Torque-free Precession: Why the Earth Wobbles,* 317; *Euler's Equations,* 320.
Note 7.1 FINITE AND INFINITESIMAL ROTATIONS 326
Note 7.2 MORE ABOUT GYROSCOPES 328
Case 1 Uniform Precession, 331; Case 2 Torque-free Precession, 331; Case 3 Nutation, 331.
PROBLEMS 334

8 NONINERTIAL SYSTEMS AND FICTITIOUS FORCES

8.1 INTRODUCTION 340
8.2 THE GALILEAN TRANSFORMATIONS 340
8.3 UNIFORMLY ACCELERATING SYSTEMS 343
8.4 THE PRINCIPLE OF EQUIVALENCE 346
8.5 PHYSICS IN A ROTATING COORDINATE SYSTEM 355
Time Derivatives and Rotating Coordinates, 356; *Acceleration Relative to Rotating Coordinates,* 358; *The Apparent Force in a Rotating Coordinate System,* 359.
Note 8.1 THE EQUIVALENCE PRINCIPLE AND THE GRAVITATIONAL RED SHIFT 369
Note 8.2 ROTATING COORDINATE TRANSFORMATION 371
PROBLEMS 372

9 CENTRAL FORCE MOTION

9.1 INTRODUCTION 378
9.2 CENTRAL FORCE MOTION AS A ONE BODY PROBLEM 378
9.3 GENERAL PROPERTIES OF CENTRAL FORCE MOTION 380
The Motion Is Confined to a Plane, 380; *The Energy and Angular Momentum Are Constants of the Motion,* 380; *The Law of Equal Areas,* 382.
9.4 FINDING THE MOTION IN REAL PROBLEMS 382
9.5 THE ENERGY EQUATION AND ENERGY DIAGRAMS 383
9.6 PLANETARY MOTION 390
9.7 KEPLER'S LAWS 400
Note 9.1 PROPERTIES OF THE ELLIPSE 403
PROBLEMS 406

10 THE HARMONIC OSCILLATOR

10.1 INTRODUCTION AND REVIEW 410
Standard Form of the Solution, 410; *Nomenclature,* 411; *Energy Considerations,* 412; *Time Average Values,* 413; *Average Energy,* 413.
10.2 THE DAMPED HARMONIC OSCILLATOR 414
Energy, 416; *The Q of an Oscillator,* 418.

10.3 THE FORCED HARMONIC OSCILLATOR 421
The Undamped Forced Oscillator, 421; *Resonance,* 423; *The Forced Damped Harmonic Oscillator,* 424; *Resonance in a Lightly Damped System: The Quality Factor Q,* 426.
10.4 RESPONSE IN TIME VERSUS RESPONSE IN FREQUENCY 432
Note 10.1 SOLUTION OF THE EQUATION OF MOTION FOR THE UNDRIVEN DAMPED OSCILLATOR 433
The Use of Complex Variables, 433; *The Damped Oscillator,* 435.
Note 10.2 SOLUTION OF THE EQUATION OF MOTION FOR THE FORCED OSCILLATOR 437
PROBLEMS 438

11 THE SPECIAL THEORY OF RELATIVITY
11.1 THE NEED FOR A NEW MODE OF THOUGHT 442
11.2 THE MICHELSON-MORLEY EXPERIMENT 445
11.3 THE POSTULATES OF SPECIAL RELATIVITY 450
The Universal Velocity, 451; *The Principle of Relativity,* 451; *The Postulates of Special Relativity,* 452.
11.4 THE GALILEAN TRANSFORMATIONS 453
11.5 THE LORENTZ TRANSFORMATIONS 455
PROBLEMS 459

12 RELATIVISTIC KINEMATICS
12.1 INTRODUCTION 462
12.2 SIMULTANEITY AND THE ORDER OF EVENTS 463
12.3 THE LORENTZ CONTRACTION AND TIME DILATION 466
The Lorentz Contraction, 466; *Time Dilation,* 468.
12.4 THE RELATIVISTIC TRANSFORMATION OF VELOCITY 472
12.5 THE DOPPLER EFFECT 475
The Doppler Shift in Sound, 475; *Relativistic Doppler Effect,* 477; *The Doppler Effect for an Observer off the Line of Motion,* 478.
12.6 THE TWIN PARADOX 480
PROBLEMS 484

13 RELATIVISTIC MOMENTUM AND ENERGY
13.1 MOMENTUM 490
13.2 ENERGY 493
13.3 MASSLESS PARTICLES 500
13.4 DOES LIGHT TRAVEL AT THE VELOCITY OF LIGHT? 508
PROBLEMS 512

14 FOUR-VECTORS AND RELATIVISTIC INVARIANCE
14.1 INTRODUCTION 516
14.2 VECTORS AND TRANSFORMATIONS 516
Rotation about the z Axis, 517; *Invariants of a Transformation,* 520; *The Transformation Properties of Physical Laws,* 520; *Scalar Invariants,* 521.
14.3 MINIKOWSKI SPACE AND FOUR-VECTORS 521
14.4 THE MOMENTUM-ENERGY FOUR-VECTOR 527
14.5 CONCLUDING REMARKS 534
PROBLEMS 536

INDEX 539

LIST OF EXAMPLES

1 VECTORS AND KINEMATICS—A FEW MATHEMATICAL PRELIMINARIES

EXAMPLES, CHAPTER 1

1.1 Law of Cosines, 5; 1.2 Work and the Dot Product, 5; 1.3 Examples of the Vector Product in Physics, 7; 1.4 Area as a Vector, 7.
1.5 Vector Algebra, 9; 1.6 Construction of a Perpendicular Vector, 10.
1.7 Finding **v** from **r**, 16; 1.8 Uniform Circular Motion, 17.
1.9 Finding Velocity from Acceleration, 20; 1.10 Motion in a Uniform Gravitational Field, 21; 1.11 Nonuniform Acceleration—The Effect of a Radio Wave on an Ionospheric Electron, 22.
1.12 Circular Motion and Rotating Vectors, 25.
1.13 Circular Motion and Straight Line Motion in Polar Coordinates, 34; 1.14 Velocity of a Bead on a Spoke, 35; 1.15 Off-center Circle, 35; 1.16 Acceleration of a Bead on a Spoke, 37; 1.17 Radial Motion without Acceleration, 38.

2 NEWTON'S LAWS—THE FOUNDATIONS OF NEWTONIAN MECHANICS

EXAMPLES, CHAPTER 2

2.1 Astronauts in Space—Inertial Systems and Fictitious Force, 60.
2.2 The Astronauts' Tug-of-war, 70; 2.3 Freight Train, 72; 2.4 Constraints, 74; 2.5 Block on String 1, 75; 2.6 Block on String 2, 76; 2.7 The Whirling Block, 76; 2.8 The Conical Pendulum, 77.
2.9 Turtle in an Elevator, 84; 2.10 Block and String 3, 87; 2.11 Dangling Rope, 88; 2.12 Whirling Rope, 89; 2.13 Pulleys, 90; 2.14 Block and Wedge with Friction, 93; 2.15 The Spinning Terror, 94; 2.16 Free Motion in a Viscous Medium, 96; 2.17 Spring and Block—The Equation for Simple Harmonic Motion, 98; 2.18 The Spring Gun—An Example Illustrating Initial Conditions, 99.

3 MOMENTUM

EXAMPLES, CHAPTER 3

3.1 The Bola, 115; 3.2 Drum Major's Baton, 117; 3.3 Center of Mass of a Nonuniform Rod, 119; 3.4 Center of Mass of a Triangular Sheet, 120; 3.5 Center of Mass Motion, 122.
3.6 Spring Gun Recoil, 123; 3.7 Earth, Moon, and Sun—A Three Body System, 125; 3.8 The Push Me–Pull You, 128.
3.9 Rubber Ball Rebound, 131; 3.10 How to Avoid Broken Ankles, 132.
3.11 Mass Flow and Momentum, 134; 3.12 Freight Car and Hopper, 135; 3.13 Leaky Freight Car, 136; 3.14 Rocket in Free Space, 138; 3.15 Rocket in a Gravitational Field, 139.
3.16 Momentum Transport to a Surface, 141; 3.17 A Dike at the Bend of a River, 143; 3.18 Pressure of a Gas, 144.

4 WORK AND ENERGY

EXAMPLES, CHAPTER 4

4.1 Mass Thrown Upward in a Uniform Gravitational Field, 154; 4.2 Solving the Equation of Simple Harmonic Motion, 154.
4.3 Vertical Motion in an Inverse Square Field, 156.
4.4 The Conical Pendulum, 161; 4.5 Escape Velocity—The General Case, 162.
4.6 The Inverted Pendulum, 164; 4.7 Work Done by a Uniform Force, 165; 4.8 Work Done by a Central Force, 167; 4.9 A Path-dependent Line Integral, 167; 4.10 Parametric Evaluation of a Line Integral, 168.

4.11 Potential Energy of a Uniform Force Field, 170; 4.12 Potential Energy of an Inverse Square Force, 171; 4.13 Bead, Hoop, and Spring, 172.
4.14 Energy and Stability—The Teeter Toy, 175.
4.15 Molecular Vibrations, 179; 4.16 Small Oscillations, 181.
4.17 Block Sliding down Inclined Plane, 183.
4.18 Elastic Collision of Two Balls, 190; 4.19 Limitations on Laboratory Scattering Angle, 193.

5 SOME MATHEMATICAL ASPECTS OF FORCE AND ENERGY

EXAMPLES, CHAPTER 5
5.1 Partial Derivatives, 203; 5.2 Applications of the Partial Derivative, 205.
5.3 Gravitational Attraction by a Particle, 208; 5.4 Uniform Gravitational Field, 209; 5.5 Gravitational Attraction by Two Point Masses, 209.
5.6 Energy Contours for a Binary Star System, 212.
5.7 The Curl of the Gravitational Force, 219; 5.8 A Nonconservative Force, 220; 5.9 A Most Unusual Force Field, 221; 5.10 Construction of the Potential Energy Function, 222; 5.11 How the Curl Got Its Name, 224.
5.12 Using Stokes' Theorem, 227.

6 ANGULAR MOMENTUM AND FIXED AXIS ROTATION

EXAMPLES, CHAPTER 6
6.1 Angular Momentum of a Sliding Block, 236; 6.2 Angular Momentum of the Conical Pendulum, 237.
6.3 Central Force Motion and the Law of Equal Areas, 240; 6.4 Capture Cross Section of a Planet, 241; 6.5 Torque on a Sliding Block, 244; 6.6 Torque on the Conical Pendulum, 245; 6.7 Torque due to Gravity, 247.
6.8 Moments of Inertia of Some Simple Objects, 250; 6.9 The Parallel Axis Theorem, 252.
6.10 Atwood's Machine with a Massive Pulley, 254.
6.11 Grandfather's Clock, 256; 6.12 Kater's Pendulum, 258; 6.13 The Doorstep, 259.
6.14 Angular Momentum of a Rolling Wheel, 262; 6.15 Disk on Ice, 264; 6.16 Drum Rolling down a Plane, 265; 6.17 Drum Rolling down a Plane: Energy Method, 268; 6.18 The Falling Stick, 269.

7 RIGID BODY MOTION AND THE CONSERVATION OF ANGULAR MOMENTUM

EXAMPLES, CHAPTER 7
7.1 Rotations through Finite Angles, 289; 7.2 Rotation in the *xy* Plane, 291; 7.3 Vector Nature of Angular Velocity, 291; 7.4 Angular Momentum of a Rotating Skew Rod, 292; 7.5 Torque on the Rotating Skew Rod, 293; 7.6 Torque on the Rotating Skew Rod (Geometric Method), 294.
7.7 Gyroscope Precession, 298; 7.8 Why a Gyroscope Precesses, 299.
7.9 Precession of the Equinoxes, 300; 7.10 The Gyrocompass Effect, 301; 7.11 Gyrocompass Motion, 302; 7.12 The Stability of Rotating Objects, 304.
7.13 Rotating Dumbbell, 310; 7.14 The Tensor of Inertia for a Rotating Skew Rod, 312; 7.15 Why Flying Saucers Make Better Spacecraft than Do Flying Cigars, 314.
7.16 Stability of Rotational Motion, 322; 7.17 The Rotating Rod, 323; 7.18 Euler's Equations and Torque-free Precession, 324.

8 NONINERTIAL SYSTEMS AND FICTITIOUS FORCES

EXAMPLES, CHAPTER 8
8.1 The Apparent Force of Gravity, 346; 8.2 Cylinder on an Accelerating Plank, 347; 8.3 Pendulum in an Accelerating Car, 347.
8.4 The Driving Force of the Tides, 350; 8.5 Equilibrium Height of the Tide, 352.
8.6 Surface of a Rotating Liquid, 362; 8.7 The Coriolis Force, 363; 8.8 Deflection of a Falling Mass, 364; 8.9 Motion on the Rotating Earth, 366; 8.10 Weather Systems, 366; 8.11 The Foucault Pendulum, 369.

9 CENTRAL FORCE MOTION

EXAMPLES, CHAPTER 9
9.1 Noninteracting Particles, 384; 9.2 The Capture of Comets, 387; 9.3 Perturbed Circular Orbit, 388.
9.4 Hyperbolic Orbits, 393; 9.5 Satellite Orbit, 396; 9.6 Satellite Maneuver, 398.
9.7 The Law of Periods, 403.

10 THE HARMONIC OSCILLATOR

EXAMPLES, CHAPTER 10
10.1 Initial Conditions and the Frictionless Harmonic Oscillator, 411.
10.2 The Q of Two Simple Oscillators, 419; 10.3 Graphical Analysis of a Damped Oscillator, 420.
10.4 Forced Harmonic Oscillator Demonstration, 424; 10.5 Vibration Eliminator, 428.

11 THE SPECIAL THEORY OF RELATIVITY

EXAMPLES, CHAPTER 11
11.1 The Galilean Transformations, 453; 11.2 A Light Pulse as Described by the Galilean Transformations, 455.

12 RELATIVISTIC KINEMATICS

EXAMPLES, CHAPTER 12
12.1 Simultaneity, 463; 12.2 An Application of the Lorentz Transformations, 464; 12.3 The Order of Events: Timelike and Spacelike Intervals, 465.
12.4 The Orientation of a Moving Rod, 467; 12.5 Time Dilation and Meson Decay, 468; 12.6 The Role of Time Dilation in an Atomic Clock, 470.
12.7 The Speed of Light in a Moving Medium, 474.
12.8 Doppler Navigation, 479.

13 RELATIVISTIC MOMENTUM AND ENERGY

EXAMPLES, CHAPTER 13
13.1 Velocity Dependence of the Electron's Mass, 492.
13.2 Relativistic Energy and Momentum in an Inelastic Collision, 496; 13.3 The Equivalence of Mass and Energy, 498.
13.4 The Photoelectric Effect, 502; 13.5 Radiation Pressure of Light, 502;

13.6 The Compton Effect, 503; 13.7 Pair Production, 505; 13.8 The Photon Picture of the Doppler Effect, 507.
13.9 The Rest Mass of the Photon, 510; 13.10 Light from a Pulsar, 510.

14 FOUR-VECTORS AND RELATIVISTIC INVARIANCE

EXAMPLES, CHAPTER 14
14.1 Transformation Properties of the Vector Product, 518; 14.2 A Non-vector, 519.
14.3 Time Dilation, 524; 14.4 Construction of a Four-vector: The Four-velocity, 525; 14.5 The Relativistic Addition of Velocities, 526.
14.6 The Doppler Effect, Once More, 530; 14.7 Relativistic Center of Mass Systems, 531; 14.8 Pair Production in Electron-electron Collisions, 533.

PREFACE

There is good reason for the tradition that students of science and engineering start college physics with the study of mechanics: mechanics is the cornerstone of pure and applied science. The concept of energy, for example, is essential for the study of the evolution of the universe, the properties of elementary particles, and the mechanisms of biochemical reactions. The concept of energy is also essential to the design of a cardiac pacemaker and to the analysis of the limits of growth of industrial society. However, there are difficulties in presenting an introductory course in mechanics which is both exciting and intellectually rewarding. Mechanics is a mature science and a satisfying discussion of its principles is easily lost in a superficial treatment. At the other extreme, attempts to "enrich" the subject by emphasizing advanced topics can produce a false sophistication which emphasizes technique rather than understanding.

This text was developed from a first-year course which we taught for a number of years at the Massachusetts Institute of Technology and, earlier, at Harvard University. We have tried to present mechanics in an engaging form which offers a strong base for future work in pure and applied science. Our approach departs from tradition more in depth and style than in the choice of topics; nevertheless, it reflects a view of mechanics held by twentieth-century physicists.

Our book is written primarily for students who come to the course knowing some calculus, enough to differentiate and integrate simple functions.[1] It has also been used successfully in courses requiring only concurrent registration in calculus. (For a course of this nature, Chapter 1 should be treated as a resource chapter, deferring the detailed discussion of vector kinematics for a time. Other suggestions are listed in To The Teacher.) Our experience has been that the principal source of difficulty for most students is in learning how to apply mathematics to physical problems, not with mathematical techniques as such. The elements of calculus can be mastered relatively easily, but the development of problem-solving ability requires careful guidance. We have provided numerous worked examples throughout the text to help supply this guidance. Some of the examples, particularly in the early chapters, are essentially pedagogical. Many examples, however, illustrate principles and techniques by application to problems of real physical interest.

The first chapter is a mathematical introduction, chiefly on vectors and kinematics. The concept of rate of change of a vector,

[1] The background provided in "Quick Calculus" by Daniel Kleppner and Norman Ramsey, John Wiley & Sons, New York, 1965, is adequate.

probably the most difficult mathematical concept in the text, plays an important role throughout mechanics. Consequently, this topic is developed with care, both analytically and geometrically. The geometrical approach, in particular, later proves to be invaluable for visualizing the dynamics of angular momentum.

Chapter 2 discusses inertial systems, Newton's laws, and some common forces. Much of the discussion centers on applying Newton's laws, since analyzing even simple problems according to general principles can be a challenging task at first. Visualizing a complex system in terms of its essentials, selecting suitable inertial coordinates, and distinguishing between forces and accelerations are all acquired skills. The numerous illustrative examples in the text have been carefully chosen to help develop these skills.

Momentum and energy are developed in the following two chapters. Chapter 3, on momentum, applies Newton's laws to extended systems. Students frequently become confused when they try to apply momentum considerations to rockets and other systems involving flow of mass. Our approach is to apply a differential method to a system defined so that no mass crosses its boundary during the chosen time interval. This ensures that no contribution to the total momentum is overlooked. The chapter concludes with a discussion of momentum flux. Chapter 4, on energy, develops the work-energy theorem and its application to conservative and nonconservative forces. The conservation laws for momentum and energy are illustrated by a discussion of collision problems.

Chapter 5 deals with some mathematical aspects of conservative forces and potential energy; this material is not needed elsewhere in the text, but it will be of interest to students who want a mathematically complete treatment of the subject.

Students usually find it difficult to grasp the properties of angular momentum and rigid body motion, partly because rotational motion lies so far from their experience that they cannot rely on intuition. As a result, introductory texts often slight these topics, despite their importance. We have found that rotational motion can be made understandable by emphasizing physical reasoning rather than mathematical formalism, by appealing to geometric arguments, and by providing numerous worked examples. In Chapter 6 angular momentum is introduced, and the dynamics of fixed axis rotation is treated. Chapter 7 develops the important features of rigid body motion by applying vector arguments to systems dominated by spin angular momentum. An elementary treatment of general rigid body motion is presented in the last sections of Chapter 7 to show how Euler's equations can be developed from simple physical arguments. This more advanced material is optional however; we do not usually treat it in our own course.

Chapter 8, on noninertial coordinate systems, completes the development of the principles of newtonian mechanics. Up to this point in the text, inertial systems have been used exclusively in order to avoid confusion between forces and accelerations. Our discussion of noninertial systems emphasizes their value as computational tools and their implications for the foundations of mechanics.

Chapters 9 and 10 treat central force motion and the harmonic oscillator, respectively. Although no new physical concepts are involved, these chapters illustrate the application of the principles of mechanics to topics of general interest and importance in physics. Much of the algebraic complexity of the harmonic oscillator is avoided by focusing the discussion on energy, and by using simple approximations.

Chapters 11 through 14 present a discussion of the principles of special relativity and some of its applications. We attempt to emphasize the harmony between relativistic and classical thought, believing, for example, that it is more valuable to show how the classical conservation laws are unified in relativity than to dwell at length on the so-called "paradoxes." Our treatment is concise and minimizes algebraic complexities. Chapter 14 shows how ideas of symmetry play a fundamental role in the formulation of relativity. Although we have kept the beginning students in mind, the concepts here are more subtle than in the previous chapters. Chapter 14 can be omitted if desired; but by illustrating how symmetry bears on the principles of mechanics, it offers an exciting mode of thought and a powerful new tool.

Physics cannot be learned passively; there is absolutely no substitute for tackling challenging problems. Here is where students gain the sense of satisfaction and involvement produced by a genuine understanding of the principles of physics. The collection of problems in this book was developed over many years of classroom use. A few problems are straightforward and intended for drill; most emphasize basic principles and require serious thought and effort. We have tried to choose problems which make this effort worthwhile in the spirit of Piet Hein's aphorism

Problems worthy
  of attack
prove their worth
  by hitting back[1]

[1] From *Grooks I*, by Piet Hein, copyrighted 1966, The M.I.T. Press.

It gives us pleasure to acknowledge the many contributions to this book from our colleagues and from our students. In particular, we thank Professors George B. Benedek and David E. Pritchard for a number of examples and problems. We should also like to thank Lynne Rieck and Mary Pat Fitzgerald for their cheerful fortitude in typing the manuscript.

Daniel Kleppner

Robert J. Kolenkow

TO THE TEACHER

The first eight chapters form a comprehensive introduction to classical mechanics and constitute the heart of a one-semester course. In a 12-week semester, we have generally covered the first 8 chapters and parts of Chapters 9 or 10. However, Chapter 5 and some of the advanced topics in Chapters 7 and 8 are usually omitted, although some students pursue them independently.

Chapters 11, 12, and 13 present a complete introduction to special relativity. Chapter 14, on transformation theory and four-vectors, provides deeper insight into the subject for interested students. We have used the chapters on relativity in a three-week short course and also as part of the second-term course in electricity and magnetism.

The problems at the end of each chapter are generally graded in difficulty. They are also cumulative; concepts and techniques from earlier chapters are repeatedly called upon in later sections of the book. The hope is that by the end of the course the student will have developed a good intuition for tackling new problems, that he will be able to make an intelligent estimate, for instance, about whether to start from the momentum approach or from the energy approach, and that he will know how to set off on a new tack if his first approach is unsuccessful. Many students report a deep sense of satisfaction from acquiring these skills.

Many of the problems require a symbolic rather than a numerical solution. This is not meant to minimize the importance of numerical work but to reinforce the habit of analyzing problems symbolically. Answers are given to some problems; in others, a numerical "answer clue" is provided to allow the student to check his symbolic result. Some of the problems are challenging and require serious thought and discussion. Since too many such problems at once can result in frustration, each assignment should have a mix of easier and harder problems.

Chapter 1 Although we would prefer to start a course in mechanics by discussing physics rather than mathematics, there are real advantages to devoting the first few lectures to the mathematics of motion. The concepts of kinematics are straightforward for the most part, and it is helpful to have them clearly in hand before tackling the much subtler problems presented by newtonian dynamics in Chapter 2. A departure from tradition in this chapter is the discussion of kinematics using polar coordinates. Many students find this topic troublesome at first, requiring serious effort. However, we feel that the effort will be amply rewarded. In the first place, by being able to use polar coordinates freely, the kinematics of rotational motion are much easier to understand;

the mystery of radial acceleration disappears. More important, this topic gives valuable insights into the nature of a time-varying vector, insights which not only simplify the dynamics of particle motion in Chapter 2 but which are invaluable to the discussion of momentum flux in Chapter 3, angular momentum in Chapters 6 and 7, and the use of noninertial coordinates in Chapter 8. Thus, the effort put into understanding the nature of time-varying vectors in Chapter 1 pays important dividends throughout the course.

If the course is intended for students who are concurrently beginning their study of calculus, we recommend that parts of Chapter 1 be deferred. Chapter 2 can be started after having covered only the first six sections of Chapter 1. Starting with Example 2.5, the kinematics of rotational motion are needed; at this point the ideas presented in Section 1.9 should be introduced. Section 1.7, on the integration of vectors, can be postponed until the class has become familiar with integrals. Occasional examples and problems involving integration will have to be omitted until that time. Section 1.8, on the geometric interpretation of vector differentiation, is essential preparation for Chapters 6 and 7 but need not be discussed earlier.

Chapter 2 The material in Chapter 2 often represents the student's first serious attempt to apply abstract principles to concrete situations. Newton's laws of motion are not self-evident; most people unconsciously follow aristotelian thought. We find that after an initial period of uncertainty, students become accustomed to analyzing problems according to principles rather than vague intuition. A common source of difficulty at first is to confuse force and acceleration. We therefore emphasize the use of inertial systems and recommend strongly that noninertial coordinate systems be reserved until Chapter 8, where their correct use is discussed. In particular, the use of centrifugal force in the early chapters can lead to endless confusion between inertial and noninertial systems and, in any case, it is not adequate for the analysis of motion in rotating coordinate systems.

Chapters 3 and 4 There are many different ways to derive the rocket equations. However, rocket problems are not the only ones in which there is a mass flow, so that it is important to adopt a method which is easily generalized. It is also desirable that the method be in harmony with the laws of conservation of momentum or, to put it more crudely, that there is no swindle involved. The differential approach used in Section 3.5 was developed to meet these requirements. The approach may not be elegant, but it is straightforward and quite general.

In Chapter 4, we attempt to emphasize the general nature of the work-energy theorem and the difference between conservative and nonconservative forces. Although the line integral is introduced and explained, only simple line integrals need to be evaluated, and general computational techniques should not be given undue attention.

Chapter 5 This chapter completes the discussion of energy and provides a useful introduction to potential theory and vector calculus. However, it is relatively advanced and will appeal only to students with an appetite for mathematics. The results are not needed elsewhere in the text, and we recommend leaving this chapter for optional use, or as a special topic.

Chapters 6 and 7 Most students find that angular momentum is the most difficult physical concept in elementary mechanics. The major conceptual hurdle is visualizing the vector properties of angular momentum. We therefore emphasize the vector nature of angular momentum repeatedly throughout these chapters. In particular, many features of rigid body motion can be understood intuitively by relying on the understanding of time-varying vectors developed in earlier chapters. It is more profitable to emphasize the qualitative features of rigid body motion than formal aspects such as the tensor of inertia. If desired, these qualitative arguments can be pressed quite far, as in the analysis of gyroscopic nutation in Note 7.2. The elementary discussion of Euler's equations in Section 7.7 is intended as optional reading only. Although Chapters 6 and 7 require hard work, many students develop a physical insight into angular momentum and rigid body motion which is seldom gained at the introductory level and which is often obscured by mathematics in advanced courses.

Chapter 8 The subject of noninertial systems offers a natural springboard to such speculative and interesting topics as transformation theory and the principle of equivalence. From a more practical point of view, the use of noninertial systems is an important technique for solving many physical problems.

Chapters 9 and 10 In these chapters the principles developed earlier are applied to two important problems, central force motion and the harmonic oscillator. Although both topics are generally treated rather formally, we have tried to simplify the mathematical development. The discussion of central force motion relies heavily on the conservation laws and on energy diagrams. The treatment of the harmonic oscillator sidesteps much of the usual algebraic complexity by focusing on the lightly damped oscillator. Applications and examples play an important role in both chapters.

Chapters 11 to 14 Special relativity offers an exciting change of pace to a course in mechanics. Our approach attempts to emphasize the connection of relativity with classical thought. We have used the Michelson-Morley experiment to motivate the discussion. Although the prominence of this experiment in Einstein's thought has been much exaggerated, this approach has the advantage of grounding the discussion on a real experiment.

We have tried to focus on the ideas of events and their transformations without emphasizing computational aids such as diagrammatic methods. This approach allows us to deemphasize many of the so-called paradoxes.

For many students, the real mystery of relativity lies not in the postulates or transformation laws but in why transformation principles should suddenly become the fundamental concept for generating new physical laws. This touches on the deepest and most provocative aspects of Einstein's thought. Chapter 14, on four-vectors, provides an introduction to transformation theory which unifies and summarizes the preceding development. The chapter is intended to be optional.

Daniel Kleppner
Robert J. Kolenkow

AN INTRODUCTION TO MECHANICS

1 VECTORS AND KINEMATICS—A FEW MATHEMATICAL PRELIMINARIES

1.1 Introduction

The goal of this book is to help you acquire a deep understanding of the principles of mechanics. The subject of mechanics is at the very heart of physics; its concepts are essential for understanding the everyday physical world as wèll as phenomena on the atomic and cosmic scales. The concepts of mechanics, such as momentum, angular momentum, and energy, play a vital role in practically every area of physics.

We shall use mathematics frequently in our discussion of physical principles, since mathematics lets us express complicated ideas quickly and transparently, and it often points the way to new insights. Furthermore, the interplay of theory and experiment in physics is based on quantitative prediction and measurement. For these reasons, we shall devote this chapter to developing some necessary mathematical tools and postpone our discussion of the principles of mechanics until Chap. 2.

1.2 Vectors

The study of vectors provides a good introduction to the role of mathematics in physics. By using vector notation, physical laws can often be written in compact and simple form. (As a matter of fact, modern vector notation was invented by a physicist, Willard Gibbs of Yale University, primarily to simplify the appearance of equations.) For example, here is how Newton's second law (which we shall discuss in the next chapter) appears in nineteenth century notation:

$$F_x = ma_x$$
$$F_y = ma_y$$
$$F_z = ma_z.$$

In vector notation, one simply writes

$$\mathbf{F} = m\mathbf{a}.$$

Our principal motivation for introducing vectors is to simplify the form of equations. However, as we shall see in the last chapter of the book, vectors have a much deeper significance. Vectors are closely related to the fundamental ideas of symmetry and their use can lead to valuable insights into the possible forms of unknown laws.

Definition of a Vector

Vectors can be approached from three points of view—geometric, analytic, and axiomatic. Although all three points of view are useful, we shall need only the geometric and analytic approaches in our discussion of mechanics.

From the geometric point of view, a vector is a *directed line segment.* In writing, we can represent a vector by an arrow and label it with a letter capped by a symbolic arrow. In print, boldfaced letters are traditionally used.

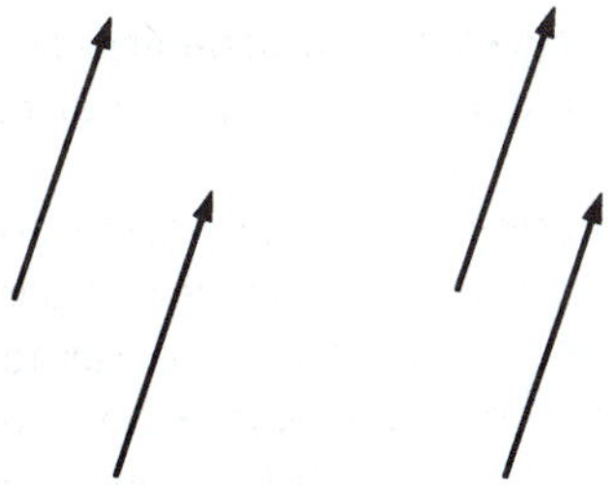

In order to describe a vector we must specify both its length and its direction. Unless indicated otherwise, we shall assume that parallel translation does not change a vector. Thus the arrows at left all represent the same vector.

If two vectors have the same length and the same direction they are equal. The vectors **B** and **C** are equal:

$$\mathbf{B} = \mathbf{C}.$$

The length of a vector is called its *magnitude.* The magnitude of a vector is indicated by vertical bars or, if no confusion will occur, by using italics. For example, the magnitude of **A** is written $|\mathbf{A}|$, or simply A. If the length of **A** is $\sqrt{2}$, then $|\mathbf{A}| = A = \sqrt{2}$.

If the length of a vector is one unit, we call it a *unit vector.* A unit vector is labeled by a caret; the vector of unit length parallel to **A** is $\hat{\mathbf{A}}$. It follows that

$$\hat{\mathbf{A}} = \frac{\mathbf{A}}{|\mathbf{A}|},$$

and conversely

$$\mathbf{A} = |\mathbf{A}|\hat{\mathbf{A}}.$$

The Algebra of Vectors

Multiplication of a Vector by a Scalar If we multiply **A** by a positive scalar b, the result is a new vector $\mathbf{C} = b\mathbf{A}$. The vector **C** is parallel to **A**, and its length is b times greater. Thus $\hat{\mathbf{C}} = \hat{\mathbf{A}}$, and $|\mathbf{C}| = b|\mathbf{A}|$.

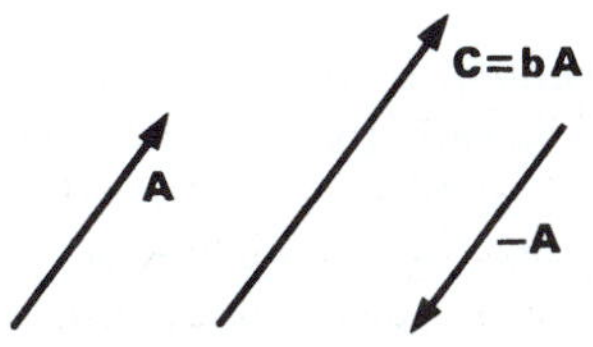

The result of multiplying a vector by -1 is a new vector opposite in direction (antiparallel) to the original vector.

Multiplication of a vector by a negative scalar evidently can change both the magnitude and the direction sense.

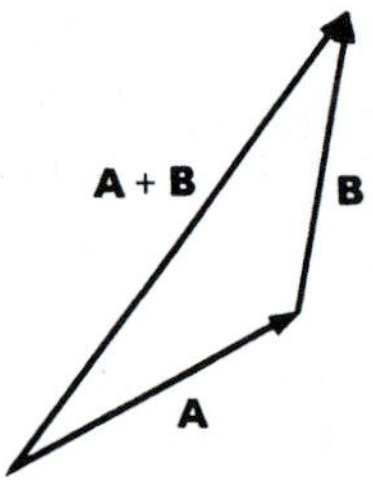

Addition of Two Vectors Addition of vectors has the simple geometrical interpretation shown by the drawing.

The rule is: To add **B** to **A**, place the tail of **B** at the head of **A**. The sum is a vector from the tail of **A** to the head of **B**.

Subtraction of Two Vectors Since $\mathbf{A} - \mathbf{B} = \mathbf{A} + (-\mathbf{B})$, in order to subtract **B** from **A** we can simply multiply it by -1 and then add. The sketches below show how.

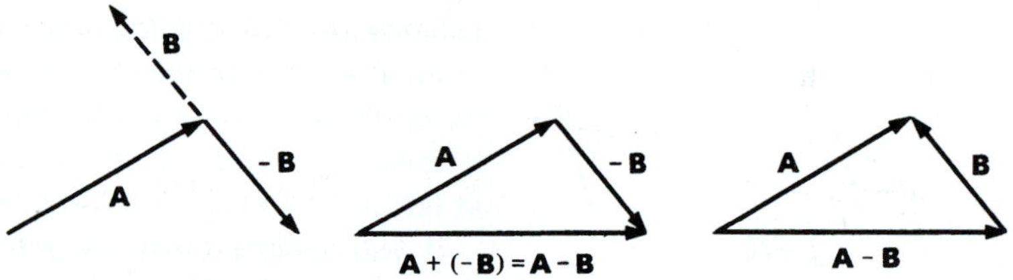

An equivalent way to construct $\mathbf{A} - \mathbf{B}$ is to place the *head* of **B** at the *head* of **A**. Then $\mathbf{A} - \mathbf{B}$ extends from the *tail* of **A** to the *tail* of **B**, as shown in the right hand drawing above.

It is not difficult to prove the following laws. We give a geometrical proof of the commutative law; try to cook up your own proofs of the others.

$$\mathbf{A} + \mathbf{B} = \mathbf{B} + \mathbf{A} \qquad \text{Commutative law}$$

$$\left.\begin{aligned} \mathbf{A} + (\mathbf{B} + \mathbf{C}) &= (\mathbf{A} + \mathbf{B}) + \mathbf{C} \\ c(d\mathbf{A}) &= (cd)\mathbf{A} \end{aligned}\right. \qquad \text{Associative law}$$

$$\left.\begin{aligned} (c + d)\mathbf{A} &= c\mathbf{A} + d\mathbf{A} \\ c(\mathbf{A} + \mathbf{B}) &= c\mathbf{A} + c\mathbf{B} \end{aligned}\right. \qquad \text{Distributive law}$$

Proof of the Commutative law of vector addition

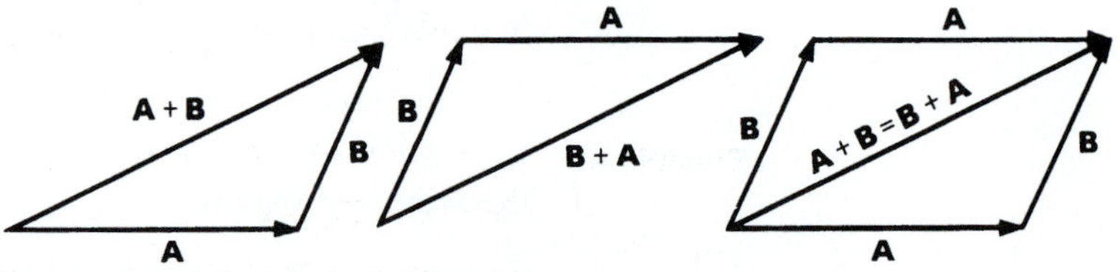

Although there is no great mystery to addition, subtraction, and multiplication of a vector by a scalar, the result of "multiplying" one vector by another is somewhat less apparent. Does multiplication yield a vector, a scalar, or some other quantity? The choice is up to us, and we shall define two types of products which are useful in our applications to physics.

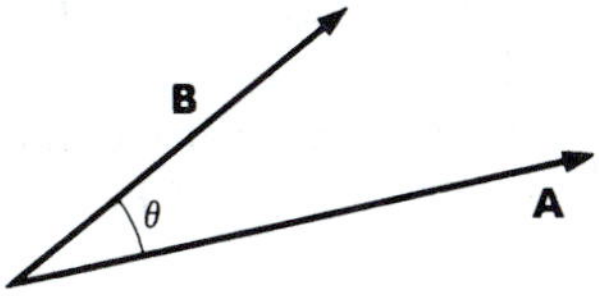

Scalar Product ("Dot" Product) The first type of product is called the *scalar* product, since it represents a way of combining two vectors to form a scalar. The scalar product of **A** and **B** is denoted by **A · B** and is often called the dot product. **A · B** is defined by

$$\mathbf{A}\cdot\mathbf{B} \equiv |\mathbf{A}|\,|\mathbf{B}|\cos\theta.$$

Here θ is the angle between **A** and **B** when they are drawn tail to tail.

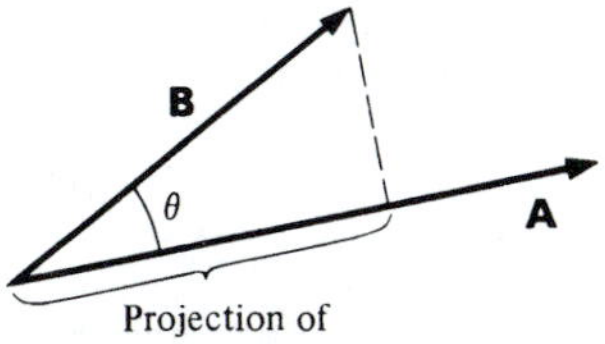

Projection of **B** on **A**

Since $|\mathbf{B}|\cos\theta$ is the projection of **B** along the direction of **A**, $\mathbf{A}\cdot\mathbf{B} = |\mathbf{A}| \times$ (projection of **B** on **A**).

Similarly,

$$\mathbf{A}\cdot\mathbf{B} = |\mathbf{B}| \times (\text{projection of } \mathbf{A} \text{ on } \mathbf{B}).$$

If $\mathbf{A}\cdot\mathbf{B} = 0$, then $|\mathbf{A}| = 0$ or $|\mathbf{B}| = 0$, or **A** is perpendicular to **B** (that is, $\cos\theta = 0$). Scalar multiplication is unusual in that the dot product of two nonzero vectors can be 0.

Note that $\mathbf{A}\cdot\mathbf{A} = |\mathbf{A}|^2$.

By way of demonstrating the usefulness of the dot product, here is an almost trivial proof of the law of cosines.

Example 1.1 **Law of Cosines**

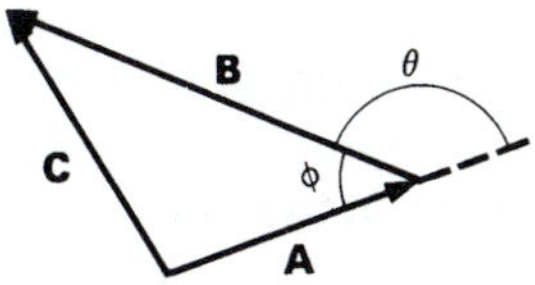

$$\mathbf{C} = \mathbf{A} + \mathbf{B}$$
$$\mathbf{C}\cdot\mathbf{C} = (\mathbf{A}+\mathbf{B})\cdot(\mathbf{A}+\mathbf{B})$$
$$|\mathbf{C}|^2 = |\mathbf{A}|^2 + |\mathbf{B}|^2 + 2|\mathbf{A}|\,|\mathbf{B}|\cos\theta$$

This result is generally expressed in terms of the angle ϕ:

$$C^2 = A^2 + B^2 - 2AB\cos\phi.$$

(We have used $\cos\theta = \cos(\pi - \phi) = -\cos\phi$.)

Example 1.2 **Work and the Dot Product**

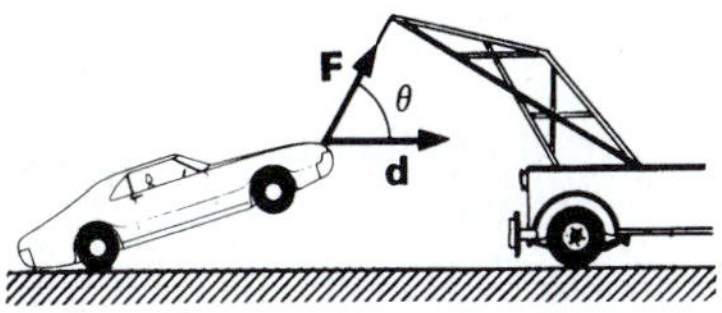

The dot product finds its most important application in the discussion of work and energy in Chap. 4. As you may already know, the work W done by a force F on an object is the displacement d of the object times the component of F along the direction of d. If the force is applied at an angle θ to the displacement,

$$W = (F\cos\theta)d.$$

Granting for the time being that force and displacement are vectors,

$$W = \mathbf{F}\cdot\mathbf{d}.$$

Vector Product ("Cross" Product) The second type of product we need is the *vector* product. In this case, two vectors **A** and **B** are combined to form a third vector **C**. The symbol for vector product is a cross:

$$\mathbf{C} = \mathbf{A} \times \mathbf{B}.$$

An alternative name is the *cross product.*

The vector product is more complicated than the scalar product because we have to specify both the magnitude and direction of **A** × **B**. The magnitude is defined as follows: if

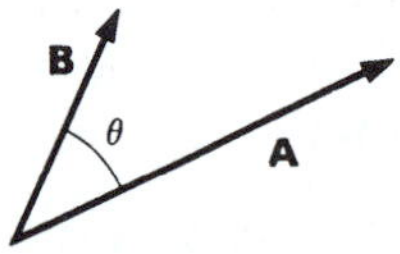

$$\mathbf{C} = \mathbf{A} \times \mathbf{B},$$

then

$$|\mathbf{C}| = |\mathbf{A}|\,|\mathbf{B}| \sin\theta,$$

where θ is the angle between **A** and **B** when they are drawn tail to tail. (To eliminate ambiguity, θ is always taken as the angle smaller than π.) Note that the vector product is zero when $\theta = 0$ or π, even if $|\mathbf{A}|$ and $|\mathbf{B}|$ are not zero.

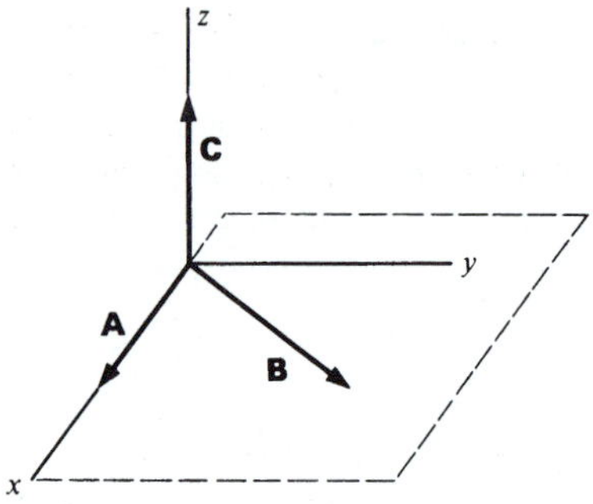

When we draw **A** and **B** tail to tail, they determine a plane. We define the direction of **C** to be perpendicular to the plane of **A** and **B**. **A**, **B**, and **C** form what is called a *right hand triple.* Imagine a right hand coordinate system with **A** and **B** in the xy plane as shown in the sketch. **A** lies on the x axis and **B** lies toward the y axis. If **A**, **B**, and **C** form a right hand triple, then **C** lies on the z axis. We shall always use right hand coordinate systems such as the one shown at left. Here is another way to determine the direction of the cross product. Think of a right hand screw with the axis perpendicular to **A** and **B**. Rotate it in the direction which swings **A** into **B**. **C** lies in the direction the screw advances. (Warning: Be sure not to use a left hand screw. Fortunately, they are rare. Hot water faucets are among the chief offenders; your honest everyday wood screw is right handed.)

(**A** is into paper)

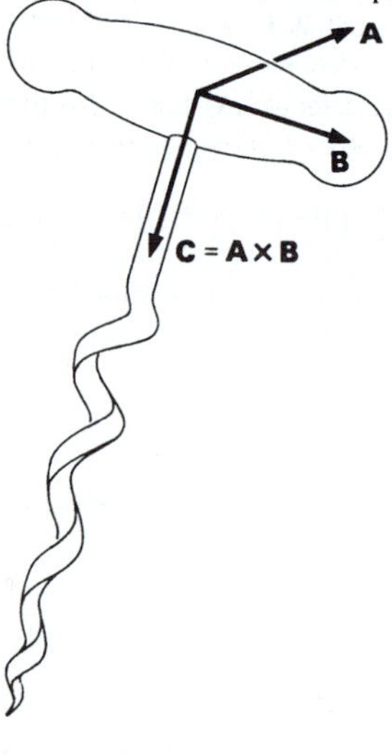

A result of our definition of the cross product is that

$$\mathbf{B} \times \mathbf{A} = -\mathbf{A} \times \mathbf{B}.$$

Here we have a case in which the order of multiplication is important. The vector product is *not* commutative. (In fact, since reversing the order reverses the sign, it is anticommutative.) We see that

$$\mathbf{A} \times \mathbf{A} = 0$$

for any vector **A**.

Example 1.3 **Examples of the Vector Product in Physics**

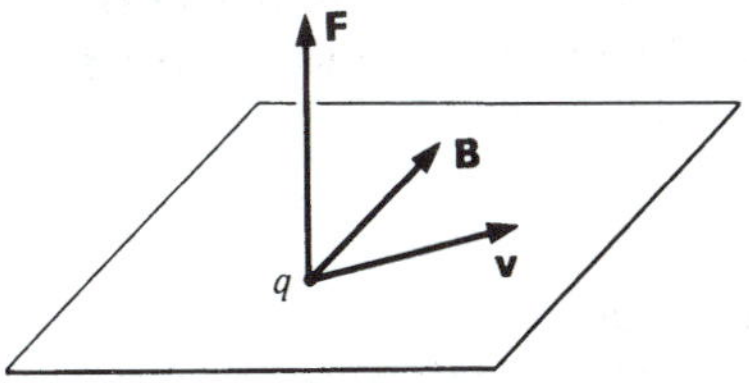

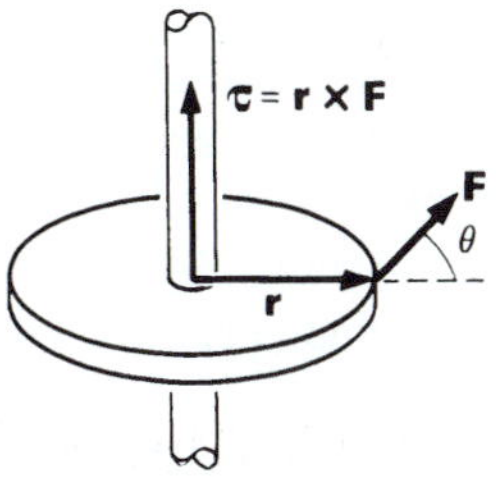

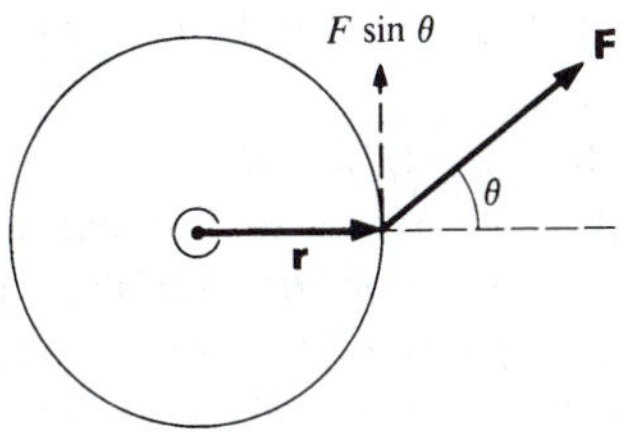

Top view

The vector product has a multitude of applications in physics. For instance, if you have learned about the interaction of a charged particle with a magnetic field, you know that the force is proportional to the charge q, the magnetic field B, and the velocity of the particle v. The force varies as the sine of the angle between v and B, and is perpendicular to the plane formed by v and B, in the direction indicated. A simpler way to give all these rules is

$$\mathbf{F} = q\mathbf{v} \times \mathbf{B}.$$

Another application is the definition of torque. We shall develop this idea later. For now we simply mention in passing that the torque $\boldsymbol{\tau}$ is defined by

$$\boldsymbol{\tau} = \mathbf{r} \times \mathbf{F},$$

where **r** is a vector from the axis about which the torque is evaluated to the point of application of the force **F**. This definition is consistent with the familiar idea that torque is a measure of the ability of an applied force to produce a twist. Note that a large force directed parallel to **r** produces no twist; it merely pulls. Only $F \sin \theta$, the component of force perpendicular to **r**, produces a torque. The torque increases as the lever arm gets larger. As you will see in Chap. 6, it is extremely useful to associate a direction with torque. The natural direction is along the axis of rotation which the torque tends to produce. All these ideas are summarized in a nutshell by the simple equation $\boldsymbol{\tau} = \mathbf{r} \times \mathbf{F}$.

Example 1.4 **Area as a Vector**

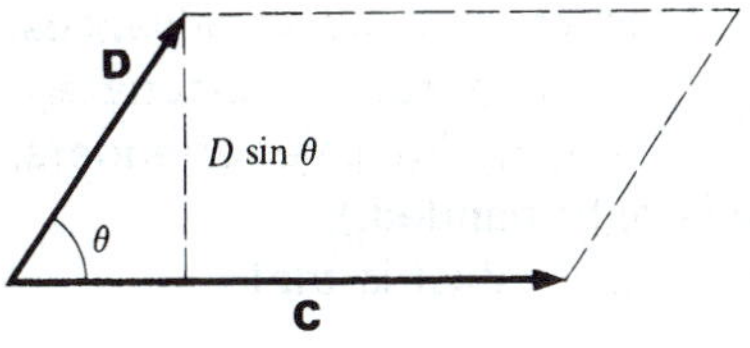

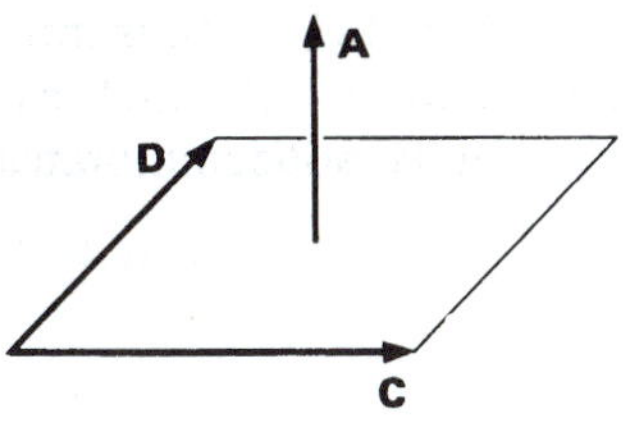

We can use the cross product to describe an area. Usually one thinks of area in terms of magnitude only. However, many applications in physics require that we also specify the orientation of the area. For example, if we wish to calculate the rate at which water in a stream flows through a wire loop of given area, it obviously makes a difference whether the plane of the loop is perpendicular or parallel to the flow. (In the latter case the flow through the loop is zero.) Here is how the vector product accomplishes this:

Consider the area of a quadrilateral formed by two vectors, **C** and **D**. The area of the parallelogram A is given by

$$\begin{aligned} A &= \text{base} \times \text{height} \\ &= CD \sin \theta \\ &= |\mathbf{C} \times \mathbf{D}|. \end{aligned}$$

If we think of A as a vector, we have

$$\mathbf{A} = \mathbf{C} \times \mathbf{D}.$$

We have already shown that the magnitude of A is the area of the parallelogram, and the vector product defines the convention for assigning a direction to the area. The direction is defined to be perpendicular to the plane of the area; that is, the direction is parallel to a *normal* to the surface. The sign of the direction is to some extent arbitrary; we could just as well have defined the area by $\mathbf{A} = \mathbf{D} \times \mathbf{C}$. However, once the sign is chosen, it is unique.

1.3 Components of a Vector

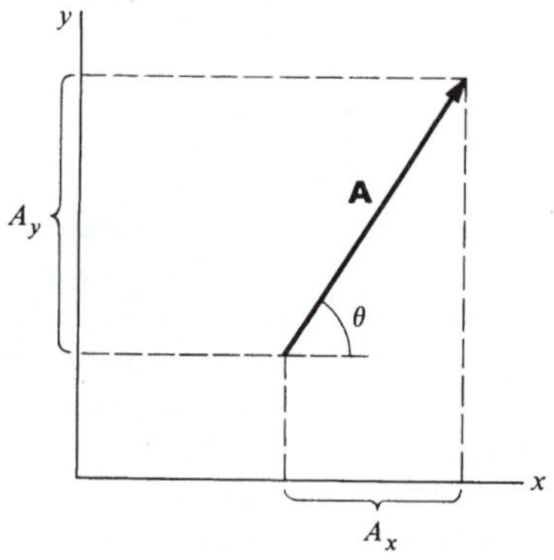

The fact that we have discussed vectors without introducing a particular coordinate system shows why vectors are so useful; vector operations are defined without reference to coordinate systems. However, eventually we have to translate our results from the abstract to the concrete, and at this point we have to choose a coordinate system in which to work.

For simplicity, let us restrict ourselves to a two-dimensional system, the familiar xy plane. The diagram shows a vector $\mathbf{A}$ in the xy plane. The projections of $\mathbf{A}$ along the two coordinate axes are called the components of $\mathbf{A}$. The components of $\mathbf{A}$ along the x and y axes are, respectively, A_x and A_y. The magnitude of $\mathbf{A}$ is $|\mathbf{A}| = (A_x{}^2 + A_y{}^2)^{\frac{1}{2}}$, and the direction of $\mathbf{A}$ is such that it makes an angle $\theta = \arctan (A_y/A_x)$ with the x axis.

Since the components of a vector define it, we can specify a vector entirely by its components. Thus

$$\mathbf{A} = (A_x,A_y)$$

or, more generally, in three dimensions,

$$\mathbf{A} = (A_x,A_y,A_z).$$

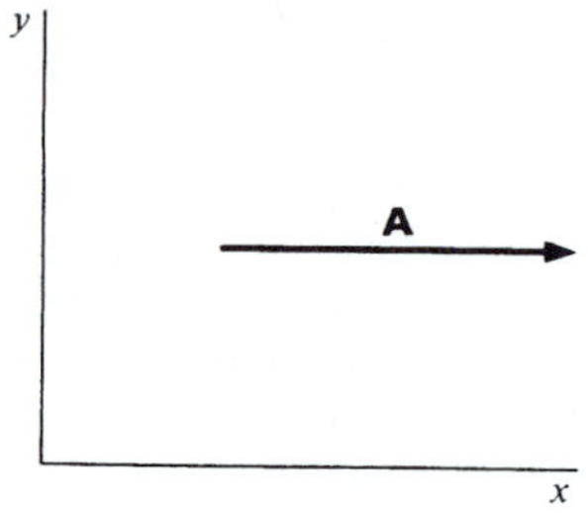

Prove for yourself that $|\mathbf{A}| = (A_x{}^2 + A_y{}^2 + A_z{}^2)^{\frac{1}{2}}$. The vector $\mathbf{A}$ has a meaning independent of any coordinate system. However, the components of $\mathbf{A}$ depend on the coordinate system being used. To illustrate this, here is a vector $\mathbf{A}$ drawn in two different coordinate systems. In the first case,

$$\mathbf{A} = (A,0) \qquad (x,y \text{ system}),$$

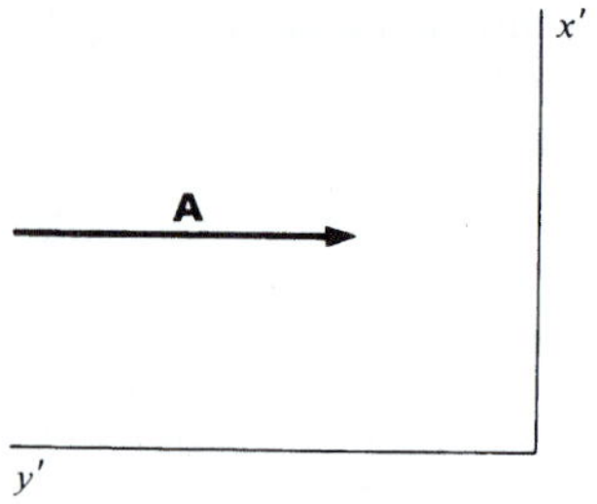

while in the second

$$\mathbf{A} = (0,-A) \qquad (x',y' \text{ system}).$$

Unless noted otherwise, we shall restrict ourselves to a single coordinate system, so that if

$$\mathbf{A} = \mathbf{B},$$

then

$$A_x = B_x \qquad A_y = B_y \qquad A_z = B_z.$$

The single vector equation $\mathbf{A} = \mathbf{B}$ symbolically represents three scalar equations.

All vector operations can be written as equations for components. For instance, multiplication by a scalar gives

$$c\mathbf{A} = (cA_x, cA_y).$$

The law for vector addition is

$$\mathbf{A} + \mathbf{B} = (A_x + B_x,\ A_y + B_y,\ A_z + B_z).$$

By writing $\mathbf{A}$ and $\mathbf{B}$ as the sums of vectors along each of the coordinate axes, you can verify that

$$\mathbf{A} \cdot \mathbf{B} = A_xB_x + A_yB_y + A_zB_z.$$

We shall defer evaluating the cross product until the next section.

Example 1.5 Vector Algebra

Let

$$\mathbf{A} = (3,5,-7)$$
$$\mathbf{B} = (2,7,1).$$

Find $\mathbf{A} + \mathbf{B}$, $\mathbf{A} - \mathbf{B}$, $|\mathbf{A}|$, $|\mathbf{B}|$, $\mathbf{A} \cdot \mathbf{B}$, and the cosine of the angle between $\mathbf{A}$ and $\mathbf{B}$.

$$\begin{aligned}
\mathbf{A} + \mathbf{B} &= (3 + 2,\ 5 + 7,\ -7 + 1) \\
&= (5,12,-6) \\
\mathbf{A} - \mathbf{B} &= (3 - 2,\ 5 - 7,\ -7 - 1) \\
&= (1,-2,-8) \\
|\mathbf{A}| &= (3^2 + 5^2 + 7^2)^{\frac{1}{2}} \\
&= \sqrt{83} \\
&= 9.11 \\
|\mathbf{B}| &= (2^2 + 7^2 + 1^2)^{\frac{1}{2}} \\
&= \sqrt{54} \\
&= 7.35 \\
\mathbf{A} \cdot \mathbf{B} &= 3 \times 2 + 5 \times 7 - 7 \times 1 \\
&= 34
\end{aligned}$$

$$\cos(\mathbf{A},\mathbf{B}) = \frac{\mathbf{A} \cdot \mathbf{B}}{|\mathbf{A}|\,|\mathbf{B}|} = \frac{34}{(9.11)(7.35)} = 0.507$$

Example 1.6 Construction of a Perpendicular Vector

Find a unit vector in the xy plane which is perpendicular to **A** = (3,5,1).

We denote the vector by **B** = (B_x,B_y,B_z). Since **B** is in the xy plane, $B_z = 0$. For **B** to be perpendicular to **A**, we have **A · B** = 0.

$$\mathbf{A}\cdot\mathbf{B} = 3B_x + 5B_y$$
$$= 0$$

Hence $B_y = -\frac{3}{5}B_x$. However, **B** is a unit vector, which means that $B_x^2 + B_y^2 = 1$. Combining these gives $B_x^2 + \frac{9}{25}B_x^2 = 1$, or $B_x = \sqrt{\frac{25}{34}} = \pm 0.857$ and $B_y = -\frac{3}{5}B_x = \mp 0.514$.

The ambiguity in sign of B_x and B_y indicates that **B** can point along a line perpendicular to **A** in either of two directions.

1.4 Base Vectors

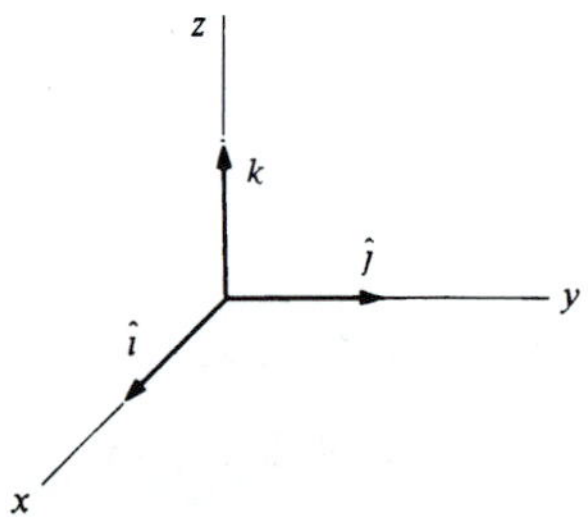

Base vectors are a set of orthogonal (perpendicular) unit vectors, one for each dimension. For example, if we are dealing with the familiar cartesian coordinate system of three dimensions, the base vectors lie along the x, y, and z axes. The x unit vector is denoted by **î**, the y unit vector by **ĵ**, and the z unit vector by $\hat{\mathbf{k}}$.

The base vectors have the following properties, as you can readily verify:

$$\hat{\mathbf{i}}\cdot\hat{\mathbf{i}} = \hat{\mathbf{j}}\cdot\hat{\mathbf{j}} = \hat{\mathbf{k}}\cdot\hat{\mathbf{k}} = 1$$
$$\hat{\mathbf{i}}\cdot\hat{\mathbf{j}} = \hat{\mathbf{j}}\cdot\hat{\mathbf{k}} = \hat{\mathbf{k}}\cdot\hat{\mathbf{i}} = 0$$
$$\hat{\mathbf{i}}\times\hat{\mathbf{j}} = \hat{\mathbf{k}}$$
$$\hat{\mathbf{j}}\times\hat{\mathbf{k}} = \hat{\mathbf{i}}$$
$$\hat{\mathbf{k}}\times\hat{\mathbf{i}} = \hat{\mathbf{j}}.$$

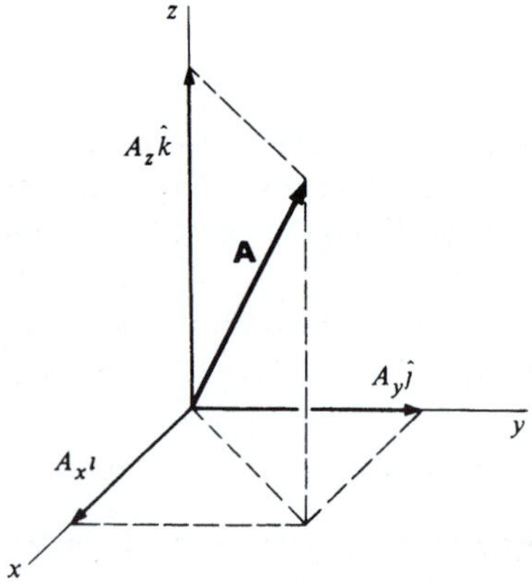

We can write any vector in terms of the base vectors.

$$\mathbf{A} = A_x\hat{\mathbf{i}} + A_y\hat{\mathbf{j}} + A_z\hat{\mathbf{k}}$$

The sketch illustrates these two representations of a vector.

To find the component of a vector in any direction, take the dot product with a unit vector in that direction. For instance,

$$A_z = \mathbf{A}\cdot\hat{\mathbf{k}}.$$

It is easy to evaluate the vector product **A × B** with the aid of the base vectors.

$$\mathbf{A}\times\mathbf{B} = (A_x\hat{\mathbf{i}} + A_y\hat{\mathbf{j}} + A_z\hat{\mathbf{k}})\times(B_x\hat{\mathbf{i}} + B_y\hat{\mathbf{j}} + B_z\hat{\mathbf{k}})$$

Consider the first term:

$$A_x\hat{\mathbf{i}} \times \mathbf{B} = A_xB_x(\hat{\mathbf{i}} \times \hat{\mathbf{i}}) + A_xB_y(\hat{\mathbf{i}} \times \hat{\mathbf{j}}) + A_xB_z(\hat{\mathbf{i}} \times \hat{\mathbf{k}}).$$

(We have assumed the associative law here.) Since $\hat{\mathbf{i}} \times \hat{\mathbf{i}} = 0$, $\hat{\mathbf{i}} \times \hat{\mathbf{j}} = \hat{\mathbf{k}}$, and $\hat{\mathbf{i}} \times \hat{\mathbf{k}} = -\hat{\mathbf{j}}$, we find

$$A_x\hat{\mathbf{i}} \times \mathbf{B} = A_x(B_y\hat{\mathbf{k}} - B_z\hat{\mathbf{j}}).$$

The same argument applied to the y and z components gives

$$A_y\hat{\mathbf{j}} \times \mathbf{B} = A_y(B_z\hat{\mathbf{i}} - B_x\hat{\mathbf{k}})$$
$$A_z\hat{\mathbf{k}} \times \mathbf{B} = A_z(B_x\hat{\mathbf{j}} - B_y\hat{\mathbf{i}}).$$

A quick way to derive these relations is to work out the first and then to obtain the others by cyclically permuting x, y, z, and $\hat{\mathbf{i}}$, $\hat{\mathbf{j}}$, $\hat{\mathbf{k}}$ (that is, $x \to y$, $y \to z$, $z \to x$, and $\hat{\mathbf{i}} \to \hat{\mathbf{j}}$, $\hat{\mathbf{j}} \to \hat{\mathbf{k}}$, $\hat{\mathbf{k}} \to \hat{\mathbf{i}}$.) A simple way to remember the result is to use the following device: write the base vectors and the components of **A** and **B** as three rows of a determinant,[1] like this

$$\mathbf{A} \times \mathbf{B} = \begin{vmatrix} \hat{\mathbf{i}} & \hat{\mathbf{j}} & \hat{\mathbf{k}} \\ A_x & A_y & A_z \\ B_x & B_y & B_z \end{vmatrix}$$
$$= \hat{\mathbf{i}}(A_yB_z - A_zB_y) - \hat{\mathbf{j}}(A_xB_z - A_zB_x) + \hat{\mathbf{k}}(A_xB_y - A_yB_x).$$

For instance, if $\mathbf{A} = \hat{\mathbf{i}} + 3\hat{\mathbf{j}} - \hat{\mathbf{k}}$ and $\mathbf{B} = 4\hat{\mathbf{i}} + \hat{\mathbf{j}} + 3\hat{\mathbf{k}}$, then

$$\mathbf{A} \times \mathbf{B} = \begin{vmatrix} \hat{\mathbf{i}} & \hat{\mathbf{j}} & \hat{\mathbf{k}} \\ 1 & 3 & -1 \\ 4 & 1 & 3 \end{vmatrix}$$
$$= 10\hat{\mathbf{i}} - 7\hat{\mathbf{j}} - 11\hat{\mathbf{k}}.$$

1.5 Displacement and the Position Vector

So far we have discussed only abstract vectors. However, the reason for introducing vectors here is concrete—they are just right for describing kinematical laws, the laws governing the geometrical properties of motion, which we need to begin our discussion of mechanics. Our first application of vectors will be to the description of position and motion in familiar three dimensional space. Although our first application of vectors is to the motion of a point in space, don't conclude that this is the only

[1] If you are unfamiliar with simple determinants, most of the books listed at the end of the chapter discuss determinants.

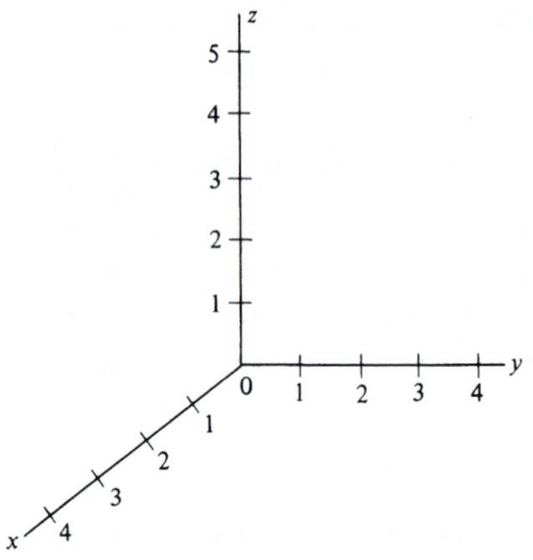

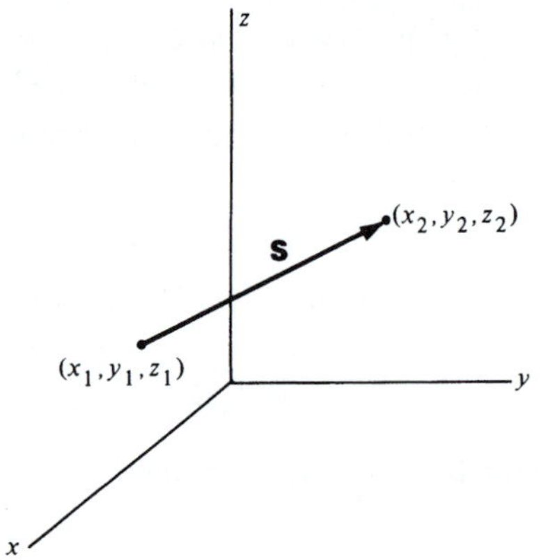

application, or even an unusually important one. Many physical quantities besides displacements are vectors. Among these are velocity, force, momentum, and gravitational and electric fields.

To locate the position of a point in space, we start by setting up a coordinate system. For convenience we choose a three dimensional cartesian system with axes x, y, and z, as shown.

In order to measure position, the axes must be marked off in some convenient unit of length—meters, for instance.

The position of the point of interest is given by listing the values of its three coordinates, x_1, y_1, z_1. These numbers do *not* represent the components of a vector according to our previous discussion. (They specify a position, not a magnitude and direction.) However, if we move the point to some new position, x_2, y_2, z_2, then the *displacement* defines a vector **S** with coordinates $S_x = x_2 - x_1$, $S_y = y_2 - y_1$, $S_z = z_2 - z_1$.

S is a vector from the initial position to the final position—it defines the displacement of a point of interest. Note, however, that **S** contains no information about the initial and final positions separately—only about the *relative* position of each. Thus, $S_z = z_2 - z_1$ depends on the *difference* between the final and initial values of the z coordinates; it does not specify z_2 or z_1 separately. **S** is a true vector; although the values of the coordinates of the initial and final points depend on the coordinate system, **S** does not, as the sketches below indicate.

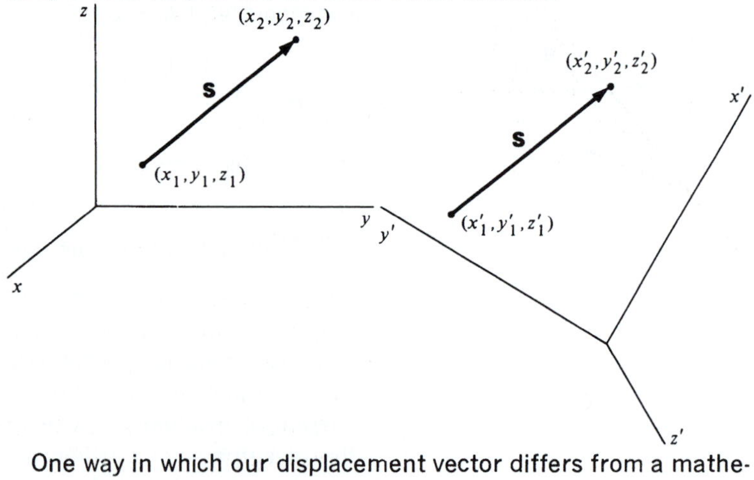

One way in which our displacement vector differs from a mathematician's vector is that his vectors are usually pure quantities, with components given by absolute numbers, whereas **S** has the physical dimension of length associated with it. We will use the convention that the magnitude of a vector has dimensions so that a unit vector is dimensionless. Thus, a displacement of 8 m (8 meters) in the x direction is $\mathbf{S} = (8\text{ m}, 0, 0)$. $|\mathbf{S}| = 8$ m, and $\hat{\mathbf{S}} = \mathbf{S}/|\mathbf{S}| = \hat{\mathbf{i}}$.

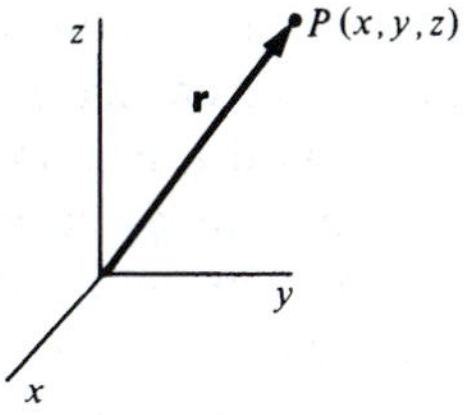

Although vectors define displacements rather than positions, it is in fact possible to describe the position of a point with respect to the origin of a given coordinate system by a special vector, known as the *position vector,* which extends from the origin to the point of interest. We shall use the symbol $\mathbf{r}$ to denote the position vector. The position of an arbitrary point P at (x,y,z) is written as

$$\mathbf{r} = (x,y,z) = x\hat{\jmath} + y\hat{\jmath} + z\hat{\mathbf{k}}.$$

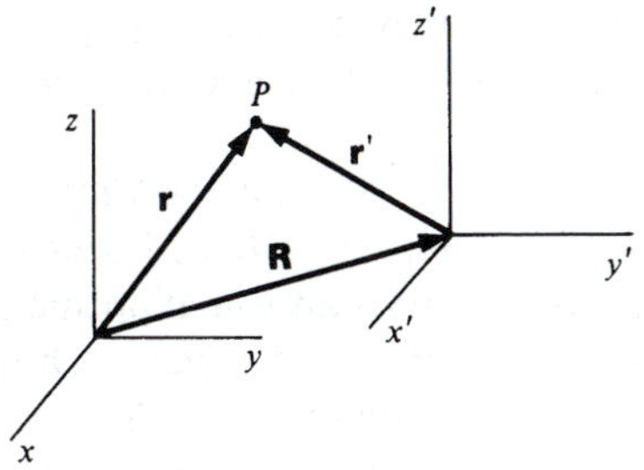

Unlike ordinary vectors, $\mathbf{r}$ depends on the coordinate system. The sketch to the left shows position vectors $\mathbf{r}$ and $\mathbf{r}'$ indicating the position of the same point in space but drawn in different coordinate systems. If $\mathbf{R}$ is the vector from the origin of the unprimed coordinate system to the origin of the primed coordinate system, we have

$$\mathbf{r}' = \mathbf{r} - \mathbf{R}.$$

In contrast, a true vector, such as a displacement $\mathbf{S}$, is independent of coordinate system. As the bottom sketch indicates,

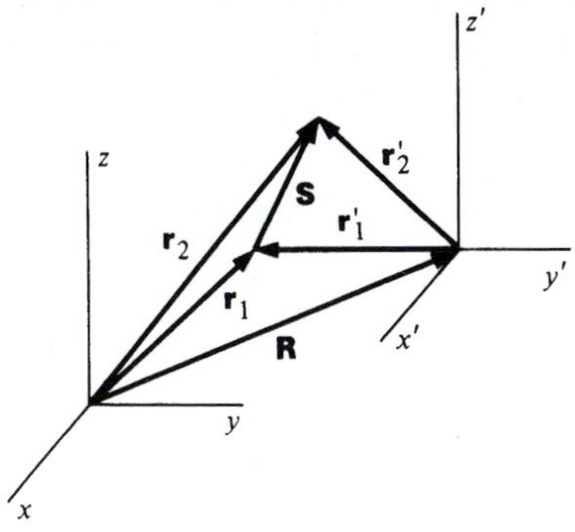

$$\begin{aligned}\mathbf{S} &= \mathbf{r}_2 - \mathbf{r}_1 \\ &= (\mathbf{r}_2' + \mathbf{R}) - (\mathbf{r}_1' + \mathbf{R}) \\ &= \mathbf{r}_2' - \mathbf{r}_1'.\end{aligned}$$

1.6 Velocity and Acceleration

Motion in One Dimension

Before applying vectors to velocity and acceleration in three dimensions, it may be helpful to review briefly the case of one dimension, motion along a straight line.

Let x be the value of the coordinate of a particle moving along a line. x is measured in some convenient unit, such as meters, and we assume that we have a continuous record of position versus time.

The *average velocity* $\bar{v}$ of the point between two times, t_1 and t_2, is defined by

$$\bar{v} = \frac{x(t_2) - x(t_1)}{t_2 - t_1}.$$

(We shall often use a bar to indicate an average of a quantity.)

The *instantaneous velocity* v is the limit of the average velocity as the time interval approaches zero.

$$v = \lim_{\Delta t \to 0} \frac{x(t + \Delta t) - x(t)}{\Delta t}.$$

The limit we have introduced in defining v is precisely that involved in the definition of a derivative. In fact, we have[1]

$$v = \frac{dx}{dt}.$$

In a similar fashion, the *instantaneous acceleration* is

$$a = \lim_{\Delta t \to 0} \frac{v(t + \Delta t) - v(t)}{\Delta t}$$
$$= \frac{dv}{dt}.$$

The concept of speed is sometimes useful. Speed s is simply the magnitude of the velocity: $s = |\mathbf{v}|$.

Motion in Several Dimensions

Our task now is to extend the ideas of velocity and acceleration to several dimensions. Consider a particle moving in a plane. As time goes on, the particle traces out a path, and we suppose that we know the particle's coordinates as a function of time. The instantaneous position of the particle at some time t_1 is

$\mathbf{r}(t_1) = [x(t_1),y(t_1)]$ or $\mathbf{r}_1 = (x_1,y_1)$,

[1] Physicists generally use the Leibnitz notation dx/dt, since this is a handy form for using differentials (see Note 1.1). Starting in Sec. 1.9 we shall use Newton's notation $\dot{x}$, but only to denote derivatives with respect to time.

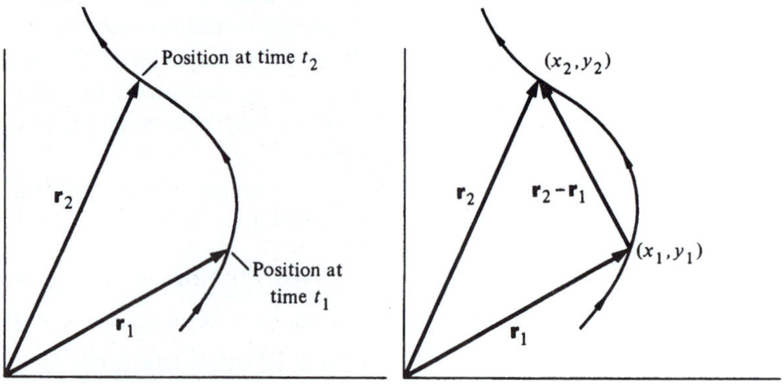

where x_1 is the value of x at $t = t_1$, and so forth. At time t_2 the position is

$$\mathbf{r}_2 = (x_2, y_2).$$

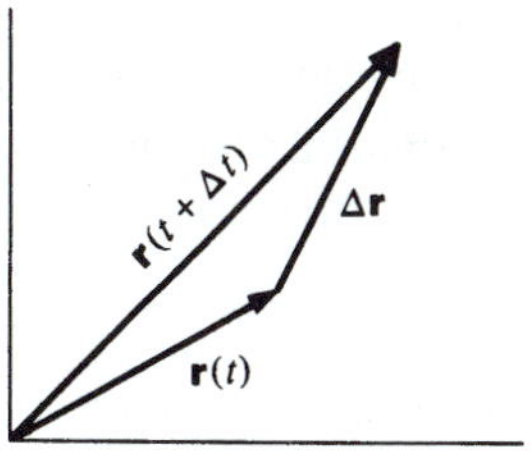

The displacement of the particle between times t_1 and t_2 is

$$\mathbf{r}_2 - \mathbf{r}_1 = (x_2 - x_1, y_2 - y_1).$$

We can generalize our example by considering the position at some time t, and at some later time $t + \Delta t$.† The displacement of the particle between these times is

$$\Delta\mathbf{r} = \mathbf{r}(t + \Delta t) - \mathbf{r}(t).$$

This vector equation is equivalent to the two scalar equations

$$\Delta x = x(t + \Delta t) - x(t)$$
$$\Delta y = y(t + \Delta t) - y(t).$$

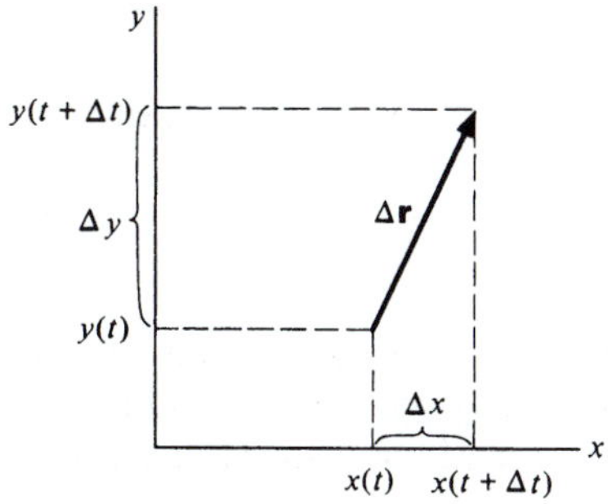

The *velocity* **v** of the particle as it moves along the path is defined to be

$$\mathbf{v} = \lim_{\Delta t \to 0} \frac{\Delta\mathbf{r}}{\Delta t} = \frac{d\mathbf{r}}{dt},$$

which is equivalent to the two scalar equations

$$v_x = \lim_{\Delta t \to 0} \frac{\Delta x}{\Delta t} = \frac{dx}{dt}$$
$$v_y = \lim_{\Delta t \to 0} \frac{\Delta y}{\Delta t} = \frac{dy}{dt}.$$

Extension of the argument to three dimensions is trivial. The third component of velocity is

$$v_z = \lim_{\Delta t \to 0} \frac{z(t + \Delta t) - z(t)}{\Delta t} = \frac{dz}{dt}.$$

Our definition of velocity as a vector is a straightforward generalization of the familiar concept of motion in a straight line. Vector notation allows us to describe motion in three dimensions with a single equation, a great economy compared with the three equations we would need otherwise. The equation $\mathbf{v} = d\mathbf{r}/dt$ expresses the results we have just found.

† We will often use the quantity Δ to denote a difference or change, as in the case here of $\Delta\mathbf{r}$ and Δt. However, this implies nothing about the size of the quantity, which may be large or small, as we please.

Alternatively, since $\mathbf{r} = x\hat{\mathbf{i}} + y\hat{\mathbf{j}} + z\hat{\mathbf{k}}$, we obtain by simple differentiation[1]

$$\frac{d\mathbf{r}}{dt} = \frac{dx}{dt}\hat{\mathbf{i}} + \frac{dy}{dt}\hat{\mathbf{j}} + \frac{dz}{dt}\hat{\mathbf{k}}$$

as before.

Let the particle undergo a displacement $\Delta\mathbf{r}$ in time Δt. In the limit $\Delta t \to 0$, $\Delta\mathbf{r}$ becomes tangent to the trajectory, as the sketch indicates. However, the relation

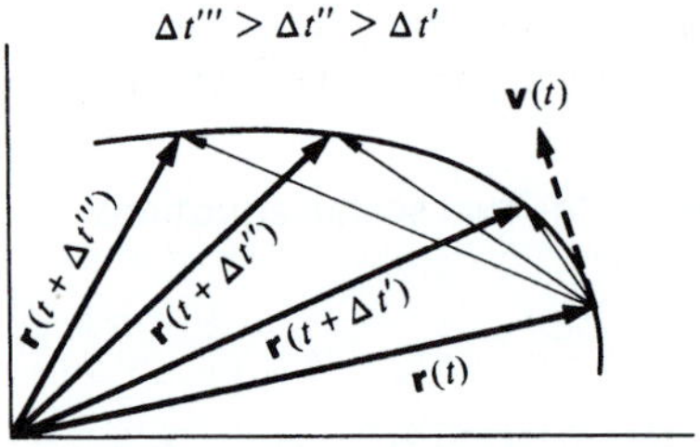

$$\Delta\mathbf{r} \approx \frac{d\mathbf{r}}{dt}\Delta t$$
$$= \mathbf{v}\,\Delta t,$$

which becomes exact in the limit $\Delta t \to 0$, shows that $\mathbf{v}$ is parallel to $\Delta\mathbf{r}$; the instantaneous velocity $\mathbf{v}$ of a particle is everywhere tangent to the trajectory.

Example 1.7 Finding v from r

The position of a particle is given by

$$\mathbf{r} = A(e^{\alpha t}\hat{\mathbf{i}} + e^{-\alpha t}\hat{\mathbf{j}}),$$

where α is a constant. Find the velocity, and sketch the trajectory.

$$\mathbf{v} = \frac{d\mathbf{r}}{dt}$$
$$= A(\alpha e^{\alpha t}\hat{\mathbf{i}} - \alpha e^{-\alpha t}\hat{\mathbf{j}})$$

or

$$v_x = A\alpha e^{\alpha t}$$
$$v_y = -A\alpha e^{-\alpha t}.$$

The magnitude of $\mathbf{v}$ is

$$v = (v_x^2 + v_y^2)^{\frac{1}{2}}$$
$$= A\alpha(e^{2\alpha t} + e^{-2\alpha t})^{\frac{1}{2}}.$$

In sketching the motion of a point, it is usually helpful to look at limiting cases. At $t = 0$, we have

$$\mathbf{r}(0) = A(\hat{\mathbf{i}} + \hat{\mathbf{j}})$$
$$v(0) = \alpha A(\hat{\mathbf{i}} - \hat{\mathbf{j}}).$$

[1] Caution: We can neglect the cartesian unit vectors when we differentiate, since their directions are fixed. Later we shall encounter unit vectors which can change direction, and then differentiation is more elaborate.

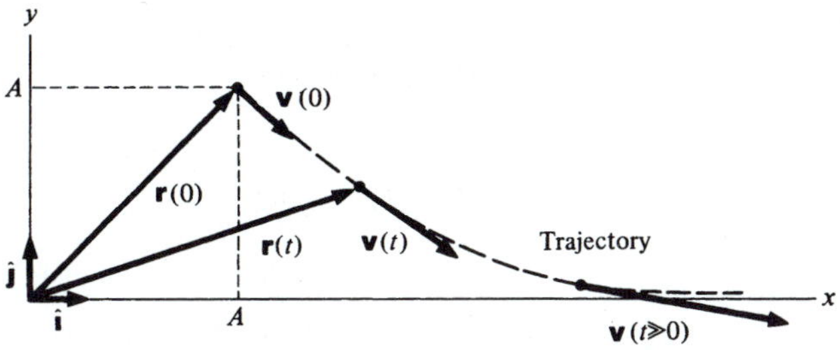

As $t \to \infty$, $e^{\alpha t} \to \infty$ and $e^{-\alpha t} \to 0$. In this limit $\mathbf{r} \to Ae^{\alpha t}\hat{\mathbf{i}}$, which is a vector along the x axis, and $\mathbf{v} \to \alpha A e^{\alpha t}\hat{\mathbf{i}}$; the speed increases without limit.

Similarly, the acceleration **a** is defined by

$$\mathbf{a} = \frac{d\mathbf{v}}{dt} = \frac{dv_x}{dt}\hat{\mathbf{i}} + \frac{dv_y}{dt}\hat{\mathbf{j}} + \frac{dv_z}{dt}\hat{\mathbf{k}}$$

$$= \frac{d^2\mathbf{r}}{dt^2}.$$

We could continue to form new vectors by taking higher derivatives of **r**, but we shall see in our study of dynamics that **r**, **v**, and **a** are of chief interest.

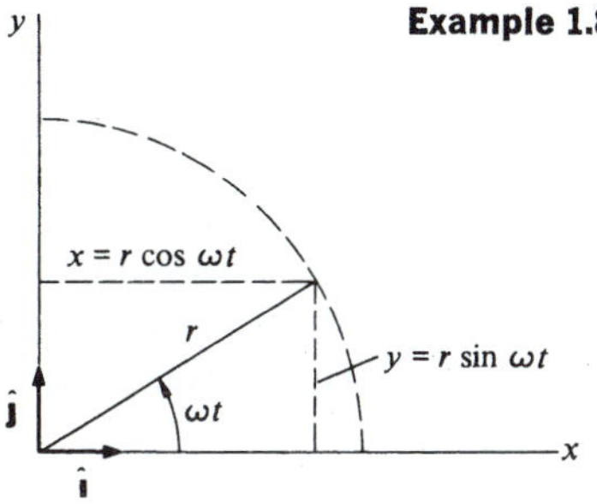

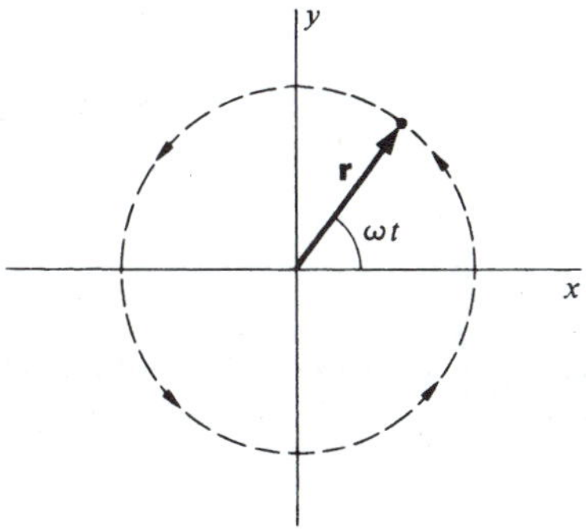

Example 1.8 **Uniform Circular Motion**

Circular motion plays an important role in physics. Here we look at the simplest and most important case—*uniform* circular motion, which is circular motion at constant speed.

Consider a particle moving in the xy plane according to $\mathbf{r} = r(\cos\omega t\hat{\mathbf{i}} + \sin\omega t\hat{\mathbf{j}})$, where r and ω are constants. Find the trajectory, the velocity, and the acceleration.

$$|\mathbf{r}| = [r^2\cos^2\omega t + r^2\sin^2\omega t]^{\frac{1}{2}}$$

Using the familiar identity $\sin^2\theta + \cos^2\theta = 1$,

$$|\mathbf{r}| = [r^2(\cos^2\omega t + \sin^2\omega t)]^{\frac{1}{2}}$$

$$= r = \text{constant}.$$

The trajectory is a circle.

The particle moves counterclockwise around the circle, starting from $(r,0)$ at $t = 0$. It traverses the circle in a time T such that $\omega T = 2\pi$. ω is called the *angular velocity* of the motion and is measured in radians

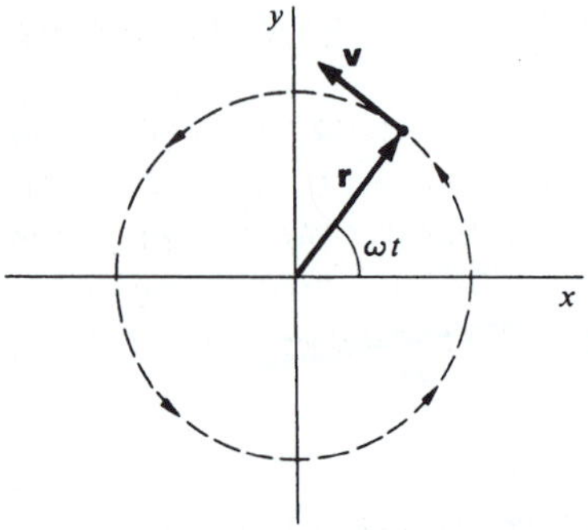

per second. T, the time required to execute one complete cycle, is called the *period.*

$$\mathbf{v} = \frac{d\mathbf{r}}{dt}$$
$$= r\omega(-\sin\omega t\hat{\mathbf{i}} + \cos\omega t\hat{\mathbf{j}})$$

We can show that $\mathbf{v}$ is tangent to the trajectory by calculating $\mathbf{v}\cdot\mathbf{r}$:

$$\mathbf{v}\cdot\mathbf{r} = r^2\omega(-\sin\omega t\cos\omega t + \cos\omega t\sin\omega t)$$
$$= 0.$$

Since $\mathbf{v}$ is perpendicular to $\mathbf{r}$, it is tangent to the circle as we expect. Incidentally, it is easy to show that $|\mathbf{v}| = r\omega =$ constant.

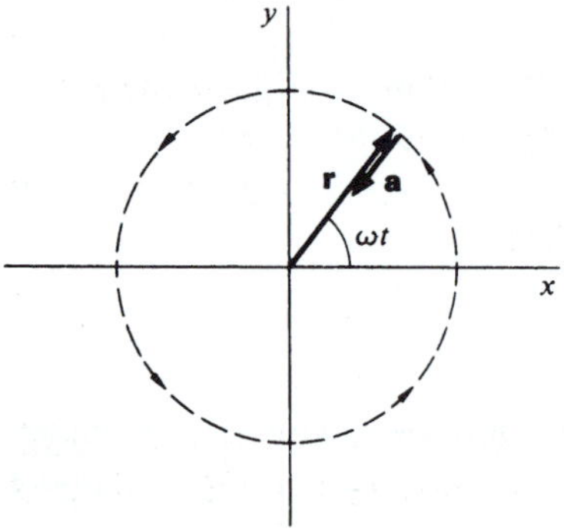

$$\mathbf{a} = \frac{d\mathbf{v}}{dt}$$
$$= r\omega^2[-\cos\omega t\hat{\mathbf{i}} - \sin\omega t\hat{\mathbf{j}}]$$
$$= -\omega^2\mathbf{r}$$

The acceleration is directed radially inward, and is known as the *centripetal acceleration.* We shall have more to say about it shortly.

A Word about Dimension and Units

Physicists call the fundamental physical units in which a quantity is measured the *dimension* of the quantity. For example, the dimension of velocity is distance/time and the dimension of acceleration is velocity/time or (distance/time)/time = distance/time^2. As we shall discuss in Chap. 2, mass, distance, and time are the fundamental physical units used in mechanics.

To introduce a system of units, we specify the standards of measurement for mass, distance, and time. Ordinarily we measure distance in meters and time in seconds. The units of velocity are then meters per second (m/s) and the units of acceleration are meters per second^2 (m/s^2).

The natural unit for measuring angle is the *radian* (rad). The angle θ in radians is S/r, where S is the arc subtended by θ in a circle of radius $\mathbf{r}$:

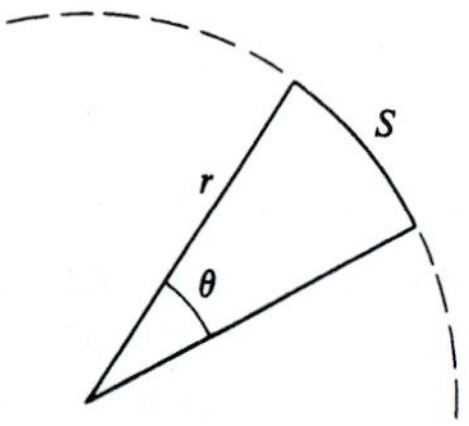

$$\theta = \frac{S}{r}.$$

2π rad = 360°. We shall always use the radian as the unit of angle, unless otherwise stated. For example, in $\sin\omega t$, ωt is in radians. ω therefore has the dimensions 1/time and the units radians per second. (The radian is dimensionless, since it is the ratio of two lengths.)

To avoid gross errors, it is a good idea to check to see that both sides of an equation have the same dimensions or units. For example, the equation $v = \alpha r e^{\alpha t}$ is dimensionally correct; since exponentials and their arguments are always dimensionless, α has the units 1/s, and the right hand side has the correct units, meters per second.

1.7 Formal Solution of Kinematical Equations

Dynamics, which we shall take up in the next chapter, enables us to find the acceleration of a body directly. Once we know the acceleration, finding the velocity and position is a simple matter of integration. Here is the formal integration procedure.

If the acceleration is known as a function of time, the velocity can be found from the defining equation

$$\frac{d\mathbf{v}(t)}{dt} = \mathbf{a}(t)$$

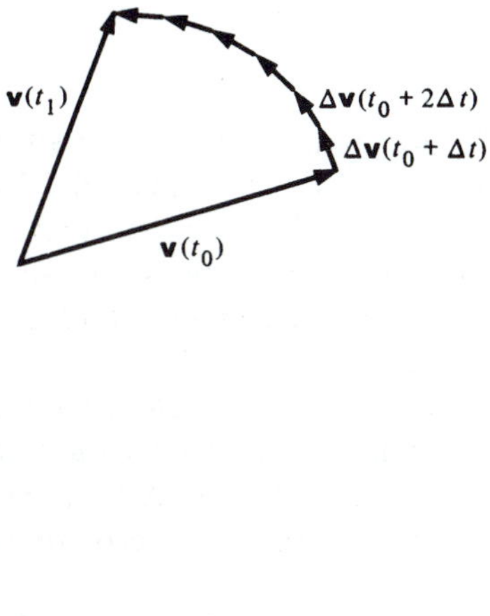

by integration with respect to time. Suppose we want to find $\mathbf{v}(t_1)$ given the initial velocity $\mathbf{v}(t_0)$ and the acceleration $\mathbf{a}(t)$. Dividing the time interval $t_1 - t_0$ into n parts $\Delta t = (t_1 - t_0)/n$,

$$\begin{aligned}\mathbf{v}(t_1) &\approx \mathbf{v}(t_0) + \Delta\mathbf{v}(t_0 + \Delta t) + \Delta\mathbf{v}(t_0 + 2\Delta t) + \cdots + \Delta\mathbf{v}(t_1)\\ &\approx \mathbf{v}(t_0) + \mathbf{a}(t_0 + \Delta t)\,\Delta t + \mathbf{a}(t_0 + 2\Delta t)\,\Delta t + \cdots + \mathbf{a}(t_1)\,\Delta t,\end{aligned}$$

since $\Delta\mathbf{v}(t) \approx \mathbf{a}(t)\,\Delta t$. Taking the x component,

$$v_x(t_1) \approx v_x(t_0) + a_x(t_0 + \Delta t)\,\Delta t + \cdots + a_x(t_1)\,\Delta t.$$

The approximation becomes exact in the limit $n \to \infty\,(\Delta t \to 0)$, and the sum becomes an integral:

$$v_x(t_1) = v_x(t_0) + \int_{t_0}^{t_1} a_x(t)\,dt.$$

The y and z components can be treated similarly. Combining the results,

$$\begin{aligned}v_x(t)_1\hat{\mathbf{i}} + v_y(t_1)\hat{\mathbf{j}} + v_z(t_1)\hat{\mathbf{k}} &= v_x(t_0)\hat{\mathbf{i}} + \int_{t_0}^{t_1} a_x(t)\,dt\,\hat{\mathbf{i}}\\ &\quad + v_y(t_0)\hat{\mathbf{j}} + \int_{t_0}^{t_1} a_y(t)\,dt\,\hat{\mathbf{j}} + v_z(t_0)\hat{\mathbf{k}} + \int_{t_0}^{t_1} a_z(t)\,dt\,\hat{\mathbf{k}}\end{aligned}$$

or

$$\mathbf{v}(t_1) = \mathbf{v}(t_0) + \int_{t_0}^{t_1} \mathbf{a}(t)\,dt.$$

This result is the same as the formal integration of $d\mathbf{v} = \mathbf{a}\, dt$.

$$\int_{t_0}^{t_1} d\mathbf{v} = \int_{t_0}^{t_1} \mathbf{a}(t)\, dt$$

$$\mathbf{v}(t_1) - \mathbf{v}(t_0) = \int_{t_0}^{t_1} \mathbf{a}(t)\, dt$$

Sometimes we need an expression for the velocity at an arbitrary time t, in which case we have

$$\mathbf{v}(t) = \mathbf{v}_0 + \int_{t_0}^{t} \mathbf{a}(t')\, dt'.$$

The dummy variable of integration has been changed from t to t' to avoid confusion with the upper limit t. We have designated the initial velocity $\mathbf{v}(t_0)$ by $\mathbf{v}_0$ to make the notation more compact. When $t = t_0$, $\mathbf{v}(t)$ reduces to $\mathbf{v}_0$, as we expect.

Example 1.9 Finding Velocity from Acceleration

A Ping-Pong ball is released near the surface of the moon with velocity $\mathbf{v}_0 = (0,5,-3)$ m/s. It accelerates (downward) with acceleration $\mathbf{a} = (0,0,-2)$ m/s^2. Find its velocity after 5 s.

The equation

$$\mathbf{v}(t) = \mathbf{v}_0 + \int_{t_0}^{t} \mathbf{a}(t')\, dt'$$

is equivalent to the three component equations

$$v_x(t) = v_{0x} + \int_0^t a_x(t')\, dt'$$

$$v_y(t) = v_{0y} + \int_0^t a_y(t')\, dt'$$

$$v_z(t) = v_{0z} + \int_0^t a_z(t')\, dt'.$$

Taking these equations in turn with the given values of $\mathbf{v}_0$ and $\mathbf{a}$, we obtain at $t = 5$ s:

$$v_x = 0 \text{ m/s}$$

$$v_y = 5 \text{ m/s}$$

$$v_z = -3 + \int_0^5 (-2)\, dt' = -13 \text{ m/s}.$$

Position is found by a second integration. Starting with

$$\frac{d\mathbf{r}(t)}{dt} = \mathbf{v}(t),$$

we find, by an argument identical to the above,

$$\mathbf{r}(t) = \mathbf{r}_0 + \int_0^t \mathbf{v}(t')\, dt'.$$

A particularly important case is that of *uniform acceleration.* If we take $\mathbf{a}$ = constant and $t_0 = 0$, we have

$$\mathbf{v}(t) = \mathbf{v}_0 + \mathbf{a}t$$

and

$$\mathbf{r}(t) = \mathbf{r}_0 + \int_0^t (\mathbf{v}_0 + \mathbf{a}t')\,dt'$$

or

$$\mathbf{r}(t) = \mathbf{r}_0 + \mathbf{v}_0 t + \tfrac{1}{2}\mathbf{a}t^2.$$

Quite likely you are already familiar with this in its one dimensional form. For instance, the x component of this equation is

$$x = x_0 + v_{0x}t + \tfrac{1}{2}a_x t^2$$

where v_{0x} is the x component of $\mathbf{v}_0$. This expression is so familiar that you may inadvertently apply it to the general case of varying acceleration. Don't! It only holds for *uniform* acceleration. In general, the full procedure described above must be used.

Example 1.10 **Motion in a Uniform Gravitational Field**

Suppose that an object moves freely under the influence of gravity so that it has a constant downward acceleration g. Choosing the z axis vertically upward, we have

$$\mathbf{a} = -g\hat{\mathbf{k}}.$$

If the object is released at $t = 0$ with initial velocity $\mathbf{v}_0$, we have

$$x = x_0 + v_{0x}t$$
$$y = y_0 + v_{0y}t$$
$$z = z_0 + v_{0z}t - \tfrac{1}{2}gt^2.$$

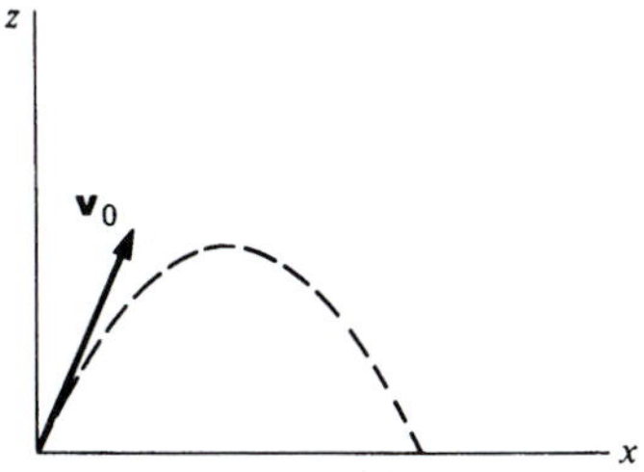

Without loss of generality, we can let $\mathbf{r}_0 = 0$, and assume that $v_{0y} = 0$. (The latter assumption simply means that we choose the coordinate system so that the initial velocity is in the xz plane.) Then

$$x = v_{0x}t$$
$$z = v_{0z}t - \tfrac{1}{2}gt^2.$$

The path of the object is shown in the sketch. We can eliminate time from the two equations for x and z to obtain the *trajectory*.

$$z = \frac{v_{0z}}{v_{0x}}x - \frac{g}{2v_{0x}^2}x^2$$

This is the well-known parabola of free fall projectile motion. However, as mentioned above, uniform acceleration is not the most general case.

Example 1.11 Nonuniform Acceleration—The Effect of a Radio Wave on an Ionospheric Electron

The ionosphere is a region of electrically neutral gas, composed of positively charged ions and negatively charged electrons, which surrounds the earth at a height of approximately 200 km (120 mi). If a radio wave passes through the ionosphere, its electric field accelerates the charged particle. Because the electric field oscillates in time, the charged particles tend to jiggle back and forth. The problem is to find the motion of an electron of charge $-e$ and mass m which is initially at rest, and which is suddenly subjected to an electric field $\mathbf{E} = \mathbf{E}_0 \sin \omega t$ (ω is the frequency of oscillation in radians per second).

The law of force for the charge in the electric field is $\mathbf{F} = -e\mathbf{E}$, and by Newton's second law we have $\mathbf{a} = \mathbf{F}/m = -e\mathbf{E}/m$. (If the reasoning behind this is a mystery to you, ignore it for now. It will be clear later. This example is meant to be a mathematical exercise—the physics is an added dividend.) We have

$$\mathbf{a} = \frac{-e\mathbf{E}}{m} = \frac{-e\mathbf{E}_0}{m} \sin \omega t.$$

$\mathbf{E}_0$ is a constant vector and we shall choose our coordinate system so that the x axis lies along it. Since there is no acceleration in the y or z directions, we need consider only the x motion. With this understanding, we can drop subscripts and write a for a_x.

$$a(t) = \frac{-eE_0}{m} \sin \omega t = a_0 \sin \omega t$$

where

$$a_0 = \frac{-eE_0}{m}.$$

Then

$$\begin{aligned} v(t) &= v_0 + \int_0^t a(t')\,dt' \\ &= v_0 + \int_0^t a_0 \sin \omega t'\,dt' \\ &= v_0 - \frac{a_0}{\omega} \cos \omega t' \Big|_0^t = v_0 - \frac{a_0}{\omega} (\cos \omega t - 1) \end{aligned}$$

and

$$\begin{aligned} x &= x_0 + \int_0^t v(t')\,dt' \\ &= x_0 + \int_0^t \left[v_0 - \frac{a_0}{\omega}(\cos \omega t' - 1) \right] dt' \\ &= x_0 + \left(v_0 + \frac{a_0}{\omega} \right) t - \frac{a_0}{\omega^2} \sin \omega t. \end{aligned}$$

We are given that $x_0 = v_0 = 0$, so we have

$$x = \frac{a_0}{\omega} t - \frac{a_0}{\omega^2} \sin \omega t.$$

The result is interesting: the second term oscillates and corresponds to the jiggling motion of the electron, which we predicted. The first term, however, corresponds to motion with uniform velocity, so in addition to the jiggling motion the electron starts to drift away. Can you see why?

1.8 More about the Derivative of a Vector

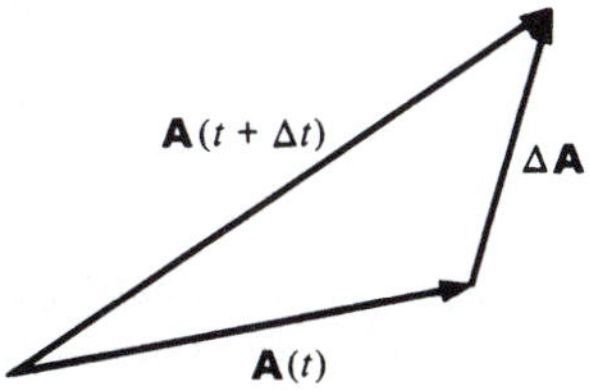

In Sec. 1.6 we demonstrated how to describe velocity and acceleration by vectors. In particular, we showed how to differentiate the vector **r** to obtain a new vector $\mathbf{v} = d\mathbf{r}/dt$. We will want to differentiate other vectors with respect to time on occasion, and so it is worthwhile generalizing our discussion.

Consider some vector $\mathbf{A}(t)$ which is a function of time. The change in **A** during the interval from t to $t + \Delta t$ is

$$\Delta\mathbf{A} = \mathbf{A}(t + \Delta t) - \mathbf{A}(t).$$

In complete analogy to the procedure we followed in differentiating **r** in Sec. 1.6, we define the time derivative of **A** by

$$\frac{d\mathbf{A}}{dt} = \lim_{\Delta t \to 0} \frac{\mathbf{A}(t + \Delta t) - \mathbf{A}(t)}{\Delta t}.$$

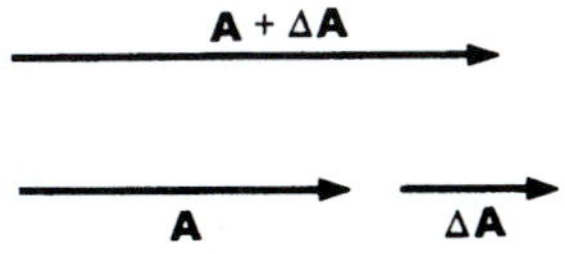

Case 1

It is important to appreciate that $d\mathbf{A}/dt$ is a new vector which can be large or small, and can point in any direction, depending on the behavior of **A**.

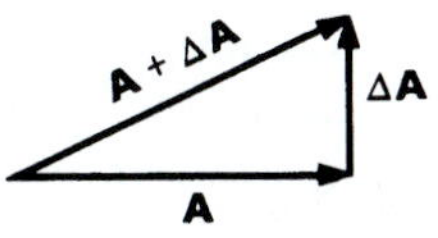

Case 2

There is one important respect in which $d\mathbf{A}/dt$ differs from the derivative of a simple scalar function. **A** can change in both *magnitude* and *direction*—a scalar function can change only in magnitude. This difference is important. The figure illustrates the addition of a small increment $\Delta\mathbf{A}$ to **A**. In the first case $\Delta\mathbf{A}$ is parallel to **A**; this leaves the direction unaltered but changes the magnitude to $|\mathbf{A}| + |\Delta\mathbf{A}|$. In the second, $\Delta\mathbf{A}$ is perpendicular to **A**. This causes a change of *direction* but leaves the magnitude practically unaltered.

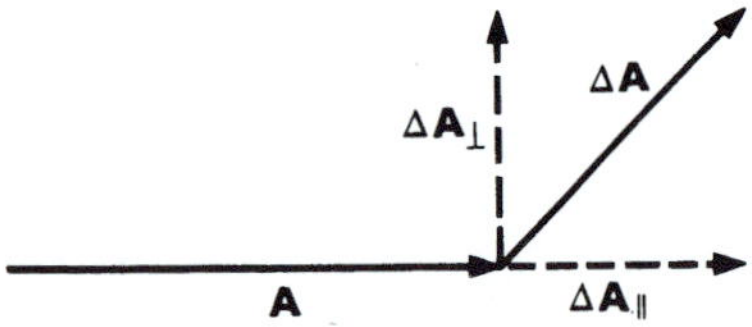

In general, **A** will change in both magnitude and direction. Even so, it is useful to visualize both types of change taking place simultaneously. In the sketch to the left we show a small increment $\Delta\mathbf{A}$ resolved into a component vector $\Delta\mathbf{A}_\parallel$ parallel to **A** and a component vector $\Delta\mathbf{A}_\perp$ perpendicular to **A**. In the limit where we take the derivative, $\Delta\mathbf{A}_\parallel$ changes the magnitude of **A** but not its direction, while $\Delta\mathbf{A}_\perp$ changes the direction of **A** but not its magnitude.

Students who do not have a clear understanding of the two ways a vector can change sometimes make an error by neglecting one of them. For instance, if $d\mathbf{A}/dt$ is always perpendicular to **A**, **A** must *rotate,* since its magnitude cannot change; its time dependence arises solely from change in direction. The illustrations below show how rotation occurs when $\Delta\mathbf{A}$ is always perpendicular to **A**. The rotational motion is made more apparent by drawing

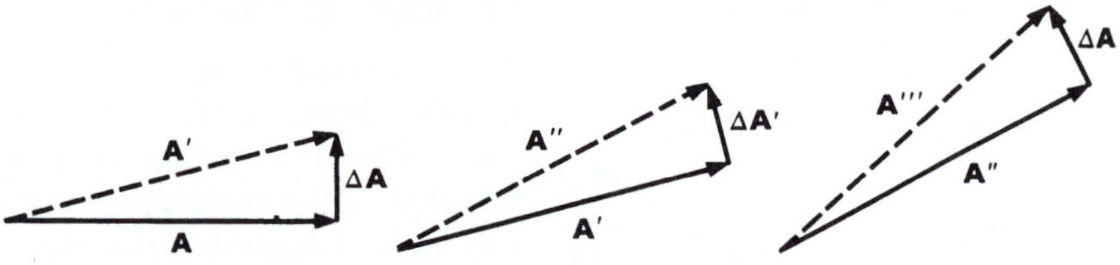

the successive vectors at a common origin.

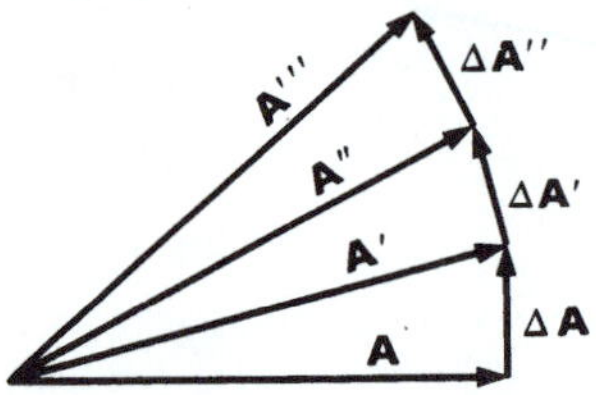

Contrast this with the case where $\Delta\mathbf{A}$ is always parallel to **A**.

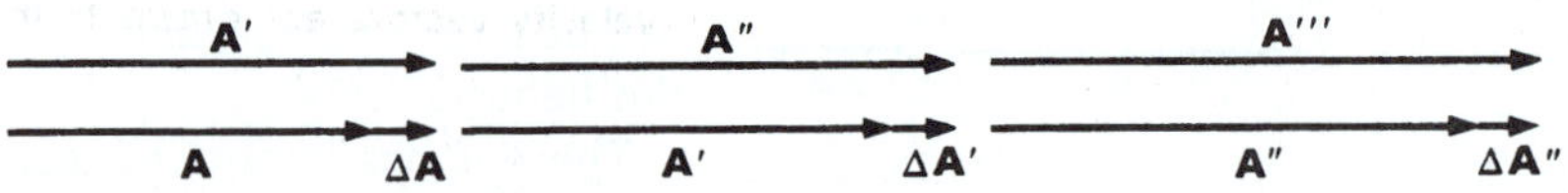

Drawn from a common origin, the vectors look like this:

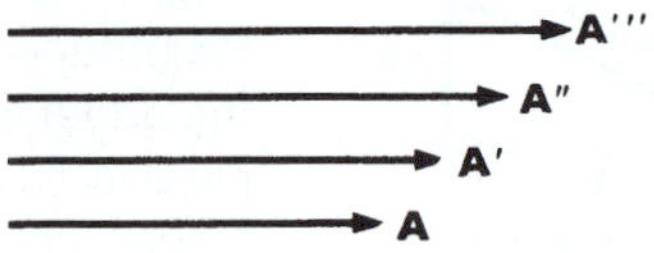

The following example relates the idea of rotating vectors to circular motion.

Example 1.12 **Circular Motion and Rotating Vectors**

In Example 1.8 we discussed the motion given by

$\mathbf{r} = r(\cos \omega t\hat{\mathbf{i}} + \sin \omega t\hat{\mathbf{j}})$.

The velocity is

$\mathbf{v} = r\omega(-\sin \omega t\hat{\mathbf{i}} + \cos \omega t\hat{\mathbf{j}})$.

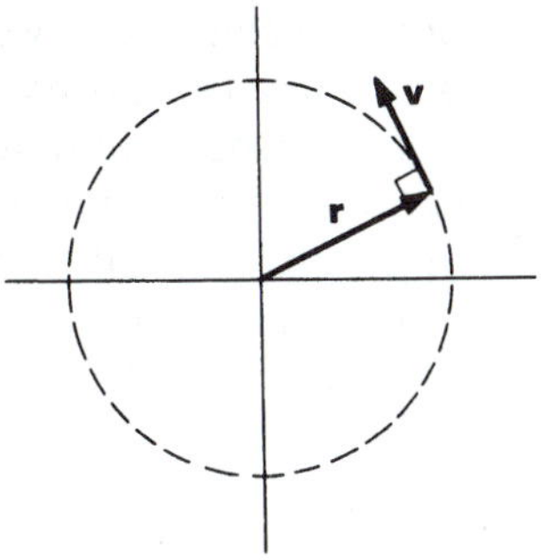

Since

$$\mathbf{r} \cdot \mathbf{v} = r^2\omega(-\cos \omega t \sin \omega t + \sin \omega t \cos \omega t)$$
$$= 0,$$

we see that $d\mathbf{r}/dt$ is perpendicular to $\mathbf{r}$. We conclude that the magnitude of $\mathbf{r}$ is constant, so that the only possible change in $\mathbf{r}$ is due to rotation. Since the trajectory is a circle, this is precisely the case: $\mathbf{r}$ rotates about the origin.

We showed earlier that $\mathbf{a} = -\omega^2\mathbf{r}$. Since $\mathbf{r} \cdot \mathbf{v} = 0$, it follows that $\mathbf{a} \cdot \mathbf{v} = -\omega^2\mathbf{r} \cdot \mathbf{v} = 0$ and $d\mathbf{v}/dt$ is perpendicular to $\mathbf{v}$. This means that the velocity vector has constant magnitude, so that it too must rotate if it is to change in time.

That $\mathbf{v}$ indeed rotates is readily seen from the sketch, which shows $\mathbf{v}$ at various positions along the trajectory. In the second sketch the same

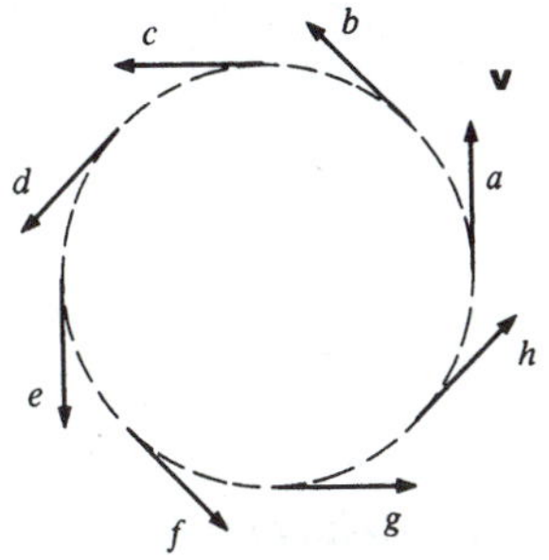

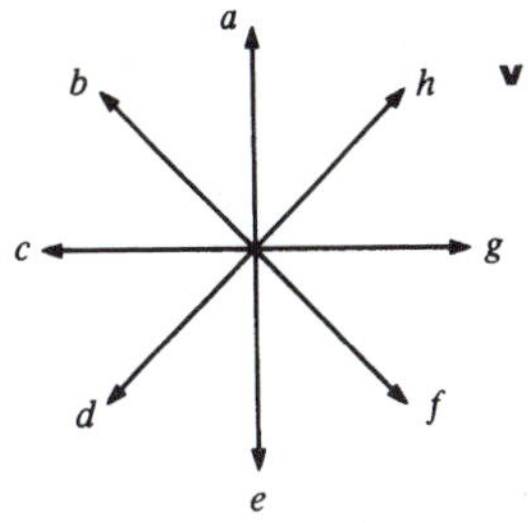

velocity vectors are drawn from a common origin. It is apparent that each time the particle completes a traversal, the velocity vector has swung around through a full circle.

Perhaps you can show that the acceleration vector also undergoes uniform rotation.

Suppose a vector $\mathbf{A}(t)$ has constant magnitude A. The only way $\mathbf{A}(t)$ can change in time is by rotating, and we shall now develop a useful expression for the time derivative $d\mathbf{A}/dt$ of such a

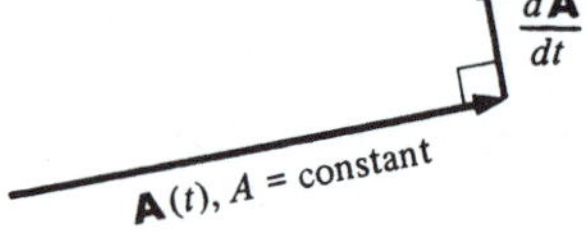

rotating vector. The direction of $d\mathbf{A}/dt$ is always perpendicular to $\mathbf{A}$. The magnitude of $d\mathbf{A}/dt$ can be found by the following geometrical argument.

The change in $\mathbf{A}$ in the time interval t to $t + \Delta t$ is

$$\Delta\mathbf{A} = \mathbf{A}(t + \Delta t) - \mathbf{A}(t).$$

Using the angle $\Delta\theta$ defined in the sketch,

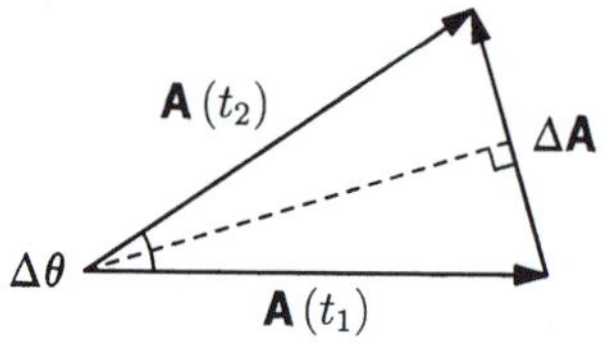

$$|\Delta\mathbf{A}| = 2A \sin\frac{\Delta\theta}{2}.$$

For $\Delta\theta \ll 1$, $\sin \Delta\theta/2 \approx \Delta\theta/2$, as discussed in Note 1.1. We have

$$|\Delta\mathbf{A}| \approx 2A\frac{\Delta\theta}{2}$$
$$= A\,\Delta\theta$$

and

$$\left|\frac{\Delta\mathbf{A}}{\Delta t}\right| = A\frac{\Delta\theta}{\Delta t}.$$

Taking the limit $\Delta t \to 0$,

$$\left|\frac{d\mathbf{A}}{dt}\right| = A\frac{d\theta}{dt}.$$

$d\theta/dt$ is called the *angular velocity* of $\mathbf{A}$.

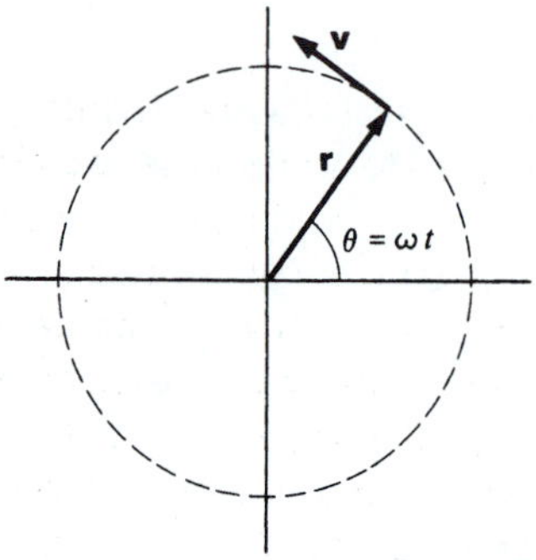

For a simple application of this result, let $\mathbf{A}$ be the rotating vector $\mathbf{r}$ discussed in Examples 1.8 and 1.12. Then $\theta = \omega t$ and

$$\left|\frac{d\mathbf{r}}{dt}\right| = r\frac{d}{dt}(\omega t) = r\omega \qquad \text{or} \qquad v = r\omega.$$

Returning now to the general case, a change in $\mathbf{A}$ is the result of a rotation *and* a change in magnitude.

$$\Delta\mathbf{A} = \Delta\mathbf{A}_\perp + \Delta\mathbf{A}_\parallel.$$

For $\Delta\theta$ sufficiently small,

$$|\Delta\mathbf{A}_\perp| = A\,\Delta\theta$$
$$|\Delta\mathbf{A}_\parallel| = \Delta A$$

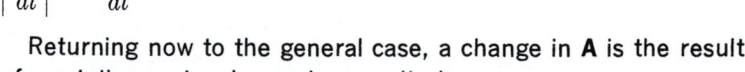

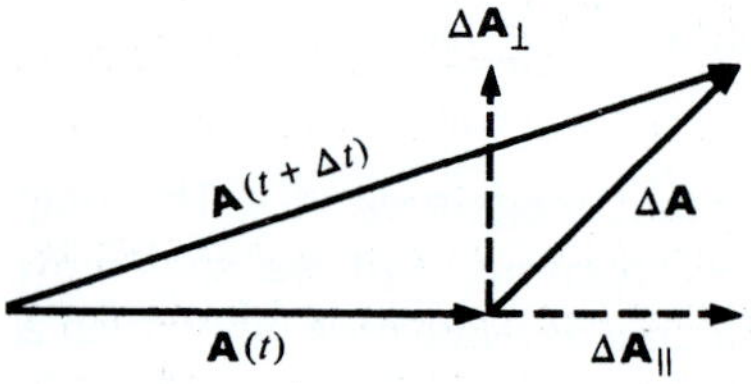

and, dividing by Δt and taking the limit,

$$\left|\frac{d\mathbf{A}_\perp}{dt}\right| = A\frac{d\theta}{dt}$$
$$\left|\frac{d\mathbf{A}_\parallel}{dt}\right| = \frac{dA}{dt}.$$

$d\mathbf{A}_\perp/dt$ is zero if **A** does not rotate ($d\theta/dt = 0$), and $d\mathbf{A}_\parallel/dt$ is zero if **A** is constant in magnitude.

We conclude this section by stating some formal identities in vector differentiation. Their proofs are left as exercises. Let the scalar c and the vectors **A** and **B** be functions of time. Then

$$\frac{d}{dt}(c\mathbf{A}) = \frac{dc}{dt}\mathbf{A} + c\frac{d\mathbf{A}}{dt}$$

$$\frac{d}{dt}(\mathbf{A}\cdot\mathbf{B}) = \frac{d\mathbf{A}}{dt}\cdot\mathbf{B} + \mathbf{A}\cdot\frac{d\mathbf{B}}{dt}$$

$$\frac{d}{dt}(\mathbf{A}\times\mathbf{B}) = \frac{d\mathbf{A}}{dt}\times\mathbf{B} + \mathbf{A}\times\frac{d\mathbf{B}}{dt}.$$

In the second relation, let $\mathbf{A} = \mathbf{B}$. Then

$$\frac{d}{dt}(A^2) = 2\mathbf{A}\cdot\frac{d\mathbf{A}}{dt},$$

and we see again that if $d\mathbf{A}/dt$ is perpendicular to **A**, the magnitude of **A** is constant.

1.9 Motion in Plane Polar Coordinates

Polar Coordinates

Rectangular, or cartesian, coordinates are well suited to describing motion in a straight line. For instance, if we orient the coordinate system so that one axis lies in the direction of motion, then only a single coordinate changes as the point moves. However, rectangular coordinates are not so useful for describing circular motion, and since circular motion plays a prominent role in physics, it is worth introducing a coordinate system more natural to it.

We should mention that although we can use any coordinate system we like, the proper choice of a coordinate system can vastly simplify a problem, so that the material in this section is very much in the spirit of more advanced physics. Quite likely some of this material will be entirely new to you. Be patient if it seems strange or even difficult at first. Once you have studied the examples and worked a few problems, it will seem much more natural.

Our new coordinate system is based on the cylindrical coordinate system. The z axis of the cylindrical system is identical to that of the cartesian system. However, position in the xy plane is

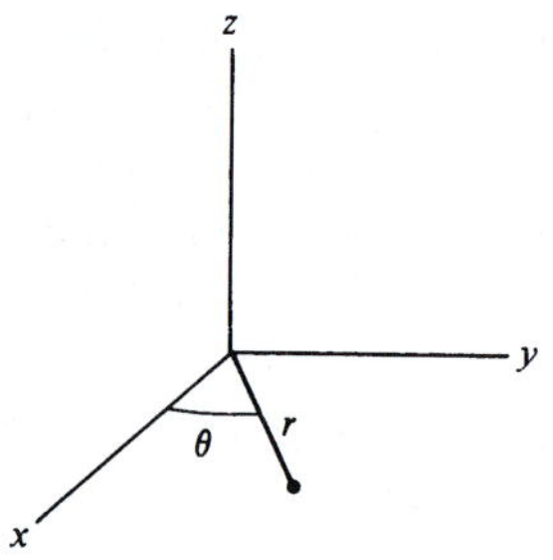

described by distance r from the z axis and the angle θ that r makes with the x axis. These coordinates are shown in the sketch. We see that

$$r = \sqrt{x^2 + y^2}$$

$$\theta = \arctan \frac{y}{x}.$$

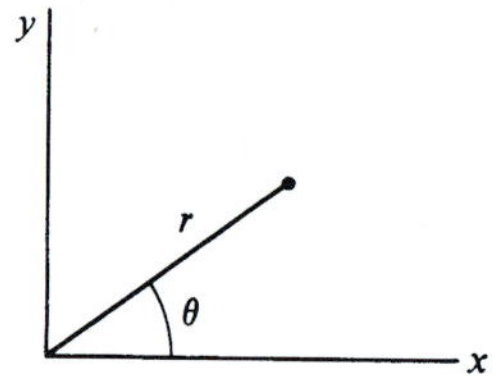

Since we shall be concerned primarily with motion in a plane, we neglect the z axis and restrict our discussion to two dimensions. The coordinates r and θ are called *plane polar* coordinates. In the following sections we shall learn to describe position, velocity, and acceleration in plane polar coordinates.

The contrast between cartesian and plane polar coordinates is readily seen by comparing drawings of constant coordinate lines for the two systems.

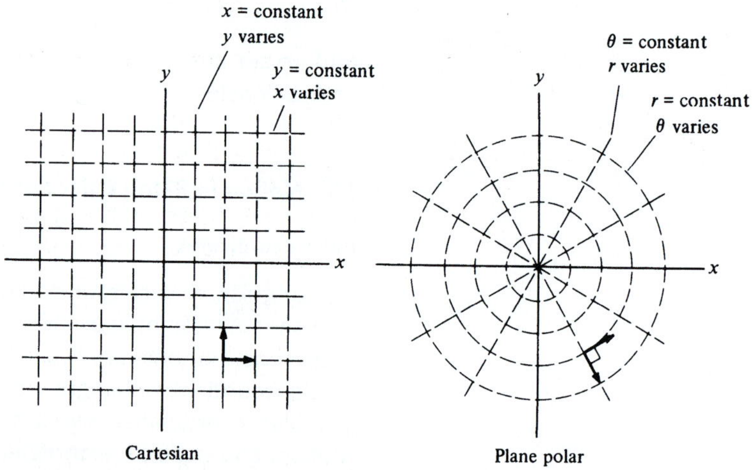

The lines of constant x and of constant y are straight and perpendicular to each other. Lines of constant θ are also straight, directed radially outward from the origin. In contrast, lines of constant r are circles concentric to the origin. Note, however, that the lines of constant θ and constant r are perpendicular wherever they intersect.

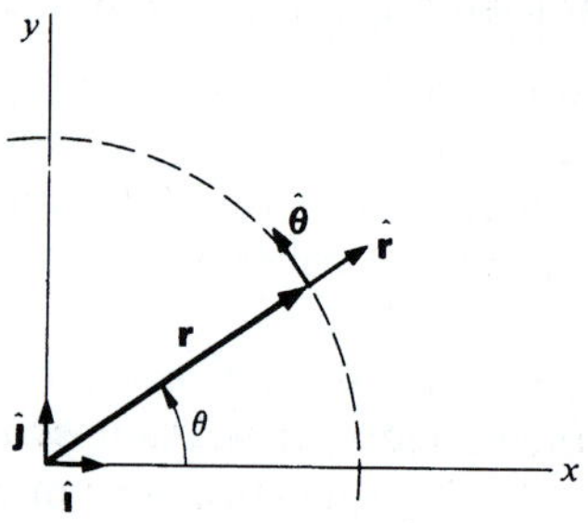

In Sec. 1.4 we introduced the base vectors $\hat{\mathbf{i}}$ and $\hat{\mathbf{j}}$ which point in the direction of increasing x and increasing y, respectively. In a similar fashion we now introduce two new unit vectors, $\hat{\mathbf{r}}$ and $\hat{\boldsymbol{\theta}}$, which point in the direction of increasing r and increasing θ. There is an important difference between these base vectors and the

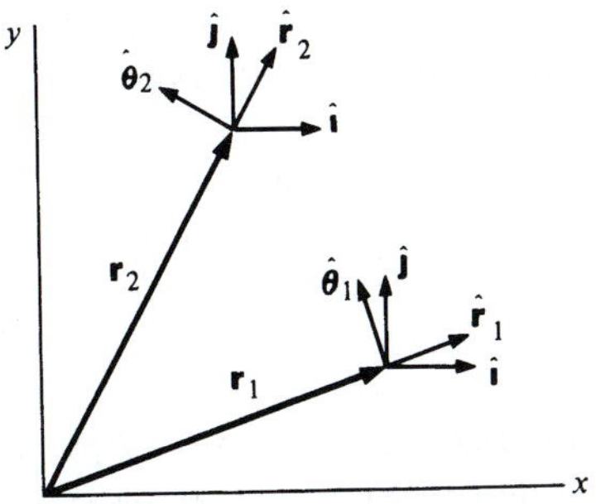

cartesian base vectors: the directions of $\hat{\mathbf{r}}$ and $\hat{\boldsymbol{\theta}}$ vary with position, whereas $\hat{\mathbf{i}}$ and $\hat{\mathbf{j}}$ have fixed directions. The drawing shows this by illustrating both sets of base vectors at two points in space. Because $\hat{\mathbf{r}}$ and $\hat{\boldsymbol{\theta}}$ vary with position, kinematical formulas can look more complicated in polar coordinates than in the cartesian system. (It is not that polar coordinates are complicated, it is simply that cartesian coordinates are simpler than they have a right to be. Cartesian coordinates are the only coordinates whose base vectors have fixed directions.)

Although $\hat{\mathbf{r}}$ and $\hat{\boldsymbol{\theta}}$ vary with position, note that they depend on θ only, not on r. We can think of $\hat{\mathbf{r}}$ and $\hat{\boldsymbol{\theta}}$ as being functionally dependent on θ.

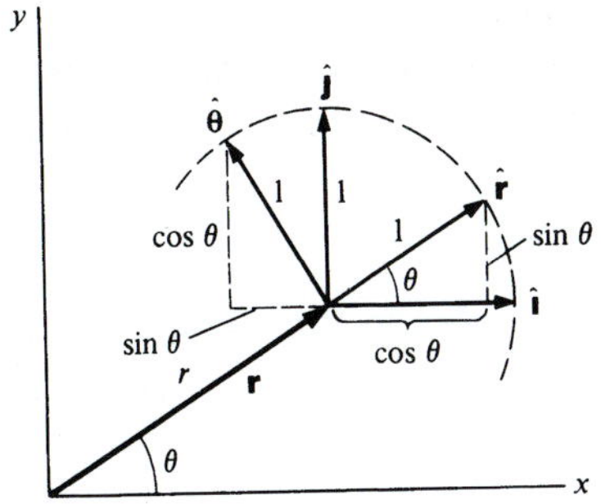

The drawing shows the unit vectors $\hat{\mathbf{i}}$, $\hat{\mathbf{j}}$ and $\hat{\mathbf{r}}$, $\hat{\boldsymbol{\theta}}$ at a point in the xy plane. We see that

$$\hat{\mathbf{r}} = \hat{\mathbf{i}} \cos \theta + \hat{\mathbf{j}} \sin \theta$$

$$\hat{\boldsymbol{\theta}} = -\hat{\mathbf{i}} \sin \theta + \hat{\mathbf{j}} \cos \theta.$$

Before proceeding, convince yourself that these expressions are reasonable by checking them at a few particularly simple points, such as $\theta = 0$, and $\pi/2$. Also verify that $\hat{\mathbf{r}}$ and $\hat{\boldsymbol{\theta}}$ are orthogonal (i.e., perpendicular) by showing that $\hat{\mathbf{r}} \cdot \hat{\boldsymbol{\theta}} = 0$.

It is easy to verify that we indeed have the same vector $\mathbf{r}$ no matter whether we describe it by cartesian or polar coordinates. In cartesian coordinates we have

$$\mathbf{r} = x\hat{\mathbf{i}} + y\hat{\mathbf{j}},$$

and in polar coordinates we have

$$\mathbf{r} = r\hat{\mathbf{r}}.$$

If we insert the above expression for $\hat{\mathbf{r}}$, we obtain

$$x\hat{\mathbf{i}} + y\hat{\mathbf{j}} = r(\hat{\mathbf{i}} \cos \theta + \hat{\mathbf{j}} \sin \theta).$$

We can separately equate the coefficients of $\hat{\mathbf{i}}$ and $\hat{\mathbf{j}}$ to obtain

$$x = r \cos \theta \qquad y = r \sin \theta,$$

as we expect.

The relation

$$\mathbf{r} = r\hat{\mathbf{r}}$$

is sometimes confusing, because the equation as written seems to make no reference to the angle θ. We know that two parameters are needed to specify a position in two dimensional space (in cartesian coordinates they are x and y), but the equation $\mathbf{r} = r\hat{\mathbf{r}}$ seems to contain only the quantity r. The answer is that $\hat{\mathbf{r}}$ is not a fixed vector and we need to know the value of θ to tell how $\hat{\mathbf{r}}$ is oriented as well as the value of r to tell how far we are from the origin. Although θ does not occur explicitly in $r\hat{\mathbf{r}}$, its value must be known to fix the direction of $\hat{\mathbf{r}}$. This would be apparent if we wrote $\mathbf{r} = r\hat{\mathbf{r}}(\theta)$ to emphasize the dependence of $\hat{\mathbf{r}}$ on θ. However, by common convention $\hat{\mathbf{r}}$ is understood to stand for $\hat{\mathbf{r}}(\theta)$.

The orthogonality of $\hat{\mathbf{r}}$ and $\hat{\boldsymbol{\theta}}$ plus the fact that they are unit vectors, $|\hat{\mathbf{r}}| = 1$, $|\hat{\boldsymbol{\theta}}| = 1$, means that we can continue to evaluate scalar products in the simple way we are accustomed to. If

$$\mathbf{A} = A_r\hat{\mathbf{r}} + A_\theta\hat{\boldsymbol{\theta}} \qquad \text{and} \qquad \mathbf{B} = B_r\hat{\mathbf{r}} + B_\theta\hat{\boldsymbol{\theta}},$$

then

$$\mathbf{A} \cdot \mathbf{B} = A_rB_r + A_\theta B_\theta.$$

Of course, the $\hat{\mathbf{r}}$'s and the $\hat{\boldsymbol{\theta}}$'s must refer to the same point in space for this simple rule to hold.

Velocity in Polar Coordinates

Now let us turn our attention to describing velocity with polar coordinates. Recall that in cartesian coordinates we have

$$\begin{aligned}\mathbf{v} &= \frac{d}{dt}(x\hat{\mathbf{i}} + y\hat{\mathbf{j}}) \\ &= \dot{x}\hat{\mathbf{i}} + \dot{y}\hat{\mathbf{j}}.\end{aligned}$$

(Remember that $\dot{x}$ stands for dx/dt.)

The same vector, $\mathbf{v}$, expressed in polar coordinates is given by

$$\begin{aligned}\mathbf{v} &= \frac{d}{dt}(r\hat{\mathbf{r}}) \\ &= \dot{r}\hat{\mathbf{r}} + r\frac{d\hat{\mathbf{r}}}{dt}.\end{aligned}$$

The first term on the right is obviously the component of the velocity directed radially outward. We suspect that the second term is the component of velocity in the tangential ($\hat{\boldsymbol{\theta}}$) direction. This is indeed the case. However to prove it we must evaluate $d\hat{\mathbf{r}}/dt$. Since this step is slightly tricky, we shall do it three different ways. Take your pick!

Evaluating $d\hat{\mathbf{r}}/dt$

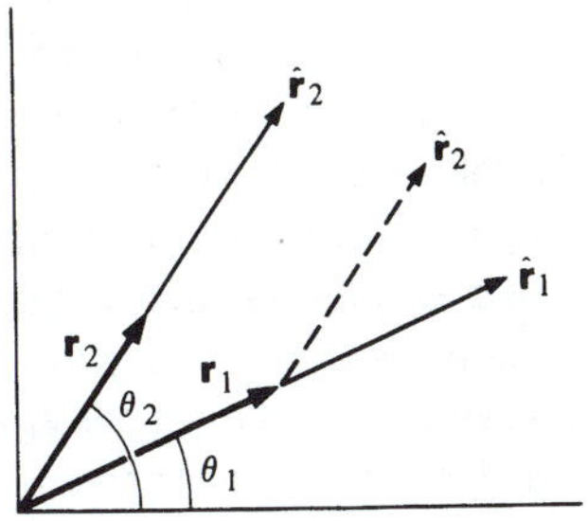

Method 1 We can invoke the ideas of the last section to find $d\hat{\mathbf{r}}/dt$. Since $\hat{\mathbf{r}}$ is a unit vector, its magnitude is constant and $d\hat{\mathbf{r}}/dt$ is perpendicular to $\hat{\mathbf{r}}$; as θ increases, $\hat{\mathbf{r}}$ rotates.

$$|\Delta\hat{\mathbf{r}}| \approx |\hat{\mathbf{r}}|\,\Delta\theta = \Delta\theta,$$

$$\frac{|\Delta\hat{\mathbf{r}}|}{\Delta t} \approx \frac{\Delta\theta}{\Delta t},$$

and, taking the limit, we obtain

$$\left|\frac{d\hat{\mathbf{r}}}{dt}\right| = \frac{d\theta}{dt}.$$

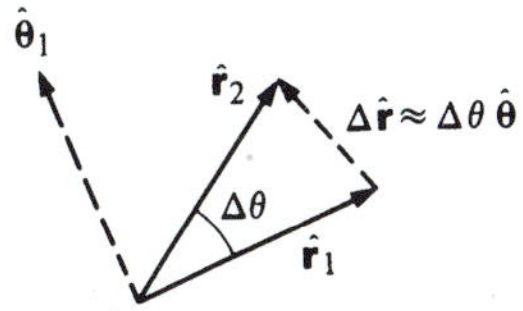

As the sketch shows, as θ increases, $\hat{\mathbf{r}}$ swings in the $\hat{\boldsymbol{\theta}}$ direction, hence

$$\frac{d\hat{\mathbf{r}}}{dt} = \dot{\theta}\hat{\boldsymbol{\theta}}.$$

If this method is too casual for your taste, you may find methods 2 or 3 more appealing.

Method 2

$$\hat{\mathbf{r}} = \hat{\mathbf{i}}\cos\theta + \hat{\mathbf{j}}\sin\theta$$

We note that $\hat{\mathbf{i}}$ and $\hat{\mathbf{j}}$ are fixed unit vectors, and thus cannot vary in time. θ, on the other hand, does vary as $\mathbf{r}$ changes. Using

$$\begin{aligned}\frac{d}{dt}(\cos\theta) &= \left(\frac{d}{d\theta}\cos\theta\right)\frac{d\theta}{dt} \\ &= -\sin\theta\,\dot{\theta}\end{aligned}$$

and

$$\begin{aligned}\frac{d}{dt}(\sin\theta) &= \left(\frac{d}{d\theta}\sin\theta\right)\frac{d\theta}{dt} \\ &= \cos\theta\,\dot{\theta},\end{aligned}$$

we obtain

$$\begin{aligned}\frac{d\hat{\mathbf{r}}}{dt} &= \hat{\mathbf{i}}\frac{d}{dt}(\cos\theta) + \hat{\mathbf{j}}\frac{d}{dt}(\sin\theta) \\ &= -\hat{\mathbf{i}}\sin\theta\,\dot{\theta} + \hat{\mathbf{j}}\cos\theta\,\dot{\theta} \\ &= (-\hat{\mathbf{i}}\sin\theta + \hat{\mathbf{j}}\cos\theta)\,\dot{\theta}.\end{aligned}$$

However, recall that $-\hat{\mathbf{i}}\sin\theta + \hat{\mathbf{j}}\cos\theta = \hat{\boldsymbol{\theta}}$. We obtain

$$\frac{d\hat{\mathbf{r}}}{dt} = \dot{\theta}\hat{\boldsymbol{\theta}}.$$

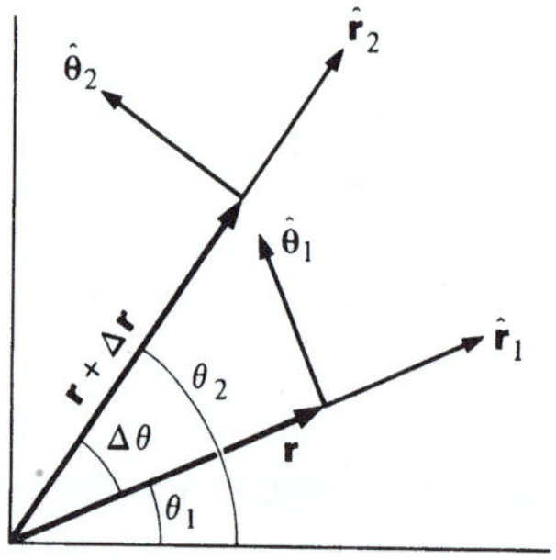

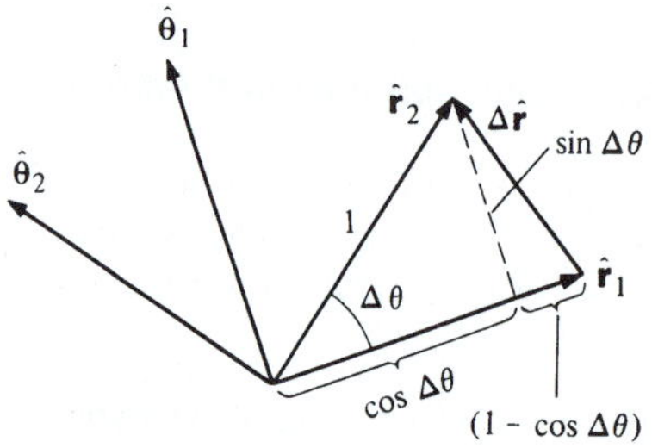

Method 3

The drawing shows **r** at two different times, t and $t + \Delta t$. The coordinates are, respectively, (r,θ) and $(r + \Delta r,\ \theta + \Delta\theta)$. Note that the angle between $\hat{\mathbf{r}}_1$ and $\hat{\mathbf{r}}_2$ is equal to the angle between $\hat{\boldsymbol{\theta}}_1$ and $\hat{\boldsymbol{\theta}}_2$; this angle is $\theta_2 - \theta_1 = \Delta\theta$.

The change in $\hat{\mathbf{r}}$ during the time Δt is illustrated by the lower drawing. We see that

$$\Delta\hat{\mathbf{r}} = \hat{\boldsymbol{\theta}}_1 \sin\Delta\theta - \hat{\mathbf{r}}_1 (1 - \cos\Delta\theta).$$

Hence

$$\frac{\Delta\hat{\mathbf{r}}}{\Delta t} = \hat{\boldsymbol{\theta}}_1 \frac{\sin\Delta\theta}{\Delta t} - \hat{\mathbf{r}}_1 \frac{(1 - \cos\Delta\theta)}{\Delta t}$$

$$= \hat{\boldsymbol{\theta}}_1 \left(\frac{\Delta\theta - \frac{1}{6}(\Delta\theta)^3 + \cdots}{\Delta t}\right) - \hat{\mathbf{r}}_1 \left(\frac{\frac{1}{2}(\Delta\theta)^2 - \frac{1}{24}(\Delta\theta)^4 + \cdots}{\Delta t}\right),$$

where we have used the series expansions discussed in Note 1.1. We need to evaluate

$$\frac{d\hat{\mathbf{r}}}{dt} = \lim_{\Delta t \to 0} \frac{\Delta\hat{\mathbf{r}}}{\Delta t}.$$

In the limit $\Delta t \to 0$, $\Delta\theta$ also approaches zero, but $\Delta\theta/\Delta t$ approaches the limit $d\theta/dt$. Therefore

$$\lim_{\Delta t \to 0} \frac{\Delta\theta}{\Delta t}(\Delta\theta)^n = 0 \qquad n > 0.$$

The term in $\hat{\mathbf{r}}$ entirely vanishes in the limit and we are left with

$$\frac{d\hat{\mathbf{r}}}{dt} = \dot{\theta}\hat{\boldsymbol{\theta}},$$

as before. We also need an expression for $d\hat{\boldsymbol{\theta}}/dt$. You can use any, or all, of the arguments above to prove for yourself that

$$\frac{d\hat{\boldsymbol{\theta}}}{dt} = -\dot{\theta}\hat{\mathbf{r}}.$$

Since you should be familiar with both results, let's summarize them together:

$$\frac{d\hat{\mathbf{r}}}{dt} = \dot{\theta}\hat{\boldsymbol{\theta}}$$

$$\frac{d\hat{\boldsymbol{\theta}}}{dt} = -\dot{\theta}\hat{\mathbf{r}}.$$

And now, we can return to our problem. On page 30 we showed that

$$\mathbf{v} = \frac{d}{dt}r\hat{\mathbf{r}} = \dot{r}\hat{\mathbf{r}} + r\frac{d\hat{\mathbf{r}}}{dt}.$$

Using the above results, we can write this as

$$\mathbf{v} = \dot{r}\hat{\mathbf{r}} + r\dot{\theta}\hat{\boldsymbol{\theta}}.$$

As we surmised, the second term is indeed in the tangential (that is, $\hat{\boldsymbol{\theta}}$) direction. We can get more insight into the meaning of each term by considering special cases where only one component varies at a time.

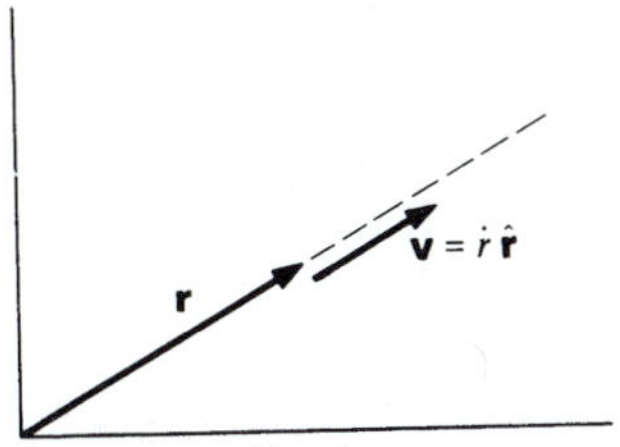

Case 1

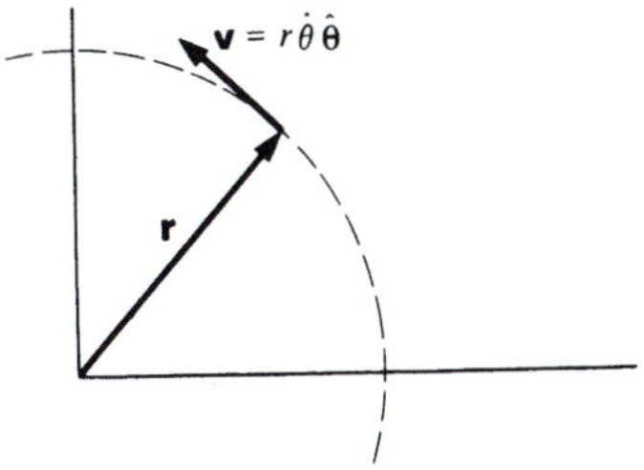

Case 2

1. θ = constant, velocity is radial. If θ is a constant, $\dot{\theta} = 0$, and $\mathbf{v} = \dot{r}\hat{\mathbf{r}}$. We have one dimensional motion in a fixed radial direction.

2. r = constant, velocity is tangential. In this case $\mathbf{v} = r\dot{\theta}\hat{\boldsymbol{\theta}}$. Since r is fixed, the motion lies on the arc of a circle. The speed of the point on the circle is $r\dot{\theta}$, and it follows that $\mathbf{v} = r\dot{\theta}\hat{\boldsymbol{\theta}}$.

For motion in general, both r and θ change in time.

The next three examples illustrate the use of polar coordinates to describe velocity.

Example 1.13 **Circular Motion and Straight Line Motion in Polar Coordinates**

A particle moves in a circle of radius b with angular velocity $\dot{\theta} = \alpha t$, where α is a constant. (α has the units radians per second².) Describe the particle's velocity in polar coordinates.

Since $r = b =$ constant, $\mathbf{v}$ is purely tangential and $\mathbf{v} = b\alpha t\hat{\boldsymbol{\theta}}$. The sketches show $\hat{\mathbf{r}}$, $\hat{\boldsymbol{\theta}}$, and $\mathbf{v}$ at a time t_1 and at a later time t_2.

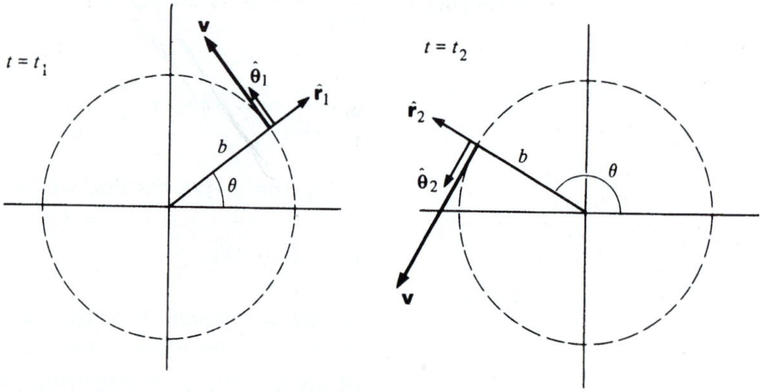

The particle is located at the position

$$r = b \qquad \theta = \theta_0 + \int_0^t \dot{\theta}\, dt = \theta_0 + \tfrac{1}{2}\alpha t^2.$$

If the particle is on the x axis at $t = 0$, $\theta_0 = 0$. The particle's position vector is $\mathbf{r} = b\hat{\mathbf{r}}$, but as the sketches indicate, θ must be given to specify the direction of $\hat{\mathbf{r}}$.

Consider a particle moving with constant velocity $\mathbf{v} = u\hat{\mathbf{i}}$ along the line $y = 2$. Describe $\mathbf{v}$ in polar coordinates.

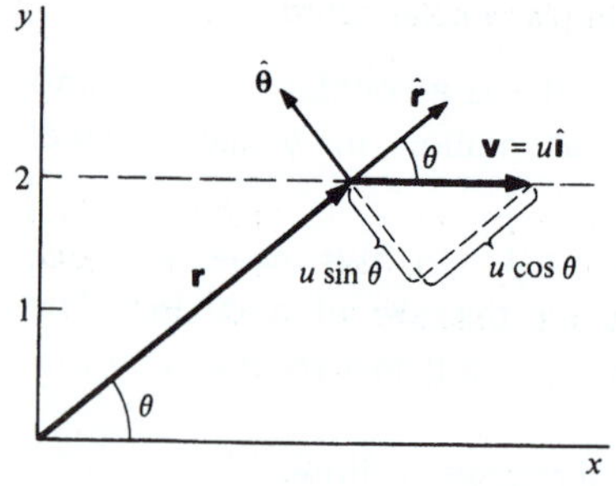

$$\mathbf{v} = v_r\hat{\mathbf{r}} + v_\theta\hat{\boldsymbol{\theta}}.$$

From the sketch,

$$v_r = u \cos \theta$$

$$v_\theta = -u \sin \theta$$

$$\mathbf{v} = u \cos \theta\hat{\mathbf{r}} - u \sin \theta\hat{\boldsymbol{\theta}}.$$

As the particle moves to the right, θ decreases and $\hat{\mathbf{r}}$ and $\hat{\boldsymbol{\theta}}$ change direction. Ordinarily, of course, we try to use coordinates that make the problem as simple as possible; polar coordinates are not well suited here.

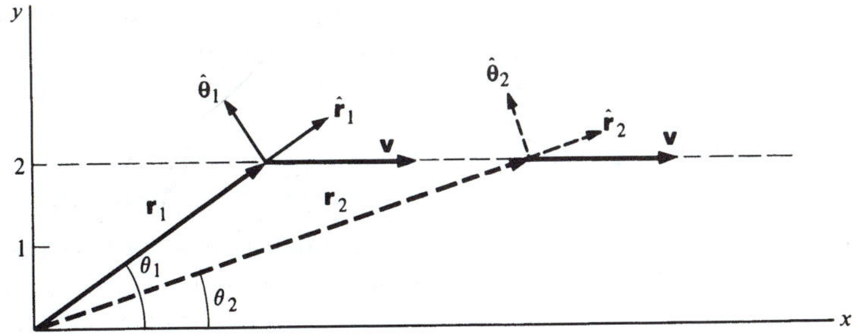

Example 1.14 **Velocity of a Bead on a Spoke**

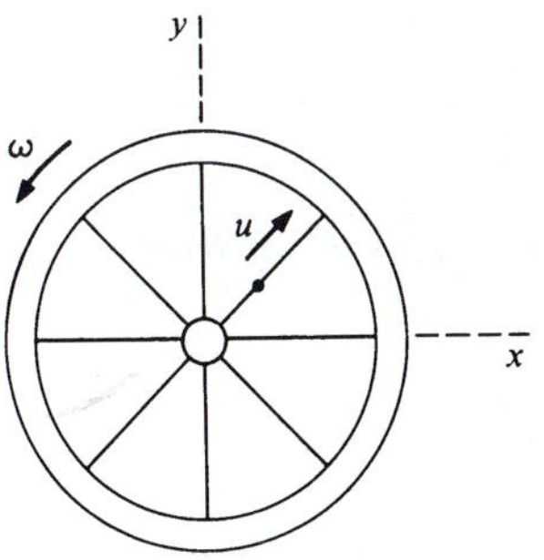

A bead moves along the spoke of a wheel at constant speed u meters per second. The wheel rotates with uniform angular velocity $\dot{\theta} = \omega$ radians per second about an axis fixed in space. At $t = 0$ the spoke is along the x axis, and the bead is at the origin. Find the velocity at time t

a. In polar coordinates

b. In cartesian coordinates.

a. We have $r = ut$, $\dot{r} = u$, $\dot{\theta} = \omega$. Hence

$$\mathbf{v} = \dot{r}\hat{\mathbf{r}} + r\dot{\theta}\hat{\boldsymbol{\theta}} = u\hat{\mathbf{r}} + ut\omega\hat{\boldsymbol{\theta}}.$$

To specify the velocity completely, we need to know the direction of $\hat{\mathbf{r}}$ and $\hat{\boldsymbol{\theta}}$. This is obtained from $\mathbf{r} = (r,\theta) = (ut,\omega t)$.

b. In cartesian coordinates, we have

$$v_x = v_r \cos\theta - v_\theta \sin\theta$$
$$v_y = v_r \sin\theta + v_\theta \cos\theta.$$

Since $v_r = u$, $v_\theta = r\omega = ut\omega$, $\theta = \omega t$, we obtain

$$\mathbf{v} = (u\cos\omega t - ut\omega\sin\omega t)\hat{\mathbf{i}} + (u\sin\omega t + ut\omega\cos\omega t)\hat{\mathbf{j}}.$$

Note how much simpler the result is in plane polar coordinates.

Example 1.15 **Off-center Circle**

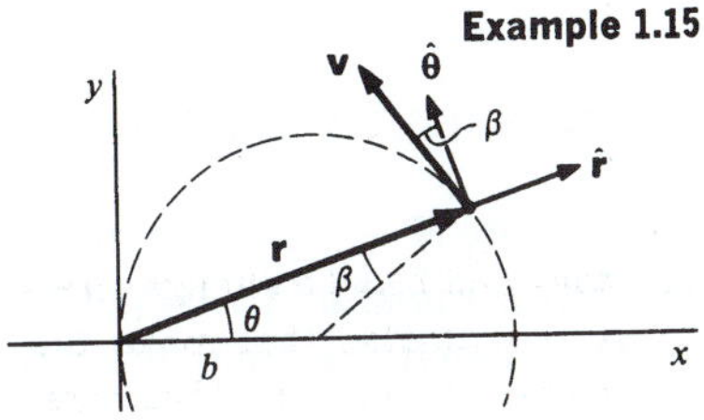

A particle moves with constant speed v around a circle of radius b. Find its velocity vector in polar coordinates using an origin lying on the circle. With this origin, $\mathbf{v}$ is no longer purely tangential, as the sketch indicates.

$$\mathbf{v} = -v\sin\beta\hat{\mathbf{r}} + v\cos\beta\hat{\boldsymbol{\theta}}$$
$$= -v\sin\theta\hat{\mathbf{r}} + v\cos\theta\hat{\boldsymbol{\theta}}.$$

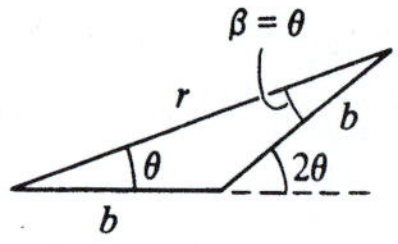

The last step follows since β and θ are the base angles of an isosceles triangle and are therefore equal. To complete the calculation, we must find θ as a function of time. By geometry, $2\theta = \omega t$ or $\theta = \omega t/2$, where $\omega = v/b$.

Acceleration in Polar Coordinates

Our final task is to find the acceleration. We differentiate $\mathbf{v}$ to obtain

$$\begin{aligned}\mathbf{a} &= \frac{d}{dt}\mathbf{v} \\ &= \frac{d}{dt}(\dot{r}\hat{\mathbf{r}} + r\dot{\theta}\hat{\boldsymbol{\theta}}) \\ &= \ddot{r}\hat{\mathbf{r}} + \dot{r}\frac{d}{dt}\hat{\mathbf{r}} + \dot{r}\dot{\theta}\hat{\boldsymbol{\theta}} + r\ddot{\theta}\hat{\boldsymbol{\theta}} + r\dot{\theta}\frac{d}{dt}\hat{\boldsymbol{\theta}}.\end{aligned}$$

If we substitute the results for $d\hat{\mathbf{r}}/dt$ and $d\hat{\boldsymbol{\theta}}/dt$ from page 33, we obtain

$$\begin{aligned}\mathbf{a} &= \ddot{r}\hat{\mathbf{r}} + \dot{r}\dot{\theta}\hat{\boldsymbol{\theta}} + \dot{r}\dot{\theta}\hat{\boldsymbol{\theta}} + r\ddot{\theta}\hat{\boldsymbol{\theta}} - r\dot{\theta}^2\hat{\mathbf{r}} \\ &= (\ddot{r} - r\dot{\theta}^2)\hat{\mathbf{r}} + (r\ddot{\theta} + 2\dot{r}\dot{\theta})\hat{\boldsymbol{\theta}}.\end{aligned}$$

The term $\ddot{r}\hat{\mathbf{r}}$ is a linear acceleration in the radial direction due to change in radial speed. Similarly, $r\ddot{\theta}\hat{\boldsymbol{\theta}}$ is a linear acceleration in the tangential direction due to change in the magnitude of the angular velocity.

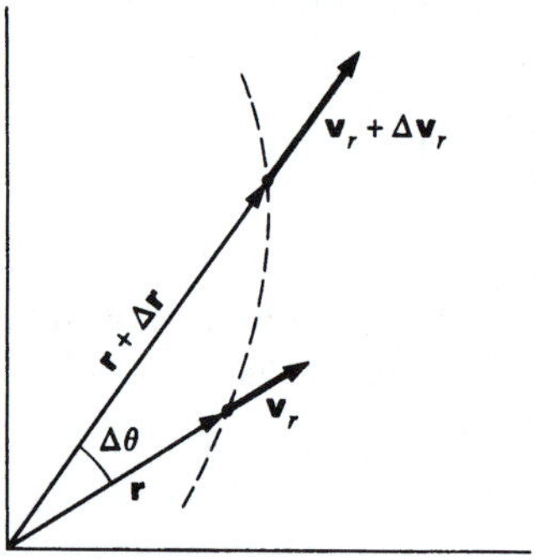

The term $-r\dot{\theta}^2\hat{\mathbf{r}}$ is the centripetal acceleration which we encountered in Example 1.8. Finally, $2\dot{r}\dot{\theta}\hat{\boldsymbol{\theta}}$ is the *Coriolis* acceleration. Perhaps you have heard of the Coriolis force, a fictitious force which appears to act in a rotating coordinate system, and which we shall study in Chap. 8. The Coriolis acceleration that we are discussing here is a real acceleration which is present when r and θ both change with time.

The expression for acceleration in polar coordinates appears complicated. However, by looking at it from the geometric point of view, we can obtain a more intuitive picture.

The instantaneous velocity is

$$\mathbf{v} = \dot{r}\hat{\mathbf{r}} + r\dot{\theta}\hat{\boldsymbol{\theta}} = v_r\hat{\mathbf{r}} + v_\theta\hat{\boldsymbol{\theta}}.$$

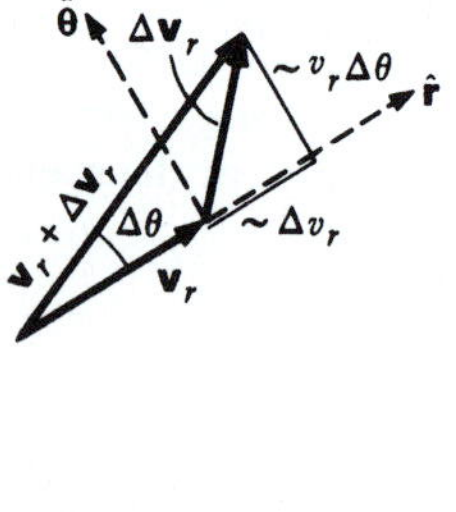

Let us look at the velocity at two different times, treating the radial and tangential terms separately.

The sketch at left shows the radial velocity $\dot{r}\hat{\mathbf{r}} = v_r\hat{\mathbf{r}}$ at two different instants. The change $\Delta\mathbf{v}_r$ has both a radial and a tangential component. As we can see from the sketch (or from the discussion at the end of Sec. 1.8), the radial component of $\Delta \mathbf{v}_r$ is $\Delta v_r \hat{\mathbf{r}}$ and the tangential component is $v_r \, \Delta\theta \hat{\boldsymbol{\theta}}$. The radial component contributes

$$\lim_{\Delta t \to 0} \left(\frac{\Delta v_r}{\Delta t} \hat{\mathbf{r}} \right) = \frac{dv_r}{dt} \hat{\mathbf{r}} = \ddot{r}\hat{\mathbf{r}}$$

to the acceleration. The tangential component contributes

$$\lim_{\Delta t \to 0} \left(v_r \frac{\Delta\theta}{\Delta t} \hat{\boldsymbol{\theta}} \right) = v_r \frac{d\theta}{dt} \hat{\boldsymbol{\theta}} = \dot{r}\dot{\theta}\hat{\boldsymbol{\theta}},$$

which is one-half the Coriolis acceleration. We see that half the Coriolis acceleration arises from the change of direction of the radial velocity.

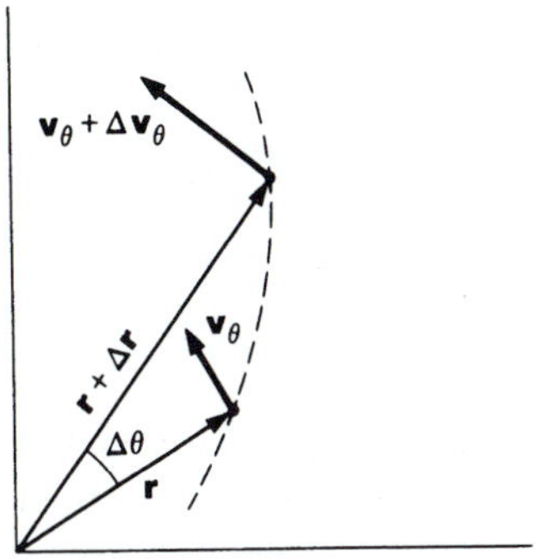

The tangential velocity $r\dot{\theta}\hat{\boldsymbol{\theta}} = v_\theta \hat{\boldsymbol{\theta}}$ can be treated similarly. The change in direction of $\hat{\boldsymbol{\theta}}$ gives $\Delta \mathbf{v}_\theta$ an inward radial component $-v_\theta \, \Delta\theta \hat{\mathbf{r}}$. This contributes

$$\lim_{\Delta t \to 0} \left(-v_\theta \frac{\Delta\theta}{\Delta t} \hat{\mathbf{r}} \right) = -v_\theta \dot{\theta} \hat{\mathbf{r}} = -r\dot{\theta}^2 \hat{\mathbf{r}},$$

which we recognize as the centripetal acceleration. Finally, the tangential component of $\Delta \mathbf{v}_\theta$ is $\Delta v_\theta \hat{\boldsymbol{\theta}}$. Since $v_\theta = r\dot{\theta}$, there are two ways the tangential speed can change. If $\dot{\theta}$ increases by $\Delta\dot{\theta}$, v_θ increases by $r \, \Delta\dot{\theta}$. Second, if r increases by Δr, v_θ increases by $\Delta r\dot{\theta}$. Hence $\Delta v_\theta = r \, \Delta\dot{\theta} + \Delta r \, \dot{\theta}$, and the contribution to the acceleration is

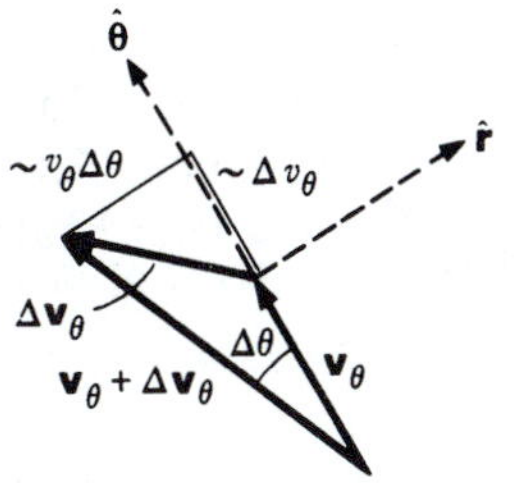

$$\lim_{\Delta t \to 0} \left(\frac{\Delta v_\theta}{\Delta t} \hat{\boldsymbol{\theta}} \right) = \lim_{\Delta t \to 0} \left(r \frac{\Delta\dot{\theta}}{\Delta t} + \frac{\Delta r}{\Delta t} \dot{\theta} \right) \hat{\boldsymbol{\theta}}$$
$$= (r\ddot{\theta} + \dot{r}\dot{\theta})\hat{\boldsymbol{\theta}}.$$

The second term is the remaining half of the Coriolis acceleration; we see that this part arises from the change in tangential speed due to the change in radial distance.

Example 1.16 **Acceleration of a Bead on a Spoke**

A bead moves outward with constant speed u along the spoke of a wheel. It starts from the center at $t = 0$. The angular position of the spoke is given by $\theta = \omega t$, where ω is a constant. Find the velocity and acceleration.

$$\mathbf{v} = \dot{r}\hat{\mathbf{r}} + r\dot{\theta}\hat{\boldsymbol{\theta}}$$

We are given that $\dot{r} = u$ and $\dot{\theta} = \omega$. The radial position is given by $r = ut$, and we have

$$\mathbf{v} = u\hat{\mathbf{r}} + ut\omega\hat{\boldsymbol{\theta}}.$$

The acceleration is

$$\mathbf{a} = (\ddot{r} - r\dot{\theta}^2)\hat{\mathbf{r}} + (r\ddot{\theta} + 2\dot{r}\dot{\theta})\hat{\boldsymbol{\theta}}$$
$$= -ut\omega^2\hat{\mathbf{r}} + 2u\omega\hat{\boldsymbol{\theta}}.$$

The velocity is shown in the sketch for several different positions of the wheel. Note that the radial velocity is constant. The tangential acceleration is also constant—can you visualize this?

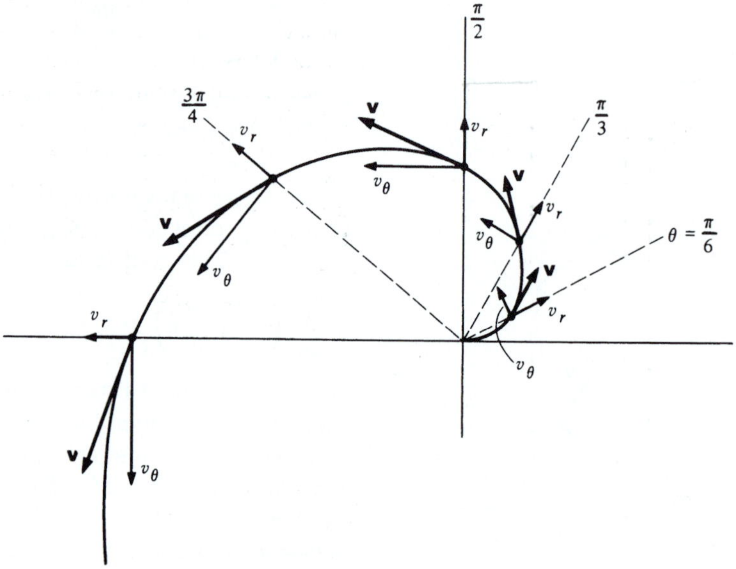

Example 1.17 Radial Motion without Acceleration

A particle moves with $\dot{\theta} = \omega =$ constant and $r = r_0e^{\beta t}$, where r_0 and β are constants. We shall show that for certain values of β, the particle moves with $a_r = 0$.

$$\mathbf{a} = (\ddot{r} - r\dot{\theta}^2)\hat{\mathbf{r}} + (r\ddot{\theta} + 2\dot{r}\dot{\theta})\hat{\boldsymbol{\theta}}$$
$$= (\beta^2r_0e^{\beta t} - r_0e^{\beta t}\omega^2)\hat{\mathbf{r}} + 2\beta r_0\omega e^{\beta t}\hat{\boldsymbol{\theta}}.$$

If $\beta = \pm\omega$, the radial part of **a** vanishes.

It is very surprising at first that when $r = r_0e^{\beta t}$ the particle moves with zero radial acceleration. The error is in thinking that $\ddot{r}$ makes the only contribution to a_r; the term $-r\dot{\theta}^2$ is also part of the radial acceleration, and cannot be neglected.

The paradox is that even though $a_r = 0$, the radial velocity $v_r = \dot{r} = r_0\omega e^{\beta t}$ is increasing rapidly with time. The answer is that we can be misled by the special case of cartesian coordinates; in polar coordinates,

$$v_r \neq \int a_r(t)\, dt,$$

because $\int a_r(t)\, dt$ does not take into account the fact that the unit vectors $\hat{\mathbf{r}}$ and $\hat{\boldsymbol{\theta}}$ are functions of time.

Note 1.1 Mathematical Approximation Methods

Occasionally in the course of solving a problem in physics you may find that you have become so involved with the mathematics that the physics is totally obscured. In such cases, it is worth stepping back for a moment to see if you cannot sidestep the mathematics by using simple approximate expressions instead of exact but complicated formulas. If you have not yet acquired the knack of using approximations, you may feel that there is something essentially wrong with the procedure of substituting inexact results for exact ones. However, this is not really the case, as the following example illustrates.

Suppose that a physicist is studying the free fall of bodies in vacuum, using a tall vertical evacuated tube. The timing apparatus is turned on when the falling body interrupts a thin horizontal ray of light located a distance L below the initial position. By measuring how long the body takes to pass through the light beam, the physicist hopes to determine the local value of g, the acceleration due to gravity. The falling body in the experiment has a height l.

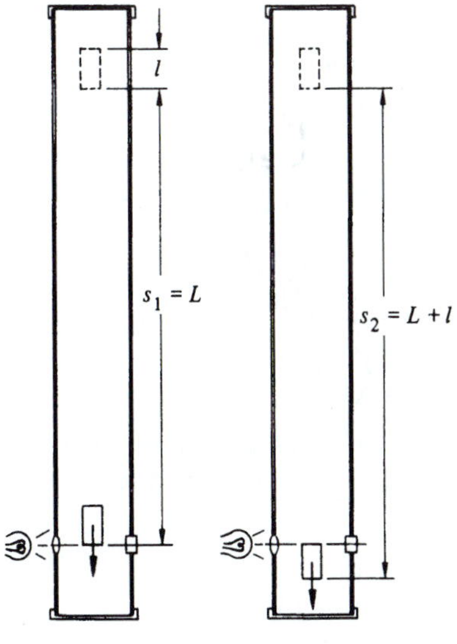

For a freely falling body starting from rest, the distance s traveled in time t is

$$s = \tfrac{1}{2}gt^2,$$

which gives

$$t = \sqrt{\frac{2}{g}}\sqrt{s}.$$

The time interval $t_2 - t_1$ required for the body to fall from $s_1 = L$ centimeters to $s_2 = (L + l)$ centimeters is

$$t_2 - t_1 = \sqrt{\frac{2}{g}}(\sqrt{s_2} - \sqrt{s_1})$$

$$= \sqrt{\frac{2}{g}}(\sqrt{L+l} - \sqrt{L}).$$

If $t_2 - t_1$ is measured experimentally, g is given by

$$g = 2\left(\frac{\sqrt{L+l} - \sqrt{L}}{(t_2 - t_1)}\right)^2$$

This formula is exact under the stated conditions, but it may not be the most useful expression for our purposes.

Consider the factor

$$\sqrt{L+l} - \sqrt{L}.$$

In practice, L will be large compared with l (typical values might be L = 100 cm, l = 1 cm). Our factor is the small difference between two large numbers and is hard to evaluate accurately by using a slide rule or ordinary mathematical tables. Here is a simple approach, known as the method of power series expansion, which enables us to evaluate the factor to any accuracy we please. As we shall discuss formally later in this Note, the quantity $\sqrt{1+x}$ can be written in the series form

$$\sqrt{1+x} = 1 + \tfrac{1}{2}x - \tfrac{1}{8}x^2 + \tfrac{1}{16}x^3 + \cdots$$

for $-1 < x < 1$. Furthermore, if we cut off the series at some point, the error we incur by this approximation is of the order of the first neglected term. We can put the factor in a form suitable for expansion by first extracting $\sqrt{L}$:

$$\sqrt{L+l} - \sqrt{L} = \sqrt{L}\left(\sqrt{1+\frac{l}{L}} - 1\right).$$

The dimensionless ratio l/L plays the part of x in our expansion. Expanding $\sqrt{1 + l/L}$ in the series form gives

$$\begin{aligned}\sqrt{L}\left(\sqrt{1+\frac{l}{L}} - 1\right) &= \sqrt{L}\left[1 + \frac{1}{2}\left(\frac{l}{L}\right) - \frac{1}{8}\left(\frac{l}{L}\right)^2 + \frac{1}{16}\left(\frac{l}{L}\right)^3 + \cdots - 1\right] \\ &= \sqrt{L}\left[\frac{1}{2}\left(\frac{l}{L}\right) - \frac{1}{8}\left(\frac{l}{L}\right)^2 + \frac{1}{16}\left(\frac{l}{L}\right)^3 + \cdots\right].\end{aligned}$$

We see that if l/L is much smaller than 1, the successive terms decrease rapidly. The first term in the bracket, $\frac{1}{2}(l/L)$, is the largest term, and extracting it from the bracket yields

$$\begin{aligned}\sqrt{L+l} - \sqrt{L} &= \sqrt{L}\,\frac{1}{2}\left(\frac{l}{L}\right)\left[1 - \frac{1}{4}\left(\frac{l}{L}\right) + \frac{1}{8}\left(\frac{l}{L}\right)^2 + \cdots\right] \\ &= \frac{l}{2\sqrt{L}}\left[1 - \frac{1}{4}\left(\frac{l}{L}\right) + \frac{1}{8}\left(\frac{l}{L}\right)^2 + \cdots\right].\end{aligned}$$

Our expansion is now in its final and most useful form. The first factor, $l/(2\sqrt{L})$, gives the dominant behavior and is a useful first approximation. Furthermore, writing the series as we have, with leading term 1, shows clearly the contributions of the successive powers of l/L. For example, if $l/L = 0.01$, the term $\frac{1}{8}(l/L)^2 = 1.2 \times 10^{-5}$ and we make a fractional error of about 1 part in 10^5 by retaining only the preceding terms. In many cases this accuracy is more than enough. For instance, if the time interval $t_2 - t_1$ in the falling body experiment can be measured to only 1 part in 1,000, we gain nothing by evaluating $\sqrt{L+l} - \sqrt{L}$ to greater accuracy than this. On the other hand, if we require greater accuracy, we can easily tell how many terms of the series should be retained.

Practicing physicists make mathematical approximations freely (when justified) and have no compunctions about discarding negligible terms. The ability to do this often makes the difference between being stymied by impenetrable algebra and arithmetic and successfully solving a problem.

Furthermore, series approximations often allow us to simplify complicated algebraic expressions to bring out the essential physical behavior.

Here are some helpful methods for making mathematical approximations.

1 THE BINOMIAL SERIES

$$(1 + x)^n = 1 + nx + \frac{n(n-1)}{2!}x^2 + \frac{n(n-1)(n-2)}{3!}x^3 + \cdots + \frac{n(n-1)\cdots(n-k+1)}{k!}x^k + \cdots$$

This series is valid for $-1 < x < 1$, and for any value of n. (If n is an integer, the series terminates, the last term being x^n.) The series is exact; the approximation enters when we truncate it. For $n = \frac{1}{2}$, as in our example,

$$(1 + x)^{\frac{1}{2}} = 1 + \tfrac{1}{2}x - \tfrac{1}{8}x^2 + \tfrac{1}{16}x^3 + \cdots \qquad -1 < x < 1.$$

If we need accuracy only to $O(x^2)$ (order of x^2), we have

$$(1 + x)^{\frac{1}{2}} = 1 + \tfrac{1}{2}x - \tfrac{1}{8}x^2 + O(x^3),$$

where the term $O(x^3)$ indicates that terms of order x^3 and higher are not being considered. As a rule of thumb, the error is approximately the size of the first term dropped.

The series can also be applied if $|x| > 1$ as follows:

$$(1 + x)^n = x^n\left(1 + \frac{1}{x}\right)^n = x^n\left[1 + n\frac{1}{x} + \frac{n(n-1)}{2!}\left(\frac{1}{x}\right)^2 + \cdots\right].$$

Examples:

1. $$\frac{1}{1+x} = (1 + x)^{-1} = 1 - x + x^2 - x^3 + \cdots \qquad -1 < x < 1$$

2. $$\frac{1}{1-x} = (1 - x)^{-1} = 1 + x + x^2 + x^3 + \cdots \qquad -1 < x < 1$$

3. $$(1{,}001)^{\frac{1}{3}} = (1{,}000 + 1)^{\frac{1}{3}} = 1{,}000^{\frac{1}{3}}(1 + 0.001)^{\frac{1}{3}} = 10[1 + 0.001(\tfrac{1}{3}) + \cdots] \approx 10(1.0003) = 10.003$$

4. $2 - \dfrac{1}{\sqrt{1+x}} - \dfrac{1}{\sqrt{1-x}}$: for small x, this expression is zero to first

approximation. However, this approximation may not be adequate. Using the binomial series, we have

$$2 - \frac{1}{\sqrt{1+x}} - \frac{1}{\sqrt{1-x}} = 2 - (1 - \tfrac{1}{2}x + \tfrac{3}{8}x^2 + \cdots) - (1 + \tfrac{1}{2}x + \tfrac{3}{8}x^2 + \cdots)$$

$$= -\tfrac{3}{4}x^2.$$

Notice that the terms linear in x also cancel. To obtain a nonvanishing result we had to go to a high enough order, in this case to order x^2. It is clear that for a correct result we have to expand all terms to the same order.

2 TAYLOR'S SERIES[1]

Analogous to the binomial series, we can try to represent an arbitrary function $f(x)$ by a power series in x:

$$f(x) = a_0 + a_1x + a_2x^2 + \cdots = \sum_{k=0}^{\infty} a_kx^k.$$

For $x = 0$ we must have

$$f(0) = a_0.$$

Assuming for the moment that it is permissible to differentiate, we have

$$\frac{df}{dx} = f'(x) = a_1 + 2a_2x + \cdots$$

Evaluating at $x = 0$ we have

$$a_1 = f'(x)\Big|_{x=0}.$$

Continuing this process, we find

$$a_k = \frac{1}{k!} f^{(k)}(x)\Big|_{x=0},$$

where $f^{(k)}(x)$ is the kth derivative of $f(x)$. For the sake of a less cumbersome notation, we often write $f^{(k)}(0)$ to stand for $f^{(k)}(x)\Big|_{x=0}$; but bear in mind that $f^{(k)}(0)$ means that we should differentiate $f(x)$ k times and *then* set x equal to 0.

The power series for $f(x)$, known as a *Taylor series,* can then be expressed formally as

$$f(x) = f(0) + f'(0)x + f''(0)\frac{x^2}{2!} + f'''(0)\frac{x^3}{3!} + \cdots.$$

This series, if it converges, allows us to find good approximations to $f(x)$ for small values of x (that is, for values of x near zero). Generalizing,

$$f(a + x) = f(a) + f'(a)x + f''(a)\frac{x^2}{2!} + \cdots$$

[1] Taylor's series is discussed in most elementary calculus texts. See the list at the end of the chapter.

gives us the behavior of the function in the neighborhood of the point a. An alternative form for this expression is

$$f(t) = f(a) + f'(a)(t - a) + f''(a)\frac{(t-a)^2}{2!} + \cdots .$$

Our formal manipulations are valid only if the series converges. The range of convergence of a Taylor series may be $-\infty < x < \infty$ for some functions (such as e^x) but quite limited for other functions. (The binomial series converges only if $-1 < x < 1$.) The range of convergence is hard to find without considering functions of a complex variable, and we shall avoid these questions by simply assuming that we are dealing with simple functions for which the range of convergence is either infinite or is readily apparent. Here are some examples:

a. The Trigonometric Functions

Let $f(x) = \sin x$, and expand about $x = 0$.

$$f(0) = \sin(0) = 0$$
$$f'(0) = \cos(0) = 1$$
$$f''(0) = -\sin(0) = 0$$
$$f'''(0) = -\cos(0) = -1, \quad \text{etc.}$$

Hence

$$\sin x = x - \frac{1}{3!}x^3 + \frac{1}{5!}x^5 - \frac{1}{7!}x^7 + \cdots .$$

Similarly

$$\cos x = 1 - \frac{1}{2!}x^2 + \frac{1}{4!}x^4 - \cdots .$$

These expansions converge for all values of x but are particularly useful for small values of x. To $O(x^2)$, $\sin x = x$, $\cos x = 1 - x^2/2$.

The figure below compares the exact value for $\sin x$ with a Taylor series in which successively higher terms are included. Note how each

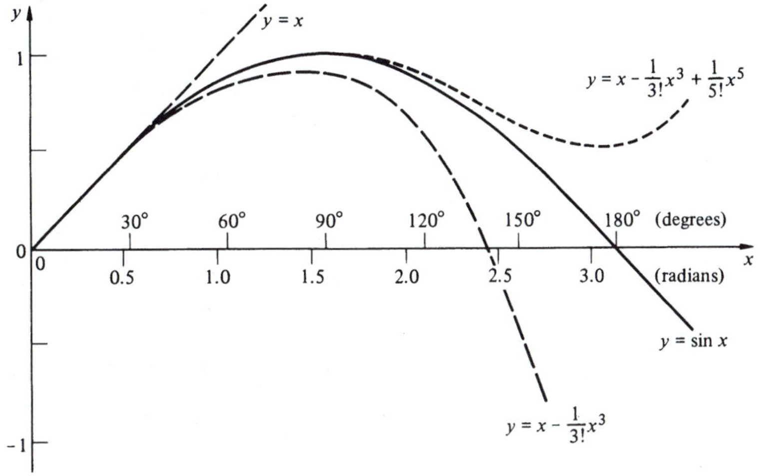

term increases the range over which the series is accurate. If an infinite number of terms are included, the Taylor series represents the function accurately everywhere.

b. The Binomial Series

We can derive the binomial series introduced in the last section by letting

$$f(x) = (1 + x)^n.$$

Then

$$\begin{aligned} f(0) &= 1 \\ f'(0) &= n(1 + 0)^n = n \\ f''(0) &= n(n - 1) \\ f^{(k)}(0) &= n(n - 1)(n - 2) \cdots (n - k + 1) \end{aligned}$$

$$(1 + x)^n = 1 + nx + \frac{1}{2!}n(n - 1)x^2 + \cdots + \cdots \frac{n(n - 1) \cdots (n - k + 1)}{k!}x^k + \cdots$$

c. The Exponential Function

If we let $f(x) = e^x$, we have $f'(x) = f(x)$, by the definition of the exponential function. Similarly $f^{(k)}(x) = f(x)$. Since $f(0) = e^0 = 1$, we have

$$e^x = 1 + x + \frac{1}{2!}x^2 + \frac{1}{3!}x^3 + \cdots.$$

This series converges for all values of x.

A useful result from the theory of the Taylor series is that if the series converges at all, it represents the function so well that we are allowed to differentiate or integrate the series any number of times. For example,

$$\begin{aligned} \frac{d}{dx}(\sin x) &= \frac{d}{dx}\left(x - \frac{1}{3!}x^3 + \frac{1}{5!}x^5 + \cdots\right) \\ &= 1 - \frac{1}{2!}x^2 + \frac{1}{4!}x^4 + \cdots \\ &= \cos x. \end{aligned}$$

Furthermore, the Taylor series for the product of two functions is the product of the individual series:

$$\begin{aligned} \sin x \cos x &= \left(x - \frac{1}{3!}x^3 + \frac{1}{5!}x^5 + \cdots\right)\left(1 - \frac{1}{2!}x^2 + \frac{1}{4!}x^4 + \cdots\right) \\ &= x - \left(\frac{1}{3!} + \frac{1}{2!}\right)x^3 + \left(\frac{1}{4!} + \frac{1}{3!2!} + \frac{1}{5!}\right)x^5 + \cdots \end{aligned}$$

$$= x - \frac{4x^3}{3!} + \frac{16x^5}{5!} + \cdots$$

$$= \frac{1}{2}\left[(2x) - \frac{(2x)^3}{3!} + \frac{(2x)^5}{5!} + \cdots\right]$$

$$= \frac{1}{2}[\sin(2x)].$$

The Taylor series sometimes comes in handy in the evaluation of integrals. To estimate

$$\int_1^{1.1} \frac{e^z}{z}\, dz,$$

let $z = 1 + x$. We then have

$$\int_1^{1.1} \frac{e^z}{z}\, dz = \int_0^{0.1} \frac{e^{(1+x)}}{1+x}\, dx$$

$$= (e)\int_0^{0.1} \frac{e^x}{1+x}\, dx$$

$$\approx (e)\int_0^{0.1} \frac{(1+x)}{(1+x)}\, dx$$

$$\approx 0.1e.$$

The approximation should be better than 1 part in 100 or so, for x always lies in the interval $0 \le x \le 0.1$. In this range, $e^x \approx 1 + x$ is a good approximation to two or three significant figures.

3 DIFFERENTIALS

Consider $f(x)$, a function of the independent variable x. Often we need to have a simple approximation for the change in $f(x)$ when x is changed to $x + \Delta x$. Let us denote the change by $\Delta f = f(x + \Delta x) - f(x)$. It is natural to turn to the Taylor series. Expanding the Taylor series for $f(x)$ about the point x gives

$$f(x + \Delta x) = f(x) + f'(x)\,\Delta x + \frac{1}{2!}f''(x)\,\Delta x^2 + \cdots,$$

where, for example, $f'(x)$ stands for df/dx evaluated at the point x. Omitting terms of order $(\Delta x)^2$ and higher yields the simple linear approximation

$$\Delta f = f(x + \Delta x) - f(x) \approx f'(x)\,\Delta x.$$

This approximation becomes increasingly accurate the smaller the size of Δx. However, for finite values of Δx, the expression

$$\Delta f \approx f'(x)\,\Delta x$$

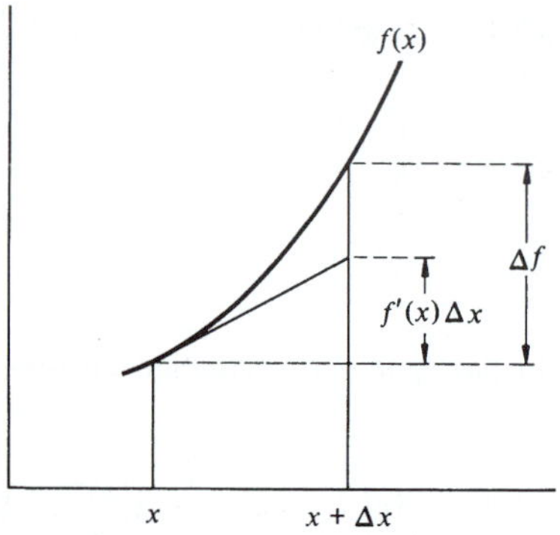

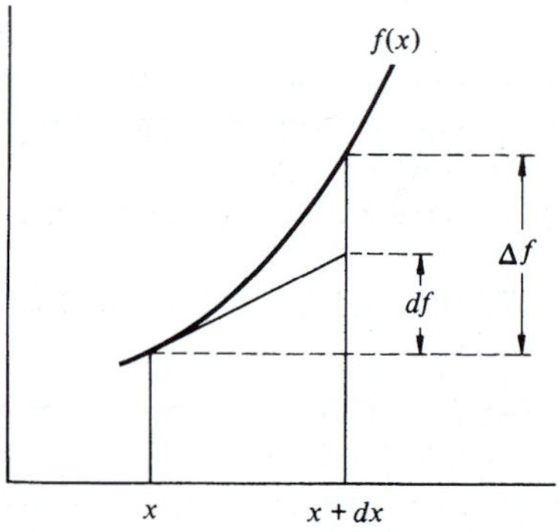

has to be considered to be an approximation. The graph at left shows a comparison of $\Delta f \equiv f(x + \Delta x) - f(x)$ with the linear extrapolation $f'(x)\,\Delta x$. It is apparent that Δf, the actual change in $f(x)$ as x is changed, is generally not exactly equal to Δf for finite Δx.

As a matter of notation, we use the symbol dx to stand for Δx, the increment in x. dx is known as the *differential* of x; it can be as large or small as we please. We *define* df, the differential of f, by

$$df \equiv f'(x)\,dx.$$

This notation is illustrated in the lower drawing. Note that dx and Δx are used interchangeably. On the other hand, df and Δf are different quantities. df is a differential defined by $df = f'(x)\,dx$, whereas Δf is the actual change $f(x + dx) - f(x)$. Nevertheless, when the linear approximation is justified in a problem, we often use df to represent Δf. We can always do this when eventually a limit will be taken. Here are some examples.

1. $d(\sin\theta) = \cos\theta\,d\theta$.
2. $d(xe^{x^2}) = (e^{x^2} + 2x^2e^{x^2})\,dx$.
3. Let V be the volume of a sphere of radius r:

$$V = \tfrac{4}{3}\pi r^3$$
$$dV = 4\pi r^2\,dr.$$

4. What is the fractional increase in the volume of the earth if its average radius, 6.4×10^6 m, increases by 1 m?

$$\frac{dV}{V} = \frac{4\pi r^2\,dr}{\tfrac{4}{3}\pi r^3} = 3\frac{dr}{r} = \frac{3}{6.4 \times 10^6} = 4.7 \times 10^{-7}.$$

One common use of differentials is in changing the variable of integration. For instance, consider the integral

$$\int_a^b xe^{x^2}\,dx.$$

A useful substitution is $t = x^2$. The procedure is first to solve for x in terms of t,

$$x = \sqrt{t},$$

and then to take differentials:

$$dx = \frac{1}{2}\frac{1}{\sqrt{t}}\,dt.$$

This result is exact, since we are effectively taking the limit. The original integral can now be written in terms of t:

$$\int_a^b xe^{x^2}\,dx = \int_{t_1}^{t_2} \sqrt{t}\, e^t \left(\frac{1}{2}\frac{1}{\sqrt{t}}\,dt\right) = \tfrac{1}{2}\int_{t_1}^{t_2} e^t\,dt$$

$$= \tfrac{1}{2}(e^{t_2} - e^{t_1}),$$

where $t_1 = a^2$ and $t_2 = b^2$.

Some References to Calculus Texts

A very popular textbook is G. B. Thomas, Jr., "Calculus and Analytic Geometry," 4th ed., Addison-Wesley Publishing Company, Inc., Reading, Mass.

The following introductory texts in calculus are also widely used:
M. H. Protter and C. B. Morrey, "Calculus with Analytic Geometry," Addison-Wesley Publishing Company, Inc., Reading, Mass.
A. E. Taylor, "Calculus with Analytic Geometry," Prentice-Hall, Inc., Englewood Cliffs, N.J.
R. E. Johnson and E. L. Keokemeister, "Calculus With Analytic Geometry," Allyn and Bacon, Inc., Boston.

A highly regarded advanced calculus text is R. Courant, "Differential and Integral Calculus," Interscience Publishing, Inc., New York.

If you need to review calculus, you may find the following helpful: Daniel Kleppner and Norman Ramsey, "Quick Calculus," John Wiley & Sons, Inc., New York.

Problems

1.1 Given two vectors, $\mathbf{A} = (2\hat{\mathbf{i}} - 3\hat{\mathbf{j}} + 7\hat{\mathbf{k}})$ and $\mathbf{B} = (5\hat{\mathbf{i}} + \hat{\mathbf{j}} + 2\hat{\mathbf{k}})$, find: (*a*) $\mathbf{A} + \mathbf{B}$; (*b*) $\mathbf{A} - \mathbf{B}$; (*c*) $\mathbf{A} \cdot \mathbf{B}$; (*d*) $\mathbf{A} \times \mathbf{B}$.

Ans. (*a*) $7\hat{\mathbf{i}} - 2\hat{\mathbf{j}} + 9\hat{\mathbf{k}}$; (*c*) 21

1.2 Find the cosine of the angle between

$\mathbf{A} = (3\hat{\mathbf{i}} + \hat{\mathbf{j}} + \hat{\mathbf{k}})$ and $\mathbf{B} = (-2\hat{\mathbf{i}} - 3\hat{\mathbf{j}} - \hat{\mathbf{k}})$.

Ans. -0.805

1.3 The direction cosines of a vector are the cosines of the angles it makes with the coordinate axes. The cosine of the angles between the vector and the x, y, and z axes are usually called, in turn α, β, and γ. Prove that $\alpha^2 + \beta^2 + \gamma^2 = 1$, using either geometry or vector algebra.

1.4 Show that if $|\mathbf{A} - \mathbf{B}| = |\mathbf{A} + \mathbf{B}|$, then $\mathbf{A}$ is perpendicular to $\mathbf{B}$.

1.5 Prove that the diagonals of an equilateral parallelogram are perpendicular.

1.6 Prove the law of sines using the cross product. It should only take a couple of lines. (*Hint*: Consider the area of a triangle formed by $\mathbf{A}$, $\mathbf{B}$, $\mathbf{C}$, where $\mathbf{A} + \mathbf{B} + \mathbf{C} = 0$.)

1.7 Let $\hat{\mathbf{a}}$ and $\hat{\mathbf{b}}$ be unit vectors in the xy plane making angles θ and ϕ with the x axis, respectively. Show that $\hat{\mathbf{a}} = \cos\theta\hat{\mathbf{i}} + \sin\theta\hat{\mathbf{j}}$, $\hat{\mathbf{b}} = \cos\phi\hat{\mathbf{i}} + \sin\phi\hat{\mathbf{j}}$, and using vector algebra prove that

$\cos(\theta - \phi) = \cos\theta\cos\phi + \sin\theta\sin\phi$.

1.8 Find a unit vector perpendicular to

$\mathbf{A} = (\hat{\mathbf{i}} + \hat{\mathbf{j}} - \hat{\mathbf{k}})$ and $\mathbf{B} = (2\hat{\mathbf{i}} - \hat{\mathbf{j}} + 3\hat{\mathbf{k}})$.

Ans. $\hat{\mathbf{n}} = \pm(2\hat{\mathbf{i}} - 5\hat{\mathbf{j}} - 3\hat{\mathbf{k}})/\sqrt{38}$

1.9 Show that the volume of a parallelepiped with edges **A**, **B**, and **C** is given by $\mathbf{A} \cdot (\mathbf{B} \times \mathbf{C})$.

1.10 Consider two points located at $\mathbf{r}_1$ and $\mathbf{r}_2$, separated by distance $r = |\mathbf{r}_1 - \mathbf{r}_2|$. Find a vector **A** from the origin to a point on the line between $\mathbf{r}_1$ and $\mathbf{r}_2$ at distance xr from the point at $\mathbf{r}_1$, where x is some number.

1.11 Let **A** be an arbitrary vector and let $\hat{\mathbf{n}}$ be a unit vector in some fixed direction. Show that $\mathbf{A} = (\mathbf{A} \cdot \hat{\mathbf{n}})\hat{\mathbf{n}} + (\hat{\mathbf{n}} \times \mathbf{A}) \times \hat{\mathbf{n}}$.

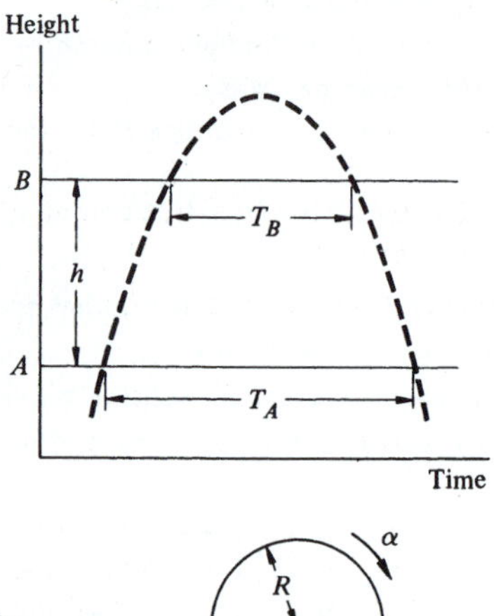

1.12 The acceleration of gravity can be measured by projecting a body upward and measuring the time that it takes to pass two given points in both directions.

Show that if the time the body takes to pass a horizontal line A in both directions is T_A, and the time to go by a second line B in both directions is T_B, then, assuming that the acceleration is constant, its magnitude is

$$g = \frac{8h}{T_A^2 - T_B^2},$$

where h is the height of line B above line A.

1.13 At $t = 0$, an elevator departs from the ground with uniform speed. At time T_1 a boy drops a marble through the floor. The marble falls with uniform acceleration $g = 9.8\ \text{m/s}^2$, and hits the ground T_2 seconds later. Find the height of the elevator at time T_1.

Ans. clue. If $T_1 = T_2 = 4$ s, $h = 39.2$ m

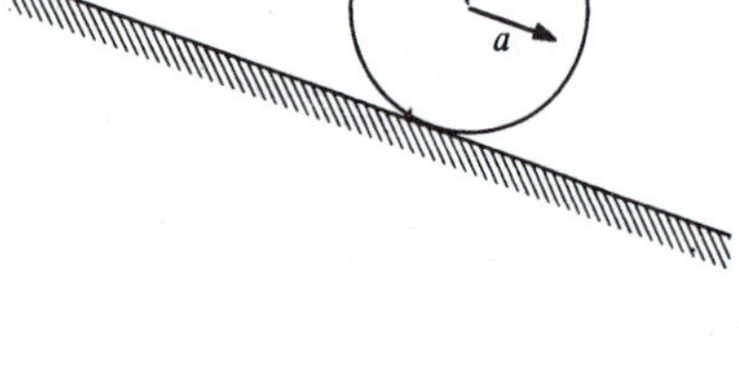

1.14 A drum of radius R rolls down a slope without slipping. Its axis has acceleration a parallel to the slope. What is the drum's angular acceleration α?

1.15 By *relative velocity* we mean velocity with respect to a specified coordinate system. (The term velocity, alone, is understood to be relative to the observer's coordinate system.)

a. A point is observed to have velocity $\mathbf{v}_A$ relative to coordinate system A. What is its velocity relative to coordinate system B, which is displaced from system A by distance **R**? (**R** can change in time.)

Ans. $\mathbf{v}_B = \mathbf{v}_A - d\mathbf{R}/dt$

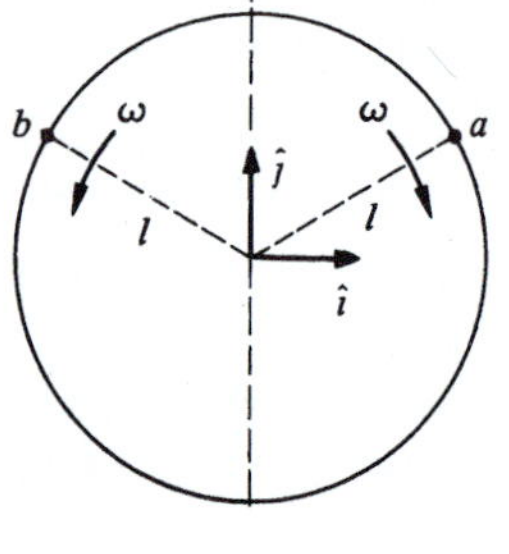

b. Particles a and b move in opposite directions around a circle with angular speed ω, as shown. At $t = 0$ they are both at the point $\mathbf{r} = l\hat{\mathbf{j}}$, where l is the radius of the circle.

Find the velocity of a relative to b.

1.16 A sportscar, Fiasco I, can accelerate uniformly to 120 mi/h in 30 s. Its *maximum* braking rate cannot exceed $0.7g$. What is the minimum time required to go $\frac{1}{2}$ mi, assuming it begins and ends at rest? (*Hint:* A graph of velocity vs. time can be helpful.)

1.17 A particle moves in a plane with constant radial velocity $\dot{r} = 4$ m/s. The angular velocity is constant and has magnitude $\dot{\theta} = 2$ rad/s. When the particle is 3 m from the origin, find the magnitude of (*a*) the velocity and (*b*) the acceleration.

Ans. (*a*) $v = \sqrt{52}$ m/s

1.18 The rate of change of acceleration is sometimes known as "jerk." Find the direction and magnitude of jerk for a particle moving in a circle of radius R at angular velocity ω. Draw a vector diagram showing the instantaneous position, velocity, acceleration, and jerk.

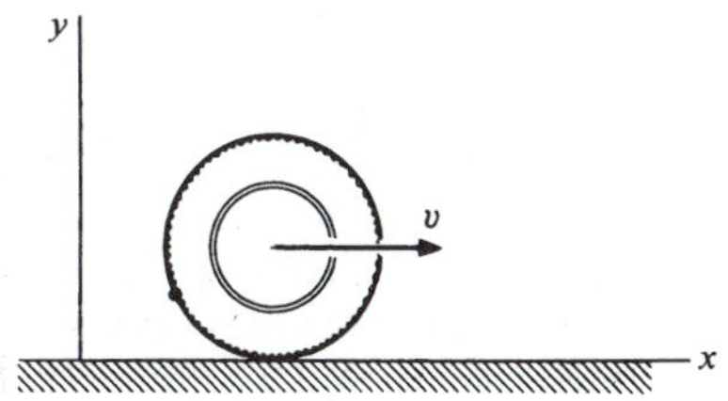

1.19 A tire of radius R rolls in a straight line without slipping. Its center moves with constant speed V. A small pebble lodged in the tread of the tire touches the road at $t = 0$. Find the pebble's position, velocity and acceleration as functions of time.

1.20 A particle moves outward along a spiral. Its trajectory is given by $r = A\theta$, where A is a constant. $A = (1/\pi)$ m/rad. θ increases in time according to $\theta = \alpha t^2/2$, where α is a constant.

a. Sketch the motion, and indicate the approximate velocity and acceleration at a few points.

b. Show that the radial acceleration is zero when $\theta = 1/\sqrt{2}$ rad.

c. At what angles do the radial and tangential accelerations have equal magnitude?

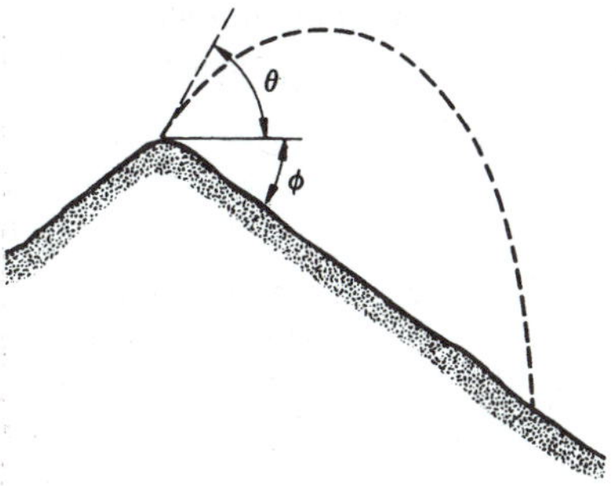

1.21 A boy stands at the peak of a hill which slopes downward uniformly at angle ϕ. At what angle θ from the horizontal should he throw a rock so that it has the greatest range?

Ans. clue. If $\phi = 60°$, $\theta = 15°$

2 NEWTON'S LAWS–THE FOUNDATIONS OF NEWTONIAN MECHANICS

2.1 Introduction

Our aim in this chapter is to understand Newton's laws of motion. From one point of view this is a modest task: Newton's laws are simple to state and involve little mathematical complexity. Their simplicity is deceptive, however. As we shall see, they combine definitions, observations from nature, partly intuitive concepts, and some unexamined assumptions on the properties of space and time. Newton's statement of the laws of motion left many of these points unclear. It was not until two hundred years after Newton that the foundations of classical mechanics were carefully examined, principally by Ernst Mach,[1] and our treatment is very much in the spirit of Mach.

Newton's laws of motion are by no means self-evident. In Aristotle's system of mechanics, a force was thought to be needed to maintain a body in uniform motion. Aristotelian mechanics was accepted for thousands of years because, superficially, it seemed intuitively correct. Careful reasoning from observation and a real effort of thought was needed to break out of the aristotelian mold. Most of us are still not accustomed to thinking in newtonian terms, and it takes both effort and practice to learn to analyze situations from the newtonian point of view. We shall spend a good deal of time in this chapter looking at applications of Newton's laws, for only in this way can we really come to understand them. However, in addition to deepening our understanding of dynamics, there is an immediate reward—we shall be able to analyze quantitatively physical phenomena which at first sight may seem incomprehensible.

Although Newton's laws provide a direct introduction to classical mechanics, it should be pointed out that there are a number of other approaches. Among these are the formulations of Lagrange and Hamilton, which take energy rather than force as the fundamental concept. However, these methods are physically equivalent to the newtonian approach, and even though we could use one of them as our point of departure, a deep understanding of Newton's laws is an invaluable asset to understanding any systematic treatment of mechanics.

A word about the validity of newtonian mechanics: possibly you already know something about modern physics—the development early in this century of relativity and quantum mechanics. If so,

[1] Mach's text, "The Science of Mechanics" (1883), translated the arguments from Newton's "Principia" into a more logically satisfying form. His analysis of the assumptions of newtonian mechanics played a major role in the development of Einstein's special theory of relativity, as we shall see in Chap. 10.

you know that there are important areas of physics in which newtonian mechanics fails, while relativity and quantum mechanics succeed. Briefly, newtonian mechanics breaks down for systems moving with a speed comparable to the speed of light, 3×10^8 m/s, and it also fails for systems of atomic dimensions or smaller where quantum effects are significant. The failure arises because of inadequacies in classical concepts of space, time, and the nature of measurement. A natural impulse might be to throw out classical physics and proceed directly to modern physics. We do not accept this point of view for several reasons. In the first place, although the more advanced theories have shown us where classical physics breaks down, they also show us where the simpler methods of classical physics give accurate results. Rather than make a blanket statement that classical physics is right or wrong, we recognize that newtonian mechanics is exceptionally useful in many areas of physics but of limited applicability in other areas. For instance, newtonian physics enables us to predict eclipses centuries in advance, but is useless for predicting the motions of electrons in atoms. It should also be recognized that because classical physics explains so many everyday phenomena, it is an essential tool for all practicing scientists and engineers. Furthermore, most of the important concepts of classical physics are preserved in modern physics, albeit in altered form.

2.2 Newton's Laws

It is important to understand which parts of Newton's laws are based on experiment and which parts are matters of definition. In discussing the laws we must also learn how to apply them, not only because this is the bread and butter of physics but also because this is essential for a real understanding of the underlying concepts.

We start by appealing directly to experiment. Unfortunately, experiments in mechanics are among the hardest in physics because motion in our everyday surroundings is complicated by forces such as gravity and friction. To see the physical essentials, we would like to eliminate all disturbances and examine very simple systems. One way to accomplish this would be to enroll as astronauts, for in the environment of space most of the everyday disturbances are negligible. However, lacking the resources to put ourselves in orbit, we settle for second best, a device known as a *linear air track,* which approximates ideal conditions, but only in one dimension. (Although it is not clear that we can learn anything about three dimensional motion from studying motion in one dimension, happily this turns out to be the case.)

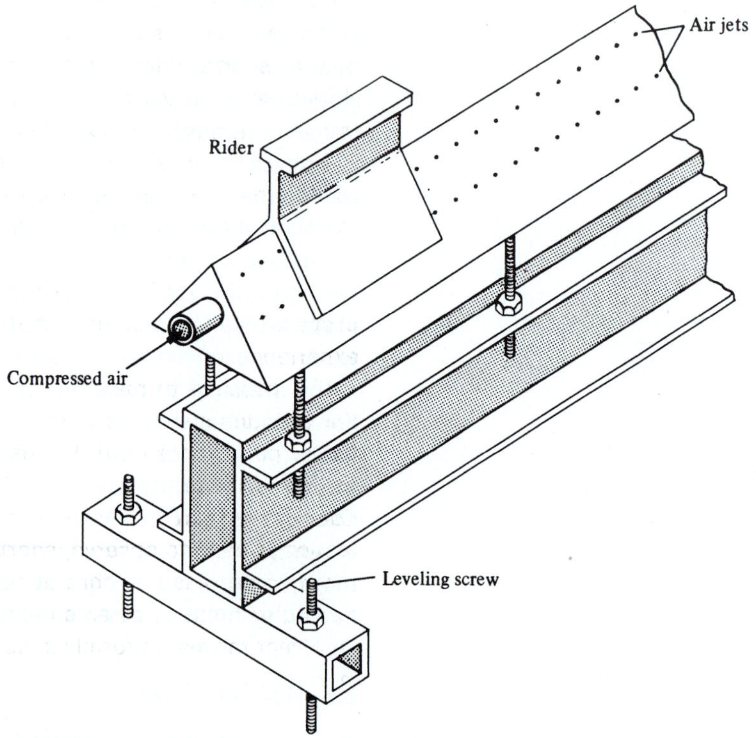

Linear air track

The linear air track is a hollow triangular beam perhaps 2 m long, pierced by many small holes which emit gentle streams of air. A rider rests on the beam, and when the air is turned on, the rider floats on a thin cushion of air. Because of the air suspension, the rider moves with negligible friction. (The reason for this is that the thin film of air has a viscosity typically 5,000 times less than a film of oil.) If the track is leveled carefully, and if we eliminate stray air currents, the rider behaves as if it were isolated in its motion along the track. The rider moves along the track free of gravity, friction, or any other detectable influences.

Now let's observe how the rider behaves. (Try these experiments yourself if possible.) Suppose that we place the rider on the track and carefully release it from rest. As we might expect, the rider stays at rest, at least until a draft hits it or somebody bumps the apparatus. (This isn't too surprising, since we leveled the track until the rider stayed put when left at rest.) Next, we give the rider a slight shove and then let it move freely. The motion seems uncanny, for the rider continues to move along slowly and evenly, neither gaining nor losing speed. This is contrary to our everyday experience that moving bodies stop moving unless we push them. The reason is that in everyday motion, friction usually plays an important role. For instance, the air track rider comes to a grinding halt if we turn off the air and let sliding friction act. Apparently the friction stops the motion. But we are getting ahead of ourselves; let us return to the properly functioning air track and try to generalize from our experience.

It is possible to make a two dimensional air table analogous to the one dimensional air track. (A smooth sheet of glass with a flat piece of dry ice on it does pretty well. The evaporating dry ice provides the gas cushion.) We find again that the undisturbed rider moves with uniform velocity. Three dimensional isolated motion is hard to observe, short of going into space, but let us for the moment assume that our experience in one and two dimensions also holds in three dimensions. We therefore surmise that an object moves uniformly in space provided there are no external influences.

Newton's First Law

In our discussion of the air track experiments, we glossed over an important point. Motion has meaning only with respect to a particular coordinate system, and in describing motion it is essential to specify the coordinate system we are using. For example, in describing motion along the air track, we implicitly used a coordinate system fixed to the track. However, we are free to choose any coordinate system we please, including systems which are moving with respect to the track. In a coordinate system moving uniformly with respect to the track, the undisturbed rider moves with constant velocity. Such a coordinate system is called an *inertial system*. Not all coordinate systems are inertial; in a coordinate system accelerating with respect to the track, the undisturbed rider does not have constant velocity. However, it is always possible to find a coordinate system with respect to which isolated bodies move uniformly. This is the essence of Newton's first law of motion.

Newton's first law of motion is the assertion that inertial systems exist.

Newton's first law is part definition and part experimental fact. Isolated bodies move uniformly in inertial systems by virtue of the definition of an inertial system. In constrast, that inertial systems exist is a statement about the physical world.

Newton's first law raises a number of questions, such as what we mean by an "isolated body," but we will defer these temporarily and go on.

Newton's Second Law

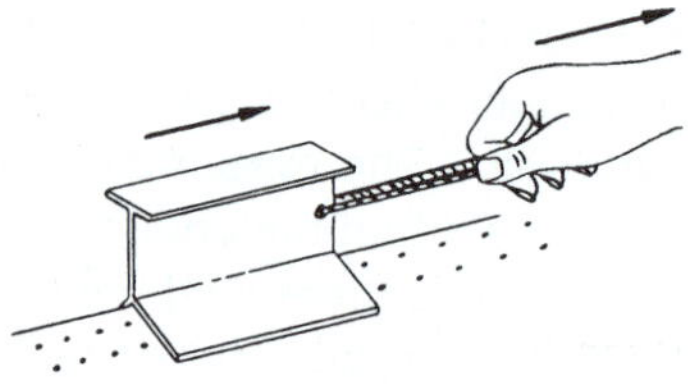

We now turn to how the rider on the air track behaves when it is no longer isolated. Suppose that we pull the rider with a rubber band. Nothing happens while the rubber band is loose, but as soon as we pull hard enough to stretch the rubber band, the rider starts to move. If we move our hand ahead of the rider so that the rubber band is always stretched to the same standard length, we find that the rider moves in a wonderfully simple way; its velocity increases uniformly with time. The rider moves with constant acceleration.

Now suppose that we try the same experiment with a different rider, perhaps one a good deal larger than the first. Again, the same rubber band stretched to the standard length produces a constant acceleration, but the acceleration is different from that in the first case. Apparently the acceleration depends not only on what we do to the object, since presumably we do the same thing in each case, but also on some property of the object, which we call *mass*.

We can use our rubber band experiment to *define* what we mean by mass. We start by arbitrarily saying that the first body has a mass m_1. (m_1 could be one unit of mass or x units of mass, where x is any number we choose.) We then *define* the mass of the second body to be

$$m_2 = m_1 \frac{a_1}{a_2},$$

where a_1 is the acceleration of the first body in our rubber band experiment and a_2 is the acceleration of the second body.

Continuing this procedure, we can assign masses to other objects by measuring their accelerations with the standard stretched rubber band. Thus

$$m_3 = m_1 \frac{a_1}{a_3}$$

$$m_4 = m_1 \frac{a_1}{a_4} \qquad \text{etc.}$$

Although this procedure is straightforward, there is no obvious reason why the quantity we define this way is particularly important. For instance, why not consider instead some other property, call it property Z, such that $Z_2 = Z_1(a_1/a_2)^2$? The reason is that mass is useful, whereas property Z (or most other quantities you try) is not. By making further experiments with the air track, for instance by using springs or magnets instead of a rubber band, we find that the ratios of accelerations, hence the mass ratios, are the same no matter how we produce the uniform accelerations, provided that we do the same thing to each body. Thus, mass so defined turns out to be independent of the source of acceleration and appears to be an inherent property of a body. Of course, the actual mass value of an individual body depends on our choice of mass unit. The important thing is that two bodies have a unique mass ratio.

Our definition of mass is an example of an *operational* definition. By operational we mean that the definition is dominantly in terms of experiments we perform and not in terms of abstract concepts, such as "mass is a measure of the resistance of bodies to a change in motion." Of course, there can be many abstract concepts hidden in apparently simple operations. For instance, when we measure acceleration, we tacitly assume that we have a clear understanding of distance and time. Although our intuitive ideas are adequate for our purposes here, we shall see when we discuss relativity that the behavior of measuring rods and clocks is itself a matter for experiment.

A second troublesome aspect of operational definitions is that they are limited to situations in which the operations can actually be performed. In practice this is usually not a problem; physics proceeds by constructing a chain of theory and experiment which allows us to employ convenient methods of measurement ultimately based on the operational definitions. For instance, the most practical way to measure the mass of a mountain is to observe its gravitational pull on a test body, such as a hanging plumb bob. According to the operational definition, we should apply a standard force and measure the mountain's acceleration. Nevertheless, the two methods are directly related conceptually.

We defined mass by experiments on laboratory obiects; we cannot say a priori whether the results are consistent on a much larger or smaller scale. In fact, one of the major goals of physics is to find the limitations of such definitions, for the limitations normally reveal new physical laws. Nevertheless, if an operational definition is to be at all useful, it must have very wide applicability. For instance, our definition of mass holds not only for everyday objects on the earth but also, to a very high degree, for planetary motion, motion on an enormously larger scale. It should not surprise us, however, if eventually we find situations in which the operations are no longer useful.

Now that we have defined mass, let us turn our attention to *force*.

We describe the operation of acting on the test mass with a stretched rubber band as "applying" a force. (Note that we have sidestepped the question of what a force is and have limited ourselves to describing how to produce it—namely, by stretching a rubber band by a given amount.) When we apply the force, the test mass accelerates at some rate, a. If we apply two standard stretched rubber bands, side by side, we find that the mass accelerates at the rate $2a$, and if we apply them in opposite directions, the acceleration is zero. The effects of the rubber bands add algebraically for the case of motion in a straight line.

We can establish a force scale by defining the unit force as the force which produces unit acceleration when applied to the unit mass. It follows from our experiments that F units of force accelerate the unit mass by F units of acceleration and, from our definition of mass, it will produce $F \times (1/m)$ units of acceleration in mass m. Hence, the acceleration produced by force F acting on mass m is $a = F/m$ or, in a more familiar order, $F = ma$. In the International System of units (SI), the unit of force is the *newton* (N), the unit of mass is the *kilogram* (kg), and acceleration is in meters per second^2 (m/s^2). Units are discussed further in Sec. 2.3.

So far we have limited our experiments to one dimension. Since acceleration is a vector, and mass, as far as we know, is a scalar, we expect that force is also a vector. It is natural to think of the force as pointing in the direction of the acceleration it produces when acting alone. This assumption appears trivial, but it is not—its justification lies in experiment. We find that forces obey the *principle of superposition:* The acceleration produced by several forces acting on a body is equal to the vector sum of the accelerations produced by each of the forces acting separately. Not only does this confirm the vector nature of force, but it also enables us to analyze problems by considering one force at a time.

Combining all these observations, we conclude that the total force $\mathbf{F}$ on a body of mass m is $\mathbf{F} = \Sigma\mathbf{F}_i$, where $\mathbf{F}_i$ is the ith applied force. If $\mathbf{a}$ is the net acceleration, and $\mathbf{a}_i$ the acceleration due to $\mathbf{F}_i$ alone, then we have

$$\begin{aligned}\mathbf{F} &= \Sigma\mathbf{F}_i \\ &= \Sigma m\mathbf{a}_i \\ &= m\Sigma\mathbf{a}_i \\ &= m\mathbf{a}\end{aligned}$$

or

$$\mathbf{F} = m\mathbf{a}.$$

This is Newton's second law of motion. It will underlie much of our subsequent discussion.

It is important to understand clearly that force is not merely a matter of definition. For instance, if the air track rider starts accelerating, it is not sufficient to claim that there is a force acting defined by $\mathbf{F} = m\mathbf{a}$. Forces always arise from *interactions* between systems, and if we ever found an acceleration without an interaction, we would be in a terrible mess. It is the interaction which is physically significant and which is responsible for the force. For this reason, when we isolate a body sufficiently from its surroundings, we expect the body to move uniformly in an inertial system. Isolation means eliminating interactions. You may question whether it is always possible to isolate a body. Fortunately, as far as we know, the answer is yes. All known interactions decrease with distance. (The forces which extend over the greatest distance are the familiar gravitational and Coulomb forces. They decrease as $1/r^2$, where r is the distance. Most forces decrease much more rapidly. For example, the force between separated atoms decreases as $1/r^7$.) By moving the test body sufficiently far from everything else, the interactions can be reduced as much as desired.

Newton's Third Law

The fact that force is necessarily the result of an interaction between two systems is made explicit by Newton's third law. The third law states that forces always appear in pairs: if body b exerts force $\mathbf{F}_a$ on body a, then there must be a force $\mathbf{F}_b$ acting on body b, due to body a, such that $\mathbf{F}_b = -\mathbf{F}_a$. There is no such thing as a lone force without a partner. As we shall see in the next chapter, the third law leads directly to the powerful law of conservation of momentum.

We have argued that a body can be isolated by removing it sufficiently far from other bodies. However, the following problem arises. Suppose that an isolated body starts to accelerate in defiance of Newton's second law. What prevents us from explaining away the difficulty by attributing the acceleration to carelessness in isolating the system? If this option is open to us, Newton's second law becomes meaningless. We need an independent way of telling whether or not there is a physical interaction on a system. Newton's third law provides such a test. If the acceleration of a body is the result of an outside force, then somewhere in the universe there must be an equal and opposite force acting on another body. If we find such a force, the dilemma is resolved; the body was not completely isolated. The interaction may be new and interesting, but as long as the forces are equal and opposite, Newton's laws are satisfied.

If an isolated body accelerates and we cannot find some external object which suffers an equal and opposite force, then we are in trouble. As far as we know this has never occurred. Thus Newton's third law is not only a vitally important dynamical tool, but it is also an important logical element in making sense of the first two laws.

Newton's second law $\mathbf{F} = m\mathbf{a}$ holds true only in inertial systems. The existence of inertial systems seems almost trivial to us, since the earth provides a reasonably good inertial reference frame for everyday observations. However, there is nothing trivial about the concept of an inertial system, as the following example shows.

Example 2.1 Astronauts in Space—Inertial Systems and Fictitious Forces

Two spaceships are moving in empty space chasing an unidentified flying object, possibly a flying saucer. The captains of the two ships, A and B, must find out if the saucer is flying freely or if it is accelerating. A, B, and the saucer are all moving along a straight line.

The captain of A sets to work and measures the distance to the saucer as a function of time. In principle, he sets up a coordinate system along the line of motion with his ship as origin and notes the position of the saucer, which he calls $x_A(t)$. (In practice he uses his radar set to measure the distance to the saucer.) From $x_A(t)$ he calculates the velocity

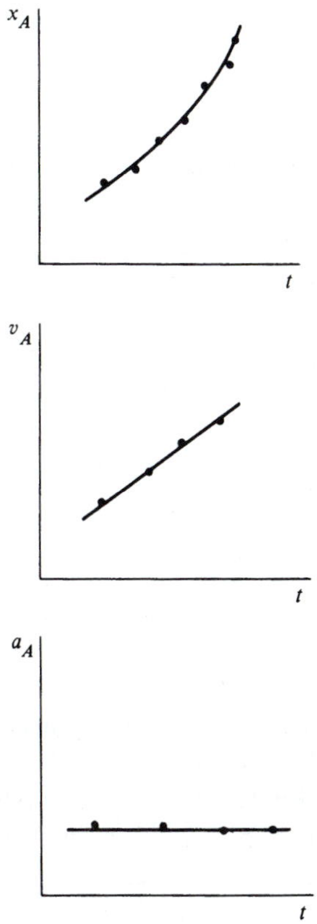

$v_A = \dot{x}_A$ and the acceleration $a_A = \ddot{x}_A$. The results are shown in the sketches. The captain of A concludes that the saucer has a positive acceleration $a_A = 1{,}000\ \text{m/s}^2$. He therefore assumes that its engines are on and that the force on the saucer is

$$\begin{aligned} F_A &= a_A M \\ &= 1{,}000M \text{ newtons,} \end{aligned}$$

where M is the saucer's mass in kilograms.

The captain of B goes through the same procedure. He finds that the acceleration is $a_B = 950\ \text{m/s}^2$ and concludes that the force on the saucer is

$$\begin{aligned} F_B &= a_B M \\ &= 950M \text{ newtons.} \end{aligned}$$

This presents a serious problem. There is nothing arbitrary about force; if different observers obtain different values for the force, at least one of them must be mistaken. The captains of A and B have confidence in the laws of mechanics, so they set about resolving the discrepancy. In particular, they recall that Newton's laws hold only in inertial systems. How can they decide whether or not their systems are inertial?

A's captain sets out by checking to see if all his engines are off. Since they are, he suspects that he is not accelerating and that his spaceship defines an inertial system. To check that this is the case, he undertakes a simple but sensitive experiment. He observes that a pencil, carefully released at rest, floats without motion. He concludes that the pencil's acceleration is negligible and that he is in an inertial system. The reasoning is as follows: as long as he holds the pencil it must have the same instantaneous velocity and acceleration as the spaceship. However, there are no forces acting on the pencil after it is released, assuming that we can neglect gravitational or electrical interactions with the spaceship, air currents, etc. The pencil, then, can be presumed to represent an isolated body. If the spaceship is itself accelerating, it will catch up with the pencil—the pencil will appear to accelerate relative to the cabin. Otherwise, the spaceship must itself define an inertial system.

The determination of the force on the saucer by the captain of A must be correct because A is in an inertial system. But what can we say about the observations made by the captain of B? To answer this problem, we look at the relation of x_A and x_B. From the sketch,

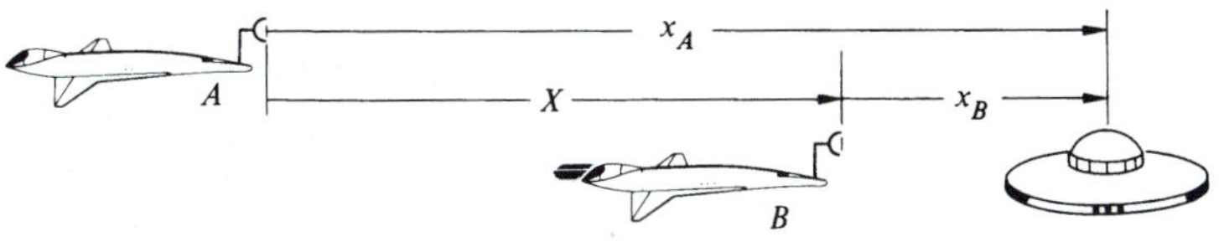

$$x_A(t) = x_B(t) + X(t),$$

where $X(t)$ is the position of B relative to A. Differentiating twice with respect to time, we have

$$\ddot{x}_A = \ddot{x}_B + \ddot{X}. \quad 1$$

Since system A is inertial, Newton's second law for the saucer is

$$F_{\text{true}} = M\ddot{x}_A \quad 2$$

where F_{true} is the true force on the saucer.

What about the observations made by the captain of B? The apparent force observed by B is

$$F_{B,\text{apparent}} = M\ddot{x}_B. \quad 3$$

Using the results of (1) and (2), we have

$$\begin{aligned} F_{B,\text{apparent}} &= M\ddot{x}_A - M\ddot{X} \\ &= F_{\text{true}} - M\ddot{X}. \end{aligned} \quad 4$$

B will not measure the true force unless $\ddot{X} = 0$. However, $\ddot{X} = 0$ only when B moves uniformly with respect to A. As we suspect, this is not the case here. The captain of B has accidently left on a rocket engine, and he is accelerating away from A at 50 m/s^2. After shutting off the engine, he obtains the same value for the force on the saucer as does A.

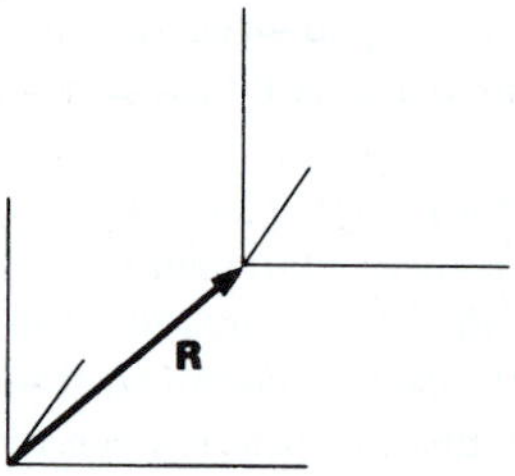

Although we considered only motion along a line in Example 2.1, it is easy to generalize the result to three dimensions. If **R** is the vector from the origin of an inertial system to the origin of another coordinate system, we have

$$\mathbf{F}_{\text{apparent}} = \mathbf{F}_{\text{true}} - M\ddot{\mathbf{R}}.$$

If $\ddot{\mathbf{R}} = 0$, then $\mathbf{F}_{\text{apparent}} = \mathbf{F}_{\text{true}}$, which means that the second coordinate system is also inertial. In fact, we have merely proven what we asserted earlier, namely, that any system moving uniformly with respect to an inertial system is also inertial.

Sometimes we would like to carry out measurements in noninertial systems. What can we do to get the correct equations of motion? The answer lies in the relation $\mathbf{F}_{\text{apparent}} = \mathbf{F}_{\text{true}} - M\ddot{\mathbf{R}}$. We can think of the last term as an additional force, which we call a *fictitious force*. (The term fictitious indicates that there is no real interaction involved.) We then write

$$\mathbf{F}_{\text{apparent}} = \mathbf{F}_{\text{true}} + \mathbf{F}_{\text{fictitious}},$$

where $\mathbf{F}_{\text{fictitious}} = -M\ddot{\mathbf{R}}$. Here M is the mass of the particle and $\ddot{\mathbf{R}}$ is the acceleration of the noninertial system with respect to any inertial system.

Fictitious forces are useful in solving certain problems, but they must be treated with care. They generally cause more confusion then they are worth at this stage of your studies, and for that reason we shall avoid them for the present and agree to use inertial systems only. Later on, in Chap. 8, we shall examine fictitious forces in detail and learn how to deal with them.

Although Newton's laws can be stated in a reasonably clear and consistent fashion, it should be realized that there are fundamental difficulties which cannot be argued away. We shall return to these in later chapters after we have had a chance to become better acquainted with the concepts of newtonian physics. Some points, however, are well to bear in mind now.

1. You have had to take our word that the experiments we used to define mass and to develop the second law of motion really give the results claimed. It should come as no surprise (although it was a considerable shock when it was first discovered) that this is not always so. For instance, the mass scale we have set up is no longer consistent when the particles are moving at high speeds. It turns out that instead of the mass we defined, called the rest mass m_0, a more useful quantity is $m = m_0/\sqrt{1 - v^2/c^2}$, where c is the speed of light and v is the speed of the particle. For the case $v \ll c$, m and m_0 differ negligibly. The reason that our tabletop experiments did not lead us to the more general expression for mass is that even for the largest everyday velocities, say the velocity of a spacecraft going around the earth, $v/c \approx 3 \times 10^{-5}$, and m and m_0 differ by only a few parts in 10^{10}.

2. Newton's laws describe the behavior of point masses. In the case where the size of the body is small compared with the interaction distance, this offers no problem. For instance, the earth and sun are so small compared with the distance between them that for many purposes their motion can be adequately described by considering the motion of point masses located at the center of each. However, the approximation that we are dealing with point masses is fortunately not essential, and if we wish to describe the motion of large bodies, we can readily generalize Newton's laws, as we shall do in the next chapter. It turns out to be not much more difficult to discuss the motion of a rigid body composed of 10^{24} atoms than the motion of a single point mass.

3. Newton's laws deal with particles and are poorly suited for describing a continuous system such as a fluid. We cannot directly apply $\mathbf{F} = m\mathbf{a}$ to a fluid, for both the force and the mass are continuously distributed. However, newtonian mechanics can be extended to deal with fluids and provides the underlying principles of fluid mechanics.

One system which is particularly troublesome for our present formulation of newtonian mechanics is the electromagnetic field. Paradoxes can arise when such a field is present. For instance, two charged bodies which interact electrically actually interact via the electric fields they create. The interaction is not instantaneously transmitted from one particle to the other but propagates at the velocity of light. During the propagation time there is an apparent breakdown of Newton's third law; the forces on the particles are not equal and opposite. Similar problems arise in considering gravitational and other interactions. However, the problem lies not so much with newtonian mechanics as with its misapplication. Simply put, fields possess mechanical properties like momentum and energy which must not be overlooked. From this point of view there is no such thing as a simple two particle system. However, for many systems the fields can be taken into account and the paradoxes can be resolved within the newtonian framework.

2.3 Standards and Units

Length, time, and mass play a fundamental role in every branch of physics. These quantities are defined in terms of certain fundamental physical standards which are agreed to by the scientific community. Since a particular standard generally does not have a convenient size for every application, a number of systems of units have come into use. For example, the centimeter, the angstrom, and the yard are all units of length, but each is defined in terms of the standard meter. There are a number of systems of units in widespread use, the choice being chiefly a matter of custom and convenience. This section presents a brief description of the current standards and summarizes the units which we shall encounter.

The Fundamental Standards

The fundamental standards play two vital roles. In the first place, the precision with which these standards can be defined and reproduced limits the ultimate accuracy of experiments. In some cases the precision is almost unbelievably high—time, for instance, can be measured to a few parts in 10^{12}. In addition, agreeing to a standard for a physical quantity simultaneously provides an operational definition for that quantity. For example, the modern view is that time is what is measured by clocks, and that the properties of time can be understood only by observing the properties of clocks. This is not a trivial point; the rates of all clocks are affected by motion and by gravity (as we shall discuss in Chaps. 8 and 12), and unless we are willing to accept the fact that time itself is altered by motion and gravity, we are led into contradictions.

Once a physical quantity has been defined in terms of a measurement procedure, we must appeal to experiment, not to preconceived notions, to understand its properties. To contrast this viewpoint with a nonoperational approach, consider, for example, Newton's definition of time: "Absolute, true, and mathematical time, of itself, and from its own nature, flows equally without relation to anything external." This may be intuitively and philosophically appealing, but it is hard to see how such a definition can be applied. The idea is metaphysical and not of much use in physics.

Once we have agreed on the operation underlying a particular physical quantity, the problem is to construct the most precise practical standard. Until recently, physical standards were man-made, in the sense that they consisted of particular objects to which all other measurements had to be referred. Thus, the unit length, the meter, was defined to be the distance between two scratches on a platinum bar. Such man-made standards have a number of disadvantages. Since the standard must be carefully preserved, actual measurements are often done with secondary standards, which causes a loss of accuracy. Furthermore, the precision of a man-made standard is intrinsically limited. In the case of the standard meter, precision was found to be limited by fuzziness in the engraved lines which defined the meter interval. When more accurate optical techniques for locating position were developed in the latter part of the nineteenth century, it was realized that the standard meter bar was no longer adequate.

Length is now defined by a natural, rather than man-made, standard. The meter is defined to be a given multiple of the wavelength of a particular spectral line. The advantage of such a unit is that anyone who has the required optical equipment can reproduce it. Also, as the instrumentation improves, the accuracy of the standard will correspondingly increase. Most of the standards of physics are now natural.

Here is a brief account of the current status of the standards of length, time, and mass.

Length The meter was intended to be one ten-millionth of the distance from the equator to the pole of the earth along the Dunkirk-Barcelona line. This cannot be measured accurately (in fact it changes due to distortions of the earth), and in 1889 it was agreed to define the meter as the separation between two scratches in a platinum-iridium bar which is preserved at the International Bureau of Weights and Measures, Sèvres, France. In 1960 the meter was redefined to be 1,650,763.73 wavelengths of the orange-red line of krypton 86. The accuracy of this standard is a few parts in 10^8.

Recent advances in laser techniques provide methods which should allow the velocity of light to be measured to better than 1 part in 10^8. It is likely that the velocity of light will replace length as a fundamental quantity. In this case the unit of length would be derived from velocity and time.

Time Time has traditionally been measured in terms of rotation of the earth. Until 1956 the basic unit, the second, was defined as 1/86,400 of the mean solar day. Unfortunately, the period of rotation of the earth is not very uniform. Variations of up to one part in 10^7 per day occur due to atmospheric tides and changes in the earth's core. The motion of the earth around the sun is not influenced by these perturbations, and until recently the mean solar year was used to define the second. Here the accuracy was a few parts in 10^9. Fortunately, time can now be measured in terms of a natural atomic frequency. In 1967 the second was defined to be the time required to execute 9,192,637,770 cycles of a hyperfine transition in cesium 133. This transition frequency can be reliably measured to a few parts in 10^{12}, which means that time is by far the most accurately determined fundamental quantity.

Mass Of the three fundamental units, only mass is defined in terms of a man-made standard. Originally, the kilogram was defined to be the mass of 1,000 cubic centimeters of water at a temperature of 4 degrees Centigrade. The definition is difficult to apply, and in 1889 the kilogram was defined to be the mass of a platinum-iridium cylinder which is maintained at the International Bureau of Weights and Measures. Secondary standards can be compared with it to an accuracy of one part in 10^9. Perhaps someday we will learn how to define the kilogram in terms of a natural unit, such as the mass of an atom. However, at present nobody knows how to count reliably the large number of atoms needed to constitute a useful sample. Perhaps you can discover a method.

Systems of Units

Although the standards for mass, length, and time are accepted by the entire scientific community, there are a variety of systems of units which differ in the scaling factors. The most widely used system of units is the International System, abbreviated SI (for Système International d'Unités). It is the legal system in most countries. The SI units are *meter*, *kilogram*, and *second;* SI replaces the former mks system. The related cgs system, based on the centimeter, gram, and second, is also commonly used. A third system, the English system of units, is used for nonscientific measurements in Britain and North America, although Britain is in the process of switching to the metric system. It is essential to know how to work problems in any system of units. We shall work chiefly with SI units, with occasional use of the cgs system and one or two lapses into the English system.

Here is a table listing the names of units in the SI, cgs, and English systems.

	SI	CGS	ENGLISH
Length	1 meter (m)	1 centimeter (cm)	1 foot (ft)
Mass	1 kilogram (kg)	1 gram (g)	1 slug
Time	1 second (s)	1 second (s)	1 second (s)
Acceleration	1 m/s^2	1 cm/s^2	1 ft/s^2
Force	1 newton (N) $= 1\ kg{\cdot}m/s^2$	1 dyne $= 1\ g{\cdot}cm/s^2$	1 pound (lb) $= 1\ slug{\cdot}ft/s^2$

Some useful relations between these units systems are:

$1\ m = 100\ cm$	$1\ in = \frac{1}{12}\ ft \approx 2.54\ cm$
$1\ kg = 1000\ g$	$1\ slug \approx 14.6\ kg$
$1\ N = 10^5\ dyne$	$1\ N \approx 0.224\ lb$

The word pound sometimes refers to a unit of mass. In this context it stands for the mass which experiences a gravitational force of one pound at the surface of the earth, approximately 0.454 kg. We shall avoid this confusing usage.

2.4 Some Applications of Newton's Laws

Newton's laws are meaningless equations until we know how to apply them. A number of steps are involved which, once learned, are so natural that the procedure becomes intuitive. Our aim in this section is to outline a method of analyzing physical problems and to illustrate it by examples. A note of reassurance lest you feel that matters are presented too dogmatically: There are many ways of attacking most problems, and the procedure we suggest is certainly not the only one. In fact, no cut-and-dried procedure can ever substitute for intelligent analytical thinking. However, the systematic method suggested here will be helpful in getting started, and we urge you to master it even if you should later resort to shortcuts or a different approach.

Here are the steps:

1. Mentally divide the system[1] into smaller systems, each of which can be treated as a point mass.

2. Draw a force diagram for each mass as follows:
   a. Represent the body by a point or simple symbol, and label it.
   b. Draw a force vector on the mass for each force acting on it.

Point 2*b* can be tricky. Draw only forces acting *on* the body, not forces exerted *by* the body. The body may be attached to strings, pushed by other bodies, etc. We replace all these physical interactions with other bodies by a system of forces; according to Newton's laws, only forces acting *on* the body influence its motion.

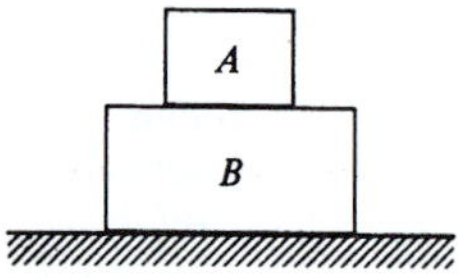

As an example, here are two blocks at rest on a table top. The force diagram for A is shown at left. F_1 is the force exerted on block A by block B, and W_A is the force of gravity on A, called the *weight*.

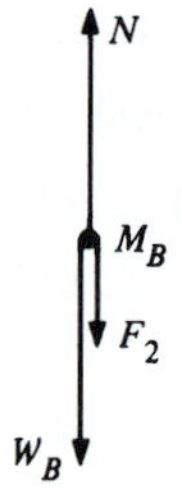

Similarly, we can draw the force diagram for block B. W_B is the force of gravity on B, N is the normal (perpendicular) force exerted by the table top on B, and F_2 is the force exerted by A on B. There are no other physical interactions that would produce a force on B.

It is important not to confuse a force with an acceleration; draw only *real* forces. Since we are using only inertial systems for the present, all the forces are associated with physical interactions. For every force you should be able to answer the question, "What

[1] We use "system" here to mean a collection of physical objects rather than a coordinate system. The meaning should be clear from the context.

exerts this force on the body?'' (We shall see how to use so-called fictitious forces in Chap. 8.[1])

3. Introduce a coordinate system. The coordinate system must be inertial—that is, it must be fixed to an inertial frame. With the force diagram as a guide, write separately the component equations of motion for each body. By equation of motion we mean an equation of the form $F_{1x} + F_{2x} + \cdot \cdot \cdot = Ma_x$, where the x component of each force on the body is represented by a term on the left hand side of the equation. The algebraic sign of each component must be consistent with the force diagram and with the choice of coordinate system.

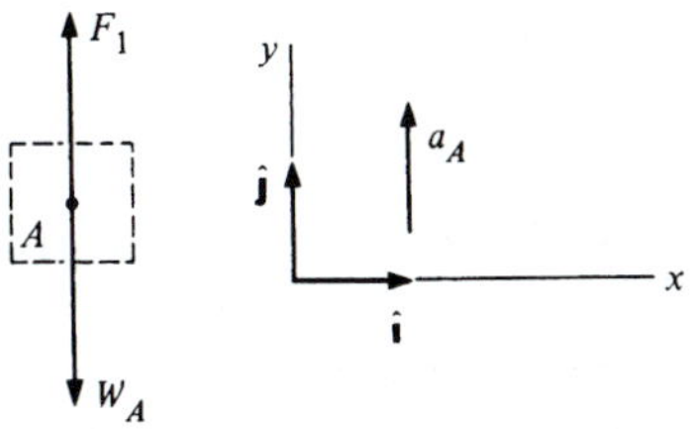

For instance, returning to the force diagram for block A, Newton's second law gives

$$\mathbf{F}_1 + \mathbf{W}_A = m_A\mathbf{a}_A.$$

Since $\mathbf{F}_1 = F_1\hat{\mathbf{j}}$, $\mathbf{W}_A = -W_A\hat{\mathbf{j}}$, we have

$$0 = m_A(\mathbf{a}_A)_x$$

and

$$F_1 - W_A = m_A(\mathbf{a}_A)_y.$$

The x equation of motion is trivial and normally we omit it, writing simply

$$F_1 - W_A = m_Aa_A.$$

The equation of motion for B is

$$N - F_2 - W_B = m_Ba_B.$$

4. If two bodies in the same system interact, the forces between them must be equal and opposite by Newton's third law. These relations should be written explicitly.

For example, in the case of the two blocks on the tabletop, $\mathbf{F}_1 = -\mathbf{F}_2$. Hence

$$F_1 = F_2.$$

Note that Newton's third law never relates two forces acting on the *same* body; forces on two different bodies must be involved.

[1] The most notorious fictitious force is the centrifugal force. Long experience has shown that using this force before one has a really solid grasp of Newton's laws invariably causes confusion. Besides, it is only one of several fictitious forces which play a role in rotating systems. For both these reasons, we shall strictly avoid centrifugal forces for the present.

5. In many problems, bodies are constrained to move along certain paths. A pendulum bob, for instance, moves in a circle, and a block sliding on a tabletop is constrained to move in a plane. Each constraint can be described by a kinematical equation known as a *constraint* equation. Write each constraint equation.

Sometimes the constraints are implicit in the statement of the problem. For the two blocks on the tabletop, there is no vertical acceleration, and the constraint equations are

$$(\mathbf{a}_A)_y = 0 \qquad (\mathbf{a}_B)_y = 0.$$

6. Keep track of which variables are known and which are unknown. The force equations and the constraint equations should provide enough relations to allow every unknown to be found. If an equation is overlooked, there will be too few equations for the unknowns.

Completing the problem of the two blocks on the table, we have

$$\left.\begin{aligned} F_1 - W_A &= m_A a_A \\ N - F_2 - W_B &= m_B a_B \end{aligned}\right\} \text{Equations of motion}$$

$$F_1 = F_2 \qquad \text{From Newton's third law}$$

$$\left.\begin{aligned} a_A &= 0 \\ a_B &= 0 \end{aligned}\right\} \text{Constraint equations}$$

All that remains is the mathematical task of solving the equations. We find

$$F_1 = F_2 = W_A$$
$$N = W_A + W_B.$$

Here are a few examples which illustrate the application of Newton's laws.

The main point of the first example is to help us distinguish between the force we apply to an object and the force it exerts on us. Physiologically, these forces are often confused. If you push a book across a table, the force you feel is not the force that makes the book move; it is the force the book exerts on you. According to Newton's third law, these two forces are always equal and opposite. If one force is limited, so is the other.

Example 2.2 The Astronauts' Tug-of-war

Two astronauts, initially at rest in free space, pull on either end of a rope. Astronaut Alex played football in high school and is stronger than astronaut Bob, whose hobby was chess. The maximum force with which

Alex can pull, F_A, is larger than the maximum force with which Bob can pull, F_B. Their masses are M_A and M_B, and the mass of the rope, M_r, is negligible. Find their motion if each pulls on the rope as hard as he can.

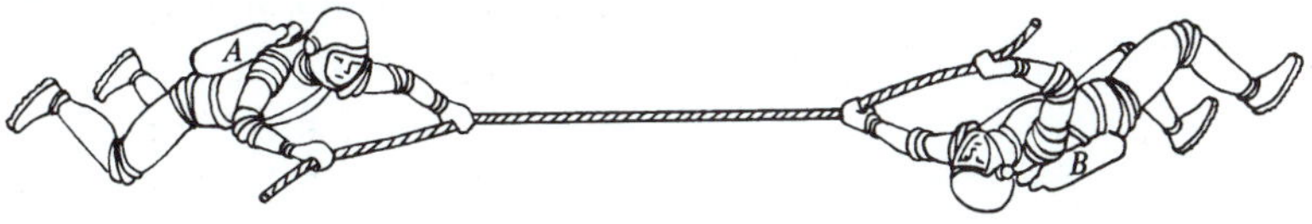

Here are the force diagrams. For clarity, we show the rope as a line.

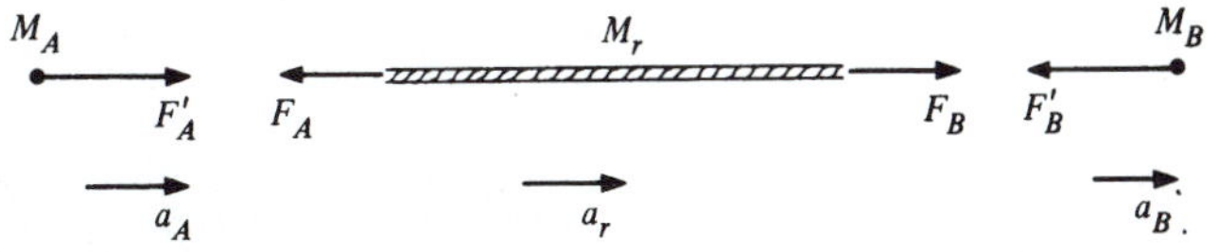

Note that the forces F_A and F_B exerted by the astronauts act on the *rope,* not on the astronauts. The forces exerted by the rope on the astronauts are $F_A{}'$ and $F_B{}'$. The diagram shows the directions of the forces and the coordinate system we have adopted; acceleration to the right is positive.

By Newton's third law,

$$F'_A = F_A$$
$$F'_B = F_B. \qquad 1$$

The equation of motion for the rope is

$$F_B - F_A = M_r a_r. \qquad 2$$

Only motion along the line of the rope is of interest, and we omit the equations of motion in the remaining two directions. There are no constraints, and we proceed to the solution.

Since the mass of the rope, M_r, is negligible, we take $M_r = 0$ in Eq. (2). This gives $F_B - F_A = 0$ or

$$F_B = F_A.$$

The total force on the rope is F_B to the right and F_A to the left. These forces are equal in magnitude, and the total force on the rope is zero. In general, the total force on any body of negligible mass must be effectively zero; a finite force acting on zero mass would produce an infinite acceleration.

Since $F_B = F_A$, Eq. (1) gives $F'_A = F_A = F_B = F'_B$. Hence

$$F'_A = F'_B.$$

The astronauts each pull with the *same* force. Physically, there is a limit to how hard Bob can grip the rope; if Alex tries to pull too hard, the rope slips through Bob's fingers. The force Alex can exert is limited by the strength of Bob's grip. If the rope were tied to Bob, Alex could exert his maximum pull.

The accelerations of the two astronauts are

$$a_A = \frac{F'_A}{M_A}$$

$$a_B = \frac{-F'_B}{M_B}$$

$$= \frac{-F'_A}{M_B}.$$

The negative sign means that a_B is to the left. In many problems the directions of some acceleration or force components are initially unknown. In writing the equations of motion, any choice is valid, provided we are consistent with the convention assumed in the force diagram. If the solution yields a negative sign, the acceleration or force is opposite to the direction assumed.

The next example shows that in order for a compound system to accelerate, there must be a net force on each part of the system.

Example 2.3 **Freight Train**

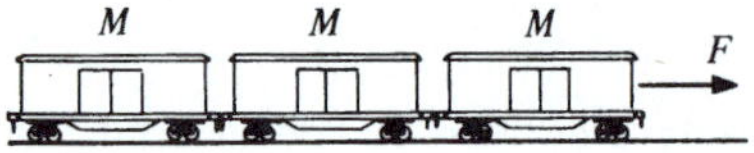

Three freight cars of mass M are pulled with force F by a locomotive. Friction is negligible. Find the forces on each car.

Before drawing the force diagram, it is worth thinking about the system as a whole. Since the cars are joined, they are constrained to have the same acceleration. Since the total mass is $3M$, the acceleration is

$$a = \frac{F}{3M}.$$

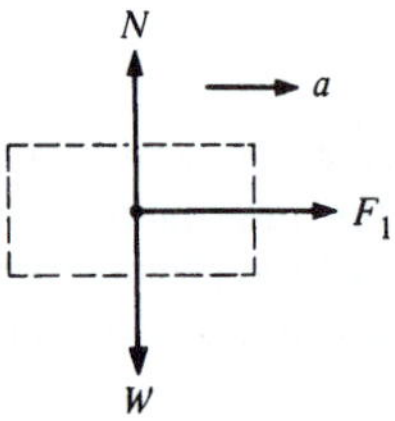

A force diagram for the last car is shown at the left. W is the weight and N is the upward force exerted by the track. The vertical acceleration is zero, so that $N = W$. F_1 is the force exerted by the next car. We have

$$F_1 = Ma$$

$$= M\left(\frac{F}{3M}\right)$$

$$= \frac{F}{3}.$$

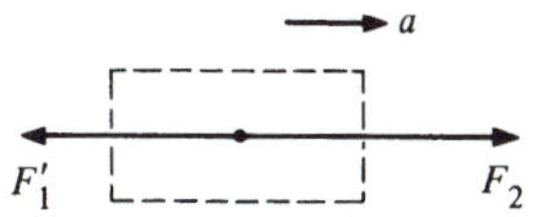

Now let us consider the middle car. The vertical forces are as before, and we omit them. F_1' is the force exerted by the last car, and F_2 is the force exerted by the first car. The equation of motion is

$$F_2 - F_1' = Ma.$$

By Newton's third law, $F_1' = F_1 = F/3$. Since $a = F/3M$, we have

$$F_2 = M\left(\frac{F}{3M}\right) + \frac{F}{3}$$
$$= \frac{2F}{3}.$$

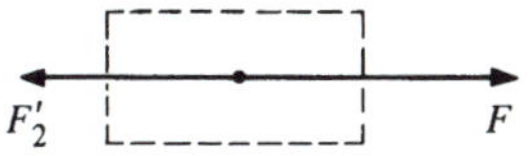

The horizontal forces on the first car are F, to the right, and

$$F_2' = F_2 = \frac{2F}{3},$$

to the left. Each car experiences a total force $F/3$ to the right.

Here is a slightly more general way to look at the problem. Consider a string of N cars, each of mass M, pulled by a force F. The accelera-

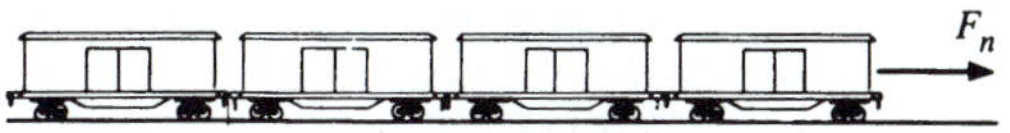

tion is $a = F/(NM)$. To find the force F_n pulling the last n cars, note that F_n must give the mass nM an acceleration $F/(NM)$. Hence

$$F_n = nM\frac{F}{NM}$$
$$= \frac{n}{N}F.$$

The force is proportional to the number of cars pulled.

In systems composed of several bodies, the accelerations are often related by constraints. The equations of constraint can sometimes be found by simple inspection, but the most general approach is to start with the coordinate geometry, as shown in the next example.

Example 2.4 Constraints

a. WEDGE AND BLOCK

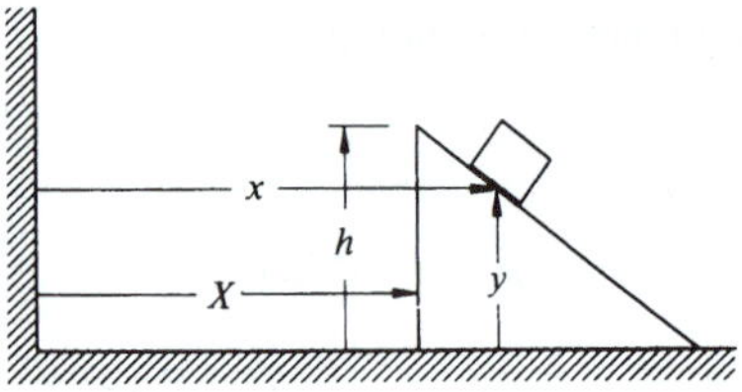

A block moves on a wedge which in turns moves on a horizontal table, as shown in the sketch. The wedge angle is θ. How are the accelerations of the block and the wedge related?

As long as the wedge is in contact with the table, we have the trivial constraint that the vertical acceleration of the wedge is zero. To find the less obvious constraint, let X be the horizontal coordinate of the end of the wedge and let x and y be the horizontal and vertical coordinates of the block, as shown. Let h be the height of the wedge.

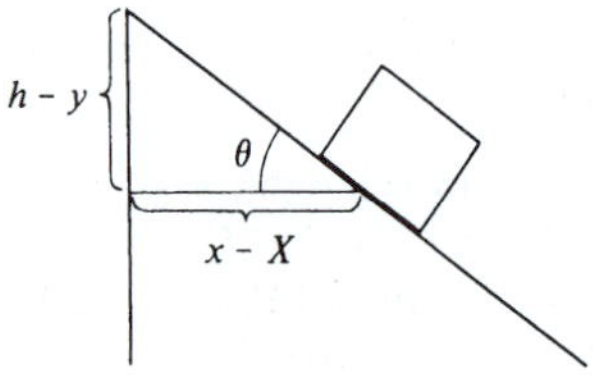

From the geometry, we see that

$$(x - X) = (h - y) \cot \theta.$$

Differentiating twice with respect to time, we obtain the equation of constraint

$$\ddot{x} - \ddot{X} = -\ddot{y} \cot \theta. \tag{1}$$

A few comments are in order. Note that the coordinates are inertial. We would have trouble using Newton's second law if we measured the position of the block with respect to the wedge; the wedge is accelerating and cannot specify an inertial system. Second, unimportant parameters, like the height of the wedge, disappear when we take time derivatives, but they can be useful in setting up the geometry. Finally, constraint equations are independent of applied forces. For example, even if friction between the block and wedge affects their accelerations, Eq. (1) is valid as long as the bodies are in contact.

b. MASSES AND PULLEY

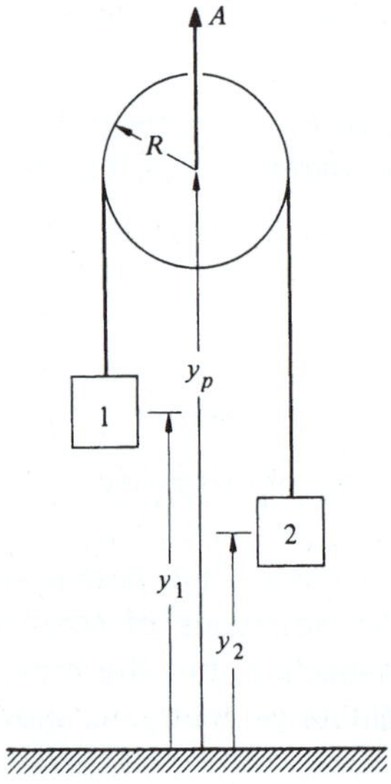

Two masses are connected by a string which passes over a pulley accelerating upward at rate A, as shown. Find how the accelerations of the bodies are related. Assume that there is no horizontal motion.

We shall use the coordinates shown in the drawing. The length of the string, l, is constant. Hence, if y_p is measured to the center of the pulley of radius R,

$$l = \pi R + (y_p - y_1) + (y_p - y_2). \tag{2}$$

Differentiating twice with respect to time, we find the constraint condition

$$0 = 2\ddot{y}_p - \ddot{y}_1 - \ddot{y}_2.$$

Using $A = \ddot{y}_p$, we have

$$A = \tfrac{1}{2}(\ddot{y}_1 + \ddot{y}_2).$$

c. PULLEY SYSTEM

The pulley system shown on the opposite page is used to hoist the block. How does the acceleration of the end of the rope compare with the

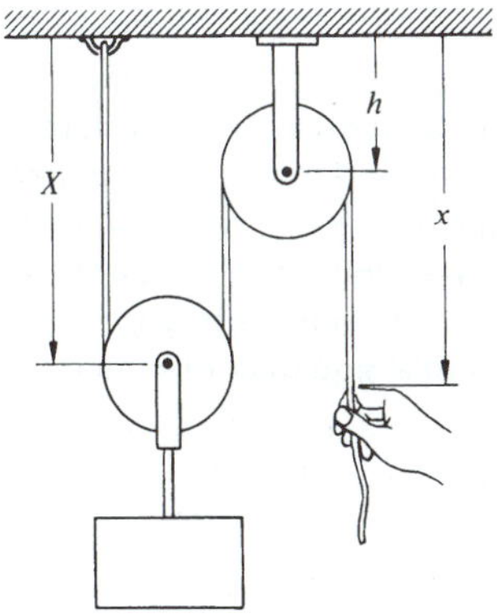

acceleration of the block? Using the coordinates indicated, the length of the rope is given by

$$l = X + \pi R + (X - h) + \pi R + (x - h),$$

where R is the radius of the pulleys. Hence

$$\ddot{X} = -\tfrac{1}{2}\ddot{x}.$$

The block accelerates half as fast as the hand, and in the opposite direction.

Our examples so far have involved linear motion only. Let us look at the dynamics of rotational motion.

A particle undergoing circular motion must have a radial acceleration. This sometimes causes confusion, since our intuitive idea of acceleration usually relates to change in speed rather than to change in direction of motion. For this reason, we start with as simple an example as possible.

Example 2.5 **Block on String 1**

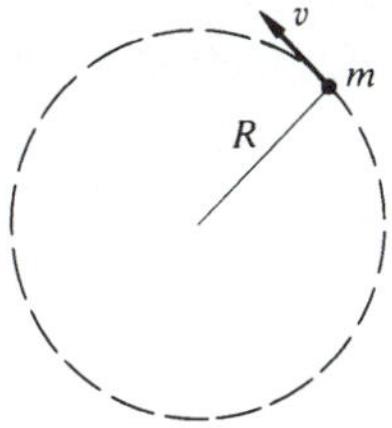

Mass m whirls with constant speed v at the end of a string of length R. Find the force on m in the absence of gravity or friction.

The only force on m is the string force T, which acts toward the center, as shown in the diagram. It is natural to use polar coordinates. Note that according to the derivation in Sec. 1.9, the radial acceleration is $a_r = \ddot{r} - r\dot{\theta}^2$, where $\dot{\theta}$ is the angular velocity. a_r is positive outward. Since **T** is directed toward the origin, $\mathbf{T} = -T\hat{\mathbf{r}}$ and the radial equation of motion is

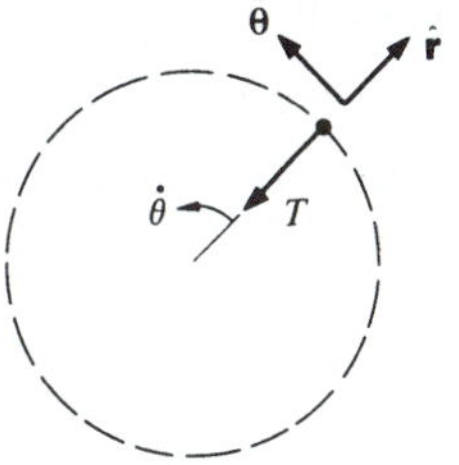

$$\begin{aligned} -T &= ma_r \\ &= m(\ddot{r} - r\dot{\theta}^2). \end{aligned}$$

$\ddot{r} = \ddot{R} = 0$ and $\dot{\theta} = v/R$. Hence $a_r = -R(v/R)^2 = -v^2/R$ and

$$T = \frac{mv^2}{R}.$$

Note that T is directed toward the origin; there is no outward force on m. If you whirl a pebble at the end of a string, *you* feel an outward force. However, the force you feel does not act on the pebble, it acts on you. This force is equal in magnitude and opposite in direction to the force with which you pull the pebble, assuming the string's mass to be negligible.

In the following example both radial and tangential acceleration play a role in circular motion.

Example 2.6 **Block on String 2**

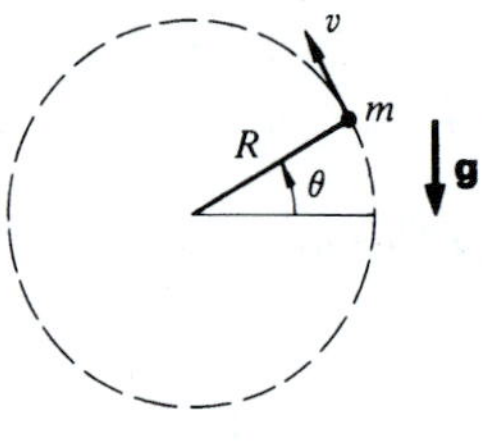

Mass m is whirled on the end of a string length R. The motion is in a vertical plane in the gravitational field of the earth. The forces on m are the weight W down, and the string force T toward the center. The instantaneous speed is v, and the string makes angle θ with the horizontal. Find T and the tangential acceleration at this instant.

The lower diagram shows the forces and unit vectors $\hat{\mathbf{r}}$ and $\hat{\boldsymbol{\theta}}$. The radial force is $-T - W \sin \theta$, so the radial equation of motion is

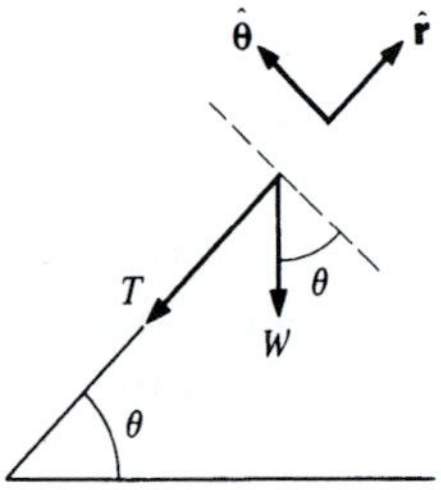

$$-(T + W \sin \theta) = ma_r$$

$$= m(\ddot{r} - r\dot{\theta}^2). \qquad 1$$

The tangential force is $-W \cos \theta$. Hence

$$-W \cos \theta = ma_\theta$$

$$= m(r\ddot{\theta} + 2\dot{r}\dot{\theta}). \qquad 2$$

Since $r = R$ = constant, $a_r = -R(\dot{\theta}^2) = -v^2/R$, and Eq. (1) gives

$$T = \frac{mv^2}{R} - W \sin \theta.$$

The string can pull but not push, so that T cannot be negative. This requires that $mv^2/R \geq W \sin \theta$. The maximum value of $W \sin \theta$ occurs when the mass is vertically up; in this case $mv^2/R > W$. If this condition is not satisfied, the mass does not follow a circular path but starts to fall; $\ddot{r}$ is no longer zero.

The tangential acceleration is given by Eq. (2). Since $\dot{r} = 0$ we have

$$a_\theta = R\ddot{\theta}$$

$$= -\frac{W \cos \theta}{m}.$$

The mass does not move with constant speed; it accelerates tangentially. On the downswing the tangential speed increases, on the upswing it decreases.

The next example involves rotational motion, translational motion, and constraints.

Example 2.7 **The Whirling Block**

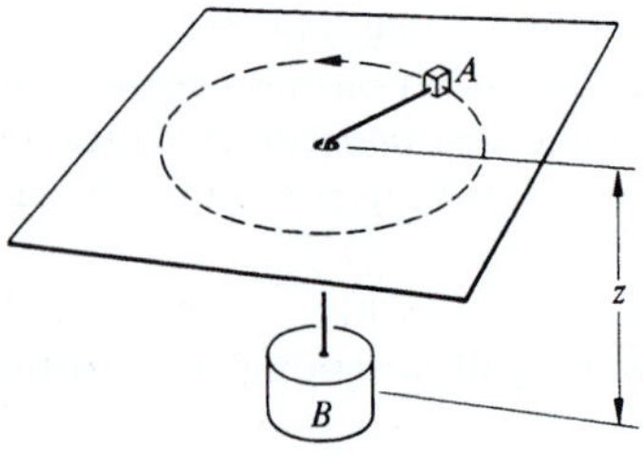

A horizontal frictionless table has a small hole in its center. Block A on the table is connected to block B hanging beneath by a string of negligible mass which passes through the hole.

Initially, B is held stationary and A rotates at constant radius r_0 with steady angular velocity ω_0. If B is released at $t = 0$, what is its acceleration immediately afterward?

The force diagrams for A and B after the moment of release are shown in the sketches.

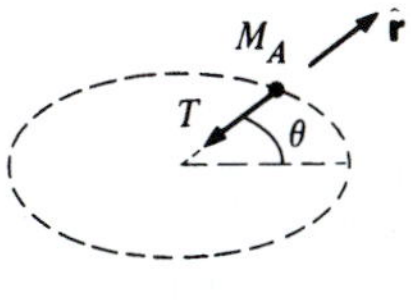

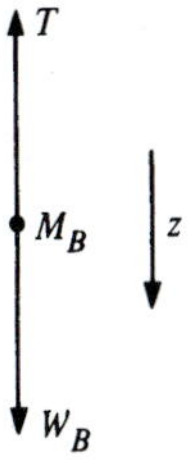

The vertical forces acting on A are in balance and we need not consider them. The only horizontal force acting on A is the string force T. The forces on B are the string force T and the weight W_B.

It is natural to use polar coordinates r, θ for A, and a single linear coordinate z for B, as shown in the force diagrams. As usual, the unit vector $\hat{\mathbf{r}}$ is radially outward. The equations of motion are

$$-T = M_A(\ddot{r} - r\dot{\theta}^2) \qquad \text{Radial} \qquad 1$$

$$0 = M_A(r\ddot{\theta} + 2\dot{r}\dot{\theta}) \qquad \text{Tangential} \qquad 2$$

$$W_B - T = M_B\ddot{z} \qquad \text{Vertical.} \qquad 3$$

Since the length of the string, l, is constant, we have

$$r + z = l. \qquad 4$$

Differentiating Eq. (4) twice with respect to time gives us the constraint equation

$$\ddot{r} = -\ddot{z}. \qquad 5$$

The negative sign means that if A moves inward, B falls. Combining Eqs. (1), (3), and (5), we find

$$\ddot{z} = \frac{W_B - M_A r\dot{\theta}^2}{M_A + M_B}.$$

It is important to realize that although acceleration can change instantaneously, velocity and position cannot. Thus immediately after B is released, $r = r_0$ and $\dot{\theta} = \omega_0$. Hence

$$\ddot{z}(0) = \frac{W_B - M_A r_0\omega_0^2}{M_A + M_B}. \qquad 6$$

$z(0)$ can be positive, negative, or zero depending on the value of the numerator in Eq. (6); if ω_0 is large enough, block B will begin to rise after release.

The apparently simple problem in the next example has some unexpected subtleties.

Example 2.8 The Conical Pendulum

Mass M hangs from a string of length l which is attached to a rod rotating at constant angular frequency ω, as shown in the drawing on the next page. The mass moves with steady speed in a circular path of constant radius. Find α, the angle the string makes with the vertical.

We start with the force diagram. T is the string force and W is the weight of the bob. (Note that there are no other forces on the bob. If this is not clear, you are most likely confusing an acceleration with a

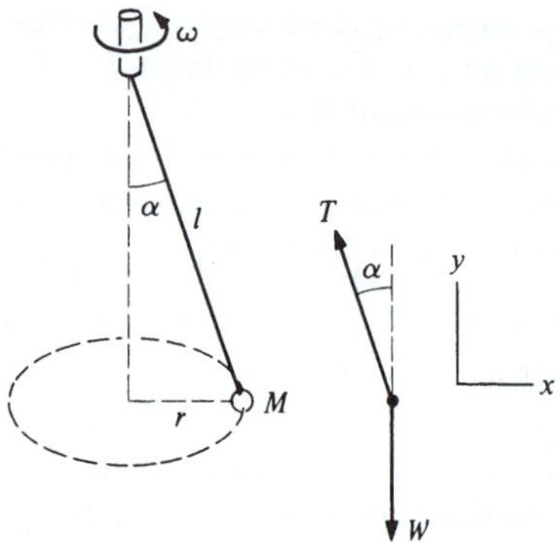

force—a serious error.) The vertical equation of motion is

$$T \cos\alpha - W = 0$$

because y is constant and $\ddot{y}$ is therefore zero.

To find the horizontal equation of motion note that the bob is accelerating in the $\hat{\mathbf{r}}$ direction at rate $a_r = -\omega^2 r$. Then

$$-T \sin\alpha = -Mr\omega^2. \qquad 2$$

Since $r = l \sin\alpha$ we have

$$T \sin\alpha = Ml\omega^2 \sin\alpha \qquad 3$$

or

$$T = Ml\omega^2. \qquad 4$$

Combining Eqs. (1) and (3) gives

$$Ml\omega^2 \cos\alpha = W.$$

As we shall discuss in Sec. 2.5, $W = Mg$, where M is the mass and g is known as the acceleration due to gravity. We obtain

$$\cos\alpha = \frac{g}{l\omega^2}.$$

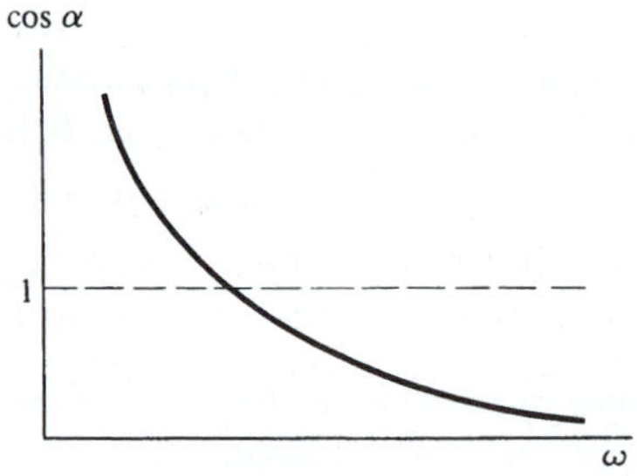

This appears to be the desired solution. For $\omega \to \infty$, $\cos\alpha \to 0$ and $\alpha \to \pi/2$. At high speeds the bob flies out until it is almost horizontal. However, at low speeds the solution does not make sense. As $\omega \to 0$, our solution predicts $\cos\alpha \to \infty$, which is nonsense since $\cos\alpha \le 1$. Something has gone wrong. Here is the trouble.

Our solution predicts $\cos\alpha > 1$ for $\omega < \sqrt{g/l}$. When $\omega = \sqrt{g/l}$, $\cos\alpha = 1$ and $\sin\alpha = 0$; the bob simply hangs vertically. In going from Eq. (2) to Eq. (3) we divided both sides of Eq. (2) by $\sin\alpha$ and, in this case we divided by 0, which is not permissible. However, we see that we have overlooked a second possible solution, namely, $\sin\alpha = 0$, $T = W$, which is true for all values of ω. The solution corresponds to the pendulum hanging straight down. Here is a plot of the complete solution.

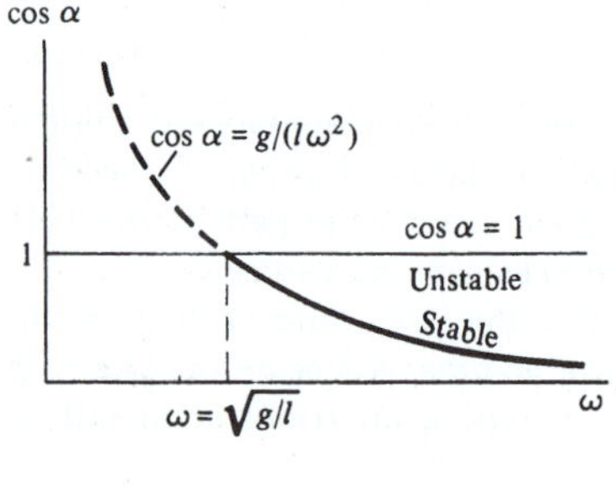

Physically, for $\omega \le \sqrt{g/l}$ the only acceptable solution is $\alpha = 0$, $\cos\alpha = 1$. For $\omega > \sqrt{g/l}$ there are two acceptable solutions:

1. $\cos\alpha = 1$
2. $\cos\alpha = \dfrac{g}{l\omega^2}$.

Solution 1 corresponds to the bob rotating rapidly but hanging vertically. Solution 2 corresponds to the bob flying around at an angle with the vertical. For $\omega > \sqrt{g/l}$, solution 1 is unstable—if the system is in that state and is slightly perturbed, it will jump outward. Can you see why this is so?

The moral of this example is that you have to be sure that the mathematics makes good physical sense.

2.5 The Everyday Forces of Physics

When a physicist sets out to design an accelerator, he uses the laws of mechanics and his knowledge of electric and magnetic forces to determine the paths that the particles will follow. Predicting motion from known forces is an important part of physics and underlies most of its applications. Equally important, however, is the converse process of deducing the physical interaction by observing the motion; this is how new laws are discovered. A classic example is Newton's deduction of the law of gravitation from Kepler's laws of planetary motion. The current attempt to understand the interactions between elementary particles from high energy scattering experiments provides a more contemporary illustration.

Unscrambling experimental observations to find the force can be difficult. In a facetious mood, Eddington once said that force is the mathematical expression we put into the left hand side of Newton's second law to obtain results that agree with observed motions. Fortunately, force has a more concrete physical reality.

Much of our effort in the following chapters will be to learn how systems behave under applied forces. If every pair of particles in the universe had its own special interaction, the task would be impossible. Fortunately, nature is kinder than this. As far as we know, there are only four fundamentally different types of interactions in the universe: gravity, electromagnetic interactions, the so-called weak interaction, and the strong interaction.

Gravity and the electromagnetic interactions can act over a long range because they decrease only as the inverse square of the distance. However, the gravitational force always attracts, whereas electrical forces can either attract or repel. In large systems, electrical attraction and repulsion cancel to a high degree, and gravity alone is left. For this reason, gravitational forces dominate the cosmic scale of our universe. In contrast, the world immediately around us is dominated by the electrical forces, since they are far stronger than gravity on the atomic scale. Electrical forces are responsible for the structure of atoms, molecules, and more complex forms of matter, as well as the existence of light.

The weak and strong interactions have such short ranges that they are important only at nuclear distances, typically 10^{-15} m.

They are negligible even at atomic distances, 10^{-10} m. As its name implies, the strong interaction is very strong, much stronger than the electromagnetic force at nuclear distances. It is the "glue" that binds the atomic nucleus, but aside from this it has little effect in the everyday world. The weak interaction plays a less dramatic role; it mediates in the creation and destruction of neutrinos—particles of no mass and no charge which are essential to our understanding of matter but which can be detected only by the most arduous experiments.

Our object in the remainder of the chapter is to become familiar with the forces which are important in everyday mechanics. Two of these, the forces of gravity and electricity, are fundamental and cannot be explained in simpler terms. The other forces we shall discuss, friction, the contact force, and the viscous force, can be understood as the macroscopic manifestation of interatomic forces.

Gravity, Weight, and the Gravitational Field

Gravity is the most familiar of the fundamental forces. It has close historical ties to the development of mechanics; Newton discovered the law of universal gravitation in 1666, the same year that he formulated his laws of motion. By calculating the motion of two gravitating particles, he was able to derive Kepler's empirical laws of planetary motion. (By accomplishing all this by age 26, Newton established a tradition which still maintains—that great advances are often made by young physicists.)

According to Newton's law of gravitation, two particles attract each other with a force directed along their line of centers. The magnitude of the force is proportional to the product of the masses and decreases as the inverse square of the distance between the particles.

In verbal form the law is bulky and hard to use. However. we can reduce it to a simple mathematical expression.

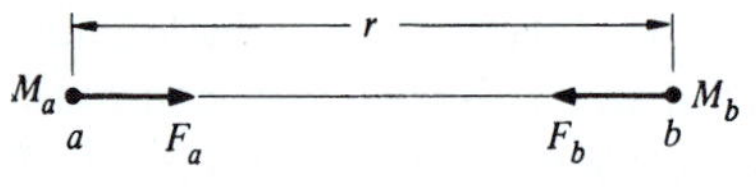

Consider two particles, a and b, with masses M_a and M_b, respectively, separated by distance r. Let $\mathbf{F}_b$ be the force exerted on particle b by particle a. Our verbal description of the magnitude of the force is summarized by

$$|\mathbf{F}_b| = \frac{GM_aM_b}{r^2}.$$

G is a constant of proportionality called the *gravitational constant.* Its value is found by measuring the force between masses in a known geometry. The first measurements were performed by Henry Cavendish in 1771 using a torsion balance. The modern value of G is 6.67×10^{-11} N·m²/kg². (G is the least accurately known of the fundamental constants. Perhaps you can devise a new way to measure it more precisely.) Experimentally, G is the same for all materials—aluminum, lead, neutrons, or what have you. For this reason, the law is called the universal law of gravitation.

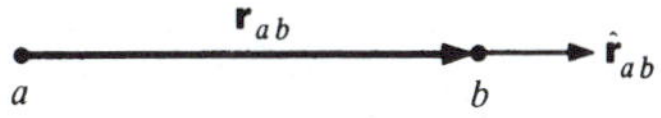

The gravitational force between two particles is *central* (along the line of centers) and attractive. The simplest way to describe these properties is to use vectors. By convention, we introduce a vector $\mathbf{r}_{ab}$ *from* the particle exerting the force, particle a in this case, *to* the particle experiencing the force, particle b. Note that $|\mathbf{r}_{ab}| = r$. Using the unit vector $\hat{\mathbf{r}}_{ab} = \mathbf{r}_{ab}/r$, we have

$$\mathbf{F}_b = -\frac{GM_aM_b}{r^2}\hat{\mathbf{r}}_{ab}.$$

The negative sign indicates that the force is attractive. The force on a due to b is

$$\mathbf{F}_a = -\frac{GM_aM_b}{r^2}\hat{\mathbf{r}}_{ba} = +\frac{GM_aM_b}{r^2}\hat{\mathbf{r}}_{ab} = -\mathbf{F}_b,$$

since $\hat{\mathbf{r}}_{ba} = -\hat{\mathbf{r}}_{ab}$. The forces are equal and opposite, and Newton's third law is automatically satisfied.

The gravitational force has a unique and mysterious property. Consider the equation of motion of particle b under the gravitational attraction of particle a.

$$\mathbf{F}_b = -\frac{GM_aM_b}{r^2}\hat{\mathbf{r}}_{ab}$$

$$= M_b\mathbf{a}_b$$

or

$$\mathbf{a}_b = -\frac{GM_a}{r^2}\hat{\mathbf{r}}_{ab}.$$

The acceleration of a particle under gravity is independent of its mass! There is a subtle point connected with our cancelation of M_b, however. The "mass" (*gravitational* mass) in the law of gravitation, which measures the strength of gravitational interaction, is operationally distinct from the "mass" (*inertial* mass) which characterizes inertia in Newton's second law. Why gravitational mass is proportional to inertial mass for all matter is one of the great mysteries of physics. However, the proportionality has been experimentally verified to very high accuracy, approximately 1 part in 10^{11}; we shall have more to say about this in Chap. 8.

The Gravitational Force of a Sphere The law of gravitation applies only to particles. How can we find the gravitational force on a particle due to an extended body like the earth? Fortunately, the gravitational force obeys the *law of superposition:* the force due to a collection of particles is the vector sum of the forces exerted by the particles individually. This allows us to mentally divide the body into a collection of small elements which can be treated as particles. Using integral calculus, we can sum the forces from all the particles. This method is applied in Note 2.1 to calculate the force between a particle of mass m and a uniform thin spherical shell of mass M and radius R. The result is

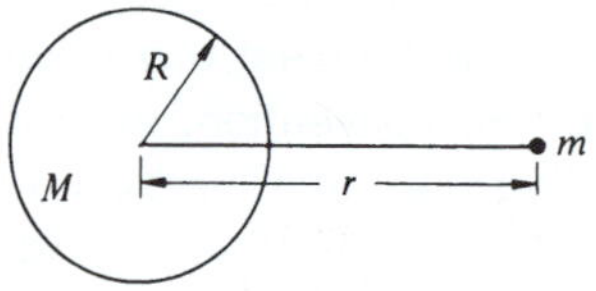

$$\mathbf{F} = -G\frac{Mm}{r^2}\hat{\mathbf{r}} \qquad r > R$$

$$\mathbf{F} = 0 \qquad r < R,$$

where r is the distance from the center of the shell to the particle. If the particle lies outside the shell, the force is the same as if all the mass of the shell were concentrated at its center.

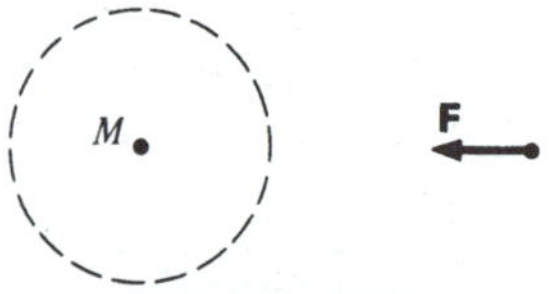

The reason the gravitational force vanishes inside the spherical shell can be seen by a simple argument due to Newton. Consider the two small mass elements marked out by a conical surface with its apex at m. The amount of mass in each element is proportional to its surface area. The area increases as (distance)^2. However, the strength of the force varies as 1/(distance)^2. Thus the forces of the two mass elements are equal and opposite, and cancel. The total force on m is zero, because we can pair up all the elements of the shell this way.

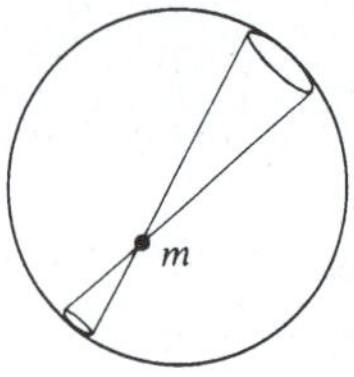

A uniform solid sphere can be regarded as a succession of thin spherical shells, and it follows that for particles outside it, a sphere behaves gravitationally as if its mass were concentrated at its center. This result also holds if the density of the sphere varies with radius, provided the mass distribution is spherically symmetric. For example, although the earth has a dense core, the mass distribution is nearly spherically symmetric, so that to good approximation the gravitational force of the earth on a mass m at distance r is

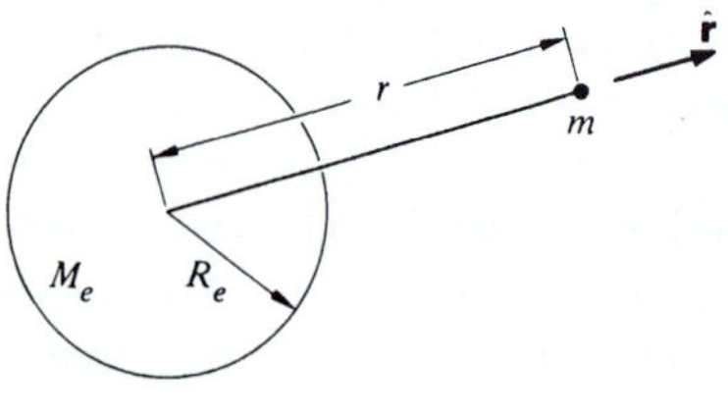

$$\mathbf{F} = -\frac{GM_e m}{r^2}\hat{\mathbf{r}} \qquad r \geq R_e,$$

where M_e is the mass of the earth and R_e is its radius.

At the surface of the earth, the gravitational force is

$$\mathbf{F} = -\frac{GM_e m}{R_e^2}\hat{\mathbf{r}},$$

and the acceleration due to gravity is

$$\mathbf{a} = \frac{\mathbf{F}}{m}$$

$$= -\frac{GM_e}{R_e^2}\hat{\mathbf{r}}.$$

As we expect, the acceleration is independent of m. GM_e/R_e^2 is usually called g. Sometimes g is written as a vector directed down. toward the center of the earth.

$$\mathbf{g} = -\frac{GM_e}{R_e^2}\hat{\mathbf{r}}$$

Numerically, $|g|$ is approximately 9.8 m/s^2 = 980 cm/s^2 $\approx$ 32 ft/s^2.

By convention, g usually stands for the downward acceleration of an object measured with respect to the earth's surface. This differs slightly from the true gravitational acceleration because of the rotation of the earth, a point we shall return to in Chap. 8. g increases by about five parts per thousand from the equator to the poles. About half this variation is due to the slight flattening of the earth about the poles, and the remainder arises from the earth's rotation. Local mass concentrations also affect g; a variation in g of ten parts per million is typical.

The acceleration due to gravity decreases with altitude. We can estimate this effect by taking differentials of the expression

$$g(r) = \frac{GM_e}{r^2}.$$

We have

$$\Delta g(r) = \frac{dg}{dr}\Delta r = -\frac{2GM_e}{r^3}\Delta r$$

$$= -\frac{2g}{r}\Delta r.$$

The fractional change in g with altitude is

$$\frac{\Delta g}{g} = -\frac{2\,\Delta r}{r}.$$

At the earth's surface, $r = 6 \times 10^6$ m, and g decreases by one part per million for an increase in altitude of 3 m.

Weight We define the weight of a body near the earth to be the gravitational force exerted on it by the earth. At the surface of the earth the weight of a mass m is

$$\mathbf{W} = -G\frac{M_e m}{R_e^2}\hat{\mathbf{r}}$$
$$= m\mathbf{g}.$$

The unit of weight is the newton (SI), dyne (cgs), or pound (English). Since $g = 9.8$ m/s², the weight of 1 kg mass is 9.8 N. An automobile which weighs 3,200 lb has mass

$$m = \frac{W}{g} = \frac{3{,}200 \text{ lb}}{32 \text{ ft/s}^2} = 100 \text{ slugs}.$$

Our definition of weight is unambiguous. According to our definition, the weight of a body is not affected by its motion. However, weight is often used in another sense. In this sense, the magnitude of the weight is the magnitude of the force which must be exerted on a body by its surroundings to keep it at rest; its direction is the direction of gravitational attraction. The next example illustrates the difference between these two definitions.

Example 2.9 **Turtle in an Elevator**

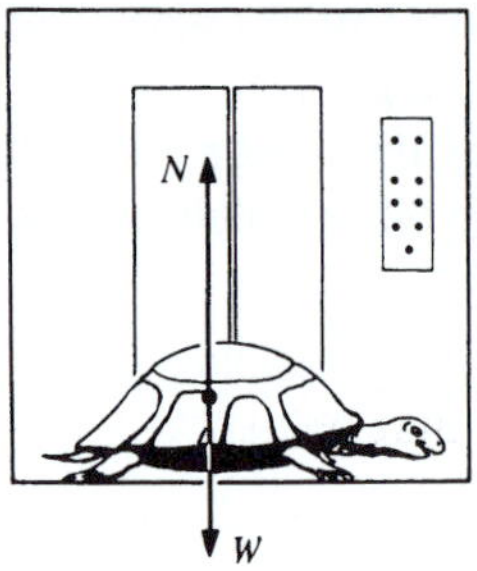

An amiable turtle of mass M stands in an elevator accelerating at rate a Find N, the force exerted on him by the floor of the elevator.

The forces acting on the turtle are N and the weight, the true gravitational force $W = Mg$. Taking up to be the positive direction, we have

$$N - W = Ma$$
$$N = Mg + Ma$$
$$= M(g + a).$$

This result illustrates the two senses in which weight is used. In the sense that weight is the gravitational force, the weight of the turtle, Mg, is independent of the motion of the elevator. In contrast, the weight of the turtle has magnitude $N = M(g + a)$, if the magnitude of the weight is taken to be the magnitude of the force exerted by the elevator on the turtle. If the turtle were standing on a scale, the scale would indicate a weight N. With this definition, the turtle's weight increases when the elevator accelerates up.

If the elevator accelerates down, a is negative and N is less than Mg. If the downward acceleration equals g, N becomes zero, and the turtle "floats" in the elevator. The turtle is then said to be in a state of weightlessness.

Although the two definitions of weight are both commonly used and are both acceptable, we shall generally consider weight to mean the true gravitational force. This is consistent with our resolve to refer all motion to inertial systems and helps us to keep the real forces on a body distinct. If the acceleration due to gravity is g, the real gravitational force on a body of mass m is $W = mg$.

Our definition of weight has one minor drawback. As we saw in the last example, a scale does not read mg in an accelerating system. As we have already pointed out, systems at rest on the earth's surface have a small acceleration due to the earth's rotation, so that the reading of a scale is not the true gravitational force on a mass. However, the effect is small, and we shall treat the surface of the earth as an inertial system for the present.

The Gravitational Field The gravitational force on particle b due to particle a is

$$\mathbf{F}_b = -\frac{GM_aM_b}{r^2}\hat{\mathbf{r}}_{ab},$$

where $\hat{\mathbf{r}}_{ab}$ is a unit vector which points from a toward b. The ratio $\mathbf{F}_b/M_b$, which is independent of M_b, is called the *gravitational field* due to M_a. Denoting the field by $\mathbf{G}_a$, we have

$$\begin{aligned}\mathbf{G}_a &= \frac{\mathbf{F}_b}{M_b}\\ &= -G\frac{M_a}{r^2}\hat{\mathbf{r}}_{ab}.\end{aligned}$$

In general, if the gravitational field at a point in space is $\mathbf{G}$, the gravitational force on mass M at that point is

$$\mathbf{F} = M\mathbf{G}.$$

The dimension of gravitation field is force/mass = acceleration. The acceleration of mass M by gravitational field $\mathbf{G}$ is given by

$$\begin{aligned}\mathbf{F} &= M\mathbf{a}\\ &= M\mathbf{G}\end{aligned}$$

or

$$\mathbf{a} = \mathbf{G}.$$

We see that the gravitational field at a point is numerically equal to the gravitational acceleration experienced by a body located there. For example, the gravitational field of the earth is **g**.

For the present we can regard the gravitational field as a mathematical convenience that allows us to focus on the source of the gravitational attraction. However, the concept of field has a broader significance in physics. Fields have important physical properties, such as the ability to store or transmit energy and momentum. Until recently, the dynamical properties of the gravitational field were chiefly of theoretical interest, since their effects were too small to be observed. However, there is now lively experimental activity in searching for such dynamical features as gravitational waves and "black holes."

The Electrostatic Force

We mention the electrostatic force only in passing since its full implications are better left to a more detailed study of electricity and magnetism. The salient feature of the electrostatic force between two particles is that the force, like gravity, is an inverse square central force. The force depends upon a fundamental property of the particle called its *electric charge* q. There are two different kinds of electric charge: like charges repel, unlike charges attract.

For the sake of convenience, we distinguish the two different kinds of charges by associating an algebraic sign with q, and for this reason we talk about negative and positive charges. The electrostatic force F_b on charge q_b due to charge q_a is given by Coulomb's law:

$$\mathbf{F}_b = k\frac{q_a q_b}{r^2}\hat{\mathbf{r}}_{ab}.$$

k is a constant of proportionality and $\hat{\mathbf{r}}_{ab}$ is a unit vector which points from a to b. If q_a and q_b are both negative or both positive, the force is repulsive, but if the charges are of different sign, $\mathbf{F}_b$ is attractive.

In the SI system, the unit of charge is the *coulomb*, abbreviated C. (The coulomb is defined in terms of electric currents and magnetic forces.) In this system, k is found by experiment to be

$$k = 8.99 \times 10^9 \text{ N}\cdot\text{m}^2/\text{C}^2.$$

In analogy with the gravitational field, we can define the electric field **E** as the electric force on a body divided by its charge. The electric field at **r** due to a charge q at the origin is

$$\mathbf{E} = k\frac{q}{r^2}\hat{\mathbf{r}}.$$

Contact Forces

By contact forces we mean the forces which are transmitted between bodies by short-range atomic or molecular interactions. Examples include the pull of a string, the surface force of sliding friction, and the force of viscosity between a moving body and a fluid. One of the achievements of twentieth century physics is that these forces can now be explained in terms of the fundamental properties of matter. However, our approach will emphasize the empirical properties of these forces and the techniques for dealing with them in physical problems, with only brief mention of their microscopic origins.

Tension—The Force of a String We have been taking the "string" force for granted, having some primitive idea of this kind of force. The following example is intended to help put these ideas into sharper focus.

Example 2.10 **Block and String 3**

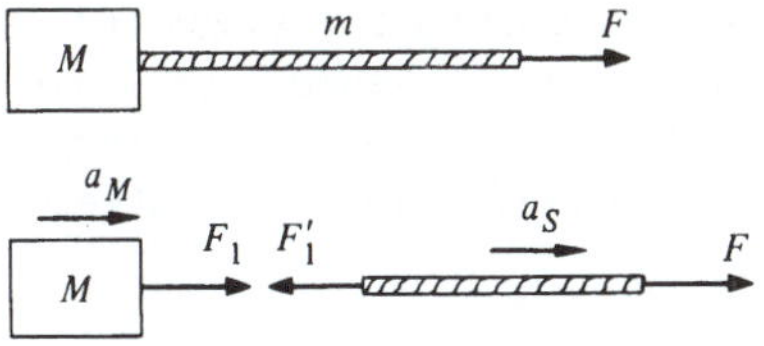

Consider a block of mass M in free space pulled by a string of mass m. A force F is applied to the string, as shown. What is the force that the string "transmits" to the block?

The sketch shows the force diagrams. F_1 is the force of the string on the block, F_1' is the force of the block on the string, a_M is the acceleration of the block, and a_S is the acceleration of the string. The equations of motion are

$$F_1 = Ma_M$$
$$F - F_1' = ma_S.$$

Assuming that the string is inextensible, it accelerates at the same rate as the block, giving the constraint equation $a_S = a_M$. Furthermore, $F_1 = F_1'$ by Newton's third law. Solving for the acceleration, we find that

$$a = \frac{F}{M + m},$$

as we expect, and

$$F_1 = F_1'$$
$$= \frac{M}{M+m}F.$$

The force on the block is less than F; the string does not transmit the full applied force. However, if the mass of the string is negligible compared with the block, $F_1 = F$ to good approximation.

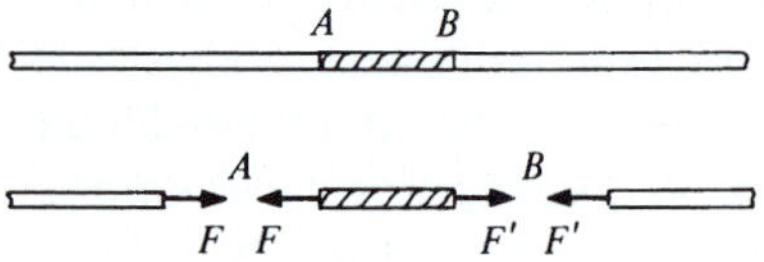

We can think of a string as composed of short sections interacting by contact forces. Each section pulls the sections to either side of it, and by Newton's third law, it is pulled by the adjacent sections. The magnitude of the force acting between adjacent sections is called *tension*. There is no direction associated with tension. In the sketch, the tension at A is F and the tension at B is F'.

Although a string may be under considerable tension (for example a string on a guitar), if the tension is uniform, the net string force on each small section is zero and the section remains at rest unless external forces act on it. If there are external forces on the section, or if the string is accelerating, the tension generally varies along the string, as Examples 2.11 and 2.12 show.

Example 2.11 **Dangling Rope**

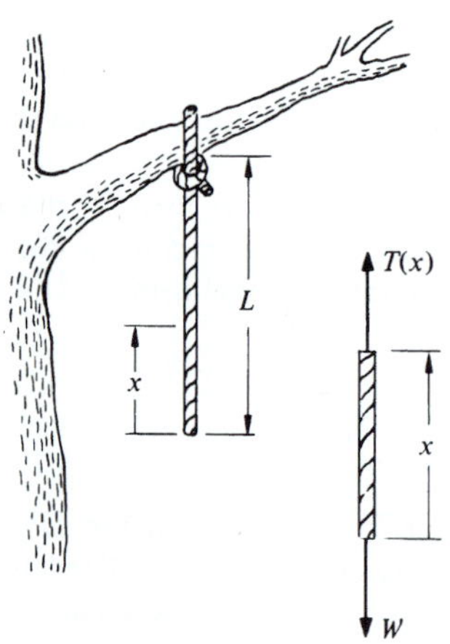

A uniform rope of mass M and length L hangs from the limb of a tree. Find the tension a distance x from the bottom.

The force diagram for the lower section of the rope is shown in the sketch. The section is pulled up by a force of magnitude $T(x)$, where $T(x)$ is the tension at x. The downward force on the rope is its weight $W = Mg(x/L)$. The total force on the section is zero since it is at rest. Hence

$$T(x) = \frac{Mg}{L}x.$$

At the bottom of the rope the tension is zero, while at the top the tension equals the total weight of the rope Mg.

The next example cannot be solved by direct application of Newton's second law. However, by treating each small section of the system as a particle, and taking the limit using calculus, we can obtain a differential equation which leads to the solution.

The technique is so useful that it is employed time and again in physics.

Example 2.12 **Whirling Rope**

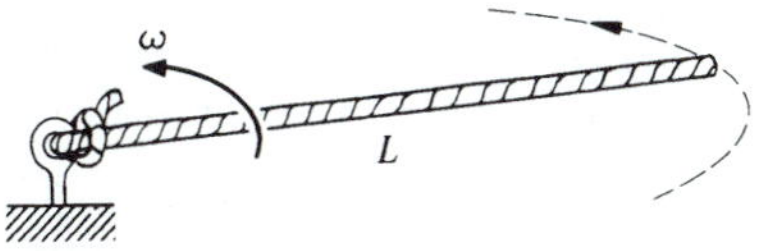

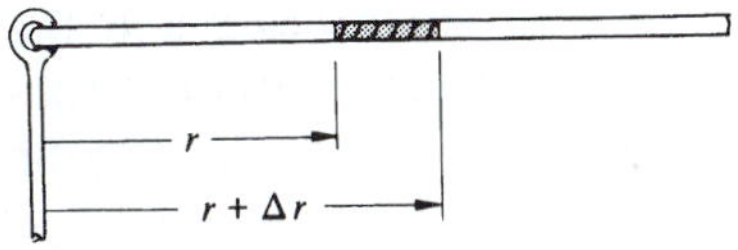

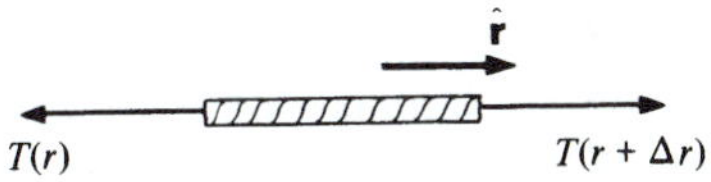

A uniform rope of mass M and length L is pivoted at one end and whirls with uniform angular velocity ω. What is the tension in the rope at distance r from the pivot? Neglect gravity.

Consider the small section of rope between r and $r + \Delta r$. The length of the section is Δr and its mass is $\Delta m = M\,\Delta r/L$. Because of its circular motion, the section has a radial acceleration. Therefore, the forces pulling either end of the section cannot be equal, and we conclude that the tension must vary with r.

The inward force on the section is $T(r)$, the tension at r, and the outward force is $T(r + \Delta r)$. Treating the section as a particle, its inward radial acceleration is $r\omega^2$. [This point can be confusing; it is just as reasonable to take the acceleration to be $(r + \Delta r)\omega^2$. However, we shall shortly take the limit $\Delta r \to 0$, and in this limit the two expressions give the same result.]

The equation of motion for the section is

$$
\begin{aligned}
T(r + \Delta r) - T(r) &= -(\Delta m) r\omega^2 \\
&= -\frac{Mr\omega^2\,\Delta r}{L}.
\end{aligned}
$$

The problem is to find $T(r)$, but we are not yet ready to do this. However, by dividing the last equation by Δr and taking the limit $\Delta r \to 0$, we can find an exact expression for dT/dr.

$$
\begin{aligned}
\frac{dT}{dr} &= \lim_{\Delta r \to 0} \frac{T(r + \Delta r) - T(r)}{\Delta r} \\
&= -\frac{Mr\omega^2}{L}
\end{aligned}
$$

To find the tension, we integrate.

$$dT = -\frac{M\omega^2}{L}\, r\, dr$$

$$\int_{T_0}^{T(r)} dT = -\int_0^r \frac{M\omega^2}{L}\, r\, dr,$$

where T_0 is the tension at $r = 0$.

$$T(r) - T_0 = -\frac{M\omega^2}{L}\frac{r^2}{2}$$

or

$$T(r) = T_0 - \frac{M\omega^2}{2L} r^2.$$

To evaluate T_0 we need one additional piece of information. Since the end of the rope at $r = L$ is free, the tension there must be zero. We have

$$T(L) = 0 = T_0 - \tfrac{1}{2}M\omega^2 L.$$

Hence, $T_0 = \frac{1}{2}M\omega^2 L$, and the final result can be written

$$T(r) = \frac{M\omega^2}{2L}(L^2 - r^2).$$

When a pulley is used to change the direction of a rope under tension, there is a reaction force on the pulley. As every sailor knows, the force on the pulley depends on the tension and the angle through which the rope is deflected. Working out this problem in detail provides another illustration of how calculus can be applied to a physical problem.

Example 2.13 Pulleys

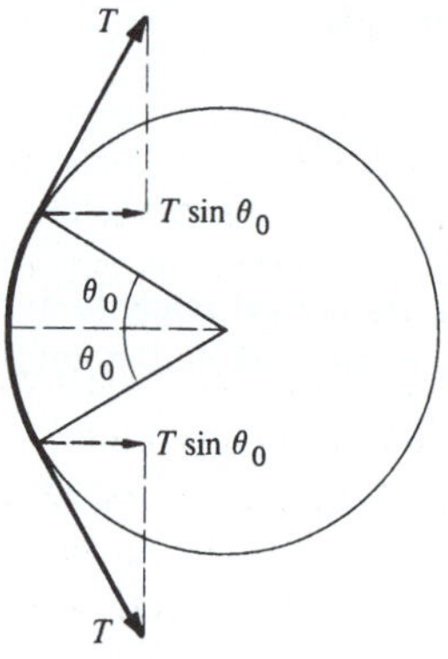

A string with constant tension T is deflected through angle $2\theta_0$ by a smooth fixed pulley. What is the force on the pulley?

Intuitively, the magnitude of the force is $2T \sin \theta_0$. To prove this result, we shall find the force due to each element of the string and then add them vectorially.

Consider the section of string between θ and $\theta + \Delta\theta$. The force diagram is drawn below, center. ΔF is the outward force due to the pulley

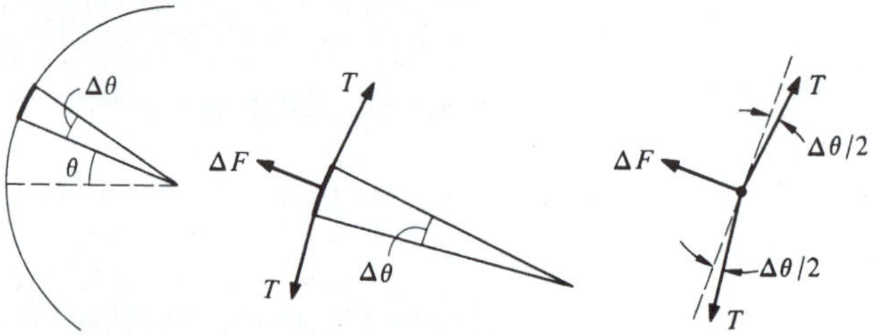

The tension in the string is constant, but the forces T at either end of the element are not parallel. Since we shall shortly take the limit $\Delta\theta \to 0$, we can treat the element like a particle. For equilibrium, the total force is zero. We have

$$\Delta F - 2T \sin \frac{\Delta\theta}{2} = 0.$$

For small $\Delta\theta$, $\sin(\Delta\theta/2) \approx \Delta\theta/2$ and

$$\Delta F = 2T\frac{\Delta\theta}{2} = T\Delta\theta.$$

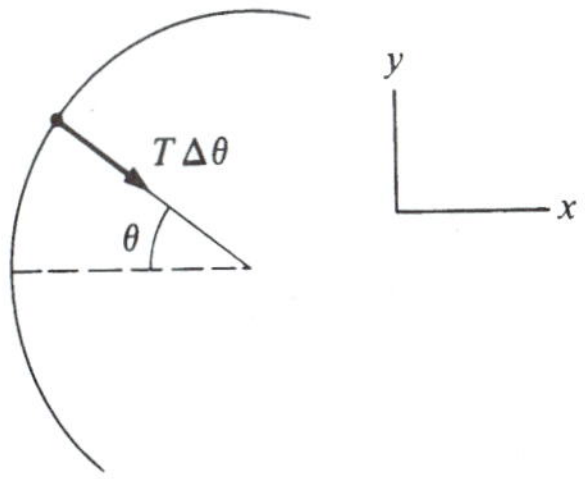

Thus the element exerts an inward radial force of magnitude $T\,\Delta\theta$ on the pulley.

The element at angle θ exerts a force in the x direction of $(T\,\Delta\theta)\cos\theta$. The total force in the x direction is $\Sigma T\cos\theta\,\Delta\theta$, where the sum is over all elements of the string which are touching the pulley. In the limit $\Delta\theta \to 0$, the sum becomes an integral. The total force in the x direction is therefore

$$\int_{-\theta_0}^{\theta_0} T\cos\theta\,d\theta = 2T\sin\theta_0.$$

Tension and Atomic Forces The force on each element of a string in equilibrium is zero. Nevertheless, the string will break if the tension is too large. We can understand this qualitatively by looking at strings from the atomic viewpoint. An idealized model of a string is a single long chain of molecules. Suppose that force F is applied to molecule 1 at the end of the string. The force diagrams for molecules 1 and 2 are shown in the sketch below. In

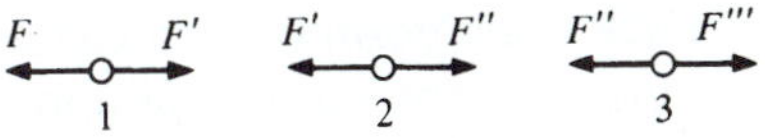

equilibrium, $F = F'$ and $F' = F''$, so that $F'' = F$. We see that the string "transmits" the force F. To understand how this comes about, we need to look at the nature of intermolecular forces.

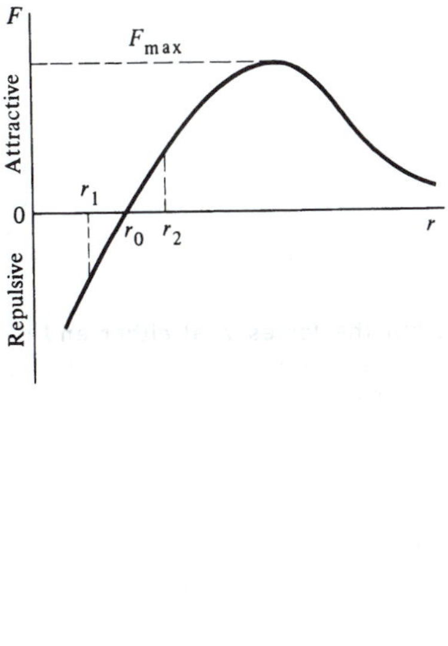

Qualitatively, the force between two molecules depends on the distance r between them, as shown in the drawing. The intermolecular force is repulsive at small distances, is zero at some separation r_0, and is attractive for $r > r_0$. For large values of r the force falls to zero. There are no scales on our sketch, but r_0 is typically a few angstroms (1 Å $= 10^{-10}$ m).

When there is no applied force, the molecules must be a distance r_0 apart; otherwise the intermolecular forces would make the string contract or expand. As we pull on the string, the molecules move apart slightly, say to $r = r_2$, where the intermolecular attractive force just balances the applied force so that the total force on each molecule is zero. If the string were stiff like a metal rod, we could push as well as pull. A push makes the molecules move slightly together, say to $r = r_1$, where the intermolecular repulsive force balances the applied force. The change in the length depends on the slope of the interatomic force curve at r_0. The steeper the curve, the less the stretch for a given pull.

The attractive intermolecular force has a maximum value F_{max}, as shown in the sketch. If the applied pull is greater than F_{max}, the intermolecular force is too weak to restore balance—the molecules continue to separate and the string breaks.

For a real string or rod, the intermolecular forces act in a three dimensional lattice work of atoms. The breaking strength of most materials is considerably less than the limit set by F_{max}. Breaks occur at points of weakness, or "defects," in the lattice, where the molecular arrangement departs from regularity. Microscopic metal "whiskers" seem to be nearly free from defects, and they exhibit breaking strengths close to the theoretical maximum.

The Normal Force The force exerted by a surface on a body in contact with it can be resolved into two components, one perpendicular to the surface and one tangential to the surface. The perpendicular component is called the *normal* force and the tangential component is called *friction.*

The origin of the normal force is similar to the origin of tension in a string. When we put a book on a table, the molecules of the book exert downward forces on the molecules of the table. The molecules composing the upper layers of the tabletop move downward until the repulsion of the molecules below balances the force applied by the book. From the atomic point of view, no surface is perfectly rigid. Although compression always occurs, it is often too slight to notice, and we shall neglect it and treat surfaces as rigid.

The normal force on a body, generally denoted by N, has the following simple property: for a body resting on a surface, N is equal and opposite to the resultant of all other forces which act on the body in a direction perpendicular to the surface. For instance, when you stand still, the normal force exerted by the ground is equal to your weight. However, when you walk, the normal force fluctuates as you accelerate up and down.

Friction Friction cannot be described by a simple formula, but macroscopic mechanics is hard to understand without some idea of the properties of friction.

Friction arises when the surface of one body moves, or tries to move, along the surface of a second body. The magnitude of the force of friction varies in a complicated way with the nature of the surfaces and their relative velocity. In fact, the only thing we can always say about friction is that it opposes the motion which would occur in its absence. For instance, suppose that we try to push a book across a table. If we push gently, the book remains at rest; the force of friction assumes a value equal and opposite to the tangential force we apply. In this case, the force of friction assumes whatever value is needed to keep the book at rest. However, the friction force cannot increase indefinitely. If we push hard enough, the book starts to slide. For many surfaces the maximum value of the friction is found to be equal to μN, where N is the normal force and μ is the *coefficient of friction.*

When a body slides across a surface, the friction force is directed opposite to the instantaneous velocity and has magnitude μN. Experimentally, the force of sliding friction decreases slightly when bodies begin to slide, but for the most part we shall neglect this effect. For two given surfaces the force of sliding friction is essentially independent of the area of contact.

It may seem strange that friction is independent of the area of contact. The reason is that the actual area of contact on an atomic scale is a minute fraction of the total surface area. Friction occurs because of the interatomic forces at these minute regions of atomic contact. The fraction of the geometric area in atomic contact is proportional to the normal force divided by the geometric area. If the normal force is doubled, the area of atomic contact is doubled and the friction force is twice as large. However, if the geometric area is doubled while the normal force remains the same, the fraction of area in atomic contact is halved and the actual area in atomic contact—hence the friction force—remains constant. (Nonrigid bodies, like automobile tires, are more complicated. A wide tire is generally better than a narrow one for good acceleration and braking.)

In summary, we take the force of friction f to behave as follows:

1. For bodies not in relative motion,

$$0 \leq f \leq \mu N.$$

f opposes the motion that would occur in its absence.

2. For bodies in relative motion,

$$f = \mu N.$$

f is directed opposite to the relative velocity.

Example 2.14 **Block and Wedge with Friction**

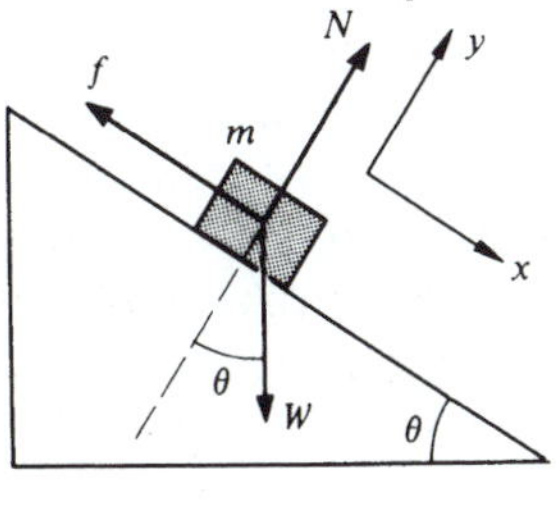

A block of mass m rests on a fixed wedge of angle θ. The coefficient of friction is μ. (For wooden blocks, μ is of the order of 0.2 to 0.5.) Find the value of θ at which the block starts to slide.

In the absence of friction, the block would slide down the plane; hence the friction force f points up the plane. With the coordinates shown, we have

$$m\ddot{x} = W \sin\theta - f$$

and

$$m\ddot{y} = N - W\cos\theta$$
$$= 0.$$

When sliding starts, f has its maximum value μN, and $\ddot{x} = 0$. The equations then give

$$W\sin\theta_{max} = \mu N$$
$$W\cos\theta_{max} = N.$$

Hence,

$$\tan\theta_{max} = \mu.$$

Notice that as the wedge angle is gradually increased from zero, the friction force grows in magnitude from zero toward its maximum value μN, since before the block begins to slide we have

$$f = W\sin\theta \qquad \theta \le \theta_{max}.$$

Example 2.15 **The Spinning Terror**

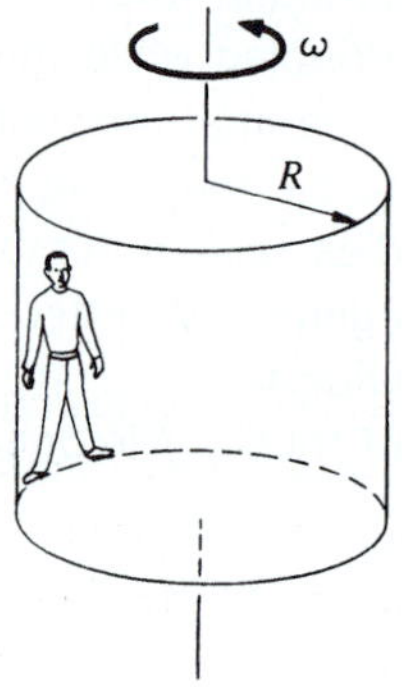

The Spinning Terror is an amusement park ride—a large vertical drum which spins so fast that everyone inside stays pinned against the wall when the floor drops away. What is the minimum steady angular velocity ω which allows the floor to be dropped away safely?

Suppose that the radius of the drum is R and the mass of the body is M. Let μ be the coefficient of friction between the drum and M. The forces on M are the weight W, the friction force f, and the normal force exerted by the wall, N, as shown below.

The radial acceleration is $R\omega^2$ toward the axis, and the radial equation of motion is

$$N = MR\omega^2.$$

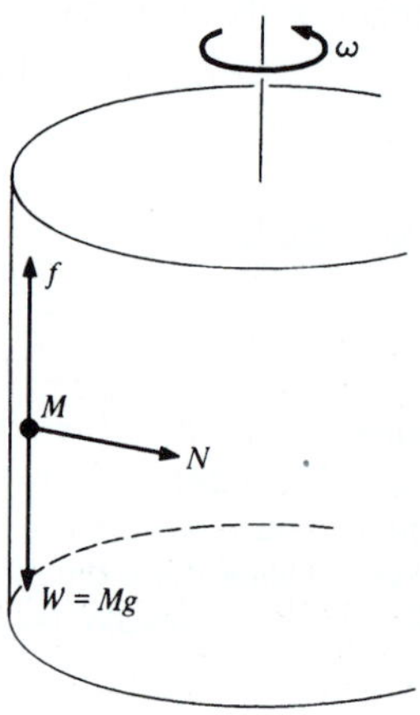

By the law of static friction,

$$f \le \mu N = \mu MR\omega^2.$$

Since we require M to be in vertical equilibrium,

$$f = Mg,$$

and we have

$$Mg \le \mu MR\omega^2$$

or

$$\omega^2 \ge \frac{g}{\mu R}.$$

The smallest value of ω that will work is

$$\omega_{\min} = \sqrt{\frac{g}{\mu R}}.$$

For cloth on wood μ is at least 0.3, and if the drum has radius 6 ft, then $\omega_{\min} = [32/(0.3 \times 6)]^{\frac{1}{2}} = 4$ rad/s. The drum must make at least $\omega/2\pi =$ 0.6 turns per second.

Viscosity

A body moving through a liquid or gas is retarded by the force of viscosity exerted on it by the fluid. Unlike the friction force between dry surfaces, the viscous force has a simple velocity dependence; it is proportional to the velocity. At high speeds other forces due to turbulence occur and the total drag force can have a complicated velocity dependence. (Sports car designers use a force proportional to the square of the speed to account for the drag forces.) However, in many practical cases viscosity is the only important drag force.

Viscosity arises because a body moving through a medium exerts forces which set the nearby fluid into motion. By Newton's third law the fluid exerts a reaction force on the body.

We can write the viscous retarding force in the form

$$\mathbf{F}_v = -C\mathbf{v},$$

where C is a constant which depends on the fluid and the geometry of the body. $\mathbf{F}_v$ is always along the line of motion, because it is proportional to $\mathbf{v}$. The negative sign assures that $\mathbf{F}_v$ opposes the motion. For objects of simple shape moving through a gas at low pressure, C can be calculated from first principles. We shall treat it as an empirical constant.

When the only force on a body is the viscous retarding force, the equation of motion is

$$-C\mathbf{v} = m\frac{d\mathbf{v}}{dt}.$$

What we have here is a differential equation for $\mathbf{v}$. Since the force is along the line of motion, only the magnitude of $\mathbf{v}$ changes[1]

[1] Formally, this is proved as follows. Since $\mathbf{v} = v\hat{\mathbf{v}}$, $d\mathbf{v}/dt = dv/dt\,\hat{\mathbf{v}} + v\,d\hat{\mathbf{v}}/dt$. The equation of motion is $-Cv\hat{\mathbf{v}} = m\,dv/dt\,\hat{\mathbf{v}} + mv\,d\hat{\mathbf{v}}/dt$. Because $\hat{\mathbf{v}}$ is a unit vector, $d\hat{\mathbf{v}}/dt$ is perpendicular to $\hat{\mathbf{v}}$. The other terms of the equation lie in the $\hat{\mathbf{v}}$ direction, so that $d\hat{\mathbf{v}}/dt$ must be zero. The same conclusion follows more directly from the simple physical argument that a force directed along the line of motion can change the speed but cannot change the direction of motion.

and the vector equation reduces to the scalar equation

$$-Cv = m\frac{dv}{dt}$$

or

$$m\frac{dv}{dt} + Cv = 0.$$

The task of solving such a differential equation occurs often in physics. A few differential equations are so simple and occur so frequently that it is helpful to be thoroughly familiar with them and their solutions. The equation of the form $m\,dv/dt + Cv = 0$ is one of the most common, and the following example should make you feel at home with it.

Example 2.16 Free Motion in a Viscous Medium

A body of mass m released with velocity v_0 in a viscous fluid is retarded by a force Cv. Find the motion, supposing that no other forces act.

The equation of motion is

$$m\frac{dv}{dt} + Cv = 0,$$

which we can rewrite in the standard form

$$\frac{dv}{dt} + \frac{C}{m}v = 0. \qquad 1$$

If you are familiar with the properties of the exponential function e^{ax}, then you know that $(d/dx)e^{ax} = ae^{ax}$, or $(d/dx)e^{ax} - ae^{ax} = 0$. This suggests that we use a trial solution $v = e^{at}$, where a is a constant to be determined. Then $dv/dt = ae^{at}$, and substituting this in Eq. (1) gives us

$$ae^{at} + \frac{C}{m}e^{at} = 0.$$

This holds true at all times if $a = -C/m$. Hence, a solution is

$$v = e^{-Ct/m}.$$

However, this cannot be the correct solution; v has the dimension of velocity whereas the exponential function is dimensionless. Let us try

$$v = Ae^{-Ct/m},$$

where A is a constant. Substituting this in Eq. (1) gives

$$-\frac{C}{m}Ae^{-Ct/m} + \frac{C}{m}Ae^{-Ct/m} = 0,$$

so that the solution is acceptable. But A can be any constant, whereas our solution must be quite specific. To evaluate A we make use of the given initial condition. An *initial condition* is a specific piece of information about the motion at some particular time. We were given that $v = v_0$ at $t = 0$. Hence

$$v(t = 0) = Ae^0 = v_0.$$

Since $e^0 = 1$, it follows that $A = v_0$, and the full solution is

$$v = v_0 e^{-Ct/m}.$$

We solved Eq. (1) by what might be called a common sense approach—we simply guessed the answer. This particular equation can also be solved by formal integration after appropriate "separation of the variables."

$$\frac{dv}{dt} + \frac{C}{m}v = 0$$

$$\frac{dv}{v} = -\frac{C}{m}dt$$

$$\int_{v_0}^{v} \frac{dv}{v} = -\int_0^t \frac{C}{m}dt$$

Note the correspondence between the limits: v is the velocity at time t and v_0 is the velocity at time 0.

$$\ln\frac{v}{v_0} = -\frac{C}{m}t$$

$$\frac{v}{v_0} = e^{(-C/m)t}$$

$$v = v_0 e^{-Ct/m}.$$

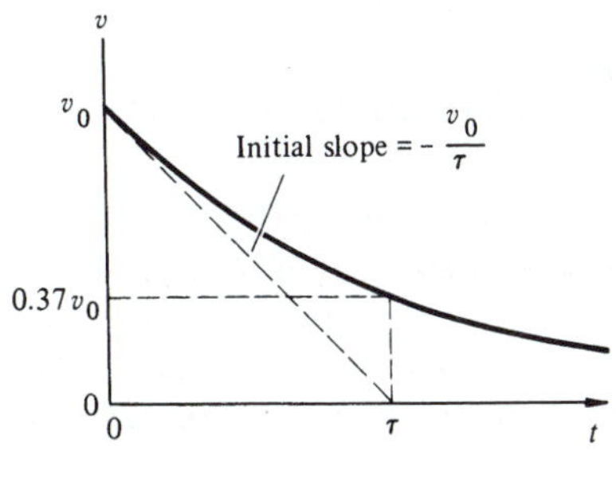

Before leaving this problem, let us look at the solution in a little more detail. The velocity decreases exponentially in time. If we let $\tau = m/c$, then we have $v = v_0 e^{-t/\tau}$. τ is a *characteristic time* for the system; it is the time for the velocity to drop to $e^{-1} \approx 0.37$ of its original velocity.

The Linear Restoring Force: Hooke's Law, the Spring, and Simple Harmonic Motion

In the mid-seventeenth century Robert Hooke discovered that the extension of a spring is proportional to the applied force, both for positive and negative displacements. The force F_S exerted by a stretched spring is given by Hooke's law

$$F_S = -kx,$$

where k is a constant called the *spring constant* and x is the displacement of the end of the spring from its equilibrium position. The magnitude of F_S increases linearly with displacement. The

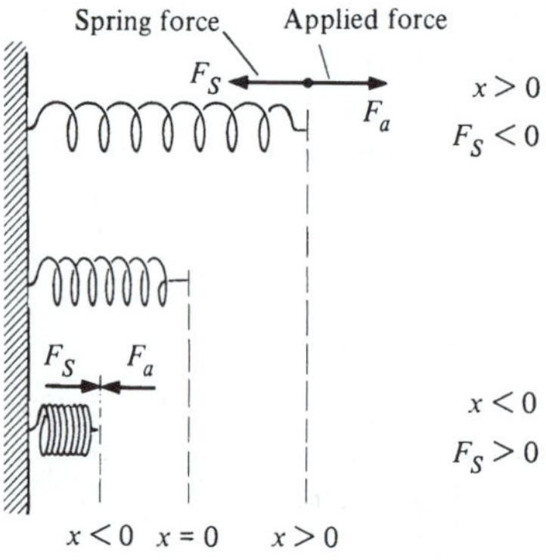

negative sign indicates that F_S is a restoring force; the spring force is always in the direction that tends to restore the spring to its equilibrium length. A force obeying Hooke's law is called a *linear restoring force.*

If the spring is stretched by an applied force F_a, then $x > 0$ and F_S is negative, directed toward the origin.

If the spring is compressed by F_a, then $x < 0$ and F_S is positive.

Hooke's law is essentially empirical and breaks down for large displacements. Taking a jaundiced view of affairs, we could rephrase Hooke's law as "extension is proportional to force, as long as it is." However, this misses the important point. For sufficiently small displacements Hooke's law is remarkably accurate, not only for springs but also for practically every system near equilibrium. Consequently, the motion of a system under a linear restoring force occurs persistently throughout physics.

By looking at the intermolecular force curve on page 91, we can see why the linear restoring force is so common. If the force curve is linear in the neighborhood of the equilibrium point, then the force is proportional to the displacement from equilibrium. This is almost always the case; a sufficiently short segment of a curve is generally linear to good approximation. Only in pathological cases does the force curve have no linear component. It is also apparent that the linear approximation necessarily breaks down for large displacements. We shall return to these considerations in Chap. 4.

In the following example we investigate simple harmonic motion —the motion of a mass under a linear restoring force. We shall again encounter a differential equation. Like the equation for viscous drag, the differential equation for simple harmonic motion occurs frequently and is well worth learning to recognize early in the game. Fortunately, the solution has a simple form.

Example 2.17 **Spring and Block—The Equation for Simple Harmonic Motion**

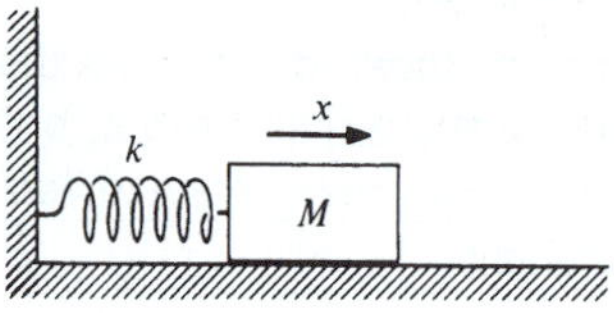

A block of mass M is attached to one end of a horizontal spring, the other end of which is fixed. The block rests on a horizontal frictionless surface. What motion is possible for the block?

Since the spring force is the only horizontal force acting on the block, the equation of motion is

$$M\ddot{x} = -kx$$

or

$$\ddot{x} + \frac{k}{M}x = 0,$$

where x is measured from the equilibrium position. It is convenient to write

$$\omega = \sqrt{\frac{k}{M}}.$$

The equation takes the standard form

$$\ddot{x} + \omega^2 x = 0.$$

You should learn to recognize the mathematical form of this equation, since it arises in many different physical contexts. It is called the equation of *simple harmonic motion* (SHM). Without going into the theory of differential equations, we simply write down the solution

$$x = A \sin \omega t + B \cos \omega t.$$

ω is known as the *angular frequency* of the motion. By substitution it is easy to show that this solution satisfies the original equation for arbitrary values of A and B. The theory of differential equations tells us that there are no further nontrivial solutions. The main point here, however, is to become familiar with the mathematical form of the SHM differential equation and the form of its solution. We shall derive the solution in Example 4.2, but this purely mathematical process does not concern us now.

As we show in the following example, the constants A and B are to be determined from the initial conditions. We shall show that A and B can be found by knowing the position and velocity at some particular time.

Example 2.18 **The Spring Gun—An Example Illustrating Initial Conditions**

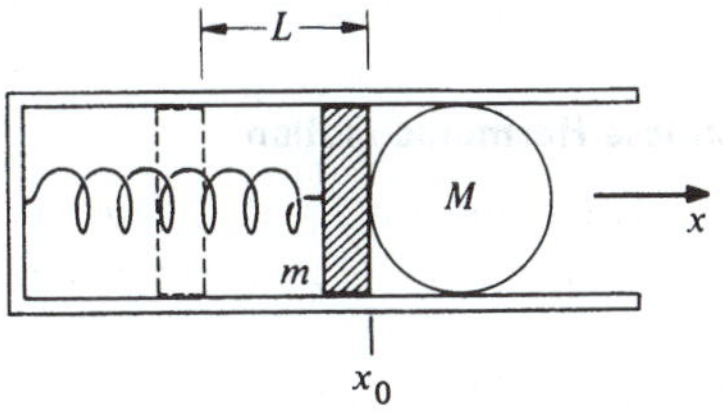

The piston of a spring gun has mass m and is attached to one end of a spring with spring constant k. The projectile is a marble of mass M. The piston and marble are pulled back a distance L from the equilibrium position and suddenly released. What is the speed of the marble as it loses contact with the piston? Neglect friction.

Let the x axis be along the direction of motion with the origin at the unstretched position. The position of the piston is given by

$$x(t) = A \sin \omega t + B \cos \omega t, \qquad 1$$

where $\omega = \sqrt{k/(m + M)}$. This equation holds up to the time the marble and piston lose contact. The velocity is

$$v(t) = \dot{x}(t)$$
$$= \omega A \cos \omega t - \omega B \sin \omega t. \qquad 2$$

There are two arbitrary constants in the solution, A and B, and to evaluate them we need two pieces of information. We know that at $t = 0$, when the spring is released, the position and velocity are given by

$$x(0) = -L$$
$$v(0) = 0.$$

Using these values in Eqs. (1) and (2), we find

$$\begin{aligned} -L &= x(0) \\ &= A \sin(0) + B \cos(0) \\ &= B, \end{aligned}$$

and

$$\begin{aligned} 0 &= v(0) \\ &= \omega A \cos(0) - \omega B \sin(0) \\ &= \omega A. \end{aligned}$$

Hence

$$B = -L$$
$$A = 0.$$

Then, from the time of release until the time when the marble leaves the piston, the motion is described by the equations

$$x(t) = -L \cos \omega t \qquad 3$$
$$v(t) = \omega L \sin \omega t. \qquad 4$$

When do the marble and piston lose contact? The piston can only push, not pull, on the marble, and when the piston begins to slow down, contact is lost and the marble moves on at a constant velocity. From Eq. (4), we see that the time t_m at which the velocity reaches a maximum is given by

$$\omega t_m = \frac{\pi}{2}.$$

Substituting this in Eq. (3), we find

$$\begin{aligned} x(t_m) &= -L \cos \frac{\pi}{2} \\ &= 0. \end{aligned}$$

The marble loses contact as the spring passes its equilibrium point, as we expect, since the spring force retards the piston for $x > 0$.

From Eq. (4), the final speed of the marble is

$$v_{\max} = v(t_m) = \omega L \sin\frac{\pi}{2} = \sqrt{\frac{k}{m+M}}\,L.$$

For the highest speeds, k and L should be large and $m + M$ should be small.

Note 2.1 The Gravitational Attraction of a Spherical Shell

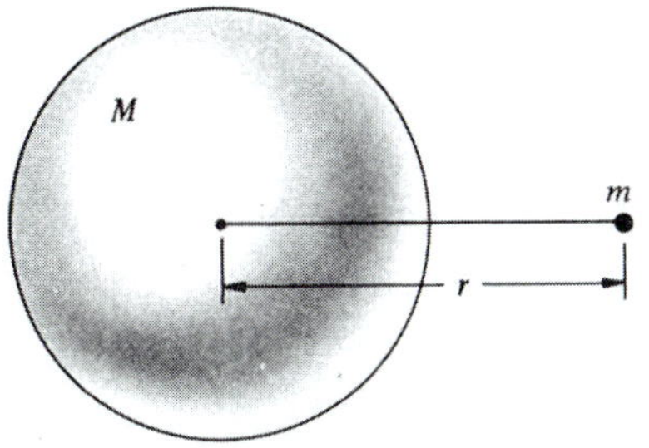

In this note we calculate the gravitational force between a uniform thin spherical shell of mass M and a particle of mass m located a distance r from its center. We shall show that the magnitude of the force is GMm/r^2 if the particle is outside the shell and zero if the particle is inside.

To attack the problem, we divide the shell into narrow rings and add their forces by using integral calculus. Let R be the radius of the shell and t its thickness, $t \ll R$. The ring at angle θ, which subtends angle $d\theta$, has circumference $2\pi R \sin\theta$, width $R\,d\theta$, and thickness t. Its volume is

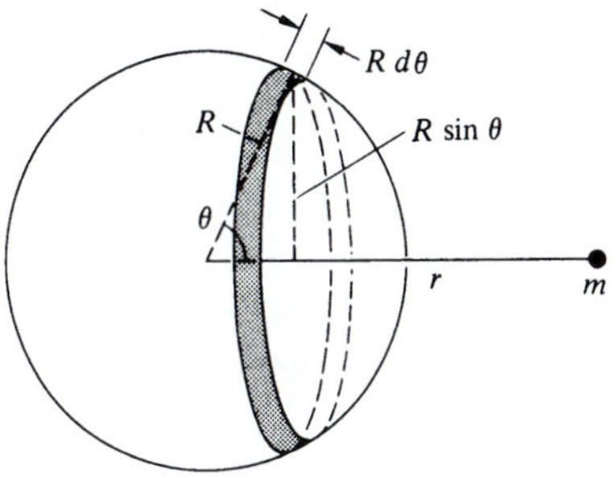

$$dV = 2\pi R^2 t \sin\theta\, d\theta$$

and its mass is

$$\rho\, dV = 2\pi R^2 t\rho \sin\theta\, d\theta = \frac{M}{2}\sin\theta\, d\theta,$$

where $\rho = M/(4\pi R^2 t)$ is the density of the shell.

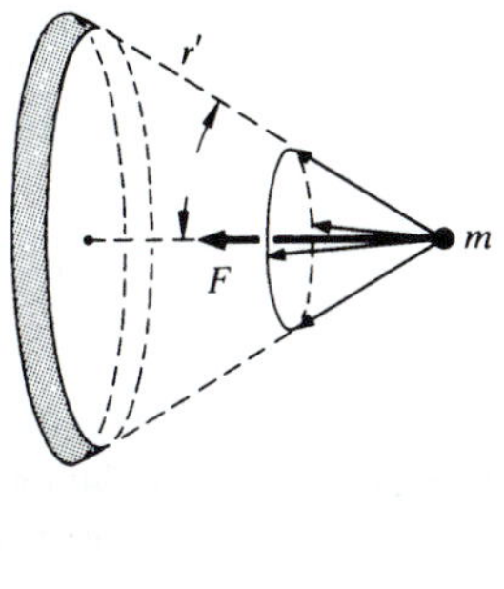

Each part of the ring is the same distance r' from m. The force on m due to a small section of the ring points toward that section. By symmetry, the transverse force components for the whole ring add vectorially to zero. Since the angle α between the force vector and the line of centers is the same for all sections of the ring, the force components along the line of centers add to give

$$dF = \frac{Gm\rho\, dV}{r'^2}\cos\alpha$$

for the whole ring.

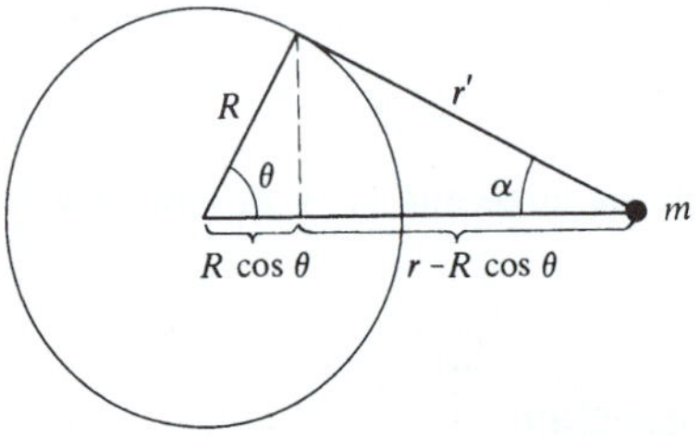

The force due to the entire shell is

$$F = \int dF$$
$$= \int \frac{Gm\rho\, dV}{r'^2} \cos \alpha.$$

The problem now is to express all the quantities in the integrand in terms of one variable, say the polar angle θ. From the sketch, $\cos \alpha = (r - R \cos \theta)/r'$, and $r' = \sqrt{r^2 + R^2 - 2rR \cos \theta}$. Since

$$\rho\, dV = M \sin \theta\, d\theta/2,$$

we have

$$F = \left(\frac{GMm}{2}\right) \int_0^\pi \frac{(r - R \cos \theta) \sin \theta\, d\theta}{(r^2 + R_2 - 2rR \cos \theta)^{3/2}}.$$

A convenient substitution for evaluating this integral is $u = r - R \cos \theta$, $du = R \sin \theta\, d\theta$. Hence

$$F = \left(\frac{GMm}{2R}\right) \int_{r-R}^{r+R} \frac{u\, du}{(R^2 - r^2 + 2ru)^{3/2}}. \qquad 1$$

This integral is listed in standard tables. The result is

$$F = \frac{GMm}{2R} \frac{1}{2r^2} \left(\sqrt{R^2 - r^2 + 2ru} - \frac{r^2 - R^2}{\sqrt{R^2 - r^2 + 2ru}} \right) \Bigg|_{r-R}^{r+R}$$
$$= \frac{GMm}{4Rr^2} \left[(r + R) - (r - R) - (r^2 - R^2) \left(\frac{1}{r + R} - \frac{1}{r - R} \right) \right]$$
$$= \frac{GMm}{r^2} \qquad r > R.$$

For $r > R$, the shell acts gravitationally as though all its mass were concentrated at its center.

There is one subtlety in our evaluation of the integral. The term $\sqrt{r^2 + R^2 - 2rR}$ is inherently positive, and we must take

$$\sqrt{r^2 + R^2 - 2rR} = r - R,$$

since $r > R$. If the particle is inside the shell, the magnitude of the force is still given by Eq. (1). However, in this case $r < R$, and we must take $\sqrt{r^2 + R^2 - 2rR} = R - r$ in the evaluation. We find

$$F = \frac{GMm}{4Rr^2} \left[(R + r) - (R - r) - (r^2 - R^2) \left(\frac{1}{R + r} - \frac{1}{R - r} \right) \right]$$
$$= 0 \qquad r < R.$$

A solid sphere can be thought of as a succession of spherical shells. It is not hard to extend our results to this case when the density of the sphere $\rho(r')$ is a function only of radial distance r' from the center of

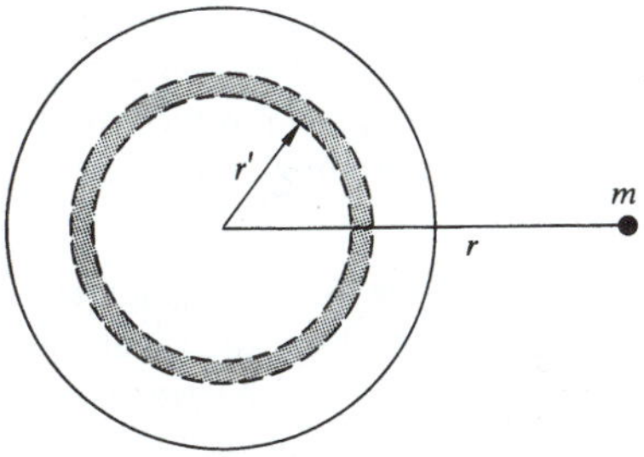

the sphere. The mass of a spherical shell of radius r' and thickness dr' is $\rho(r')4\pi r'^2\,dr'$. The force it exerts on m is

$$dF = \frac{GM}{r^2}\,\rho(r')4\pi r'^2\,dr'.$$

Since the force exerted by every shell is directed toward the center of the sphere, the total force is

$$F = \frac{Gm}{r^2}\int_0^R \rho(r')4\pi r'^2\,dr'.$$

However, the integral is simply the total mass of the sphere, and we find that for $r > R$, the force between m and the sphere is identical to the force between two particles separated a distance r.

Problems

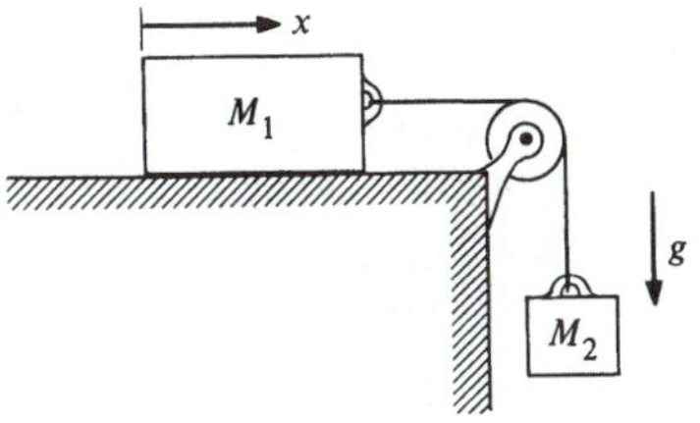

2.1 A 5-kg mass moves under the influence of a force $\mathbf{F} = (4t^2\hat{\mathbf{i}} - 3t\hat{\mathbf{j}})$ N, where t is the time in seconds (1 N = 1 newton). It starts at rest from the origin at $t = 0$. Find: (a) its velocity; (b) its position; and (c) $\mathbf{r} \times \mathbf{v}$, for any later time.

Ans. clue. (c) If $t = 1$ s, $\mathbf{r} \times \mathbf{v} = 6.7 \times 10^{-3}\hat{\mathbf{k}}$ m²/s

2.2 The two blocks shown in the sketch are connected by a string of negligible mass. If the system is released from rest, find how far block M_1 slides in time t. Neglect friction.

Ans. clue. If $M_1 = M_2$, $x = gt^2/4$

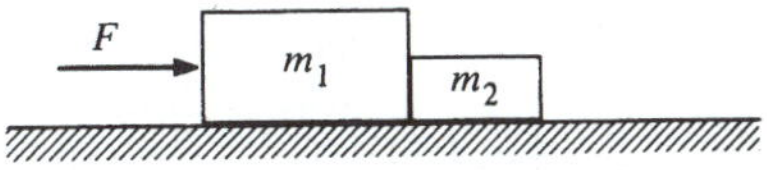

2.3 Two blocks are in contact on a horizontal table. A horizontal force is applied to one of the blocks, as shown in the drawing. If $m_1 = 2$ kg, $m_2 = 1$ kg, and $F = 3$ N, find the force of contact between the two blocks.

2.4 Two particles of mass m and M undergo uniform circular motion about each other at a separation R under the influence of an attractive force F. The angular velocity is ω radians per second. Show that $R = (F/\omega^2)(1/m + 1/M)$.

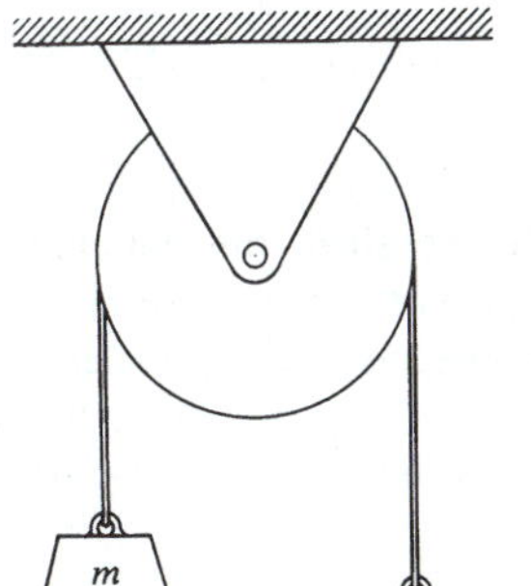

2.5 The Atwood's machine shown in the drawing has a pulley of negligible mass. Find the tension in the rope and the acceleration of M.

Ans. clue. If $M = 2m$, $T = \frac{4}{3}Mg$, $A = \frac{1}{3}g$

2.6 In a concrete mixer, cement, gravel, and water are mixed by tumbling action in a slowly rotating drum. If the drum spins too fast the ingredients stick to the drum wall instead of mixing.

Assume that the drum of a mixer has radius R and that it is mounted with its axle horizontal. What is the fastest the drum can rotate without the ingredients sticking to the wall all the time? Assume $g = 32$ ft/s².

Ans. clue. If $R = 2$ ft, $\omega_{max} = 4$ rad/s $\approx$ 38 rotations per minute

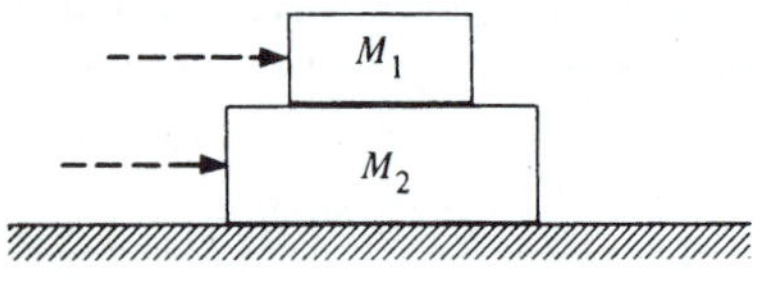

2.7 A block of mass M_1 rests on a block of mass M_2 which lies on a frictionless table. The coefficient of friction between the blocks is μ. What is the maximum horizontal force which can be applied to the blocks for them to accelerate without slipping on one another if the force is applied to (*a*) block 1 and (*b*) block 2?

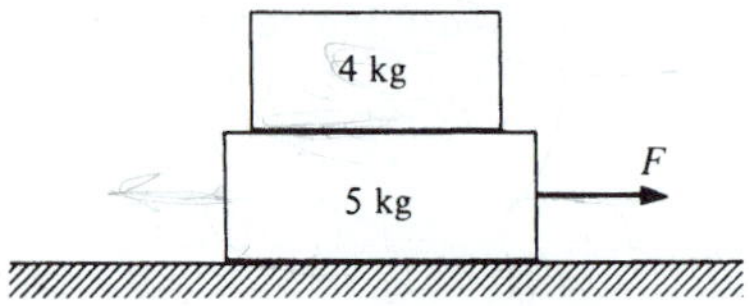

2.8 A 4-kg block rests on top of a 5-kg block, which rests on a frictionless table. The coefficient of friction between the two blocks is such that the blocks start to slip when the horizontal force F applied to the lower block is 27 N. Suppose that a horizontal force is now applied only to the upper block. What is its maximum value for the blocks to slide without slipping relative to each other?

Ans. $F = 21.6$ N

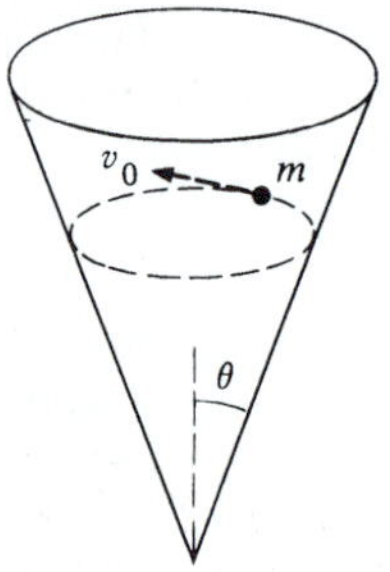

2.9 A particle of mass m slides without friction on the inside of a cone. The axis of the cone is vertical, and gravity is directed downward. The apex half-angle of the cone is θ, as shown.

The path of the particle happens to be a circle in a horizontal plane. The speed of the particle is v_0.

Draw a force diagram and find the radius of the circular path in terms of v_0, g, and θ.

2.10 Find the radius of the orbit of a synchronous satellite which circles the earth. (A synchronous satellite goes around the earth once every 24 h, so that its position appears stationary with respect to a ground station.) The simplest way to find the answer and give your results is by expressing all distances in terms of the earth's radius.

Ans. $6.6R_e$

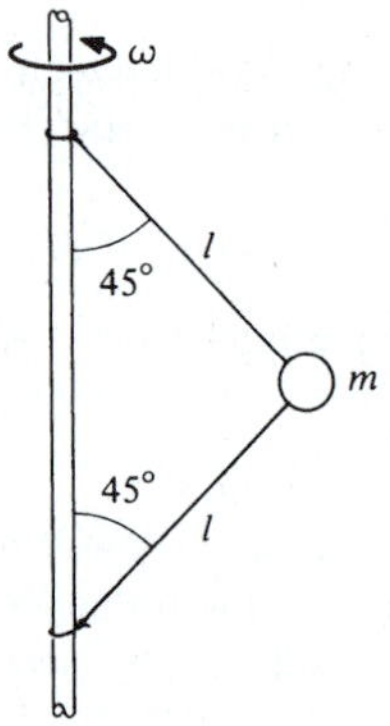

2.11 A mass m is connected to a vertical revolving axle by two strings of length l, each making an angle of 45° with the axle, as shown. Both the axle and mass are revolving with angular velocity ω. Gravity is directed downward.

a. Draw a clear force diagram for m.

b. Find the tension in the upper string, T_{up}, and lower string, T_{low}.

Ans. clue. If $l\omega^2 = \sqrt{2}\,g$, $T_{up} = \sqrt{2}\,mg$

2.12 If you have courage and a tight grip, you can yank a tablecloth out from under the dishes on a table. What is the longest time in which the cloth can be pulled out so that a glass 6 in from the edge comes to rest before falling off the table? Assume that the coefficient of friction of the glass sliding on the tablecloth or sliding on the tabletop is 0.5. (For the trick to be effective the cloth should be pulled out so rapidly that the glass does not move appreciably.)

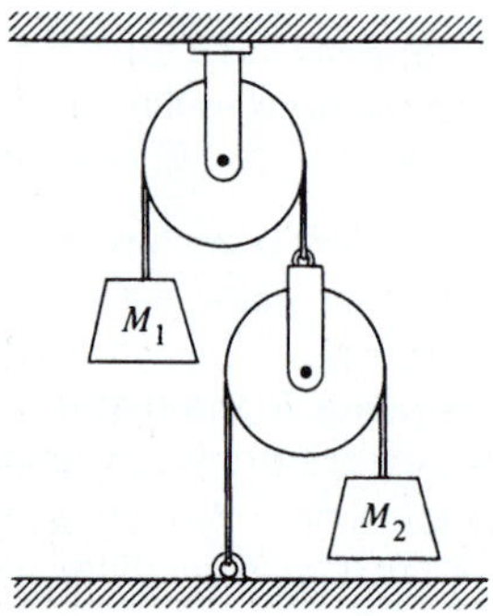

2.13 Masses M_1 and M_2 are connected to a system of strings and pulleys as shown. The strings are massless and inextensible, and the pulleys are massless and frictionless. Find the acceleration of M_1.

Ans. clue. If $M_1 = M_2$, $\ddot{x}_1 = g/5$

2.14 Two masses, A and B, lie on a frictionless table (see below left). They are attached to either end of a light rope of length l which passes around a pulley of negligible mass. The pulley is attached to a rope connected to a hanging mass, C. Find the acceleration of each mass. (You can check whether or not your answer is reasonable by considering special cases—for instance, the cases $M_A = 0$, or $M_A = M_B = M_C$.)

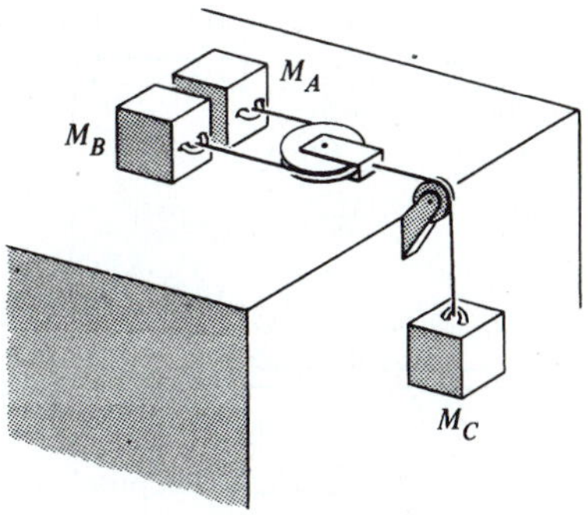

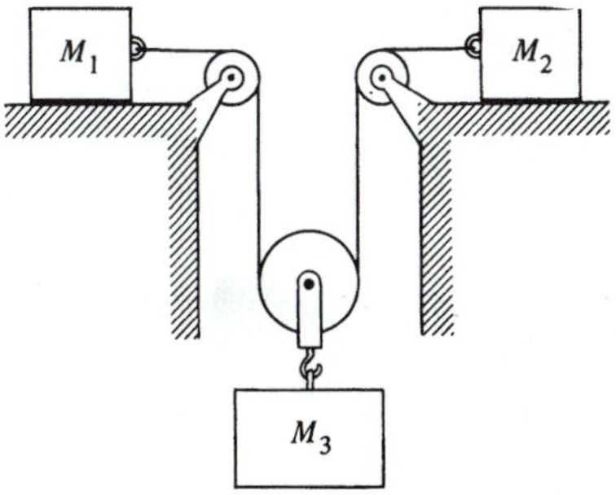

2.15 The system on the right above uses massless pulleys and rope. The coefficient of friction between the masses and horizontal surfaces is μ. Assume that M_1 and M_2 are sliding. Gravity is directed downward.

a. Draw force diagrams, and show all relevant coordinates.

b. How are the accelerations related?

c. Find the tension in the rope, T.

Ans. $T = (\mu + 1)g/[2/M_3 + 1/(2M_1) + 1/(2M_2)]$

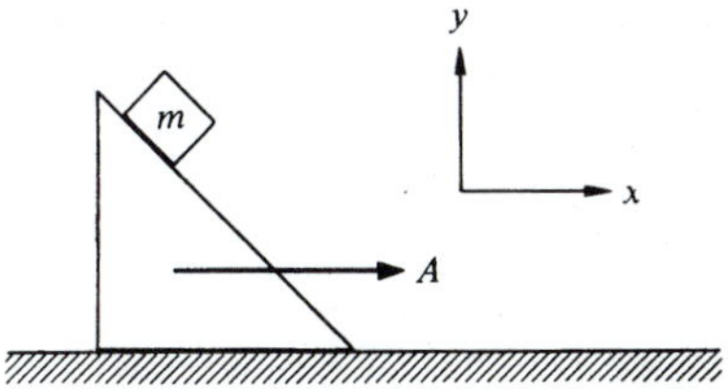

2.16 A 45° wedge is pushed along a table with constant acceleration A. A block of mass m slides without friction on the wedge. Find its acceleration. (Gravity is directed down.)

Ans. clue. If $A = 3g$, $\ddot{y} = g$

2.17 A block rests on a wedge inclined at angle θ. The coefficient of friction between the block and plane is μ.

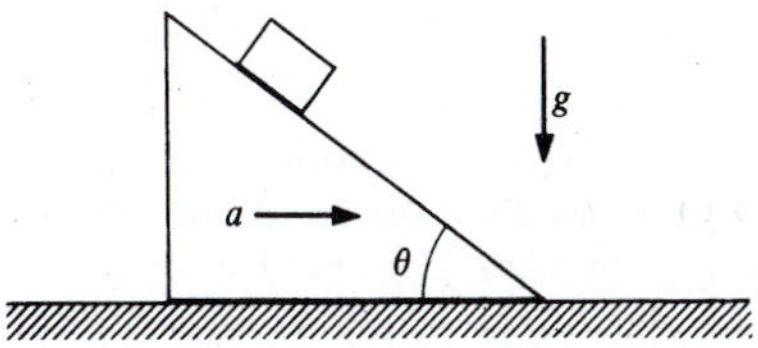

a. Find the maximum value of θ for the block to remain motionless on the wedge when the wedge is fixed in position.

Ans. $\tan \theta = \mu$

b. The wedge is given horizontal acceleration a, as shown. Assuming that $\tan \theta > \mu$, find the minimum acceleration for the block to remain on the wedge without sliding.

Ans. clue. If $\theta = \pi/4$, $a_{\min} = g(1 - \mu)/(1 + \mu)$

c. Repeat part b, but find the maximum value of the acceleration.

Ans. clue. If $\theta = \pi/4$, $a_{\max} = g(1 + \mu)/(1 - \mu)$

2.18 A painter of mass M stands on a platform of mass m and pulls himself up by two ropes which hang over pulleys, as shown. He pulls each rope with force F and accelerates upward with a uniform acceleration a. Find a—neglecting the fact that no one could do this for long.

Ans. clue. If $M = m$ and $F = Mg$, $a = g$

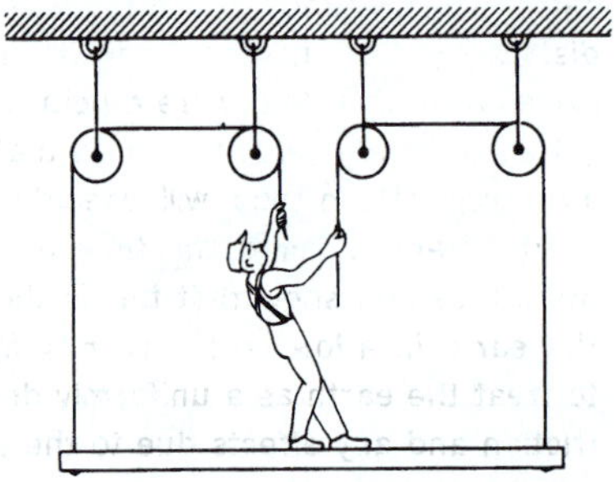

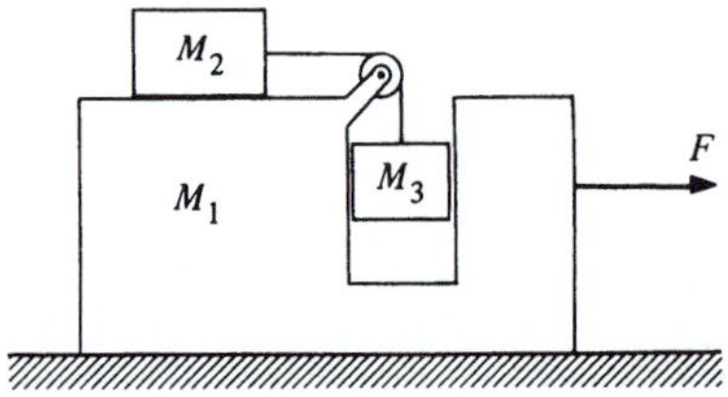

2.19 A "Pedagogical Machine" is illustrated in the sketch above. Al surfaces are frictionless. What force F must be applied to M_1 to keep M_3 from rising or falling?

Ans. clue. For equal masses, $F = 3Mg$

2.20 Consider the "Pedagogical Machine" of the last problem in the case where F is zero. Find the acceleration of M_1.

Ans. $a_1 = -M_2M_3g/(M_1M_2 + M_1M_3 + 2M_2M_3 + M_3{}^2)$

2.21 A uniform rope of mass m and length l is attached to a block of mass M. The rope is pulled with force F. Find the tension at distance x from the end of the rope. Neglect gravity.

2.22 A uniform rope of weight W hangs between two trees. The ends of the rope are the same height, and they each make angle θ with the trees. Find

a. The tension at either end of the rope

b. The tension in the middle of the rope

Ans. clue. If $\theta = 45°$, $T_{\text{end}} = W/\sqrt{2}$, $T_{\text{middle}} = W/2$

2.23 A piece of string of length l and mass M is fastened into a circular loop and set spinning about the center of a circle with uniform angular velocity ω. Find the tension in the string. Suggestion: Draw a force diagram for a small piece of the loop subtending a small angle, $\Delta\theta$.

Ans. $T = M\omega^2 l/(2\pi)^2$

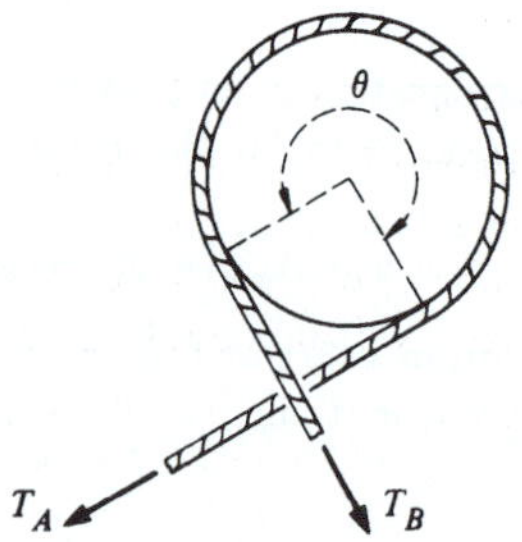

2.24 A device called a capstan is used aboard ships in order to control a rope which is under great tension. The rope is wrapped around a fixed drum, usually for several turns (the drawing shows about three-fourths turn). The load on the rope pulls it with a force T_A, and the sailor holds it with a much smaller force T_B. Can you show that $T_B = T_Ae^{-\mu\theta}$, where μ is the coefficient of friction and θ is the total angle subtended by the rope on the drum?

2.25 Find the shortest possible period of revolution of two identical gravitating solid spheres which are in circular orbit in free space about a point midway between them. (You can imagine the spheres fabricated from any material obtainable by man.)

2.26 The gravitational force on a body located at distance R from the center of a uniform spherical mass is due solely to the mass lying at distance $r \leq R$, measured from the center of the sphere. This mass exerts a force as if it were a point mass at the origin.

Use the above result to show that if you drill a hole through the earth and then fall in, you will execute simple harmonic motion about the earth's center. Find the time it takes you to return to your point of departure and show that this is the time needed for a satellite to circle the earth in a low orbit with $r \approx R_e$. In deriving this result, you need to treat the earth as a uniformly dense sphere, and you must neglect all friction and any effects due to the earth's rotation.

2.27 As a variation of the last problem, show that you will also execute simple harmonic motion with the same period even if the straight hole passes far from the earth's center.

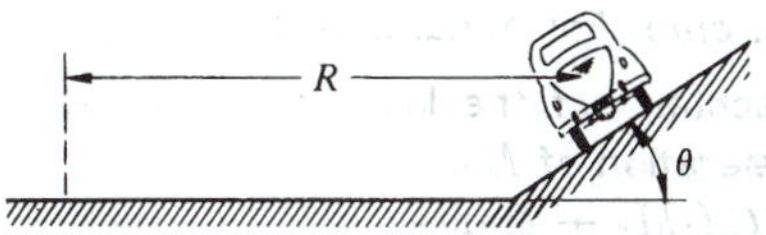

2.28 An automobile enters a turn whose radius is R. The road is banked at angle θ, and the coefficient of friction between wheels and road is μ. Find the maximum and minimum speeds for the car to stay on the road without skidding sideways.

Ans. clue. If $\mu = 1$ and $\theta = \pi/4$, all speeds are possible

2.29 A car is driven on a large revolving platform which rotates with constant angular speed ω. At $t = 0$ a driver leaves the origin and follows a line painted radially outward on the platform with constant speed v_0. The total weight of the car is W, and the coefficient of friction between the car and stage is μ.

a. Find the acceleration of the car as a function of time using polar coordinates. Draw a clear vector diagram showing the components of *acceleration* at some time $t > 0$.

b. Find the time at which the car just starts to skid.

c. Find the direction of the friction force with respect to the instantaneous position vector **r** just before the car starts to skid. Show your result on a clear diagram.

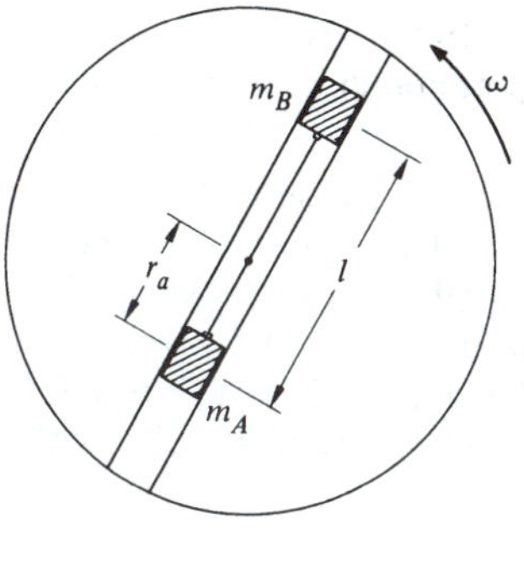

2.30 A disk rotates with constant angular velocity ω, as shown. Two masses, m_A and m_B, slide without friction in a groove passing through the center of the disk. They are connected by a light string of length l, and are initially held in position by a catch, with mass m_A at distance r_A from the center. Neglect gravity. At $t = 0$ the catch is removed and the masses are free to slide.

Find $\ddot{r}_A$ immediately after the catch is removed in terms of m_A, m_B, l, r_A, and ω.

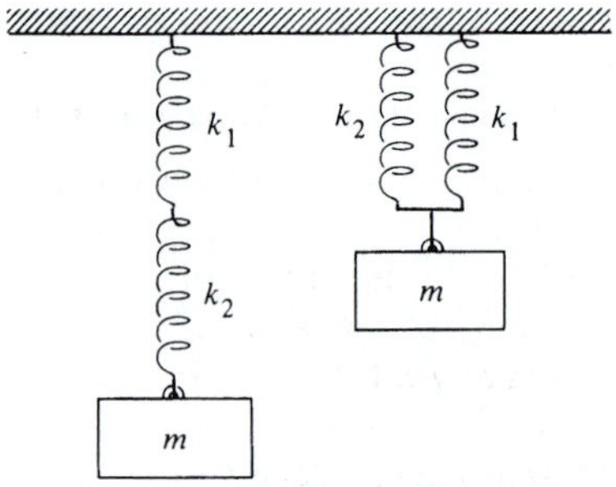

(a) (b)

2.31 Find the frequency of oscillation of mass m suspended by two springs having constants k_1 and k_2, in each of the configurations shown.

Ans. clue. If $k_1 = k_2 = k$, $\omega_a = \sqrt{k/2m}$, $\omega_b = \sqrt{2k/m}$

2.32 A wheel of radius R rolls along the ground with velocity V. A pebble is carefully released on top of the wheel so that it is instantaneously at rest on the wheel.

a. Show that the pebble will immediately fly off the wheel if $V > \sqrt{Rg}$.

b. Show that in the case where $V < \sqrt{Rg}$, and the coefficient of friction is $\mu = 1$, the pebble starts to slide when it has rotated through an angle given by $\theta = \arccos [(1/\sqrt{2})(V^2/Rg)] - \pi/4$.

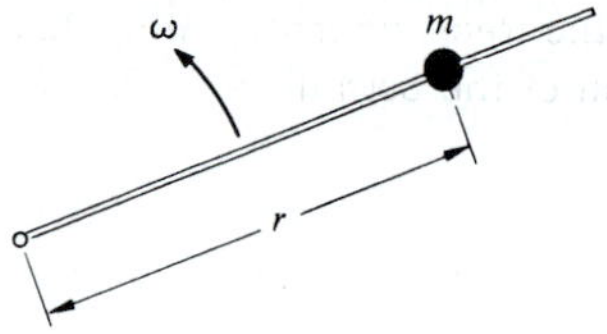

2.33 A particle of mass m is free to slide on a thin rod. The rod rotates in a plane about one end at constant angular velocity ω. Show that the motion is given by $r = Ae^{-\gamma t} + Be^{+\gamma t}$, where γ is a constant which you must find and A and B are arbitrary constants. Neglect gravity.

Show that for a particular choice of initial conditions [that is, $r(t = 0)$ and $v(t = 0)$], it is possible to obtain a solution such that r decreases continually in time, but that for any other choice r will eventually increase. (Exclude cases where the bead hits the origin.)

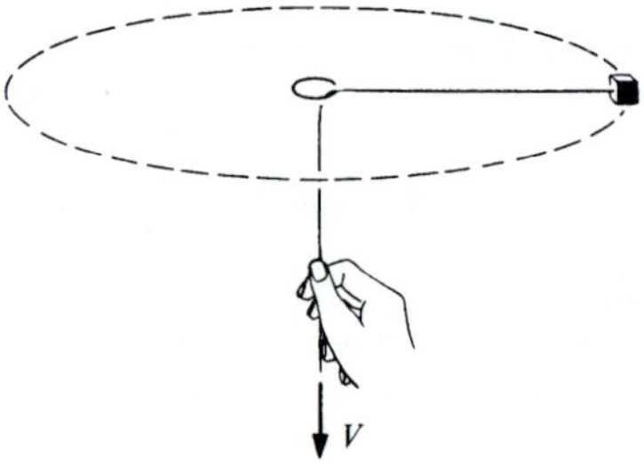

2.34. A mass m whirls around on a string which passes through a ring, as shown. Neglect gravity. Initially the mass is distance r_0 from the center and is revolving at angular velocity ω_0. The string is pulled with constant velocity V starting at $t = 0$ so that the radial distance to the mass decreases. Draw a force diagram and obtain a differential equation for ω. This equation is quite simple and can be solved either by inspection or by formal integration. Find

a. $\omega(t)$.

Ans. clue. For $Vt = r_0/2$, $\omega = 4\omega_0$

b. The force needed to pull the string.

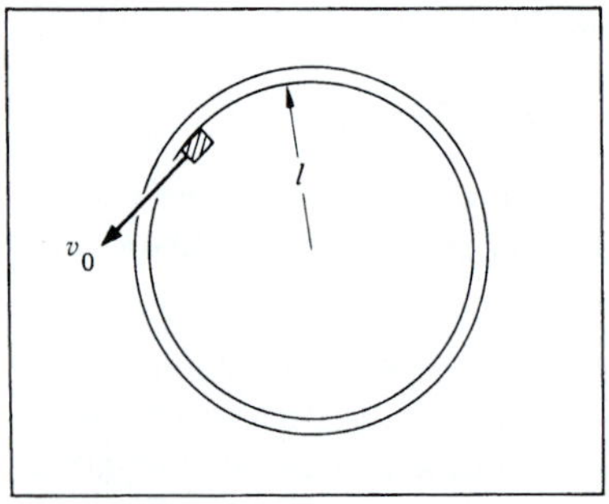

2.35 This problem involves solving a simple differential equation.

A block of mass m slides on a frictionless table. It is constrained to move inside a ring of radius l which is fixed to the table. At $t = 0$, the block is moving along the inside of the ring (i.e., in the tangential direction) with velocity v_0. The coefficient of friction between the block and the ring is μ.

a. Find the velocity of the block at later times.

Ans. $v_0/[1 + (\mu v_0 t/l)]$

b. Find the position of the block at later times.

2.36 This problem involves a simple differential equation. You should be able to integrate it after a little "playing around."

A particle of mass m moving along a straight line is acted on by a retarding force (one always directed against the motion) $F = be^{av}$, where

b and α are constants and v is the velocity. At $t = 0$ it is moving with velocity v_0. Find the velocity at later times.

Ans. $v(t) = (1/\alpha) \ln [1/(\alpha bt/m + e^{-\alpha v_0})]$

2.37 The Eureka Hovercraft Corporation wanted to hold hovercraft races as an advertising stunt. The hovercraft supports itself by blowing air downward, and has a big fixed propeller on the top deck for forward propulsion. Unfortunately, it has no steering equipment, so that the pilots found that making high speed turns was very difficult. The company decided to overcome this problem by designing a bowl shaped track in which the hovercraft, once up to speed, would coast along in a circular path with no need to steer. They hired an engineer to design and build the track, and when he finished, he hastily left the country. When the company held their first race, they found to their dismay that the craft took exactly the same time T to circle the track, no matter what its speed. Find the equation for the cross section of the bowl in terms of T.

3 MOMENTUM

3.1 Introduction

In the last chapter we made a gross simplification by treating nature as if it were composed of point particles rather than real, *extended* bodies. Sometimes this simplification is justified—as in the study of planetary motion, where the size of the planets is of little consequence compared with the vast distances which characterize our solar system, or in the case of elementary particles moving through an accelerator, where the size of the particles, about 10^{-15} m, is minute compared with the size of the machine. However, these cases are unusual. Much of the time we deal with large bodies which may have elaborate structure. For instance, consider the landing of a spacecraft on the moon. Even if we could calculate the gravitational field of such an irregular and inhomogeneous body as the moon, the spacecraft itself is certainly not a point particle—it has spiderlike legs, gawky antennas, and a lumpy body.

Furthermore, the methods of the last chapter fail us when we try to analyze systems such as rockets in which there is a flow of mass. Rockets accelerate forward by ejecting mass backward; it is hard to see how to apply $\mathbf{F} = M\mathbf{a}$ to such a system.

In this chapter we shall generalize the laws of motion to overcome these difficulties. We begin by restating Newton's second law in a slightly modified form. In Chap. 2 we wrote the law in the familiar form

$$\mathbf{F} = M\mathbf{a}. \qquad 3.1$$

This is not quite the way Newton wrote it. He chose to write

$$\mathbf{F} = \frac{d}{dt}M\mathbf{v}. \qquad 3.2$$

For a particle in newtonian mechanics, M is a constant and $(d/dt)(M\mathbf{v}) = M(d\mathbf{v}/dt) = M\mathbf{a}$, as before. The quantity $M\mathbf{v}$, which plays a prominent role in mechanics, is called *momentum*. Momentum is the product of a vector $\mathbf{v}$ and a scalar M. Denoting momentum by $\mathbf{p}$, Newton's second law becomes

$$\mathbf{F} = \frac{d\mathbf{p}}{dt}. \qquad 3.3$$

This form is preferable to $\mathbf{F} = M\mathbf{a}$ because it is readily generalized to complex systems, as we shall soon see, and because momentum turns out to be more fundamental than mass or velocity separately.

3.2 Dynamics of a System of Particles

Consider a system of interacting particles. One example of such a system is the sun and planets, which are so far apart compared with their diameters that they can be treated as simple particles to good approximation. All particles in the solar system interact via gravitational attraction; the chief interaction is with the sun, although the interaction of the planets with each other also influences their motion. In addition, the entire solar system is attracted by far off matter.

At the other extreme, the system could be a billiard ball resting on a table. Here the particles are atoms (disregarding for now the fact that atoms are not point particles but are themselves composed of smaller particles) and the interactions are primarily interatomic electric forces. The external forces on the billiard ball include the gravitational force of the earth and the contact force of the tabletop.

We shall now prove some simple properties of physical systems. We are free to choose the boundaries of the system as we please, but once the choice is made, we must be consistent about which particles are included in the system and which are not. We suppose that the particles in the system interact with particles outside the system as well as with each other. To make the argument general, consider a system of N interacting particles with masses $m_1, m_2, m_3, \ldots, m_N$. The position of the jth particle is $\mathbf{r}_j$, the force on it is $\mathbf{f}_j$, and its momentum is $\mathbf{p}_j = m_j\dot{\mathbf{r}}_j$. The equation of motion for the jth particle is

$$\mathbf{f}_j = \frac{d\mathbf{p}_j}{dt}. \qquad 3.4$$

The force on particle j can be split into two terms:

$$\mathbf{f}_j = \mathbf{f}_j^{\text{int}} + \mathbf{f}_j^{\text{ext}}. \qquad 3.5$$

Here $\mathbf{f}_j^{\text{int}}$, the *internal* force on particle j, is the force due to all other particles in the system, and $\mathbf{f}_j^{\text{ext}}$, the *external* force on particle j, is the force due to sources outside the system. The equation of motion becomes

$$\mathbf{f}_j^{\text{int}} + \mathbf{f}_j^{\text{ext}} = \frac{d\mathbf{p}_j}{dt}. \qquad 3.6$$

Now let us focus on the system as a whole by the following stratagem: add all the equations of motion of all the particles in the system.

$$\mathbf{f}_1^{\text{int}} + \mathbf{f}_1^{\text{ext}} = \frac{d\mathbf{p}_1}{dt}$$

$$\cdots\cdots\cdots\cdots$$

$$\mathbf{f}_j^{\text{int}} + \mathbf{f}_j^{\text{ext}} = \frac{d\mathbf{p}_j}{dt} \qquad 3.7$$

$$\cdots\cdots\cdots\cdots$$

$$\mathbf{f}_N^{\text{int}} + \mathbf{f}_N^{\text{ext}} = \frac{d\mathbf{p}_N}{dt}.$$

The result of adding these equations can be written

$$\Sigma\mathbf{f}_j^{\text{int}} + \Sigma\mathbf{f}_j^{\text{ext}} = \sum \frac{d\mathbf{p}_j}{dt}. \qquad 3.8$$

The summations extend over all particles, $j = 1, \ldots, N$.

The second term, $\Sigma\mathbf{f}_j^{\text{ext}}$, is the sum of all external forces acting on all the particles. It is the *total external force* acting on the system, $\mathbf{F}_{\text{ext}}$.

$$\Sigma\mathbf{f}_j^{\text{ext}} \equiv \mathbf{F}_{\text{ext}}.$$

The first term in Eq. (3.8), $\Sigma\mathbf{f}_j^{\text{int}}$, is the sum of all internal forces acting on all the particles. According to Newton's third law, the forces between any two particles are equal and opposite so that their sum is zero. It follows that the sum of all the forces between all the particles is also zero; the internal forces cancel in pairs. Hence

$$\Sigma\mathbf{f}_j^{\text{int}} = 0.$$

Equation (3.8) then simplifies to

$$\mathbf{F}_{\text{ext}} = \sum \frac{d\mathbf{p}_j}{dt}. \qquad 3.9$$

The right hand side can be written $\Sigma(d\mathbf{p}_j/dt) = (d/dt)\Sigma\mathbf{p}_j$, since the derivative of a sum is the sum of the derivatives. $\Sigma\mathbf{p}_j$ is the *total momentum* of the system, which we designate by $\mathbf{P}$.

$$\mathbf{P} \equiv \Sigma\mathbf{p}_j. \qquad 3.10$$

With this substitution, Eq. (3.9) becomes

$$\mathbf{F}_{\text{ext}} = \frac{d\mathbf{P}}{dt}. \tag{3.11}$$

In words, the total external force applied to a system equals the rate of change of the system's momentum. This is true irrespective of the details of the interaction; $\mathbf{F}_{\text{ext}}$ could be a single force acting on a single particle, or it could be the resultant of many tiny interactions involving each particle of the system.

Example 3.1 **The Bola**

The bola is a weapon used by gauchos for entangling animals. It consists of three balls of stone or iron connected by thongs. The gaucho whirls the bola in the air and hurls it at the animal. What can we say about its motion?

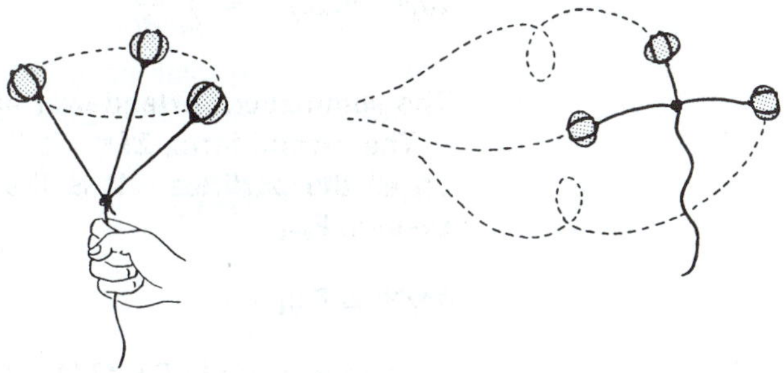

Consider a bola with masses m_1, m_2, and m_3. The balls are pulled by the binding thong and by gravity. (We neglect air resistance.) Since the constraining forces depend on the instantaneous positions of all three balls, it is a real problem even to write the equation of motion of one ball. However, the total momentum obeys the simple equation

$$\frac{d\mathbf{P}}{dt} = \mathbf{F}_{\text{ext}} = \mathbf{f}_1^{\text{ext}} + \mathbf{f}_2^{\text{ext}} + \mathbf{f}_3^{\text{ext}}$$

$$= m_1\mathbf{g} + m_2\mathbf{g} + m_3\mathbf{g}$$

or

$$\frac{d\mathbf{P}}{dt} = M\mathbf{g},$$

where M is the total mass. This equation represents an important first step in finding the detailed motion. The equation is identical to that of a single particle of mass M with momentum $\mathbf{P}$. This is a familiar fact to the gaucho who forgets that he has a complicated system when he hurls the bola; he instinctively aims it like a single mass.

Center of Mass

According to Eq. (3.11),

$$\mathbf{F} = \frac{d\mathbf{P}}{dt}, \tag{3.12}$$

where we have dropped the subscript *ext* with the understanding that **F** stands for the external force. This result is identical to the equation of motion of a single particle, although in fact it refers to a system of particles. It is tempting to push the analogy between Eq. (3.12) and single particle motion even further by writing

$$\mathbf{F} = M\ddot{\mathbf{R}}, \tag{3.13}$$

where M is the total mass of the system and **R** is a vector yet to be defined. Since $\mathbf{P} = \Sigma m_j\dot{\mathbf{r}}_j$, Eq. (3.12) and (3.13) give

$$M\ddot{\mathbf{R}} = \frac{d\mathbf{P}}{dt} = \Sigma m_j\ddot{\mathbf{r}}_j,$$

which is true if

$$\mathbf{R} = \frac{1}{M}\Sigma m_j\mathbf{r}_j. \tag{3.14}$$

R is a vector from the origin to the point called the *center of mass.* The system behaves as if all the mass is concentrated at the center of mass and all the external forces act at that point.

We are often interested in the motion of comparatively rigid bodies like baseballs or automobiles. Such a body is merely a system of particles which are fixed relative to each other by strong internal forces; Eq. (3.13) shows that with respect to external forces, the body behaves as if it were a point particle. In Chap. 2, we casually treated every body as if it were a particle; we see now that this is justified provided that we focus attention on the center of mass.

You may wonder whether this description of center of mass motion isn't a gross oversimplification—experience tells us that an extended body like a plank behaves differently from a compact body like a rock, even if the masses are the same and we apply the same force. We are indeed oversimplifying. The relation $\mathbf{F} = M\ddot{\mathbf{R}}$ describes only the translation of the body (the motion of its center of mass); it does not describe the body's orientation in space. In Chaps. 6 and 7 we shall investigate the rotation of extended bodies, and it will turn out that the rotational motion of a body depends both on its shape and the point where the forces are applied. Nevertheless, as far as translation of the center of mass is concerned, $\mathbf{F} = M\ddot{\mathbf{R}}$ tells the whole story. This result is true for any system of particles, not just for those fixed in rigid objects, as long as the forces between the particles obey Newton's third law. It is immaterial whether or not the particles move relative to each other and whether or not there happens to be any matter at the center of mass.

Example 3.2 **Drum Major's Baton**

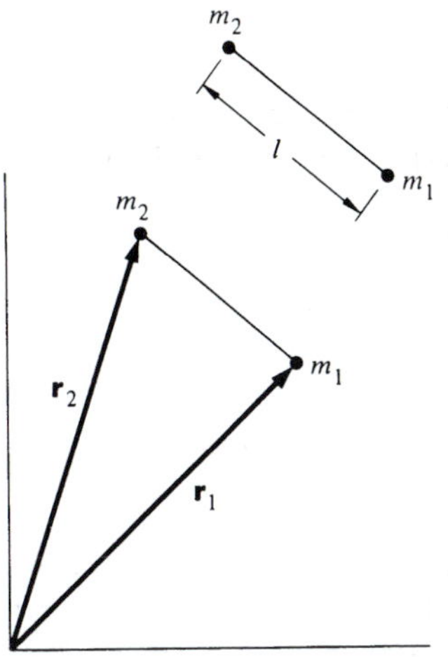

A drum major's baton consists of two masses m_1 and m_2 separated by a thin rod of length l. The baton is thrown into the air. The problem is to find the baton's center of mass and the equation of motion for the center of mass.

Let the position vectors of m_1 and m_2 be $\mathbf{r}_1$ and $\mathbf{r}_2$. The position vector of the center of mass, measured from the same origin, is

$$\mathbf{R} = \frac{m_1\mathbf{r}_1 + m_2\mathbf{r}_2}{m_1 + m_2}, \qquad 1$$

where we have neglected the mass of the thin rod. The center of mass lies on the line joining m_1 and m_2. To show this, suppose first that the tip of $\mathbf{R}$ does not lie on the line, and consider the vectors $\mathbf{r}_1'$, $\mathbf{r}_2'$ from the tip of $\mathbf{R}$ to m_1 and m_2. From the sketch we see that

$$\mathbf{r}_1' = \mathbf{r}_1 - \mathbf{R}$$
$$\mathbf{r}_2' = \mathbf{r}_2 - \mathbf{R}.$$

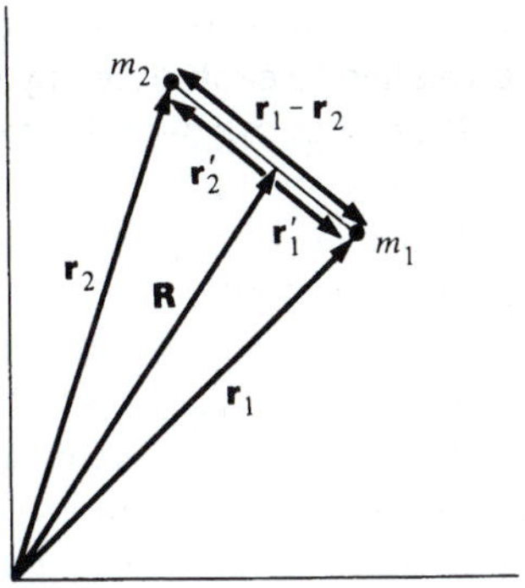

Using Eq. (1) gives

$$\mathbf{r}_1' = \mathbf{r}_1 - \frac{m_1\mathbf{r}_1}{m_1 + m_2} - \frac{m_2\mathbf{r}_2}{m_1 + m_2}$$
$$= \frac{m_2}{m_1 + m_2}(\mathbf{r}_1 - \mathbf{r}_2)$$
$$\mathbf{r}_2' = \mathbf{r}_2 - \frac{m_1\mathbf{r}_1}{m_1 + m_2} - \frac{m_2\mathbf{r}_2}{m_1 + m_2}$$
$$= -\left(\frac{m_1}{m_1 + m_2}\right)(\mathbf{r}_1 - \mathbf{r}_2).$$

$\mathbf{r}_1'$ and $\mathbf{r}_2'$ are proportional to $\mathbf{r}_1 - \mathbf{r}_2$, the vector from m_1 to m_2. Hence $\mathbf{r}_1'$ and $\mathbf{r}_2'$ lie along the line joining m_1 and m_2, as shown. Furthermore,

$$r_1' = \frac{m_2}{m_1 + m_2}|\mathbf{r}_1 - \mathbf{r}_2|$$

$$= \frac{m_2}{m_1 + m_2}l$$

and

$$r_2' = \frac{m_1}{m_1 + m_2}|\mathbf{r}_1 - \mathbf{r}_2|$$

$$= \frac{m_1}{m_1 + m_2}l.$$

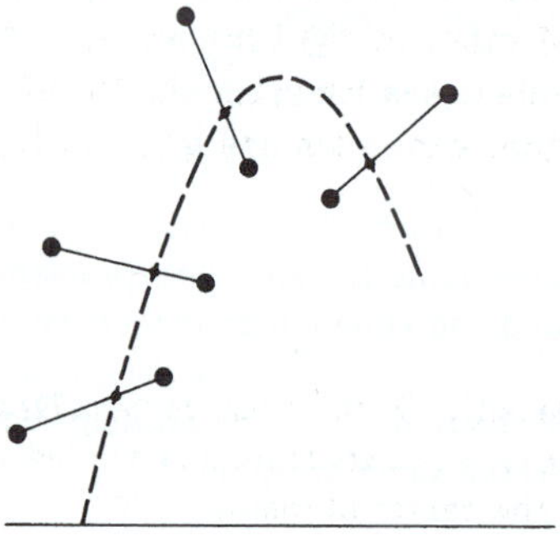

Assuming that friction is negligible, the external force on the baton is $\mathbf{F} = m_1\mathbf{g} + m_2\mathbf{g}$.

The equation of motion of the center of mass is

$$(m_1 + m_2)\ddot{\mathbf{R}} = (m_1 + m_2)\mathbf{g}$$

or

$$\ddot{\mathbf{R}} = \mathbf{g}.$$

The center of mass follows the parabolic trajectory of a single mass in a uniform gravitational field. With the methods developed in Chap. 6, we shall be able to find the motion of m_1 and m_2 about the center of mass, completing the solution to the problem.

Although it is a simple matter to find the center of mass of a system of particles, the procedure for locating the center of mass of an extended body is not so apparent. However, it is a straightforward task with the help of calculus. We proceed by dividing the body into N mass elements. If $\mathbf{r}_j$ is the position of the jth element, and m_j is its mass, then

$$\mathbf{R} = \frac{1}{M}\sum_{j=1}^{N} m_j\mathbf{r}_j.$$

The result is not rigorous, since the mass elements are not true particles. However, in the limit where N approaches infinity, the size of each element approaches zero and the approximation becomes exact.

$$\mathbf{R} = \lim_{N\to\infty} \frac{1}{M}\sum_{j=1}^{N} m_j\mathbf{r}_j.$$

This limiting process defines an integral. Formally

$$\lim_{N\to\infty} \sum_{j=1}^{\infty} m_j\mathbf{r}_j = \int \mathbf{r}\,dm,$$

where dm is a differential mass element. Then

$$\mathbf{R} = \frac{1}{M}\int \mathbf{r}\, dm. \tag{3.15}$$

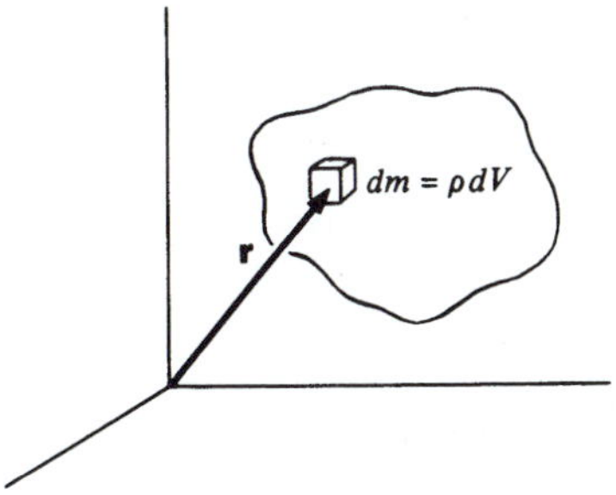

To visualize this integral, think of dm as the mass in an element of volume dV located at position $\mathbf{r}$. If the mass density at the element is ρ, then $dm = \rho\, dV$ and

$$\mathbf{R} = \frac{1}{M}\int \mathbf{r}\rho\, dV.$$

This integral is called a volume integral. Although it is important to know how to find the center of mass of rigid bodies, we shall only be concerned with a few simple cases here, as illustrated by the following two examples. Further examples are given in Note 3.1 at the end of the chapter.

Example 3.3 **Center of Mass of a Nonuniform Rod**

A rod of length L has a nonuniform density. λ, the mass per unit length of the rod, varies as $\lambda = \lambda_0(s/L)$, where λ_0 is a constant and s is the distance from the end marked 0. Find the center of mass.

It is apparent that $\mathbf{R}$ lies on the rod. Let the origin of the coordinate system coincide with the end of the rod, 0, and let the x axis lie along the rod so that $s = x$. The mass in an element of length dx is $dm = \lambda\, dx = \lambda_0 x\, dx/L$. The rod extends from $x = 0$ to $x = L$ and the total mass is

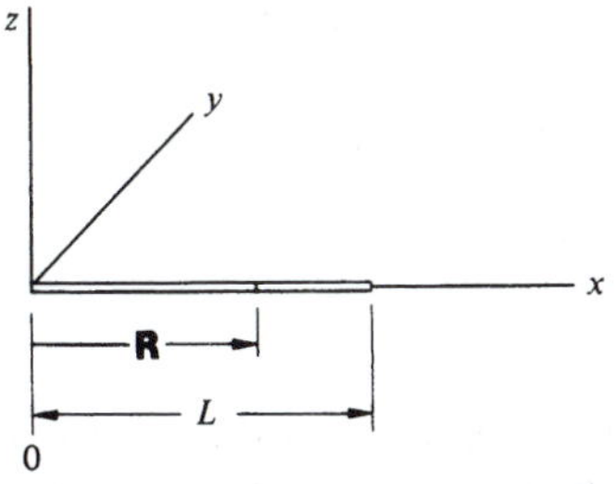

$$\begin{aligned} M &= \int dm \\ &= \int_0^L \lambda\, dx \\ &= \int_0^L \frac{\lambda_0 x\, dx}{L} \\ &= \tfrac{1}{2}\lambda_0 L. \end{aligned}$$

The center of mass is at

$$\begin{aligned} \mathbf{R} &= \frac{1}{M}\int \mathbf{r}\lambda\, dM \\ &= \frac{2}{\lambda_0 L}\int_0^L (x\hat{\mathbf{i}} + 0\hat{\mathbf{j}} + 0\hat{\mathbf{k}})\frac{\lambda_0 x\, dx}{L} \\ &= \frac{2}{L^2}\frac{\hat{\mathbf{i}}}{3}x^3\Big|_0^L \\ &= \tfrac{2}{3}L\hat{\mathbf{i}}. \end{aligned}$$

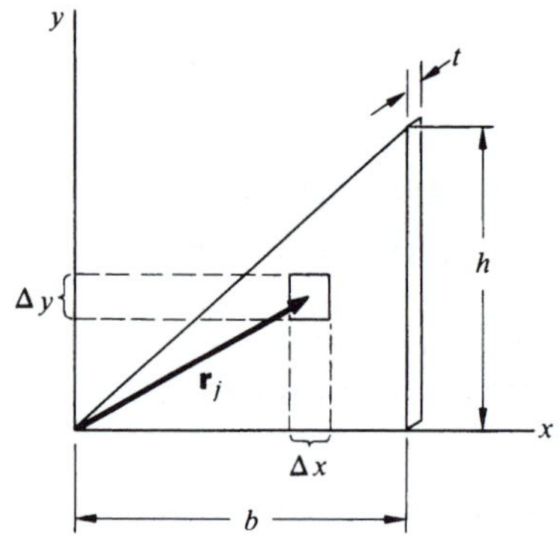

Example 3.4 Center of Mass of a Triangular Sheet

Consider the two dimensional case of a uniform right triangular sheet of mass M, base b, height h, and small thickness t. If we divide the sheet into small rectangular areas of side Δx and Δy, as shown, then the volume of each element is $\Delta V = t\,\Delta x\,\Delta y$, and

$$\mathbf{R} \approx \frac{\Sigma m_j \mathbf{r}_j}{M} = \frac{\Sigma \rho_j t\,\Delta x\,\Delta y \mathbf{r}_j}{M},$$

where j is the label of one of the volume elements and ρ_j is the density. Because the sheet is uniform,

$$\rho_j = \text{constant} = \frac{M}{V} = \frac{M}{At},$$

where A is the area of the sheet.

We can carry out the sum by summing first over the Δx's and then over the Δy's, instead of over the single index j. This gives a double sum which can be converted to a double integral by taking the limit, as follows:

$$\mathbf{R} = \lim_{\substack{\Delta x \to 0 \\ \Delta y \to 0}} \left(\frac{M}{At}\right)\left(\frac{t}{M}\right) \Sigma\Sigma \mathbf{r}_j\,\Delta x\,\Delta y = \frac{1}{A}\iint \mathbf{r}\,dx\,dy.$$

Let $\mathbf{r} = x\hat{\mathbf{i}} + y\hat{\mathbf{j}}$ be the position vector of an element $dx\,dy$. Then, writing $\mathbf{R} = X\hat{\mathbf{i}} + Y\hat{\mathbf{j}}$, we have

$$\begin{aligned}\mathbf{R} &= X\hat{\mathbf{i}} + Y\hat{\mathbf{j}} \\ &= \frac{1}{A}\iint (x\hat{\mathbf{i}} + y\hat{\mathbf{j}})\,dx\,dy \\ &= \frac{1}{A}\left(\iint x\,dx\,dy\right)\hat{\mathbf{i}} + \frac{1}{A}\left(\iint y\,dx\,dy\right)\hat{\mathbf{j}}.\end{aligned}$$

Hence the coordinates of the center of mass are given by

$$X = \frac{1}{A}\iint x\,dx\,dy$$

$$Y = \frac{1}{A}\iint y\,dx\,dy.$$

The double integrals may look strange, but they are easily evaluated. Consider first the double integral

$$X = \frac{1}{A}\iint x\,dx\,dy.$$

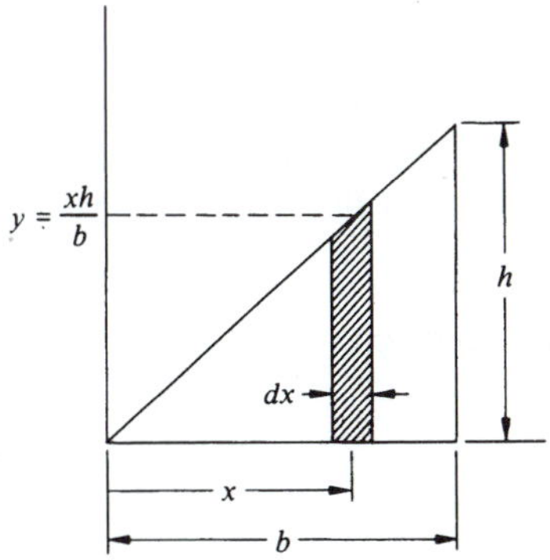

This integral instructs us to take each element, multiply its area by its x coordinate, and sum the results. We can do this in stages by first considering the elements in a strip parallel to the y axis. The strip runs from $y = 0$ to $y = xh/b$. Each element in the strip has the same x coordinate, and the contribution of the strip to the double integral is

$$\frac{1}{A}x\,dx\int_0^{xh/b} dy = \frac{h}{bA}x^2\,dx.$$

Finally, we sum the contributions of all such strips $x = 0$ to $x = b$ to find

$$X = \frac{h}{bA}\int_0^b x^2\,dx = \frac{h}{bA}\frac{b^3}{3}$$
$$= \frac{hb^2}{3A}.$$

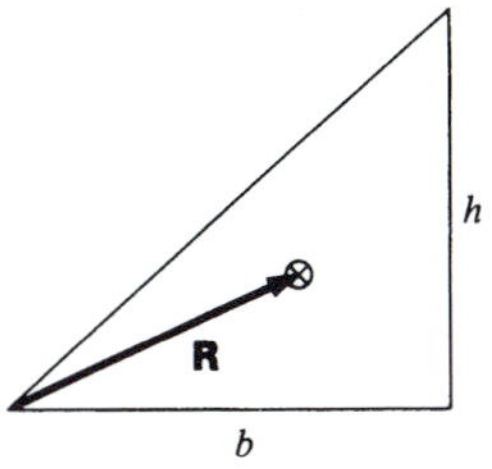

Since $A = \frac{1}{2}bh$,

$$X = \tfrac{2}{3}b.$$

Similarly,

$$Y = \frac{1}{A}\int_0^b\left(\int_0^{xh/b} y\,dy\right)dx$$
$$= \frac{h^2}{2Ab^2}\int_0^b x^2\,dx = \frac{h^2b}{6A}$$
$$= \tfrac{1}{3}h.$$

Hence

$$\mathbf{R} = \tfrac{2}{3}b\hat{\mathbf{i}} + \tfrac{1}{3}h\hat{\mathbf{j}}.$$

Although the coordinates of **R** depend on the particular coordinate system we choose, the position of the center of mass with respect to the triangular plate is, of course, independent of the coordinate system.

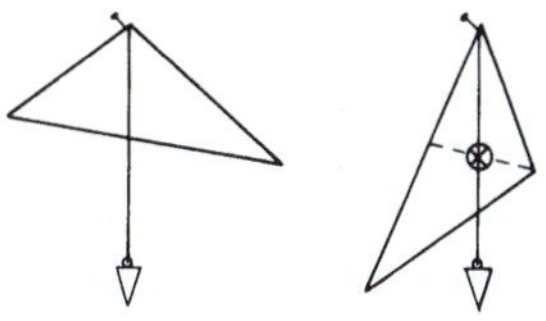

Often physical arguments are more useful than mathematical analysis. For instance, to find the center of mass of an irregular plane object, let it hang from a pivot and draw a plumb line from the pivot. The center of mass will hang directly below the pivot (this may be intuitively be obvious, and it can easily be proved with the methods of Chap. 6), and it is somewhere on the plumb line. Repeat the procedure with a different pivot point. The two lines intersect at the center of mass.

Example 3.5 Center of Mass Motion

A rectangular box is held with one corner resting on a frictionless table and is gently released. It falls in a complex tumbling motion, which we are not yet prepared to solve because it involves rotation. However, there is no difficulty in finding the trajectory of the center of mass.

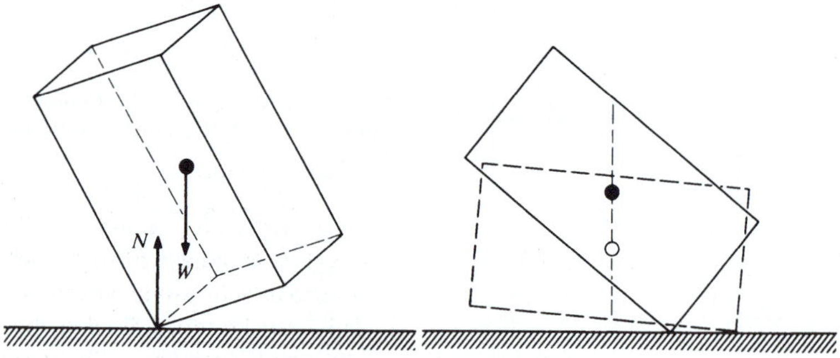

The external forces acting on the box are gravity and the normal force of the table. Neither of these has a horizontal component, and so the center of mass must accelerate vertically. For a uniform box, the center of mass is at the geometrical center. If the box is released from rest, then its center falls straight down.

3.3 Conservation of Momentum

In the last section we found that the total external force **F** acting on a system is related to the total momentum **P** of the system by

$$\mathbf{F} = \frac{d\mathbf{P}}{dt}.$$

Consider the implications of this for an isolated system, that is, a system which does not interact with its surroundings. In this case $\mathbf{F} = 0$, and $d\mathbf{P}/dt = 0$. The total momentum is constant; no matter how strong the interactions among an isolated system of particles, and no matter how complicated the motions, the total momentum of an isolated system is constant. This is the law of conservation of momentum. As we shall show, this apparently simple law can provide powerful insights into complicated systems.

Example 3.6 **Spring Gun Recoil**

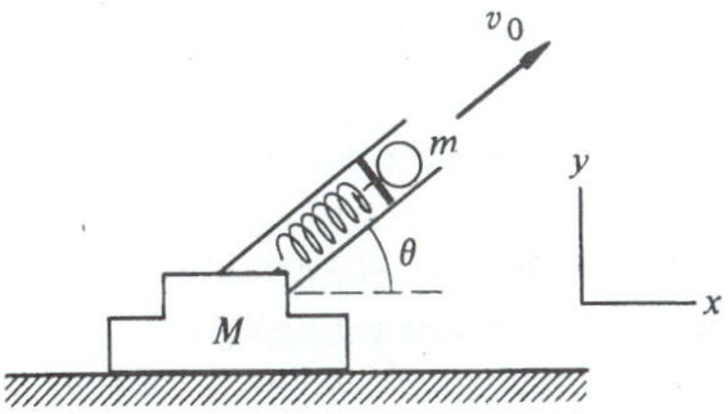

A loaded spring gun, initially at rest on a horizontal frictionless surface, fires a marble at angle of elevation θ. The mass of the gun is M, the mass of the marble is m, and the muzzle velocity of the marble is v_0. What is the final motion of the gun?

Take the physical system to be the gun and marble. Gravity and the normal force of the table act on the system. Both these forces are vertical. Since there are no horizontal external forces, the x component of the vector equation $\mathbf{F} = d\mathbf{P}/dt$ is

$$0 = \frac{dP_x}{dt}. \qquad 1$$

According to Eq. (1), P_x is conserved:

$$P_{x,\text{initial}} = P_{x,\text{final}}. \qquad 2$$

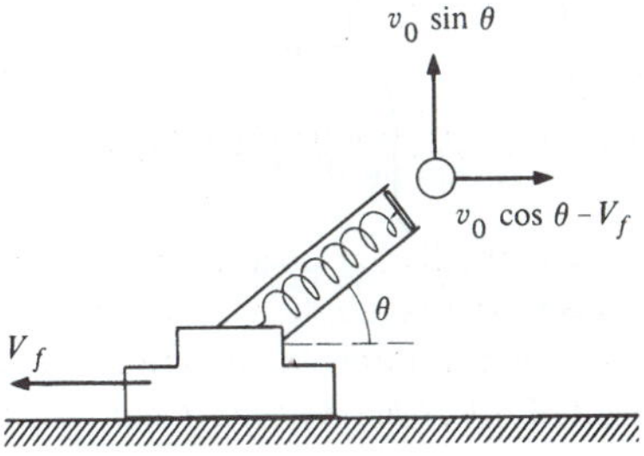

Let the initial time be prior to firing the gun. Then $P_{x,\text{initial}} = 0$, since the system is initially at rest. After the marble has left the muzzle, the gun recoils with some speed V_f, and its final horizontal momentum is MV_f, to the left. Finding the final velocity of the marble involves a subtle point, however. Physically, the marble's acceleration is due to the force of the gun, and the gun's recoil is due to the reaction force of the marble. The gun stops accelerating once the marble leaves the barrel, so that at the instant the marble and the gun part company, the gun has its final speed V_f. At that same instant the speed of the marble *relative to the gun* is v_0. Hence, the final horizontal speed of the marble relative to the table is $v_0 \cos \theta - V_f$. By conservation of horizontal momentum, we therefore have

$$0 = m(v_0 \cos \theta - V_f) - MV_f$$

or

$$V_f = \frac{mv_0 \cos \theta}{M + m}.$$

By using conservation of momentum we found the final motion of the system in a few steps. To show the advantage of this method, let us repeat the problem using Newton's laws directly.

Let $\mathbf{v}(t)$ be the velocity of marble at time t and let $\mathbf{V}(t)$ be the velocity of the gun. While the marble is being fired, it is acted on by the spring, by gravity, and by friction forces with the muzzle wall. Let the net force on the marble be $\mathbf{f}(t)$. The x equation of motion for the marble is

$$m \frac{dv_x}{dt} = f_x(t). \qquad 3$$

Formal integration of Eq. (3) gives

$$mv_x(t) = mv_x(0) + \int_0^t f_x\,dt. \qquad 4$$

The external forces are all vertical, and therefore the horizontal force f_x on the marble is due entirely to the gun. By Newton's third law, there is a reaction force $-f_x$ on the gun due to the marble. No other horizontal forces act on the gun, and the horizontal equation of motion for the gun is therefore

$$M\frac{dV_x}{dt} = -f_x(t),$$

which can be integrated to give

$$MV_x(t) = MV_x(0) - \int_0^t f_x\,dt. \qquad 5$$

We can eliminate the integral by combining Eqs. (4) and (5):

$$MV_x(t) + mv_x(t) = MV_x(0) + mv_x(0). \qquad 6$$

We have rediscovered that the horizontal component of momentum is conserved.

What about the motion of the center of mass? Its horizontal velocity is

$$\dot{R}_x(t) = \frac{MV_x(t) + mv_x(t)}{M + m}.$$

Using Eq. (6), the numerator can be rewritten to give

$$\dot{R}_x(t) = \frac{MV_x(0) + mv_x(0)}{M + m} = 0,$$

since the system is initially at rest. R_x is constant, as we expect.

We did not include the small force of air friction. Would the center of mass remain at rest if we had included it?

The essential step in our derivation of the law of conservation of momentum was to use Newton's third law. Thus, conservation of momentum appears to be a natural consequence of newtonian mechanics. It has been found, however, that conservation of momentum holds true even in areas where newtonian mechanics proves inadequate, including the realms of quantum mechanics and relativity. In addition, conservation of momentum can be generalized to apply to systems like the electromagnetic field, which possess momentum but not mass. For these reasons, conservation of momentum is generally regarded as being more fundamental than newtonian mechanics. From this point of view, Newton's third law is a simple consequence of conservation of momentum for interacting particles. For our present purposes it is purely a matter of taste whether we wish to regard Newton's third law or conservation of momentum as more fundamental.

Example 3.7 Earth, Moon, and Sun—a Three Body System

Newton was the first to calculate the motion of two gravitating bodies. As we shall discuss in Chap. 9, two bodies of mass M_1 and M_2 bound by gravity move so that $\mathbf{r}_{12}$ traces out an ellipse. The sketch shows the motion in a frame in which the center of mass is at rest. (Note that the center of mass of two particles lies on the line joining them.)

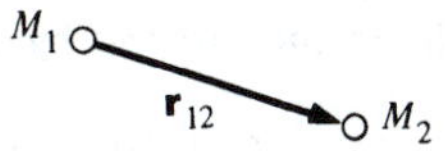

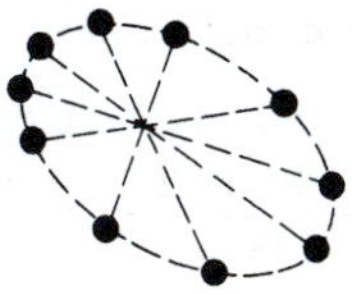

There is no general analytical solution for the motion of three gravitating bodies, however. In spite of this, we can explain many of the important features of the motion with the help of the concept of center of mass.

At first glance, the motion of the earth-moon-sun system appears to be quite complex. In the absence of the sun, the earth and moon would execute elliptical motion about their center of mass. As we shall now show, that center of mass orbits the sun like a single planet, to good approximation. The total motion is the simple result of two simultaneous elliptical orbits.

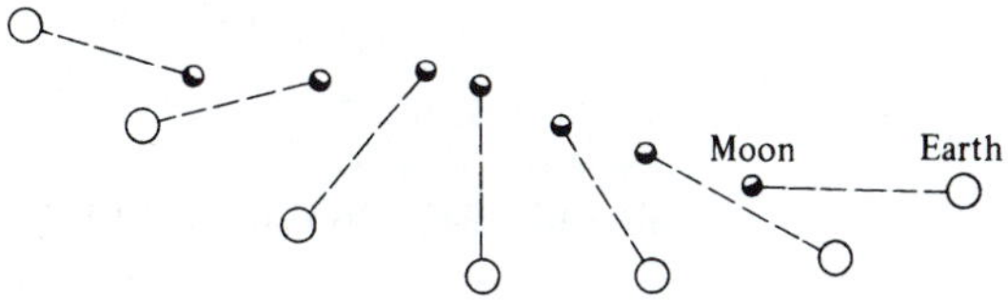

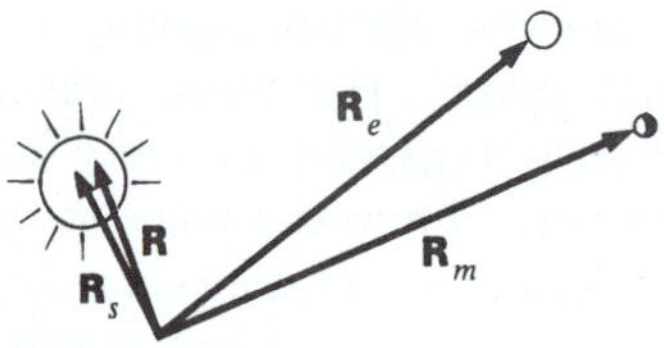

The center of mass of the earth-moon-sun system lies at

$$\mathbf{R} = \frac{M_e\mathbf{R}_e + M_m\mathbf{R}_m + M_s\mathbf{R}_s}{M_e + M_m + M_s},$$

where M_e, M_m, and M_s are the masses of the earth, moon, and sun, respectively. The sun's mass is so large compared with the mass of the earth or the moon that $\mathbf{R}_0 \approx \mathbf{R}_s$, and to good approximation the center of mass of the three body system lies at the center of the sun. Since external forces are negligible, the sun is effectively at rest in an inertial frame and it is natural to use a coordinate system with its origin at the center of the sun so that $\mathbf{R} = 0$.

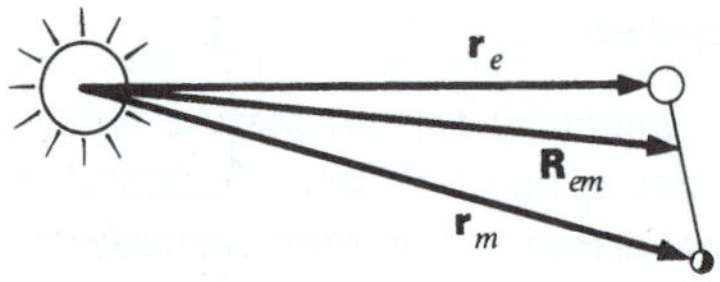

Let r_e and r_m be the positions of the earth and moon with respect to the sun, and let us focus for the moment on the system composed of the earth and moon. Their center of mass lies at

$$\mathbf{R}_{em} = \frac{M_e\mathbf{r}_e + M_m\mathbf{r}_m}{M_e + M_m}.$$

The external force on the earth-moon system is the gravitational pull of the sun:

$$\mathbf{F} = -GM_s\left(\frac{M_e}{r_e^2}\hat{\mathbf{r}}_e + \frac{M_m}{r_m^2}\hat{\mathbf{r}}_m\right).$$

The equation of motion of the center of mass is

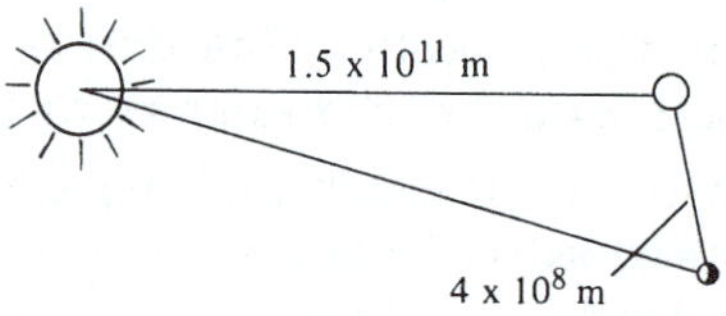

$$(M_e + M_m)\ddot{\mathbf{R}}_{em} = \mathbf{F}.$$

The earth and moon are so close compared with their distance from the sun that we shall not make a large error if we assume $r_e \approx r_m \approx R_{em}$. With this approximation,

$$(M_e + M_m)\ddot{\mathbf{R}}_{em} \approx \frac{-GM_s}{R^2}(M_e\hat{\mathbf{r}}_e + M_m\hat{\mathbf{r}}_m)$$

$$= \frac{-GM_s(M_e + M_m)\hat{\mathbf{R}}_{em}}{R^2}.$$

The center of mass of the earth and moon moves like a planet of mass $M_e + M_m$ about the sun. The total motion is the combination of this elliptical motion and the elliptical motion of the earth and moon about their center of mass, as illustrated on the opposite page. (The drawing is not to scale: the center of mass of the earth-moon system lies within the earth, and the moon's orbit is always concave toward the sun. Also, the plane of the moon's orbit is inclined by 5° with respect to the earth's orbit around the sun.)

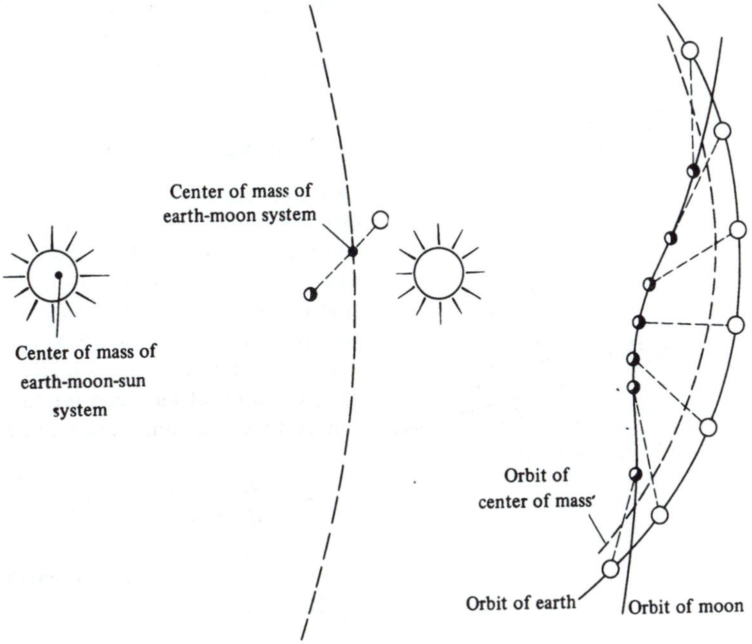

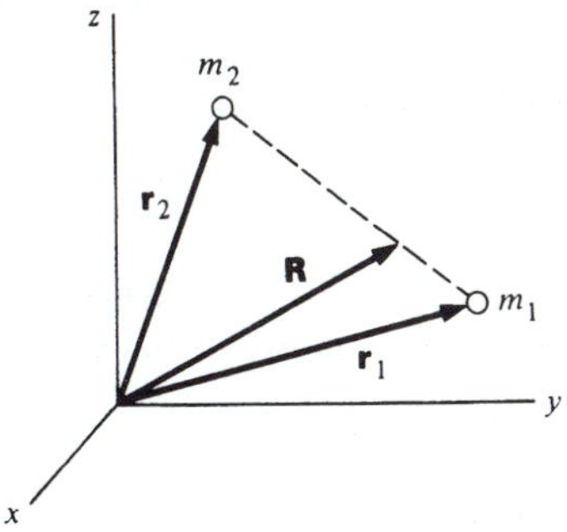

Center of Mass Coordinates

Often a problem can be simplified by the right choice of coordinates. The center of mass coordinate system, in which the origin lies at the center of mass, is particularly useful. The drawing illustrates the case of a two particle system with masses m_1 and m_2. In the initial coordinate system, x, y, z, the particles are located at $\mathbf{r}_1$ and $\mathbf{r}_2$ and their center of mass is at

$$\mathbf{R} = \frac{m_1\mathbf{r}_1 + m_2\mathbf{r}_2}{m_1 + m_2}.$$

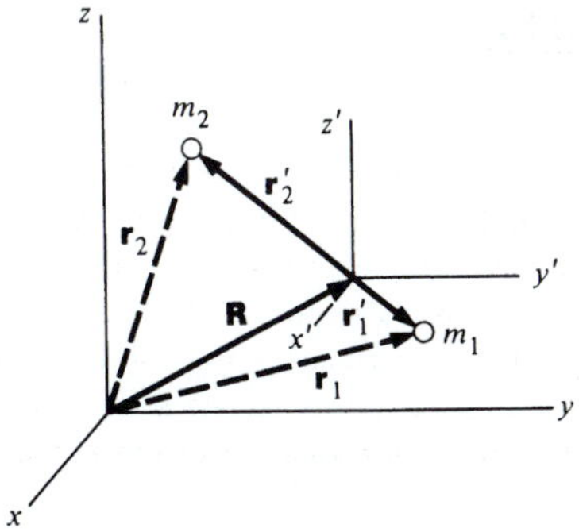

We now set up the center of mass coordinate system, x', y', z', with its origin at the center of mass. The origins of the old and new system are displaced by **R**. The center of mass coordinates of the two particles are

$$\mathbf{r}'_1 = \mathbf{r}_1 - \mathbf{R}$$
$$\mathbf{r}'_2 = \mathbf{r}_2 - \mathbf{R}.$$

Center of mass coordinates are the natural coordinates for an isolated two body system. For such a system the motion of the center of mass is trivial—it moves uniformly. Furthermore,

$m_1\mathbf{r}_1' + m_2\mathbf{r}_2' = 0$ by the definition of center of mass, so that if the motion of one particle is known, the motion of the other particle follows directly. Here is an example.

Example 3.8 **The Push Me–Pull You**

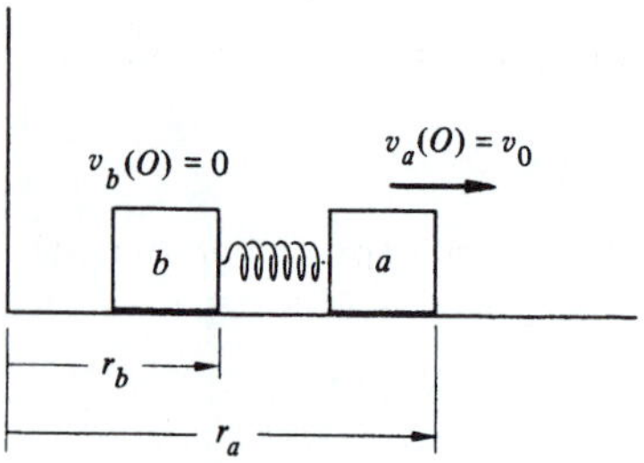

Two identical blocks a and b both of mass m slide without friction on a straight track. They are attached by a spring of length l and spring constant k. Initially they are at rest. At $t = 0$, block a is hit sharply, giving it an instantaneous velocity v_0 to the right. Find the velocities for subsequent times. (Try this yourself if there is a linear air track available—the motion is quite unexpected.)

Since the system slides freely after the collision, the center of mass moves uniformly and therefore defines an inertial frame.

Let us transform to center of mass coordinates. The center of mass lies at

$$R = \frac{mr_a + mr_b}{m + m} = \frac{1}{2}(r_a + r_b).$$

As expected, R is always halfway between a and b. The center of mass coordinates of a and b are

$$r_a' = r_a - R = \tfrac{1}{2}(r_a - r_b)$$

$$r_b' = r_b - R = -\tfrac{1}{2}(r_a - r_b) = -r_a'.$$

The sketch below shows these coordinates.

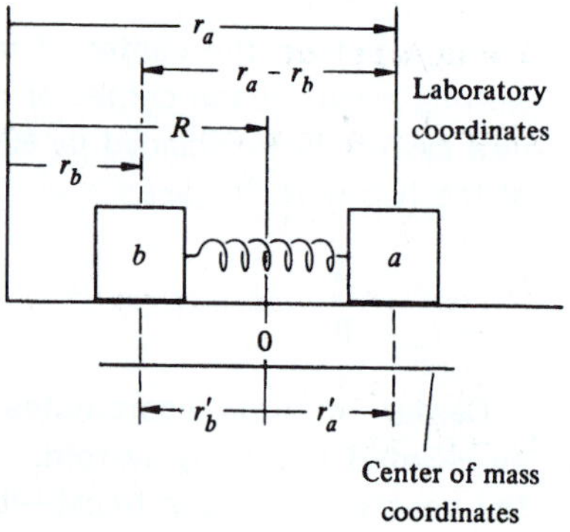

The instantaneous length of the spring is $r_a - r_b = r_a' - r_b'$. The instantaneous departure of the spring from its equilibrium length is $r_a - r_b - l = r_a' - r_b' - l$, where l is the unstretched length of the spring. The equations of motion in the center of mass system are

$$m\ddot{r}_a' = -k(r_a' - r_b' - l)$$
$$m\ddot{r}_b' = +k(r_a' - r_b' - l).$$

The form of these equations suggests that we subtract them, obtaining

$$m(\ddot{r}_a' - \ddot{r}_b') = -2k(r_a' - r_b' - l).$$

It is natural to introduce the departure of the spring from its equilibrium length as a variable. Letting $u = r_a' - r_b' - l$, we have

$$m\ddot{u} + 2ku = 0.$$

This is the equation for simple harmonic motion which we discussed in Example 2.14. The solution is

$$u = A \sin \omega t + B \cos \omega t,$$

where $\omega = \sqrt{2k/m}$. Since the spring is unstretched at $t = 0$, $u(0) = 0$ which requires $B = 0$. Furthermore, since $u = r_a' - r_b' - l = r_a - r_b - l$, we have at $t = 0$

$$\begin{aligned} \dot{u}(0) &= v_a(0) - v_b(0) \\ &= A\omega \cos (0) \\ &= v_0, \end{aligned}$$

so that

$$A = v_0/\omega$$

and

$$u = (v_0/\omega) \sin \omega t.$$

Since $v_a' - v_b' = \dot{u}$, and $v_a' = -v_b'$, we have

$$v_a' = -v_b' = \tfrac{1}{2}v_0 \cos \omega t.$$

The laboratory velocities are

$$v_a = \dot{R} + v_a'$$
$$v_b = \dot{R} + v_b'.$$

Since $\dot{R}$ is constant, it is always equal to its initial value

$$\dot{R} = \tfrac{1}{2}[v_a(0) + v_b(0)]$$
$$= \tfrac{1}{2}v_0.$$

Putting these together gives

$$v_a = \frac{v_0}{2}(1 + \cos \omega t)$$

$$v_b = \frac{v_0}{2}(1 - \cos \omega t).$$

The masses move to the right on the average, but they alternately come to rest in a push me–pull you fashion.

3.4 Impulse and a Restatement of the Momentum Relation

The relation between force and momentum is

$$\mathbf{F} = \frac{d\mathbf{P}}{dt}. \qquad 3.16$$

As a general rule, any law of physics which can be expressed in terms of derivatives can also be written in an integral form. The integral form of the force-momentum relationship is

$$\int_0^t \mathbf{F}\, dt = \mathbf{P}(t) - \mathbf{P}(0). \qquad 3.17$$

The change in momentum of a system is given by the integral of force with respect to time. This form contains essentially the same physical information as Eq. (3.16), but it gives a new way of looking at the effect of a force: the change in momentum is the time integral of the force. To produce a given change in the momentum in time interval t requires only that $\int_0^t \mathbf{F}\, dt$ have the appropriate value; we can use a small force acting for much of the time or a large force acting for only part of the interval. The integral $\int_0^t \mathbf{F}\, dt$ is called the *impulse*. The word impulse calls to mind a short, sharp shock, as in Example 3.8, where we talked of giving a blow to a mass at rest so that its final velocity was v_0. However, the physical definition of impulse can just as well be applied to a weak force acting for a long time. Change of momentum depends only on $\int \mathbf{F}\, dt$, independent of the detailed time dependence of the force.

Here are two examples involving impulse.

Example 3.9 **Rubber Ball Rebound**

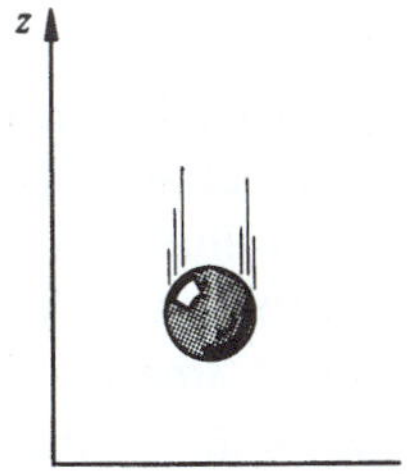

A rubber ball of mass 0.2 kg falls to the floor. The ball hits with a speed of 8 m/s and rebounds with approximately the same speed. High speed photographs show that the ball is in contact with the floor for 10^{-3} s. What can we say about the force exerted on the ball by the floor?

The momentum of the ball just before it hits the floor is $\mathbf{P}_a = -1.6\hat{\mathbf{k}}$ kg·m/s and its momentum 10^{-3} s later is $\mathbf{P}_b = +1.6\hat{\mathbf{k}}$ kg·m/s. Since $\int_{t_a}^{t_b} \mathbf{F}\,dt = \mathbf{P}_b - \mathbf{P}_a$, $\int_{t_a}^{t_b} \mathbf{F}\,dt = 1.6\hat{\mathbf{k}} - (-1.6\hat{\mathbf{k}}) = 3.2\hat{\mathbf{k}}$ kg·m/s. Although the exact variation of $\mathbf{F}$ with time is not known, it is easy to find the average force exerted by the floor on the ball. If the collision time is $\Delta t = t_b - t_a$, the average force $\mathbf{F}_{\text{av}}$ acting during the collision is

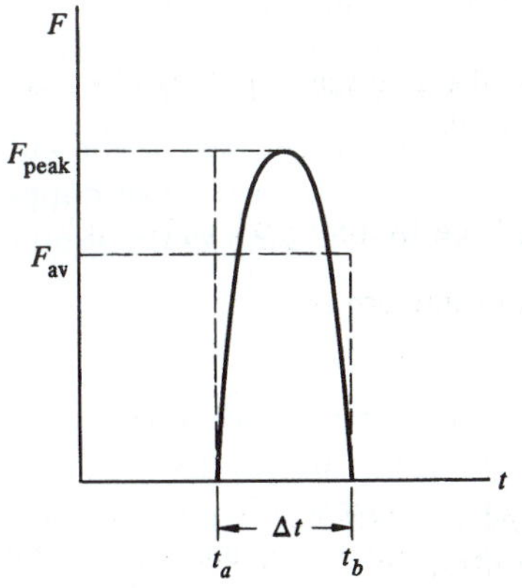

$$\mathbf{F}_{\text{av}}\,\Delta t = \int_{t_a}^{t_a+\Delta t} \mathbf{F}\,dt.$$

Since $\Delta t = 10^{-3}$ s,

$$\mathbf{F}_{\text{av}} = \frac{3.2\hat{\mathbf{k}}\ \text{kg·m/s}}{10^{-3}\ \text{s}} = 3{,}200\hat{\mathbf{k}}\ \text{N}.$$

The average force is directed upward, as we expect. In more familiar units, 3,200 N $\approx$ 720 lb—a sizable force. The instantaneous force on the ball is even larger at the peak, as the sketch shows. If the ball hits a softer surface, the collision time is longer and the peak force is less.

Actually, there is a weakness in our treatment of the rubber ball rebound. In calculating the impulse $\int \mathbf{F}\,dt$, $\mathbf{F}$ is the total force. This includes the gravitational force, which we have neglected. Proceeding more carefully, we write

$$\begin{aligned}\mathbf{F} &= \mathbf{F}_{\text{floor}} + \mathbf{F}_{\text{grav}} \\ &= \mathbf{F}_{\text{floor}} - Mg\hat{\mathbf{k}}.\end{aligned}$$

The impulse equation then becomes

$$\int_0^{10^{-3}} \mathbf{F}_{\text{floor}}\,dt - \int_0^{10^{-3}} Mg\hat{\mathbf{k}}\,dt = 3.2\hat{\mathbf{k}}\ \text{kg·m/s}.$$

The impulse due to the gravitational force is

$$\begin{aligned}-\int_0^{10^{-3}} Mg\hat{\mathbf{k}}\,dt = -Mg\hat{\mathbf{k}}\int_0^{10^{-3}} dt &= -(0.2)(9.8)(10^{-3})\hat{\mathbf{k}} \\ &= -1.96 \times 10^{-3}\hat{\mathbf{k}}\ \text{kg·m/s}.\end{aligned}$$

This is less than one-thousandth of the total impulse, and we can neglect it with little error. Over a long period of time, gravity can produce a large change in the ball's momentum (the ball gains speed as it falls, for example). In the short time of contact, however, gravity contributes little momentum change compared with the tremendous force exerted by the floor. Contact forces during a short collision are generally so huge that we can neglect the impulse due to other forces of moderate strength, such as gravity or friction.

The last example reveals why a quick collision is more violent than a slow collision, even when the initial and final velocities are identical. This is the reason that a hammer can produce a force far greater than the carpenter could produce on his own; the hard hammerhead rebounds in a very short time compared with the time of the hammer swing, and the force driving the hammer is correspondingly amplified. Many devices to prevent bodily injury in accidents are based on the same considerations, but applied in reverse—they essentially prolong the time of the collision. This is the rationale for the hockey player's helmet, as well as the automobile seat belt. The following example shows what can happen in even a relatively mild collision, as when you jump to the ground.

Example 3.10 How to Avoid Broken Ankles

Animals, including humans, instinctively reduce the force of impact with the ground by flexing while running or jumping. Consider what happens to someone who hits the ground with his legs rigid.

Suppose a man of mass M jumps to the ground from height h, and that his center of mass moves downward a distance s during the time of collision with the ground. The average force during the collision is

$$F = \frac{Mv_0}{t}, \qquad 1$$

where t is the time of the collision and v_0 is the velocity with which he hits the ground. As a reasonable approximation, we can take his acceleration due to the force of impact to be constant, so that the man comes uniformly to rest. In this case the collision time is given by $v_0 = 2s/t$, or

$$t = \frac{2s}{v_0}.$$

Inserting this in Eq. (1) gives

$$F = \frac{Mv_0^2}{2s}. \qquad 2$$

For a body in free fall for distance h,

$$v_0^2 = 2gh.$$

Inserting this in Eq. (2) gives

$$F = Mg\frac{h}{s}.$$

If the man hits the ground rigidly in a vertical position, his center of mass will not move far during the collision. Suppose that his center of mass moves 1 cm, which roughly means that his height momentarily decreases by approximately 2 cm. If he jumps from a height of 2 m, the force is 200 times his weight!

Consider the force on a 90-kg ($\approx$200-lb) man jumping from a height of 2 m. The force is

$$\begin{aligned} F &= 90 \text{ kg} \times 9.8 \text{ m/s}^2 \times 200 \\ &= 1.8 \times 10^5 \text{ N}. \end{aligned}$$

Where is a bone fracture most likely to occur? The force is a maximum at the feet, since the mass above a horizontal plane through the man decreases with height. Thus his ankles will break, not his neck. If the area of contact of bone at each ankle is 5 cm^2, then the force per unit area is

$$\begin{aligned} \frac{F}{A} &= \frac{1.8 \times 10^5 \text{ N}}{10 \text{ cm}^2} \\ &= 1.8 \times 10^4 \text{ N/cm}^2. \end{aligned}$$

This is approximately the compressive strength of human bone, and so there is a good probability that his ankles will snap.

Of course, no one would be so rash as to jump rigidly. We instinctively cushion the impact when jumping by flexing as we hit the ground, in the extreme case collapsing to the ground. If the man's center of mass drops 50 cm, instead of 1 cm, during the collision, the force is only one-fiftieth as much as we calculated, and there is no danger of compressive fracture.

3.5 Momentum and the Flow of Mass

Analyzing the forces on a system in which there is a flow of mass becomes terribly confusing if we try to apply Newton's laws blindly. A rocket provides the most dramatic example of such a system, although there are many other everyday problems where the same considerations apply—for instance, the problem of calculating the reaction force on a fire hose, or of calculating the acceleration of a snowball which grows larger as it rolls downhill.

There is no fundamental difficulty in handling any of these problems provided that we keep clearly in mind exactly what is included in the system. Recall that $\mathbf{F} = d\mathbf{P}/dt$ [Eq. (3.12)] was established for a system composed of a certain set of particles. When we apply this equation in the integral form,

$$\int_{t_a}^{t_b} \mathbf{F}\, dt = \mathbf{P}(t_b) - \mathbf{P}(t_a),$$

it is essential to deal with the same set of particles throughout the time interval t_a to t_b; we must keep track of all the particles that were originally in the system. Consequently, the mass of the system cannot change during the time of interest.

Example 3.11 Mass Flow and Momentum

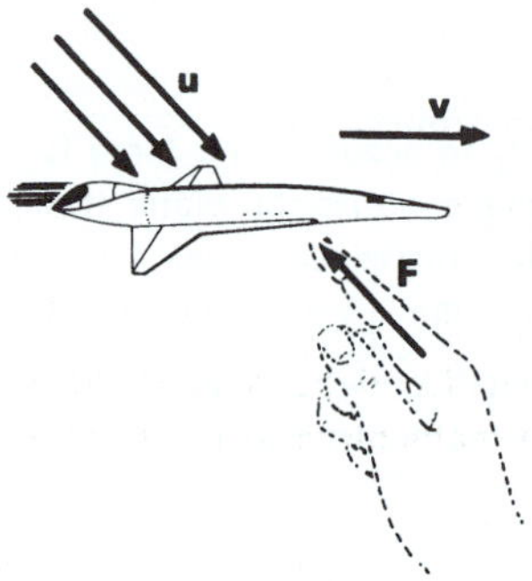

A spacecraft moves through space with constant velocity **v**. The spacecraft encounters a stream of dust particles which embed themselves in it at rate dm/dt. The dust has velocity **u** just before it hits. At time t the total mass of the spacecraft is $M(t)$. The problem is to find the external force **F** necessary to keep the spacecraft moving uniformly. (In practice, **F** would most likely come from the spacecraft's own rocket engines. For simplicity, we can visualize the source **F** to be completely external—an invisible hand, so to speak.)

Let us focus on the short time interval between t and $t + \Delta t$. The drawings below show the system at the beginning and end of the interval.

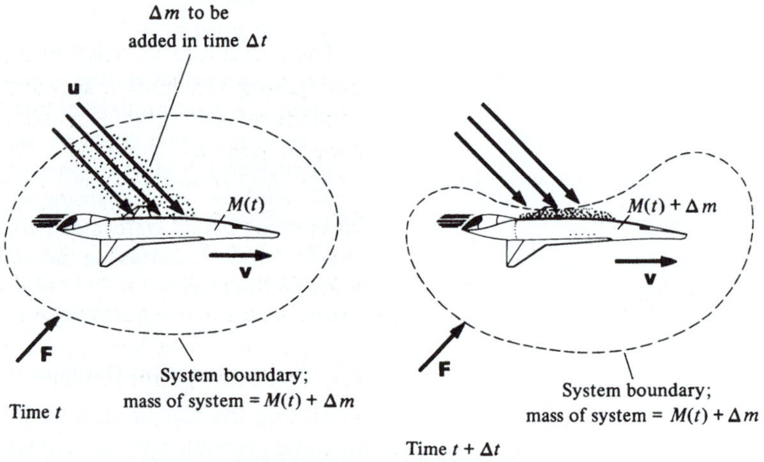

Let Δm denote the mass added to the satellite during Δt. The system consists of $M(t)$ and Δm. The initial momentum is

$$\mathbf{P}(t) = M(t)\mathbf{v} + (\Delta m)\mathbf{u}.$$

The final momentum is

$$\mathbf{P}(t + \Delta t) = M(t)\mathbf{v} + (\Delta m)\mathbf{v}.$$

The change in momentum is

$$\begin{aligned}\Delta\mathbf{P} &= \mathbf{P}(t + \Delta t) - \mathbf{P}(t)\\ &= (\mathbf{v} - \mathbf{u})\,\Delta m.\end{aligned}$$

The rate of change of momentum is approximately

$$\frac{\Delta\mathbf{P}}{\Delta t} = (\mathbf{v} - \mathbf{u})\frac{\Delta m}{\Delta t}.$$

In the limit $\Delta t \to 0$, we have the exact result

$$\frac{d\mathbf{P}}{dt} = (\mathbf{v} - \mathbf{u})\frac{dm}{dt}.$$

Since $\mathbf{F} = d\mathbf{P}/dt$, the required external force is

$$\mathbf{F} = (\mathbf{v} - \mathbf{u})\frac{dm}{dt}.$$

Note that $\mathbf{F}$ can be either positive or negative, depending on the direction of the stream of mass. If $\mathbf{u} = \mathbf{v}$, the momentum of the system is constant, and $\mathbf{F} = 0$.

The procedure of isolating the system, focusing on differentials, and taking the limit may appear a trifle formal. However, the procedure is helpful in avoiding errors in a subject where it is easy to become confused. For instance, a frequent error is to argue that $\mathbf{F} = (d/dt)(m\mathbf{v}) = m(d\mathbf{v}/dt) + \mathbf{v}(dm/dt)$. In the last example $\mathbf{v}$ is constant, and the result would be $\mathbf{F} = \mathbf{v}(dm/dt)$ rather than $(\mathbf{v} - \mathbf{u})(dm/dt)$. The difficulty arises from the fact that there are several contributions to the momentum, so that the expression for the momentum of a single particle, $\mathbf{p} = m\mathbf{v}$, is not appropriate. The limiting procedure illustrated in the last example avoids such ambiguities.

Example 3.12 Freight Car and Hopper

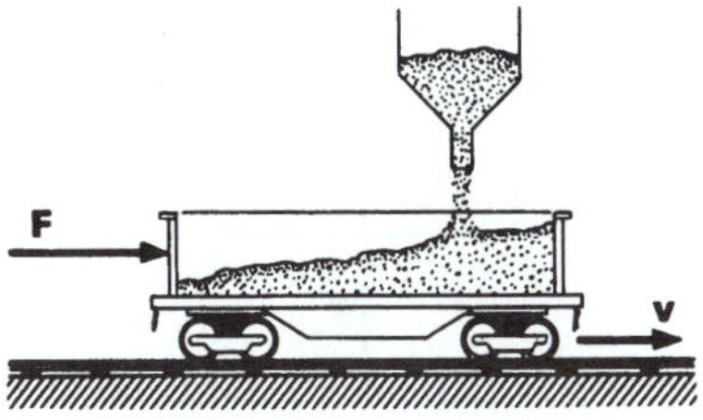

Sand falls from a stationary hopper onto a freight car which is moving with uniform velocity v. The sand falls at the rate dm/dt. How much force is needed to keep the freight car moving at the speed v?

In this case, the initial speed of the sand is 0, and

$$\frac{dP}{dt} = (v - u)\left(\frac{dm}{dt}\right) = v\frac{dm}{dt}.$$

The required force is $F = v\,dm/dt$. We can understand why this force is needed by considering in detail just what happens to a sand grain as it lands on the surface of the freight car. What would happen if the surface of the freight car were slippery?

Example 3.13 Leaky Freight Car

Now consider a related case. The same freight car is leaking sand at the rate dm/dt; what force is needed to keep the freight car moving uniformly with speed $\mathbf{v}$?

Here the mass is decreasing. However, the velocity of the sand after leaving the freight car is identical to its initial velocity, and its momentum does not change. Since $dP/dt = 0$, no force is required. (The sand does change its momentum when it hits the ground, and there is a resulting force on the ground, but that does not affect the motion of the freight car.)

The concept of momentum is invaluable in understanding the motion of a rocket. A rocket accelerates by expelling gas at a high velocity; the reaction force of the gas on the rocket accelerates the rocket in the opposite direction. The mechanism is illustrated by the drawings of the cubical chamber containing gas at high pressure.

The gas presses outward on each wall with the force $\mathbf{F}_a$. (We show only four walls for clarity.) The vector sum of the $\mathbf{F}_a$'s is zero, giving zero net force on the chamber. Similarly each wall of the chamber exerts a force on the gas $\mathbf{F}_b = -\mathbf{F}_a$; the net force on the gas is also zero. In the right hand drawings below, one wall

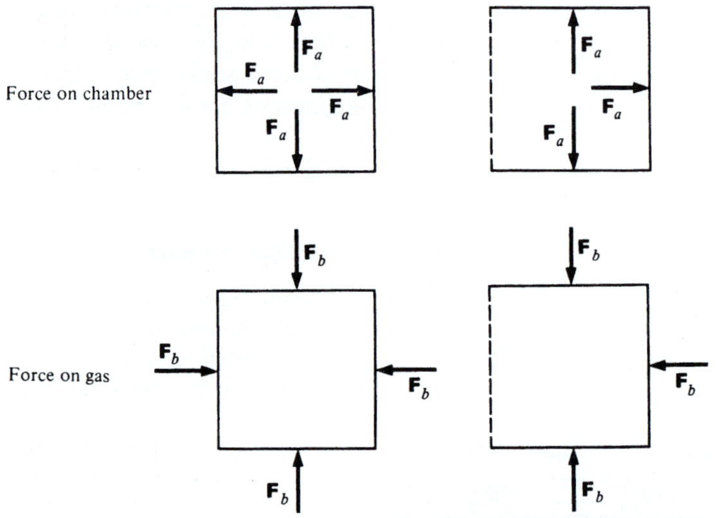

has been removed. The net force on the chamber is F_a, to the right. The net force on the gas is F_b, to the left. Hence the gas accelerates to the left, and the chamber accelerates to the right.

To analyze the motion of the rocket in detail, we must equate the external force on the system, $\mathbf{F}$, with the rate of change of momentum, $d\mathbf{P}/dt$. Consider the rocket at time t. Between t and $t + \Delta t$ a mass of fuel Δm is burned and expelled as gas with velocity $\mathbf{u}$ *relative to the rocket.*. The exhaust velocity $\mathbf{u}$ is determined by the nature of the propellants, the throttling of the engine, etc., but it is independent of the velocity of the rocket.

The sketches below show the system at time t and at time

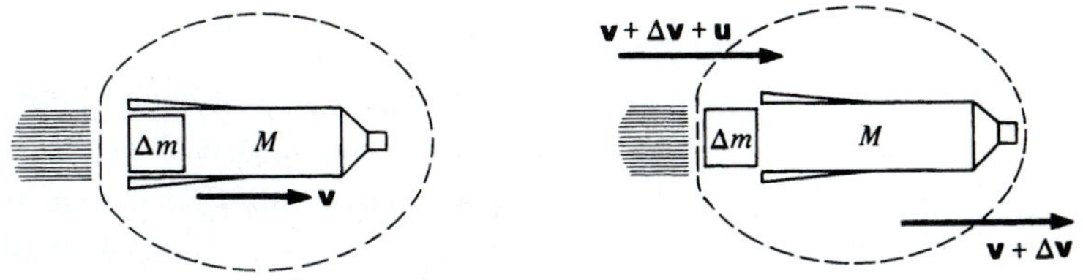

Time t Time $t + \Delta t$

$t + \Delta t$. The system consists of Δm plus the remaining mass of the rocket M. Hence the total mass is $M + \Delta m$.

The velocity of the rocket at time t is $\mathbf{v}(t)$, and at $t + \Delta t$, it is $\mathbf{v} + \Delta\mathbf{v}$. The initial momentum is

$$\mathbf{P}(t) = (M + \Delta m)\mathbf{v}$$

and the final momentum is

$$\mathbf{P}(t + \Delta t) = M(\mathbf{v} + \Delta\mathbf{v}) + \Delta m(\mathbf{v} + \Delta\mathbf{v} + \mathbf{u}).$$

The change in momentum is

$$\begin{aligned} \Delta\mathbf{P} &= \mathbf{P}(t + \Delta t) - \mathbf{P}(t) \\ &= M\,\Delta\mathbf{v} + (\Delta m)\mathbf{u} + \Delta m\,\Delta\mathbf{v} \end{aligned}$$

Therefore,

$$\begin{aligned} \frac{d\mathbf{P}}{dt} &= \lim_{\Delta t \to 0} \frac{\Delta\mathbf{P}}{\Delta t} = \lim_{\Delta t \to 0} \left(M\frac{\Delta\mathbf{v}}{\Delta t} + \mathbf{u}\frac{\Delta m}{\Delta t} + \frac{\Delta m \Delta\mathbf{v}}{\Delta t} \right) \\ &= M\frac{d\mathbf{v}}{dt} + \mathbf{u}\frac{dm}{dt}. \end{aligned} \qquad 3.18$$

Note that we have defined $\mathbf{u}$ to be positive in the direction of $\mathbf{v}$. In most rocket applications, $\mathbf{u}$ is negative, opposite to $\mathbf{v}$. It is inconvenient to have both m and M in the equation. dm/dt is the rate of increase of the exhaust mass. Since this mass comes from the rocket,

$$\frac{dm}{dt} = -\frac{dM}{dt}.$$

Using this in Eq. (3.18), and equating the external force to $d\mathbf{P}/dt$, we obtain the fundamental rocket equation

$$\mathbf{F} = M\frac{d\mathbf{v}}{dt} - \mathbf{u}\frac{dM}{dt}. \qquad 3.19$$

It may be useful to point out two minor subtleties in our development. The first is that the velocities have been expressed with respect to an inertial frame, not a frame attached to the rocket. The second is that we took the final velocity of the element of exhaust gas to be $\mathbf{v} + \Delta\mathbf{v} + \mathbf{u}$ rather than $\mathbf{v} + \mathbf{u}$. This is correct (consult Example 3.6 on spring gun recoil if you need help in seeing the reason), but actually it makes no difference here, since either expression yields the same final result when the limit is taken. Here are two examples on rockets.

Example 3.14 Rocket in Free Space

If there is no external force on a rocket, $\mathbf{F} = 0$ and its motion is given by

$$M\frac{d\mathbf{v}}{dt} = \mathbf{u}\frac{dM}{dt}$$

or

$$\frac{d\mathbf{v}}{dt} = \frac{\mathbf{u}}{M}\frac{dM}{dt}.$$

Generally the exhaust velocity $\mathbf{u}$ is constant, in which case it is easy to integrate the equation of motion.

$$\int_{t_0}^{t_f}\frac{d\mathbf{v}}{dt}\,dt = \mathbf{u}\int_{t_0}^{t_f}\frac{1}{M}\frac{dM}{dt}\,dt$$

$$= \mathbf{u}\int_{M_0}^{M_f}\frac{dM}{M}$$

or

$$\mathbf{v}_f - \mathbf{v}_0 = \mathbf{u}\ln\frac{M_f}{M_0}$$

$$= -\mathbf{u}\ln\frac{M_0}{M_f}.$$

If $\mathbf{v}_0 = 0$, then

$$\mathbf{v}_f = -\mathbf{u} \ln \frac{M_0}{M_f}.$$

The final velocity is independent of how the mass is released—the fuel can be expended rapidly or slowly without affecting $\mathbf{v}_f$. The only important quantities are the exhaust velocity and the ratio of initial to final mass.

The situation is quite different if a gravitational field is present, as shown by the next example.

Example 3.15 Rocket in a Gravitational Field

If a rocket takes off in a constant gravitational field, Eq. (3.19) becomes

$$M\mathbf{g} = M\frac{d\mathbf{v}}{dt} - \mathbf{u}\frac{dM}{dt},$$

where $\mathbf{u}$ and $\mathbf{g}$ are directed down and are assumed to be constant.

$$\frac{d\mathbf{v}}{dt} = \frac{\mathbf{u}}{M}\frac{dM}{dt} + \mathbf{g}.$$

Integrating with respect to time, we obtain

$$\mathbf{v}_f - \mathbf{v}_0 = \mathbf{u} \ln \left(\frac{M_f}{M_0}\right) + \mathbf{g}(t_f - t_0).$$

Let $\mathbf{v}_0 = 0$, $t_0 = 0$, and take velocity positive upward.

$$v_f = u \ln \left(\frac{M_0}{M_f}\right) - gt_f.$$

Now there is a premium attached to burning the fuel rapidly. The shorter the burn time, the greater the velocity. This is why the takeoff of a large rocket is so spectacular—it is essential to burn the fuel as quickly as possible.

3.6 Momentum Transport

Nearly everyone has at one time or another been on the receiving end of a stream of water from a hose. You feel a push. If the stream is intense, as in the case of a fire hose, the push can be dramatic—a jet of high pressure water can be used to break through the wall of a burning building.

The push of a water stream arises from the momentum it transfers to you. Unless another external force gives you equal momentum in the opposite direction, off you go. How can a column of water flying through the air exert a force which is every bit as real as a force transmitted by a rigid steel rod? The reason is easy to see if we picture the stream of water as a series of small uniform droplets of mass m, traveling with velocity v_0. Let the droplets be distance l apart and suppose that the stream is directed against your hand. Assume that the drops collide without rebound and simply run down your arm. Consider the force exerted by your hand on the stream. As each drop hits there is a large force for a short time. Although we do not know the instantaneous force, we can find the impulse I_{droplet} on each drop due to your hand.

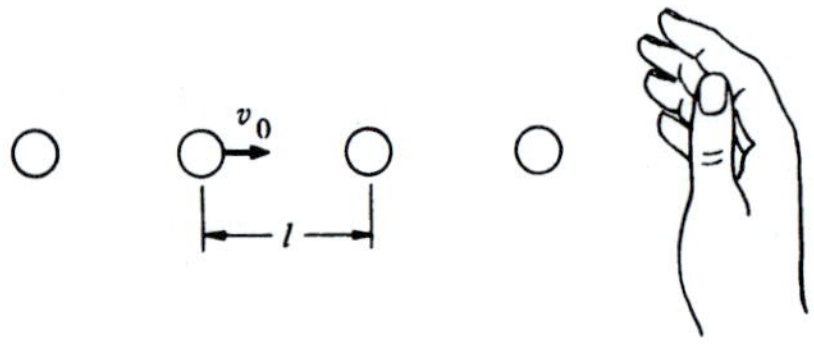

$$
\begin{aligned}
I_{\text{droplet}} &= \int_{1\ \text{collision}} F\,dt \\
&= \Delta p \\
&= m(v_f - v_0) \\
&= -mv_0.
\end{aligned}
$$

The impulse on your hand is equal and opposite.

$$I_{\text{hand}} = mv_0.$$

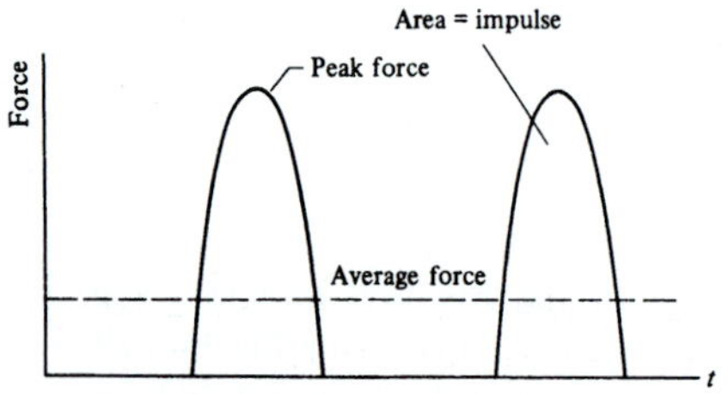

The positive sign means that the impulse on the hand is in the same direction as the velocity of the drop. The impulse equals the area under one of the peaks shown in the drawing. If there are many collisions per second, you do not feel the shock of each drop. Rather, you feel the average force F_{av} indicated by the dashed line in the drawing. The area under F_{av} during one collision period T (the time between collisions) is identical to the impulse due to one drop.

$$F_{\text{av}}T = \int_{1\ \text{collision}} F\,dt$$

Since $T = l/v_0$ and $\int F\,dt = mv_0$, the average force is

$$
\begin{aligned}
F_{\text{av}} &= \frac{mv_0}{T} \\
&= \frac{m}{l}\,v_0^2.
\end{aligned}
$$

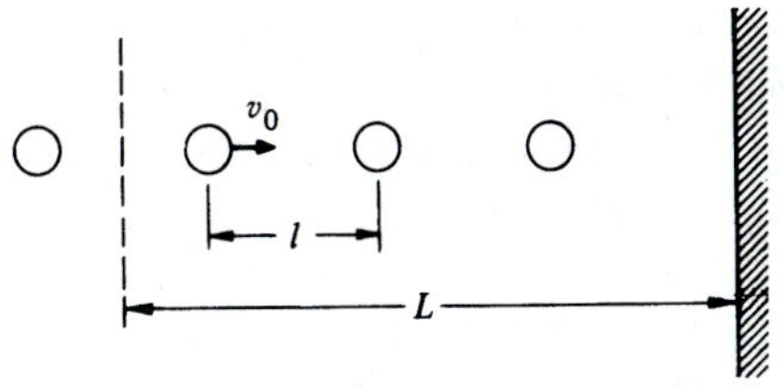

Here is another way to find the average force. Consider length L of the stream just about to hit the surface. The number of drops in L is L/l, and since each drop has momentum mv_0, the total momentum is

$$\Delta p = \frac{L}{l} mv_0.$$

All these drops will strike the wall in time

$$\Delta t = \frac{L}{v_0}.$$

The average force is

$$F_{\text{av}} = \frac{\Delta p}{\Delta t}$$
$$= \frac{m}{l} v_0^2.$$

To apply this model to a fluid, consider a stream moving with speed v. If the mass per unit length is $m/l \equiv \lambda$, the momentum per unit length is λv and the rate at which the stream transports momentum to the surface is

$$\frac{dp}{dt} = \lambda v^2. \tag{3.20}$$

If the stream comes to rest at the surface, the force on the surface is

$$F = \lambda v^2. \tag{3.21}$$

Example 3.16 **Momentum Transport to a Surface**

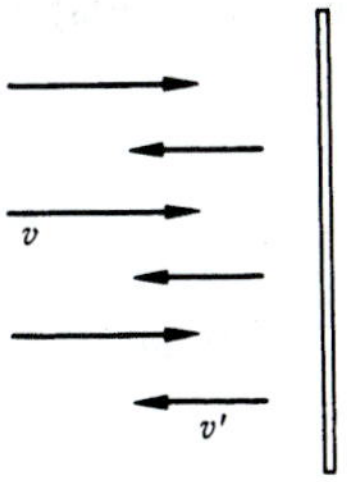

A stream of particles of mass m and separation l hits a perpendicular surface with velocity v. The stream rebounds along the original line of motion with velocity v'. The mass per unit length of the incident stream is $\lambda = m/l$. What is the force on the surface?

The incident stream transfers momentum to the surface at the rate λv^2. However, the reflected stream does not carry it away at the rate $\lambda v'^2$, since the density of the stream must change at the surface. The number of particles incident on the surface in time Δt is $v\,\Delta t/l$ and their total mass is $\Delta m = mv\,\Delta t/l$. Hence, the rate at which mass arrives at the surface is

$$\frac{dm}{dt} = \frac{m}{l} v = \lambda v.$$

The rate at which mass is carried away from the surface is $\lambda' v'$. Since mass does not accumulate on the surface, these rates must be equal. Hence $\lambda' v' = \lambda v$, and the force on the surface is

$$F = \frac{dp'}{dt} + \frac{dp}{dt} = \lambda' v'^2 + \lambda v^2$$
$$= \lambda v(v' + v).$$

If the stream collides without rebound, then $v' = 0$ and $F = \lambda v^2$, in agreement with our previous result. If the particles undergo perfect reflection, then $v' = v$, and $F = 2\lambda v^2$. The actual force lies somewhere between these extremes.

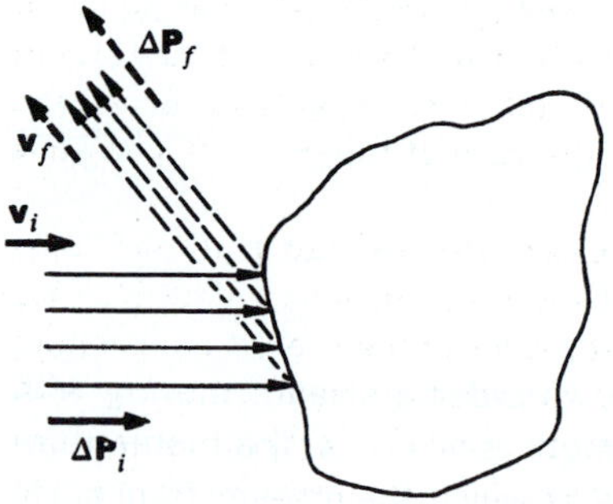

We can generalize the idea of momentum transport to three dimensions. Consider a stream of fluid which strikes an object and rebounds in some arbitrary direction. For simplicity we assume that the incident stream is uniform and that in time Δt it transports momentum $\Delta \mathbf{P}_i$. The direction of $\Delta \mathbf{P}_i$ is parallel to the initial velocity $\mathbf{v}_i$ and $\Delta P_i = \lambda_i v_i^2\,\Delta t$. During the same interval Δt the rebounding stream carries away momentum $\Delta \mathbf{P}_f$, where $\Delta P_f = \lambda_f v_f^2\,\Delta t$; the direction of $\Delta \mathbf{P}_f$ is parallel to the final velocity $\mathbf{v}_f$. The vectors are shown in the sketch.

The net momentum change of the fluid in Δt is

$$\Delta \mathbf{P}_{\text{fluid}} = \Delta \mathbf{P}_f - \Delta \mathbf{P}_i.$$

The rate of change of the fluid's momentum is

$$\left(\frac{d\mathbf{P}}{dt}\right)_{\text{fluid}} = \left(\frac{d\mathbf{P}}{dt}\right)_f - \left(\frac{d\mathbf{P}}{dt}\right)_i.$$

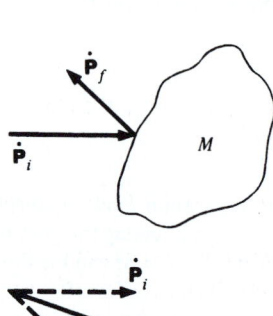

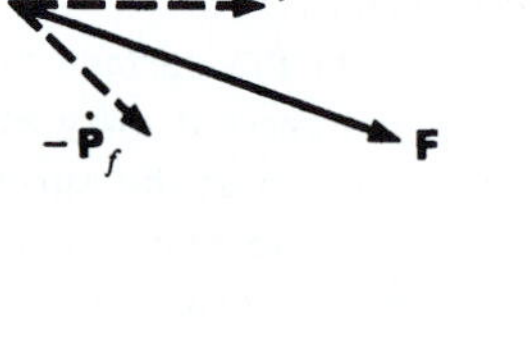

By Newton's second law, $(d\mathbf{P}/dt)_{\text{fluid}}$ equals the force on the fluid due to the object. By Newton's third law, the force on the object due to the fluid is

$$\mathbf{F} = -\left(\frac{d\mathbf{P}}{dt}\right)_{\text{fluid}}$$
$$= \left(\frac{d\mathbf{P}}{dt}\right)_i - \left(\frac{d\mathbf{P}}{dt}\right)_f$$
$$= \dot{\mathbf{P}}_i - \dot{\mathbf{P}}_f. \qquad 3.22$$

The sketches illustrate this result.

Unless there is some opposing force, the object will begin to accelerate. If $\dot{\mathbf{P}}_f = \dot{\mathbf{P}}_i$, the stream transfers no momentum and $\mathbf{F} = 0$.

The force on a moving airplane or boat can be found by considering the effect of a multitude of streams hitting the surface, each with its own velocity. Although the mathematical formalism for analyzing this would lead us too far afield, the physical principle is the same: momentum transport.

Example 3.17 **A Dike at the Bend of a River**

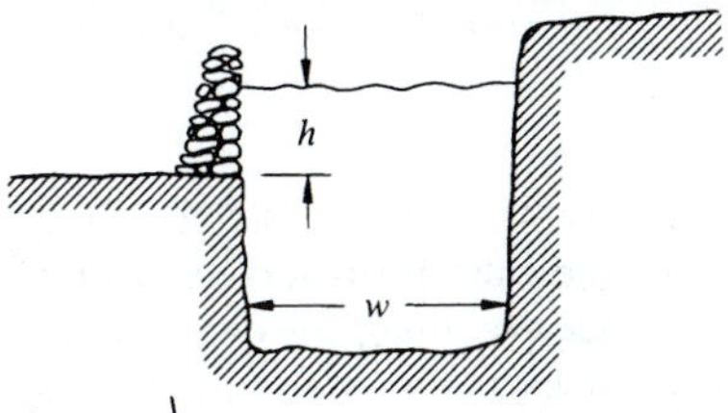

The problem is to build a dike at the bend of a river to prevent flooding when the river rises. Obviously the dike has to be strong enough to withstand the static pressure of the river $\rho g h$, where ρ is the density of the water and h is the height from the base of the dike to the surface of the water. However, because of the bend there is an additional pressure, the dynamic pressure due to the rush of water. How does this compare with the static pressure?

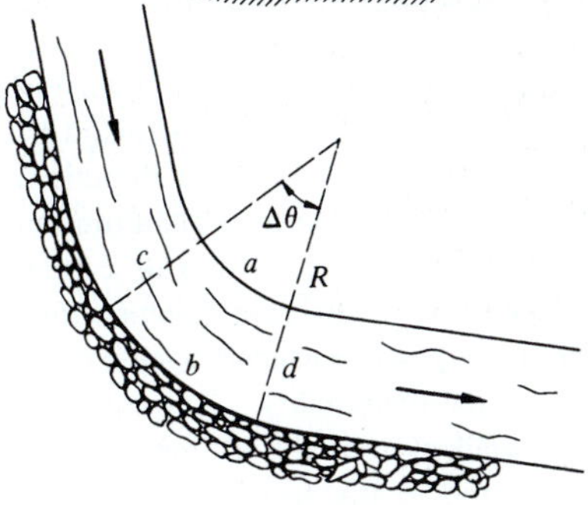

We approximate the bend by a circular curve with radius R, and focus our attention on a short length of the curve subtending angle $\Delta\theta$. We need only concern ourselves with that section of the river above the base of the dike, and we consider the volume of the river bounded by the bank a, the dike b, and two imaginary surfaces c and d. Momentum is transferred into the volume through surface c and out through surface d at rate $\dot{P} = \lambda v^2 = \rho A v^2$. Here A is the cross sectional area of the river lying above the base of the dike, $A = hw$. (Note that $\rho A = \lambda =$ mass per unit length of the river.)

However, surfaces c and d are not parallel. The rate of change of the stream's momentum is

$$\dot{\mathbf{P}} = \dot{\mathbf{P}}_d - \dot{\mathbf{P}}_c.$$

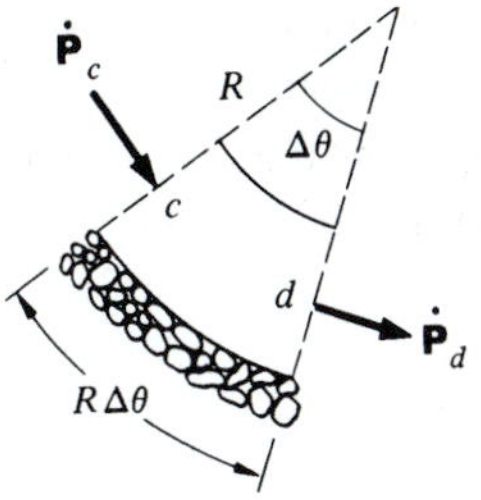

As we can see from the vector drawing below, $\dot{\mathbf{P}}$ is radially inward and has magnitude

$$|\dot{\mathbf{P}}| = \dot{P}\,\Delta\theta.$$

The dynamic force on the dike is radially outward, and has the same magnitude, $\dot{P}\,\Delta\theta$. The force is exerted over the area $(R\,\Delta\theta)h$, and the dynamic pressure is therefore

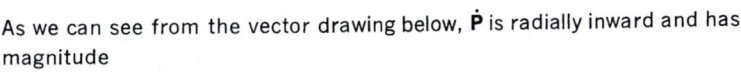

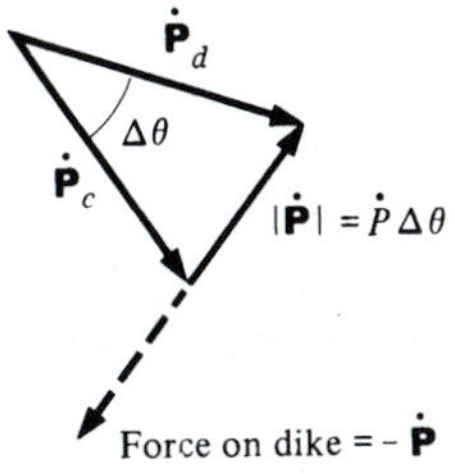

$$\begin{aligned}\text{pressure} &= \frac{\dot{P}\,\Delta\theta}{R\,\Delta\theta h}\\ &= \frac{\rho A v^2}{Rh}\\ &= \frac{\rho w v^2}{R}.\end{aligned}$$

The ratio of dynamic to static pressure is

$$\frac{\text{dynamic pressure}}{\text{static pressure}} = \frac{\rho w v^2}{R}\frac{1}{\rho g h} = \frac{w}{h}\frac{v^2}{Rg}$$

$$= \frac{\text{width}}{\text{depth}} \times \frac{\text{centripetal acceleration}}{g}.$$

For a river in flood with a speed of 10 mi/h (approximately 14 ft/s), a radius of 2,000 ft, a flood height of 3 ft, and a width of 200 ft, the ratio is 0.22, so that the dynamic pressure is by no means negligible. The ratio is even larger near the surface of the river where the static pressure is small.

Example 3.18 Pressure of a Gas

As a further application of the idea of momentum transport, let us find the pressure exerted by a gas. Although our argument will be somewhat simpleminded, it exhibits the essential ideas and gives the same result as more refined arguments.

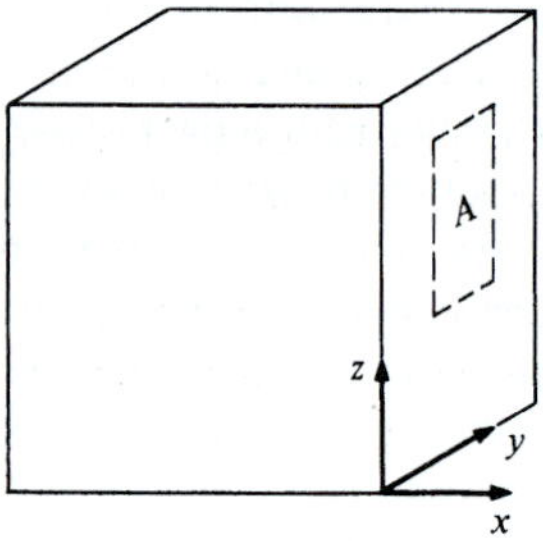

Assume that there are n atoms per unit volume of the gas, each having mass m, and that they move randomly. Let us find the force exerted on an area A in the yz plane due to motion of the atoms in the x direction. We make the plausible assumption that it is permissible to neglect motion in the y and z direction, and treat only motion parallel to the x axis. Suppose that all atoms have the same speed, v_x. The rate at which they hit the surface is $\frac{1}{2}nAv_x$, where the factor of $\frac{1}{2}$ is introduced because the atoms can move in either direction with equal probability. The momentum carried by each atom is mv_x. It is unlikely that the atoms come to rest after the collision; this would correspond to the freezing of the gas on the walls. On the average, they must leave at the same rate as they arrive, which means that the average change in momentum is $2mv_x$. Hence, the rate at which momentum changes due to collisions with area A is

$$\frac{dp}{dt} = \left(\frac{1}{2}nAv_x\right)(2mv_x)$$

$$= mnAv_x^2.$$

The force is

$$F = \frac{dp}{dt}$$

$$= mnAv_x^2$$

and the pressure P_x on the x surface is

$$P_x = \frac{F}{A}$$

$$= mnv_x^2.$$

The assumption that v_x has a fixed value is actually unnecessary. If the atoms have many different instantaneous speeds, then it can be shown that v_x^2 should be replaced by its average $\overline{v_x^2}$, and $P_x = nm\overline{v_x^2}$. By an identical argument we have $P_y = mn\overline{v_y^2}$ and $P_z = nm\overline{v_z^2}$. However, since the pressure of a gas should not depend on direction, we have $P_x = P_y = P_z$, which implies that $\overline{v_x^2} = \overline{v_y^2} = \overline{v_z^2}$. The mean squared velocity is $\overline{v^2} = \overline{v_x^2} + \overline{v_y^2} + \overline{v_z^2}$, so that $\overline{v_x^2} = \frac{1}{3}\overline{v^2}$ and the pressure is

$$P = \tfrac{1}{3}nm\overline{v^2}.$$

This is a famous result of the kinetic theory of gas, and it is a crucial point in the argument connecting heat and kinetic energy.

Note 3.1 Center of Mass

In this Note we shall find the center of mass of some nonsymmetrical objects. These examples are trivial if you have had experience evaluating two or three dimensional integrals. Otherwise, read on.

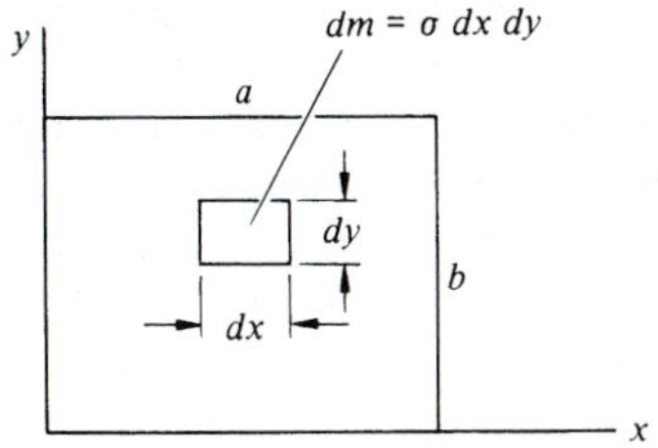

1. Find the center of mass of a thin rectangular plate with sides of length a and b, whose mass per unit area σ varies in the following fashion: $\sigma = \sigma_0(xy/ab)$, where σ_0 is a constant.

$$\mathbf{R} = \frac{1}{M}\iint (x\hat{\mathbf{i}} + y\hat{\mathbf{j}})\sigma\, dx\, dy$$

We find M, the mass of the plate, as follows:

$$\begin{aligned} M &= \int_0^b \int_0^a \sigma\, dx\, dy \\ &= \int_0^b \int_0^a \sigma_0 \frac{x}{a}\frac{y}{b}\, dx\, dy. \end{aligned}$$

We first integrate over x, treating y as a constant.

$$\begin{aligned} M &= \int_0^b \left(\int_0^a \sigma_0 \frac{x}{a}\frac{y}{b}\, dx\right) dy \\ &= \int_0^b \left(\sigma_0 \frac{y}{b}\frac{x^2}{2a}\Bigg|_{x=0}^{x=a}\right) dy \\ &= \int_0^b \sigma_0 \frac{y}{b}\frac{a}{2}\, dy \\ &= \frac{\sigma_0 a}{2}\frac{y^2}{2b}\Bigg|_{y=0}^{y=b} = \frac{1}{4}\sigma_0 ab. \end{aligned}$$

The x component of **R** is

$$X = \frac{1}{M}\iint x\sigma\, dx\, dy$$
$$= \frac{1}{M}\int_0^b \left(\int_0^a x\sigma_0 \frac{xy}{ab}\, dx\right) dy$$
$$= \frac{1}{M}\int_0^b \left(\frac{\sigma_0 y}{ab}\frac{x^3}{3}\Big|_0^a\right) dy$$
$$= \frac{1}{M}\frac{\sigma_0}{ab}\int_0^b \frac{ya^3}{3}\, dy$$
$$= \frac{1}{M}\frac{\sigma_0}{ab}\frac{a^3}{3}\frac{b^2}{2}$$
$$= \frac{4}{\sigma_0 ab}\frac{\sigma_0 a^2 b}{6}$$
$$= \frac{2}{3}a.$$

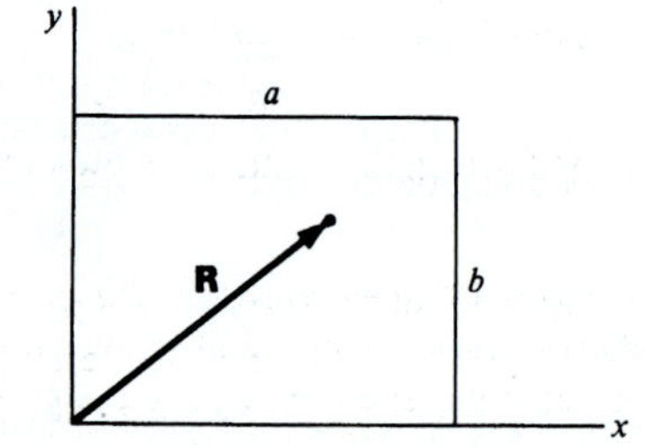

Similarly, $Y = \frac{2}{3}b$.

2. Find the center of mass of a uniform solid hemisphere of radius R and mass M.

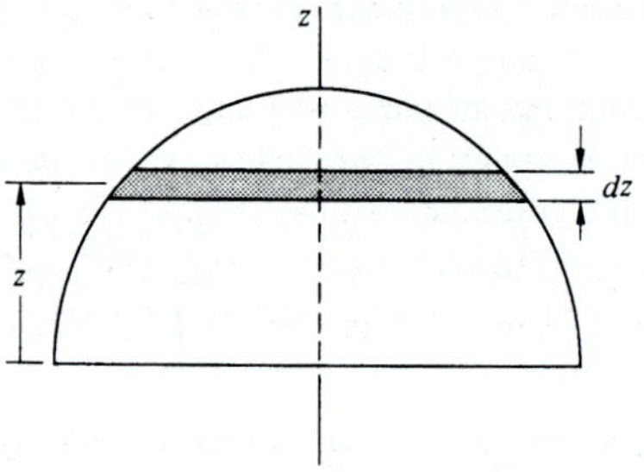

From symmetry it is apparent that the center of mass lies on the z axis, as illustrated. Its height above the equatorial plane is

$$Z = \frac{1}{M}\int z\, dM.$$

The integral is over three dimensions, but the symmetry of the situation lets us treat it as a one dimensional integral. We mentally subdivide the hemisphere into a pile of thin disks. Consider the circular disk of radius r and thickness dz. Its volume is $dV = \pi r^2\, dz$, and its mass is $dM = \rho\, dV = (M/V)(dV)$, where $V = \frac{2}{3}\pi R^3$. Hence,

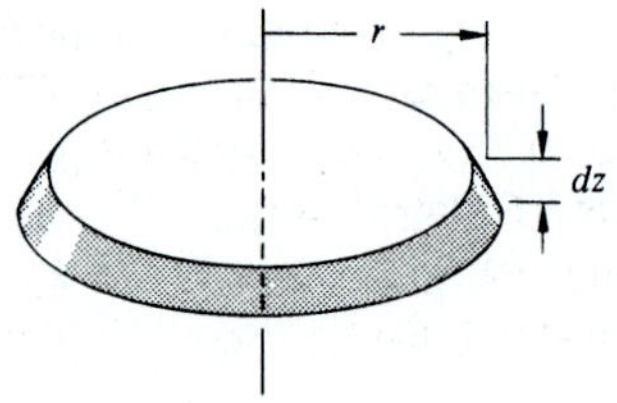

$$Z = \frac{1}{M}\int \frac{M}{V} z\, dV$$
$$= \frac{1}{V}\int_{z=0}^{R} \pi r^2 z\, dz.$$

To evaluate the integral we need to find r in terms of z. Since $r^2 = R^2 - z^2$, we have

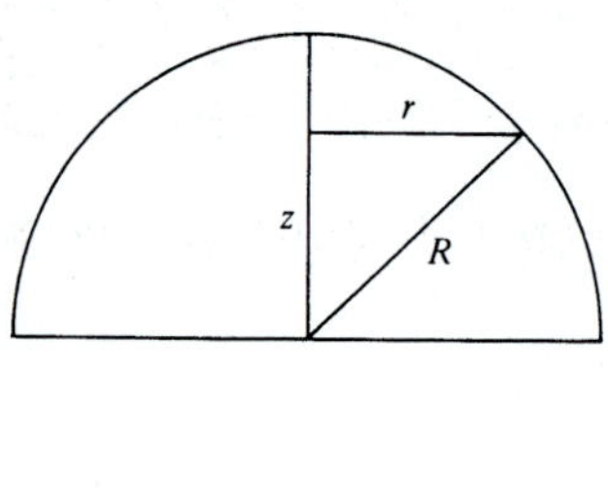

$$Z = \frac{\pi}{V}\int_0^R z(R^2 - z^2)\, dz$$
$$= \frac{\pi}{V}\left(\frac{1}{2}z^2R^2 - \frac{1}{4}z^4\right)\Big|_0^R$$

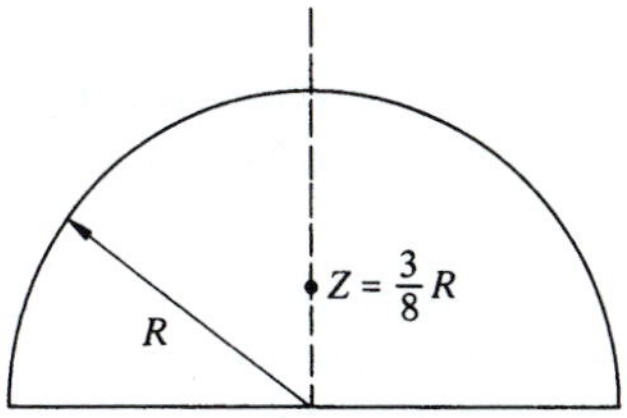

$$= \frac{\pi}{V}\left(\frac{1}{2}R^4 - \frac{1}{4}R^4\right)$$

$$= \frac{\frac{1}{4}\pi R^4}{\frac{2}{3}\pi R^3}$$

$$= \frac{3}{8}R.$$

Problems

3.1 The density of a thin rod of length l varies with the distance x from one end as $\rho = \rho_0 x^2/l^2$. Find the position of the center of mass.

Ans. $X = 3l/4$

3.2 Find the center of mass of a thin uniform plate in the shape of an equilateral triangle with sides a.

3.3 Suppose that a system consists of several bodies, and that the position of the center of mass of each body is known. Prove that the center of mass of the system can be found by treating each body as a particle concentrated at its center of mass.

3.4 An instrument-carrying projectile accidentally explodes at the top of its trajectory. The horizontal distance between the launch point and the point of explosion is L. The projectile breaks into two pieces which fly apart horizontally. The larger piece has three times the mass of the smaller piece. To the surprise of the scientist in charge, the smaller piece returns to earth at the launching station. How far away does the larger piece land? Neglect air resistance and effects due to the earth's curvature.

3.5 A circus acrobat of mass M leaps straight up with initial velocity v_0 from a trampoline. As he rises up, he takes a trained monkey of mass m off a perch at a height h above the trampoline.

What is the maximum height attained by the pair?

3.6 A light plane weighing 2,500 lb makes an emergency landing on a short runway. With its engine off, it lands on the runway at 120 ft/s. A hook on the plane snags a cable attached to a 250-lb sandbag and drags the sandbag along. If the coefficient of friction between the sandbag and the runway is 0.4, and if the plane's brakes give an additional retarding force of 300 lb, how far does the plane go before it comes to a stop?

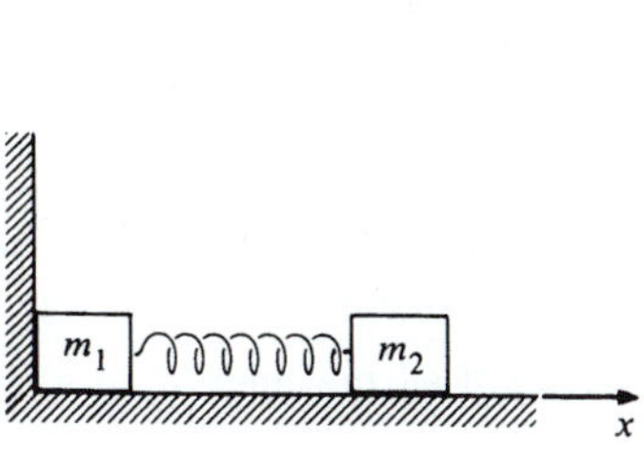

3.7 A system is composed of two blocks of mass m_1 and m_2 connected by a massless spring with spring constant k. The blocks slide on a frictionless plane. The unstretched length of the spring is l. Initially m_2 is held so that the spring is compressed to $l/2$ and m_1 is forced against a stop, as shown. m_2 is released at $t = 0$.

Find the motion of the center of mass of the system as a function of time.

3.8 A 50-kg woman jumps straight into the air, rising 0.8 m from the ground. What impulse does she receive from the ground to attain this height?

3.9 A freight car of mass M contains a mass of sand m. At $t = 0$ a constant horizontal force F is applied in the direction of rolling and at the same time a port in the bottom is opened to let the sand flow out at constant rate dm/dt. Find the speed of the freight car when all the sand is gone. Assume the freight car is at rest at $t = 0$.

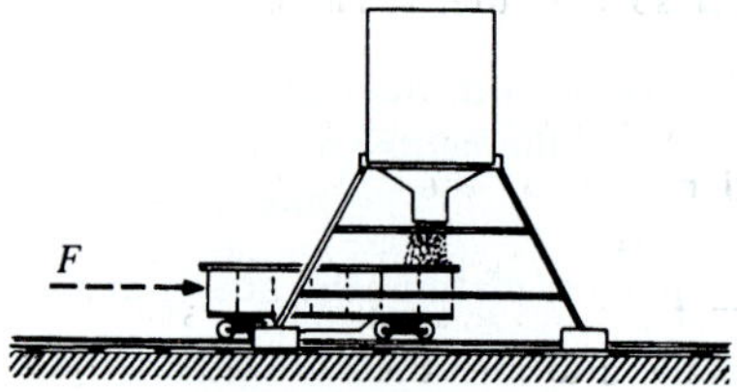

3.10 An empty freight car of mass M starts from rest under an applied force F. At the same time, sand begins to run into the car at steady rate b from a hopper at rest along the track.

Find the speed when a mass of sand, m, has been transferred. (*Hint:* There is a way to do this problem in one or two lines.)

Ans. clue. If $M = 500$ kg, $b = 20$ kg/s, $F = 100$ N, then $v = 1.4$ m/s at $t = 10$ s

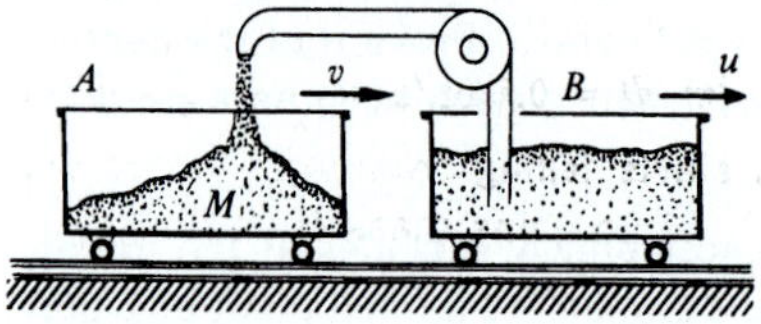

3.11 Material is blown into cart A from cart B at a rate b kilograms per second. The material leaves the chute vertically downward, so that it has the same horizontal velocity as cart B, u. At the moment of interest, cart A has mass M and velocity v, as shown. Find dv/dt, the instantaneous acceleration of A.

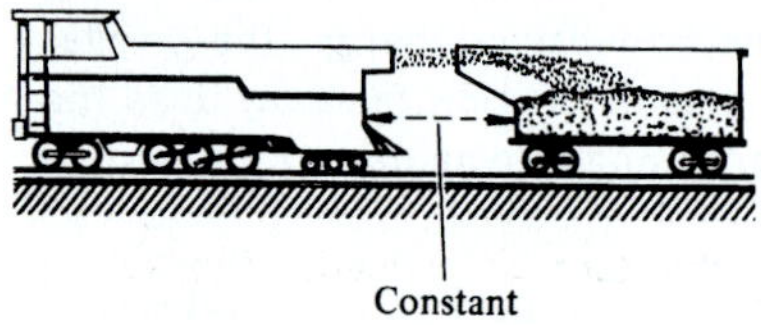

3.12 A sand-spraying locomotive sprays sand horizontally into a freight car as shown in the sketch. The locomotive and freight car are not attached. The engineer in the locomotive maintains his speed so that the distance to the freight car is constant. The sand is transferred at a rate $dm/dt = 10$ kg/s with a velocity of 5 m/s relative to the locomotive. The car starts from rest with an initial mass of 2,000 kg. Find its speed after 100 s.

3.13 A ski tow consists of a long belt of rope around two pulleys, one at the bottom of a slope and the other at the top. The pulleys are driven by a husky electric motor so that the rope moves at a steady speed of 1.5 m/s. The pulleys are separated by a distance of 100 m, and the angle of the slope is 20°.

Skiers take hold of the rope and are pulled up to the top, where they release the rope and glide off. If a skier of mass 70 kg takes the tow every 5 s on the average, what is the average force required to pull the rope? Neglect friction between the skis and the snow.

3.14 N men, each with mass m, stand on a railway flatcar of mass M. They jump off one end of the flatcar with velocity u relative to the car. The car rolls in the opposite direction without friction.

a. What is the final velocity of the flatcar if all the men jump at the same time?

b. What is the final velocity of the flatcar if they jump off one at a time? (The answer can be left in the form of a sum of terms.)

c. Does case *a* or case *b* yield the largest final velocity of the flat car? Can you give a simple physical explanation for your answer?

3.15 A rope of mass M and length l lies on a frictionless table, with a short portion, l_0, hanging through a hole. Initially the rope is at rest.

a. Find a general equation for $x(t)$, the length of rope through the hole.

Ans. $x = Ae^{\gamma t} + Be^{-\gamma t}$, $\gamma^2 = g/l$

b. Evaluate the constants A and B so that the initial conditions are satisfied.

3.16 Water shoots out of a fire hydrant having nozzle diameter D with nozzle speed V_0. What is the reaction force on the hydrant?

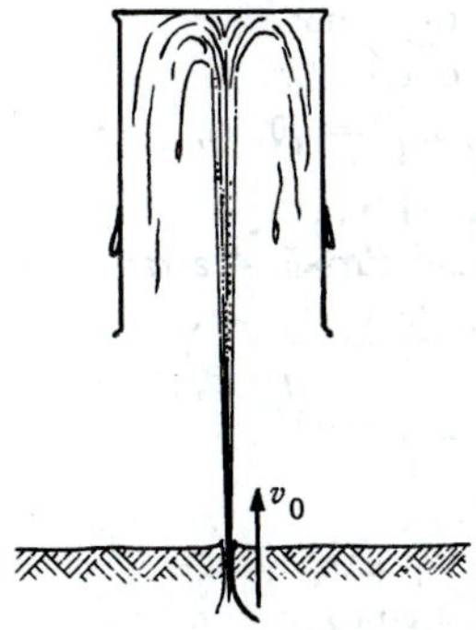

3.17 An inverted garbage can of weight W is suspended in air by water from a geyser. The water shoots up from the ground with a speed v_0, at a constant rate dm/dt. The problem is to find the maximum height at which the garbage can rides. What assumption must be fulfilled for the maximum height to be reached?

Ans. clue. If $v_0 = 20$ m/s, $W = 8.2$ N, $dm/dt = 0.5$ kg/s, then $h_{max} \approx 15$ m

3.18 A raindrop of initial mass M_0 starts falling from rest under the influence of gravity. Assume that the drop gains mass from the cloud at a rate proportional to the product of its instantaneous mass and its instantaneous velocity:

$$\frac{dM}{dt} = kMV,$$

where k is a constant.

Show that the speed of the drop eventually becomes effectively constant, and give an expression for the terminal speed. Neglect air resistance.

3.19 A bowl full of water is sitting out in a pouring rainstorm. Its surface area is 500 cm^2. The rain is coming straight down at 5 m/s at a rate of 10^{-3} g/cm^2·s. If the excess water drips out of the bowl with negligible velocity, find the force on the bowl due to the falling rain.

What is the force if the bowl is moving uniformly upward at 2 m/s?

3.20 A rocket ascends from rest in a uniform gravitational field by ejecting exhaust with constant speed u. Assume that the rate at which mass is expelled is given by $dm/dt = \gamma m$, where m is the instantaneous mass of the rocket and γ is a constant, and that the rocket is retarded by air resistance with a force bv, where b is a constant. Find the velocity of the rocket as a function of time.

Ans. clue. The terminal velocity is $(\gamma u - g)/b$.

4 WORK AND ENERGY

4.1 Introduction

In this chapter we make another attack on the fundamental problem of classical mechanics—predicting the motion of a system under known interactions. We shall encounter two important new concepts, work and energy, which first appear to be mere computational aids, mathematical crutches so to speak, but which turn out to have very real physical significance.

As first glance there seems to be no problem in finding the motion of a particle if we know the force; starting with Newton's second law, we obtain the acceleration, and by integrating we can find first the velocity and then the position. It sounds simple, but there is a problem; in order to carry out these calculations we must know the force as a function of time, whereas force is usually known as a function of position as, for example, the spring force or the gravitational force. The problem is serious because physicists are generally interested in interactions between systems, which means knowing how the force varies with position, not how it varies with time.

The task, then, is to find $\mathbf{v}(t)$ from the equation

$$m\frac{d\mathbf{v}}{dt} = \mathbf{F}(\mathbf{r}), \qquad 4.1$$

where the notation emphasizes that $\mathbf{F}$ is a known function of position. A physicist with a penchant for mathematical formalism might stop at this point and say that what we are dealing with is a problem in differential equations and that what we ought to do now is study the schemes available, including numerical methods, for solving such equations. From the strict calculational point of view, he is right. However, such an approach is too narrow and affords too little physical understanding.

Fortunately, the solution to Eq. (4.1) is simple for the important case of one dimensional motion in a single variable. The general case is more complex, but we shall see that it is not too difficult to integrate Eq. (4.1) for three dimensional motion provided that we are content with less than a complete solution. By way of compensation we shall obtain a very helpful physical relation, the work-energy theorem; its generalization, the law of conservation of energy, is among the most useful conservation laws in physics.

Let's consider the one dimensional problem before tackling the general case.

4.2 Integrating the Equation of Motion in One Dimension

A large class of important problems involves only a single variable to describe the motion. The one dimensional harmonic oscillator provides a good example. For such problems the equation of motion reduces to

$$m\frac{d^2x}{dt^2} = F(x)$$

or

$$m\frac{dv}{dt} = F(x). \qquad 4.2$$

We can solve this equation for v by a mathematical trick. First, formally integrate $m\,dv/dt = F(x)$ with respect to x:

$$m\int_{x_a}^{x_b}\frac{dv}{dt}\,dx = \int_{x_a}^{x_b}F(x)\,dx.$$

The integral on the right can be evaluated by standard methods since $F(x)$ is known. The integral on the left is intractable as it stands, but it can be integrated by changing the variable from x to t. The trick is to use[1]

$$\begin{aligned} dx &= \left(\frac{dx}{dt}\right)dt \\ &= v\,dt. \end{aligned}$$

Then

$$\begin{aligned} m\int_{x_a}^{x_b}\frac{dv}{dt}\,dx &= m\int_{t_a}^{t_b}\frac{dv}{dt}\,v\,dt \\ &= m\int_{t_a}^{t_b}\frac{d}{dt}\left(\frac{1}{2}v^2\right)dt \\ &= \frac{1}{2}mv^2\bigg|_{t_a}^{t_b} \\ &= \tfrac{1}{2}mv_b^2 - \tfrac{1}{2}mv_a^2, \end{aligned}$$

where $x_a \equiv x(t_a)$, $v_a \equiv v(t_a)$, etc.

Putting these results together yields

$$\tfrac{1}{2}mv_b^2 - \tfrac{1}{2}mv_a^2 = \int_{x_a}^{x_b}F(x)\,dx. \qquad 4.3$$

[1] Change of variables using differentials is discussed in Note 1.1.

Alternatively, we can use indefinite upper limits in Eq. (4.3):

$$\tfrac{1}{2}mv^2 - \tfrac{1}{2}mv_a{}^2 = \int_{x_a}^{x} F(x)\,dx, \qquad 4.4$$

where v is the speed of the particle when it is at position x. Equation (4.4) gives us v as a function of x. Since $v = dx/dt$, we could solve Eq. (4.4) for dx/dt and integrate again to find $x(t)$. Rather than write out the general formula, it is easier to see the method by studying a few examples.

Example 4.1 **Mass Thrown Upward in a Uniform Gravitational Field**

A mass m is thrown vertically upward with initial speed v_0. How high does it rise, assuming the gravitational force to be constant, and neglecting air friction?

Taking the z axis to be directed vertically upward,

$$F = -mg.$$

Equation (4.3) gives

$$\begin{aligned}\tfrac{1}{2}mv_1{}^2 - \tfrac{1}{2}mv_0{}^2 &= \int_{z_0}^{z_1} F\,dz \\ &= -mg\int_{z_0}^{z_1} dz \\ &= -mg(z_1 - z_0).\end{aligned}$$

At the peak, $v_1 = 0$ and we obtain the answer

$$z_1 = z_0 + \frac{v_0{}^2}{2g}.$$

It is interesting to note that the solution makes no reference to time at all. We could have solved the problem by applying Newton's second law, but we would have had to eliminate t to obtain the result.

Here is an example that is not easy to solve by direct application of Newton's second law.

Example 4.2 **Solving the Equation of Simple Harmonic Motion**

In Example 2.17 we discussed the equation of simple harmonic motion and pulled the solution out of a hat without proof. Now we shall derive the solution using Eq. (4.4).

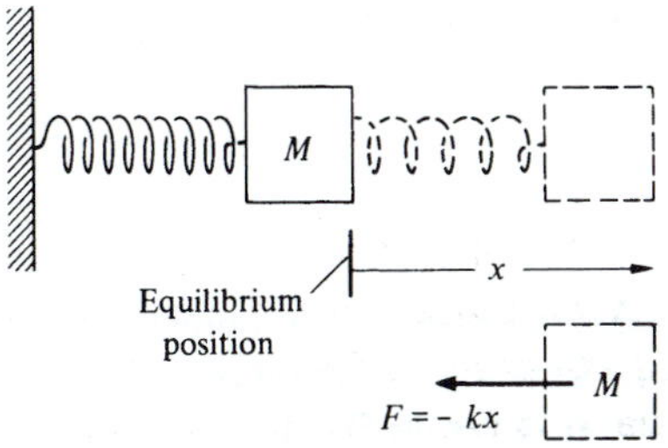

Consider a mass M attached to a spring. Using the coordinate x measured from the equilibrium point, the spring force is $F = -kx$. Then Eq. (4.4) becomes

$$\tfrac{1}{2}Mv^2 - \tfrac{1}{2}Mv_0^2 = -k\int_{x_0}^{x} x\,dx$$
$$= -\tfrac{1}{2}kx^2 + \tfrac{1}{2}kx_0^2.$$

The initial coordinates are labeled by the subscript 0.

In order to find x and v, we must know their values at some time t_0. Physically, this arises because the equation of motion by itself cannot completely specify the motion; we also need to know a set of initial conditions, in this case the initial position and velocity.[1] We are free to choose any initial conditions we wish. Let us consider the case where at $t = 0$ the mass is released from rest, $v_0 = 0$, at a distance x_0 from the origin. Then

$$v^2 = -\frac{k}{M}x^2 + \frac{k}{M}x_0^2$$

and

$$\frac{dx}{dt} = v$$
$$= \sqrt{\frac{k}{M}}\sqrt{x_0^2 - x^2}.$$

Separating the variables gives

$$\int_{x_0}^{x} \frac{dx}{\sqrt{x_0^2 - x^2}} = \sqrt{\frac{k}{M}}\int_0^t dt$$
$$= \sqrt{\frac{k}{M}}\,t.$$

The integral on the left hand side is arcsin (x/x_0). (The integral is listed in standard tables. Consulting a table of integrals is just as respectable for a physicist as consulting a dictionary is for a writer. Of course, in both cases one hopes that experience gradually reduces dependence.) Denoting $\sqrt{k/M}$ by ω, we obtain

$$\arcsin\left(\frac{x}{x_0}\right)\Bigg|_{x_0}^{x} = \omega t$$

or

$$\arcsin\left(\frac{x}{x_0}\right) - \arcsin 1 = \omega t.$$

[1] In the language of differential equations, Newton's second law is a "second order" equation in the position; the highest order derivative it involves is the acceleration, which is the second derivative of the position with respect to time. The theory of differential equations shows that the complete solution of a differential equation of *n*th order must involve *n* initial conditions.

Since arcsin 1 = $\pi/2$, we obtain

$$x = x_0 \sin\left(\omega t + \frac{\pi}{2}\right)$$
$$= x_0 \cos \omega t.$$

Note that the solution indeed satisfies the given initial conditions: at $t = 0$, $x = x_0 \cos 0 = x_0$, and $\dot{x} = x_0\omega \sin 0 = 0$. For these conditions our result agrees with the general solution given in Example 2.14.

4.3 The Work-energy Theorem in One Dimension

In Sec. 4.2 we demonstrated the formal procedure for integrating Newton's second law with respect to position. The result was

$$\tfrac{1}{2}mv_b{}^2 - \tfrac{1}{2}mv_a{}^2 = \int_{x_a}^{x_b} F(x)\,dx,$$

which we now wish to interpret in physical terms.

The quantity $\frac{1}{2}mv^2$ is called the *kinetic energy* K, and the left hand side can be written $K_b - K_a$. The integral $\int_{x_a}^{x_b} F(x)\,dx$ is called the *work* W_{ba} done *by* the force F *on* the particle as the particle moves from a to b. Our relation now takes the form

$$W_{ba} = K_b - K_a. \qquad 4.5$$

This result is known as the work-energy theorem or, more precisely, the work-energy theorem in one dimension. (We shall shortly see a more general statement.) The unit of work and energy in the SI system is the *joule* (J):

$$1\text{ J} = 1\text{ kg}\cdot\text{m}^2/\text{s}^2.$$

The unit of work and energy in the cgs system is the *erg*:

$$1\text{ erg} = 1\text{ gm}\cdot\text{cm}^2/\text{s}^2$$
$$= 10^{-7}\text{ J}.$$

The unit work in the English system is the *foot-pound*:

$$1\text{ ft}\cdot\text{lb} \approx 1.336\text{ J}.$$

Example 4.3 Vertical Motion in an Inverse Square Field

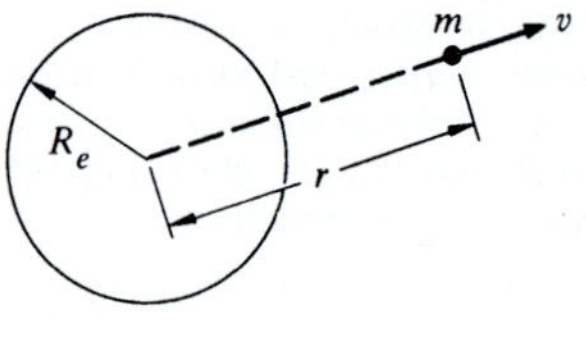

A mass m is shot vertically upward from the surface of the earth with initial speed v_0. Assuming that the only force is gravity, find its maximum altitude and the minimum value of v_0 for the mass to escape the earth completely.

The force on m is

$$F = -\frac{GM_em}{r^2}.$$

The problem is one dimensional in the variable r, and it is simple to find the kinetic energy at distance r by the work-energy theorem.

Let the particle start at $r = R_e$ with initial velocity v_0.

$$K(r) - K(R_e) = \int_{R_e}^{r} F(r)\,dr$$

$$= -GM_em \int_{R_e}^{r} \frac{dr}{r^2}$$

or

$$\tfrac{1}{2}mv(r)^2 - \tfrac{1}{2}mv_0{}^2 = GM_em\left(\frac{1}{r} - \frac{1}{R_e}\right).$$

We can immediately find the maximum height of m. At the highest point, $v(r) = 0$ and we have

$$v_0{}^2 = 2GM_e\left(\frac{1}{R_e} - \frac{1}{r_{\text{max}}}\right).$$

It is a good idea to introduce known familiar constants whenever possible. For example, since $g = GM_e/R_e{}^2$, we can write

$$v_0{}^2 = 2gR_e{}^2\left(\frac{1}{R_e} - \frac{1}{r_{\text{max}}}\right)$$

$$= 2gR_e\left(1 - \frac{R_e}{r_{\text{max}}}\right)$$

or

$$r_{\text{max}} = \frac{R_e}{1 - \dfrac{v_0{}^2}{2gR_e}}.$$

The escape velocity from the earth is the initial velocity needed to move r_{max} to infinity. The escape velocity is therefore

$$v_{\text{escape}} = \sqrt{2gR_e}$$

$$= \sqrt{2 \times 9.8 \times 6.4 \times 10^6}$$

$$= 1.1 \times 10^4 \text{ m/s}.$$

The energy needed to eject a 50-kg spacecraft from the surface of the earth is

$$W = \tfrac{1}{2}Mv_{\text{escape}}^2$$

$$= \tfrac{1}{2}(50)(1.1 \times 10^4)^2 = 3.0 \times 10^9 \text{ J}.$$

4.4 Integrating the Equation of Motion in Several Dimensions

Returning to the central problem of this chapter, let us try to integrate the equation of motion of a particle acted on by a force which depends on position.

$$\mathbf{F}(\mathbf{r}) = m\frac{d\mathbf{v}}{dt}. \qquad 4.6$$

In the case of one dimensional motion we integrated with respect to position. To generalize this, consider what happens when the particle moves a short distance $\Delta\mathbf{r}$.

We assume that $\Delta\mathbf{r}$ is so small that $\mathbf{F}$ is effectively constant over this displacement. If we take the scalar product of Eq. (4.6) with $\Delta\mathbf{r}$, we obtain

$$\mathbf{F}\cdot\Delta\mathbf{r} = m\frac{d\mathbf{v}}{dt}\cdot\Delta\mathbf{r}. \qquad 4.7$$

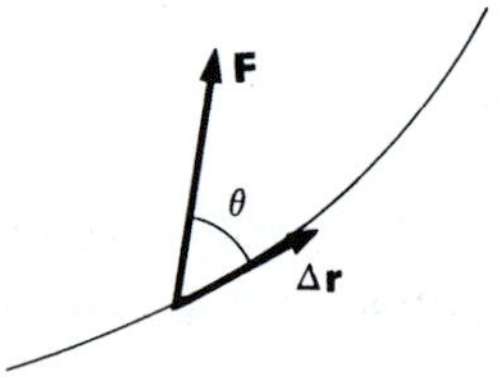

The sketch shows the trajectory and the force at some point along the trajectory. At this point,

$$\mathbf{F}\cdot\Delta\mathbf{r} = F\,\Delta r\cos\theta.$$

Perhaps you are wondering how we know $\Delta\mathbf{r}$, since this requires knowing the trajectory, which is what we are trying to find. Let us overlook this problem for a few moments and pretend we know the trajectory.

Now consider the right hand side of Eq. (4.7), $m(d\mathbf{v}/dt)\cdot\Delta\mathbf{r}$. We can transform this by noting that $\mathbf{v}$ and $\Delta\mathbf{r}$ are not independent; for a sufficiently short length of path, $\mathbf{v}$ is approximately constant. Hence $\Delta\mathbf{r} = \mathbf{v}\,\Delta t$, where Δt is the time the particle requires to travel $\Delta\mathbf{r}$, and therefore

$$m\frac{d\mathbf{v}}{dt}\cdot\Delta\mathbf{r} = m\frac{d\mathbf{v}}{dt}\cdot\mathbf{v}\,\Delta t. \qquad 4.8$$

We can transform Eq. (4.7) with the vector identity[1]

$$\mathbf{v}\cdot\frac{d\mathbf{v}}{dt} = \frac{1}{2}\frac{d}{dt}(v^2).$$

[1] The identity $\mathbf{A}\cdot(d\mathbf{A}/dt) = \frac{1}{2}(d/dt)(A^2)$ is easily proved:

$$\begin{aligned}\frac{1}{2}\frac{d}{dt}(A^2) &= \frac{1}{2}\frac{d}{dt}(\mathbf{A}\cdot\mathbf{A})\\ &= \frac{1}{2}\left(\mathbf{A}\cdot\frac{d\mathbf{A}}{dt}+\frac{d\mathbf{A}}{dt}\cdot\mathbf{A}\right)\\ &= \mathbf{A}\cdot\frac{d\mathbf{A}}{dt}.\end{aligned}$$

Equation (4.7) becomes

$$\mathbf{F}\cdot\Delta\mathbf{r} = \frac{m}{2}\frac{d}{dt}(v^2)\,\Delta t. \qquad 4.9$$

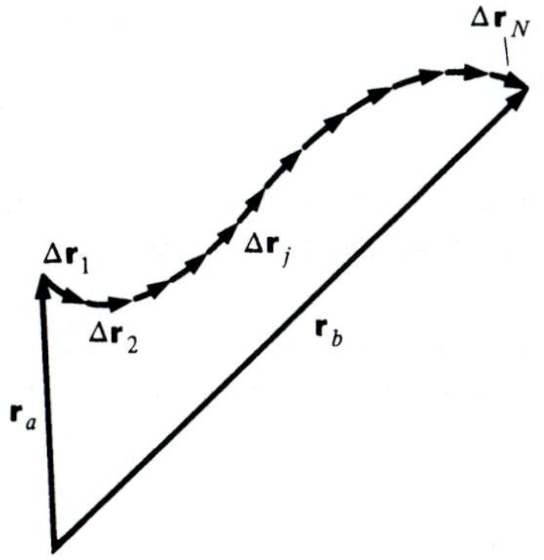

The next step is to divide the entire trajectory from the initial position $\mathbf{r}_a$ to the final position $\mathbf{r}_b$ into N short segments of length $\Delta\mathbf{r}_j$, where j is an index numbering the segments. (It makes no difference whether all the pieces have the same length.) For each segment we can write a relation similar to Eq. (4.9):

$$\mathbf{F}(\mathbf{r}_j)\cdot\Delta\mathbf{r}_j = \frac{m}{2}\frac{d}{dt}(v_j^2)\,\Delta t_j, \qquad 4.10$$

where $\mathbf{r}_j$ is the location of segment j, $\mathbf{v}_j$ is the velocity the particle has there, and Δt_j is the time it spends in traversing it. If we add together the equations of all the segments, we have

$$\sum_{j=1}^{N}\mathbf{F}(\mathbf{r}_j)\cdot\Delta\mathbf{r}_j = \sum_{j=1}^{N}\frac{m}{2}\frac{d}{dt}(v_j^2)\,\Delta t_j. \qquad 4.11$$

Next we take the limiting process where the length of each segment approaches zero, and the number of segments approaches infinity. We have

$$\int_{\mathbf{r}_a}^{\mathbf{r}_b}\mathbf{F}(\mathbf{r})\cdot d\mathbf{r} = \int_{t_a}^{t_b}\frac{m}{2}\frac{d}{dt}(v^2)\,dt, \qquad 4.12$$

where t_a and t_b are the times corresponding to $\mathbf{r}_a$ and $\mathbf{r}_b$. In converting the sum to an integral, we have dropped the numerical index j and have indicated the location of the first segment $\Delta\mathbf{r}_1$ by $\mathbf{r}_a$, and the location of the last section $\Delta\mathbf{r}_N$ by $\mathbf{r}_b$.

The integral on the right in Eq. (4.12) is

$$\frac{m}{2}\int_{t_a}^{t_b}\frac{d}{dt}(v^2)\,dt = \tfrac{1}{2}mv^2\Big|_{t_a}^{t_b}$$
$$= \tfrac{1}{2}mv_b^2 - \tfrac{1}{2}mv_a^2.$$

This represents a simple generalization of the result we found for one dimension. Here, however, $v^2 = v_x^2 + v_y^2 + v_z^2$, whereas for the one dimensional case we had $v^2 = v_x^2$.

Equation (4.12) becomes

$$\int_{\mathbf{r}_a}^{\mathbf{r}_b}\mathbf{F}\cdot d\mathbf{r} = \tfrac{1}{2}mv_b^2 - \tfrac{1}{2}mv_a^2. \qquad 4.13$$

The integral on the left is called a *line integral*. We shall see how to evaluate line integrals in the next two sections, and we shall also see how to interpret Eq. (4.13) physically. However, before proceeding, let's pause for a moment to summarize.

Our starting point was $\mathbf{F}(\mathbf{r}) = m\, d\mathbf{v}/dt$. All we have done is to integrate this equation with respect to distance, but because we described each step carefully, it looks like many operations are involved. This is not really the case; the whole argument can be stated in a few lines as follows:

$$\mathbf{F} = m\frac{d\mathbf{v}}{dt}$$

$$\begin{aligned}\int_a^b \mathbf{F}\cdot d\mathbf{r} &= \int_a^b m\frac{d\mathbf{v}}{dt}\cdot d\mathbf{r}\\ &= \int_a^b m\frac{d\mathbf{v}}{dt}\cdot \mathbf{v}\, dt\\ &= \int_a^b \frac{m}{2}\frac{d}{dt}(v^2)\, dt\\ &= \tfrac{1}{2}mv_b^2 - \tfrac{1}{2}mv_a^2.\end{aligned}$$

4.5 The Work-energy Theorem

We now want to interpret Eq. (4.13) in physical terms. The quantity $\frac{1}{2}mv^2$ is called the *kinetic energy* K, and the right hand side of Eq. (4.13) can be written as $K_b - K_a$. The integral $\int_{\mathbf{r}_a}^{\mathbf{r}_b} \mathbf{F}\cdot d\mathbf{r}$ is called the *work* W_{ba} done by the force $\mathbf{F}$ on the particle as the particle moves from a to b. Equation (4.13) now takes the form

$$W_{ba} = K_b - K_a. \qquad 4.14$$

This result is the general statement of the *work-energy theorem* which we met in restricted form in our discussion of one dimensional motion.

The work ΔW done by a force $\mathbf{F}$ in a small displacement $\Delta\mathbf{r}$ is

$$\Delta W = \mathbf{F}\cdot\Delta\mathbf{r} = F\cos\theta\,\Delta r = F_\parallel\,\Delta r,$$

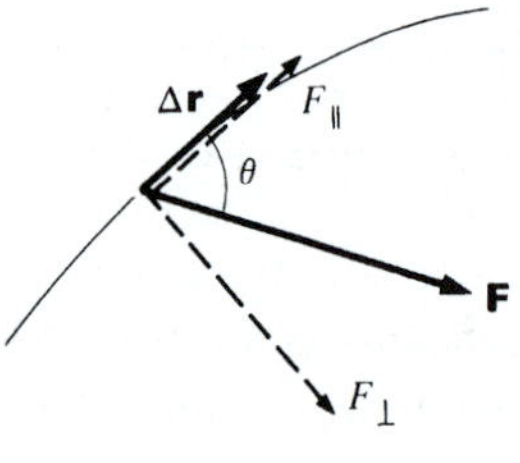

where $F_\parallel = F\cos\theta$ is the component of $\mathbf{F}$ along the direction of $\Delta\mathbf{r}$. The component of $\mathbf{F}$ perpendicular to $\Delta\mathbf{r}$ does no work. For a finite displacement from $\mathbf{r}_a$ to $\mathbf{r}_b$, the work on the particle, $\int_a^b \mathbf{F}\cdot d\mathbf{r}$, is the sum of the contributions $\Delta W = F_\parallel\,\Delta r$ from each segment of the path, in the limit where the size of each segment approaches zero.

In the work-energy theorem, $W_{ba} = K_b - K_a$, W_{ba} is the work done on the particle by the total force **F**. If **F** is the sum of several forces $\mathbf{F} = \Sigma \mathbf{F}_i$, we can write

$$W_{ba} = \sum_i (W_i)_{ba}$$
$$= K_b - K_a,$$

where

$$(W_i)_{ba} = \int_{\mathbf{r}_a}^{\mathbf{r}_b} \mathbf{F}_i \cdot d\mathbf{r}$$

is the work done by the ith force $\mathbf{F}_i$.

Our discussion so far has been restricted to the case of a single particle. However, we showed in Chap. 3 that the center of mass of an extended system moves according to the equation of motion

$$\mathbf{F} = M\ddot{\mathbf{R}}$$
$$= M\frac{d\mathbf{V}}{dt}, \qquad 4.15$$

where $\mathbf{V} = \dot{\mathbf{R}}$ is the velocity of the center of mass. Integrating Eq. (4.15) with respect to position gives

$$\int_{\mathbf{R}_a}^{\mathbf{R}_b} \mathbf{F} \cdot d\mathbf{R} = \tfrac{1}{2}MV_b^2 - \tfrac{1}{2}MV_a^2, \qquad 4.16$$

where $d\mathbf{R} = \mathbf{V}\,dt$ is the displacement of the center of mass in time dt. Equation (4.16) is the work-energy theorem for the translational motion of an extended system; in Chaps. 6 and 7 we shall extend the ideas of work and kinetic energy to include rotational motion. Note, however, that Eq. (4.16) holds regardless of the rotational motion of the system.

Example 4.4 The Conical Pendulum

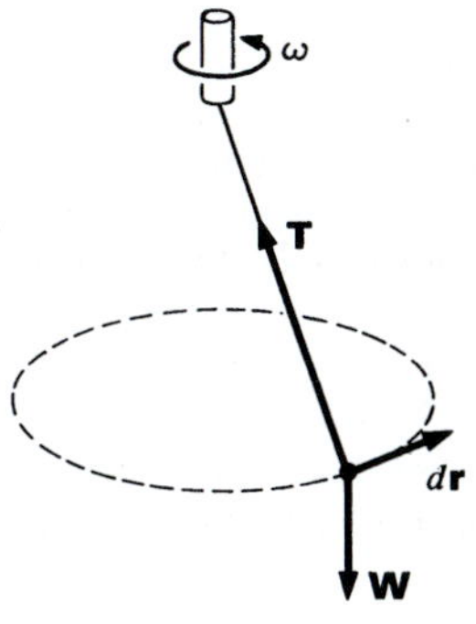

We discussed the motion of the conical pendulum in Example 2.8. Since the mass moves with constant angular velocity ω in a circle of constant radius R, the kinetic energy of the mass, $\frac{1}{2}mR\omega^2$, is constant. The work-energy theorem then tells us that no net work is being done on the mass.

Furthermore, in the conical pendulum the string force and the weight force separately do no work, since each of these forces is perpendicular to the path of the particle, making the integrand of the work integral zero.

It is important to realize that in the work integral $\int \mathbf{F} \cdot d\mathbf{r}$, the vector $d\mathbf{r}$ is along the path of the particle. Since $\mathbf{v} = d\mathbf{r}/dt$, $d\mathbf{r} = \mathbf{v}\,dt$ and $d\mathbf{r}$ is always parallel to $\mathbf{v}$.

Example 4.5 **Escape Velocity—the General Case**

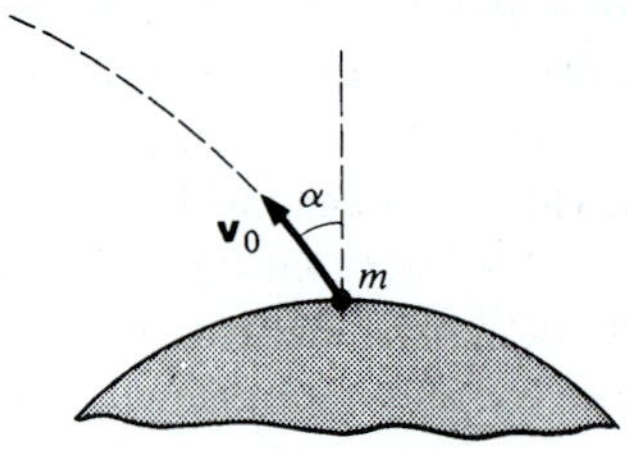

In Example 4.3 we discussed the one dimensional motion of a mass m projected vertically upward from the earth. We found that if the initial speed is greater than $v_0 = \sqrt{2gR_e}$, the mass will escape from the earth. Suppose that we look at the problem once again, but now allow the mass to be projected at angle α from the vertical.

The force on m, neglecting air resistance, is

$$\mathbf{F} = -\frac{GM_e m}{r^2}\hat{\mathbf{r}}$$

$$= -mg\frac{R_e^2}{r^2}\hat{\mathbf{r}},$$

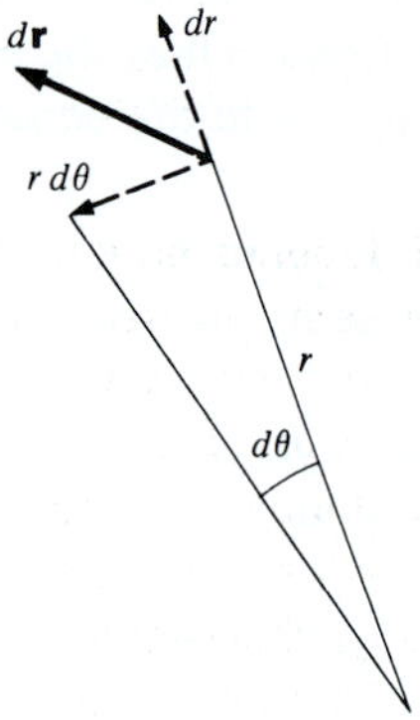

where $g = GM_e/R_e^2$ is the acceleration due to gravity at the earth's surface. We do not know the trajectory of the particle without solving the problem in detail. However, any element of the path $d\mathbf{r}$ can be written

$$d\mathbf{r} = dr\,\hat{\mathbf{r}} + r\,d\theta\,\hat{\boldsymbol{\theta}}.$$

Hence

$$\mathbf{F}\cdot d\mathbf{r} = -mg\frac{R_e^2}{r^2}\hat{\mathbf{r}}\cdot(dr\,\hat{\mathbf{r}} + r\,d\theta\,\hat{\boldsymbol{\theta}})$$

$$= -mg\frac{R_e^2}{r^2}\,dr.$$

The work-energy theorem becomes

$$\tfrac{1}{2}mv^2 - \tfrac{1}{2}mv_0^2 = -mgR_e^2\int_{R_e}^{r}\frac{dr}{r^2}$$

$$= -mgR_e^2\left(\frac{1}{r} - \frac{1}{R_e}\right).$$

The escape velocity is the value of v_0 for which $r = \infty$, $v = 0$. We find

$$v_0 = \sqrt{2gR_e}$$

$$= 1.1 \times 10^4\ \text{m/s},$$

as before. The escape velocity is independent of the launch direction.

We have neglected the earth's rotation in our analysis. In the absence of air resistance the projectile should be fired horizontally to the east, since the rotational speed of the earth's surface is then added to the launch velocity.

4.6 Applying the Work-energy Theorem

In the last section we derived the work-energy theorem

$$W_{ba} = K_b - K_a \qquad 4.17$$

and applied it to a few simple cases. In this section we shall use it to tackle more complicated problems. However, a few comments on the properties of the theorem are in order first.

To begin, we should emphasize that the work-energy theorem is a mathematical consequence of Newton's second law; we have introduced no new physical ideas. The work-energy theorem is merely the statement that the change in kinetic energy is equal to the net work done. This should not be confused with the general law of conservation of energy, an independent physical law which we shall discuss in Sec. 4.12.

Possibly you are troubled by the following problem: to apply the work-energy theorem, we have to evaluate the line integral for work[1]

$$W_{ba} = \oint_a^b \mathbf{F} \cdot d\mathbf{r}$$

and the evaluation of this integral depends on knowing what path the particle actually follows. We seem to need to know everything about the motion even before we use the work-energy theorem, and it is hard to see what use the theorem would be.

In the most general case, the work integral depends on the path followed, and since we don't know the path without completely solving the problem, the work-energy theorem is useless. There are, fortunately, two special cases of considerable practical importance. For many forces of interest, the work integral does not depend on the particular path but only on the end points. Such forces, which include most of the important forces in physics, are called *conservative* forces. As we shall discuss later in this chapter, the work-energy theorem can be put in a very simple form when the forces are conservative.

The work-energy theorem is also useful in cases where the path is known because the motion is *constrained*. By constrained motion, we mean motion in which external constraints act to keep the particle on a predetermined trajectory. The roller coaster is a perfect example. Except in cases of calamity, the roller coaster follows the track because it is held on by wheels both below and above the track. There are many other examples of constrained motion which come readily to mind—the conical pendulum is one (here the constraint is that the length of the string is fixed)—but all have one feature in common—the constraining force does no work. To see this, note that the effect of the constraint force is

[1] The C through the integral sign reminds us that the integral is to be evaluated along some specific curve.

to assure that the direction of the velocity is always tangential to the predetermined path. Hence, constraint forces change only the direction of **v** and do no work.[1]

Example 4.6 **The Inverted Pendulum**

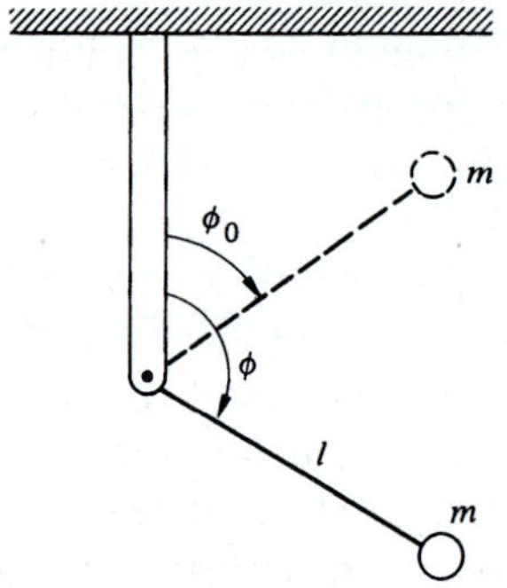

A pendulum consists of a light rigid rod of length l, pivoted at one end and with mass m attached at the other end. The pendulum is released from rest at angle ϕ_0, as shown. What is the velocity of m when the rod is at angle ϕ?

The work-energy theorem gives

$$\tfrac{1}{2}mv(\phi)^2 - \tfrac{1}{2}mv_0^2 = W_{\phi,\phi_0}.$$

Since $v_0 = 0$, we have

$$v(\phi) = \left(\frac{2W_{\phi,\phi_0}}{m}\right)^{\frac{1}{2}}$$

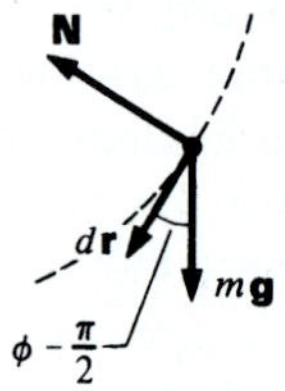

To evaluate W_{ϕ,ϕ_0}, the work done as the bob swings from ϕ_0 to ϕ, we examine the force diagram. $d\mathbf{r}$ lies along the circle of radius l. The forces acting are gravity, directed down, and the force of the rod, **N**. Since **N** lies along the radius, $\mathbf{N} \cdot d\mathbf{r} = 0$, and **N** does no work. The work done by gravity is

$$\begin{aligned} m\mathbf{g} \cdot d\mathbf{r} &= mgl \cos\left(\phi - \frac{\pi}{2}\right) d\phi \\ &= mgl \sin\phi \, d\phi \end{aligned}$$

where we have used $|d\mathbf{r}| = l\,d\phi$.

$$\begin{aligned} W_{\phi,\phi_0} &= \int_{\phi_0}^{\phi} mgl \sin\phi \, d\phi \\ &= -mgl \cos\phi \Big|_{\phi_0}^{\phi} \\ &= mgl(\cos\phi_0 - \cos\phi). \end{aligned}$$

The speed at ϕ is

$$v(\phi) = [2gl(\cos\phi_0 - \cos\phi)]^{\frac{1}{2}}.$$

The maximum velocity is obtained by letting the pendulum fall from the top, $\phi_0 = 0$, to the bottom, $\phi = \pi$:

$$v_{max} = 2(gl)^{\frac{1}{2}}.$$

[1] We can prove that constraint forces do no work as follows. Suppose that the constraint force $\mathbf{F}_{constraint}$ changes the velocity by an amount $\Delta\mathbf{v}_c$ in time Δt. $\Delta\mathbf{v}_c$ is perpendicular to the instantaneous velocity **v**. The work done by $\mathbf{F}_{constraint}$ is $\mathbf{F}_{constraint} \cdot \Delta\mathbf{r} = m(\Delta\mathbf{v}_c/\Delta t) \cdot (\mathbf{v}\,\Delta t) = m\Delta\mathbf{v}_c \cdot \mathbf{v} = 0$.

This is the same speed attained by a mass falling through the same vertical distance $2l$. However, the mass on the pendulum is not traveling vertically at the bottom of its path, it is traveling horizontally.

If you doubt the utility of the work-energy theorem, try solving the last example by integrating the equation of motion. However, the example also illustrates one of the shortcomings of the method: we found a simple solution for the speed of the mass at any point on the circle—we have no information on *when* the mass gets there. For instance, if the pendulum is released at $\phi_0 = 0$, in principle it balances there forever, never reaching the bottom. Fortunately, in many problems we are not interested in time, and even when time is important, the work-energy theorem provides a valuable first step toward obtaining a complete solution.

Next we turn to the general problem of evaluating work done by a known force over a given path, the problem of evaluating line integrals. We start by looking at the case of a constant force.

Example 4.7 **Work Done by a Uniform Force**

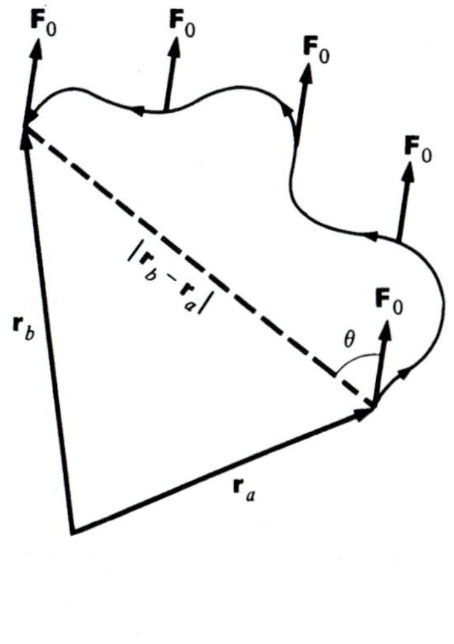

The case of a uniform force is particularly simple. Here is how to find the work done by a force, $\mathbf{F} = F_0\hat{\mathbf{n}}$, where F_0 is a constant and $\hat{\mathbf{n}}$ is a unit vector in some direction, as the particle moves from $\mathbf{r}_a$ to $\mathbf{r}_b$ along some arbitrary path. All the steps are put in to make the procedure clear, but with any practice this problem can be solved by inspection.

$$
\begin{aligned}
W_{ba} &= \oint_{\mathbf{r}_a}^{\mathbf{r}_b} \mathbf{F} \cdot d\mathbf{r} \\
&= \oint_{\mathbf{r}_a}^{\mathbf{r}_b} F_0\hat{\mathbf{n}} \cdot d\mathbf{r} \\
&= F_0\hat{\mathbf{n}} \cdot \oint_{\mathbf{r}_a}^{\mathbf{r}_b} d\mathbf{r} \\
&= F_0\hat{\mathbf{n}} \cdot \left(\hat{\mathbf{i}} \int_{x_a,y_a,z_a}^{x_b,y_b,z_b} dx + \hat{\mathbf{j}} \int_{x_a,y_a,z_a}^{x_b,y_b,z_b} dy + \hat{\mathbf{k}} \int_{x_a,y_a,z_a}^{x_b,y_b,z_b} dz\right) \\
&= F_0\hat{\mathbf{n}} \cdot [\hat{\mathbf{i}}(x_b - x_a) + \hat{\mathbf{j}}(y_b - y_a) + \hat{\mathbf{k}}(z_b - z_a)] \\
&= F_0\hat{\mathbf{n}} \cdot (\mathbf{r}_b - \mathbf{r}_a) \\
&= F_0 \cos\theta\, |\mathbf{r}_b - \mathbf{r}_a|
\end{aligned}
$$

For a constant force the work depends only on the net displacement, $\mathbf{r}_b - \mathbf{r}_a$, not on the path followed. This is not generally the case, but it holds true for an important group of forces, including central forces, as the next example shows.

Example 4.8 **Work Done by a Central Force**

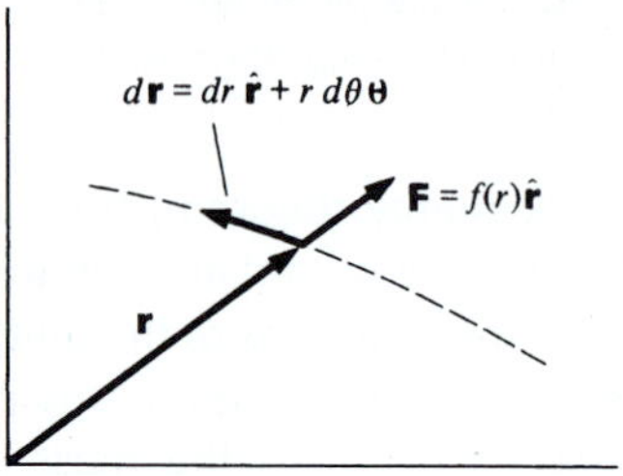

A *central force* is a radial force which depends only on the distance from the origin. Let us find the work done by the central force $\mathbf{F} = f(r)\hat{\mathbf{r}}$ on a particle which moves from $\mathbf{r}_a$ to $\mathbf{r}_b$. For simplicity we shall consider motion in a plane, for which $d\mathbf{r} = dr\,\hat{\mathbf{r}} + r\,d\theta\,\hat{\boldsymbol{\theta}}$. Then

$$\begin{aligned} W_{ba} &= \oint_a^b \mathbf{F} \cdot d\mathbf{r} \\ &= \oint_a^b f(r)\hat{\mathbf{r}} \cdot (dr\,\hat{\mathbf{r}} + r\,d\theta\,\hat{\boldsymbol{\theta}}) \\ &= \int_a^b f(r)\,dr. \end{aligned}$$

The work is given by a simple one dimensional integral over the variable r. Since θ has disappeared from the problem, it should be obvious that the work depends only on the initial and final radial distances [and, of course, on the particular form of $f(r)$], not on the particular path.

For some forces, the work is different for different paths between the initial and final points. One familiar example is work done by the force of sliding friction. Here the force always opposes the motion, so that the work done by friction in moving through distance dS is $dW = -f\,dS$, where f is the magnitude of the friction force. If we assume that f is constant, then the work done by friction in going from $\mathbf{r}_a$ to $\mathbf{r}_b$ along some path is

$$\begin{aligned} W_{ba} &= -\int_{\mathbf{r}_a}^{\mathbf{r}_b} f\,dS \\ &= -fS, \end{aligned}$$

where S is the total length of the path. The work is negative because the force always retards the particle. W_{ba} is never smaller in magnitude than fS_0, where S_0 is the distance between the two points, but by choosing a sufficiently devious route, S can be made arbitrarily large.

Example 4.9 **A Path-dependent Line Integral**

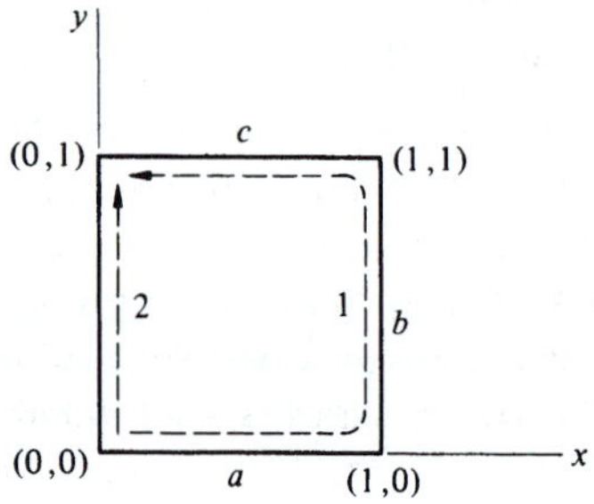

Here is a second example of a path-dependent line integral. Let $\mathbf{F} = A(xy\hat{\mathbf{i}} + y^2\hat{\mathbf{j}})$, and consider the integral from (0,0) to (0,1), first along path 1 and then along path 2, as shown in the figure. The force $\mathbf{F}$ has no physical significance, but the example illustrates the properties of nonconservative forces. Since the segments of each path lie along a coordinate axis, it is particularly simple to evaluate the integrals. For path 1 we have

$$\oint_1 \mathbf{F} \cdot d\mathbf{r} = \int_a \mathbf{F} \cdot d\mathbf{r} + \int_b \mathbf{F} \cdot d\mathbf{r} + \int_c \mathbf{F} \cdot d\mathbf{r}.$$

Along segment a, $d\mathbf{r} = dx\,\hat{\mathbf{i}}$, $\mathbf{F}\cdot d\mathbf{r} = F_x\,dx = Axy\,dx$. Since $y = 0$ along the line of this integration, $\int_a \mathbf{F}\cdot d\mathbf{r} = 0$. Similarly, for path b,

$$\int_b \mathbf{F}\cdot d\mathbf{r} = A\int_{x=1,y=0}^{x=1,y=1} y^2\,dy$$
$$= \frac{A}{3},$$

while for path c,

$$\int_c \mathbf{F}\cdot d\mathbf{r} = A\int_{x=1,y=1}^{x=0,y=1} xy\,dx$$
$$= A\int_1^0 x\,dx = -\frac{A}{2}.$$

Thus

$$\oint_1 \mathbf{F}\cdot d\mathbf{r} = \frac{A}{3} - \frac{A}{2}$$
$$= -\frac{A}{6}.$$

Along path 2 we have

$$\oint_2 \mathbf{F}\cdot d\mathbf{r} = A\int_{0,0}^{0,1} y^2\,dy$$
$$= \frac{A}{3}$$
$$\neq \oint_1 \mathbf{F}\cdot d\mathbf{r}.$$

The work done by the applied force is different for the two paths.

Usually the path of a line integral does not lie conveniently along the coordinate axes but along some arbitrary curve. The following method of evaluating a line integral in such a case is quite general; use it if all else fails.

For simplicity we again consider motion in a plane. Generalization to three dimensions is straightforward.

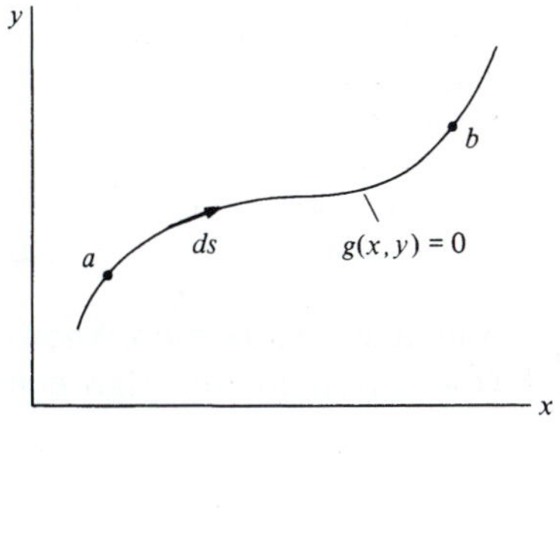

The problem is to evaluate $\oint_a^b \mathbf{F}\cdot d\mathbf{r}$ along a specified path. The path can be characterized by an equation of the form $g(x,y) = 0$. For example, if the path is a unit circle about the origin, then all points on the path obey $x^2 + y^2 - 1 = 0$.

We can characterize every point on the path by a parameter s which in practical problems could be (for example) distance along the path, or angle—anything just as long as each point on the path is associated with a value of s so that we can write

$x = x(s)$, $y = y(s)$. If we move along the path a short way, so that s changes by the amount ds, then the change in x is $dx = (dx/ds)\,ds$, and the change in y is $dy = (dy/ds)\,ds$. Since both x and y are determined by s, so are F_x and F_y. Hence, we can write $\mathbf{F} = F_x(s)\hat{\mathbf{i}} + F_y(s)\hat{\mathbf{j}}$, and we have

$$\oint_a^b \mathbf{F}\cdot d\mathbf{r} = \int_a^b (F_x\,dx + F_y\,dy)$$
$$= \int_{s_a}^{s_b}\left[F_x(s)\frac{dx}{ds} + F_y(s)\frac{dy}{ds}\right]ds.$$

We have reduced the problem to the more familiar problem of evaluating a one dimensional definite integral. The calculation is much simpler in practice than in theory. Here is an example.

Example 4.10 Parametric Evaluation of a Line Integral

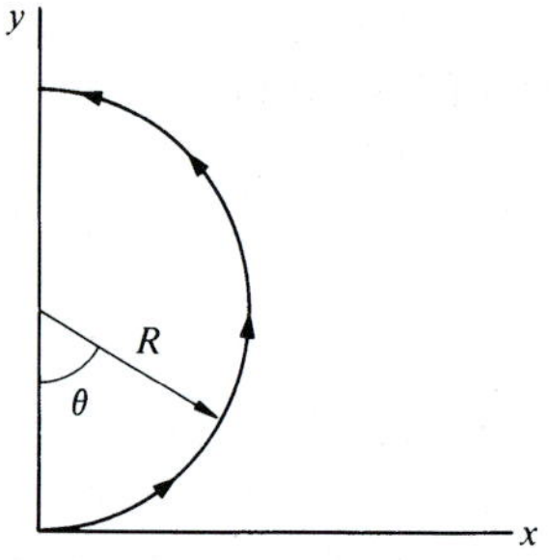

Evaluate the line integral of $\mathbf{F} = A(x^3\hat{\mathbf{i}} + xy^2\hat{\mathbf{j}})$ from $(x = 0,\ y = 0)$ to $(x = 0,\ y = 2R)$ along the semicircle shown.

The natural parameter to use here is θ, since as θ varies from 0 to π, the radius vector sweeps out the semicircle. We have

$$x = R\sin\theta \qquad dx = R\cos\theta\,d\theta \qquad F_x = AR^3\sin^3\theta$$
$$y = R(1-\cos\theta) \qquad dy = R\sin\theta\,d\theta \qquad F_y = AR^3\sin\theta(1-\cos\theta)^2$$
$$\oint \mathbf{F}\cdot d\mathbf{r} = A\int_0^\pi [(R\sin\theta)^3 R\cos\theta + R^3\sin\theta\,(1-\cos\theta)^2 R\sin\theta]\,d\theta$$
$$= R^4A\int_0^\pi [\sin^3\theta\cos\theta + \sin^2\theta(1-\cos\theta)^2]\,d\theta.$$

Evaluation of the integral is straightforward. If you are interested in carrying it through, try substituting $u = \cos\theta$.

4.7 Potential Energy

We introduced the idea of a conservative force in the last section. The work done by a conservative force on a particle as it moves from one point to another depends only on the end points, not on the path between them. Hence, for a conservative force,

$$\int_{\mathbf{r}_a}^{\mathbf{r}_b} \mathbf{F}\cdot d\mathbf{r} = \text{function of } (\mathbf{r}_b) - \text{function of } (\mathbf{r}_a)$$

or

$$\int_{\mathbf{r}_a}^{\mathbf{r}_b} \mathbf{F}\cdot d\mathbf{r} = -U(\mathbf{r}_b) + U(\mathbf{r}_a), \qquad 4.18$$

where $U(\mathbf{r})$ is a function, defined by the above expression, known as the *potential energy* function. (The reason for the sign convention will be clear in a moment.) Note that we have not proven that $U(\mathbf{r})$ exists. However, we have already seen several cases where the work is indeed path-independent, so that we can assume that U exists for at least a few forces.

The work-energy theorem $W_{ba} = K_b - K_a$ now becomes

$$\begin{aligned} W_{ba} &= -U_b + U_a \\ &= K_b - K_a \end{aligned}$$

or, rearranging,

$$K_a + U_a = K_b + U_b. \tag{4.19}$$

The left hand side of this equation, $K_a + U_a$, depends on the speed of the particle and its potential energy at $\mathbf{r}_a$; it makes no reference to $\mathbf{r}_b$. Similarly, the right hand side depends on the speed and potential energy at $\mathbf{r}_b$; it makes no reference to $\mathbf{r}_a$. This can be true only if each side of the equation equals a constant, since $\mathbf{r}_a$ and $\mathbf{r}_b$ are arbitrary and not specially chosen points. Denoting this constant by E, we have

$$K_a + U_a = K_b + U_b = E. \tag{4.20}$$

E is called the *total mechanical energy* of the particle, or, somewhat less precisely, the total energy. We have shown that if the force is conservative, the total energy is independent of the position of the particle—it remains constant, or, in the language of physics, the energy is *conserved.* Although the conservation of mechanical energy is a derived law, which means that it has basically no new physical content, it presents such a different way of looking at a physical process compared with applying Newton's laws that we have what amounts to a completely new tool. Furthermore, although the conservation of mechanical energy follows directly from Newton's laws, it is an important key to understanding the more general law of conservation of energy, which is independent of Newton's laws and which vastly increases our understanding of nature. When we discuss this in greater detail in Sec. 4.12, we shall see that the conservation law for mechanical energy turns out to be a special case of the more general law.

A peculiar property of energy is that the value of E is to a certain extent arbitrary; only changes in E have physical significance. This comes about because the equation

$$U_b - U_a = -\int_a^b \mathbf{F} \cdot d\mathbf{r}$$

defines only the difference in potential energy between a and b and not the potential energy itself. We could add a constant to U_b and the same constant to U_a and still satisfy the defining equation. However, since $E = K + U$, adding a constant to U increases E by the same amount.

Illustrations of Potential Energy

We have already seen that for a uniform force or a central force the work is path-independent. There are many other conservative forces, but by way of illustrating potential energy, here are two examples involving these forces.

Example 4.11 **Potential Energy of a Uniform Force Field**

From Example 4.7, the work done by a uniform force is $W_{ba} = \mathbf{F}_0 \cdot (\mathbf{r}_b - \mathbf{r}_a)$ For instance, the force on a particle of mass m due to a uniform gravitational field is $-mg\hat{\mathbf{k}}$, so that if the particle moves from $\mathbf{r}_a$ to $\mathbf{r}_b$, the change in potential energy is

$$\begin{aligned} U_b - U_a &= -\int_{z_a}^{z_b} (-mg)\,dz \\ &= mg(z_b - z_a). \end{aligned}$$

If we adopt the convention $U = 0$ at ground level where $z = 0$, then $U(h) = mgh$, where h is the height above the ground. However, a potential energy of the form $mgh + C$, where C is any constant, is just as suitable.

In Example 4.1 we considered the problem of a mass projected upward with a given initial velocity in a region of constant gravity. Here is how to solve the same problem by using conservation of energy.

Suppose that a mass is projected upward with initial velocity $\mathbf{v}_0 = v_{0x}\hat{\mathbf{i}} + v_{0y}\hat{\mathbf{j}} + v_{0z}\hat{\mathbf{k}}$. Find the speed at height h.

$$\begin{aligned} K_0 + U_0 &= K(h) + U(h) \\ \tfrac{1}{2}mv_0^2 + 0 &= \tfrac{1}{2}mv(h)^2 + mgh \end{aligned}$$

or

$$v(h) = \sqrt{v_0^2 - 2gh}.$$

Example 4.11 is trivial, since motion in a uniform force field is easily found from $\mathbf{F} = m\mathbf{a}$. However, it does illustrate the ease with which the energy method handles the problem. For instance, motion in all three directions is handled at once, whereas Newton's law involves one equation for each component of motion.

Example 4.12 **Potential Energy of an Inverse Square Force**

Frequently we encounter central forces $\mathbf{F} = f(r)\hat{\mathbf{r}}$, where $f(r)$ is some function of the distance to the origin. For instance, in the case of the Coulomb electrostatic force, $\mathbf{F} \propto (q_1q_2/r^2)\hat{\mathbf{r}}$, where q_1 and q_2 are the charges of two interacting particles. The gravitational force between two particles provides another example.

The potential energy of a particle in a central force $\mathbf{F} = f(r)\hat{\mathbf{r}}$ obeys

$$U_b - U_a = -\int_{\mathbf{r}_a}^{\mathbf{r}_b} \mathbf{F} \cdot d\mathbf{r}$$
$$= -\int_{r_a}^{r_b} f(r)\, dr.$$

For an inverse square force, $f(r) = A/r^2$, and we have

$$U_b - U_a = -\int_{r_a}^{r_b} \frac{A}{r^2}\, dr$$
$$= \frac{A}{r_b} - \frac{A}{r_a}.$$

To obtain the general potential energy function, we replace r_b by the radial variable r. Then

$$U(r) = \frac{A}{r} + \left(U_a - \frac{A}{r_a}\right)$$
$$= \frac{A}{r} + C.$$

The constant C has no physical meaning, since only changes in U are significant. We are free to give C any value we like. A convenient choice in this case is $C = 0$, which corresponds to taking $U(\infty) = 0$. With this convention we have

$$U(r) = \frac{A}{r}.$$

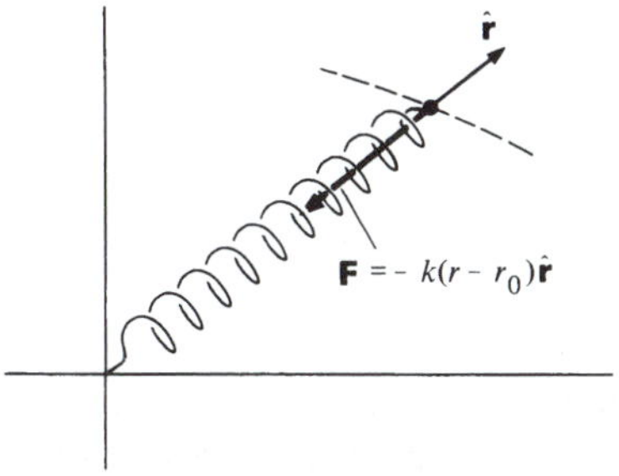

One of the most important forces in physics is the linear restoring force, the spring force. To show that the spring force is conservative, consider a spring of equilibrium length r_0 with one end attached at the origin. If the spring is stretched to length r along direction $\hat{\mathbf{r}}$, it exerts a force

$$\mathbf{F}(r) = -k(r - r_0)\hat{\mathbf{r}}.$$

Since the force is central, it is conservative. The potential energy is given by

$$U(r) - U(a) = -\int_a^r (-k)(r - r_0)\, dr$$
$$= \tfrac{1}{2}k(r - r_0)^2 \Big|_a^r.$$

Hence

$$U(r) = \tfrac{1}{2}k(r - r_0)^2 + C.$$

Conventionally, we choose the potential energy to be zero at equilibrium: $U(r_0) = 0$. This gives

$$U(r) = \tfrac{1}{2}k(r - r_0)^2. \qquad 4.21$$

When several conservative forces act on a particle, the potential energy is the sum of the potential energies for each force. In the next example, two conservative forces act.

Example 4.13 **Bead, Hoop, and Spring**

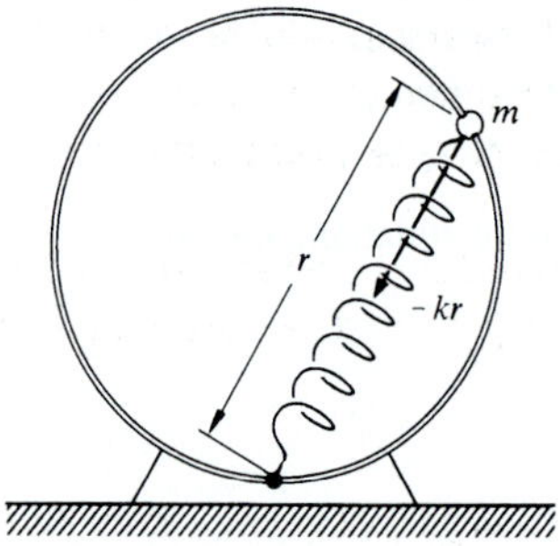

A bead of mass m slides without friction on a vertical hoop of radius R. The bead moves under the combined action of gravity and a spring attached to the bottom of the hoop. For simplicity, we assume that the equilibrium length of the spring is zero, so that the force due to the spring is $-kr$, where r is the instantaneous length of the spring, as shown.

The bead is released at the top of the hoop with negligible speed. How fast is the bead moving at the bottom of the hoop?

At the top of the hoop, the gravitational potential energy of the bead is $mg(2R)$ and the potential energy due to the spring is $\frac{1}{2}k(2R)^2 = 2kR^2$. Hence the initial potential energy is

$$U_i = 2mgR + 2kR^2.$$

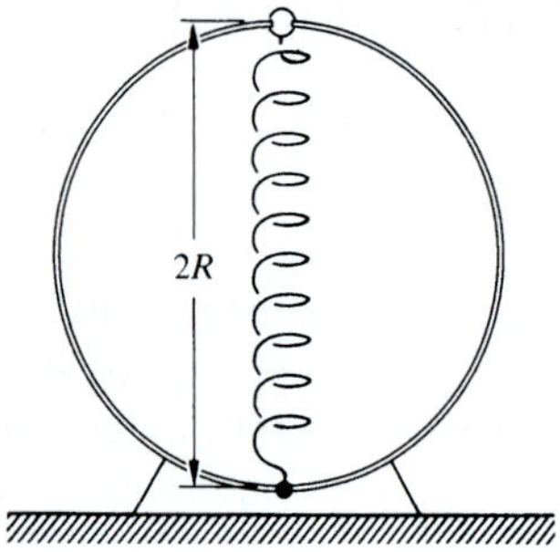

The potential energy at the bottom of the hoop is

$$U_f = 0.$$

Since all the forces are conservative, the mechanical energy is constant and we have

$$K_i + U_i = K_f + U_f.$$

The initial kinetic energy is zero and we obtain

$$K_f = U_i - U_f$$

or

$$\tfrac{1}{2}mv_f^2 = 2mgR + 2kR^2.$$

Hence

$$v_f = 2\sqrt{gR + \frac{kR^2}{m}}.$$

4.8 What Potential Energy Tells Us about Force

If we are given a conservative force, it is a straightforward matter to find the potential energy from the defining equation

$$U_b - U_a = -\int_a^b \mathbf{F} \cdot d\mathbf{r},$$

where the integral is over any path from $\mathbf{r}_a$ to $\mathbf{r}_b$. However, in many cases it is easier to characterize a force by giving its potential energy function rather than by specifying each of its components. In such cases we would like to use our knowledge of the potential energy to determine what force is acting. The procedure for finding the force turns out to be simple. In this section we shall learn how to find the force from the potential energy in a one dimensional system. The general case of three dimensions can be treated by a straightforward extension of the method developed here, but since it involves some new notation which is more readily introduced in the next chapter, let us defer the three dimensional case until then.

Suppose that we have a one dimensional system, such as a mass on a spring, in which the force is $F(x)$ and the potential energy is

$$U_b - U_a = -\int_{x_a}^{x_b} F(x)\,dx.$$

Consider the change in potential energy ΔU as the particle moves from some point x to $x + \Delta x$.

$$\begin{aligned} U(x + \Delta x) - U(x) &\equiv \Delta U \\ &= -\int_x^{x+\Delta x} F(x)\,dx. \end{aligned}$$

For Δx sufficiently small, $F(x)$ can be considered constant over the range of integration and we have

$$\begin{aligned} \Delta U &\approx -F(x)(x + \Delta x - x) \\ &= -F(x)\,\Delta x \end{aligned}$$

or

$$F(x) \approx -\frac{\Delta U}{\Delta x}.$$

In the limit $\Delta x \to 0$ we have

$$F(x) = -\frac{dU}{dx}. \qquad 4.22$$

The result is quite reasonable: potential energy is the negative integral of the force, and it follows that force is the negative derivative of the potential energy.

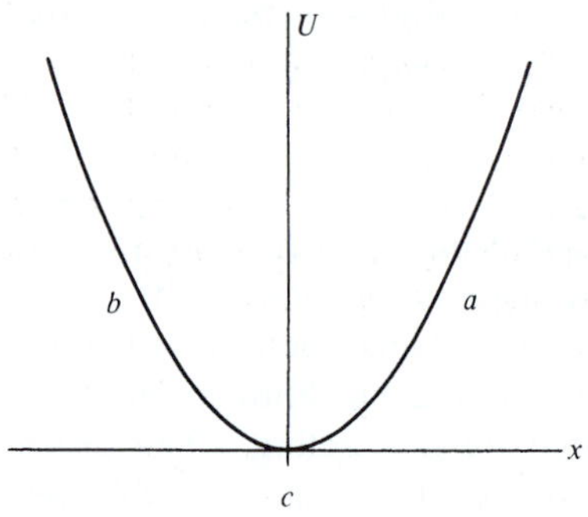

Stability

The result $F = -dU/dx$ is useful not only for computing the force but also for visualizing the stability of a system from a diagram of the potential energy. For instance, in the case of a harmonic oscillator the potential energy $U = kx^2/2$ is described by a parabola.

At point a, $dU/dx > 0$ and so the force is negative. At point b, $dU/dx < 0$ and the force is positive. At c, $dU/dx = 0$ and the force is zero. The force is directed toward the origin no matter which way the particle is displaced, and the force vanishes only when the particle is at the origin. The minimum of the potential energy curve coincides with the equilibrium position of the system. Evidently this is a stable equilibrium, since any displacement of the system produces a force which tends to push the particle toward its resting point.

Whenever $dU/dx = 0$, a system is in equilibrium. However, if this occurs at a maximum of U, the equilibrium is not stable, since a positive displacement produces a positive force, which tends to increase the displacement, and a negative displacement produces a negative force, which again causes the displacement to become larger. A pendulum of length l supporting mass m offers a good illustration of this. If we take the potential energy to be zero at the bottom of its swing, we see that

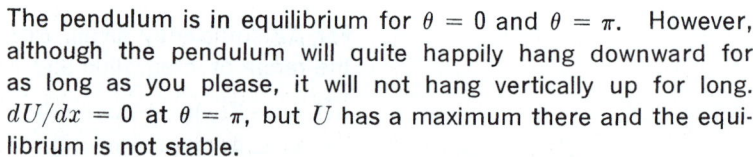

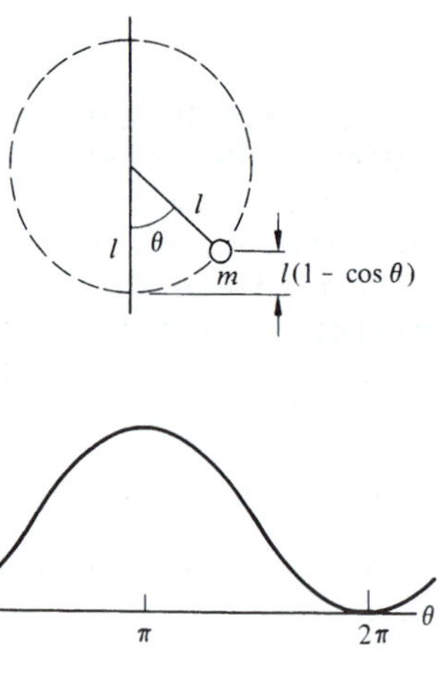

$$U(\theta) = mgz$$
$$= mgl(1 - \cos\theta).$$

The pendulum is in equilibrium for $\theta = 0$ and $\theta = \pi$. However, although the pendulum will quite happily hang downward for as long as you please, it will not hang vertically up for long. $dU/dx = 0$ at $\theta = \pi$, but U has a maximum there and the equilibrium is not stable.

The sketch of a potential energy function makes the idea of stability almost intuitively obvious. A minimum of a potential energy curve is a point of stable equilibrium, and a maximum is a point of unstable equilibrium. In more descriptive terms, the system is stable at the bottom of a potential energy "valley," and unstable at the top of a potential energy "hill."

Alternatively, we can use a simple mathematical test to determine whether or not an equilibrium point is stable. Let $U(x)$ be the potential energy function for a particle. As we have shown, the force on the particle is $F = -dU/dx$, and the system is in equilibrium where $dU/dx = 0$. Suppose that this occurs at some point x_0. To test for stability we must determine whether U has a minimum or a maximum at x_0. To accomplish this we need to examine d^2U/dx^2 at x_0. If the second derivative is positive, the equilibrium is stable; if it is negative, the system is unstable. If $d^2U/dx^2 = 0$, we must look at higher derivatives. If all derivatives vanish so that U is constant in a region about x_0, the system is said to be in a condition of neutral stability—no force results from a displacement; the particle is effectively free.

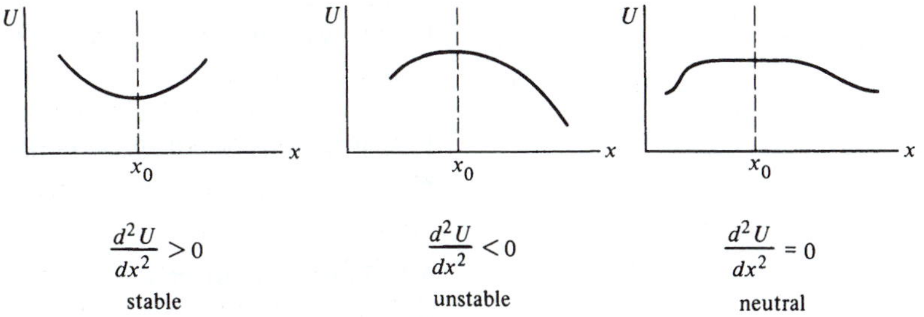

Example 4.14 **Energy and Stability—The Teeter Toy**

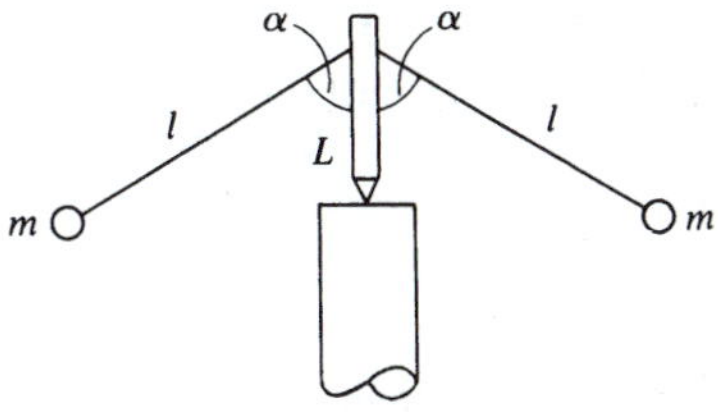

The teeter toy consists of two identical weights which hang from a peg on drooping arms, as shown. The arrangement is unexpectedly stable—the toy can be spun or rocked with little danger of toppling over. We can see why this is so by looking at its potential energy. For simplicity, we shall consider only rocking motion in the vertical plane.

Let us evaluate the potential energy when the teeter toy is cocked at angle θ, as shown in the sketch. If we take the zero of gravitational potential at the pivot, we have

$$U(\theta) = mg[L\cos\theta - l\cos(\alpha + \theta)] + mg[L\cos\theta - l\cos(\alpha - \theta)].$$

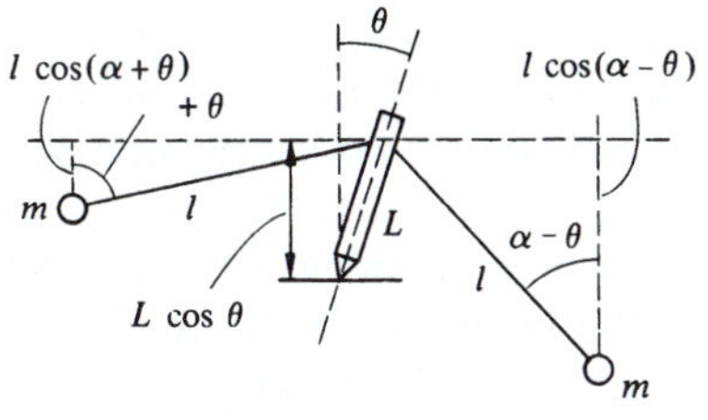

Using the identity $\cos(\alpha \pm \theta) = \cos\alpha\cos\theta \mp \sin\alpha\sin\theta$, we can rewrite $U(\theta)$ as

$$U(\theta) = 2mg\cos\theta(L - l\cos\alpha).$$

Equilibrium occurs when

$$\frac{dU}{d\theta} = -2mg\sin\theta(L - l\cos\alpha)$$
$$= 0.$$

The solution is $\theta = 0$, as we expect from symmetry. (We reject the solution $\theta = \pi$ on the grounds that θ must be limited to values less than

$\pi/2$.) To investigate the stability of the equilibrium position, we must examine the second derivative of the potential energy. We have

$$\frac{d^2U}{d\theta^2} = -2mg \cos \theta(L - l \cos \alpha).$$

At equilibrium,

$$\left.\frac{d^2U}{d\theta^2}\right|_{\theta=0} = -2mg(L - l \cos \alpha).$$

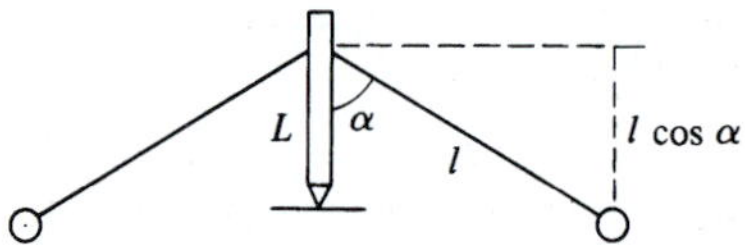

For the second derivative to be positive, we require $L - l \cos \alpha < 0$, or

$L < l \cos \alpha$.

In order for the teeter toy to be stable, the weights must hang below the pivot.

4.9 Energy Diagrams

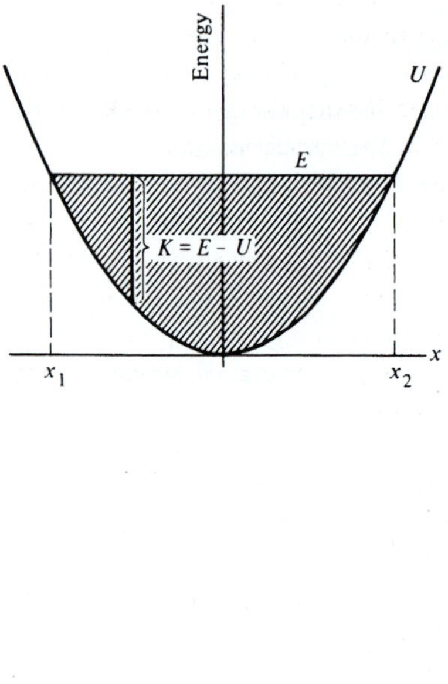

We can often find the most interesting features of the motion of a one dimensional system by using an *energy diagram*, in which the total energy E and the potential energy U are plotted as functions of position. The kinetic energy $K = E - U$ is easily found by inspection. Since kinetic energy can never be negative, the motion of the system is constrained to regions where $U \leq E$.

Here is the energy diagram for a harmonic oscillator. The potential energy $U = kx^2/2$ is a parabola centered at the origin. Since the total energy is constant for a conservative system, E is represented by a horizontal straight line. Motion is limited to the shaded region where $E \geq U$; the limits of the motion, x_1 and x_2 in the sketch, are sometimes called the turning points.

Here is what the diagram tells us. The kinetic energy, $K = E - U$, is greatest at the origin. As the particle flies past the origin in either direction, it is slowed by the spring and comes to a complete rest at one of the turning points x_1, x_2. The particle then moves toward the origin with increasing kinetic energy, and the cycle is repeated.

The harmonic oscillator provides a good example of bounded motion. As E increases, the turning points move farther and farther off, but the particle can never move away freely. If E is decreased, the amplitude of motion decreases, until finally for $E = 0$ the particle lies at rest at $x = 0$.

Quite a different behavior occurs if U does not increase indefinitely with distance. For instance, consider the case of a particle constrained to a radial line and acted on by a repulsive inverse

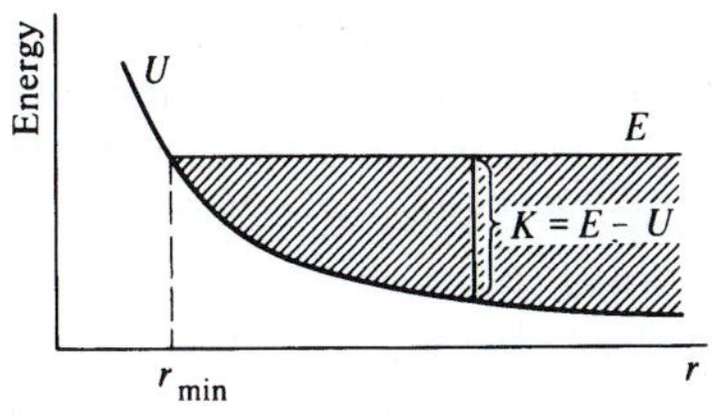

square law force $A\hat{\mathbf{r}}/r^2$. Here $U = A/r$, where A is positive. There is a distance of closest approach, $r_{\min}$, as shown in the diagram, but the motion is not bounded for large r since U decreases with distance. If the particle is shot toward the origin, it gradually loses kinetic energy until it comes momentarily to rest at $r_{\min}$. The motion then reverses and the particle moves out toward infinity. The final and initial speeds at any point are identical; the collision merely reverses the velocity.

With some potentials, either bounded or unbounded motion can occur depending upon the energy. For instance, consider the interaction between two atoms. At large separations, the atoms attract each other weakly with the van der Waals force, which varies as $1/r^7$. As the atoms approach, the electron clouds begin to overlap, producing strong forces. In this intermediate region the force is either attractive or repulsive depending on the details of the electron configuration. If the force is attractive, the potential energy decreases with decreasing r. At very short distances the atoms always repel each other strongly, so that U increases rapidly as r becomes small.

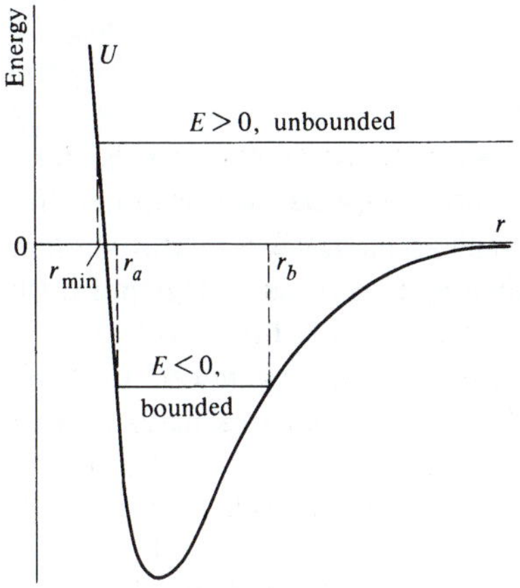

The energy diagram for a typical attractive two atom system is shown in the sketch. For positive energy, $E > 0$, the motion is unbounded, and the atoms are free to fly apart. As the diagram indicates, the distance of closest approach, $r_{\min}$, does not change appreciably as E is increased. The steep slope of the potential energy curve at small r means that the atoms behave like hard spheres—$r_{\min}$ is not sensitive to the energy of collision.

The situation is quite different if E is negative. Then the motion is bounded for both small and large separations; the atoms never approach closer than r_a or move farther apart than r_b. A bound system of two atoms is, of course, a molecule, and our sketch represents a typical diatomic molecule energy diagram. If two atoms collide with positive energy, they cannot form a molecule unless some means is available for losing enough energy to make E negative. In general, a third body is necessary to carry off the excess energy. Sometimes the third body is a surface, which is the reason surface catalysts are used to speed certain reactions. For instance, atomic hydrogen is quite stable in the gas phase even though the hydrogen molecule is tightly bound. However, if a piece of platinum is inserted in the hydrogen, the atoms immediately join to form molecules. What happens is that hydrogen atoms tightly adhere to the surface of the platinum, and if a collision occurs between two atoms on the surface, the excess energy is released to the surface, and the molecule, which is not strongly attracted to the surface, leaves. The energy delivered to the surface is so large that the platinum glows brightly. A third atom can also carry off the excess energy, but for this to happen the two atoms must collide when a third atom is nearby. This is a rare event at low pressures, but it becomes increasingly important at higher pressures. Another possibility is for the two atoms to lose energy by the emission of light. However, this occurs so rarely that it is usually not important.

4.10 Small Oscillations in a Bound System

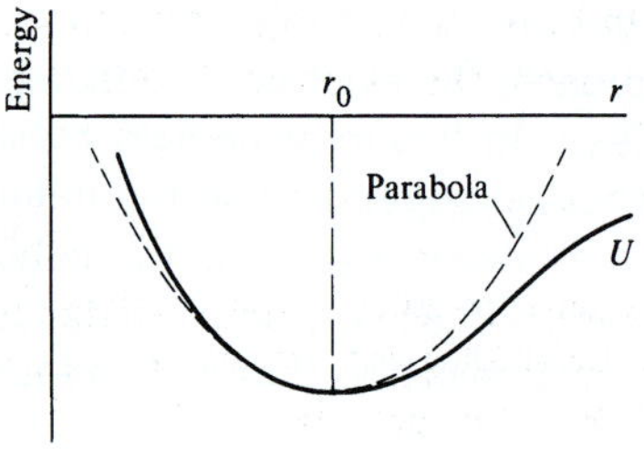

The interatomic potential we discussed in the last section illustrates an important feature of all bound systems; at equilibrium the potential energy has a minimum. As a result, nearly every bound system oscillates like a harmonic oscillator if it is slightly perturbed from its equilibrium position. This is suggested by the appearance of the energy diagram near the minimum—U has the parabolic shape of a harmonic oscillator potential. If the total energy is low enough so that the motion is restricted to the region where the curve is nearly parabolic, as illustrated in the sketch, the system must behave like a harmonic oscillator. It is not difficult to prove this.

As we have discussed in Note 1.1, any "well behaved" function $f(x)$ can be expanded in a Taylor's series about a point x_0. Thus

$$f(x) = f(x_0) + (x - x_0)f'(x_0) + \tfrac{1}{2}(x - x_0)^2 f''(x_0) + \cdots.$$

Suppose that we expand $U(r)$ about r_0, the position of the potential minimum. Then

$$U(r) = U(r_0) + (r - r_0)\left.\frac{dU}{dr}\right|_{r_0} + \frac{1}{2}(r - r_0)^2 \left.\frac{d^2U}{dr^2}\right|_{r_0} + \cdots.$$

However, since U is a minimum at r_0, $(dU/dr)\,|_{r_0} = 0$. Furthermore, for sufficiently small displacements, we can neglect the terms beyond the third in the power series. In this case,

$$U(r) = U(r_0) + \frac{1}{2}(r - r_0)^2 \left.\frac{d^2U}{dr^2}\right|_{r_0}$$

This is the potential energy of a harmonic oscillator,

$$U(x) = \text{constant} + \frac{kx^2}{2}.$$

We can even identify the effective spring constant:

$$k = \left.\frac{d^2U}{dr^2}\right|_{r_0} \tag{4.23}$$

Example 4.15 **Molecular Vibrations**

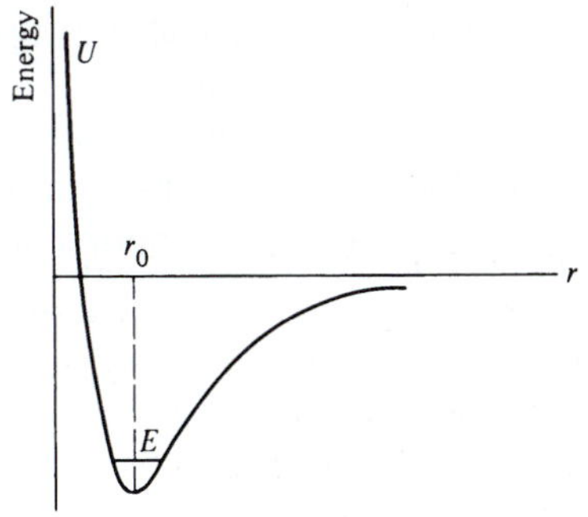

Suppose that two atoms of masses m_1 and m_2 are bound together in a molecule with energy so low that their separation is always close to the equilibrium value r_0. With the parabola approximation, the effective spring constant is $k = (d^2U/dr^2)|_{r_0}$. How can we find the vibration frequency of the molecule?

Consider the two atoms connected by a spring of equilibrium length r_0 and spring constant k, as shown below. The equations of motion are

$$m_1\ddot{r}_1 = k(r - r_0)$$
$$m_2\ddot{r}_2 = -k(r - r_0),$$

where $r = r_2 - r_1$ is the instantaneous separation of the atoms. We can find the equation of motion for r by dividing the first equation by m_1 and the second by m_2, and subtracting. The result is

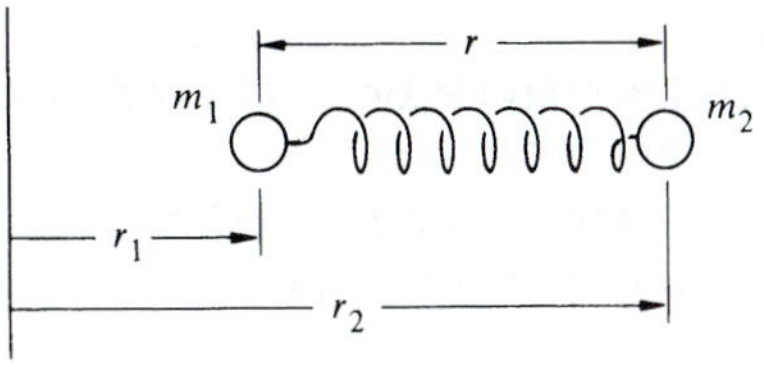

$$\ddot{r}_2 - \ddot{r}_1 = \ddot{r} = -k\left(\frac{1}{m_1} + \frac{1}{m_2}\right)(r - r_0)$$

or

$$\ddot{r} = -\frac{k}{\mu}(r - r_0),$$

where $\mu = m_1m_2/(m_1 + m_2)$. μ has the dimension of mass and is called the *reduced mass.*

By analogy with the harmonic oscillator equation $\ddot{x} = -(k/m)(x - x_0)$ for which the frequency of oscillation is $\omega = \sqrt{k/m}$, the vibrational frequency of the molecule is

$$\omega = \sqrt{\frac{k}{\mu}}$$
$$= \sqrt{\left.\frac{d^2U}{dr^2}\right|_{r_0}\frac{1}{\mu}}.$$

This vibrational motion, characteristic of all molecules, can be identified by the light the molecule radiates. The vibrational frequencies typically lie in the near infrared (3×10^{13} Hz), and by measuring the frequency we can find the value of d^2U/dr^2 at the potential energy minimum. For the HCl molecule, the effective spring constant turns out to be 5×10^5 dynes/cm = 500 N/m (roughly 3 lb/in). For large amplitudes the higher order terms in the Taylor's series start to play a role, and these lead to slight departures of the oscillator from its ideal behavior. The slight "anharmonicities" introduced by this give further details on the shape of the potential energy curve.

Since all bound systems have a potential energy minimum at equilibrium, we naturally expect that all bound systems behave like harmonic oscillators for small displacements (unless the minimum is so flat that the second derivative vanishes there also). The harmonic oscillator approximation therefore has a wide range of applicability, even down to internal motions in nuclei.

Once we have identified the kinetic and potential energies of a bound system, we can find the frequency of small oscillations by inspection. For the elementary case of a mass on a spring we have

$$U = \tfrac{1}{2}kx^2$$
$$K = \tfrac{1}{2}m\dot{x}^2$$

and

$$\omega = \sqrt{\frac{k}{m}}.$$

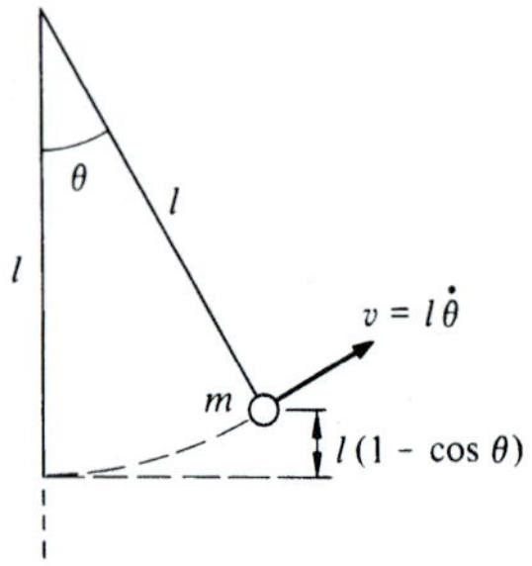

In many problems, however, it is more natural to write the energies in terms of a variable other than linear displacement. For instance, the energies of a pendulum are

$$U = mgl(1 - \cos\theta) \approx \tfrac{1}{2}mgl\theta^2$$
$$K = \tfrac{1}{2}ml^2\dot{\theta}^2.$$

More generally, the energies may have the form

$$U = \tfrac{1}{2}Aq^2 + \text{constant}$$
$$K = \tfrac{1}{2}B\dot{q}^2, \qquad 4.24$$

where q represents a variable appropriate to the problem. By analogy with the mass on a spring, we expect that the frequency of motion of the oscillator is

$$\omega = \sqrt{\frac{A}{B}}. \qquad 4.25$$

To show explicitly that any system whose energy has the form of Eq. (4.24) oscillates harmonically with a frequency $\sqrt{A/B}$, note that the total energy of the system is

$$E = K + U$$
$$= \tfrac{1}{2}B\dot{q}^2 + \tfrac{1}{2}Aq^2 + \text{constant}.$$

Since the system is conservative, E is constant. Differentiating the energy equation with respect to time gives

$$\frac{dE}{dt} = B\dot{q}\ddot{q} + Aq\dot{q}$$
$$= 0$$

or

$$\ddot{q} + \frac{A}{B}q = 0.$$

Hence q undergoes harmonic motion with frequency $\sqrt{A/B}$.

Example 4.16 **Small Oscillations**

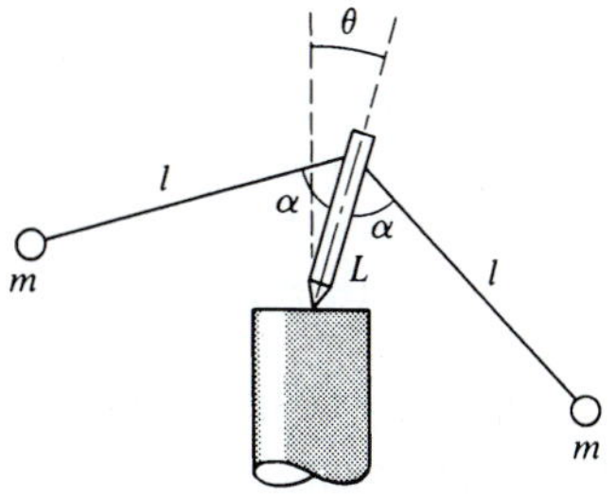

In Example 4.14 we determined the stability criterion for a teeter toy. In this example we shall find the period of oscillation of the toy when it is rocking from side to side.

From Example 4.14, the potential energy of the teeter toy is

$$U(\theta) = -A\cos\theta,$$

where $A = 2mg(l\cos\alpha - L)$. For stability, $A > 0$. If we expand $U(\theta)$ about $\theta = 0$, we have

$$U(\theta) = -A\left(1 - \frac{\theta^2}{2} + \cdots\right),$$

since $\cos\theta \approx 1 - \theta^2/2 + \cdots$. Thus,

$$U(\theta) = -A + \tfrac{1}{2}A\theta^2.$$

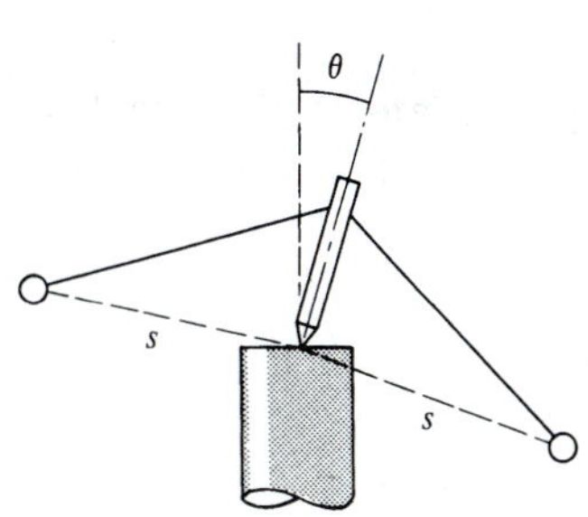

To find the kinetic energy, let s be the distance of each mass from the pivot, as shown in the sketch. If the toy rocks with angular speed $\dot{\theta}$, the speed of each mass is $s\dot{\theta}$, and the total kinetic energy is

$$K = \tfrac{1}{2}(2m)s^2\dot{\theta}^2$$
$$= \tfrac{1}{2}B\dot{\theta}^2,$$

where $B = 2ms^2$.

Hence the frequency of oscillation is

$$\omega = \sqrt{\frac{A}{B}}$$
$$= \sqrt{\frac{g(l\cos\alpha - L)}{s^2}}.$$ 1

We found in Example 4.14 that for stability $l \cos \alpha - L > 0$. Equation (1) shows that as $l \cos \alpha - L$ approaches zero, ω approaches zero, and the period of oscillation becomes infinite. In the limit $l \cos \alpha - L = 0$, the system is in neutral equilibrium, and if $l \cos \alpha - L < 0$, the system becomes unstable. Thus, a low frequency of oscillation is associated with the system operating near the threshold of stability. This is a general property of stable systems, because a low frequency of oscillation corresponds to a weak restoring force. For instance, a ship rolled by a wave oscillates about equilibrium. For comfort the period of the roll should be long. This can be accomplished by designing the hull so that its center of gravity is as high as possible consistent with stability. Lowering the center of gravity makes the system "stiffer." The roll becomes quicker and less comfortable, but the ship becomes intrinsically more stable.

4.11 Nonconservative Forces

We have stressed conservative forces and potential energy in this chapter because they play an important role in physics. However, in many physical processes nonconservative forces like friction are present. Let's see how to extend the work-energy theorem to include nonconservative forces.

Often both conservative and nonconservative forces act on the same system. For instance, an object falling through the air experiences the conservative gravitational force and the nonconservative force of air friction. We can write the total force $\mathbf{F}$ as

$$\mathbf{F} = \mathbf{F}^{c} + \mathbf{F}^{nc}$$

where $\mathbf{F}^{c}$ and $\mathbf{F}^{nc}$ are the conservative and the nonconservative forces respectively. Since the work-energy theorem is true whether or not the forces are conservative, the total work done by $\mathbf{F}$ as the particle moves from a to b is

$$\begin{aligned} W_{ba}{}^{\text{total}} &= \oint_a^b \mathbf{F} \cdot d\mathbf{r} \\ &= \oint_a^b \mathbf{F}^{c} \cdot d\mathbf{r} + \oint_a^b \mathbf{F}^{nc} \cdot d\mathbf{r} \\ &= -U_b + U_a + W_{ba}{}^{\text{nc}}. \end{aligned}$$

Here U is the potential energy associated with the conservative force and $W_{ba}{}^{\text{nc}}$ is the work done by the nonconservative force. The work-energy theorem, $W_{ba}{}^{\text{total}} = K_b - K_a$, now has the form

$$-U_b + U_a + W_{ba}{}^{\text{nc}} = K_b - K_a$$

or

$$K_b + U_b - (K_a + U_a) = W_{ba}^{\mathrm{nc}}. \qquad 4.26$$

If we define the total mechanical energy by $E = K + U$, as before, then E is no longer a constant but instead depends on the state of the system. We have

$$E_b - E_a = W_{ba}^{\mathrm{nc}}. \qquad 4.27$$

This result is a generalization of the statement of conservation of mechanical energy which we discussed in Sec. 4.7. If nonconservative forces do no work, $E_b = E_a$, and mechanical energy is conserved. However, this is a special case, since nonconservative forces are often present. Nevertheless, energy methods continue to be useful; we simply must be careful not to omit the work done by the nonconservative forces, W_{ba}^{nc}. Here is an example.

Example 4.17 Block Sliding down Inclined Plane

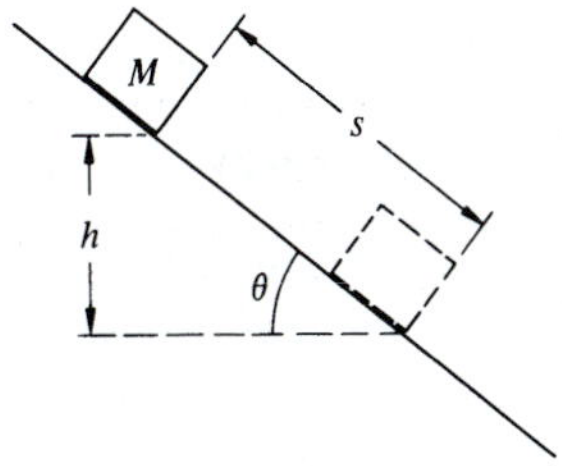

A block of mass M slides down a plane of angle θ. The problem is to find the speed of the block after it has descended through height h, assuming that it starts from rest and that the coefficient of friction μ is constant.

Initially the block is at rest at height h; finally the block is moving with speed v at height 0. Hence

$$U_a = Mgh \qquad U_b = 0$$
$$K_a = 0 \qquad K_b = \tfrac{1}{2}Mv^2$$
$$E_a = Mgh \qquad E_b = \tfrac{1}{2}Mv^2.$$

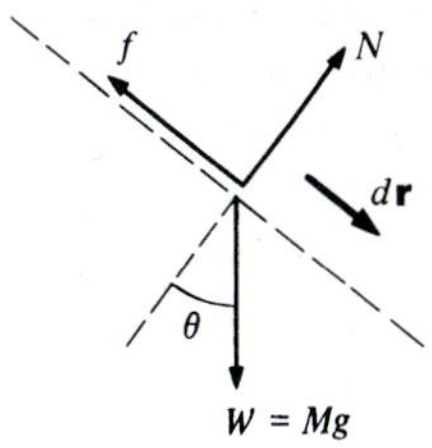

The nonconservative force is $f = \mu N = \mu Mg \cos \theta$. Hence, the nonconservative work is

$$\begin{aligned} W_{ba}^{\mathrm{nc}} &= \int_a^b \mathbf{f} \cdot d\mathbf{r} \\ &= -fs, \end{aligned}$$

where s is the distance the block slides. The negative sign arises because the direction of $\mathbf{f}$ is always opposite to the displacement, so that $\mathbf{f} \cdot d\mathbf{r} = -f\,dr$. Using $s = h/\sin \theta$, we have

$$\begin{aligned} W_{ba}^{\mathrm{nc}} &= -\mu Mg \cos \theta \frac{h}{\sin \theta} \\ &= -\mu \cot \theta\, Mgh. \end{aligned}$$

The energy equation $E_b - E_a = W_{ba}^{nc}$ becomes

$$\tfrac{1}{2}Mv^2 - Mgh = -\mu \cot\theta\, Mgh,$$

which gives

$$v = [2(1 - \mu \cot\theta)gh]^{\frac{1}{2}}.$$

Since all the forces acting on the block are constant, the expression for v could easily be found by applying our results for motion under uniform acceleration; the energy method does not represent much of a shortcut here. The power of the energy method lies in its generality. For instance, suppose that the coefficient of friction varies along the surface so that the friction force is $f = \mu(x)Mg\cos\theta$. The work done by friction is

$$W_{ba}^{nc} = -Mg\cos\theta \int_a^b \mu(x)\,dx,$$

and the final speed is easily found. In contrast, there is no simple way to find the speed by integrating the acceleration with respect to time.

4.12 The General Law of Conservation of Energy

As far as we know, the basic forces of nature, such as the force of gravity and the forces of electric and magnetic interactions, are conservative. This leads to a puzzle; if fundamental forces are conservative, how can nonconservative forces arise? The resolution of this problem lies in the point of view we adopt in describing a physical system, and in our willingness to broaden the concept of energy.

Consider friction, the most familiar nonconservative force. Mechanical energy is lost by friction when a block slides across a table, but something else occurs: the block and the table get warmer. However, there was no reference to temperature in our development of the concept of mechanical energy; a block of mass M moving with speed v has kinetic energy $\frac{1}{2}Mv^2$, whether the block is hot or cold. The fact that a block sliding across a table warms up does not affect our conclusion that mechanical energy is lost. Nevertheless, if we look carefully, we find that the heating of the system bears a definite relation to the energy dissipated. The British physicist James Prescott Joule was the first to appreciate that heat itself represents a form of energy.

By a series of meticulous experiments on the heating of water by a paddle wheel driven by a falling weight, he showed that the loss of mechanical energy by friction is accompanied by the appearance of an equivalent amount of heat. Joule concluded that heat must be a form of energy and that the sum of the mechanical energy and the heat energy of a system is conserved.

We now have a more detailed picture of heat energy than was available to Joule. We know that solids are composed of atoms held together by strong interatomic forces. Each atom can oscillate about its equilibrium position and has mechanical energy in the form of kinetic and potential energies. As the solid is heated, the amplitude of oscillation increases and the average energy of each atom grows larger. The heat energy of a solid is the mechanical energy of the random vibrations of the atoms.

There is a fundamental difference between mechanical energy on the atomic level and that on the level of everyday events. The atomic vibrations in a solid are random; at any instant there are atoms moving in all possible directions, and the center of mass of the block has no tendency to move on the average. Kinetic energy of the block represents a collective motion; when the block moves with velocity $\mathbf{v}$, each atom has, on the average, the same velocity $\mathbf{v}$.

Mechanical energy is turned into heat energy by friction, but the reverse process is never observed. No one has ever seen a hot block at rest on a table suddenly cool off and start moving, although this would not violate conservation of energy. The reason is that collective motion can easily become randomized. For instance, when a block hits an obstacle, the collective translational motion ceases and, under the impact, the atoms start to jitter more violently. Kinetic energy has been transformed to heat energy. The reverse process where the random motion of the atoms suddenly turns to collective motion is so improbable that for all practical purposes it never occurs. It is for this reason that we can distinguish between the heat energy and the mechanical energy of a chunk of matter even though on the atomic scale the distinction vanishes.

We now recognize that in addition to mechanical energy and heat there are many other forms of energy. These include the radiant energy of light, the energy of nuclear forces, and, as we shall discuss in Chap. 13, the energy associated with mass. It is apparent that the concept of energy is much wider than the simple idea of kinetic and potential energy of a mechanical system. We believe that the total energy of a system is conserved if all forms of energy are taken into account.

4.13 Power

Power is the time rate of doing work. If a force $\mathbf{F}$ acts on a body which undergoes a displacement $d\mathbf{r}$, the work is $dW = \mathbf{F} \cdot d\mathbf{r}$ and the power delivered by the force is

$$P = \frac{dW}{dt} = \mathbf{F} \cdot \frac{d\mathbf{r}}{dt}$$
$$= \mathbf{F} \cdot \mathbf{v}.$$

The unit of power in the SI system is the watt (W).

$$1 \text{ W} = 1 \text{ J/s}.$$

In the cgs system, the unit of power is the erg/s $= 10^{-7}$ W; it has no special name. The unit of power in the English system is the horsepower (hp). The horsepower is most commonly defined as 550 ft·lb/s, but slightly different definitions are sometimes encountered. The relation between the horsepower and the watt is

$$1 \text{ hp} \approx 746 \text{ W}.$$

This is a discouraging number for builders of electric cars; the average power obtainable from an ordinary automobile storage battery is only about 350 W.

The power rating of an engine is a useful indicator of its performance. For instance, a small motor with a system of reduction gears can raise a large mass M any given height, but the process will take a long time; the average power delivered is low. The power required is Mgv, where v is the weight's upward speed. To raise the mass rapidly the power must be large.

A human being in good condition can develop between $\frac{1}{2}$ to 1 hp for 30 s or so, for example while running upstairs. Over a period of 8 hours (h), however, a husky man can do work only at the rate of about 0.2 hp = 150 W. The total work done in 8 h is then $(150)(8)(3{,}600) = 4.3 \times 10^6$ J $\approx$ 1,000 kcal. The kilocalorie, approximately equal to 4,200 J, is often used to express the energy available from food. A normally active person requires 2,000 to 3,000 kcal/d. (In dietetic work the kilocalorie is sometimes called the "large" calorie, but more often simply the calorie.)

The power production of modern industrialized nations corresponds to several thousand watts per person (United States: 6,000 W per person; India: 300 W per person). The energy comes primarily from the burning of fossil fuels, which are the chief source of energy at present. In principle, we could use the sun's energy directly. When the sun is overhead, it supplies approximately 1,000 W/m^2 ($\approx$ 1 hp/yd^2) to the earth's surface. Unfortunately, present solar cells are costly and inefficient, and there is no economical way of storing the energy for later use.

4.14 Conservation Laws and Particle Collisions

Much of our knowledge of atoms, nuclei, and elementary particles has come from scattering experiments. Perhaps the most dramatic of these was the experiment performed in 1911 by Ernest Rutherford in which alpha particles (doubly ionized helium atoms) were scattered from atoms of gold in a thin foil. By studying how the number of scattered alpha particles varied with the deflection angle, Rutherford was led to the nuclear model of the atom. The techniques of experimental physics have advanced considerably since Rutherford's time. A high energy particle accelerator several miles long may appear to have little in common with Rutherford's tabletop apparatus, but its purpose is the same—to discover the interaction forces between particles by studying how they scatter.

Finding the interaction force from a scattering experiment is a difficult task. Furthermore, the detailed description of collisions on the atomic scale generally requires the use of quantum mechanics. Nevertheless, there are constraints on the motion arising from the conservation laws of momentum and energy which are so strong that they are solely responsible for many of the features of scattering. Since the conservation laws can be applied without knowing the interactions, they play a vital part in the analysis of collision phenomena.

In this section we shall see how to apply the conservation laws of momentum and energy to scattering experiments. No new physical principles are involved; the discussion is intended to illustrate ideas we have already introduced.

Collisions and Conservation Laws

The drawings below show three stages during the collision of two particles. In (*a*), long before the collision, each particle is effectively free, since the interaction forces are generally important only at very small separations. As the particles approach, (*b*), the momentum and energy of each particle change due to the interaction forces. Finally, long after the collision, (*c*), the particles are again free and move along straight lines with new directions and velocities. Experimentally, we usually know the initial velocities $\mathbf{v}_1$ and $\mathbf{v}_2$; often one particle is initially at rest in a target and is bombarded by particles of known energy. The experiment might consist of measuring the final velocities $\mathbf{v}_1'$ and $\mathbf{v}_2'$ with suitable particle detectors.

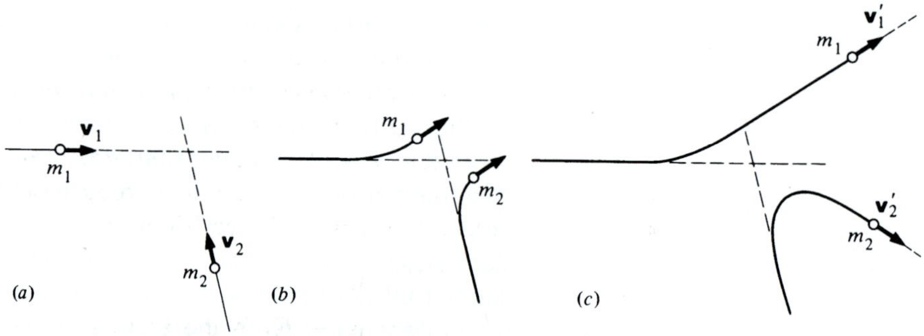

Since external forces are usually negligible, the total momentum is conserved and we have

$$\mathbf{P}_i = \mathbf{P}_f. \tag{4.28}$$

For a two body collision, this becomes

$$m_1\mathbf{v}_1 + m_2\mathbf{v}_2 = m_1\mathbf{v}_1' + m_2\mathbf{v}_2'. \tag{4.29}$$

Equation (4.29) is equivalent to three scalar equations. We have, however, six unknowns, the components of $\mathbf{v}_1'$ and $\mathbf{v}_2'$. The energy equation provides an additional relation between the velocities, as we now show.

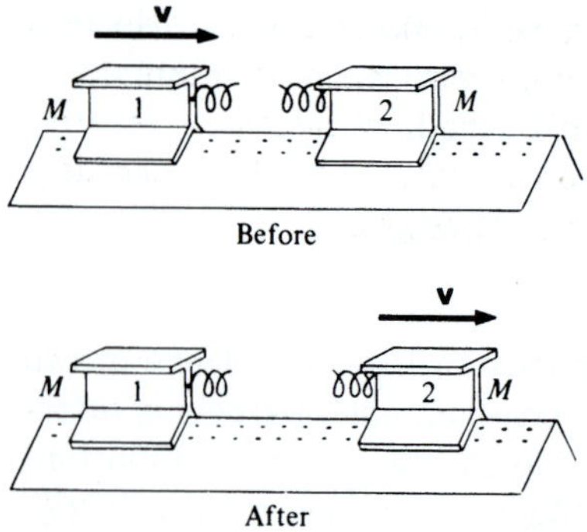

Elastic and Inelastic Collisions

Consider a collision on a linear air track between two riders of equal mass which interact via good coil springs. Suppose that initially rider 1 has speed v as shown and rider 2 is at rest. After the collision, 1 is at rest and 2 moves to the right with speed v. It is clear that momentum has been conserved and that the total kinetic energy of the two bodies, $Mv^2/2$, is the same before and after the collision. A collision in which the total kinetic energy is unchanged is called an *elastic* collision. A collision is elastic if the interaction forces are conservative, like the spring force in our example.

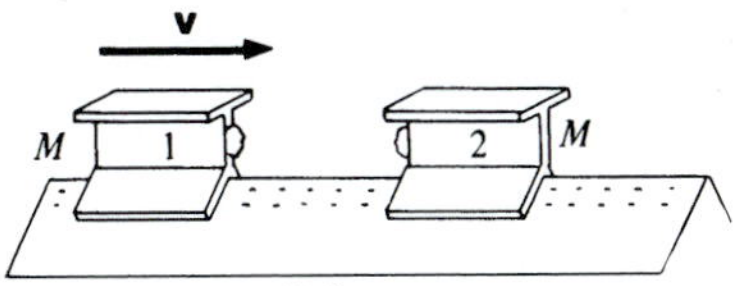

Before

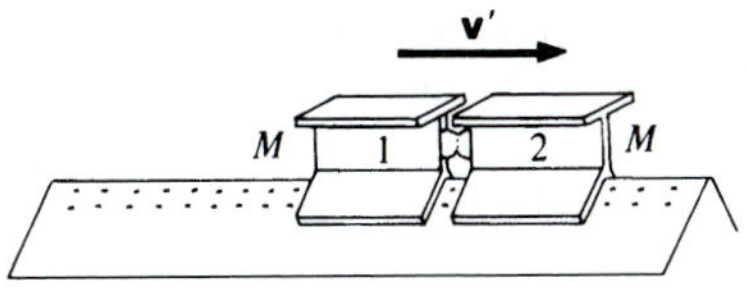

After

As a second experiment, take the same two riders and replace the springs by lumps of sticky putty. Let 2 be initially at rest. After the collision, the riders stick together and move off with speed v'. By conservation of momentum, $Mv = 2Mv'$, so that $v' = v/2$. The initial kinetic energy of the system is $Mv^2/2$, but the final kinetic energy is $(2M)v'^2/2 = Mv^2/4$. Evidently in this collision the kinetic energy is only half as much after the collision as before. The kinetic energy has changed because the interaction forces were nonconservative. Part of the energy of the collective motion was transformed to random heat energy in the putty during the collision. A collision in which the total kinetic energy is not conserved is called an *inelastic* collision.

Although the total energy of the system is always conserved in collisions, part of the kinetic energy may be converted to some other form. To take this into account, we write the conservation of energy equation for collisions as

$$K_i = K_f + Q, \tag{4.30}$$

where $Q = K_i - K_f$ is the amount of kinetic energy converted to another form. For a two body collision, Eq. (4.30) becomes

$$\tfrac{1}{2}m_1v_1^2 + \tfrac{1}{2}m_2v_2^2 = \tfrac{1}{2}m_1v_1'^2 + \tfrac{1}{2}m_2v_2'^2 + Q. \tag{4.31}$$

In most collisions on the everyday scale, kinetic energy is lost and Q is positive. However, Q can be negative if internal energy of the system is converted to kinetic energy in the collision. Such collisions are sometimes called *superelastic*, and they are important in atomic and nuclear physics. Superelastic collisions are rarely encountered in the everyday world, but one example would be the collision of two cocked mousetraps.

Collisions in One Dimension

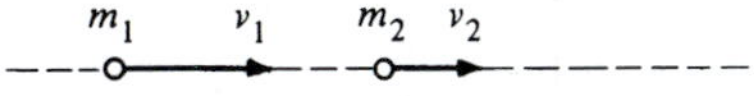

Before

After

If we have a two body collision in which the particles are constrained to move along a straight line, the conservation laws, Eqs. (4.29) and (4.31), completely determine the final velocities, regardless of the nature of the interaction forces. With the velocities shown in the sketch, the conservation laws give

Momentum:

$$m_1v_1 + m_2v_2 = m_1v_1' + m_2v_2'. \tag{4.32a}$$

Energy:

$$\tfrac{1}{2}m_1v_1^2 + \tfrac{1}{2}m_2v_2^2 = \tfrac{1}{2}m_1v_1'^2 + \tfrac{1}{2}m_2v_2'^2 + Q. \tag{4.32b}$$

These equations can be solved for v_1' and v_2' in terms of m_1, m_2, v_1, v_2, and Q. The next example illustrates the process.

Example 4.18 **Elastic Collision of Two Balls**

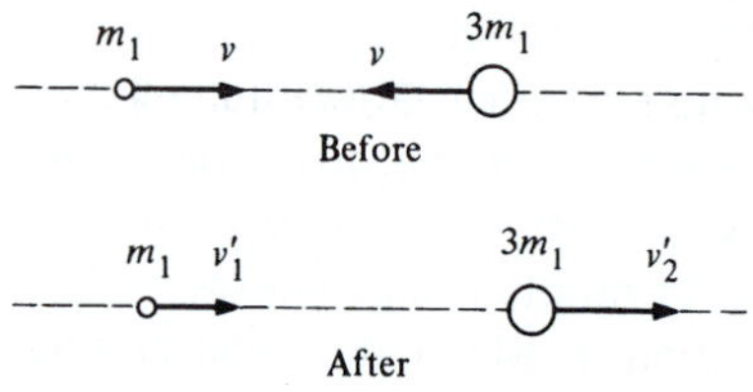

Consider the one dimensional elastic collision of two balls of masses m_1 and m_2, with $m_2 = 3m_1$. Suppose that the balls have equal and opposite velocities **v** before the collision; the problem is to find the final velocities. The conservation laws yield

$$m_1v - 3m_1v = m_1v_1' + 3m_1v_2' \quad 1$$

$$\tfrac{1}{2}m_1v^2 + \tfrac{1}{2}(3m_1)v^2 = \tfrac{1}{2}m_1v_1'^2 + \tfrac{1}{2}(3m_1)v_2'^2. \quad 2$$

We can eliminate v_1' using Eq. (1):

$$v_1' = -2v - 3v_2'. \quad 3$$

Inserting this in Eq. (2) gives

$$\begin{aligned} 4v^2 &= (-2v - 3v_2')^2 + 3v_2'^2 \\ &= 4v^2 + 12vv_2' + 12v_2'^2 \end{aligned}$$

or

$$0 = 12vv_2' + 12v_2'^2. \quad 4$$

Equation (4) has two solutions: $v_2' = -v$ and $v_2' = 0$. The corresponding values of v_1' can be found from Eq. (3).

Solution 1:

$v_1' = v$

$v_2' = -v.$

Solution 2:

$v_1' = -2v$

$v_2' = 0.$

We recognize that solution 1 simply restates the initial conditions: we always obtain such a "solution" in this type of problem because the initial velocities evidently satisfy the conservation law equations.

Solution 2 is the interesting one. It shows that after the collision, m_1 is moving to the left with twice its original speed and the heavier ball is at rest.

Collisions and Center of Mass Coordinates

It is almost always simpler to treat three dimensional collision problems in the center of mass (C) coordinate system than in the laboratory (L) system.

Consider two particles of masses m_1 and m_2, and velocities $\mathbf{v}_1$ and $\mathbf{v}_2$. The center of mass velocity is

$$\mathbf{V} = \frac{m_1\mathbf{v}_1 + m_2\mathbf{v}_2}{m_1 + m_2}.$$

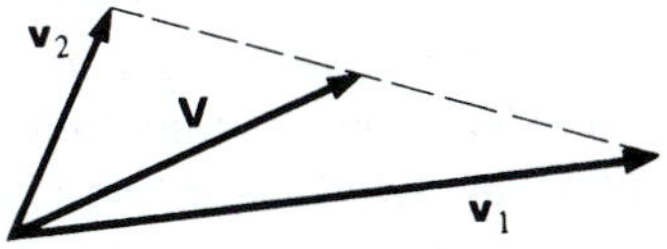

As shown in the velocity diagram at left, $\mathbf{V}$ lies on the line joining $\mathbf{v}_1$ and $\mathbf{v}_2$.

The velocities in the C system are

$$\begin{aligned}\mathbf{v}_{1c} &= \mathbf{v}_1 - \mathbf{V}\\ &= \frac{m_2}{m_1 + m_2}(\mathbf{v}_1 - \mathbf{v}_2),\end{aligned}$$

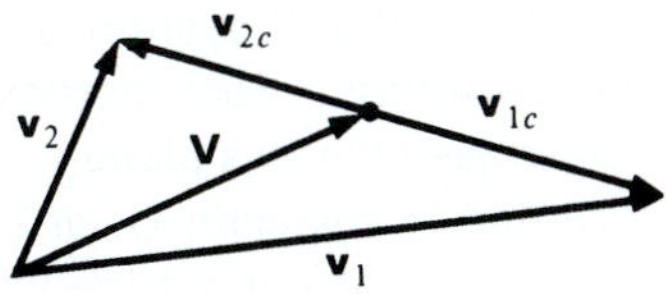

and

$$\begin{aligned}\mathbf{v}_{2c} &= \mathbf{v}_2 - \mathbf{V}\\ &= \frac{-m_1}{m_1 + m_2}(\mathbf{v}_1 - \mathbf{v}_2).\end{aligned}$$

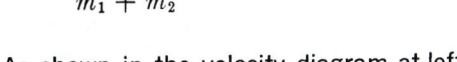

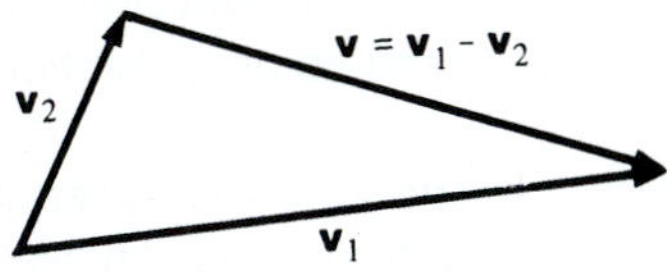

$\mathbf{v}_{1c}$ and $\mathbf{v}_{2c}$ lie back to back along the relative velocity vector $\mathbf{v} = \mathbf{v}_1 - \mathbf{v}_2$.

The momenta in the C system are

$$\begin{aligned}\mathbf{p}_{1c} &= m_1\mathbf{v}_{1c}\\ &= \frac{m_1m_2}{m_1 + m_2}(\mathbf{v}_1 - \mathbf{v}_2)\\ &= \mu\mathbf{v}\\ \mathbf{p}_{2c} &= m_2\mathbf{v}_{2c}\\ &= \frac{-m_1m_2}{m_1 + m_2}(\mathbf{v}_1 - \mathbf{v}_2)\\ &= -\mu\mathbf{v}.\end{aligned}$$

Here $\mu = m_1m_2/(m_1 + m_2)$ is the reduced mass of the system. We encountered the reduced mass for the first time in Example 4.15. As we shall see in Chap. 9, it is the natural unit of mass in a two particle system. The total momentum in the C system is zero, as we expect.

The total momentum in the L system is

$$m_1\mathbf{v}_1 + m_2\mathbf{v}_2 = (m_1 + m_2)\mathbf{V}$$

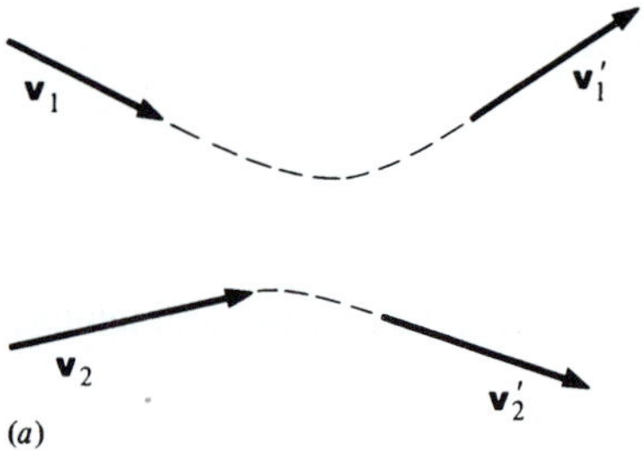

(a)

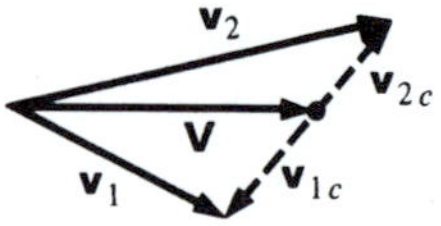

(b)

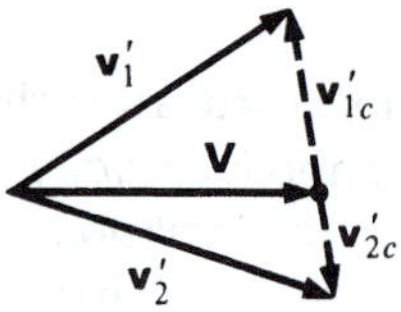

(c)

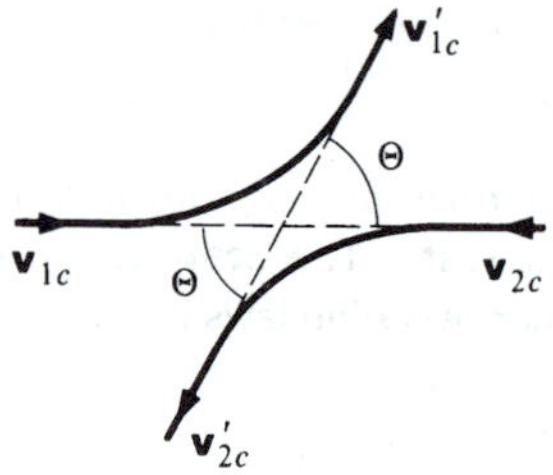

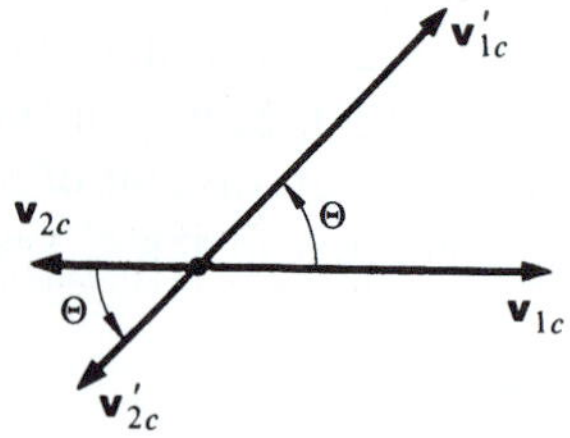

and since total momentum is conserved in any collision, **V** is constant. We can use this result to help visualize the velocity vectors before and after the collision.

Sketch (*a*) shows the trajectories and velocities of two colliding particles. In sketch (*b*) we show the initial velocities in the L and C systems. All the vectors lie in the same plane. $\mathbf{v}_{1c}$ and $\mathbf{v}_{2c}$ must be back to back since the total momentum in the C system is zero. After the collision, sketch (*c*), the velocities in the C system are again back to back. This sketch also shows the final velocities in the lab system. Note that the plane of sketch c is not necessarily the plane of sketch a. Evidently the geometrical relation between initial and final velocities in the L system is quite complicated. Fortunately, the situation in the C system is much simpler. The initial and final velocities in the C system determine a plane known as the plane of scattering. Each particle is deflected through the same scattering angle Θ in this plane. The interaction force must be known in order to calculate Θ, or conversely, by measuring the deflection we can learn about the interaction force. However, we shall defer these considerations and simply assume that the interaction has caused some deflection in the C system.

An important simplification occurs if the collision is elastic. Conservation of energy applied to the C system gives, for elastic collisions,

$$\tfrac{1}{2}m_1v_{1c}^2 + \tfrac{1}{2}m_2v_{2c}^2 = \tfrac{1}{2}m_1v_{1c}'^2 + \tfrac{1}{2}m_2v_{2c}'^2.$$

Since momentum is zero in the C system, we have

$$m_1v_{1c} - m_2v_{2c} = 0$$

$$m_1v_{1c}' - m_2v_{2c}' = 0.$$

Eliminating v_{2c} and v_{2c}' from the energy equation gives

$$\frac{1}{2}\left(m_1 + \frac{m_1^2}{m_2}\right)v_{1c}^2 = \frac{1}{2}\left(m_1 + \frac{m_1^2}{m_2}\right)v_{1c}'^2$$

or

$$v_{1c} = v_{1c}'.$$

Similarly,

$$v_{2c} = v_{2c}'.$$

In an elastic collision, the speed of each particle in the C system is the same before and after the collision. Thus, the velocity vectors simply rotate in the scattering plane.

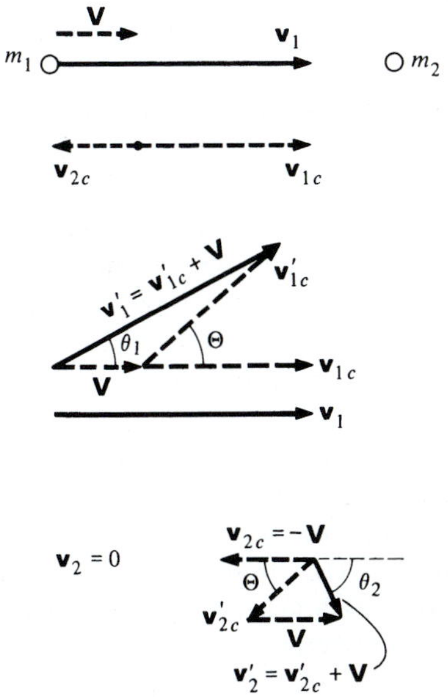

In many experiments, one of the particles, say m_2, is initially at rest in the laboratory. In this case

$$\mathbf{V} = \frac{m_1}{m_1 + m_2}\mathbf{v}$$

and

$$\mathbf{v}_{1c} = \mathbf{v}_1 - \mathbf{V}$$

$$= \frac{m_2}{m_1 + m_2}\mathbf{v}_1$$

$$\mathbf{v}_{2c} = -\mathbf{V}$$

$$= -\frac{m_1}{m_1 + m_2}\mathbf{v}_1.$$

The sketches show $\mathbf{v}_1$ and $\mathbf{v}_2$ before and after the collision in the C and L systems. θ_1 and θ_2 are the laboratory angles of the trajectories of the two particles after the collision. The velocity diagrams can be used to relate θ_1 and θ_2 to the scattering angle Θ.

Example 4.19 Limitations on Laboratory Scattering Angle

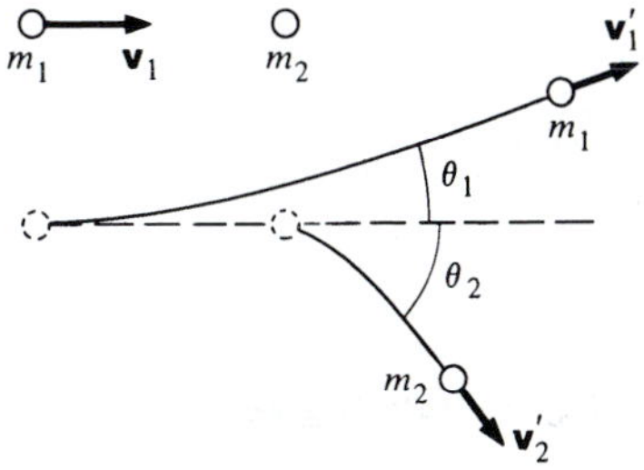

Consider the elastic scattering of a particle of mass m_1 and velocity $\mathbf{v}_1$ from a second particle of mass m_2 at rest. The scattering angle Θ in the C system is unrestricted, but the conservation laws impose limitations on the laboratory angles, as we shall show.

The center of mass velocity has magnitude

$$V = \frac{m_1 v_1}{m_1 + m_2} \qquad 1$$

and is parallel to $\mathbf{v}_1$. The initial velocities in the C system are

$$\mathbf{v}_{1c} = \frac{m_2}{m_1 + m_2}\mathbf{v}_1$$

$$\mathbf{v}_{2c} = -\frac{m_1}{m_1 + m_2}\mathbf{v}_1. \qquad 2$$

Suppose m_1 is scattered through angle Θ in the C system.

From the velocity diagram we see that the laboratory scattering angle of the incident particle is given by

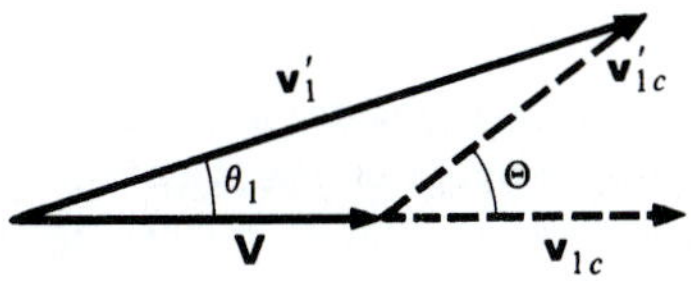

$$\tan\theta_1 = \frac{v_{1c}'\sin\Theta}{V + v_{1c}'\cos\Theta}.$$

Since the scattering is elastic, $v'_{1c} = v_{1c}$. Hence

$$\tan \theta_1 = \frac{v_{1c} \sin \Theta}{V + v_{1c} \cos \Theta} = \frac{\sin \Theta}{(V/v_{1c}) + \cos \Theta}.$$

From Eqs. (1) and (2), $V/v_{1c} = m_1/m_2$. Therefore

$$\tan \theta_1 = \frac{\sin \Theta}{(m_1/m_2) + \cos \Theta}. \qquad 3$$

The scattering angle Θ depends on the details of the interaction, but in general it can assume any value. If $m_1 < m_2$, it follows from Eq. (3) or the geometric construction in sketch (*a*) that θ_1 is unrestricted. However, the situation is quite different if $m_1 > m_2$. In this case θ_1 is never greater than a certain angle $\theta_{1,\text{max}}$. As sketch (*b*) shows, the maximum value of θ_1 occurs when $\mathbf{v}'_1$ and $\mathbf{v}'_{1c}$ are both perpendicular. In this case $\sin \theta_{1,\text{max}} = v_{1c}/V = m_2/m_1$. If $m_1 \gg m_2$, $\theta_{1,\text{max}} \approx m_2/m_1$ and the maximum scattering angle approaches zero.

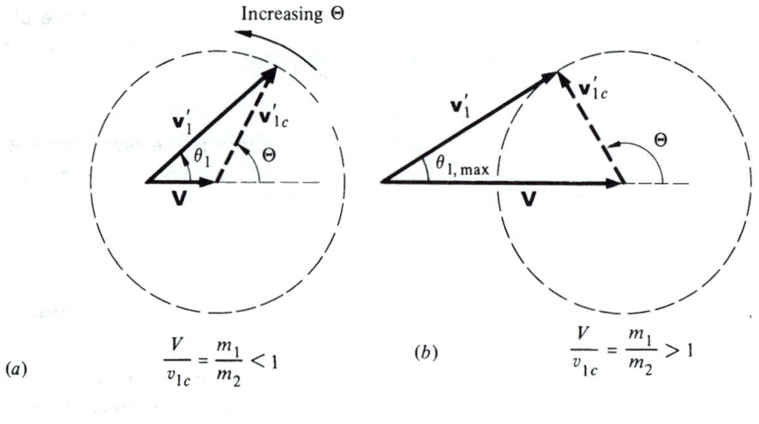

Physically, a light particle at rest cannot appreciably deflect a massive particle. The incident particle tends to continue in its forward direction no matter how the light target particle recoils.

Problems

4.1 A small block of mass m starts from rest and slides along a frictionless loop-the-loop as shown in the left-hand figure on the top of the next page. What should be the initial height z, so that m pushes against the top of the track (at a) with a force equal to its weight?

Ans. $z = 3R$

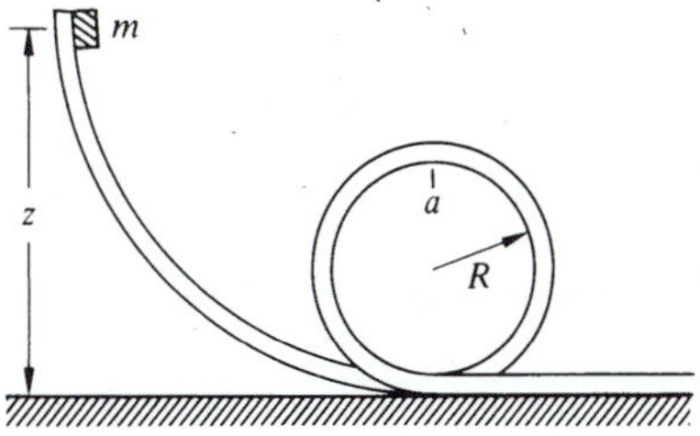

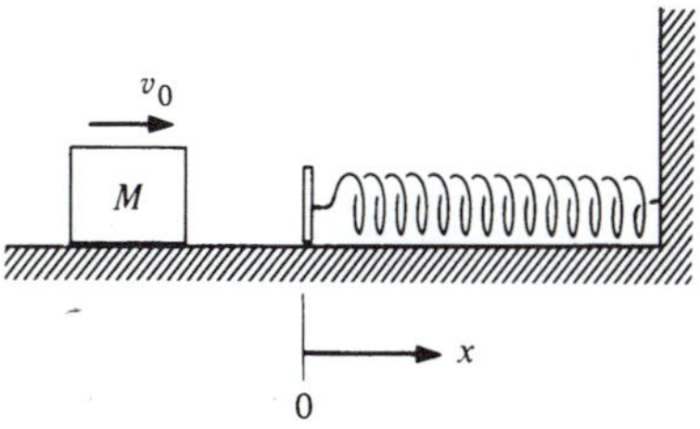

4.2 A block of mass M slides along a horizontal table with speed v_0. At $x = 0$ it hits a spring with spring constant k and begins to experience a friction force (see figure above right). The coefficient of friction is variable and is given by $\mu = bx$, where b is a constant. Find the loss in mechanical energy when the block has first come momentarily to rest.

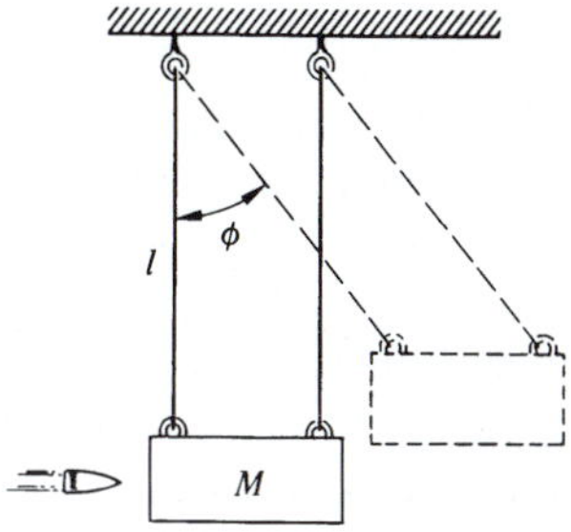

4.3 A simple way to measure the speed of a bullet is with a *ballistic pendulum*. As illustrated, this consists of a wooden block of mass M into which the bullet is shot. The block is suspended from cables of length l, and the impact of the bullet causes it to swing through a maximum angle ϕ, as shown. The initial speed of the bullet is v, and its mass is m.

a. How fast is the block moving immediately after the bullet comes to rest? (Assume that this happens quickly.)

b. Show how to find the velocity of the bullet by measuring m, M, l, and ϕ.

Ans. (*b*) $v = [(m + M)/m]\sqrt{2gl(1 - \cos\phi)}$

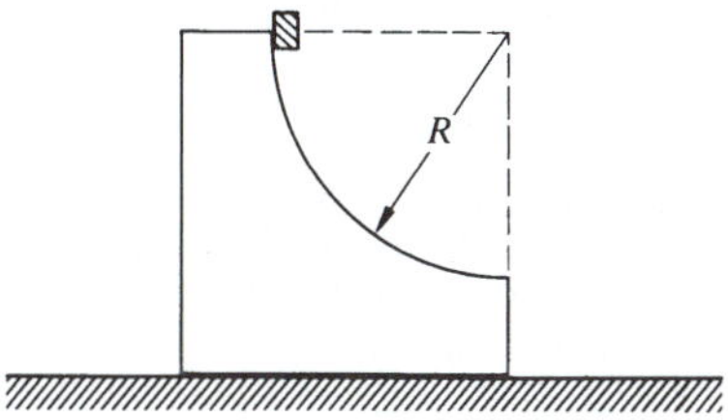

4.4 A small cube of mass m slides down a circular path of radius R cut into a large block of mass M, as shown at left. M rests on a table, and both blocks move without friction. The blocks are initially at rest, and m starts from the top of the path.

Find the velocity v of the cube as it leaves the block.

Ans. clue. If $m = M$, $v = \sqrt{gR}$

4.5 Mass m whirls on a frictionless table, held to circular motion by a string which passes through a hole in the table. The string is slowly pulled through the hole so that the radius of the circle changes from l_1 to l_2. Show that the work done in pulling the string equals the increase in kinetic energy of the mass.

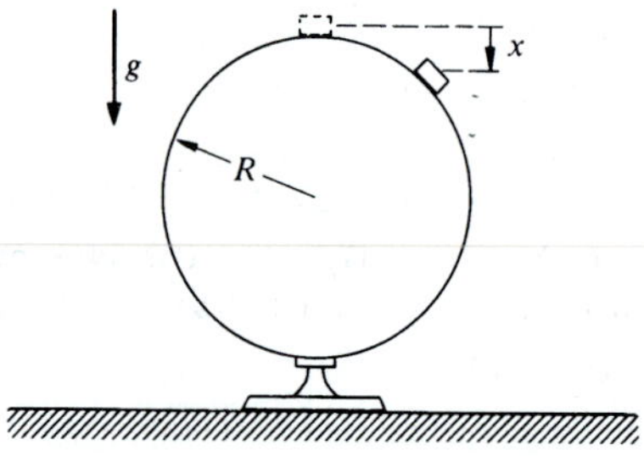

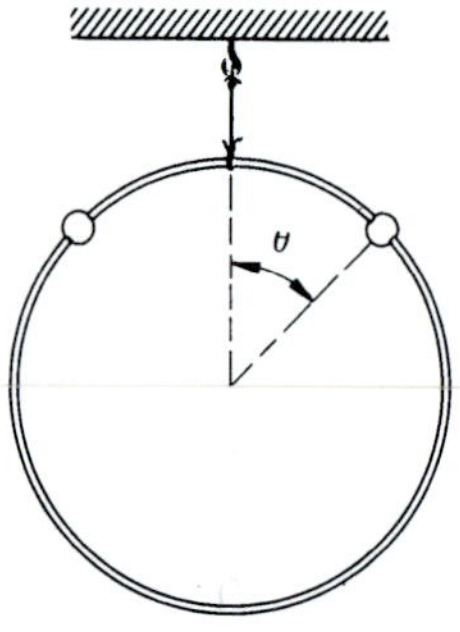

4.6 A small block slides from rest from the top of a frictionless sphere of radius R (see above left). How far below the top x does it lose contact with the sphere? The sphere does not move. *Ans.* $R/3$

4.7 A ring of mass M hangs from a thread, and two beads of mass m slide on it without friction (see above right). The beads are released simultaneously from the top of the ring and slide down opposite sides. Show that the ring will start to rise if $m > 3M/2$, and find the angle at which this occurs. *Ans. clue.* If $M = 0$, $\theta = \arccos \frac{2}{3}$

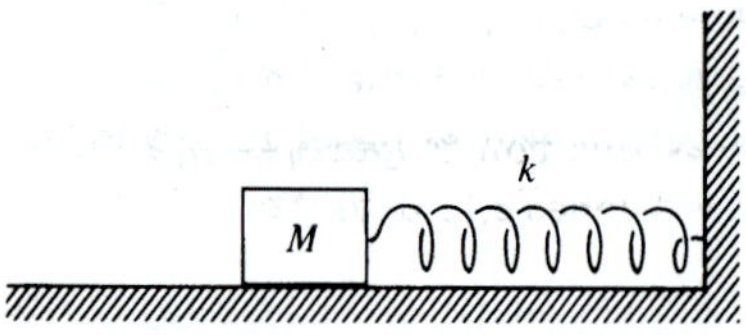

4.8 The block shown in the drawing is acted on by a spring with spring constant k and a weak friction force of constant magnitude f. The block is pulled distance x_0 from equilibrium and released. It oscillates many times and eventually comes to rest.

a. Show that the decrease of amplitude is the same for each cycle of oscillation.

b. Find the number of cycles n the mass oscillates before coming to rest. *Ans.* $n = \frac{1}{4}[(kx_0/f) - 1] \approx kx_0/4f$

4.9 A simple and very violent chemical reaction is $H + H \rightarrow H_2 + 5$ eV. (1 eV $= 1.6 \times 10^{-19}$ J, a healthy amount of energy on the atomic scale.) However, when hydrogen atoms collide in free space they simply bounce apart! The reason is that it is impossible to satisfy the laws of conservation of momentum and conservation of energy in a simple two body collision which releases energy. Can you prove this? You might start by writing the statements of conservation of momentum and energy. (Be sure to include the energy of reaction in the energy equation, and get the sign right.) By eliminating the final momentum of the molecule from the pair of equations, you should be able to show that the initial momenta would have to satisfy an impossible condition.

4.10 A block of mass M on a horizontal frictionless table is connected to a spring (spring constant k), as shown.

The block is set in motion so that it oscillates about its equilibrium point with a certain amplitude A_0. The period of motion is $T_0 = 2\pi\sqrt{M/k}$.

a. A lump of sticky putty of mass m is dropped onto the block. The putty sticks without bouncing. The putty hits M at the instant when the velocity of M is zero. Find

(1) The new period
(2) The new amplitude
(3) The change in the mechanical energy of the system

b. Repeat part *a*, but this time assume that the sticky putty hits M at the instant when M has its maximum velocity.

4.11 A chain of mass M and length l is suspended vertically with its lowest end touching a scale. The chain is released and falls onto the scale.

What is the reading of the scale when a length of chain, x, has fallen? (Neglect the size of individual links.)

Ans. clue. The maximum reading is $3Mg$

4.12 During the Second World War the Russians, lacking sufficient parachutes for airborne operations, occasionally dropped soldiers inside bales of hay onto snow. The human body can survive an average pressure on impact of 30 lb/in^2.

Suppose that the lead plane drops a dummy bale equal in weight to a loaded one from an altitude of 150 ft, and that the pilot observes that it sinks about 2 ft into the snow. If the weight of an average soldier is 144 lb and his effective area is 5 ft^2, is it safe to drop the men?

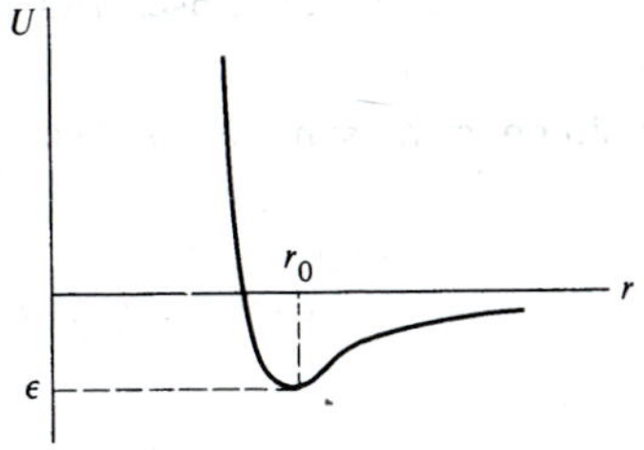

4.13 A commonly used potential energy function to describe the interaction between two atoms is the Lennard-Jones 6,12 potential

$$U = \epsilon\left[\left(\frac{r_0}{r}\right)^{12} - 2\left(\frac{r_0}{r}\right)^{6}\right].$$

a. Show that the radius at the potential minimum is r_0, and that the depth of the potential well is ϵ.

b. Find the frequency of small oscillations about equilibrium for 2 identical atoms of mass m bound to each other by the Lennard-Jones interaction.

Ans. $\omega = 12\sqrt{\epsilon/r_0^2 m}$

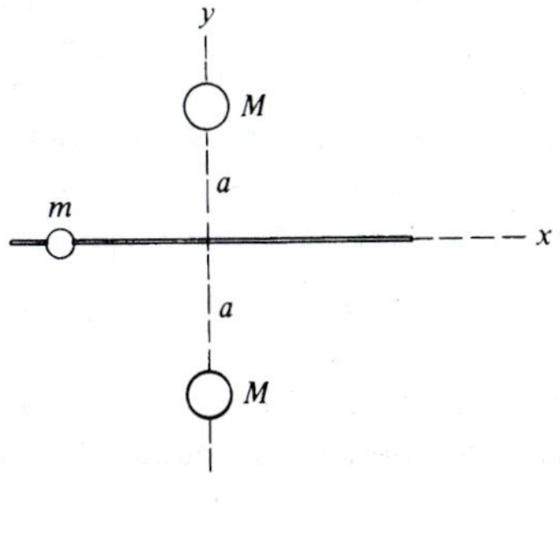

4.14 A bead of mass m slides without friction on a smooth rod along the x axis. The rod is equidistant between two spheres of mass M. The spheres are located at $x = 0$, $y = \pm a$ as shown, and attract the bead gravitationally.

a. Find the potential energy of the bead.

b. The bead is released at $x = 3a$ with velocity v_0 toward the origin. Find the speed as it passes the origin.

c. Find the frequency of small oscillations of the bead about the origin.

4.15 A particle of mass m moves in one dimension along the positive x axis. It is acted on by a constant force directed toward the origin with magnitude B, and an inverse square law repulsive force with magnitude A/x^2.

a. Find the potential energy function $U(x)$.

b. Sketch the energy diagram for the system when the maximum kinetic energy is $K_0 = \frac{1}{2}mv_0^2$.

c. Find the equilibrium position, x_0.

d. What is the frequency of small oscillations about x_0?

4.16 An 1,800-lb sportscar accelerates to 60 mi/h in 8 s. What is the average power that the engine delivers to the car's motion during this period?

4.17 A snowmobile climbs a hill at 15 mi/hr. The hill has a grade of 1 ft rise for every 40 ft. The resistive force due to the snow is 5 percent of the vehicle's weight. How fast will the snowmobile move downhill, assuming its engine delivers the same power?

Ans. 45 mi/h

4.18 A 160-lb man leaps into the air from a crouching position. His center of gravity rises 1.5 ft before he leaves the ground, and it then rises 3 ft to the top of his leap. What power does he develop assuming that he pushes the ground with constant force?

Ans. clue. More than 1 hp, less than 10 hp

4.19 The man in the preceding problem again leaps into the air, but this time the force he applies decreases from a maximum at the beginning of the leap to zero at the moment he leaves the ground. As a reasonable approximation, take the force to be $F = F_0 \cos \omega t$, where F_0 is the peak force, and contact with the ground ends when $\omega t = \pi/2$. Find the peak power the man develops during the jump.

4.20 Sand runs from a hopper at constant rate dm/dt onto a horizontal conveyor belt driven at constant speed V by a motor.

a. Find the power needed to drive the belt.

b. Compare the answer to *a* with the rate of change of kinetic energy of the sand. Can you account for the difference?

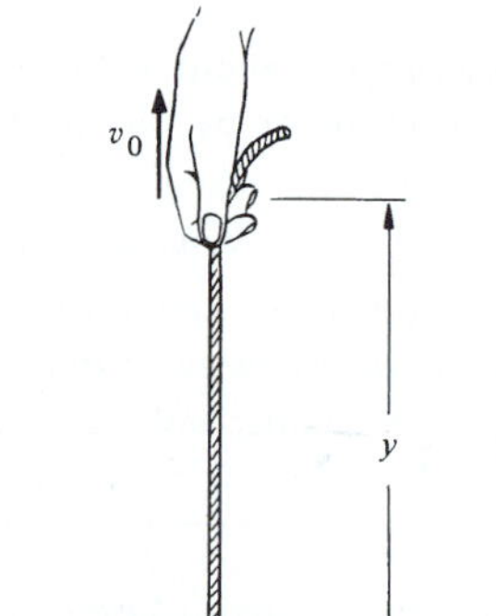

4.21 A uniform rope of mass λ per unit length is coiled on a smooth horizontal table. One end is pulled straight up with constant speed v_0.

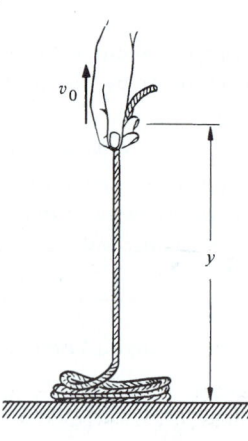

a. Find the force exerted on the end of the rope as a function of height y.

b. Compare the power delivered to the rope with the rate of change of the rope's total mechanical energy.

4.22 A ball drops to the floor and bounces, eventually coming to rest. Collisions between the ball and floor are inelastic; the speed after each collision is e times the speed before the collision where $e < 1$, (e is called the *coefficient of restitution.*) If the speed just before the first bounce is v_0, find the time to come to rest.

Ans. clue. If $v_0 = 5$ m/s, $e = 0.5$, then $T \approx 1$ s

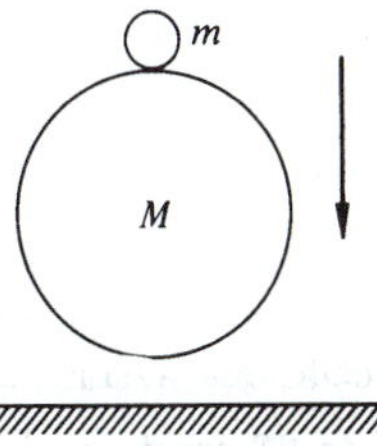

4.23 A small ball of mass m is placed on top of a "superball" of mass M, and the two balls are dropped to the floor from height h. How high does the small ball rise after the collision? Assume that collisions with the superball are elastic, and that $m \ll M$. To help visualize the problem, assume that the balls are slightly separated when the superball hits the floor. (If you are surprised at the result, try demonstrating the problem with a marble and a superball.)

4.24 Cars B and C are at rest with their brakes off. Car A plows into B at high speed, pushing B into C. If the collisions are completely inelastic, what fraction of the initial energy is dissipated in car C? Initially the cars are identical.

4.25 A proton makes a head-on collision with an unknown particle at rest. The proton rebounds straight back with $\frac{4}{9}$ of its initial kinetic energy.

Find the ratio of the mass of the unknown particle to the mass of the proton, assuming that the collision is elastic.

4.26 A particle of mass m and initial velocity v_0 collides elastically with a particle of unknown mass M coming from the opposite direction as shown at left below. After the collision m has velocity $v_0/2$ at right angles to the incident direction, and M moves off in the direction shown in the sketch. Find the ratio M/m.

4.27 Particle A of mass m has initial velocity v_0. After colliding with particle B of mass $2m$ initially at rest, the particles follow the paths shown in the sketch at right below. Find θ.

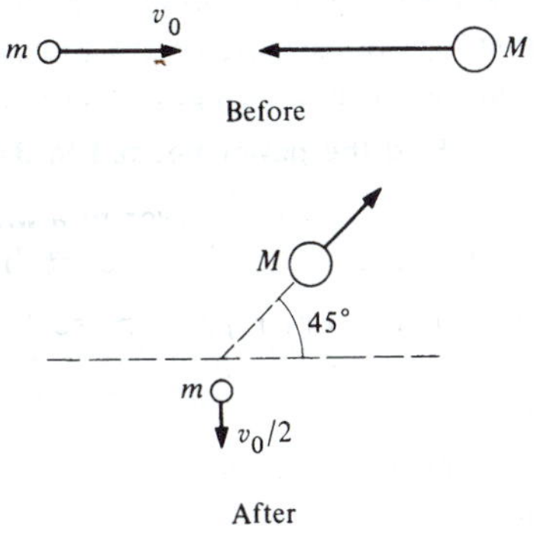

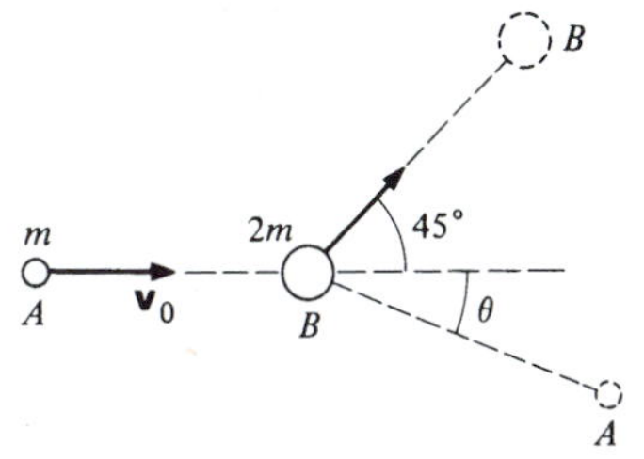

4.28 A thin target of lithium is bombarded by helium nuclei of energy E_0. The lithium nuclei are initially at rest in the target but are essentially unbound. When a helium nucleus enters a lithium nucleus, a nuclear reaction can occur in which the compound nucleus splits apart

into a boron nucleus and a neutron. The collision is inelastic, and the final kinetic energy is less than E_0 by 2.8 MeV. (1 MeV = 10^6 eV = 1.6×10^{-13} J). The relative masses of the particles are: helium, mass 4; lithium, mass 7; boron, mass 10; neutron, mass 1. The reaction can be symbolized

$$^7\text{Li} + {}^4\text{He} \rightarrow {}^{10}\text{B} + {}^1\text{n} - 2.8 \text{ MeV}.$$

a. What is $E_{0,\text{threshold}}$, the minimum value of E_0 for which neutrons can be produced? What is the energy of the neutrons at this threshold?

Ans. Neutron energy = 0.15 MeV

b. Show that if the incident energy falls in the range $E_{0,\text{threshold}} < E_0 < E_{0,\text{threshold}} + 0.27$ MeV, the neutrons ejected in the forward direction do not all have the same energy but must have either one or the other of two possible energies. (You can understand the origin of the two groups by looking at the reaction in the center of mass system.)

4.29 A "superball" of mass m bounces back and forth between two surfaces with speed v_0. Gravity is neglected and the collisions are perfectly elastic.

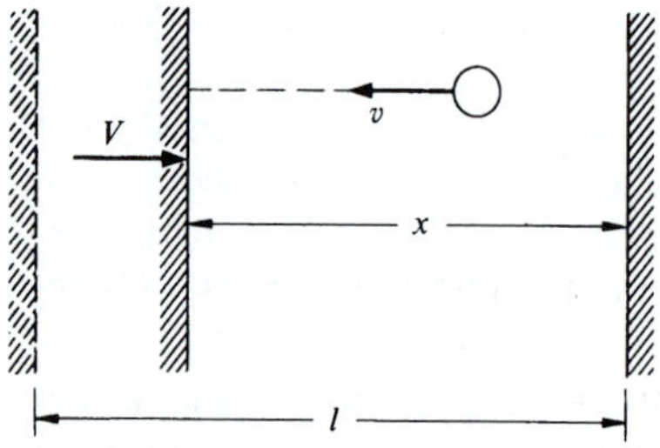

a. Find the average force F on each wall.

Ans. $F = mv_0^2/l$

b. If one surface is slowly moved toward the other with speed $V \ll v$, the bounce rate will increase due to the shorter distance between collisions, and because the ball's speed increases when it bounces from the moving surface. Find F in terms of the separation of the surfaces, x. (*Hint:* Find the average rate at which the ball's speed increases as the surface moves.)

Ans. $F = (mv_0^2/l)(l/x)^3$

c. Show that the work needed to push the surface from l to x equals the gain in kinetic energy of the ball. (This problem illustrates the mechanism which causes a gas to heat up as it is compressed.)

4.30 A particle of mass m and velocity v_0 collides elastically with a particle of mass M initially at rest and is scattered through angle Θ in the center of mass system.

a. Find the final velocity of m in the laboratory system.

Ans. $v_f = [v_0/(m + M)](m^2 + M^2 + 2mM \cos \Theta)^{\frac{1}{2}}$

b. Find the fractional loss of kinetic energy of m.

Ans. clue. If $m = M$, $(K_0 - K_f)/K_0 = (1 - \cos \Theta)/2$

5 SOME MATHEMATICAL ASPECTS OF FORCE AND ENERGY

5.1 Introduction

The last chapter introduced quite a few new physical concepts—work, potential energy, kinetic energy, the work-energy theorem, conservative and nonconservative forces, and the conservation of energy.

In this chapter there are no new physical ideas; this chapter is on mathematics. We are going to introduce several mathematical techniques which will help express the ideas of the last chapter in a more revealing manner. The rationale for this is partly that mathematical elegance can be a source of pleasure, but chiefly that the results developed here will be useful in other areas of physics, particularly in the study of electricity and magnetism. We shall find how to tell whether or not a force is conservative and how to relate the potential energy to the force.

A word of reassurance: Don't be alarmed if the mathematics looks formidable at first. Once you have a little practice with the new techniques, they will seem quite straightforward. In any case, you will probably see the same techniques presented from a different point of view in your study of calculus.

In this chapter we must deal with functions of several variables, such as a potential energy function which depends on x, y, and z. Our first task is to learn how to take derivatives and find differentials of such functions. If you are already familiar with partial differentiation the next section can be skipped. Otherwise, read on.

5.2 Partial Derivatives

We start by reviewing briefly the concept of the differential of a function $f(x)$ which depends on the single variable x. (Differentials are discussed in greater detail in Note 1.1.)

Consider the value of $f(x)$ at any point x. Let dx be an increment in x, known as the differential of x, which can be any size we please. The differential df of f is defined to be

$$df \equiv \left(\frac{df}{dx}\right) dx.$$

Note that (df/dx) stands for the derivative

$$\frac{df}{dx} = \lim_{\Delta x \to 0} \frac{\Delta f}{\Delta x}.$$

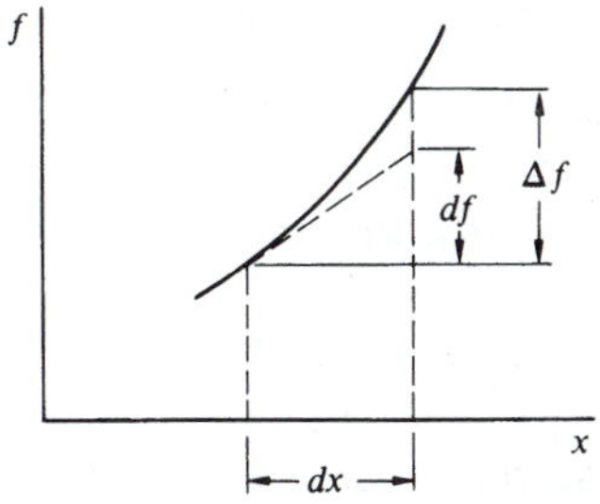

The actual change in f is $\Delta f = f(x + dx) - f(x)$. Δf differs from df, as the sketch indicates, but if the limit $dx \to 0$ is to be taken, the difference can be neglected,[1] and we can use df and Δf interchangeably.

Now let us consider a function $f(x,y)$ which depends on two variables x and y. For instance, f could be the area of a rectangle of length x and width y. If we keep the variable y fixed and let the variable x change by dx, the differential of f in this case is

$$df = \left[\lim_{\Delta x \to 0} \frac{f(x + \Delta x, y) - f(x,y)}{\Delta x}\right] dx.$$

The quantity in the bracket looks like a derivative. However, f depends on two variables and since we are differentiating with respect to only one variable, the quantity in the bracket is called a *partial derivative*. The partial derivative is denoted by $\partial f/\partial x$. (Calculus texts sometimes use f_x, but we shall avoid this notation to prevent confusion with vector components.) $\partial f/\partial x$ is read "the partial derivative of f with respect to x" or "the partial of f with respect to x." If we want to indicate that the partial derivative is to be evaluated at some particular point x_0, y_0, we can write

$$\frac{\partial f(x_0,y_0)}{\partial x} \quad \text{or} \quad \left.\frac{\partial f}{\partial x}\right|_{x_0,y_0}.$$

The procedure for evaluating partial derivatives is straightforward; in evaluating $\partial f/\partial x$, for example, all variables but x are treated as constants.

Example 5.1 Partial Derivatives

Let

$$f = x^2 \sin y.$$

Then

$$\frac{\partial f}{\partial x} = 2x \sin y,$$

$$\frac{\partial f}{\partial y} = x^2 \cos y.$$

[1] Specifically, $(\Delta f - df)$ is of order $(dx)^2$, so that $\lim_{\Delta x \to 0} [(\Delta f - df)/\Delta x] = 0$.

We can generalize the procedure to any number of variables. For instance, let

$$f = y + e^{xz}.$$

Then

$$\frac{\partial f}{\partial x} = ze^{xz},$$

$$\frac{\partial f}{\partial y} = 1,$$

$$\frac{\partial f}{\partial z} = xe^{xz}.$$

Let us consider what happens to $f(x,y)$ if x and y both vary. Let x change by dx and y change by dy. The change in f is

$$\Delta f = f(x + dx, y + dy) - f(x,y).$$

The right hand side can be written as follows:

$$f(x + dx, y + dy) - f(x,y) = [f(x + dx, y + dy) - f(x, y + dy)] + [f(x, y + dy) - f(x,y)].$$

The first term on the right is the change in f due to dx; this is given approximately by

$$(\Delta f)_{\text{due to } x} \approx \frac{\partial f(x, y + dy)}{\partial x} \Delta x.$$

The second term on the right is

$$(\Delta f)_{\text{due to } y} \approx \frac{\partial f(x,y)}{\partial y} \Delta y.$$

The total change is

$$\Delta f \approx \frac{\partial f(x, y + dy)}{\partial x} dx + \frac{\partial f(x,y)}{\partial y} dy.$$

We define the differential of f to be

$$df \equiv \frac{\partial f(x,y)}{\partial x} dx + \frac{\partial f(x,y)}{\partial y} dy. \qquad 5.1$$

If we take the limit $dx \to 0$, $dy \to 0$, Δf approaches df. In applications where we are going to take the limit, we can use Δf and df interchangeably. Furthermore, even if we do not take the limit, the differential gives a good approximation to the actual value of the change in f if dx and dy are small, as the following example illustrates.

Example 5.2 **Applications of the Partial Derivative**

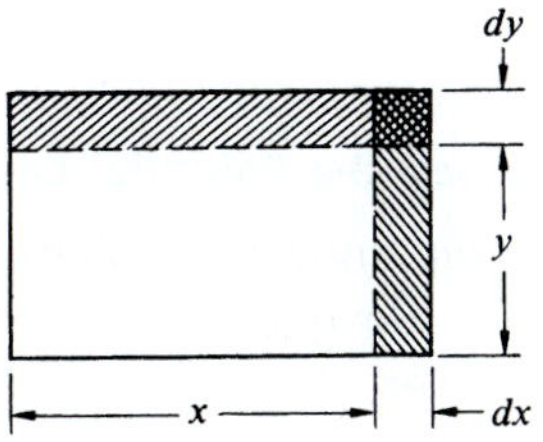

A. Suppose that f is the area of a rectangle of length x and width y. Then $f = xy$. The change in area if x increases by dx and y increases by dy is

$$\begin{aligned}\Delta f &= f(x + dx, y + dy) - f(x,y)\\ &= (x + dx)(y + dy) - xy\\ &= y\,dx + x\,dy + (dx)(dy).\end{aligned}$$

The differential of f is

$$\begin{aligned}df &= \frac{\partial(xy)}{\partial x}\,dx + \frac{\partial(xy)}{\partial y}\,dy\\ &= y\,dx + x\,dy.\end{aligned}$$

We see that

$$\Delta f - df = (dx)(dy).$$

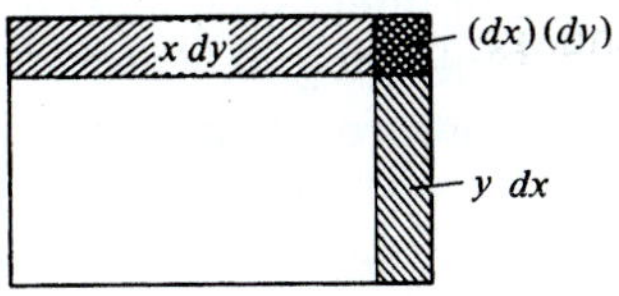

$(dx)(dy)$ is the area of the small rectangle in the figure. As $dx \to 0$ and $dy \to 0$, the area $(dx)(dy)$ becomes negligible compared with the area of the strips $x\,dy$ and $y\,dx$, and we can use the differential df as an accurate approximation to the actual change, Δf.

B. Consider the function

$$f(x,y) = y^3 e^x.$$

At $x = 0$, $y = 1$ we have $f(0,1) = 1$. What is the value of $f(0.03,1.01)$? Approximating the change in f by df we have

$$\begin{aligned}\Delta f &\approx df\\ &= \frac{\partial f}{\partial x}\,dx + \frac{\partial f}{\partial y}\,dy.\end{aligned}$$

The partial derivatives are easily evaluated.

$$\begin{aligned}\left.\frac{\partial f}{\partial x}\right|_{0,1} &= y^3 e^x\Big|_{0,1}\\ &= 1\end{aligned}$$

$$\begin{aligned}\left.\frac{\partial f}{\partial y}\right|_{0,1} &= 3y^2 e^x\Big|_{0,1}\\ &= 3\end{aligned}$$

Taking $dx = 0.03$, $dy = 0.01$, we find

$$df = (1)(0.03) + 3(0.01)$$
$$= 0.06.$$

The actual value, to four significant figures, is

$$\Delta f = 0.0617.$$

5.3 How To Find the Force if You Know the Potential Energy

Our problem is this—suppose that we know the potential energy function $U(\mathbf{r})$; how do we find $\mathbf{F}(\mathbf{r})$? For one dimensional motion we already know the answer from Sec. 4.8: $F_x = -dU/dx$. It isn't difficult to generalize this result to three dimensions.

Our starting point is the definition of potential energy:

$$U_b - U_a = -\oint_{\mathbf{r}_a}^{\mathbf{r}_b} \mathbf{F} \cdot d\mathbf{r}. \qquad 5.2$$

Let us consider the change in potential energy when a particle acted on by $\mathbf{F}$ undergoes a displacement $\Delta\mathbf{r}$.

$$U(\mathbf{r} + \Delta\mathbf{r}) - U(\mathbf{r}) = -\oint_{\mathbf{r}}^{\mathbf{r}+\Delta\mathbf{r}} \mathbf{F}(\mathbf{r}') \cdot d\mathbf{r}', \qquad 5.3$$

(We have labeled the dummy variable of integration by $\mathbf{r}'$ to avoid confusion with the end points of the line integral, $\mathbf{r}$ and $\mathbf{r} + \Delta\mathbf{r}$.) The left hand side of Eq. (5.3) is the difference in U at the two ends of the path. Let us call this ΔU. If $\Delta\mathbf{r}$ is so small that $\mathbf{F}$ does not vary appreciably over the path, the integral on the right is approximately $\mathbf{F} \cdot \Delta\mathbf{r}$. Therefore

$$\Delta U \approx -\mathbf{F} \cdot \Delta\mathbf{r}$$
$$= -(F_x\,\Delta x + F_y\,\Delta y + F_z\,\Delta z). \qquad 5.4$$

We can obtain an alternative expression for ΔU by using the results of the last section. If we approximate ΔU by the differential of U, we have from Eq. (5.1)

$$\Delta U \approx \frac{\partial U}{\partial x}\Delta x + \frac{\partial U}{\partial y}\Delta y + \frac{\partial U}{\partial z}\Delta z. \qquad 5.5$$

Combining Eq. (5.4) and (5.5) yields

$$\frac{\partial U}{\partial x}\Delta x + \frac{\partial U}{\partial y}\Delta y + \frac{\partial U}{\partial z}\Delta z \approx -F_x\,\Delta x - F_y\,\Delta y - F_z\,\Delta z. \qquad 5.6$$

When we take the limit $(\Delta x, \Delta y, \Delta z) \rightarrow 0$, the approximation becomes exact. Since Δx, Δy, and Δz are independent, Eq. (5.6) remains valid even if we choose Δy and Δz to be zero. This requires that the coefficients of Δx on either side of the equation be equal. We conclude that

$$\frac{\partial U}{\partial x} = -F_x$$
$$\frac{\partial U}{\partial y} = -F_y \qquad 5.7$$
$$\frac{\partial U}{\partial z} = -F_z.$$

We have the answer to the problem set at the beginning of this section—how to find the force from the potential energy function. However, as we shall see in the next section, there is a much neater way of expressing Eq. (5.7).

5.4 The Gradient Operator

Equation (5.7) is really a vector equation. We can write it explicitly in vector form:

$$\begin{aligned}\mathbf{F} &= \hat{\mathbf{i}}F_x + \hat{\mathbf{j}}F_y + \hat{\mathbf{k}}F_z \\ &= -\hat{\mathbf{i}}\frac{\partial U}{\partial x} - \hat{\mathbf{j}}\frac{\partial U}{\partial y} - \hat{\mathbf{k}}\frac{\partial U}{\partial z}.\end{aligned} \qquad 5.8$$

A shorthand way to symbolize this result is

$$\mathbf{F} = -\nabla U, \qquad 5.9$$

where

$$\nabla U \equiv \hat{\mathbf{i}}\frac{\partial U}{\partial x} + \hat{\mathbf{j}}\frac{\partial U}{\partial y} + \hat{\mathbf{k}}\frac{\partial U}{\partial z}. \qquad 5.10$$

Equation (5.10) is a definition, so if the notation looks strange, it is not because you have missed something. Let's see what ∇U means.

∇U is a vector called the *gradient of U* or *grad U*. The symbol ∇ (called "del") can be written in vector form as follows:

$$\nabla = \hat{\mathbf{i}}\frac{\partial}{\partial x} + \hat{\mathbf{j}}\frac{\partial}{\partial y} + \hat{\mathbf{k}}\frac{\partial}{\partial z}. \qquad 5.11$$

Obviously ∇ is not really a vector; it is a *vector operator*. This means that when ∇ operates on a scalar function (the potential energy function in our case), it forms a vector.

The relation $\mathbf{F} = -\nabla U$ is a generalization of the one dimensional case. For example, suppose that U depends only on x. Then

$$\nabla U = \frac{\partial U(x)}{\partial x}\hat{\mathbf{i}}$$

and

$$F_x = -\frac{\partial U}{\partial x}.$$

However, for a function of a single variable the partial derivative is identical to the familiar total derivative. We have

$$F_x = -\frac{dU}{dx}.$$

Here are a few more examples.

Example 5.3 Gravitational Attraction by a Particle

If a particle of mass M is at the origin, the potential energy of mass m a distance r from the origin is

$$U(x,y,z) = -\frac{GMm}{r}.$$

Then

$$\begin{aligned}\mathbf{F} &= -\nabla U \\ &= +GMm\nabla\frac{1}{r}.\end{aligned}$$

Consider the x component of $\nabla(1/r)$. Since $r = \sqrt{x^2 + y^2 + z^2}$, we have

$$\begin{aligned}\frac{\partial}{\partial x}\frac{1}{(x^2 + y^2 + z^2)^{\frac{1}{2}}} &= \frac{-x}{(x^2 + y^2 + z^2)^{\frac{3}{2}}} \\ &= -\frac{x}{r^3}.\end{aligned}$$

By symmetry the y and z terms are $-y/r^3$ and $-z/r^3$, respectively. Hence

$$\begin{aligned}\mathbf{F} &= GMm\left(\hat{\mathbf{i}}\frac{-x}{r^3} + \hat{\mathbf{j}}\frac{-y}{r^3} + \hat{\mathbf{k}}\frac{-z}{r^3}\right) \\ &= GMm\left[\frac{-\mathbf{r}}{r^3}\right] \\ &= -GMm\frac{\hat{\mathbf{r}}}{r^2}.\end{aligned}$$

We have recovered the familiar expression for the force of gravity between two particles.

Example 5.4 **Uniform Gravitational Field**

From the last chapter we know that the potential energy of mass m in a uniform gravitational field directed downward is

$$U(x,y,z) = mgz,$$

where z is the height above ground. The corresponding force is

$$\begin{aligned}\mathbf{F} &= -\nabla U \\ &= -mg\left(\hat{\mathbf{i}}\frac{\partial}{\partial x} + \hat{\mathbf{j}}\frac{\partial}{\partial y} + \hat{\mathbf{k}}\frac{\partial}{\partial z}\right)z \\ &= -mg\hat{\mathbf{k}}.\end{aligned}$$

Example 5.5 **Gravitational Attraction by Two Point Masses**

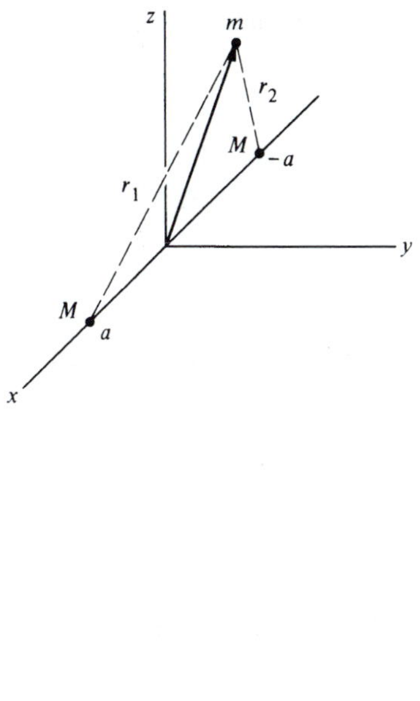

The previous examples were trivial, since the forces were obvious by inspection. Here is a more complicated case in which the energy method gives a helpful shortcut.

Two particles, each of mass M, lie on the x axis at $x = a$ and $x = -a$, respectively. Find the force on a particle of mass m located at **r**.

We start by considering the potential energy of m due to the particle at $x = a$. The distance is $\sqrt{(x-a)^2 + y^2 + z^2}$, and the potential energy is $-GMm/\sqrt{(x-a)^2 + y^2 + z^2} = -GMm/r_1$. Similarly, the potential energy due to the mass at $x = -a$ is $-GMm/\sqrt{(x+a)^2 + y^2 + z^2} = -GMm/r_2$. The total potential energy is the sum of these terms. This illustrates a major advantage of working with energy rather than force. Energy is a scalar and is simply additive, whereas forces must be added vectorially.

We have $u = -GMm/r_1 - GMm/r_2$, or

$$U = -GMm\left\{\frac{1}{[(x-a)^2 + y^2 + z^2]^{\frac{1}{2}}} + \frac{1}{[(x+a)^2 + y^2 + z^2]^{\frac{1}{2}}}\right\}.$$

The force components are easily found by differentiation.

$$\begin{aligned}F_x(x,y,z) &= -\frac{\partial U}{\partial x} \\ &= -GMm\left\{\frac{(x-a)}{[(x-a)^2 + y^2 + z^2]^{\frac{3}{2}}} + \frac{(x+a)}{[(x+a)^2 + y^2 + z^2]^{\frac{3}{2}}}\right\} \\ &= -GMm\left(\frac{x-a}{r_1^3} + \frac{x+a}{r_2^3}\right)\end{aligned}$$

Similarly,

$$F_y(x,y,z) = -\frac{\partial U}{\partial y}$$

$$= -GMm\left(\frac{y}{r_1^3} + \frac{y}{r_2^3}\right)$$

$$F_z(x,y,z) = -\frac{\partial U}{\partial z}$$

$$= -GMm\left(\frac{z}{r_1^3} + \frac{z}{r_2^3}\right).$$

If m is far from the other two masses so that $|x| \gg a$, we have $r_1 \approx r$, $r_2 \approx r$. In this case

$$F_x \approx -\frac{2GMm}{r^2}\frac{x}{r}$$

$$F_y \approx -\frac{2GMm}{r^2}\frac{y}{r}$$

$$F_z \approx -\frac{2GMm}{r^2}\frac{z}{r}.$$

At large distances the force on m is like the force $(-2GMm/r^2)\hat{\mathbf{r}}$ that would be exerted by a single mass $2M$ located at the origin.

Perhaps these examples suggest something of the convenience of the energy method. Potential energy is much simpler to manipulate than force. If force is needed, we can obtain it from $\mathbf{F} = -\boldsymbol{\nabla} U$. However, only conservative forces have potential energy functions associated with them. Nonconservative forces cannot be expressed as the gradient of a scalar function. Fortunately, most of the important forces of physics are conservative. In Sec. 5.6 we shall develop a simple means for telling whether a force is conservative or not.

We next turn to a discussion of the physical meaning of the gradient.

5.5 The Physical Meaning of the Gradient

Consider a particle moving under conservative forces with potential energy $U(x,y,z)$. As the particle moves from the point (x,y,z) to

$(x + dx, y + dy, z + dz)$, its potential energy changes by

$$U(x + dx, y + dy, z + dz) - U(x,y,z).$$

As explained in the last section, when we intend to take the limit $dx \to 0$, $dy \to 0$, $dz \to 0$, we can represent the change in U by the differential

$$dU = \frac{\partial U}{\partial x}dx + \frac{\partial U}{\partial y}dy + \frac{\partial U}{\partial z}dz.$$

The displacement is $d\mathbf{r} = dx\,\hat{\mathbf{i}} + dy\,\hat{\mathbf{j}} + dz\,\hat{\mathbf{k}}$ and we can write

$$dU = \nabla U \cdot d\mathbf{r} \qquad 5.12$$

where ∇U, the gradient of U, is

$$\nabla U = \frac{\partial U}{\partial x}\hat{\mathbf{i}} + \frac{\partial U}{\partial y}\hat{\mathbf{j}} + \frac{\partial U}{\partial z}\hat{\mathbf{k}}.$$

Equation (5.12) expresses the fundamental property of the gradient. The gradient allows us to find the change in a function induced by a change in its variables. In fact, Eq. (5.12) is actually the definition of gradient. Like a vector, the gradient operator is defined without reference to a particular coordinate system.

To develop physical insight into the meaning of ∇U, it is helpful to adopt a pictorial representation of potential energy. So let us make a brief digression.

Constant Energy Surfaces and Contour Lines

The equation $U(x,y,z) = \text{constant} = C$ defines for each value of C a surface known as a *constant energy* surface. A particle constrained to move on such a surface has constant potential energy. For example, the gravitational potential energy of a particle m at distance $r = \sqrt{x^2 + y^2 + z^2}$ from particle M is $U = -GMm/r$. The surfaces of constant energy are given by

$$-\frac{GMm}{r} = C$$

or

$$r = -\frac{GMm}{C}.$$

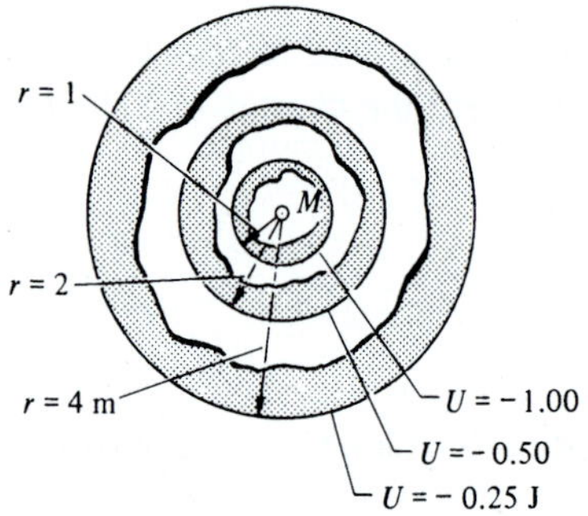

The constant energy surfaces are spheres centered on M, as shown in the drawing. (We have taken $GMm = 1$ N·m² for convenience.)

Constant energy surfaces are usually difficult to draw, and for this reason it is generally easier to visualize U by considering the lines of intersection of the constant energy surfaces with a plane. These lines are sometimes referred to as constant energy lines or, more simply, contour lines. For spherical energy surfaces the contour lines are circles. The next example discusses contour lines for a more complicated situation.

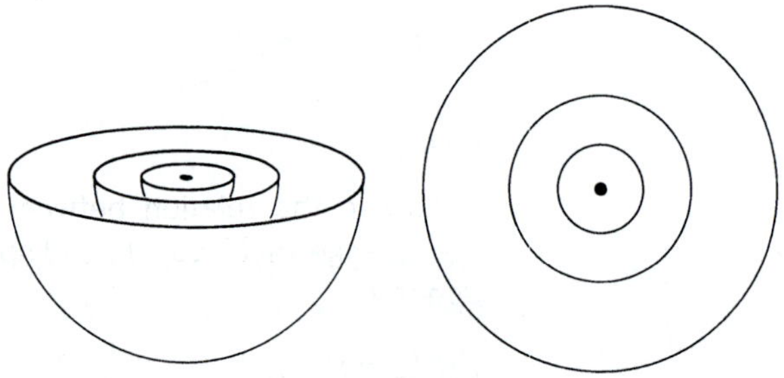

Example 5.6 Energy Contours for a Binary Star System

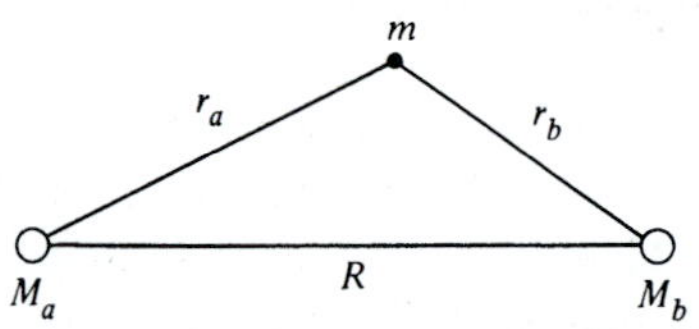

Consider a satellite of mass m in the gravitational field of a binary star system. The stars have masses M_a and M_b and are separated by distance R. The potential energy of the satellite is

$$U = -\frac{GmM_a}{r_a} - \frac{GmM_b}{r_b},$$

where r_a and r_b are its distances from the two stars. Consider the contour lines in a plane through the axis of the stars. Near star a, where $r_a \ll r_b$, we have

$$U \approx -\frac{GmM_a}{r_a}.$$

Here the contour lines are effectively circles. Near star b, where $r_b \ll r_a$, the contour lines are also effectively circles.

In the intermediate region between the two stars the effects of both bodies are important. The contour lines in the drawing opposite were calculated numerically, with $GmM_b/R = 1$, and $M_b/M_a = \frac{1}{4}$.

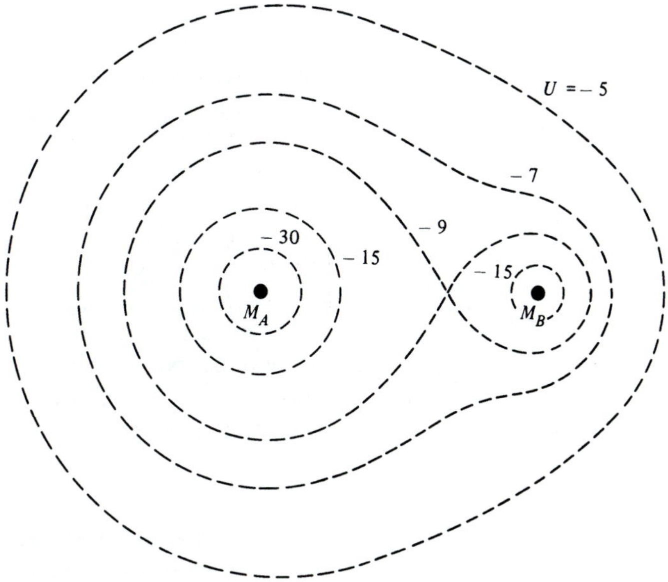

To see the relation between ∇U and contour lines, consider the change in U due to a displacement $d\mathbf{r}$ along a contour. In general

$$dU = \nabla U \cdot d\mathbf{r}.$$

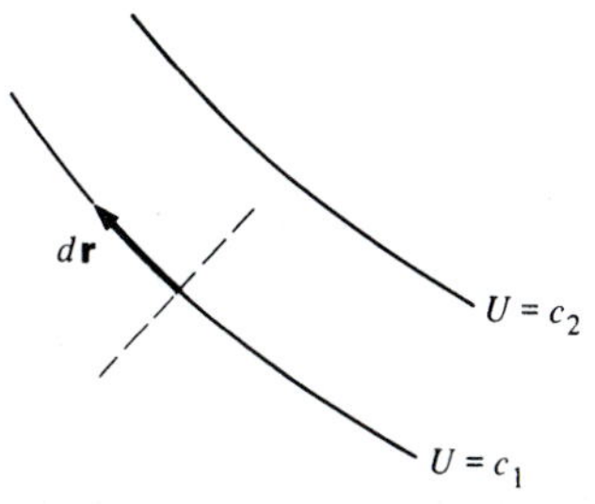

However, on a contour line, U is constant and $dU = 0$. Hence

$$\nabla U \cdot d\mathbf{r} = 0 \qquad (d\mathbf{r} \text{ along contour line}).$$

Since ∇U and $d\mathbf{r}$ are not zero, we see that the vector ∇U must be perpendicular to $d\mathbf{r}$. More generally, ∇U is perpendicular to any displacement $d\mathbf{r}$ on a constant energy surface. Hence, at every point in space, ∇U is perpendicular to the constant energy surface passing through that point.

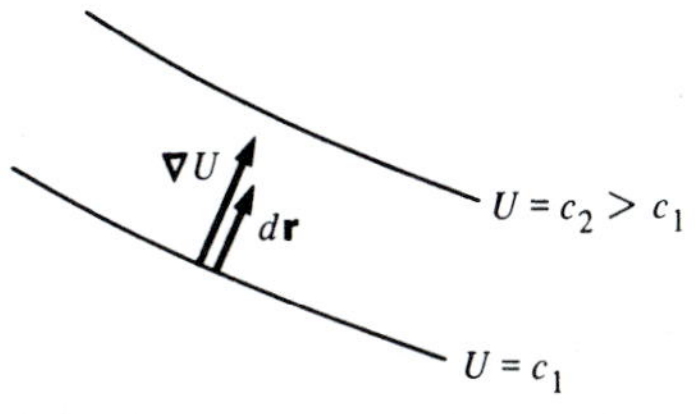

It is not hard to show that ∇U points from lower to higher potential energy. Consider a displacement $d\mathbf{r}$ pointing in the direction of increasing potential energy. For this displacement $dU > 0$, and since $dU = \nabla U \cdot d\mathbf{r} > 0$, we see that ∇U points from lower to higher potential energy. Hence the direction of ∇U is the direction in which U is increasing most rapidly.

Since $\nabla U = -\mathbf{F}$, we conclude that $\mathbf{F}$ is everywhere perpendicular to the constant energy surfaces and points from higher to lower potential energy.

Given the contour lines, it is easy to sketch the force. For the gravitational interaction of a particle with a mass located at the origin, the contour lines are circles. The force points radially inward from higher to lower potential energy, as we expect.

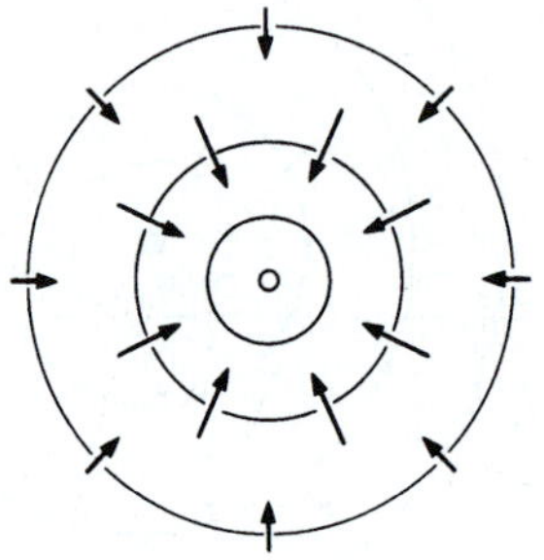

The drawing below shows the force at various points along the contour lines of the binary star system of Example 5.6. We can

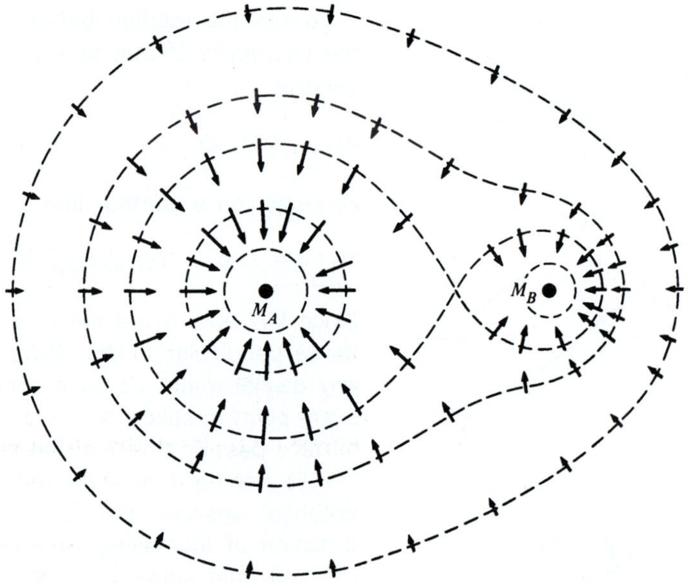

extend the arrows to form a curve everywhere parallel to **F**. These lines show the direction of the force everywhere in space and provide a simple map of the force field. Note that the force lines are perpendicular to the energy contours everywhere. Point P, where

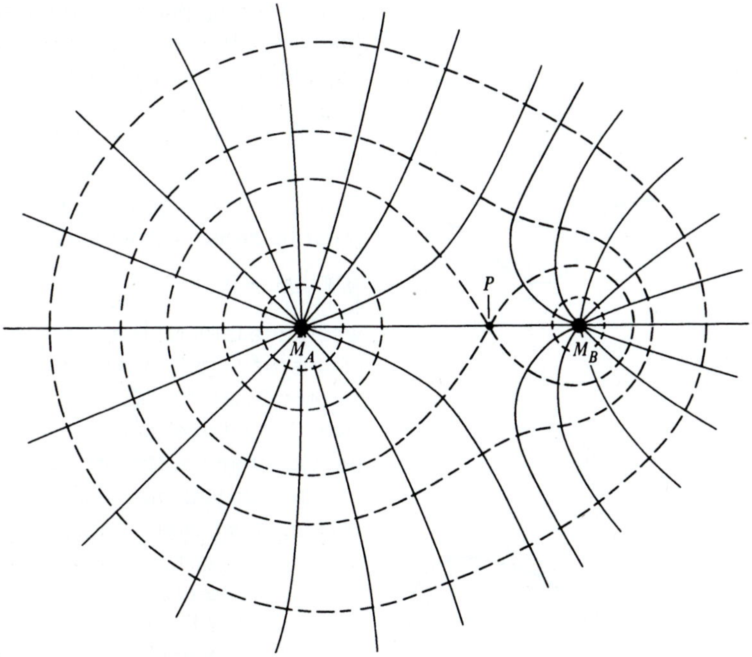

two energy contours intersect, presents a problem. How can the force point in two directions at once? The answer is that point P is the equilibrium point between the two stars where the force vanishes.

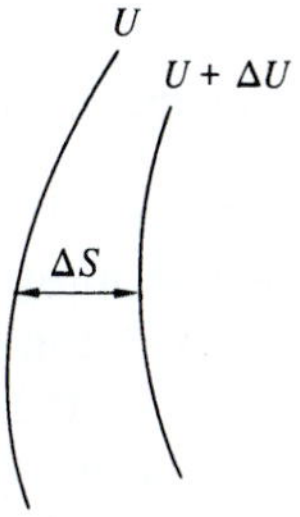

If two adjacent energy surfaces differ in energy by ΔU, then where the separation is ΔS,

$$|\nabla U| \approx \frac{\Delta U}{\Delta S}.$$

Hence, the closer the surfaces, the larger the gradient. More physically, the force is large where the potential energy is changing rapidly.

5.6 How to Find Out if a Force Is Conservative

Although we have seen numerous examples of conservative forces, we have no general test to tell us whether a given force **F(r)** is conservative. Let us now attack this problem.

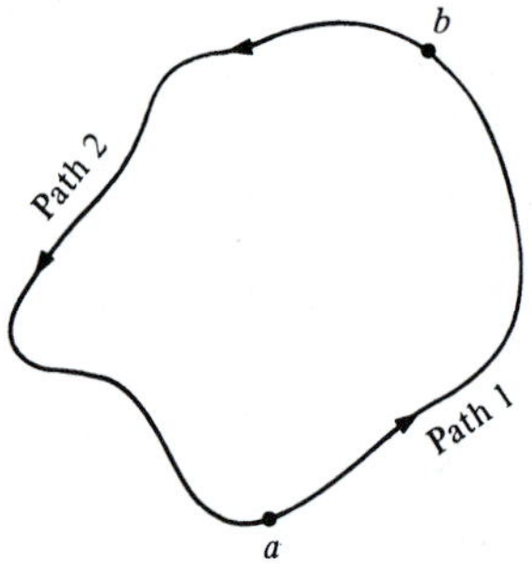

Our starting point is the observation that if $\mathbf{F}(\mathbf{r})$ is conservative, the work done on a particle by force $\mathbf{F}$ as it moves from a to b and back to a around a closed path is

$$\oint_{\substack{a \\ \text{Path 1}}}^{b} \mathbf{F}\cdot d\mathbf{r} + \oint_{\substack{b \\ \text{Path 2}}}^{a} \mathbf{F}\cdot d\mathbf{r} = (-U_b + U_a) + (-U_a + U_b) = 0.$$

Thus, the work done by a conservative force around a closed path must be zero. Symbolically,

$$\oint \mathbf{F}\cdot d\mathbf{r} = 0, \qquad 5.13$$

where the integral is a line integral taken around any closed path. (The symbol $\oint$ indicates that the path is closed.) Conversely, if a force $\mathbf{F}$ satisfies Eq. (5.13) for *all* paths (not just for a special path), the force must be conservative. Hence, Eq. (5.13) is a necessary and sufficient condition for a force to be conservative.

Although you may think that the problem is now more complicated than when we began, the fact is that we have taken a big step forward. However, in order to proceed we must further transform the problem.

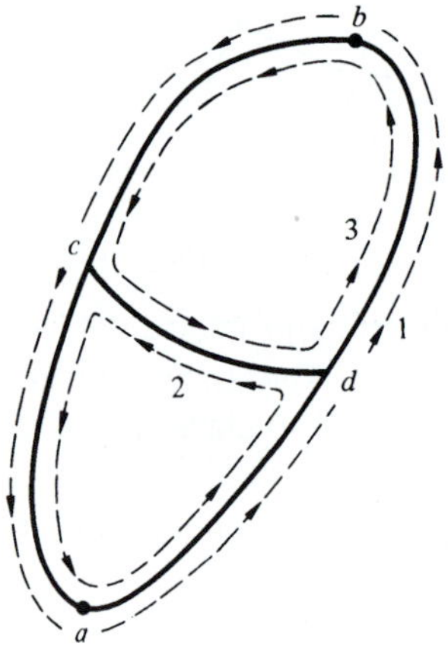

Consider $\oint \mathbf{F}\cdot d\mathbf{r}$, where the integral is around loop 1. If we break the integral into two integrals, via the "shortcut" cd, we have

$$\oint_{1} \mathbf{F}\cdot d\mathbf{r} = \oint_{2} \mathbf{F}\cdot d\mathbf{r} + \oint_{3} \mathbf{F}\cdot d\mathbf{r}.$$

This identity follows because the contribution to $\oint_{2} \mathbf{F}\cdot d\mathbf{r}$ from the line segment cd is exactly canceled by the contribution from the segment dc to $\oint_{3} \mathbf{F}\cdot d\mathbf{r}$. Traversing the same line in two directions gives zero net contribution to the total work.

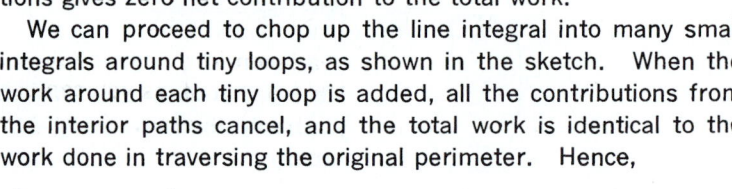

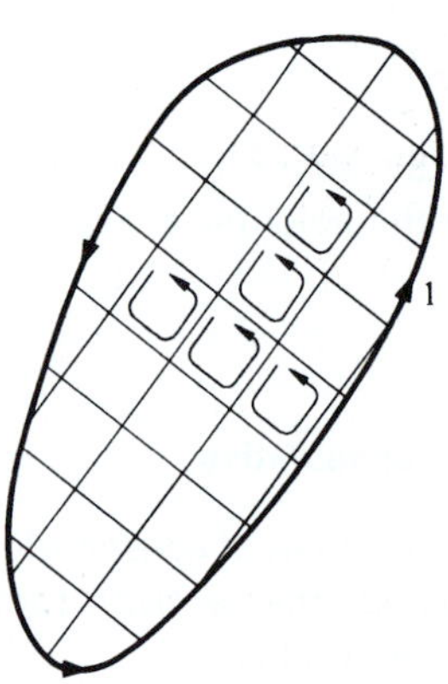

We can proceed to chop up the line integral into many small integrals around tiny loops, as shown in the sketch. When the work around each tiny loop is added, all the contributions from the interior paths cancel, and the total work is identical to the work done in traversing the original perimeter. Hence,

$$\oint_{1} \mathbf{F}\cdot d\mathbf{r} = \sum_{i} \oint_{i} \mathbf{F}\cdot d\mathbf{r} \qquad 5.14$$

where $\oint_{i} \mathbf{F}\cdot d\mathbf{r}$ is the work done in circling the ith tiny loop.

If you are wondering where this is leading, the answer is that by focusing our attention on one of the tiny paths we can convert

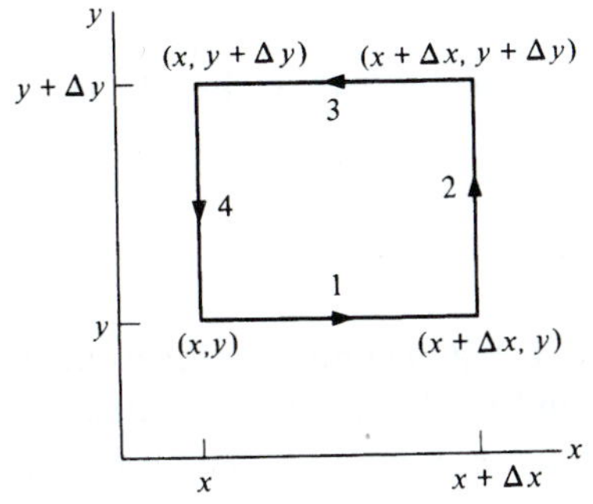

the original problem, which involves an integral over a large area, into a problem involving quantities at a single point in space. To do this, we must evaluate the line integral around one of the tiny loops. Let us consider a rectangular loop lying in the xy plane with sides of length Δx and Δy. The integral around the loop is

$$\oint \mathbf{F} \cdot d\mathbf{r} = \oint_1 \mathbf{F} \cdot d\mathbf{r} + \oint_2 \mathbf{F} \cdot d\mathbf{r} + \oint_3 \mathbf{F} \cdot d\mathbf{r} + \oint_4 \mathbf{F} \cdot d\mathbf{r}.$$

Integrals 1 and 3 both involve paths in the x direction, so let us consider them together. Integral 1 is

$$\oint_1 \mathbf{F} \cdot d\mathbf{r} = \int_{x,y}^{x+\Delta x,\, y} F_x(x,y)\, dx. \qquad 5.15$$

If Δx is small,

$$\oint_1 \mathbf{F} \cdot d\mathbf{r} \approx F_x(x,y)\, \Delta x.$$

Similarly, the integral along path 3 is

$$\oint_3 \mathbf{F} \cdot d\mathbf{r} \approx -F_x(x,\, y + \Delta y)\, \Delta x.$$

The integrals along paths 1 and 3 almost cancel. However, the small difference in y between the two paths is important. We have

$$\begin{aligned} \oint_1 \mathbf{F} \cdot d\mathbf{r} + \oint_3 \mathbf{F} \cdot d\mathbf{r} &\approx F_x(x,y)\, \Delta x - F_x(x,\, y + \Delta y)\, \Delta x \\ &= -[F_x(x,\, y + \Delta y) - F_x(x,y)]\, \Delta x. \end{aligned} \qquad 5.16$$

You may be puzzled by the fact that we are allowing for the fact that y is different between the two paths but are ignoring the variation of x along each of the paths. The reason is simply that the variation in y has an effect in first order, whereas the variation in x does not, as you can verify for yourself.

We shall eventually take the limit $\Delta x \to 0$, $\Delta y \to 0$, and from the discussion of differentials in Sec. 5.2, we have

$$F_x(x,\, y + \Delta y) - F_x(x,y) = \frac{\partial F_x}{\partial y} \Delta y.$$

Hence Eq. (5.16) can be written

$$\oint_1 \mathbf{F} \cdot d\mathbf{r} + \oint_3 \mathbf{F} \cdot d\mathbf{r} = -\frac{\partial F_x}{\partial y} \Delta x\, \Delta y.$$

Applying the same argument to paths 2 and 4 gives

$$\oint_2 \mathbf{F} \cdot d\mathbf{r} + \oint_4 \mathbf{F} \cdot d\mathbf{r} = \frac{\partial F_y}{\partial x} \Delta x \, \Delta y.$$

The line integral around the tiny rectangular loop in the xy plane is therefore

$$\oint_{xy \text{ plane}} \mathbf{F} \cdot d\mathbf{r} = \left(\frac{\partial F_y}{\partial x} - \frac{\partial F_x}{\partial y} \right) \Delta x \, \Delta y. \qquad 5.17a$$

Although we shall not stop to prove it, this result holds for a small loop of any shape if $\Delta x \, \Delta y$ is replaced by the actual area ΔA.

The line integral around a tiny loop in the yz plane can be found by simply cycling the variables, $x \to y$, $y \to z$, $z \to x$. We find

$$\oint_{yz \text{ plane}} \mathbf{F} \cdot d\mathbf{r} = \left(\frac{\partial F_z}{\partial y} - \frac{\partial F_y}{\partial z} \right) \Delta y \, \Delta z. \qquad 5.17b$$

Similarly, for a loop in the xz plane,

$$\oint_{xz \text{ plane}} \mathbf{F} \cdot d\mathbf{r} = \left(\frac{\partial F_x}{\partial z} - \frac{\partial F_z}{\partial x} \right) \Delta x \, \Delta z. \qquad 5.17c$$

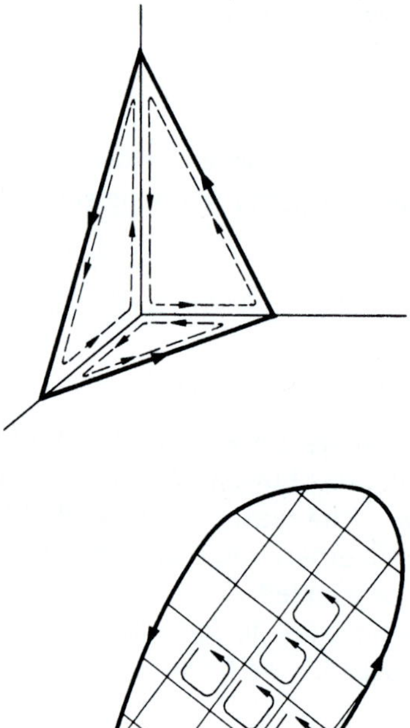

The line integral around a tiny loop in an arbitrary orientation can be decomposed into line integrals in the three coordinate planes, as the sketch suggests.

Accordingly, the line integral around any tiny loop will vanish provided

$$\begin{aligned} \frac{\partial F_y}{\partial x} - \frac{\partial F_x}{\partial y} &= 0 \\ \frac{\partial F_z}{\partial y} - \frac{\partial F_y}{\partial z} &= 0 \\ \frac{\partial F_x}{\partial z} - \frac{\partial F_z}{\partial x} &= 0. \end{aligned} \qquad 5.18$$

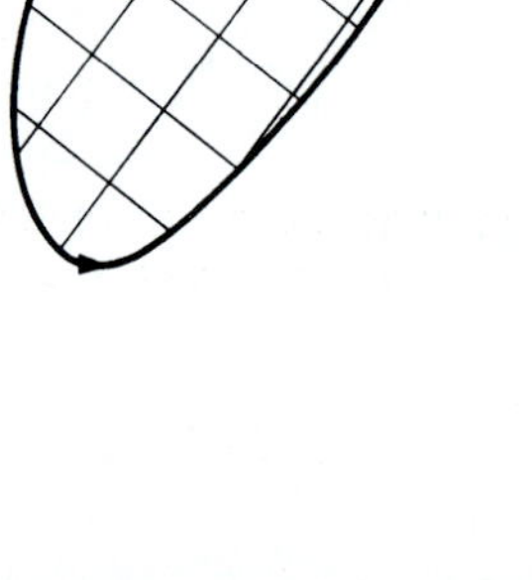

If Eq. (5.18) is satisfied everywhere, the line integral around any tiny loop vanishes and it follows that $\oint \mathbf{F} \cdot d\mathbf{r} = 0$ for any closed path. Hence, a force satisfying Eq. (5.18) is conservative.

We have achieved our goal of finding a mathematical test for whether or not a given force is conservative. However, Eq. (5.18) is rather cumbersome as it stands. Fortunately, we can summarize it in simple vector notation. If we use the familiar rules of evaluating the cross product (Sec. 1.4) and treat the vector operator ∇ as if it were a vector, then

$$\nabla \times \mathbf{F} = \begin{vmatrix} \hat{\mathbf{i}} & \hat{\mathbf{j}} & \hat{\mathbf{k}} \\ \frac{\partial}{\partial x} & \frac{\partial}{\partial y} & \frac{\partial}{\partial z} \\ F_x & F_y & F_z \end{vmatrix}$$
$$= \hat{\mathbf{i}}\left(\frac{\partial F_z}{\partial y} - \frac{\partial F_y}{\partial z}\right) + \hat{\mathbf{j}}\left(\frac{\partial F_x}{\partial z} - \frac{\partial F_z}{\partial x}\right) + \hat{\mathbf{k}}\left(\frac{\partial F_y}{\partial x} - \frac{\partial F_x}{\partial y}\right). \qquad 5.19$$

$\nabla \times \mathbf{F}$ is called the *curl* of $\mathbf{F}$.

Example 5.7 The Curl of the Gravitational Force

We know that the gravitational force is conservative since it possesses a potential energy function. However, for purposes of illustration, let us prove that the force of gravity is conservative by showing that its curl is zero.

For the gravitational force between two particles we have

$$\mathbf{F} = \frac{A}{r^2}\hat{\mathbf{r}}$$
$$= A\frac{\mathbf{r}}{r^3} = A\frac{x\hat{\mathbf{i}} + y\hat{\mathbf{j}} + z\hat{\mathbf{k}}}{r^3}.$$

$$(\nabla \times \mathbf{F})_x = \frac{\partial F_z}{\partial y} - \frac{\partial F_y}{\partial z}$$
$$= \frac{\partial}{\partial y}\left(\frac{Az}{r^3}\right) - \frac{\partial}{\partial z}\left(\frac{Ay}{r^3}\right).$$

The first term on the right hand side is

$$\frac{\partial}{\partial y} Az(x^2 + y^2 + z^2)^{-\frac{3}{2}} = Az(-\tfrac{3}{2})(x^2 + y^2 + z^2)^{-\frac{5}{2}}(2y)$$
$$= -3A\frac{zy}{r^5}.$$

Similarly,

$$\frac{\partial}{\partial z}\frac{Ay}{r^3} = -3A\frac{yz}{r^5}.$$

Hence,

$$(\nabla \times \mathbf{F})_x = -3A\frac{zy}{r^5} + 3A\frac{yz}{r^5} = 0.$$

By cycling the coordinates, we see that the other components of $\nabla \times \mathbf{F}$ are also zero. Hence $\nabla \times \mathbf{F} = 0$ and the gravitational force is conservative.

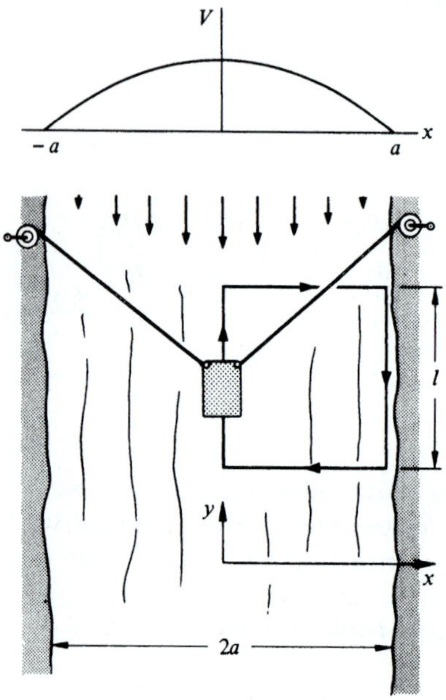

Example 5.8 A Nonconservative Force

Here is an example of a nonconservative force: consider a river with a current whose velocity **V** is maximum at the center and drops to zero at either bank.

$$\mathbf{V} = -V_0\left(1 - \frac{x^2}{a^2}\right)\hat{\mathbf{j}}$$

The width of the river is $2a$, and the coordinates are shown in the sketch.

Suppose that a barge in the stream is hauled around the path shown, by winches on the banks. The barge is pulled slowly and we shall assume that the force exerted on it by the current is

$$\mathbf{F}_{\text{river}} = b\mathbf{V},$$

where b is a constant. The barge is effectively in equilibrium, so that the force exerted by the winches is

$$\begin{aligned}\mathbf{F} &= -\mathbf{F}_{\text{river}} = -b\mathbf{V}\\ &= bV_0\left(1 - \frac{x^2}{a^2}\right)\hat{\mathbf{j}}.\end{aligned}$$

Let us evaluate $\nabla \times \mathbf{F}$ to determine whether or not the force is conservative. We have

$$\begin{aligned}(\nabla \times \mathbf{F})_x &= \frac{\partial F_z}{\partial y} - \frac{\partial F_y}{\partial z}\\ &= 0\\ (\nabla \times \mathbf{F})_y &= \frac{\partial F_x}{\partial z} - \frac{\partial F_z}{\partial x}\\ &= 0\\ (\nabla \times \mathbf{F})_z &= \frac{\partial F_y}{\partial x} - \frac{\partial F_x}{\partial y}\\ &= \frac{\partial}{\partial x} bV_0\left(1 - \frac{x^2}{a^2}\right)\\ &= -\frac{2bV_0}{a^2}x.\end{aligned}$$

Since the curl does not vanish, the force is nonconservative and the winches must do work to pull the barge around the closed path. The work done going upstream is $F(x = 0)l$, and the work done going downstream is $-F(x = a)l$. (In this idealized problem no work is needed to move the barge cross stream.) Since $F(x) = bV_0(1 - x^2/a^2)$, the total work done by the winches is

$$\begin{aligned}W &= bV_0l - bV_0l\left(1 - \frac{a^2}{a^2}\right)\\ &= bV_0l.\end{aligned}$$

Example 5.9 **A Most Unusual Force Field**

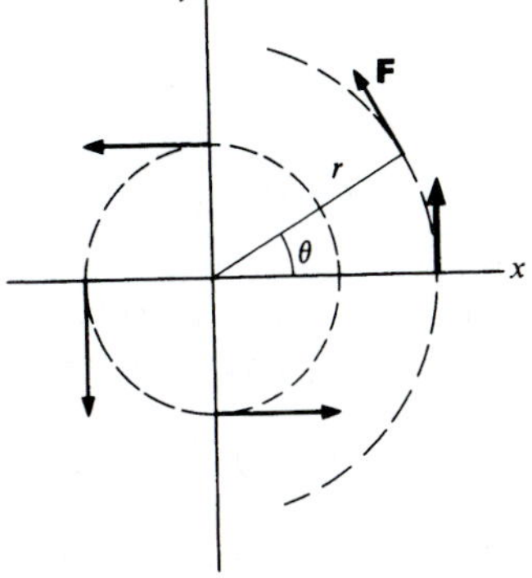

The field described in this example has some very surprising properties.

Consider a particle moving in the xy plane under the force

$$\mathbf{F}(r) = \frac{A}{r}\,\hat{\boldsymbol{\theta}},$$

where A is a constant. The force decreases as $1/r$, and is directed tangentially about the origin, as shown.

The work done as the particle travels through $d\mathbf{r} = dr\,\hat{\mathbf{r}} + r\,d\theta\,\hat{\boldsymbol{\theta}}$ is

$$\begin{aligned} dW &= \mathbf{F}\cdot d\mathbf{r} \\ &= \frac{A}{r}\,r\,d\theta \\ &= A\,d\theta. \end{aligned}$$

Surprisingly, the work does not depend on r, but only on the angle subtended.

Offhand, $\mathbf{F}$ may seem to be conservative, since the work done in going from $\mathbf{r}_1$ to $\mathbf{r}_2$ in the drawing below, left, appears to be independent of path:

$$\begin{aligned} W &= \oint_{\mathbf{r}_1}^{\mathbf{r}_2} A\,d\theta \\ &= A(\theta_2 - \theta_1). \end{aligned}$$

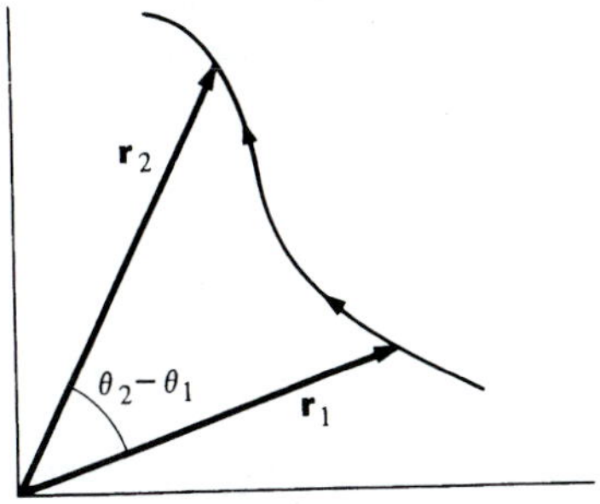

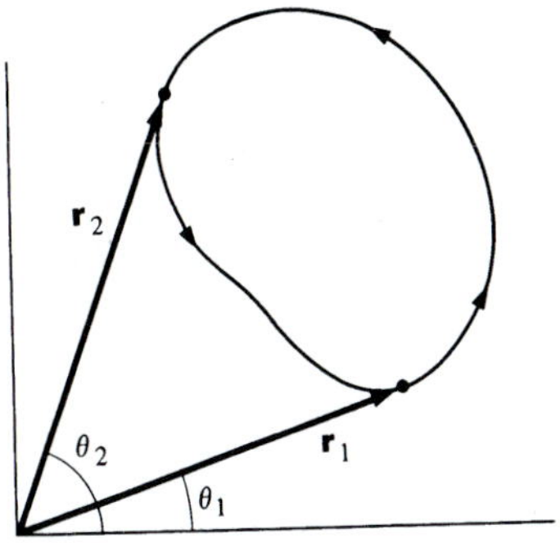

For instance, for the closed path shown above right,

$$\begin{aligned} W &= \oint_{\mathbf{r}_1}^{\mathbf{r}_2} A\,d\theta + \oint_{\mathbf{r}_2}^{\mathbf{r}_1} A\,d\theta \\ &= A(\theta_2 - \theta_1) + A(\theta_1 - \theta_2) \\ &= 0, \end{aligned}$$

as we expect for a conservative force.

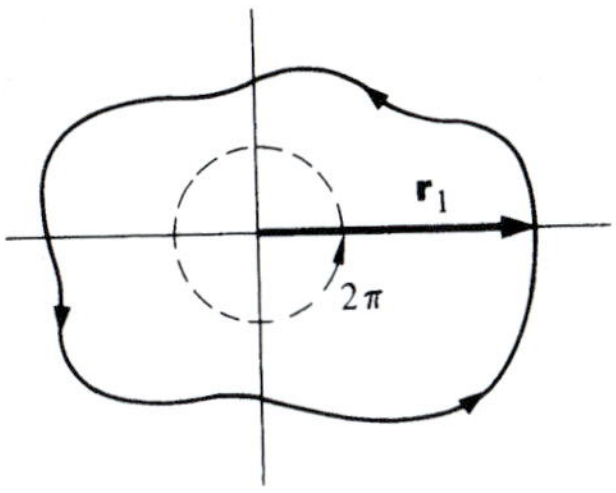

However, consider the work done along a closed path which encloses the origin as in the drawing at the left. Since $\theta_1 = 0$ and $\theta_2 = 2\pi$, the work $W = 2\pi A$. Evidently, $\mathbf{F}$ is not conservative.

Every time the particle makes a complete trip around the origin, the force does work $2\pi A$, but for a closed path that does not encircle the origin, $W = 0$. The force appears conservative provided that the path does not enclose the origin.

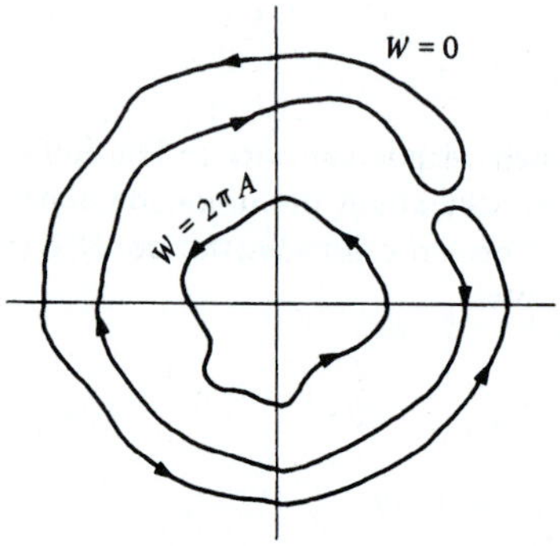

If you evaluate $\nabla \times \mathbf{F}$, you will find that it is zero everywhere except at the origin, where it has a singularity. It is this singularity which gives the force such peculiar properties. For the line integral of a force to vanish around a closed path, the curl must be zero everywhere inside the path. In this example, $\nabla \times \mathbf{F}$ is zero everywhere except at the origin.

If a force is conservative, it is always possible to find a potential energy function U such that $\mathbf{F} = -\nabla U$. The following example shows how this is done.

Example 5.10 Construction of the Potential Energy Function

In this example we shall find the potential energy function associated with the force

$$\mathbf{F} = A(x^2\hat{\mathbf{i}} + y\hat{\mathbf{j}}). \qquad 1$$

The first thing is to ascertain that $\nabla \times \mathbf{F} = 0$, for otherwise U does not exist. Since you can easily verify this for yourself, we proceed to determine U. U must obey

$$-\frac{\partial U}{\partial x} = F_x \qquad 2$$

$$= Ax^2$$

and

$$-\frac{\partial U}{\partial y} = F_y \qquad 3$$

$$= Ay.$$

We can integrate Eq. (2) to obtain

$$U(x,y) = -\frac{A}{3}x^3 + f(y). \qquad 4$$

Equation (4) needs some explanation. If U depended only on x, then integrating Eq. (2) would yield $U(x) = (-A/3)x^3 + C$, where C is a constant. However, U also depends on y. As far as partial differentiation with respect to x is concerned, $f(y)$ is a constant, since $\partial f(y)/\partial x = 0$.

Equation (4) is the most general solution of Eq. (2), and we can proceed to find the solution to Eq. (3). By substituting Eq. (4) into Eq. (3), we obtain

$$-\frac{\partial}{\partial y}\left[-\frac{A}{3}x^3 + f(y)\right] = Ay$$

or

$$-\frac{\partial f(y)}{\partial y} = -\frac{df(y)}{dy}$$
$$= Ay.$$

This can be integrated to give

$$f(y) = -\frac{A}{2}y^2 + C,$$

where C is a constant. [Since $f(y)$ is a function of the single variable y, the constant of integration cannot involve x.]

The potential energy is

$$U = -\frac{A}{3}x^3 - \frac{A}{2}y^2 + C.$$

Suppose that we try to apply this method to a nonconservative force. For instance, consider

$$\mathbf{F} = A(xy\hat{\mathbf{i}} + y^2\hat{\mathbf{j}}).$$

The curl of **F** is not zero. Nevertheless, we can attempt to solve the equations

$$-\frac{\partial U}{\partial x} = F_x$$
$$= Axy \qquad 5$$
$$-\frac{\partial U}{\partial y} = F_y$$
$$= Ay^2. \qquad 6$$

The general solution of Eq. (5) is

$$U = -\frac{A}{2}x^2y + f(y).$$

If we substitute this into Eq. (6), we have

$$\frac{A}{2}x^2 - \frac{\partial f(y)}{\partial y} = Ay^2$$

or

$$\frac{\partial f(y)}{\partial y} = -\frac{A}{2}x^2 - Ay^2.$$

But $f(y)$ cannot depend on x, so that this equation has no solution. Hence, it is impossible to construct a potential energy function for this force.

In closing this section, let's take a brief look at the physical meaning of the curl.

Example 5.11 **How the Curl Got Its Name**

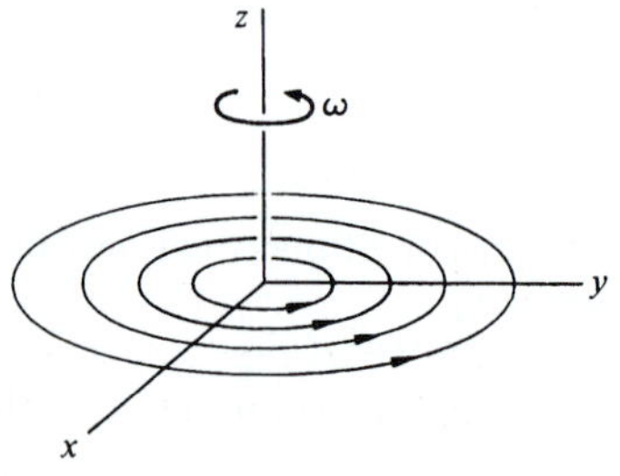

The curl was invented to help describe the properties of moving fluids. To see how the curl is connected with "curliness" or rotation, consider an idealized whirlpool turning with constant angular velocity ω about the z axis. The velocity of the fluid at **r** is

$$\mathbf{v} = r\omega\hat{\boldsymbol{\theta}},$$

where $\hat{\boldsymbol{\theta}}$ is the unit vector in the tangential direction. In cartesian coordinates,

$$\begin{aligned}\mathbf{v} &= r\omega(-\sin\omega t\,\hat{\mathbf{i}} + \cos\omega t\,\hat{\mathbf{j}})\\ &= r\omega\left(-\frac{y}{r}\hat{\mathbf{i}} + \frac{x}{r}\hat{\mathbf{j}}\right)\\ &= -\omega y\hat{\mathbf{i}} + \omega x\hat{\mathbf{j}}.\end{aligned}$$

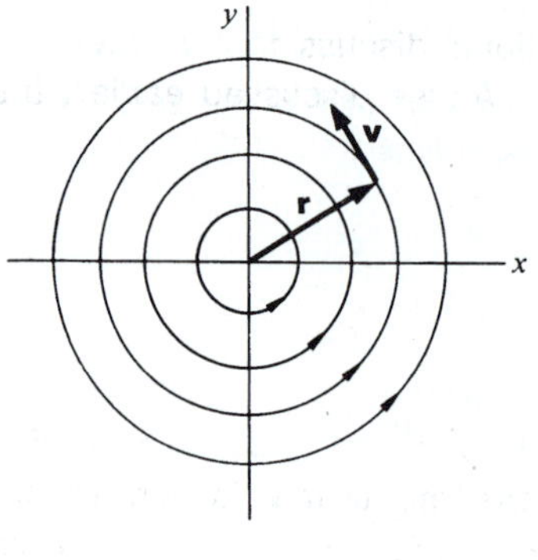

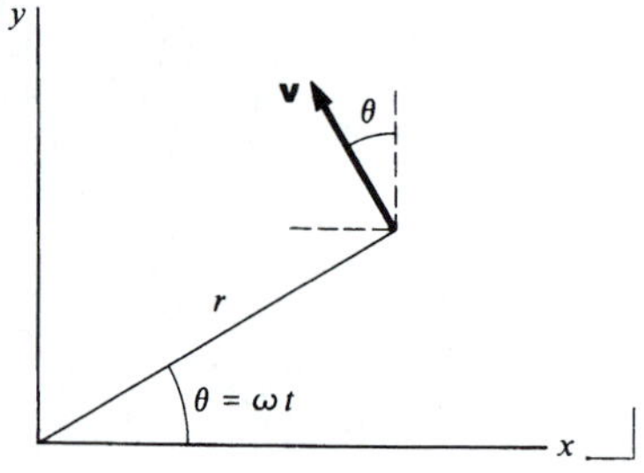

The curl of **v** is

$$\nabla \times \mathbf{v} = \begin{vmatrix} \hat{\mathbf{i}} & \hat{\mathbf{j}} & \hat{\mathbf{k}} \\ \dfrac{\partial}{\partial x} & \dfrac{\partial}{\partial y} & \dfrac{\partial}{\partial z} \\ -\omega y & \omega x & 0 \end{vmatrix}$$

$$= \hat{\mathbf{k}} \left[\frac{\partial}{\partial x}(\omega x) + \frac{\partial}{\partial y}(\omega y) \right]$$

$$= 2\omega \hat{\mathbf{k}}.$$

If a paddle wheel is placed in the liquid, it will start to rotate. The rotation will be a maximum when the axis of the wheel points along the z axis parallel to $\nabla \times \mathbf{v}$. In Europe, curl is often called "rot" (for rotation). A vector field with zero curl gives no impression of rotation, as the sketches illustrate.

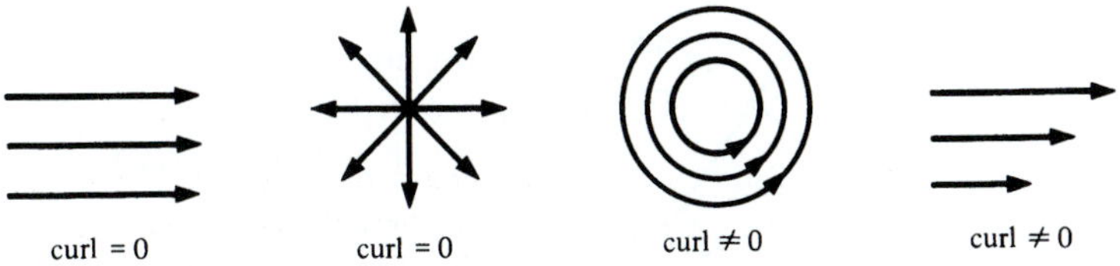

5.7 Stokes' Theorem

In Sec. 5.6 we stopped short of proving a remarkable result, known as Stokes' theorem, which relates the line integral of a vector field around a closed path to an integral over an area bounded by the path. Although Stokes' theorem is indispensible to the study of electricity and magnetism, we shall have little further use for it in our study of mechanics. Nevertheless, we have already developed most of the ideas involved in its proof, and only a brief additional discussion is needed.

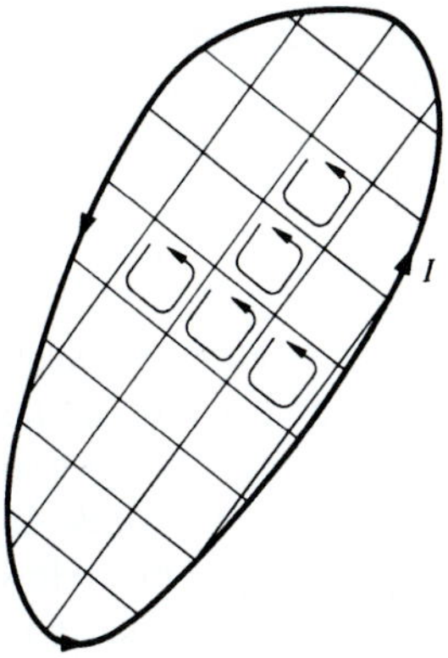

As we discussed earlier, the line integral of **F** around a closed path I can be written as the sum of the line integrals around each tiny loop.

$$\oint_{\mathrm{I}} \mathbf{F} \cdot d\mathbf{r} = \sum_i \oint_i \mathbf{F} \cdot d\mathbf{r}$$

This result holds whether **F** is conservative or not; we shall not assume that **F** is conservative in this proof. Stokes' theorem contains no physics—it is a purely mathematical result.

Our starting point is Eq. (5.17). For a tiny rectangular loop in the xy plane,

$$\oint_i \mathbf{F}\cdot d\mathbf{r} = \left(\frac{\partial F_y}{\partial x} - \frac{\partial F_x}{\partial y}\right)_i (\Delta x\,\Delta y)_i.$$

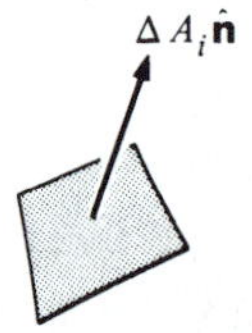

As we have pointed out, the result is independent of the shape of the loop provided that we replace $(\Delta x\,\Delta y)_i$ by the loop's area ΔA_i. We can write the area element as a vector $\Delta\mathbf{A}_i = \Delta A_i\hat{\mathbf{n}}$, where $\hat{\mathbf{n}}$ is normal to the plane of the loop. (Example 1.4 discusses the use of vectors to represent areas.) For a loop in the xy plane. $\Delta\mathbf{A} = \Delta A_z\hat{\mathbf{k}}$ and we have

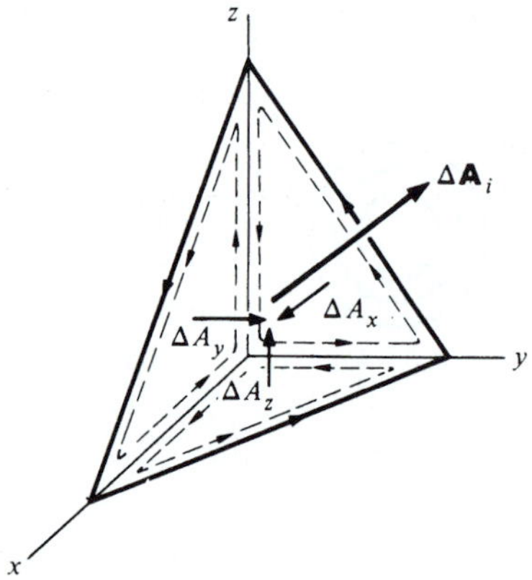

$$\begin{aligned}\oint_i \mathbf{F}\cdot d\mathbf{r} &= \left(\frac{\partial F_y}{\partial x} - \frac{\partial F_x}{\partial y}\right)_i (\Delta A_z)_i \\ &= [(\nabla\times\mathbf{F})_z\,\Delta A_z]_i. \end{aligned} \qquad 5.20$$

If the tiny loop is at an arbitrary orientation, it is plausible that

$$\begin{aligned}\oint_i \mathbf{F}\cdot d\mathbf{r} &= [(\text{curl }\mathbf{F})_x\,\Delta A_x + (\text{curl }\mathbf{F})_y\,\Delta A_y + (\text{curl }\mathbf{F})_z\,\Delta A_z]_i \\ &= [\text{curl }\mathbf{F}\cdot\Delta\mathbf{A}]_i.\end{aligned}$$

The line integral of $\mathbf{F}$ around path I is therefore

$$\begin{aligned}\oint \mathbf{F}\cdot d\mathbf{r} &= \sum_i \oint_i \mathbf{F}\cdot d\mathbf{r} \\ &= \sum_i (\text{curl }\mathbf{F}\cdot\Delta\mathbf{A})_i.\end{aligned} \qquad 5.21$$

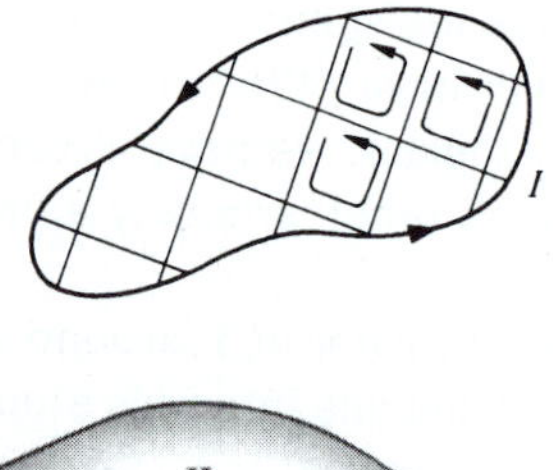

In words, the line integral is equal to the result of taking the scalar product of each vector area element with the curl of $\mathbf{F}$ at that element and summing over all elements bounded by the curve. In the limit $\Delta\mathbf{A}_i \to 0$, the number of area elements approaches infinity and the sum in Eq. (5.21) becomes an integral. We then have Stokes' theorem

$$\oint \mathbf{F}\cdot d\mathbf{r} = \int \text{curl }\mathbf{F}\cdot d\mathbf{A}. \qquad 5.22$$

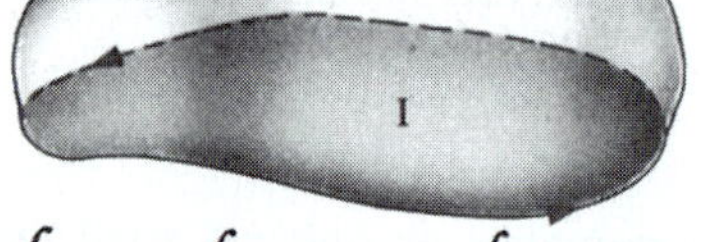

$$\oint \mathbf{F}\cdot d\mathbf{r} = \int_{\substack{\text{area}\\ \text{I}}} \text{curl }\mathbf{F}\cdot d\mathbf{A} = \int_{\substack{\text{area}\\ \text{II}}} \text{curl }\mathbf{F}\cdot d\mathbf{A}$$

Two important remarks should be made about Stokes' theorem, Eq. (5.22). First, the area of integration on the right hand side can be *any* area bounded by the closed path. Second, there is an apparent ambiguity to the direction of $d\mathbf{A}$, since the normal can be out from either side of the area element. However, Eq. (5.17) was deduced using a counterclockwise circulation about the loop, and in defining the vector associated with the area element, we automatically set up the convention that the direction of $d\mathbf{A}$ is given by the right hand rule. If the circulation is counterclockwise as seen from above, the correct direction of $d\mathbf{A}$ is the one that tends to point "up."

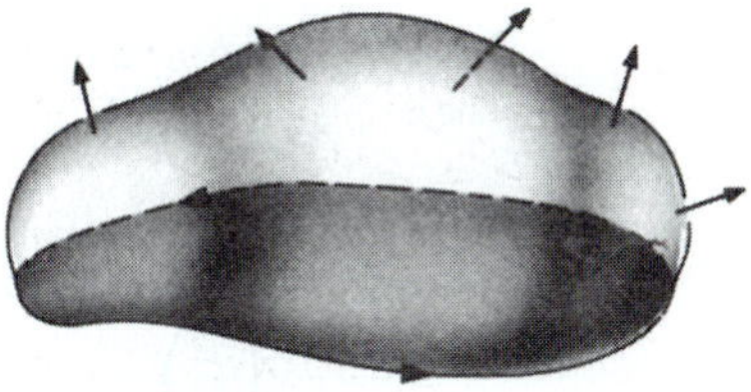

Example 5.12 **Using Stokes' Theorem**

In Example 5.8 we discussed a barge being towed against the current. We found the work done in going around the path in the sketch by evaluating the line integral $\oint \mathbf{F} \cdot d\mathbf{r} = W$. In this example we shall find the work by using Stokes' theorem

$W = \int (\nabla \times \mathbf{F}) \cdot d\mathbf{A}.$

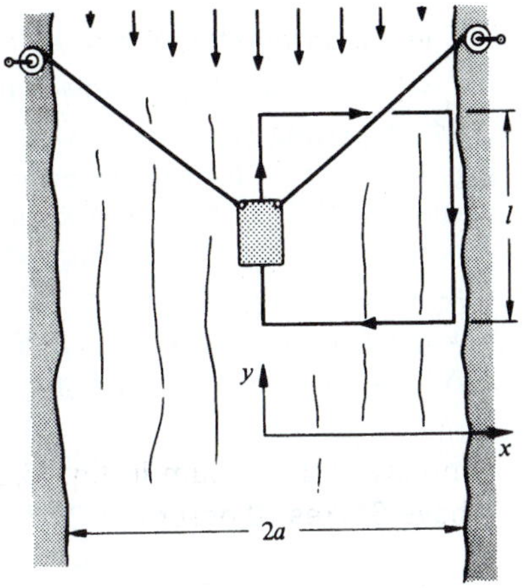

It is natural to integrate over the surface in the xy plane, as shown in the drawing above right. Since the direction of circulation is clockwise, $d\mathbf{A} = -dA\,\hat{\mathbf{k}}$, and we have $W = -\int (\nabla \times \mathbf{F})_z\, dA$.
From Example 5.8, the force is

$$\mathbf{F} = bV_0\left(1 - \frac{x^2}{a^2}\right)\hat{\mathbf{j}}$$

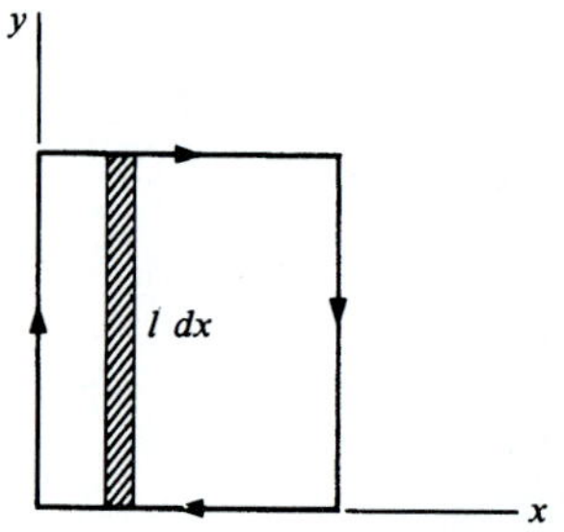

and

$$(\nabla \times \mathbf{F})_z = \frac{\partial F_y}{\partial x} - \frac{\partial F_x}{\partial y}$$
$$= -\frac{2bV_0x}{a^2}.$$

Since the integrand does not involve y, it is convenient to take $dA = l\,dx$. Then

$$W = \int_0^a \frac{2bV_0l}{a^2}\,x\,dx$$
$$= \frac{2bV_0l}{a^2}\left(\frac{a^2}{2}\right)$$
$$= bV_0l,$$

as we found previously by evaluating the line integral.

Problems 5.1 Find the forces for the following potential energies.

a. $U = Ax^2 + By^2 + Cz^2$

b. $U = A\ \ln(x^2 + y^2 + z^2)$ $(\ln = \log_e)(x^2 + y^2 + z^2)$

c. $U = A\cos\theta/r^2$ (plane polar coordinates)

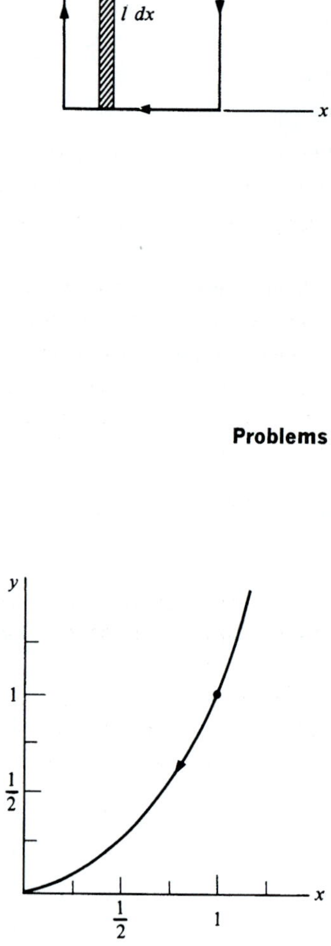

5.2 A particle of mass m moves in a horizontal plane along the parabola $y = x^2$. At $t = 0$ it is at the point (1,1) moving in the direction shown with speed v_0. Aside from the force of constraint holding it to the path, it is acted upon by the following external forces:

A radial force $\mathbf{F}_a = -Ar^3\hat{\mathbf{r}}$
A force given by $\mathbf{F}_b = B(y^2\hat{\mathbf{i}} - x^2\hat{\mathbf{j}})$

where A and B are constants.

a. Are the forces conservative?

b. What is the speed v_f of the particle when it arrives at the origin?

Ans. $v_f = (v_0^2 + A/2m + 3B/5m)^{\frac{1}{2}}$

5.3 Decide whether the following forces are conservative.

a. $\mathbf{F} = \mathbf{F}_0 \sin at$, where $\mathbf{F}_0$ is a constant vector.

b. $F = A\theta\hat{\mathbf{r}}$, A = constant and $0 \leq \theta < 2\pi$. ($\mathbf{F}$ is limited to the xy plane.)

c. A force which depends on the velocity of a particle but which is always perpendicular to the velocity.

5.4 Determine whether each of the following forces is conservative. Find the potential energy function if it exists. A, α, β are constants.

a. $\mathbf{F} = A(3\hat{\mathbf{i}} + z\hat{\mathbf{j}} + y\hat{\mathbf{k}})$

b. $\mathbf{F} = Axyz(\hat{\mathbf{i}} + \hat{\mathbf{j}} + \hat{\mathbf{k}})$

c. $F_x = 3Ax^2y^5e^{\alpha z}$, $F_y = 5Ax^3y^4e^{\alpha z}$, $F_z = \alpha Ax^3y^5e^{\alpha z}$

d. $F_x = A \sin(\alpha y) \cos(\beta z)$, $F_y = -Ax\alpha \cos(\alpha y) \cos(\beta z)$, and $F_z = Ax \sin(\alpha y) \sin(\beta z)$

5.5 The potential energy function for a particular two dimensional force field is given by $U = Cxe^{-y}$, where C is a constant.

a. Sketch the constant energy lines.

b. Show that if a point is displaced by a short distance dx along a constant energy line, then its total displacement must be $d\mathbf{r} = dx(\hat{\mathbf{i}} + \hat{\mathbf{j}}/x)$.

c. Using the result of b, show explicitly that ∇U is perpendicular to the constant energy line.

5.6 If $\mathbf{A}(\mathbf{r})$ is a vector function of $\mathbf{r}$ which everywhere satisfies $\nabla \times \mathbf{A} = 0$, show that $\mathbf{A}$ can be expressed by $\mathbf{A}(\mathbf{r}) = \nabla \phi(\mathbf{r})$, where $\phi(\mathbf{r})$ is some scalar function. (*Hint*: The result follows directly from physical arguments.)

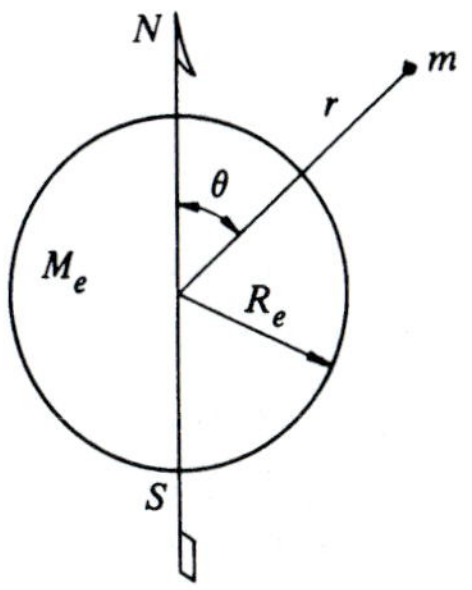

5.7 When the flattening of the earth at the poles is taken into account, it is found that the gravitational potential energy of a mass m a distance r from the center of the earth is approximately

$$U = -\frac{GM_em}{r}\left[1 - 5.4 \times 10^{-4}\left(\frac{R_e}{r}\right)^2 (3\cos^2\theta - 1)\right],$$

where θ is measured from the pole.

Show that there is a small tangential gravitational force on m except above the poles or the equator. Find the ratio of this force to GM_em/r^2 for $\theta = 45°$ and $r = R_e$.

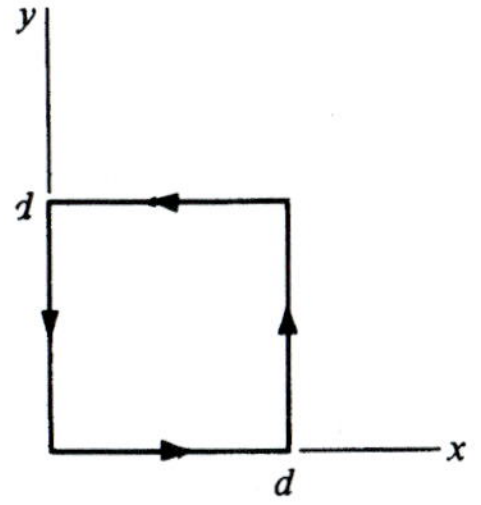

5.8 How much work is done around the path that is shown by the force $\mathbf{F} = A(y^2\hat{\mathbf{i}} + 2x^2\hat{\mathbf{j}})$, where A is a constant and x and y are in meters? Find the answer by evaluating the line integral, and also by using Stokes' theorem.

Ans. $W = Ad^3$

6 ANGULAR MOMENTUM AND FIXED AXIS ROTATION

6.1 Introduction

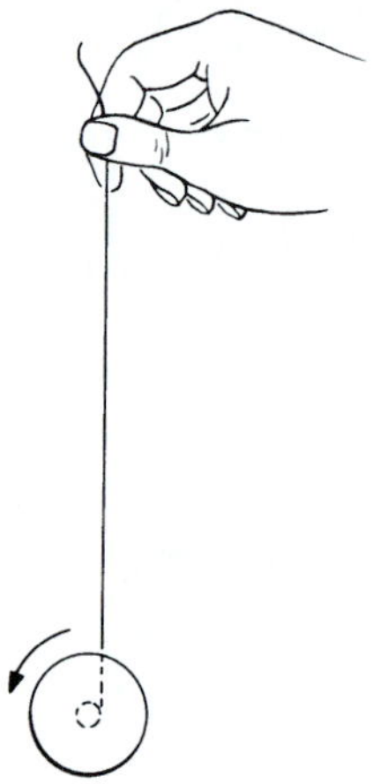

Our development of the principles of mechanics in the past five chapters is lacking in one important respect: we have not developed techniques to handle the rotational motion of solid bodies. For example, consider the common Yo-Yo running up and down its string as the spool winds and unwinds. In principle we already know how to analyze the motion: each particle of the Yo-Yo moves according to Newton's laws. Unfortunately, analyzing rotational problems on a particle-by-particle basis is an impossible task. What we need is a simple method for treating the rotational motion of an extended body as a whole. The goal of this chapter is to develop such a method. In attacking the problem of translational motion, we needed the concepts of force, linear momentum, and center of mass; in this chapter we shall develop for rotational motion the analogous concepts of torque, angular momentum, and moment of inertia.

Our aim, of course, is more ambitious than merely to understand Yo-Yos; our aim is to find a way of analyzing the general motion of a rigid body under any combination of applied forces. Fortunately this problem can be divided into two simpler problems —finding the center of mass motion, a problem we have already solved, and finding the rotational motion about the center of mass, the task at hand. The justification for this is a theorem of rigid body motion which asserts that any displacement of a rigid body can be decomposed into two independent motions: a translation of the center of mass and a rotation about the center

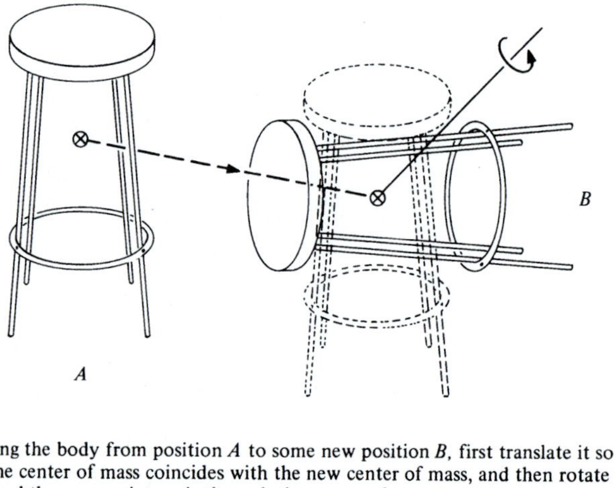

To bring the body from position A to some new position B, first translate it so that the center of mass coincides with the new center of mass, and then rotate it around the appropriate axis through the center of mass until the body is in the desired position.

of mass. A few minutes spent playing with a rigid body such as a book or a chair should convince you that the theorem is plausible. Note that the theorem does not say that this is the *only* way to represent a general displacement—merely that it is one possible way of doing so. The general proof of this theorem[1] is presented in Note 6.1 at the end of the chapter. However, detailed attention to a formal proof is not necessary at this point. What is important is being able to visualize any displacement as the combination of a single translation and a single rotation.

Leaving aside extended bodies for a time, we start in the best tradition of physics by considering the simplest possible system—a particle. Since a particle has no size, its orientation in space is of no consequence, and we need concern ourselves only with translational motion. In spite of this, particle motion is useful for introducing the concepts of angular momentum and torque. We shall then move to progressively more complex systems, culminating, in Chap. 7, with a treatment of the general motion of a rigid body.

6.2 Angular Momentum of a Particle

Here is the formal definition of the *angular momentum* **L** of a particle which has momentum **p** and position vector **r** with respect to a given coordinate system.

$$\mathbf{L} = \mathbf{r} \times \mathbf{p} \qquad 6.1$$

The unit of angular momentum is $kg{\cdot}m^2/s$ in the SI system or $g{\cdot}cm^2/s$ in cgs. There are no special names for these units.

Angular momentum is our first physical quantity to involve the cross product. (See Secs. 1.2 and 1.4 if you need to review the cross product.) Because angular momentum is so different from anything we have yet encountered, we shall discuss it in great detail at first.

Possibly the strangest aspect of angular momentum is its direction. The vectors **r** and **p** determine a plane (sometimes known as the plane of motion), and by the properties of the cross product, **L** is perpendicular to this plane. There is nothing particularly "natural" about the definition of angular momentum. However, **L** obeys a very simple dynamical equation, as we shall see, and therein lies its usefulness.

[1] Euler proved that the general displacement of a rigid body with one point fixed is a rotation about some axis; the theorem quoted in the text, called Chasle's theorem, follows directly from this.

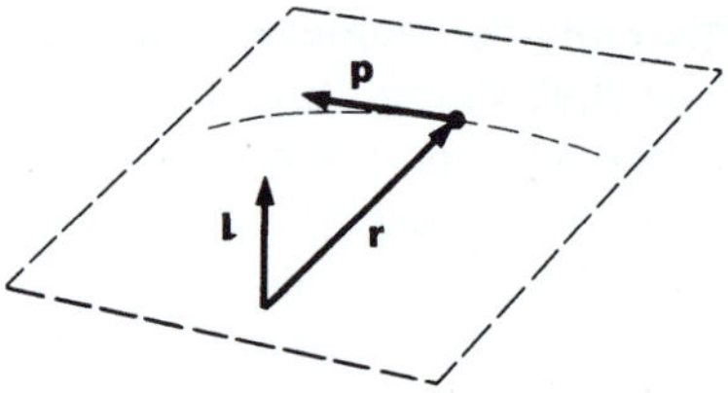

The diagram at left shows the trajectory and instantaneous position and momentum of a particle. $\mathbf{L} = \mathbf{r} \times \mathbf{p}$ is perpendicular to the plane of **r** and **p**, and points in the direction dictated by the right hand rule for vector multiplication. Although **L** has been drawn through the origin, this location has no significance. Only the direction and magnitude of **L** are important.

If **r** and **p** lie in the xy plane, then **L** is in the z direction. **L** is in the positive z direction if the "sense of rotation" of the point about the origin is counterclockwise, and in the negative z direction if the sense of rotation is clockwise. Note that the sense of rotation is well defined even if the trajectory is a straight line. The only exception is when the trajectory aims at the origin, in which case **r** and **p** are along the same line so that **L** is 0 anyway.

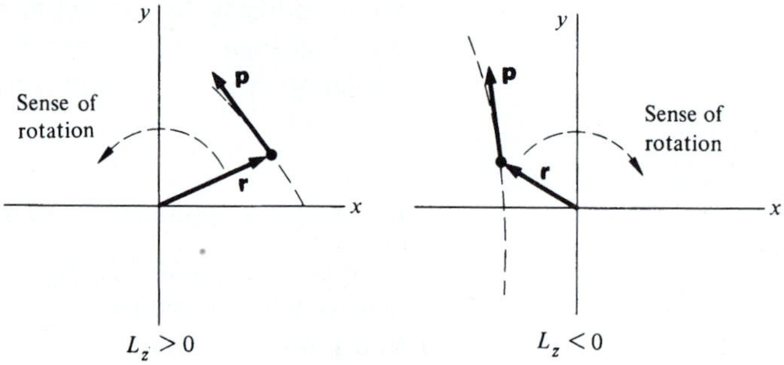

There are various methods for visualizing and calculating angular momentum. Here are three ways to calculate the angular momentum of a particle moving in the xy plane.

Method 1

$$\mathbf{L} = \mathbf{r} \times \mathbf{p}$$
$$= rp \sin \phi \hat{\mathbf{k}}$$

or

$$L_z = rp \sin \phi.$$

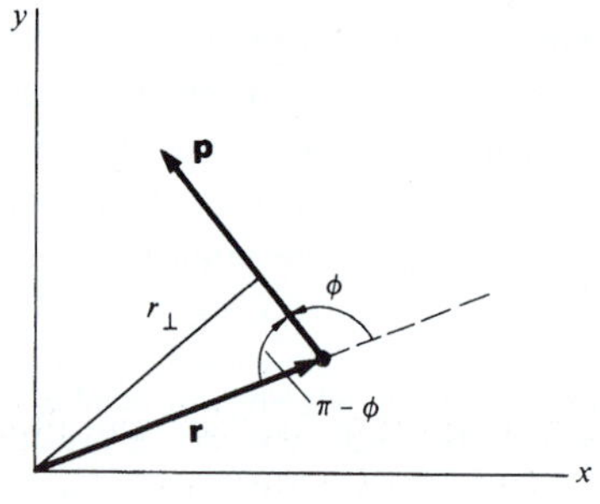

For motion in the xy plane, **L** lies in the z direction. Its magnitude has a simple geometrical interpretation: the line $r_\perp$ has length $r_\perp = r \sin (\pi - \phi) = r \sin \phi$. Therefore,

$$L_z = r_\perp p,$$

where $r_\perp$ is the perpendicular distance between the origin and the line of **p**. This result illustrates that angular momentum is proportional to the distance from the origin to the line of motion.

As the sketches show, an alternative way of writing L_z is

$$L_z = rp_\perp,$$

where $p_\perp$ is the component of **p** perpendicular to **r**.

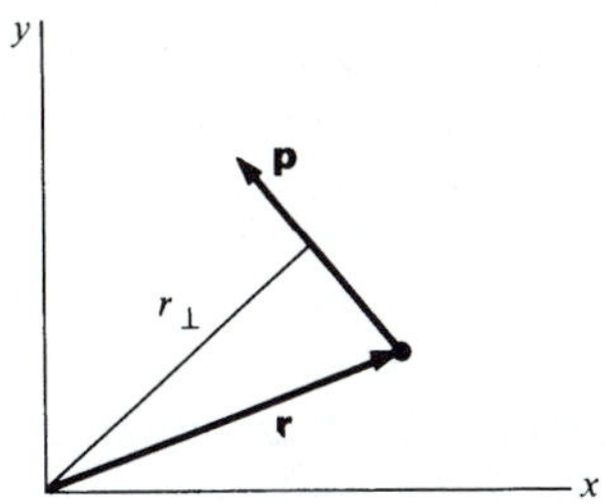

$L_z = r_\perp p$

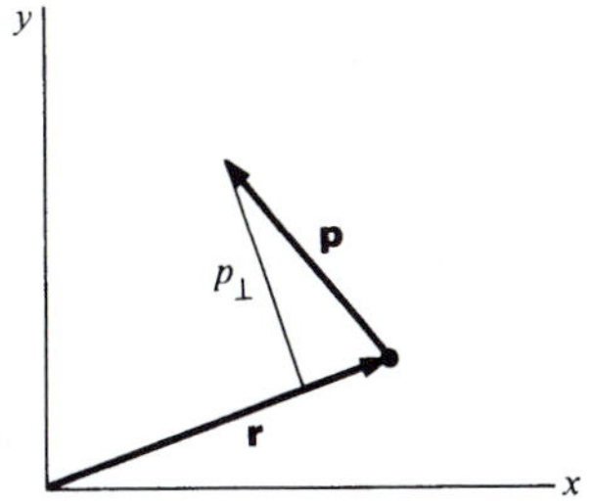

$L_z = rp_\perp$

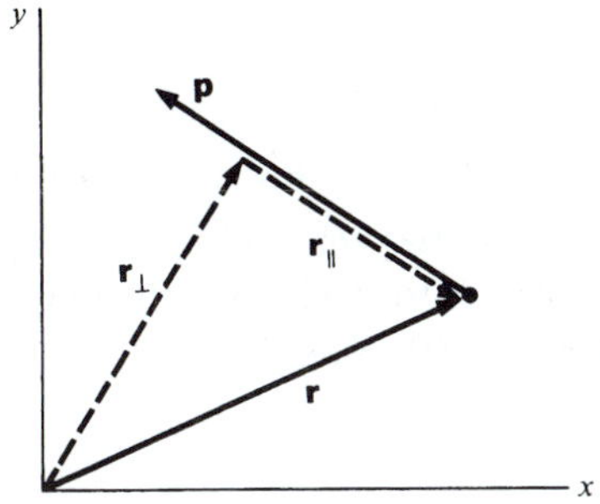

Method 2

Resolve **r** into two vectors $\mathbf{r}_\perp$ and $\mathbf{r}_\parallel$,

$$\mathbf{r} = \mathbf{r}_\perp + \mathbf{r}_\parallel,$$

such that $\mathbf{r}_\perp$ is perpendicular to **p**, and $\mathbf{r}_\parallel$ is parallel to **p**. Then

$$\begin{aligned}\mathbf{L} = \mathbf{r} \times \mathbf{p} &= (\mathbf{r}_\perp + \mathbf{r}_\parallel) \times \mathbf{p} \\ &= (\mathbf{r}_\perp \times \mathbf{p}) + (\mathbf{r}_\parallel \times \mathbf{p}) \\ &= \mathbf{r}_\perp \times \mathbf{p},\end{aligned}$$

since $\mathbf{r}_\parallel \times \mathbf{p} = 0$. (Parallel vectors have zero cross product.) Evaluating the cross product $\mathbf{r}_\perp \times \mathbf{p}$ is trivial because the vectors are perpendicular by construction. We have

$$L_z = |\mathbf{r}_\perp|\,|\mathbf{p}|$$

as before. By a similar argument,

$$L_z = |\mathbf{r}|\,|\mathbf{p}_\perp|.$$

Method 3

Consider motion in the xy plane, first in the x direction and then in the y direction, as in drawings a and b on the next page.

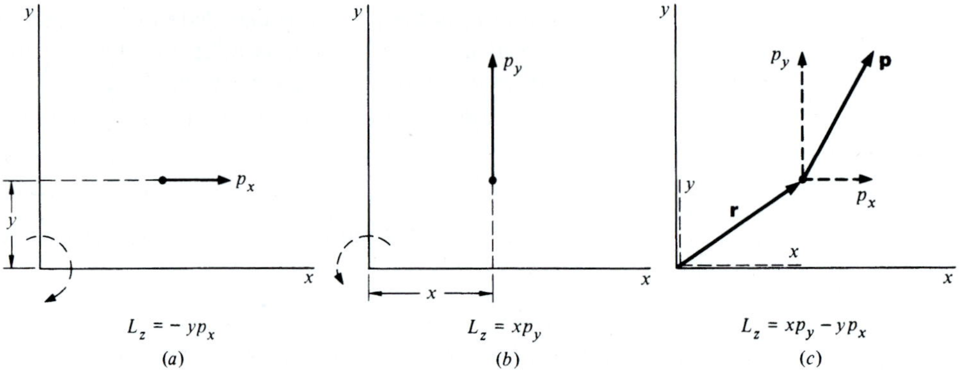

The most general case involves both these motions simultaneously, as drawings above show.

Hence $L_z = xp_y - yp_x$, as you can verify by inspection or by evaluating the cross product as follows. Using $\mathbf{r} = (x,y,0)$ and $\mathbf{p} = (p_x,p_y,0)$, we have

$$\begin{aligned}\mathbf{L} &= \mathbf{r} \times \mathbf{p} \\ &= \begin{vmatrix} \hat{\mathbf{i}} & \hat{\mathbf{j}} & \hat{\mathbf{k}} \\ x & y & 0 \\ p_x & p_y & 0 \end{vmatrix} \\ &= (xp_y - yp_x)\hat{\mathbf{k}}.\end{aligned}$$

We have limited our illustrations to motion in the xy plane where the angular momentum lies entirely along the z axis. There is, however, no difficulty applying any of these methods to the general case where $\mathbf{L}$ has components along all three axes.

Example 6.1 Angular Momentum of a Sliding Block

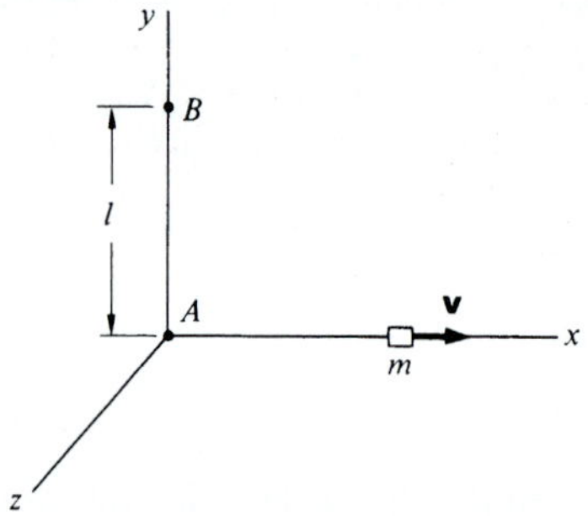

Consider a block of mass m and negligible dimensions sliding freely in the x direction with velocity $\mathbf{v} = v\hat{\mathbf{i}}$, as shown in the sketch. What is its angular momentum $\mathbf{L}_A$ about origin A and its angular momentum $\mathbf{L}_B$ about the origin B?

As shown in the drawing on the top of page 237, the vector from origin A to the block is

$$\mathbf{r}_A = x\hat{\mathbf{i}}.$$

Since $\mathbf{r}_A$ is parallel to $\mathbf{v}$, their cross product is zero and

$$\begin{aligned}\mathbf{L}_A &= m\mathbf{r}_A \times \mathbf{v} \\ &= 0.\end{aligned}$$

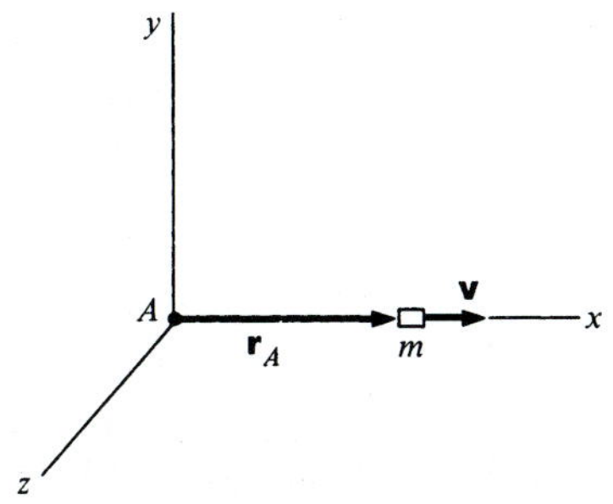

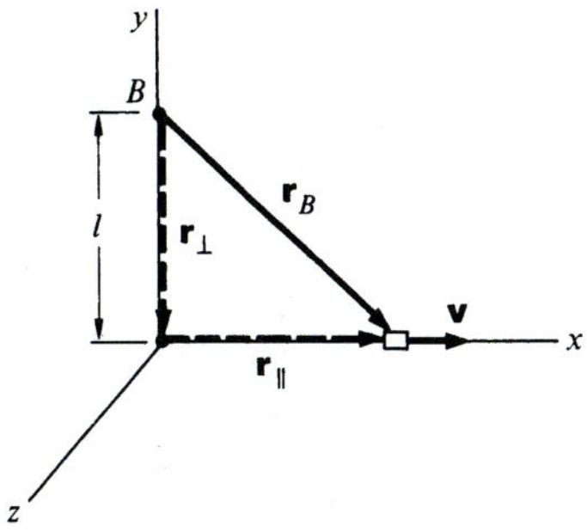

Taking origin B, we can resolve the position vector $\mathbf{r}_B$ into a component $\mathbf{r}_\parallel$ parallel to $\mathbf{v}$ and a component $\mathbf{r}_\perp$ perpendicular to $\mathbf{v}$. Since $\mathbf{r}_\parallel \times \mathbf{v} = 0$, only $\mathbf{r}_\perp$ gives a contribution to $\mathbf{L}_B$. We have $|\mathbf{r}_\perp \times \mathbf{v}| = lv$ and

$$\begin{aligned}\mathbf{L}_B &= m\mathbf{r}_B \times \mathbf{v} \\ &= mlv\hat{\mathbf{k}}.\end{aligned}$$

$\mathbf{L}_B$ lies in the positive z direction because the sense of rotation is counterclockwise about the z axis.

To calculate $\mathbf{L}_B$ formally we can write $\mathbf{r}_B = x\hat{\mathbf{i}} - l\hat{\mathbf{j}}$ and evaluate $\mathbf{r}_B \times \mathbf{v}$ using our determinantal form.

$$\begin{aligned}\mathbf{L}_B &= m\mathbf{r}_B \times \mathbf{v} \\ &= m\begin{vmatrix}\hat{\mathbf{i}} & \hat{\mathbf{j}} & \hat{\mathbf{k}} \\ x & -l & 0 \\ v & 0 & 0\end{vmatrix} \\ &= mlv\hat{\mathbf{k}}\end{aligned}$$

as before.

The following example shows in a striking way how $\mathbf{L}$ depends on our choice of origin.

Example 6.2 **Angular Momentum of the Conical Pendulum**

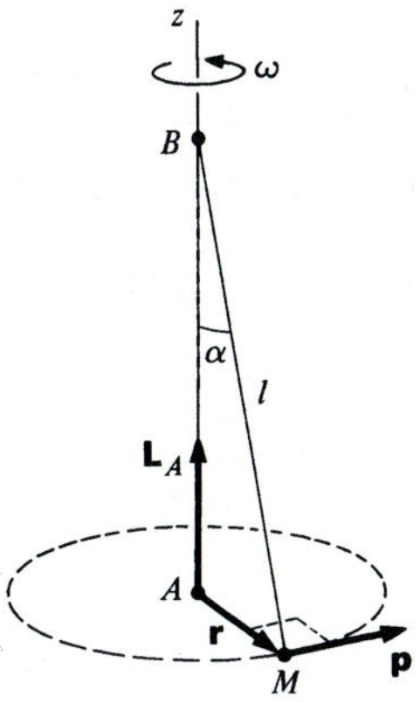

Let us return to the conical pendulum, which we encountered in Example 2.8, to illustrate some features of angular momentum. Assume that the pendulum is in steady circular motion with constant angular velocity ω.

We begin by evaluating $\mathbf{L}_A$, the angular momentum about origin A. From the sketch we see that $\mathbf{L}_A$ lies in the positive z direction. It has magnitude $|\mathbf{r}_\perp|\,|\mathbf{p}| = |\mathbf{r}|\,|\mathbf{p}| = rp$, where r is the radius of the circular motion. Since

$$\begin{aligned}|\mathbf{p}| &= Mv \\ &= Mr\omega,\end{aligned}$$

we have

$$\mathbf{L}_A = Mr^2\omega\hat{\mathbf{k}}.$$

Note that $\mathbf{L}_A$ is constant, both in magnitude and direction.

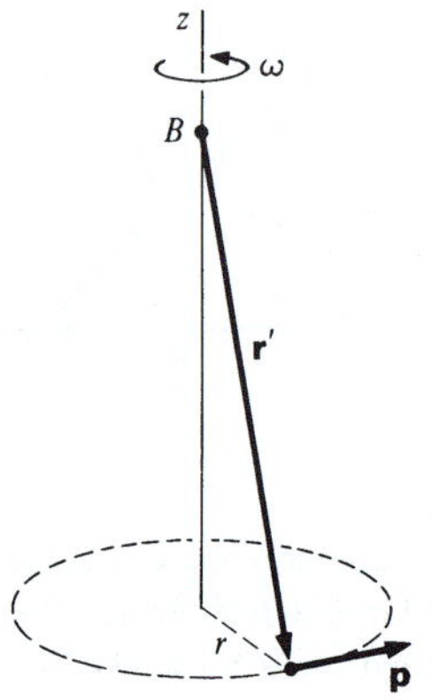

Now let us evaluate the angular momentum about the origin B located at the pivot. The magnitude of $\mathbf{L}_B$ is

$$
\begin{aligned}
|\mathbf{L}_B| &= |\mathbf{r}' \times \mathbf{p}| \\
&= |\mathbf{r}'|\,|\mathbf{p}| = l|\mathbf{p}| \\
&= Mlr\omega,
\end{aligned}
$$

where $|\mathbf{r}'| = l$, the length of the string. It is apparent that the magnitude of $\mathbf{L}$ depends on the origin we choose.

Unlike $\mathbf{L}_A$, the direction of $\mathbf{L}_B$ is not constant. $\mathbf{L}_B$ is perpendicular to both $\mathbf{r}'$ and $\mathbf{p}$, and the sketches below show $\mathbf{L}_B$ at different times. Two sketches are given to emphasize that only the magnitude and direction of $\mathbf{L}$ are important, not the position at which we choose to draw it. The magnitude of $\mathbf{L}_B$ is constant, but its *direction* is obviously not constant; as the bob swings around, $\mathbf{L}_B$ sweeps out the shaded cone shown in the sketch at the right. The z component of $\mathbf{L}_B$ is constant, but the horizontal component travels around the circle with the bob. We shall see the dynamical consequences of this in Example 6.6.

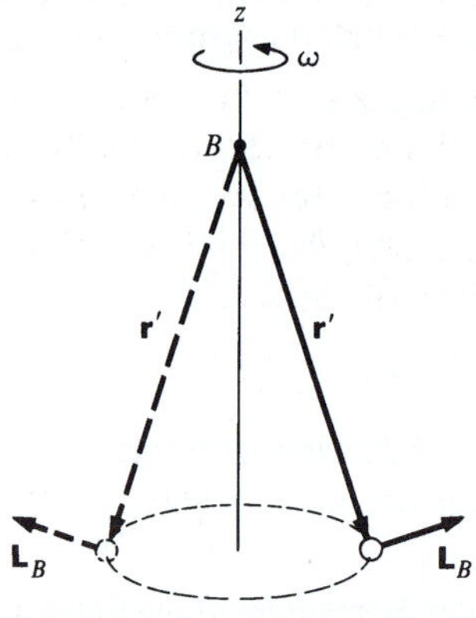

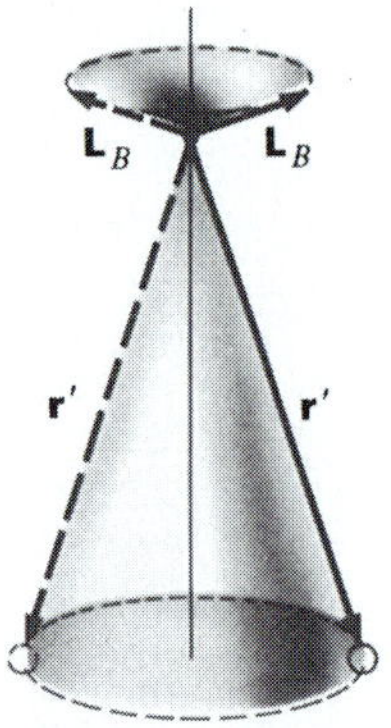

6.3 Torque

To continue our development of rotational motion we must introduce a new quantity *torque* $\boldsymbol{\tau}$. The torque due to force $\mathbf{F}$ which acts on a particle at position $\mathbf{r}$ is defined by

$$\boldsymbol{\tau} = \mathbf{r} \times \mathbf{F}. \qquad 6.2$$

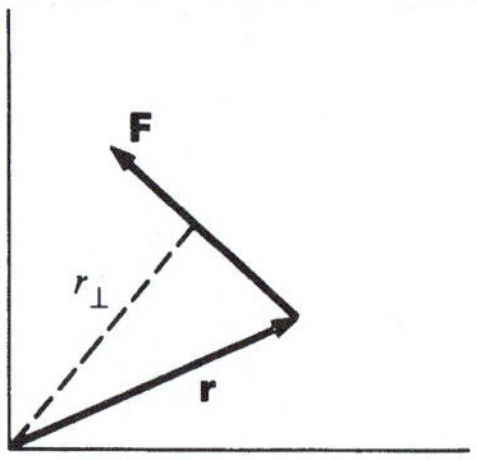

In the last section we discussed several ways of evaluating angular momentum, $\mathbf{r} \times \mathbf{p}$. The mathematical methods we developed for calculating the cross product can also be applied to torque $\mathbf{r} \times \mathbf{F}$. For example, we have

$$|\boldsymbol{\tau}| = |\mathbf{r}_\perp|\,|\mathbf{F}|$$

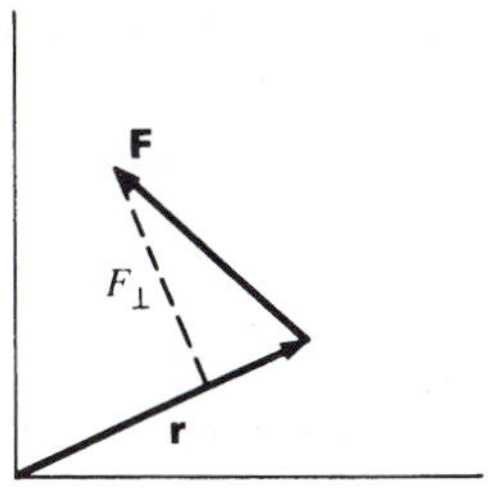

or

$$|\boldsymbol{\tau}| = |\mathbf{r}|\,|\mathbf{F}_\perp|$$

or, formally,

$$\boldsymbol{\tau} = \begin{vmatrix} \hat{\mathbf{i}} & \hat{\mathbf{j}} & \hat{\mathbf{k}} \\ x & y & z \\ F_x & F_y & F_z \end{vmatrix}.$$

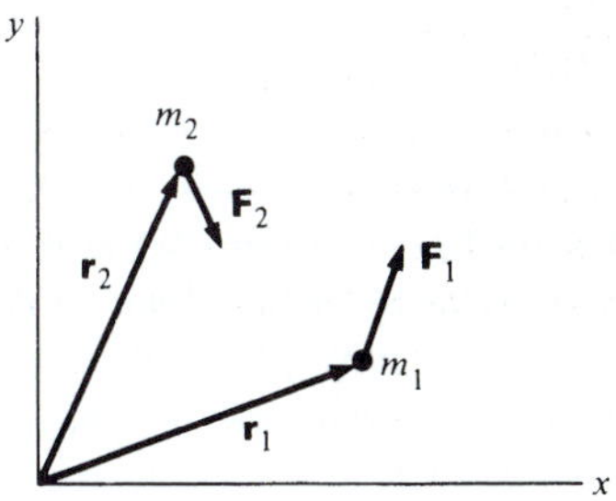

We can also associate a "sense of rotation" using **r** and **F**. Assume in the sketch that all the vectors are in the xy plane. The torque on m_1 due to $\mathbf{F}_1$ is along the positive z axis (out of the paper) and the torque on m_2 due to $\mathbf{F}_2$ is along the negative z axis (into the paper).

It is important to realize that torque and force are entirely different quantities. For one thing, torque depends on the origin we choose but force does not. For another, we see from the definition $\boldsymbol{\tau} = \mathbf{r} \times \mathbf{F}$ that $\boldsymbol{\tau}$ and **F** are always mutually perpendicular. There can be a torque on a system with zero net force, and there can be force with zero net torque. In general, there will be both torque and force. These three cases are illustrated in the sketches below. (The torques are evaluated about the centers of the disks.)

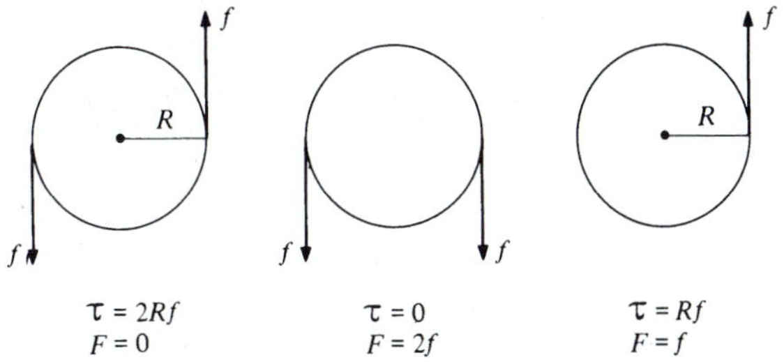

Torque is important because it is intimately related to the rate of change of angular momentum:

$$\frac{d\mathbf{L}}{dt} = \frac{d}{dt}(\mathbf{r} \times \mathbf{p})$$
$$= \left(\frac{d\mathbf{r}}{dt} \times \mathbf{p}\right) + \left(\mathbf{r} \times \frac{d\mathbf{p}}{dt}\right).$$

But $(d\mathbf{r}/dt) \times \mathbf{p} = \mathbf{v} \times m\mathbf{v} = 0$, since the cross product of two parallel vectors is zero. Also, $d\mathbf{p}/dt = \mathbf{F}$, by Newton's second law. Hence, the second term is $\mathbf{r} \times \mathbf{F} = \boldsymbol{\tau}$, and we have

$$\boldsymbol{\tau} = \frac{d\mathbf{L}}{dt}. \qquad 6.3$$

Equation (6.3) shows that if the torque is zero, $\mathbf{L}$ = constant and the angular momentum is conserved. As you may already realize from our work with linear momentum and energy, conservation laws are powerful tools. However, because we have considered only the angular momentum of a single particle, the conservation law for angular momentum has not been presented in much generality. In fact, Eq. (6.3) follows directly from Newton's second law—only when we talk about extended systems does angular momentum assume its proper role as a new physical concept. Nevertheless, even in its present context, considerations of angular momentum lead to some surprising simplifications, as the next two examples show.

Example 6.3 Central Force Motion and the Law of Equal Areas

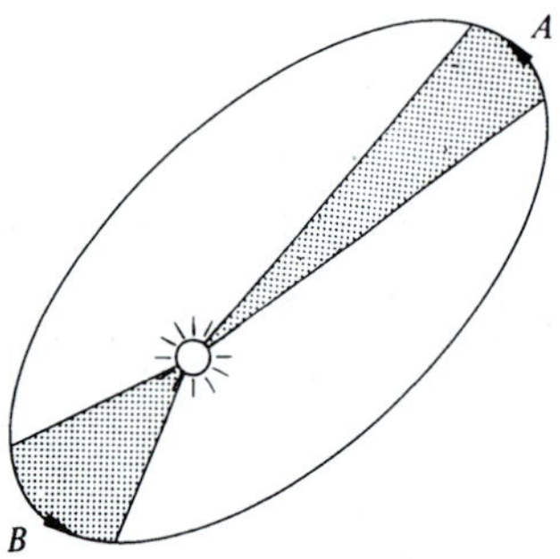

In 1609 Kepler announced his second law of planetary motion, the law of equal areas: that is, the area swept out by the radius vector from the sun to a planet in a given time is the same for any location of the planet in its orbit. The sketch (not to scale) shows the areas swept out by the earth during a month at two different seasons. The shorter radius vector at B is compensated by the greater speed of the earth when it is nearer the sun. We shall now show that the law of equal areas follows directly from considerations of angular momentum, and that it holds not only for motion under the gravitational force but also for motion under any central force.

Consider a particle moving under a central force, $\mathbf{F}(\mathbf{r}) = f(r)\hat{\mathbf{r}}$, where $f(r)$ has any dependence on r we care to choose. The torque on the particle about the origin is $\boldsymbol{\tau} = \mathbf{r} \times \mathbf{F}(\mathbf{r}) = \mathbf{r} \times f(r)\hat{\mathbf{r}} = 0$. Hence, the angular momentum of the particle $\mathbf{L} = \mathbf{r} \times \mathbf{p}$ is constant both in magnitude and direction. An immediate consequence is that the motion is confined to a plane; otherwise the direction of $\mathbf{L}$ would change with time. We shall now prove that the rate at which area is swept out is constant, a result that leads directly to the law of equal areas.

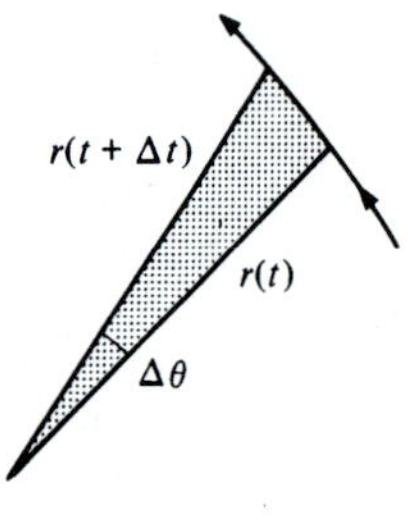

Consider the position of the particle at t and $t + \Delta t$, when its polar coordinates are, respectively, (r,θ) and $(r + \Delta r,\ \theta + \Delta\theta)$. The area swept out is shown shaded in the drawing at left.

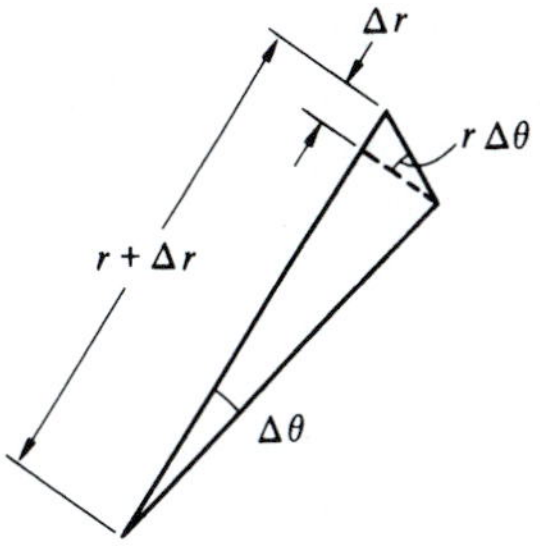

For small values of $\Delta\theta$, the area ΔA is approximately equal to the area of a triangle with base $r + \Delta r$ and altitude $r\,\Delta\theta$, as shown.

$$\Delta A \approx \tfrac{1}{2}(r + \Delta r)(r\,\Delta\theta)$$
$$= \tfrac{1}{2}r^2\,\Delta\theta + \tfrac{1}{2}r\,\Delta r\,\Delta\theta$$

The rate at which area is swept out is

$$\frac{dA}{dt} = \lim_{\Delta t\to 0}\frac{\Delta A}{\Delta t}$$
$$= \lim_{\Delta t\to 0}\frac{1}{2}\left(r^2\frac{\Delta\theta}{\Delta t} + r\frac{\Delta\theta\,\Delta r}{\Delta t}\right)$$
$$= \frac{1}{2}r^2\frac{d\theta}{dt}.$$

(The small triangle with sides $r\,\Delta\theta$ and Δr makes no contribution in the limit.)

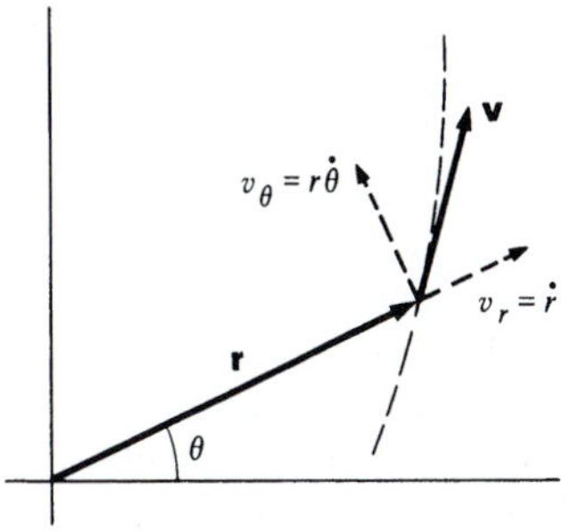

In polar coordinates the velocity of the particle is $\mathbf{v} = \dot{r}\hat{\mathbf{r}} + r\dot{\theta}\hat{\boldsymbol{\theta}}$. Its angular momentum is

$$\mathbf{L} = (\mathbf{r} \times m\mathbf{v}) = r\hat{\mathbf{r}} \times m(\dot{r}\hat{\mathbf{r}} + r\dot{\theta}\hat{\boldsymbol{\theta}}) = mr^2\dot{\theta}\hat{\mathbf{k}}.$$

(Note that $\hat{\mathbf{r}} \times \hat{\boldsymbol{\theta}} = \hat{\mathbf{k}}$). Hence,

$$\frac{dA}{dt} = \frac{1}{2}r^2\dot{\theta}$$
$$= \frac{L_z}{2m}.$$

Since L_z is constant for any central force, it follows that dA/dt is constant also.

Here is another way to prove the law of equal areas. For a central force, $F_\theta = 0$, so that $a_\theta = 0$. It follows that $ra_\theta = 0$, but $ra_\theta = r(2\dot{r}\dot{\theta} + r\ddot{\theta}) = (d/dt)(r^2\dot{\theta}) = 2(d/dt)(dA/dt)$. Hence, $dA/dt =$ constant.

Example 6.4 Capture Cross Section of a Planet

This example concerns the problem of aiming an unpowered spacecraft to hit a far-off planet. If you have ever looked at a planet through a telescope, you know that it appears to have the shape of a disk. The area of the disk is πR^2, where R is the planet's radius. If gravity played no role, we would have to aim the spacecraft to head for this area in order to assure a hit. However, the situation is more favorable than this because of the gravitational attraction of the spacecraft by the planet. Gravity tends to deflect the spacecraft toward the planet, so that some trajectories which are aimed outside the planetary disk nevertheless end in a hit. Consequently, the effective area for a hit A_e is greater than the geometrical area $A_g = \pi R^2$. Our problem is to find A_e.

We shall neglect effects of the sun and other planets here, although they would obviously have to be taken into account for a real space mission.

One approach to the problem would be to work out the full solution for the orbit of the spacecraft in the gravitational field of the planet. This involves a lengthy calculation which is not really necessary; by using conservation of energy and angular momentum, we can find the answer in a few short steps.

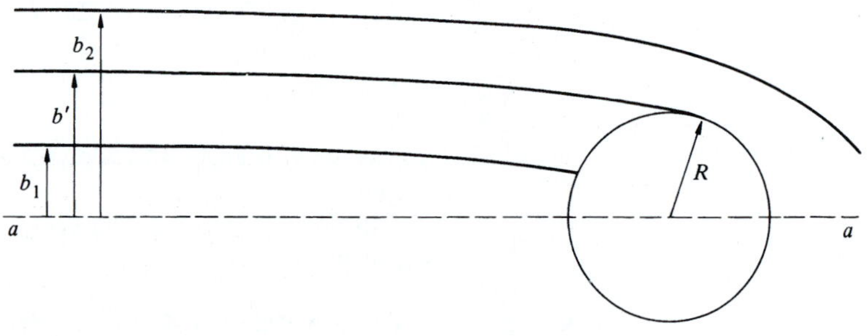

The sketch shows several possible trajectories of the spacecraft. The distance between the launch point and the target planet is assumed to be extremely large compared with R, so that the different trajectories are effectively parallel before the gravitational force of the planet becomes important. The line aa is parallel to the initial trajectories and passes through the center of the planet. The distance b between the initial trajectory and line aa is called the *impact parameter* of the trajectory. The largest value of b for which the trajectory hits the planet is indicated by b' in the sketch. The area through which the trajectory must pass to assure a hit is $A_e = \pi(b')^2$. (If there were no attraction, the trajectories would be straight lines. In this case, $b' = R$ and $A_e = \pi R^2 = A_g$.)

To find b', we note that both the energy and angular momentum of the spacecraft are conserved. (Linear momentum of the spacecraft is not conserved. Do you see why?)

The kinetic energy is $\frac{1}{2}mv^2$, and the potential energy is $-mMG/r$. The total energy $E = K + U$ is

$$E = \frac{1}{2}mv^2 - mMG\frac{1}{r}.$$

The angular momentum about the center of the planet is

$$L = -mrv \sin \phi.$$

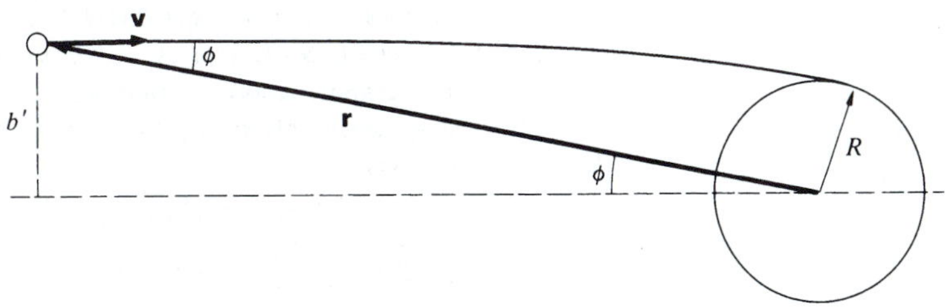

Initially, $r \to \infty$, $v = v_0$, and $r \sin \phi = b'$. Hence,

$$L = -mb'v_0,$$

$$E = \frac{1}{2} mv_0^2.$$

The point of collision occurs at the distance of closest approach of the orbit, $r = R$; otherwise the trajectory would not "just graze" the planet. At the distance of closest approach, **r** and **v** are perpendicular. If $v(R)$ is the speed at this point,

$$L = -mRv(R)$$

$$E = \frac{1}{2} mv(R)^2 - \frac{mMG}{R}.$$

Since L and E are conserved, their values at $r = R$ must be the same as their values at $r = \infty$. Hence

$$-mb'v_0 = -mRv(R) \qquad 1$$

$$\frac{1}{2} mv_0^2 = \frac{1}{2} mv(R)^2 - \frac{mMG}{R}. \qquad 2$$

Equation (1) gives $v(R) = v_0b'/R$, and by substituting this in Eq. (2) we obtain

$$(b')^2 = R^2\left(1 + \frac{mMG/R}{mv_0^2/2}\right).$$

The effective area is

$$A_e = \pi(b')^2$$

$$= \pi R^2\left(1 + \frac{mMG/R}{mv_0^2/2}\right).$$

As we expect, the effective area is greater than the geometrical area. Since $mMG/R = -U(R)$, and $mv_0^2/2 = E$, we have

$$A_e = A_g\left(1 - \frac{U(R)}{E}\right).$$

If we "turn off" gravity, $U(R) \to 0$ and $A_e \to A_g$, as we require. Furthermore, as $E \to 0$, $A_e \to \infty$, which means that it is impossible to miss the planet, provided that you start from rest. For $E = 0$, the spacecraft inevitably falls into the planet.

If there is a torque on a system the angular momentum must change according to $\boldsymbol{\tau} = d\mathbf{L}/dt$, as the following examples illustrate.

Example 6.5 **Torque on a Sliding Block**

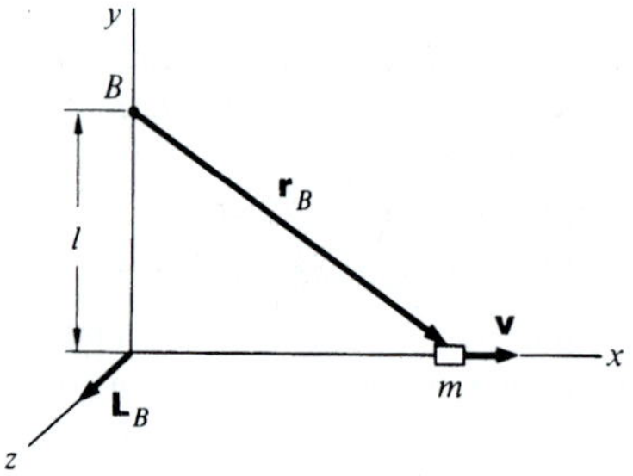

For a simple illustration of the relation $\boldsymbol{\tau} = d\mathbf{L}/dt$, consider a small block of mass m sliding in the x direction with velocity $\mathbf{v} = v\hat{\mathbf{i}}$. The angular momentum of the block about origin B is

$$\begin{aligned}\mathbf{L}_B &= m\mathbf{r}_B \times \mathbf{v} \\ &= mlv\hat{\mathbf{k}},\end{aligned} \qquad 1$$

as we discussed in Example 6.1. If the block is sliding freely, $\mathbf{v}$ does not change, and $\mathbf{L}_B$ is therefore constant, as we expect, since there is no torque acting on the block.

Suppose now that the block slows down because of a friction force $\mathbf{f} = -f\hat{\mathbf{i}}$. The torque on the block about origin B is

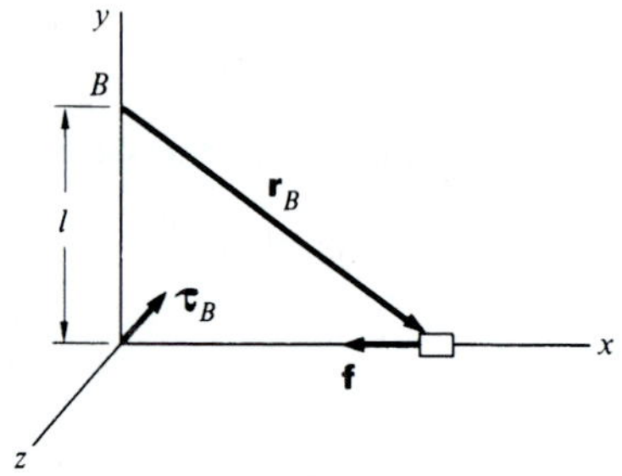

$$\begin{aligned}\boldsymbol{\tau}_B &= \mathbf{r}_B \times \mathbf{f} \\ &= -lf\hat{\mathbf{k}}.\end{aligned} \qquad 2$$

We see from Eq. (1) that as the block slows, $\mathbf{L}_B$ remains along the positive z direction but its magnitude decreases. Therefore, the change $\Delta\mathbf{L}_B$ in $\mathbf{L}_B$ points in the negative z direction, as shown in the lower sketch. The direction of $\Delta\mathbf{L}_B$ is the same as the direction of $\boldsymbol{\tau}_B$. Since $\boldsymbol{\tau} = d\mathbf{L}/dt$ in general, the vectors $\boldsymbol{\tau}$ and $\Delta\mathbf{L}$ are always parallel.

From Eq. (1),

$$\Delta\mathbf{L}_B = ml\,\Delta v\,\hat{\mathbf{k}}, \qquad 3$$

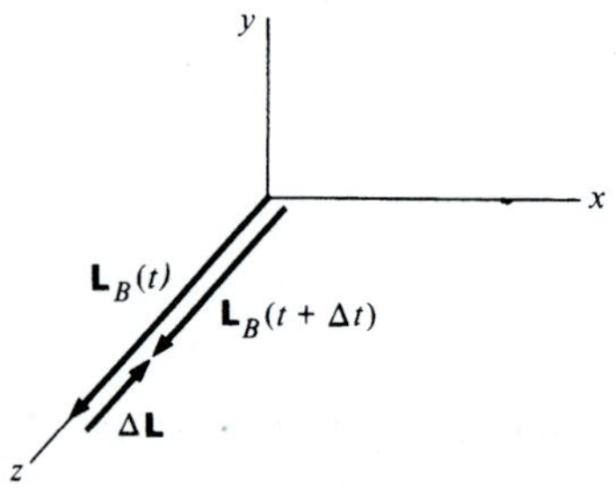

where $\Delta v < 0$. Dividing Eq. (3) by Δt and taking the limit $\Delta t \to 0$, we have

$$\frac{d\mathbf{L}_B}{dt} = ml\frac{dv}{dt}\hat{\mathbf{k}}. \qquad 4$$

By Newton's second law, $m\,dv/dt = -f$ and Eq. (4) becomes

$$\begin{aligned}\frac{d\mathbf{L}_B}{dt} &= -lf\hat{\mathbf{k}} \\ &= \boldsymbol{\tau}_B,\end{aligned}$$

as we expect.

It is important to keep in mind that since $\boldsymbol{\tau}$ and $\mathbf{L}$ depend on the choice of origin, the same origin must be used for both when applying the relation $\boldsymbol{\tau} = d\mathbf{L}/dt$, as we were careful to do in this problem.

The angular momentum of the block in this example changed only in magnitude and not in direction, since $\boldsymbol{\tau}$ and $\mathbf{L}$ happened to be along the same line. In the next example we return to the conical pendulum to study a case in which the angular momentum is constant in magnitude but changes direction due to an applied torque.

Example 6.6 **Torque on the Conical Pendulum**

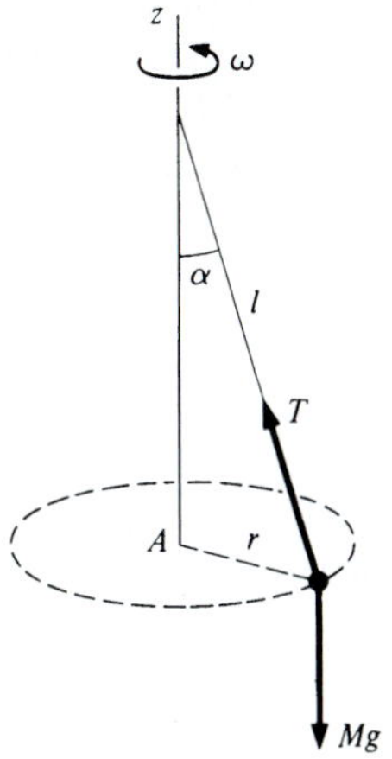

In Example 6.2, we calculated the angular momentum of a conical pendulum about two different origins. Now we shall complete the analysis by showing that the relation $\boldsymbol{\tau} = d\mathbf{L}/dt$ is satisfied.

The sketch illustrates the forces on the bob. T is the tension in the string. For uniform circular motion there is no vertical acceleration, and consequently

$$T \cos \alpha - Mg = 0. \qquad 1$$

The total force $\mathbf{F}$ on the bob is radially inward: $\mathbf{F} = -T \sin \alpha \hat{\mathbf{r}}$. The torque on M about A is

$$\begin{aligned}\boldsymbol{\tau}_A &= \mathbf{r}_A \times \mathbf{F}\\ &= 0,\end{aligned}$$

since $\mathbf{r}_A$ and $\mathbf{F}$ are both in the $\hat{\mathbf{r}}$ direction. Hence

$$\frac{d\mathbf{L}_A}{dt} = 0$$

and we have the result

$$\mathbf{L}_A = \text{constant}$$

as we already know from Example 6.2.

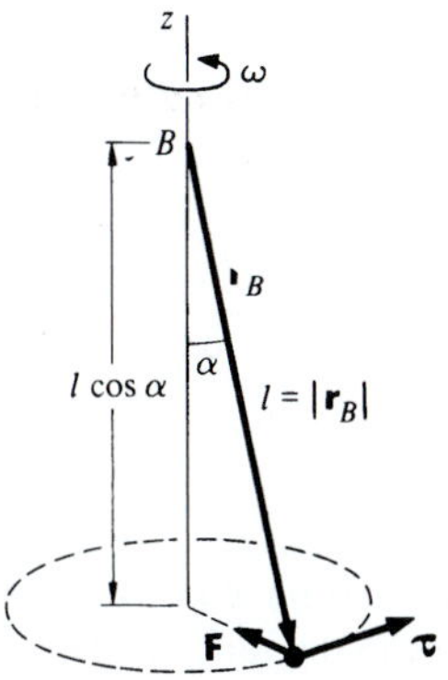

The problem looks entirely different if we take the origin at B. The torque $\boldsymbol{\tau}_B$ is

$$\boldsymbol{\tau}_B = \mathbf{r}_B \times \mathbf{F}.$$

Hence

$$\begin{aligned}|\boldsymbol{\tau}_B| &= l \cos \alpha F = l \cos \alpha\, T \sin \alpha\\ &= Mgl \sin \alpha,\end{aligned}$$

where we have used Eq. (1), $T \cos \alpha = Mg$. The direction of $\boldsymbol{\tau}_B$ is tangential to the line of motion of M:

$$\boldsymbol{\tau}_B = Mgl \sin \alpha \hat{\boldsymbol{\theta}}, \qquad 2$$

where $\hat{\boldsymbol{\theta}}$ is the unit tangential vector in the plane of motion.

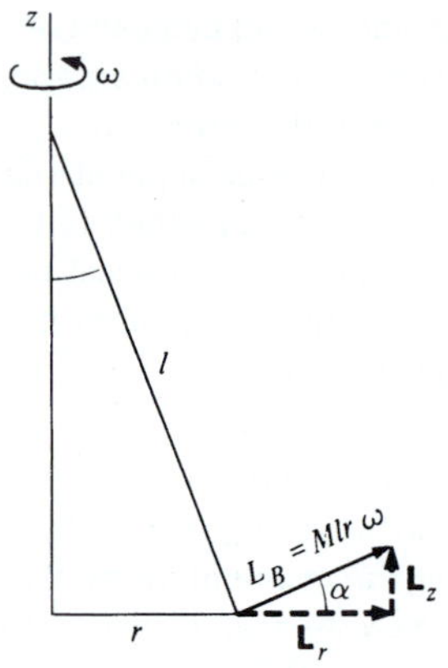

Our problem is to show that the relation

$$\boldsymbol{\tau}_B = \frac{d\mathbf{L}_B}{dt} \qquad 3$$

is satisfied. From Example 6.2, we know that $\mathbf{L}_B$ has constant magnitude $Mlr\omega$. As the diagram at left shows, $\mathbf{L}_B$ has a vertical component $L_z = Mlr\omega \sin\alpha$ and a horizontal radial component $L_r = Mlr\omega \cos\alpha$. Writing $\mathbf{L}_B = \mathbf{L}_z + \mathbf{L}_r$, we see that $\mathbf{L}_z$ is constant, as we expect, since $\boldsymbol{\tau}_B$ has no vertical component. $\mathbf{L}_r$ is not constant; it changes direction as the bob swings around. However, the magnitude of $\mathbf{L}_r$ is constant. We encountered such a situation in Sec. 1.8, where we showed that the only way a vector $\mathbf{A}$ of constant magnitude can change in time is to rotate, and that if its instantaneous rate of rotation is $d\theta/dt$, then $|d\mathbf{A}/dt| = A\,d\theta/dt$. We can employ this relation directly to obtain

$$\left|\frac{d\mathbf{L}_r}{dt}\right| = L_r\omega.$$

However, since we shall invoke this result frequently, let us take a moment to rederive it geometrically.

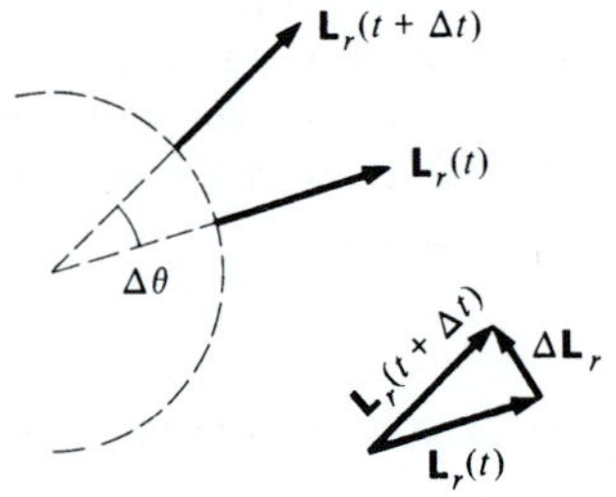

The vector diagrams show $\mathbf{L}_r$ at some time t and at $t + \Delta t$. During the interval Δt, the bob swings through angle $\Delta\theta = \omega\,\Delta t$, and $\mathbf{L}_r$ rotates through the same angle. The magnitude of the vector difference $\Delta\mathbf{L}_r = \mathbf{L}_r(t + \Delta t) - \mathbf{L}_r(t)$ is given approximately by

$$|\Delta\mathbf{L}_r| \approx L_r\,\Delta\theta.$$

In the limit $\Delta t \to 0$, we have

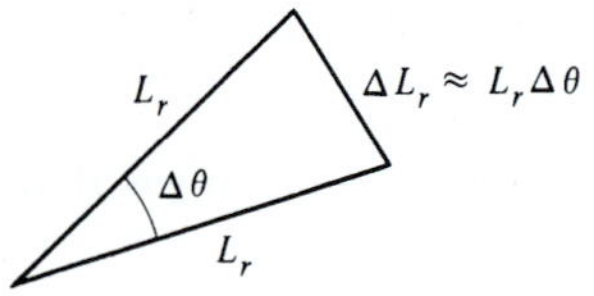

$$\frac{dL_r}{dt} = L_r\frac{d\theta}{dt}$$
$$= L_r\omega.$$

Since $L_r = Mlr\omega\cos\alpha$, we have

$$\frac{dL_r}{dt} = Mlr\omega^2\cos\alpha.$$

$Mr\omega^2$ is the radial force, $T\sin\alpha$, and since $T\cos\alpha = Mg$, we have

$$\frac{dL_r}{dt} = Mgl\sin\alpha,$$

which agrees with the magnitude of $\boldsymbol{\tau}_B$ from Eq. (2). Furthermore, as the vector drawings indicate, $d\mathbf{L}_r/dt$ lies in the tangential direction, parallel to $\boldsymbol{\tau}_B$, as we expect.

Another way to calculate $d\mathbf{L}_B/dt$ is to write $\mathbf{L}_B$ in vector form and then differentiate:

$$\mathbf{L}_B = (Mlr\omega \sin\alpha)\hat{\mathbf{k}} + (Mlr\omega \cos\alpha)\hat{\mathbf{r}}.$$

$$\frac{d\mathbf{L}_B}{dt} = Mlr\omega \cos\alpha \frac{d\hat{\mathbf{r}}}{dt}$$

$$= Mlr\omega^2 \cos\alpha\hat{\boldsymbol{\theta}},$$

where we have used $d\hat{\mathbf{r}}/dt = \omega\hat{\boldsymbol{\theta}}$.

It is important to be able to visualize angular momentum as a vector which can rotate in space. This type of reasoning occurs often in analyzing the motion of rigid bodies; we shall find it particularly helpful in understanding gyroscope motion in Chap. 7.

Example 6.7 Torque due to Gravity

We often encounter systems in which there is a torque exerted by gravity. Examples include a pendulum, a child's top, and a falling chimney. In the usual case of a uniform gravitational field, the torque on a body about any point is $\mathbf{R} \times \mathbf{W}$, where $\mathbf{R}$ is a vector from the point to the center of mass and $\mathbf{W}$ is the weight. Here is the proof.

The problem is to find the torque on a body of mass M about origin A when the applied force is due to a uniform gravitational field $\mathbf{g}$. We can regard the body as a collection of particles. The torque $\boldsymbol{\tau}_j$ on the jth particle is

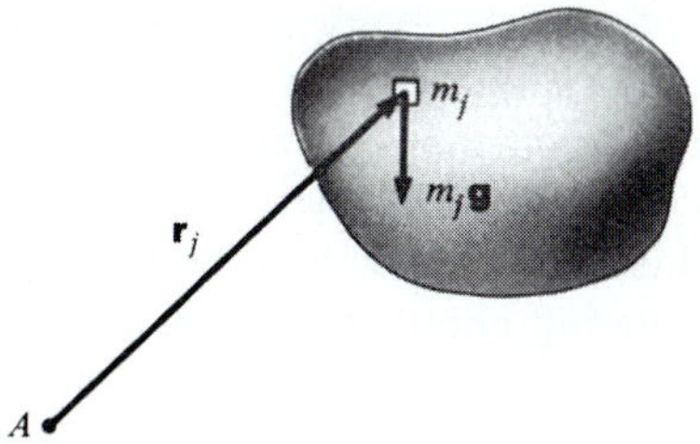

$$\boldsymbol{\tau}_j = \mathbf{r}_j \times m_j\mathbf{g},$$

where $\mathbf{r}_j$ is the position vector of the jth particle from origin A, and m_j is its mass.

The total torque is

$$\boldsymbol{\tau} = \Sigma\boldsymbol{\tau}_j$$

$$= \Sigma\mathbf{r}_j \times m_j\mathbf{g}$$

$$= (\Sigma m_j\mathbf{r}_j) \times \mathbf{g}.$$

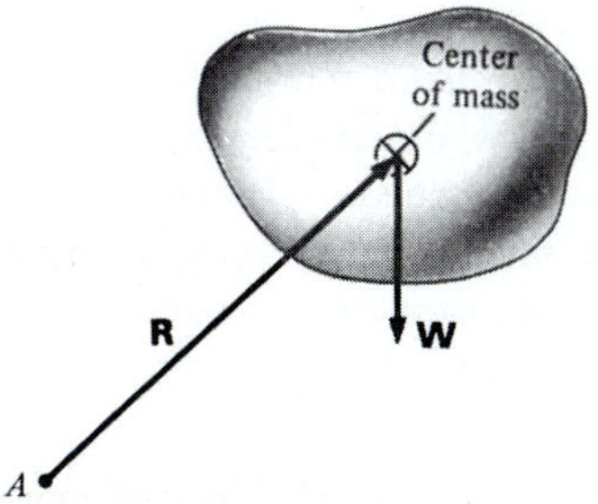

By definition of center of mass,

$$\Sigma m_j\mathbf{r}_j = M\mathbf{R},$$

where $\mathbf{R}$ is the position vector of the center of mass. Hence

$$\boldsymbol{\tau} = M\mathbf{R} \times \mathbf{g}$$

$$= \mathbf{R} \times M\mathbf{g}$$

$$= \mathbf{R} \times \mathbf{W}.$$

A corollary to this result is that in order to balance an object, the pivot point must be at the center of mass.

6.4 Angular Momentum and Fixed Axis Rotation

The most prominent application of angular momentum in classical mechanics is to the analysis of the motion of rigid bodies. The general case of rigid body motion involves free rotation about any axis—for instance, the motion of a baseball bat flung spinning and tumbling into the air. Analysis of the general case involves a number of mathematical complexities which we are going to postpone for a chapter, and in this chapter we restrict ourselves to a special, but important, case—rotation about a fixed axis. By fixed axis we mean that the *direction* of the axis of rotation is always along the same line; the axis itself may translate. For example, a car wheel attached to an axle undergoes fixed axis rotation as long as the car drives straight ahead. If the car turns, the wheel must rotate about a vertical axis while simultaneously spinning on the axle; the motion is no longer fixed axis rotation. If the wheel flies off the axle and wobbles down the road, the motion is definitely not rotation about a fixed axis.

We can choose the axis of rotation to be in the z direction, without loss of generality. The rotating object can be a wheel or a baseball bat, or anything we choose, the only restriction being that it is rigid—which is to say that its shape does not change as it rotates.

When a rigid body rotates about an axis, every particle in the body remains at a fixed distance from the axis. If we choose a coordinate system with its origin lying on the axis, then for each particle in the body, $|\mathbf{r}|$ = constant. The only way that $\mathbf{r}$ can change while $|\mathbf{r}|$ remains constant is for the velocity to be perpendicular to $\mathbf{r}$. Hence, for a body rotating about the z axis,

$$|\mathbf{v}_j| = |\dot{\mathbf{r}}_j| \qquad 6.4$$
$$= \omega\rho_j,$$

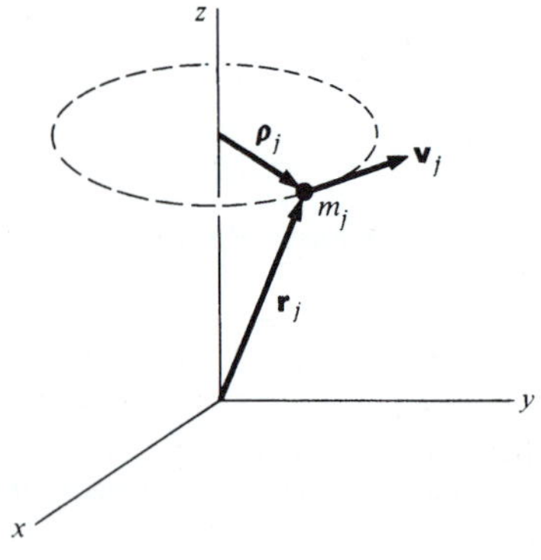

where ρ_j is the perpendicular distance from the axis of rotation to particle m_j of the rigid body and $\boldsymbol{\rho}_j$ is the corresponding vector. ω is the rate of rotation, or angular velocity. Since the axis of rotation lies in the z direction, we have $\rho_j = (x_j^2 + y_j^2)^{\frac{1}{2}}$. [In this chapter and the next we shall use the symbol ρ to denote perpendicular distance to the axis of rotation. Note that r stands for the distance to the origin: $r = (x^2 + y^2 + z^2)^{\frac{1}{2}}$.]

The angular momentum of the jth particle of the body, $\mathbf{L}(j)$, is.

$$\mathbf{L}(j) = \mathbf{r}_j \times m_j\mathbf{v}_j.$$

In this chapter we are concerned only with L_z, the component of angular momentum along the axis of rotation. Since $\mathbf{v}_j$ lies in the xy plane,

$$L_z(j) = m_j v_j \times (\text{distance to } z \text{ axis}) = m_j v_j \rho_j.$$

Using Eq. (6.4), $v_j = \omega \rho_j$, we have

$$L_z(j) = m_j \rho_j^2 \omega.$$

The z component of the total angular momentum of the body L_z is the sum of the individual z components:

$$\begin{aligned} L_z &= \sum_j L_z(j) \\ &= \Sigma m_j \rho_j^2 \omega, \end{aligned} \qquad 6.5$$

where the sum is over all particles of the body. We have taken ω to be constant throughout the body; can you see why this must be so?

Equation (6.5) can be written as

$$L_z = I\omega, \qquad 6.6$$

where

$$I \equiv \sum_j m_j \rho_j^2. \qquad 6.7$$

I is a geometrical quantity called the *moment of inertia*. I depends on both the distribution of mass in the body and the location of the axis of rotation. (We shall give a more general definition for I in the next chapter when we talk about unrestricted rigid body motion.) For continuously distributed matter we can replace the sum over mass particles by an integral over differential mass elements. In this case

$$\sum_j m_j \rho_j^2 \rightarrow \int \rho^2\, dm,$$

and

$$\begin{aligned} I &= \int \rho^2\, dm \\ &= \int (x^2 + y^2)\, dm. \end{aligned}$$

To evaluate such an integral we generally replace the mass element dm by the product of the density (mass per unit volume) w at the position of dm and the volume dV occupied by dm:

$$dm = w\, dV.$$

(Often ρ is used to denote density, but that would cause confusion here.) We can write

$$I = \int \rho^2 dm$$
$$= \int (x^2 + y^2)w\, dV.$$

For simple shapes with a high degree of symmetry, calculation of the moment of inertia is straightforward, as the following examples show.

Example 6.8 Moments of Inertia of Some Simple Objects

a. UNIFORM THIN HOOP OF MASS M AND RADIUS R, AXIS THROUGH THE CENTER AND PERPENDICULAR TO THE PLANE OF THE HOOP

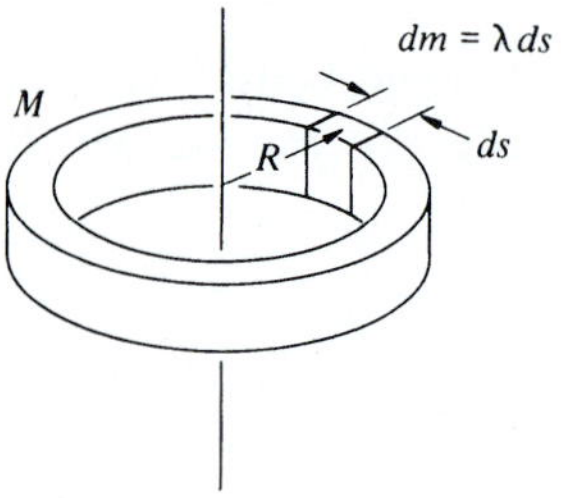

The moment of inertia about the axis is given by

$$I = \int \rho^2\, dm.$$

Since the hoop is thin, $dm = \lambda ds$, where $\lambda = M/2\pi R$ is the mass per unit length of the hoop. All points on the hoop are distance R from the axis so that $\rho = R$, and we have

$$I = \int_0^{2\pi R} R^2 \lambda\, ds$$
$$= R^2 \left(\frac{M}{2\pi R}\right) s \Big|_0^{2\pi R}$$
$$= MR^2.$$

b. UNIFORM DISK OF MASS M, RADIUS R, AXIS THROUGH THE CENTER AND PERPENDICULAR TO THE PLANE OF THE DISK

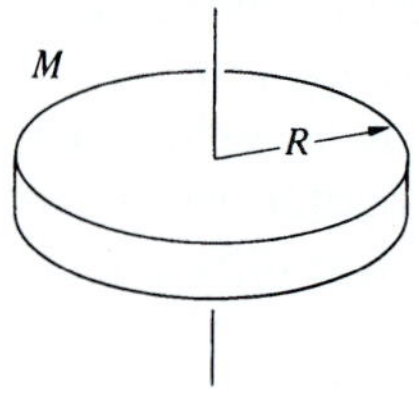

We can subdivide the disk into a series of thin hoops with radius ρ width $d\rho$, and moment of inertia dI. Then $I = \int dI$.

The area of one of the thin hoops is $dA = 2\pi\rho\, d\rho$, and its mass is

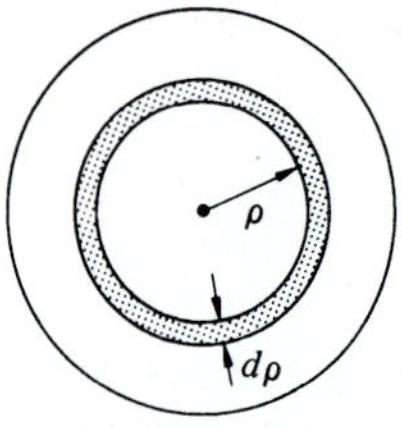

$$dm = M\frac{dA}{A} = \frac{M 2\pi\rho\, d\rho}{\pi R^2}$$
$$= \frac{2M\rho\, d\rho}{R^2}.$$

$$dI = \rho^2\, dm = \frac{2M\rho^3\, d\rho}{R^2}$$

$$I = \int_0^R \frac{2M\rho^3\, d\rho}{R^2}$$
$$= \frac{1}{2} MR^2.$$

Let us also solve this problem by double integration to illustrate the most general approach.

$$I = \int \rho^2\, dm$$
$$= \int \rho^2 \sigma\, dS,$$

where σ is the mass per unit area. For the uniform disk, $\sigma = M/\pi R^2$, Polar coordinates are the obvious choice. In plane polar coordinates,

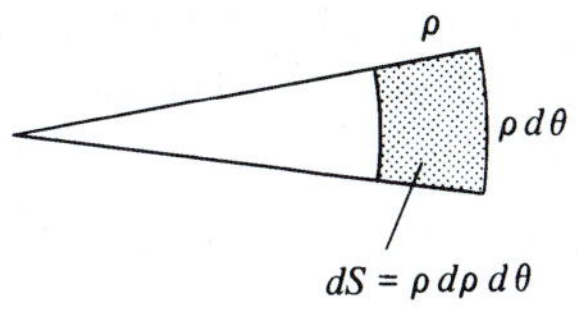

$$dS = \rho\, d\rho\, d\theta.$$

Then

$$I = \int \rho^2\, \sigma dS$$
$$= \left(\frac{M}{\pi R^2}\right) \int \rho^2\, dS$$
$$= \left(\frac{M}{\pi R^2}\right) \int_0^R \int_0^{2\pi} \rho^2 \rho\, d\rho\, d\theta$$
$$= \left(\frac{2M}{R^2}\right) \int_0^R \rho^3\, d\rho$$
$$= \tfrac{1}{2} MR^2,$$

as before.

c. UNIFORM THIN STICK OF MASS M, LENGTH L, AXIS THROUGH THE MIDPOINT AND PERPENDICULAR TO THE STICK

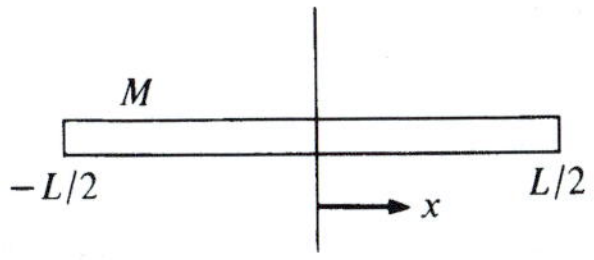

$$I = \int_{-L/2}^{+L/2} x^2\, dm$$
$$= \frac{M}{L} \int_{-L/2}^{+L/2} x^2\, dx$$
$$= \frac{M}{L} \frac{1}{3} x^3 \Big|_{-L/2}^{+L/2}$$
$$= \tfrac{1}{12} ML^2$$

d. UNIFORM THIN STICK, AXIS AT ONE END AND PERPENDICULAR TO THE STICK

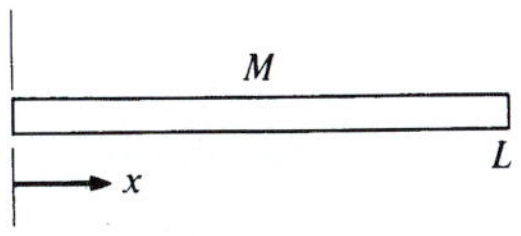

$$I = \frac{M}{L} \int_0^L x^2\, dx$$
$$= \tfrac{1}{3} ML^2.$$

e. UNIFORM SPHERE OF MASS M, RADIUS R, AXIS THROUGH CENTER

We quote this result without proof—perhaps you can derive it for yourself.

$$I = \tfrac{2}{5} MR^2.$$

Example 6.9 The Parallel Axis Theorem

This handy theorem tells us I, the moment of inertia about any axis, provided that we know I_0, the moment of inertia about a parallel axis through the center of mass. If the mass of the body is M and the distance between the axes is l, the theorem states that

$$I = I_0 + Ml^2.$$

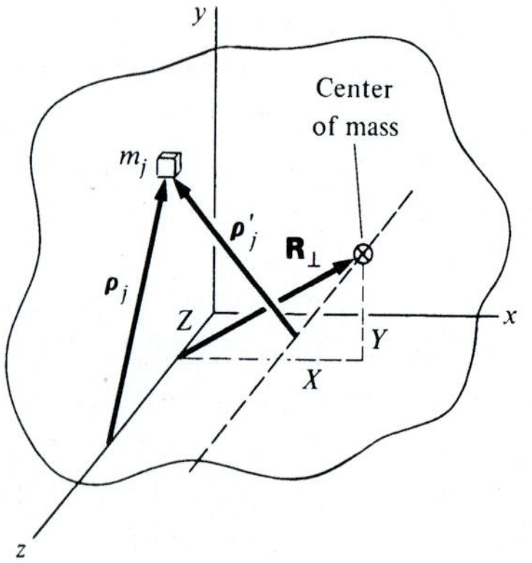

To prove this, consider the moment of inertia of the body about an axis which we choose to have lie in the z direction. The vector from the z axis to particle j is

$$\boldsymbol{\rho}_j = x_j\hat{\mathbf{i}} + y_j\hat{\mathbf{k}},$$

and

$$I = \Sigma m_j \rho_j^2.$$

If the center of mass is at $\mathbf{R} = X\hat{\mathbf{i}} + Y\hat{\mathbf{j}} + Z\hat{\mathbf{k}}$, the vector perpendicular from the z axis to the center of mass is

$$\mathbf{R}_\perp = X\hat{\mathbf{i}} + Y\hat{\mathbf{j}}.$$

If the vector from the axis through the center of mass to particle j is $\boldsymbol{\rho}'_j$, then the moment of inertia about the center of mass is

$$I_0 = \Sigma m_j \rho_j'^2.$$

From the diagram we see that

$$\boldsymbol{\rho}_j = \boldsymbol{\rho}'_j + \mathbf{R}_\perp,$$

so that

$$\begin{aligned} I &= \Sigma m_j \rho_j^2 \\ &= \Sigma m_j(\boldsymbol{\rho}'_j + \mathbf{R}_\perp)^2 \\ &= \Sigma m_j(\rho_j'^2 + 2\boldsymbol{\rho}'_j \cdot \mathbf{R}_\perp + R_\perp^2). \end{aligned}$$

The middle term vanishes, since

$$\begin{aligned} \Sigma m_j \boldsymbol{\rho}'_j &= \Sigma m_j(\boldsymbol{\rho}_j - \mathbf{R}_\perp) = M(\mathbf{R}_\perp - \mathbf{R}_\perp) \\ &= 0. \end{aligned}$$

If we designate the magnitude of $\mathbf{R}_\perp$ by l, then

$$I = I_0 + Ml^2.$$

For example, in Example 6.8c we showed that the moment of inertia of a stick about its midpoint is $ML^2/12$. The moment of inertia about its end, which is $L/2$ away from the center of mass, is therefore

$$\begin{aligned} I_a &= \frac{ML^2}{12} + M\left(\frac{L}{2}\right)^2 \\ &= \frac{ML^2}{3}, \end{aligned}$$

which is the result we found in Example 6.8d.

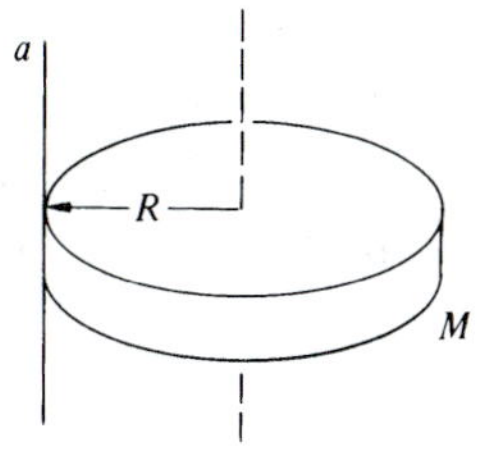

Similarly, the moment of inertia of a disk about an axis at the rim, perpendicular to the plane of the disk, is

$$I_a = \frac{MR^2}{2} + MR^2 = \frac{3MR^2}{2}.$$

6.5 Dynamics of Pure Rotation about an Axis

In Chap. 3 we showed that the motion of a system of particles is simple to describe if we distinguish between external forces and internal forces acting on the particles. The internal forces cancel by Newton's third law, and the momentum changes only because of external forces. This leads to the law of conservation of momentum: the momentum of an isolated system is constant. In describing rotational motion we are tempted to follow the same procedure and to distinguish between external and internal *torques*. Unfortunately, there is no way to prove from Newton's laws that the internal torques add to zero. However, it is an experimental fact that they always do cancel, since the angular momentum of an isolated system has never been observed to change. We shall discuss this more fully in Sec. 7.5 and for the remainder of this chapter simply assume that only external torques change the angular momentum of a rigid body.

In this section we consider fixed axis rotation with no translation of the axis, as, for instance, the motion of a door on its hinges or the spinning of a fan blade. Motion like this, where there is an axis of rotation at rest, is called *pure rotation*. Pure rotation is important because it is simple and because it is frequently encountered.

Consider a body rotating with angular velocity ω about the z axis. From Eq. (6.6) the z component of angular momentum is

$$L_z = I\omega.$$

Since $\boldsymbol{\tau} = d\mathbf{L}/dt$, where $\boldsymbol{\tau}$ is the external torque, we have

$$\begin{aligned}\tau_z &= \frac{d}{dt}(I\omega)\\ &= I\frac{d\omega}{dt}\\ &= I\alpha,\end{aligned}$$

where $\alpha = d\omega/dt$ is called the *angular acceleration*. In this chapter we are concerned with rotation only about the z axis, so we drop the subscript z and write

$$\tau = I\alpha. \qquad 6.8$$

Equation (6.8) is reminiscent of $\mathbf{F} = m\mathbf{a}$, and in fact there is a close analogy between linear and rotational motion. We can develop this further by evaluating the kinetic energy of a body undergoing pure rotation:

$$\begin{aligned} K &= \Sigma \tfrac{1}{2} m_j v_j^2 \\ &= \Sigma \tfrac{1}{2} m_j \rho_j^2 \omega^2 \\ &= \tfrac{1}{2} I \omega^2, \end{aligned}$$

where we have used $v_j = \rho_j \omega$ and $I = \Sigma m_j \rho_j^2$.

The method of handling problems involving rotation under applied torques is a straightforward extension of the familiar procedure for treating translational motion under applied forces, as the following example illustrates.

Example 6.10 **Atwood's Machine with a Massive Pulley**

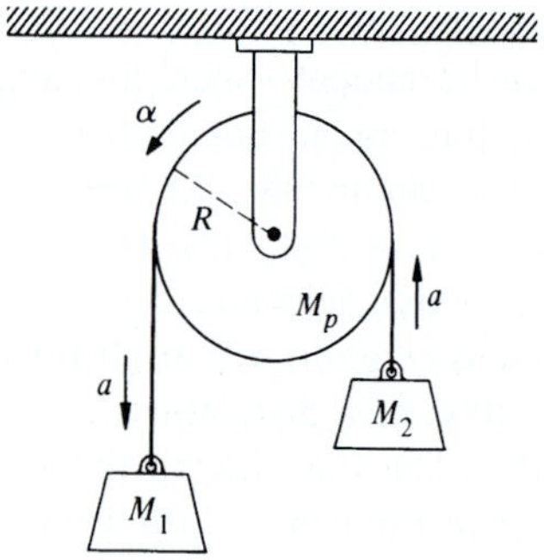

The problem is to find the acceleration a for the arrangement shown in the sketch. The effect of the pulley is to be included.

Force diagrams for the three masses are shown below left. The points of application of the forces on the pulley are shown; this is necessary whenever we need to calculate torques. The pulley evidently undergoes pure rotation about its axle, so we take the axis of rotation to be the axle.

The equations of motion are

$$\left.\begin{aligned} W_1 - T_1 &= M_1 a \\ T_2 - W_2 &= M_2 a \end{aligned}\right\} \text{Masses}$$

$$\left.\begin{aligned} \tau = T_1 R - T_2 R &= I\alpha \\ N - T_1 - T_2 - W_p &= 0 \end{aligned}\right\} \text{Pulley}$$

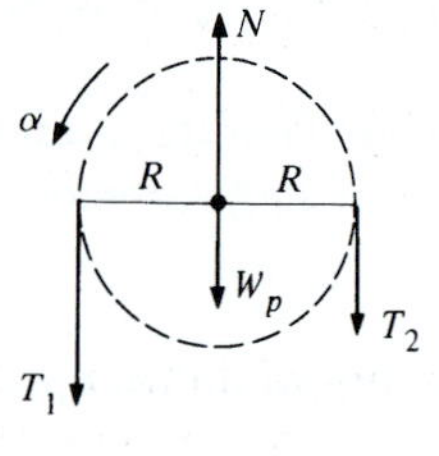

Note that in the torque equation, α must be positive counterclockwise to correspond to our convention that torque out of the paper is positive. N is the force on the axle, and the last equation simply assures that the pulley does not fall. Since we don't need to know N, it does not contribute to the solution.

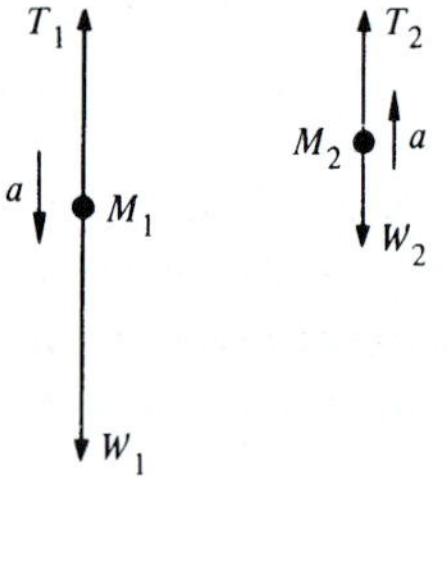

There is a constraint relating a and α, assuming that the rope does not slip. The velocity of the rope is the velocity of a point on the surface of the wheel, $v = \omega R$, from which it follows that

$$a = \alpha R.$$

We can now eliminate T_1, T_2, and α;

$$W_1 - W_2 - (T_1 - T_2) = (M_1 + M_2)a$$

$$T_1 - T_2 = \frac{I\alpha}{R} = \frac{Ia}{R^2}$$

$$W_1 - W_2 - \frac{Ia}{R^2} = (M_1 + M_2)a.$$

If the pulley is a simple disk, we have

$$I = \frac{M_p R^2}{2}$$

and it follows that

$$a = \frac{(M_1 - M_2)g}{M_1 + M_2 + M_p/2}.$$

The pulley increases the total inertial mass of the system, but in comparison with the hanging weights, the effective mass of the pulley is only one-half its real mass.

6.6 The Physical Pendulum

A mass hanging from a string is a *simple pendulum* if we assume that the mass has negligible size and the mass of the string is zero. We shall review its behavior as an introduction to the more realistic object, the *physical pendulum*, for which we do not need to make these assumptions.

The Simple Pendulum

At the left is a sketch of a simple pendulum and the force diagram. The tangential force is $-W \sin \phi$, and we obtain

$$ml\ddot{\phi} = -W \sin \phi.$$

(Incidentally, we get the same result by considering pure rotation about the point of suspension: $I = ml^2$, $\alpha = \ddot{\phi}$, and $\tau = -Wl \sin \phi$, so $ml^2\ddot{\phi} = -Wl \sin \phi$.) We can rewrite the equation of motion as

$$l\ddot{\phi} + g \sin \phi = 0.$$

This equation cannot be solved in terms of familiar functions. However, if the pendulum never swings far from the vertical, then $\phi \ll 1$, and we can use the approximation $\sin \phi \approx \phi$. Then

$$l\ddot{\phi} + g\phi = 0.$$

This is the equation for simple harmonic motion. (See Example 2.14.) The solution is $\phi = A \sin \omega t + B \cos \omega t$, where $\omega = \sqrt{g/l}$ and A and B are constants. If the pendulum starts from rest at angle ϕ_0, the solution is

$$\phi = \phi_0 \cos \omega t.$$

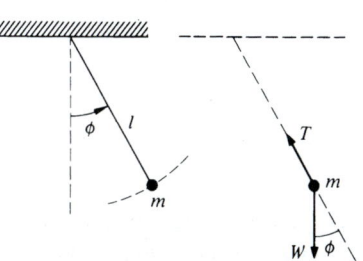

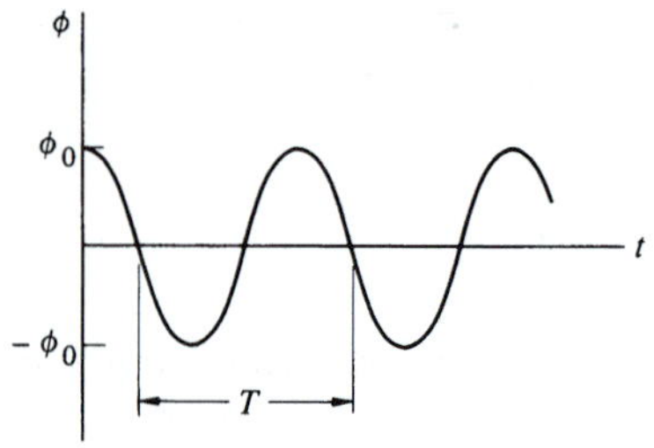

The motion is *periodic*, which means it occurs identically over and over again. The *period* T, the time between successive repetitions of the motion, is given by $\omega T = 2\pi$, or

$$T = \frac{2\pi}{\sqrt{g/l}}$$
$$= 2\pi\sqrt{\frac{l}{g}}.$$

The maximum angle ϕ_0 is called the *amplitude* of the motion. The period is independent of the amplitude, which is why the pendulum is so well suited to regulating the rate of a clock. However, this feature of the motion is a consequence of the approximation $\sin\phi \approx \phi$. The exact solution, which is developed in Note 6.2 at the end of the chapter, shows that the period lengthens slightly with increasing amplitude. The following example illustrates the consequence of this.

Example 6.11 Grandfather's Clock

As shown in Eq. (7) of Note 6.2, for small amplitudes the period of a pendulum is given by

$$T = T_0(1 + \tfrac{1}{16}\phi_0^2 + \cdots). \qquad 1$$

where

$$T_0 = 2\pi\sqrt{\frac{l}{g}}.$$

For $\phi_0 \approx 0$ we have our previous result, $T = 2\pi\sqrt{l/g}$. The correction term, $\frac{1}{16}\phi_0^2$ is surprisingly small: Consider a grandfather's clock with $T_0 = 2$ s and $l \approx 1$ m. If the pendulum swings 4 cm to either side, then $\phi_0 = 4 \times 10^{-2}$ rad and the correction term is $\phi_0^2/16 = 10^{-4}$. This by itself is of no consequence, since the length of the pendulum can be adjusted to make the clock run at any desired rate. However, the amplitude may vary slightly due to friction and other effects. Suppose that the amplitude changes by an amount $d\phi$. Taking differentials of Eq. (1) gives

$$dT = \tfrac{1}{8}T_0\phi_0\,d\phi.$$

The fractional change in T is

$$\frac{dT}{T_0} = \frac{1}{8}\phi_0\,d\phi.$$

If the amplitude changes by 10 percent, then $d\phi = 0.1\phi_0 = 4 \times 10^{-3}$ rad, and $dT/T_0 = 2 \times 10^{-5}$, giving an error of about 2 seconds per day.

The Physical Pendulum

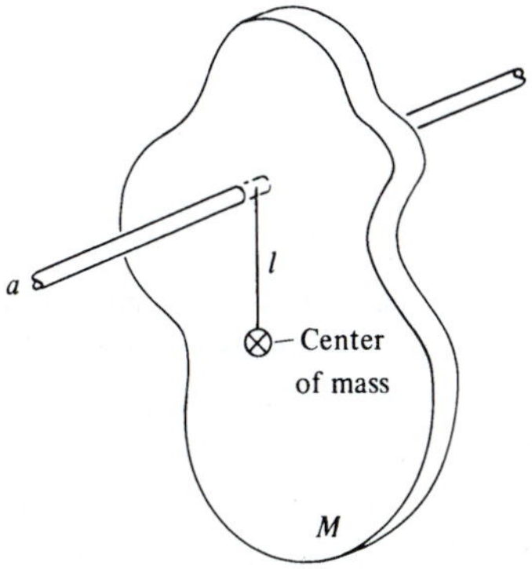

Now let us turn to the physical pendulum such as the one in the sketch. The swinging object can have any shape. Its mass is M, and its center of mass is at distance l from the pivot. One other quantity we need is the moment of inertia about the pivot, I_a. The motion is pure rotation about the pivot. Choosing the axis of rotation through the pivot, we find that the only torque is that due to gravity, and we have

$$-lW \sin \phi = I_a \ddot{\phi}.$$

Making the small angle approximation,

$$I_a \ddot{\phi} + Mlg\phi = 0.$$

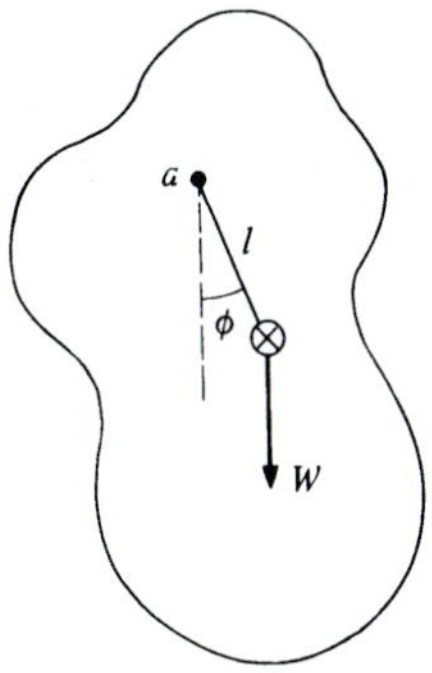

This is again the equation of simple harmonic motion with the solution

$$\phi = A \cos \omega t + B \sin \omega t,$$

where $\omega = \sqrt{Mlg/I_a}$.

We can write this result in a simpler form if we introduce the *radius of gyration*. If the moment of inertia of an object about its center of mass is I_0, the radius of gyration k is defined as

$$k = \sqrt{\frac{I_0}{M}} \qquad \text{or} \qquad I_0 = Mk^2.$$

For instance, for a hoop of radius R, $k = R$; for a disk, $k = \sqrt{\frac{1}{2}}\,R$; and for a solid sphere, $k = \sqrt{\frac{2}{5}}\,R$.

By the parallel axis theorem we have

$$\begin{aligned} I_a &= I_0 + Ml^2 \\ &= M(k^2 + l^2), \end{aligned}$$

so that

$$\omega = \sqrt{\frac{gl}{k^2 + l^2}}.$$

The simple pendulum corresponds to $k = 0$, and in this case we obtain $\omega = \sqrt{g/l}$, as before.

Example 6.12 Kater's Pendulum

Between the sixteenth and twentieth centuries, the most accurate measurements of g were obtained from experiments with pendulums. The method is attractive because the only quantities needed are the period of the pendulum, which can be determined to great accuracy by counting many swings, and the pendulum's dimensions. For very precise measurements, the limiting feature turns out to be the precision with which the center of mass of the pendulum and its radius of gyration can be determined. A clever invention, named after the nineteenth century English physicist, surveyor, and inventor Henry Kater, overcomes this difficulty.

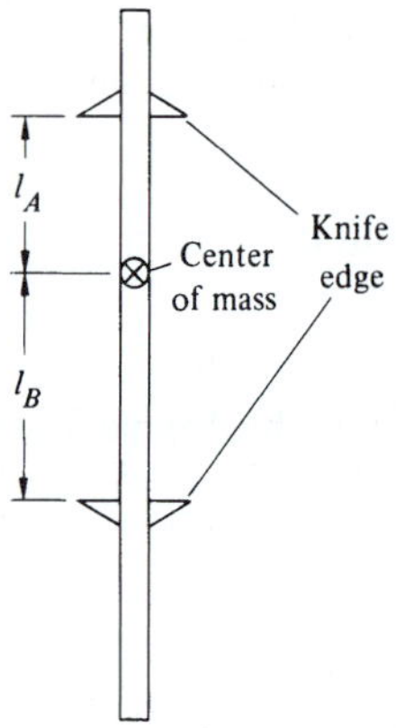

Kater's pendulum has two knife edges; the pendulum can be suspended from either. If the knife edges are distances l_A and l_B from the center of mass, then the period for small oscillations from each of these is, respectively,

$$T_A = 2\pi\left(\frac{k^2 + l_A{}^2}{gl_A}\right)^{\frac{1}{2}}$$

$$T_B = 2\pi\left(\frac{k^2 + l_B{}^2}{gl_B}\right)^{\frac{1}{2}}.$$

l_A or l_B is adjusted until the periods are identical: $T_A = T_B = T$. We can then eliminate T and solve for k^2:

$$k^2 = \frac{l_A l_B{}^2 - l_B l_A{}^2}{l_B - l_A} = l_A l_B.$$

Then

$$T = 2\pi\left(\frac{l_A l_B + l_A{}^2}{gl_A}\right)^{\frac{1}{2}} = 2\pi\left(\frac{l_A + l_B}{g}\right)^{\frac{1}{2}}$$

or

$$g = 4\pi^2\left(\frac{l_A + l_B}{T^2}\right)$$

The beauty of Kater's invention is that the only geometrical quantity needed is $l_A + l_B$, the distance between the knife edges, which can be measured to great accuracy. The position of the center of mass need not be known.

Example 6.13 **The Doorstop**

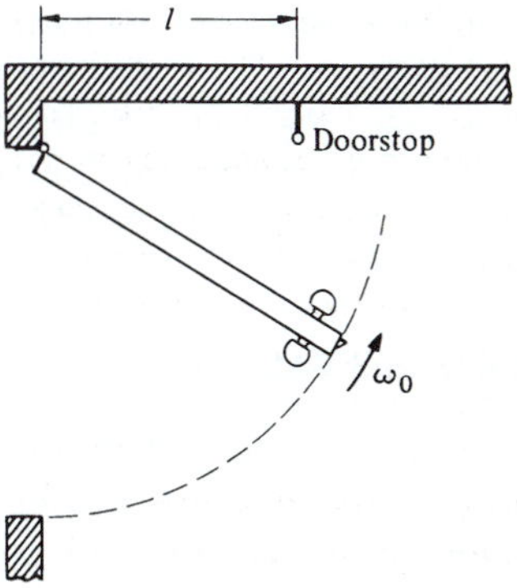

The banging of a door against its stop can tear loose the hinges. However, by the proper choice of l, the impact forces on the hinge can be made to vanish.

The forces on the door during impact are F_d, due to the stop, and F' and F'' due to the hinge. F'' is the small radial force which provides the centripetal acceleration of the swinging door. F' and F_d are the large impact forces which bring the door to rest when it bangs against the stop. The force on the hinges is equal and opposite to F' and F'' To minimize the stress on the hinges, we must make F' as small as possible.

To derive an expression for F', we shall consider in turn the angular momentum of the door about the hinges and the linear momentum of the center of mass.

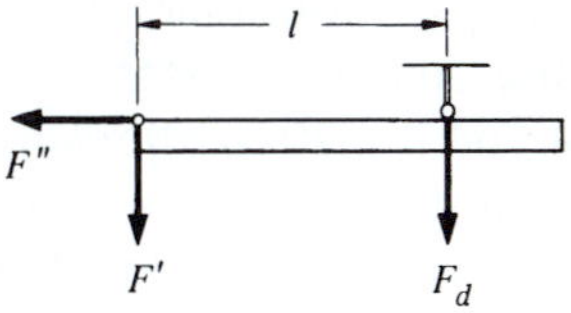

Since $dL = \tau dt$, we have

$$L_{\text{final}} - L_{\text{initial}} = \int_{t_i}^{t_f} \tau \, dt.$$

The initial angular momentum of the door is $I\omega_0$, where I is the moment of inertia about the hinges. Since the door comes to rest, $L_{\text{final}} = 0$. The torque on the door during the collision is $\tau = -lF_d$, and we obtain

$$I\omega_0 = l \int F_d \, dt, \qquad 1$$

where the integral is over the duration of the collision.

The center of mass motion obeys

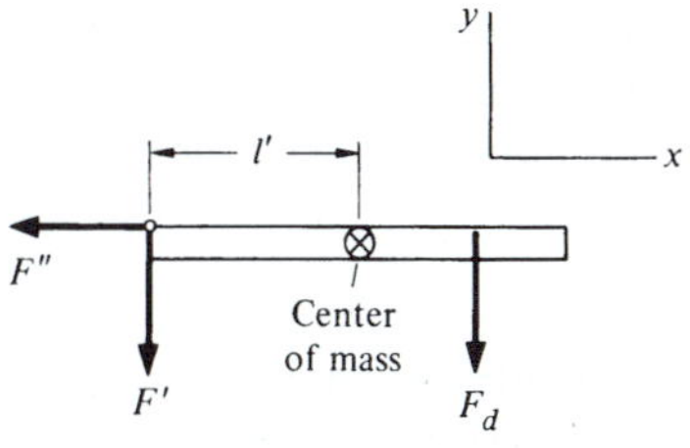

$$\mathbf{P}_{\text{final}} - \mathbf{P}_{\text{initial}} = \int \mathbf{F} \, dt,$$

where $\mathbf{F}$ is the total force. The momentum in the y direction immediately before the collision is $MV_y = Ml'\omega_0$, where l' is the distance from the hinge to the center of mass of the door. $P_{\text{final}} = 0$, and the y component of $\mathbf{F}$ is $F_y = -(F' + F_d)$. Hence,

$$Ml'\omega_0 = \int (F' + F_d) \, dt. \qquad 2$$

According to Eq. (1), $\int F_d \, dt = I\omega_0/l$, and substituting this in Eq. (2) gives

$$\int F' \, dt = \left(Ml' - \frac{I}{l} \right) \omega_0.$$

By choosing

$$l = \frac{I}{Ml'}, \qquad 3$$

the impact force is made zero. If the door is uniform, and of width w, then $I = Mw^2/3$ and $l' = w/2$. In this case $l = \frac{2}{3}w$.

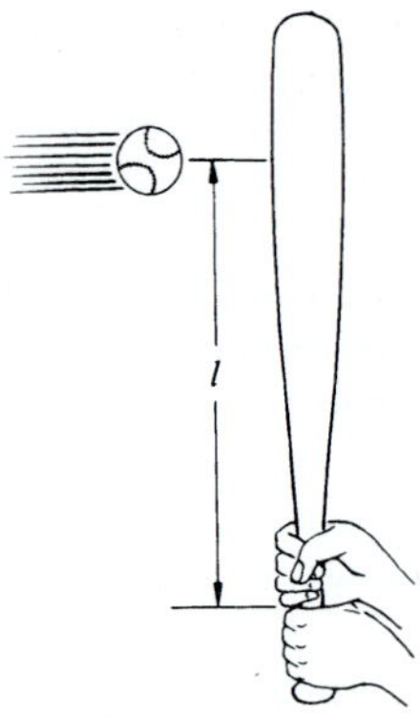

Incidentally, the stop must be at the height of the center of mass rather than at floor level. Otherwise the impact forces will not be identic l on the two hinges and the door will tend to rotate about a horizontal axis, an effect we have not taken into account.

The distance l specified by Eq. (3) is called the *center of percussion.* In batting a baseball it is important to hit the ball at the bat's center of percussion to avoid a reaction on the batter's hands and a painful sting.

6.7 Motion Involving Both Translation and Rotation

Often translation and rotation occur simultaneously, as in the case of a rolling drum. There is no obvious axis as there was in Sec. 6.5 when we analyzed pure rotation, and the problem seems confusing until we recall the theorem in Sec. 6.1—that one possible way to describe a general motion is by a translation of the center of mass plus a rotation about the center of mass. By using center of mass coordinates we will find it a straightforward matter to obtain simple expressions for both the angular momentum and the torque and to find the dynamical equation connecting them.

As before, we shall consider only motion for which the axis of rotation remains parallel to the z axis. We shall show that L_z, the z component of the angular momentum of the body, can be written as the sum of two terms. L_z is the angular momentum $I_0\omega$ due to rotation of the body about its center of mass, plus the angular momentum $(\mathbf{R} \times M\mathbf{V})_z$ due to motion of the center of mass with respect to the origin of the inertial coordinate system:

$$L_z = I_0\omega + (\mathbf{R} \times M\mathbf{V})_z,$$

where $\mathbf{R}$ is the position vector of the center of mass and $\mathbf{V} = \dot{\mathbf{R}}$.

To find the angular momentum, we start by considering the body to be an aggregation of N particles with masses $m_j(j = 1, \ldots, N)$ and position vectors $\mathbf{r}_j$ with respect to an inertial coordinate system. The angular momentum of the body can be written

$$\mathbf{L} = \sum_{j=1}^{N} (\mathbf{r}_j \times m_j\dot{\mathbf{r}}_j). \qquad 6.9$$

The center of mass of the body has position vector $\mathbf{R}$:

$$\mathbf{R} = \frac{\Sigma m_j\mathbf{r}_j}{M}, \qquad 6.10$$

where M is the total mass. The center of mass coordinates $\mathbf{r}_j'$ can be introduced as we did in Sec. 3.3:

$$\mathbf{r}_j = \mathbf{R} + \mathbf{r}_j'.$$

Eliminating $\mathbf{r}_j$ from Eq. (6.9) gives

$$\begin{aligned}\mathbf{L} &= \Sigma(\mathbf{r}_j \times m_j\dot{\mathbf{r}}_j)\\ &= \Sigma(\mathbf{R} + \mathbf{r}_j') \times m_j(\dot{\mathbf{R}} + \dot{\mathbf{r}}_j')\\ &= \mathbf{R} \times \Sigma m_j\dot{\mathbf{R}} + \Sigma m_j\mathbf{r}_j' \times \dot{\mathbf{R}} + \mathbf{R} \times \Sigma m_j\dot{\mathbf{r}}_j' + \Sigma m_j\mathbf{r}_j' \times \dot{\mathbf{r}}_j'.\end{aligned}$$

This expression looks cumbersome, but we can show that the middle two terms are identically zero and that the first and last terms have simple physical interpretations. Starting with the second term, we have

$$\begin{aligned}\Sigma m_j\mathbf{r}_j' &= \Sigma m_j(\mathbf{r}_j - \mathbf{R})\\ &= \Sigma m_j\mathbf{r}_j - M\mathbf{R}\\ &= 0.\end{aligned}$$

by Eq. (6.10). The third term is also zero; since $\Sigma m_j\mathbf{r}_j'$ is identically zero, its time derivative $\Sigma m_j\dot{\mathbf{r}}_j' = 0$ as well.

The first term is

$$\begin{aligned}\mathbf{R} \times \Sigma m_j\dot{\mathbf{R}} &= \mathbf{R} \times M\dot{\mathbf{R}}\\ &= \mathbf{R} \times M\mathbf{V},\end{aligned}$$

where $\mathbf{V} \equiv \dot{\mathbf{R}}$ is the velocity of the center of mass with respect to the inertial system. The expression for $\mathbf{L}$ then becomes

$$\mathbf{L} = \mathbf{R} \times M\mathbf{V} + \Sigma\mathbf{r}_j' \times m_j\dot{\mathbf{r}}_j'. \qquad 6.11$$

The first term of Eq. (6.11) represents the angular momentum due to the center of mass motion. The second term represents angular momentum due to motion around the center of mass. The only way for the particles of a rigid body to move with respect to the center of mass is for the body as a whole to rotate. We shall evaluate the second term for an arbitrary axis of rotation in the next chapter. In this chapter, however, we are restricting ourselves to fixed axis rotation about the z axis. Taking the z component of Eq. (6.11) gives

$$L_z = (\mathbf{R} \times M\mathbf{V})_z + (\Sigma\mathbf{r}_j' \times m_j\dot{\mathbf{r}}_j')_z. \qquad 6.12$$

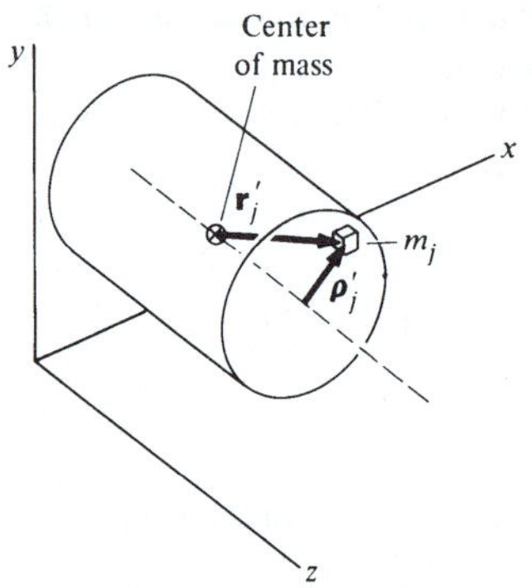

For rotation about the z axis, the second term $(\Sigma\mathbf{r}'_j \times m_j\dot{\mathbf{r}}'_j)_z$ can be simplified by recognizing that we dealt with this kind of expression before, in Sec. 6.4. The body has angular velocity $\omega\hat{\mathbf{k}}$ about its center of mass, and since the origin of $\mathbf{r}'_j$ is the center of mass, the second term is identical in form to the case of pure rotation we treated in Sec. 6.4.

$$(\Sigma m_j\mathbf{r}'_j \times \dot{\mathbf{r}}'_j)_z = (\Sigma m_j\boldsymbol{\rho}'_j \times \dot{\boldsymbol{\rho}}'_j)_z$$
$$= \Sigma m_j\rho_j'^2\omega = I_0\omega,$$

where $\boldsymbol{\rho}'_j$ is the vector to m_j perpendicular from an axis in the z direction through the center of mass. $I_0 = \Sigma m_j\rho_j'^2$ is the moment of inertia of the body about this axis.

Collecting our results, we have

$$L_z = I_0\omega + (\mathbf{R} \times M\mathbf{V})_z. \qquad 6.13$$

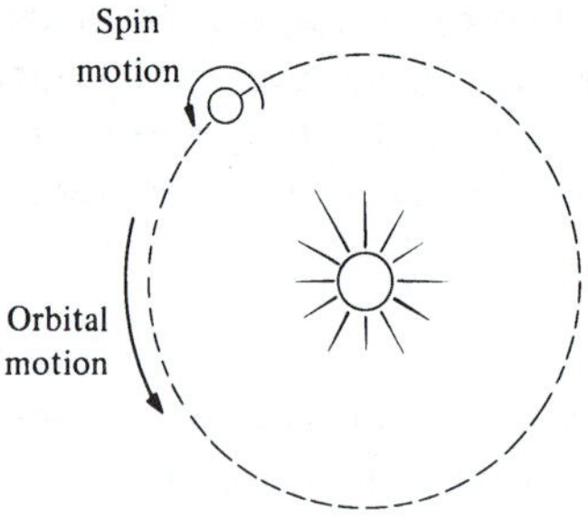

We have proven the result stated at the beginning of this section. The angular momentum of a rigid object is the sum of the angular momentum about its center of mass and the angular momentum of the center of mass about the origin. These two terms are often referred to as the *spin* and *orbital* terms, respectively. The earth illustrates them nicely. The daily rotation of the earth about its axis gives rise to the earth's spin angular momentum, and its annual revolution about the sun gives rise to the earth's orbital angular momentum about the sun. An important feature of the spin angular momentum is that it is independent of the coordinate system. In this sense it is intrinsic to the body; no change in coordinate system can eliminate spin, whereas orbital angular momentum disappears if the origin is along the line of motion.

It should be kept in mind that Eq. (6.13) is valid even when the center of mass is accelerating, since $\mathbf{L}$ was calculated with respect to an inertial coordinate system.

Example 6.14 Angular Momentum of a Rolling Wheel

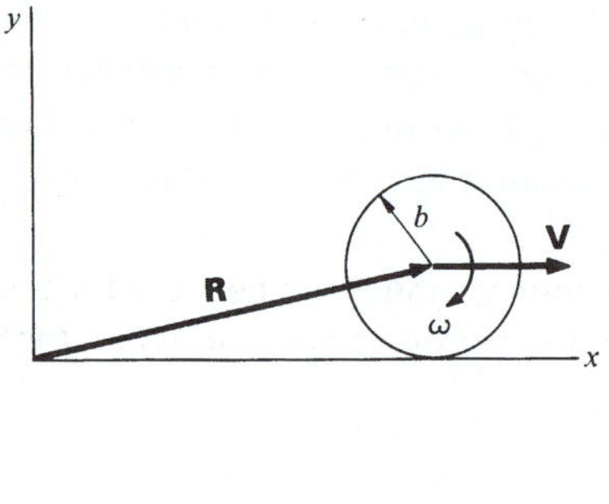

In this example we apply Eq. (6.13) to the calculation of the angular momentum of a uniform wheel of mass M and radius b which rolls uniformly and without slipping. The moment of inertia of the wheel about its center of mass is $I_0 = \frac{1}{2}Mb^2$ and its angular momentum about the center of mass is

$$L_0 = -I_0\omega$$
$$= -\tfrac{1}{2}Mb^2\omega.$$

L_0 is parallel to the z axis. The minus sign indicates that L_0 is directed into the paper, in the negative z direction.

If we calculate the angular momentum of the center of mass of the wheel with respect to the origin, we have

$$(\mathbf{R} \times M\mathbf{V})_z = -MbV.$$

The total angular momentum about the origin is then

$$\begin{aligned} L_z &= -\tfrac{1}{2}Mb^2\omega - MbV \\ &= -\tfrac{1}{2}Mb^2\omega - Mb^2\omega \\ &= -\tfrac{3}{2}Mb^2\omega, \end{aligned}$$

where we have used the result $V = b\omega$, which holds for a wheel that rolls without slipping.

Torque also naturally divides itself into two components. The torque on a body is

$$\begin{aligned} \boldsymbol{\tau} &= \Sigma \mathbf{r}_j \times \mathbf{f}_j \\ &= \Sigma(\mathbf{r}'_j + \mathbf{R}) \times \mathbf{f}_j \\ &= \Sigma(\mathbf{r}'_j \times \mathbf{f}_j) + \mathbf{R} \times \mathbf{F}, \end{aligned} \qquad 6.14$$

where $\mathbf{F} = \Sigma \mathbf{f}_j$ is the total applied force. The first term in Eq. (6.14) is the torque about the center of mass due to the various external forces, and the second term is the torque due to the total external force acting at the center of mass. For fixed axis rotation $\boldsymbol{\omega} = \omega\hat{\mathbf{k}}$, and Eq. (6.14) can be written

$$\tau_z = \tau_0 + (\mathbf{R} \times \mathbf{F})_z, \qquad 6.15$$

where τ_0 is the z component of the torque about the center of mass. But from Eq. (6.13) for L_z we have

$$\begin{aligned} \frac{dL_z}{dt} &= I_0 \frac{d\omega}{dt} + \frac{d}{dt}(\mathbf{R} \times M\mathbf{V})_z \\ &= I_0\alpha + (\mathbf{R} \times M\mathbf{a})_z. \end{aligned} \qquad 6.16$$

Using $\tau_z = dL_z/dt$, Eq. (6.15) and (6.16) yield

$$\begin{aligned} \tau_0 + (\mathbf{R} \times \mathbf{F})_z &= I_0\alpha + (\mathbf{R} \times M\mathbf{a})_z \\ &= I_0\alpha + (\mathbf{R} \times \mathbf{F})_z, \end{aligned}$$

since $\mathbf{F} = M\mathbf{a}$. Hence,

$$\tau_0 = I_0\alpha. \qquad 6.17$$

According to Eq. (6.17), rotational motion about the center of mass depends only on the torque about the center of mass, independent of the translational motion. In other words, Eq. (6.17) is correct even if the axis is accelerating.

These relations will seem quite natural when we use them. Before doing so, we complete the development by examining the kinetic energy.

$$\begin{aligned} K &= \tfrac{1}{2}\Sigma m_j v_j^2 \\ &= \tfrac{1}{2}\Sigma m_j(\dot{\boldsymbol{\rho}}_j' + \mathbf{V})^2 \\ &= \tfrac{1}{2}\Sigma m_j \dot{\rho}_j'^2 + \Sigma m_j \dot{\boldsymbol{\rho}}_j' \cdot \mathbf{V} + \tfrac{1}{2}\Sigma m_j V^2 \\ &= \tfrac{1}{2}I_0\omega^2 + \tfrac{1}{2}MV^2 \end{aligned} \qquad 6.18$$

The first term corresponds to the kinetic energy of spin, while the last term arises from the orbital center of mass motion.

Here is a summary of these results.

TABLE 6.1
Summary of Dynamical Formulas for Fixed Axis Motion

a Pure rotation about an axis—no translation.

$$L = I\omega$$
$$\tau = I\alpha$$
$$K = \tfrac{1}{2}I\omega^2$$

b Rotation and translation (subscript 0 refers to center of mass)

$$L_z = I_0\omega + (\mathbf{R} \times M\mathbf{V})_z$$
$$\tau_z = \tau_0 + (\mathbf{R} \times \mathbf{F})_z$$
$$\tau_0 = I_0\alpha$$
$$K = \tfrac{1}{2}I_0\omega^2 + \tfrac{1}{2}MV^2$$

Example 6.15 Disk on Ice

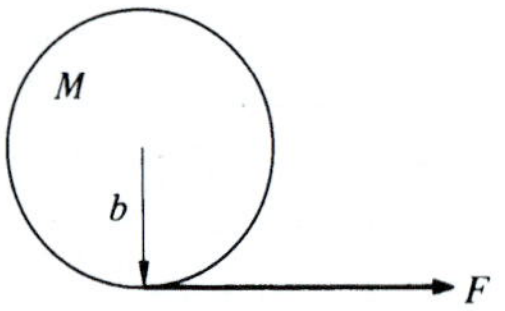

A disk of mass M and radius b is pulled with constant force F by a thin tape wound around its circumference. The disk slides on ice without friction. What is its motion?

We shall solve the problem by two different methods.

METHOD 1
Analyzing the motion about the center of mass we have

$$\begin{aligned} \tau_0 &= bF \\ &= I_0\alpha \end{aligned}$$

or

$$\alpha = \frac{bF}{I_0}.$$

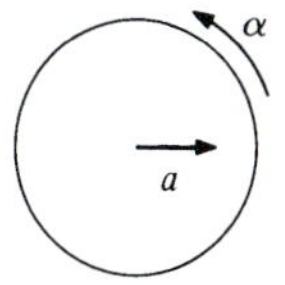

The acceleration of the center of mass is

$$a = \frac{F}{M}.$$

METHOD 2

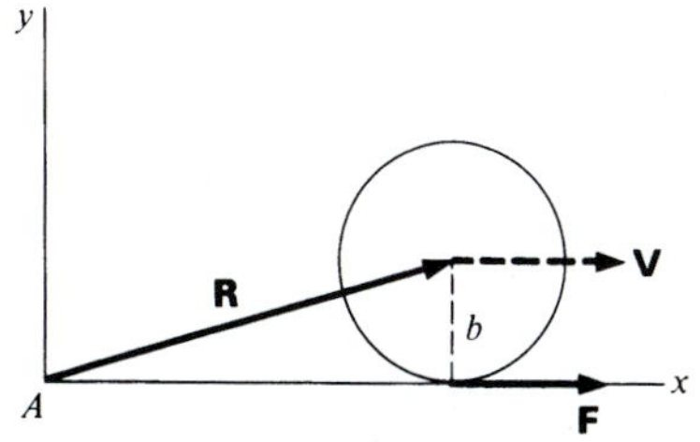

We choose a coordinate system whose origin A is along the line of **F**. The torque about A is, from Table 6.1b,

$$\tau_z = \tau_0 + (\mathbf{R} \times \mathbf{F})_z$$
$$= bF - bF = 0.$$

The torque is zero, as we expect, and angular momentum about the origin is conserved. The angular momentum about A is, from Table 6.1b,

$$L_z = I_0\omega + (\mathbf{R} \times M\mathbf{V})_z$$
$$= I_0\omega - bMV.$$

Since $dL_z/dt = 0$, we have

$$0 = I_0\alpha - bMa$$

or

$$\alpha = \frac{bMa}{I_0} = \frac{bF}{I_0},$$

as before.

Example 6.16 Drum Rolling down a Plane

A uniform drum of radius b and mass M rolls without slipping down a plane inclined at angle θ. Find its acceleration along the plane. The moment of inertia of the drum about its axis is $I_0 = Mb^2/2$.

METHOD 1

The forces acting on the drum are shown in the diagram. f is the force of friction. The translation of the center of mass along the plane is given by

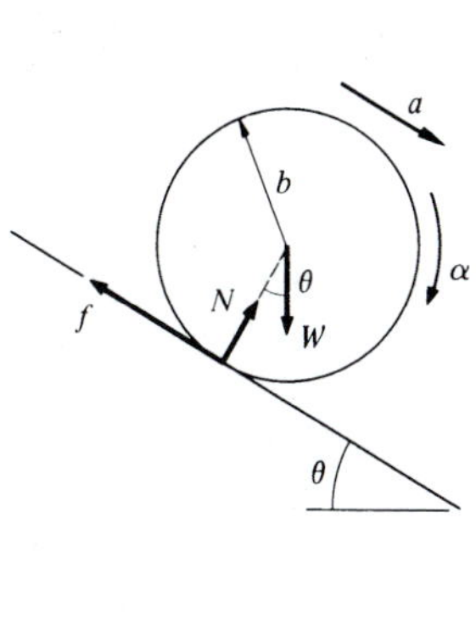

$$W \sin\theta - f = Ma$$

and the rotation about the center of mass by

$$bf = I_0\alpha.$$

For rolling without slipping, we also have

$$a = b\alpha.$$

If we eliminate f, we obtain

$$W \sin\theta - I_0 \frac{\alpha}{b} = Ma.$$

Using $I_0 = Mb^2/2$, and $\alpha = a/b$, we obtain

$$Mg \sin\theta - \frac{Ma}{2} = Ma,$$

or

$$a = \tfrac{2}{3} g \sin\theta.$$

METHOD 2

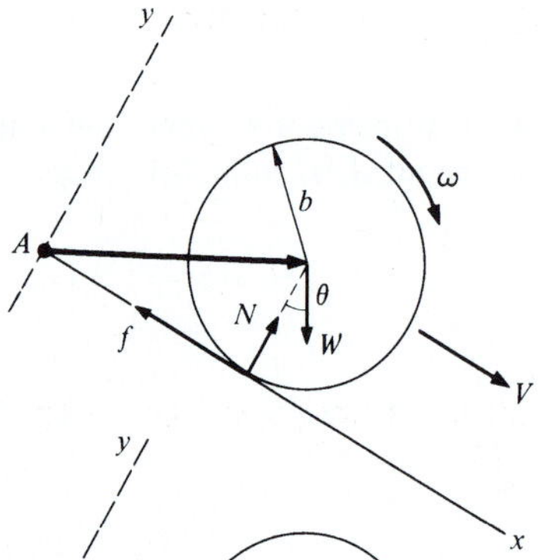

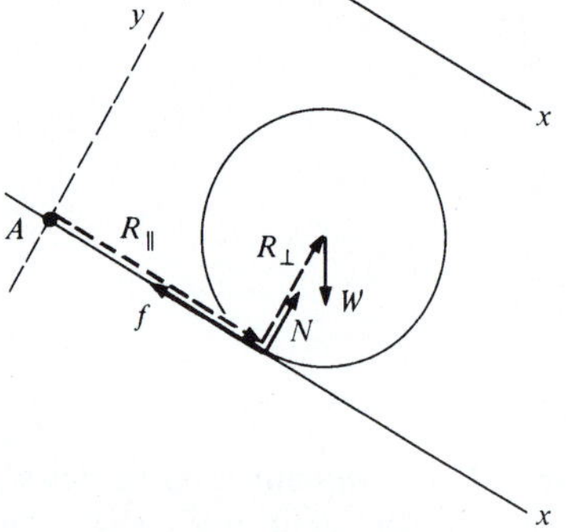

Choose a coordinate system whose origin A is on the plane. The torque about A is

$$\begin{aligned} \tau_s &= \tau_0 + (\mathbf{R} \times \mathbf{F})_z \\ &= -R_\perp f + R_\perp(f - W \sin\theta) + R_\parallel(N - W\cos\theta) \\ &= -bW \sin\theta, \end{aligned}$$

since $R_\perp = b$ and $W\cos\theta = N$. The angular momentum about A is

$$\begin{aligned} L_z &= -I_0\omega + (\mathbf{R} \times M\mathbf{V})_z \\ &= -\tfrac{1}{2}Mb^2\omega - Mb^2\omega \\ &= -\tfrac{3}{2}Mb^2\omega, \end{aligned}$$

where $(\mathbf{R} \times M\mathbf{V})_z = -Mb^2\omega$, as in Example 6.14. Since $\tau_z = dL_z/dt$, we have

$$bW \sin\theta = \frac{3}{2} Mb^2\alpha,$$

or

$$\alpha = \frac{2}{3}\frac{W}{Mb}\sin\theta = \frac{2}{3}\frac{g\sin\theta}{b}.$$

For rolling without slipping, $a = b\alpha$ and

$$a = \tfrac{2}{3} g \sin\theta.$$

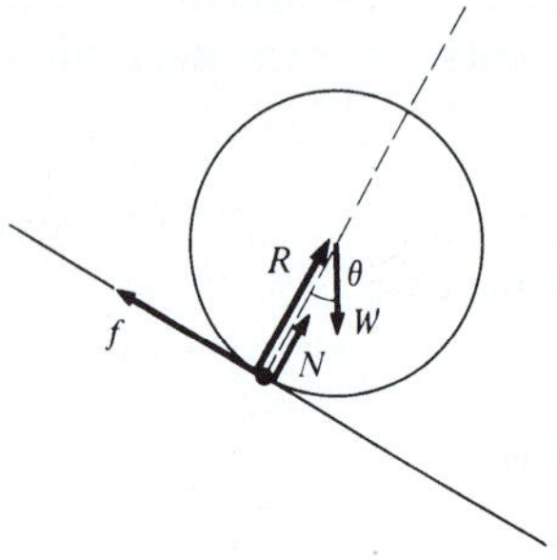

Note that the analysis would have been even more direct if we had chosen the origin at the point of contact. In this case we can calculate τ_z directly from

$$\tau_z = \Sigma(\mathbf{r}_j \times \mathbf{f}_j)_z.$$

Since **f** and **N** act at the origin, the torque is due only to W, and

$$\tau_z = -bW \sin\theta$$

as we obtained above. With this origin, however, the unknown forces **f** and **N** do not appear.

The Work-energy Theorem

In Chap. 4 we derived the work-energy theorem for a particle

$$K_b - K_a = W_{ba}$$

where

$$W_{ba} = \oint_{\mathbf{r}_a}^{\mathbf{r}_b} \mathbf{F} \cdot d\mathbf{r}.$$

We can generalize this for a rigid body and show that the work-energy theorem divides naturally into two parts, one dealing with translational energy and one dealing with rotational energy.

To derive the translational part, we start with the equation of motion for the center of mass.

$$\begin{aligned}\mathbf{F} &= M\frac{d^2\mathbf{R}}{dt^2}\\ &= M\frac{d\mathbf{V}}{dt}.\end{aligned}$$

The work done when the center of mass is displaced by $d\mathbf{R} = \mathbf{V}\,dt$ is

$$\begin{aligned}\mathbf{F} \cdot d\mathbf{R} &= M\frac{d\mathbf{V}}{dt} \cdot \mathbf{V}\,dt\\ &= d(\tfrac{1}{2}MV^2).\end{aligned}$$

Integrating, we obtain

$$\oint_{\mathbf{R}_a}^{\mathbf{R}_b} \mathbf{F} \cdot d\mathbf{R} = \tfrac{1}{2}MV_b^2 - \tfrac{1}{2}MV_a^2. \qquad 6.19$$

Now let us evaluate the work associated with the rotational kinetic energy. The equation of motion for fixed axis rotation about the center of mass is

$$\begin{aligned}\tau_0 &= I_0\alpha\\ &= I_0\frac{d\omega}{dt}.\end{aligned}$$

Rotational kinetic energy has the form $\frac{1}{2}I_0\omega^2$, which suggests that we multiply the equation of motion by $d\theta = \omega\,dt$:

$$\begin{aligned}\tau_0\,d\theta &= I_0\frac{d\omega}{dt}\,\omega\,dt\\ &= d(\tfrac{1}{2}I_0\omega^2).\end{aligned}$$

Integrating, we find that

$$\int_{\theta_a}^{\theta_b} \tau_0 \, d\theta = \tfrac{1}{2}I_0\omega_b^2 - \tfrac{1}{2}I_0\omega_a^2. \qquad 6.20$$

The integral on the left evidently represents the work done by the applied torque.

The general work-energy theorem for a rigid body is therefore

$$K_b - K_a = W_{ba},$$

where $K = \frac{1}{2}MV^2 + \frac{1}{2}I_0\omega^2$ and W_{ba} is the total work done on the body as it moves from position a to position b. We see from Eqs. (6.19) and (6.20) that the work-energy theorem is composed of two independent theorems, one for translation and one for rotation. In many problems these theorems can be applied separately, as the following example shows.

Example 6.17 Drum Rolling down a Plane: Energy Method

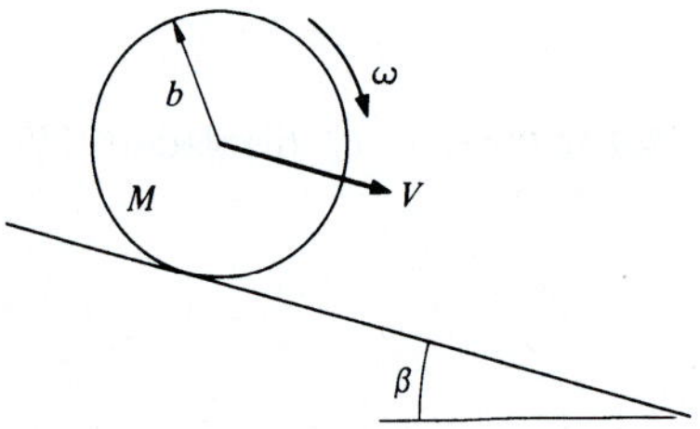

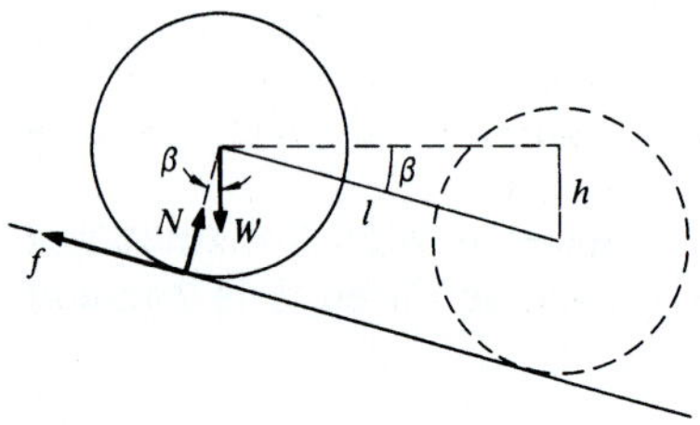

Consider once again a uniform drum of radius b, mass M, and moment of inertia $I_0 = Mb^2/2$ on a plane of angle β. If the drum starts from rest and rolls without slipping, find the speed of its center of mass, V, after it has descended a height h.

The forces on the drum are shown in the sketch. The energy equation for the translational motion is

$$\oint_a^b \mathbf{F} \cdot d\mathbf{r} = \tfrac{1}{2}MV_b^2 - \tfrac{1}{2}MV_a^2$$

or

$$(W \sin \beta - f)l = \tfrac{1}{2}MV^2, \qquad 1$$

where $l = h/\sin \beta$ is the displacement of the center of mass as the drum descends height h.

The energy equation for the rotational motion is

$$\int_{\theta_a}^{\theta_b} \tau \, d\theta = \tfrac{1}{2}I_0\omega_b^2 - \tfrac{1}{2}I_0\omega_a^2$$

or

$$fb\theta = \tfrac{1}{2}I_0\omega^2,$$

where θ is the rotation angle about the center of mass. For rolling without slipping, $b\theta = l$. Hence,

$$fl = \tfrac{1}{2}I_0\omega^2. \qquad 2$$

We also have $\omega = V/b$, so that

$$fl = \frac{1}{2}\frac{I_0V^2}{b^2}.$$

Using this in Eq. (1) to eliminate f gives

$$Wh = \frac{1}{2}\left(\frac{I_0}{b^2} + M\right)V^2$$
$$= \frac{1}{2}\left(\frac{M}{2} + M\right)V^2$$
$$= \tfrac{3}{4}MV^2$$

or

$$V = \sqrt{\frac{4gh}{3}}.$$

An interesting point in this example is that the friction force is not dissipative. From Eq. (1), friction decreases the translational energy by an amount fl. However, from Eq. (2), the torque exerted by friction increases the rotational energy by the same amount. In this motion, friction simply transforms mechanical energy from one mode to another. If slipping occurs, this is no longer the case and some of the mechanical energy is dissipated as heat.

We conclude this section with an example involving constraints which is easily handled by energy methods.

Example 6.18 **The Falling Stick**

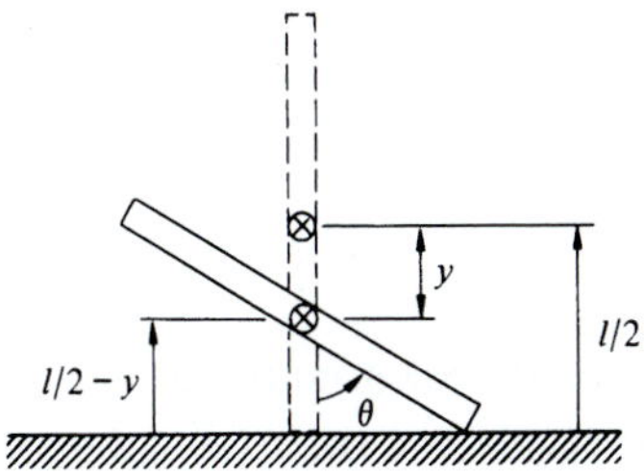

A stick of length l and mass M, initially upright on a frictionless table, starts falling. The problem is to find the speed of the center of mass as a function of position.

The key lies in realizing that since there are no horizontal forces, the center of mass must fall straight down. Since we must find velocity as a function of position, it is natural to apply energy methods.

The sketch shows the stick after it has rotated through angle θ and the center of mass has fallen distance y. The initial energy is

$$E = K_0 + U_0$$
$$= \frac{Mgl}{2}.$$

The kinetic energy at a later time is

$$K = \tfrac{1}{2}I_0\dot{\theta}^2 + \tfrac{1}{2}M\dot{y}^2$$

and the corresponding potential energy is

$$U = Mg\left(\frac{l}{2} - y\right).$$

Since there are no dissipative forces, mechanical energy is conserved and $K + U = K_0 + U_0 = Mgl/2$. Hence,

$$\tfrac{1}{2}M\dot{y}^2 + \tfrac{1}{2}I_0\dot{\theta}^2 + Mg\left(\frac{l}{2} - y\right) = Mg\frac{l}{2}.$$

We can eliminate $\dot{\theta}$ by turning to the constraint equation. From the sketch we see that

$$y = \frac{l}{2}(1 - \cos\theta).$$

Hence,

$$\dot{y} = \frac{l}{2}\sin\theta\,\dot{\theta}$$

and

$$\dot{\theta} = \frac{2}{l\sin\theta}\dot{y}.$$

Since $I_0 = M(l^2/12)$, we obtain

$$\tfrac{1}{2}M\dot{y}^2 + \tfrac{1}{2}M\frac{l^2}{12}\left(\frac{2}{l\sin\theta}\right)^2\dot{y}^2 + Mg\left(\frac{l}{2} - y\right) = Mg\frac{l}{2}$$

or

$$\dot{y}^2 = \frac{2gy}{[1 + 1/(3\sin^2\theta)]},$$

$$\dot{y} = \left[\frac{6gy\sin^2\theta}{3\sin^2\theta + 1}\right]^{1/2}.$$

6.8 The Bohr Atom

We conclude this chapter with an historical account of the Bohr theory of the hydrogen atom. Although this material represents an interesting application of the principles we have encountered, it is not essential to our development of classical mechanics.

The Bohr theory of the hydrogen atom is the major link between classical physics and quantum mechanics. We present here a brief outline of the Bohr theory as an exciting example of the application of concepts we have studied, particularly energy and angular momentum. Our description is similar, though not identical, to Bohr's original paper which he published in 1913 at the age of 26. Although this brief account cannot deal adequately with the background to the Bohr theory, it may give some of the flavor of one of the great chapters in physics.

The development of optical spectroscopy in the nineteenth century made available a great deal of experimental data on the structure of atoms. The light from atoms excited by an electric discharge is radiated only at certain discrete wavelengths characteristic of the element involved, and the last half of the nineteenth century saw tremendous effort in the measurement and interpretation of these line spectra. The wavelength measurements represented a notable experimental achievement; some were made to an accuracy of better than a part in a million. Interpretation, on the other hand, was a dismal failure; aside from certain empirical rules which gave no insight into the underlying physical laws, there was no progress.

The most celebrated empirical formula was discovered in 1886 by the Swiss high school art teacher Joseph Balmer. He found that the wavelengths of the optical spectrum of atomic hydrogen are given within experimental accuracy by the formula

$$\frac{1}{\lambda} = Ry\left(\frac{1}{2^2} - \frac{1}{n^2}\right) \qquad n = 3, 4, 5, \ldots,$$

where λ is the wavelength of a particular spectral line, and Ry is a constant, named the *Rydberg constant* after the Swedish spectroscopist who modified Balmer's formula to apply to certain other spectra. Numerically, $Ry = 109{,}700\ \text{cm}^{-1}$. (In this section we shall follow the tradition of atomic physics by using cgs units.)

Not only did Balmer's formula account for the known lines of hydrogen, $n = 3$ through $n = 6$, it predicted other lines, $n = 7, 8, \ldots$, which were quickly found. Furthermore, Balmer suggested that there might be other lines given by

$$\lambda = Ry\left(\frac{1}{m^2} - \frac{1}{n^2}\right) \qquad m = 3, 4, 5, \ldots \qquad n = m + 1, m + 2, \ldots \qquad 6.21$$

and these, too, were found. (Balmer overlooked the series with $m = 1$, lying in the ultraviolet, which was found in 1916.)

Undoubtedly the Balmer formula contained the key to the structure of hydrogen. Yet no one was able to create a model for an atom which could radiate such a spectrum.

Bohr was familiar with the Balmer formula. He was also familiar with ideas of atomic structure current at the time, ideas based on the experimental researches of J. J. Thomson and Ernest Rutherford. Thomson, working in the Cavendish physical laboratory at Cambridge University, surmised the existence of electrons in 1897. This first indication of the divisibility of the atom stimulated further work, and in 1911 Ernest Rutherford's[1] alpha scattering experiments at the University of Manchester showed that atoms have a charged core which contains most of the mass. Each atom has an integral number of electrons and an equal number of positive charges on the massive core.

A further development in physics which played an essential role in Bohr's theory was Einstein's theory of the photoelectric effect. In 1905, the same year that he published the special theory of relativity, Einstein proposed that the energy transmitted by light consists of discrete "packages," or quanta. The quantum of light is called a *photon*, and Einstein asserted that the energy of a photon is $E = h\nu$, where ν is the frequency of the light and $h = 6.62 \times 10^{-27}$ erg · s is Planck's constant.[2]

Bohr made the following postulates:

1. Atoms cannot possess arbitrary amounts of energy but must exist only in certain *stationary states*. While in a stationary state, an atom does not radiate.

2. An atom can pass from one stationary state a to a lower state by emitting a photon with energy $E_a - E_b$. The frequency of the emitted photon is

$$\nu = \frac{E_a - E_b}{h}. \qquad 6.22$$

3. While in a stationary state, the motion of the atom is given accurately by classical physics.

4. The angular momentum of the atom is $nh/2\pi$, where n is an integer.

Assumption 1, the most drastic, was absolutely necessary to account for the fact that atoms are stable. According to classical theory, an orbiting electron would continuously lose energy by radiation and spiral into the nucleus.

In view of the fact that assumption 1 breaks completely with classical physics, assumption 3 hardly seems justified. Bohr recognized this difficulty and justified the assumption on the ground that the electrodynamical forces connected with the emission of radiation would be very small in comparison with the

[1] Rutherford had earlier been a student of J. J. Thomson and in 1919 succeeded Thomson as director of the Cavendish laboratory. Bohr in turn studied with Rutherford while working out the Bohr theory.

[2] Max Planck had introduced h in 1901 in his theory of radiation from hot bodies.

electrostatic attraction of the charged particles. Possibly the real reason that Bohr continued to apply classical physics to this nonclassical situation was that he felt that at least some of the fundamental concepts of classical physics should carry over into the new physics, and that they should not be discarded until proven to be unworkable.

Bohr did not utilize postulate 4, known as the quantization of angular momentum, in his original work, although he pointed out the possibility of doing so. It has become traditional to treat this postulate as a fundamental assumption.

Let us apply these four postulates to hydrogen. The hydrogen atom consists of a single electron of charge $-e$ and mass m_0, and a nucleus of charge $+e$ and mass M. We assume that the massive nucleus is essentially at rest and that the electron is in a circular orbit of radius r with velocity v. The radial equation of motion is

$$-\frac{m_0 v^2}{r} = -\frac{e^2}{r^2}, \qquad 6.23$$

where $-e^2/r^2$ is the attractive Coulomb force between the charges The energy is

$$E = K + U = \tfrac{1}{2}m_0 v^2 - \frac{e^2}{r}. \qquad 6.24$$

Equations (6.23) and (6.24) yield

$$E = -\frac{1}{2}\frac{e^2}{r}. \qquad 6.25$$

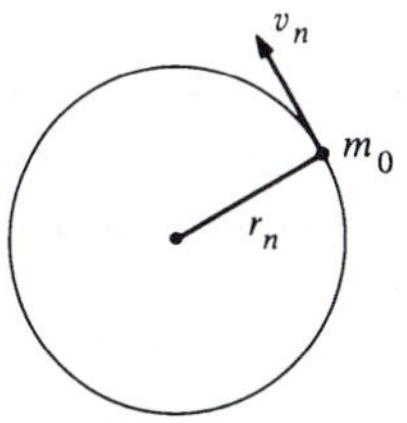

By postulate 4, the angular momentum is $nh/2\pi$, where n is an integer. Labeling the orbit parameters by n, we have

$$\frac{nh}{2\pi} = m_0 r_n v_n. \qquad 6.26$$

Equations (6.26) and (6.23) yield

$$r_n = \frac{n^2h^2}{m_0 e^2}\frac{1}{(2\pi)^2}, \qquad 6.27$$

and Eq. (6.25) gives

$$E_n = -\frac{1}{2}\frac{(2\pi)^2 m_0 e^4}{n^2h^2}. \qquad 6.28$$

If the electron makes a transition from state n to state m, the emitted photon has frequency

$$\nu = \frac{E_n - E_m}{h}$$

$$= \frac{(2\pi)^2}{2}\frac{m_0e^4}{h^3}\left(\frac{1}{m^2} - \frac{1}{n^2}\right). \qquad 6.29$$

The wavelength of the radiation is given by

$$\frac{1}{\lambda} = \frac{\nu}{c}$$

$$= \frac{2\pi^2}{c}\frac{m_0e^4}{h^3}\left(\frac{1}{m^2} - \frac{1}{n^2}\right). \qquad 6.30$$

This is identical in form to the Balmer formula, Eq. 6.21. What is even more impressive is that the numerical coefficients agree extemely well; Bohr was able to calculate the Rydberg constant from the fundamental atomic constants.

The Bohr theory, with its strong flavor of elementary classical mechanics, formed an important bridge between classical physics and present-day atomic theory. Although the Bohr theory was unsuccessful in explaining more complicated atoms, the impetus provided by Bohr's work led to the development of modern quantum mechanics in the 1920s.

Note 6.1 Chasles' Theorem

Chasles' theorem asserts that is always possible to represent an arbitrary displacement of a rigid body by a translation of its center of mass plus a rotation about its center of mass. This appendix is rather detailed and an understanding of it is not necessary for following the development of the text. However, the result is interesting and its proof provides a nice exercise in vector methods for those interested.

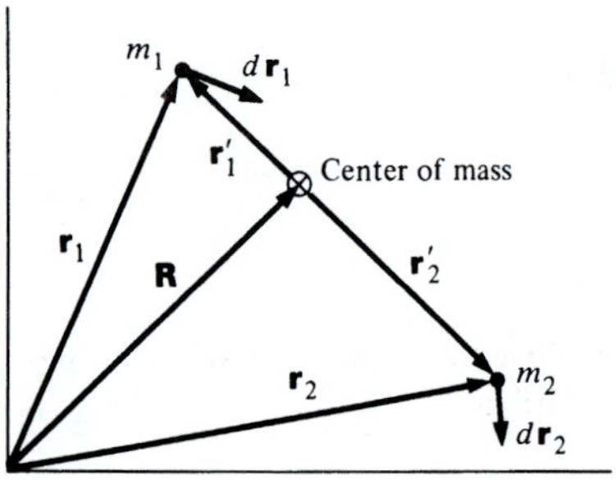

To avoid algebraic complexities, we consider here a simple rigid body consisting of two masses m_1 and m_2 joined by a rigid rod of length l. The position vectors of m_1 and m_2 are $\mathbf{r}_1$ and $\mathbf{r}_2$, respectively, as shown in the sketch. The position vector of the center of mass of the body is $\mathbf{R}$, and $\mathbf{r}_1'$ and $\mathbf{r}_2'$ are the position vectors of m_1 and m_2 with respect to the center of mass. The vectors $\mathbf{r}_1'$ and $\mathbf{r}_2'$ are back to back along the same line.

In an arbitrary displacement of the body, m_1 is displaced by $d\mathbf{r}_1$ and m_2 is displaced by $d\mathbf{r}_2$. Because the body is rigid, $d\mathbf{r}_1$ and $d\mathbf{r}_2$ are not independent, and we begin our analysis by finding their relation. The distance between m_1 and m_2 is fixed and of length l. Therefore,

$$|\mathbf{r}_1 - \mathbf{r}_2| = l$$

or

$$(\mathbf{r}_1 - \mathbf{r}_2) \cdot (\mathbf{r}_1 - \mathbf{r}_2) = l^2. \qquad 1$$

Taking differentials of Eq. (1),[1]

$$(\mathbf{r}_1 - \mathbf{r}_2) \cdot (d\mathbf{r}_1 - d\mathbf{r}_2) = 0. \qquad 2$$

Equation (2) is the "rigid body condition" we seek. There are evidently two ways of satisfying Eq. (2): either $d\mathbf{r}_1 = d\mathbf{r}_2$, or $(d\mathbf{r}_1 - d\mathbf{r}_2)$ is perpendicular to $(\mathbf{r}_1 - \mathbf{r}_2)$.

We now turn to the translational motion of the center of mass. By definition,

$$\mathbf{R} = \frac{m_1\mathbf{r}_1 + m_2\mathbf{r}_2}{m_1 + m_2}.$$

Therefore, the displacement $d\mathbf{R}$ of the center of mass is

$$d\mathbf{R} = \frac{m_1 d\mathbf{r}_1 + m_2 d\mathbf{r}_2}{m_1 + m_2}. \qquad 3$$

If we subtract this translational displacement from $d\mathbf{r}_1$ and $d\mathbf{r}_2$, the residual displacements $d\mathbf{r}_1 - d\mathbf{R}$ and $d\mathbf{r}_2 - d\mathbf{R}$ should give a pure rotation about the center of mass. Before investigating this point, we notice that since

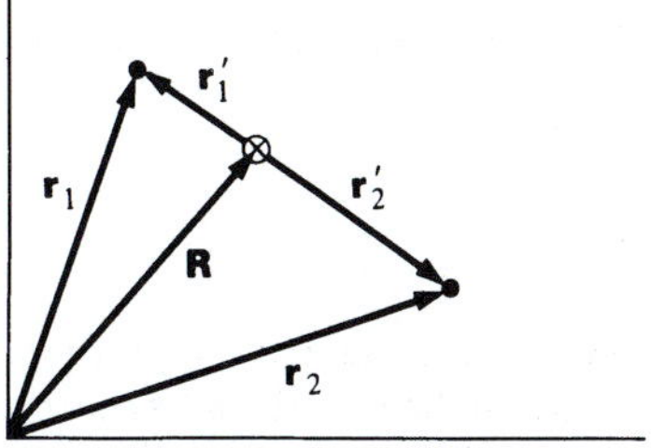

$$\mathbf{r}_1 - \mathbf{R} = \mathbf{r}'_1$$
$$\mathbf{r}_2 - \mathbf{R} = \mathbf{r}'_2,$$

the residual displacements are

$$d\mathbf{r}_1 - d\mathbf{R} = d\mathbf{r}'_1$$
$$d\mathbf{r}_2 - d\mathbf{R} = d\mathbf{r}'_2. \qquad 4$$

Using Eq. (3) in Eq. (4) we have

$$d\mathbf{r}'_1 = d\mathbf{r}_1 - d\mathbf{R}$$
$$= \left(\frac{m_2}{m_1 + m_2}\right)(d\mathbf{r}_1 - d\mathbf{r}_2) \qquad 5$$

and

$$d\mathbf{r}'_2 = d\mathbf{r}_2 - d\mathbf{R}$$
$$= -\left(\frac{m_1}{m_1 + m_2}\right)(d\mathbf{r}_1 - d\mathbf{r}_2). \qquad 6$$

Note that if $d\mathbf{r}_1 = d\mathbf{r}_2$, the residual displacements $d\mathbf{r}'_1$ and $d\mathbf{r}'_2$ are zero and the rigid body translates without rotating.

[1] Remember that $d(\mathbf{A} \cdot \mathbf{A}) = 2\mathbf{A} \cdot d\mathbf{A}$.

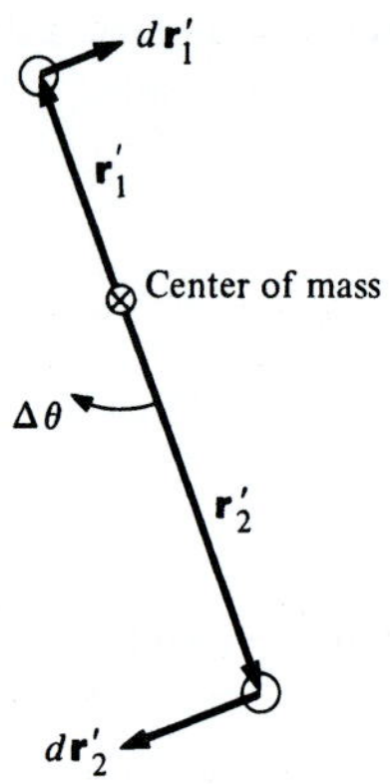

We must show that the residual displacements represent a pure rotation about the center of mass to complete the theorem. The sketch shows what a pure rotation would look like. First we show that $d\mathbf{r}_1'$ and $d\mathbf{r}_2'$ are perpendicular to the line $\mathbf{r}_1' - \mathbf{r}_2'$.

$$\begin{aligned} d\mathbf{r}_1' \cdot (\mathbf{r}_1' - \mathbf{r}_2') &= d\mathbf{r}_1' \cdot (\mathbf{r}_1 - \mathbf{r}_2) \\ &= \left(\frac{m_2}{m_1 + m_2}\right)(d\mathbf{r}_1 - d\mathbf{r}_2) \cdot (\mathbf{r}_1 - \mathbf{r}_2) \\ &= 0, \end{aligned}$$

where we have used Eq. (5) and the rigid body condition, Eq. (2). Similarly,

$$d\mathbf{r}_2' \cdot (\mathbf{r}_1' - \mathbf{r}_2') = 0.$$

Finally, we require that the residual displacements correspond to rotation through the same angle, $\Delta\theta$. With reference to our sketch, this condition in vector form is

$$\frac{d\mathbf{r}_1'}{r_1'} = -\frac{d\mathbf{r}_2'}{r_2'}.$$

Keeping in mind that

$$\frac{r_1'}{r_2'} = \frac{m_2}{m_1}$$

by definition of center of mass, and using Eq. (5) and (6), we have

$$\begin{aligned} \frac{d\mathbf{r}_1'}{r_1'} &= \left(\frac{m_2}{m_1 + m_2}\right)\frac{(d\mathbf{r}_1 - d\mathbf{r}_2)}{r_1'} \\ &= \left(\frac{m_1}{m_1 + m_2}\right)\frac{(d\mathbf{r}_1 - d\mathbf{r}_2)}{r_2'} \\ &= -\frac{d\mathbf{r}_2'}{r_2'}, \end{aligned}$$

completing the proof.

Note 6.2 Pendulum Motion

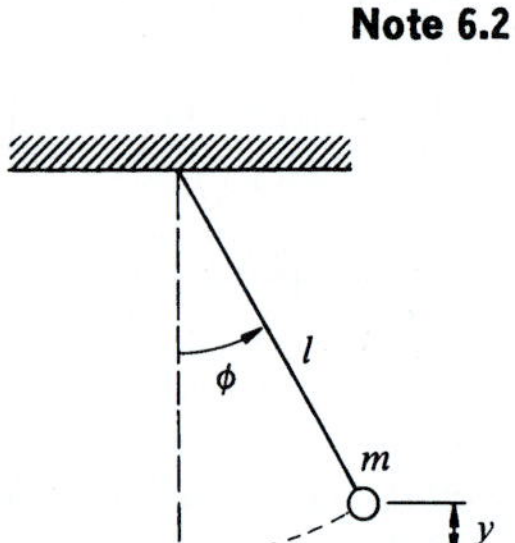

The motion of a body moving under conservative forces can always be solved formally by energy methods, and it is natural to use this approach to find the motion of a pendulum.

The total energy of the pendulum is

$$\begin{aligned} E &= K + U \\ &= \tfrac{1}{2}ml^2\dot{\phi}^2 + mgy, \end{aligned}$$

where l is the length of the pendulum and y is the vertical distance from the lowest point. From the sketch we have $y = l(1 - \cos\phi)$.

At the end of the swing, $\phi = \phi_0$ and $\dot{\phi} = 0$. The total energy is

$E = mgl(1 - \cos\phi_0)$.

The energy equation is

$$\tfrac{1}{2}ml^2\dot{\phi}^2 + mgl(1 - \cos\phi) = mgl(1 - \cos\phi_0),$$

$$\frac{d\phi}{dt} = \sqrt{\frac{2g}{l}(\cos\phi - \cos\phi_0)},$$

and

$$\int \frac{d\phi}{\sqrt{\cos\phi - \cos\phi_0}} = \sqrt{\frac{2g}{l}} \int dt. \qquad 1$$

Before looking at the general solution, let us find the solution for the case of small amplitudes. With the approximation $\cos\phi \approx 1 - \frac{1}{2}\phi^2$, we have

$$\int \frac{d\phi}{\sqrt{\frac{1}{2}}\sqrt{\phi_0^2 - \phi^2}} = \sqrt{\frac{2g}{l}} \int dt.$$

Let us integrate over one-fourth of the swing, from $\phi = 0$ to $\phi = \phi_0$. The time varies between $t = 0$ and $t = T/4$, where T is the period. We have

$$\int_0^{\phi_0} \frac{d\phi}{\sqrt{\frac{1}{2}}\,\phi_0\sqrt{1 - (\phi/\phi_0)^2}} = \sqrt{\frac{2g}{l}} \int_0^{T/4} dt$$

or

$$\arcsin \frac{\phi}{\phi_0}\Big|_0^{\phi_0} = \sqrt{\frac{g}{l}}\,\frac{T}{4}$$

$$\frac{\pi}{2} - 0 = \sqrt{\frac{g}{l}}\,\frac{T}{4}$$

$$T = 2\pi\sqrt{\frac{l}{g}},$$

as we found in the text.

To obtain a more accurate solution to Eq. (1), it is helpful to use the identity $\cos\phi = 1 - 2\sin^2(\phi/2)$. Then

$$\cos\phi - \cos\phi_0 = 2[\sin^2(\phi_0/2) - \sin^2(\phi/2)]. \qquad 2$$

Introducing Eq. (2) in Eq. (1) gives

$$\int \frac{d\phi}{\sqrt{2}\sqrt{\sin^2(\phi_0/2) - \sin^2(\phi/2)}} = \sqrt{\frac{2g}{l}} \int dt. \qquad 3$$

Now let us change variables as follows:

$$\sin u = \frac{\sin(\phi/2)}{\sin(\phi_0/2)}. \qquad 4$$

As the pendulum swings through a cycle, ϕ varies between $-\phi_0$ and $+\phi_0$. At the same time, u varies between $-\pi$ and $+\pi$. If we let

$$K = \sin\frac{\phi_0}{2},$$

then

$$\sin\frac{\phi}{2} = K \sin u$$

$$\frac{1}{2}\cos\frac{\phi}{2}\,d\phi = K\cos u\,du$$

and

$$d\phi = \left(\frac{1 - \sin^2 u}{1 - K^2\sin^2 u}\right)^{\frac{1}{2}} 2K\,du. \qquad 5$$

Substituting Eqs. (4) and (5) in Eq. (3) gives

$$\int \frac{du}{\sqrt{1 - K^2\sin^2 u}} = \sqrt{\frac{g}{l}}\int dt.$$

Let us take the integral over one period. The limits on u are 0 and 2π, while t ranges from 0 to T. We have

$$\int_0^{2\pi} \frac{du}{\sqrt{1 - K^2\sin^2 u}} = \sqrt{\frac{g}{l}}\,T. \qquad 6$$

The integral on the left is an *elliptic integral*: specifically, it is a complete elliptic integral of the first kind. Values for this function are available from computed tables. However, for our purposes it is more convenient to expand the integrand:

$$(1 - K^2\sin^2 u)^{-\frac{1}{2}} = 1 + \tfrac{1}{2}K^2\sin^2 u + \cdots$$

and

$$T = \sqrt{\frac{l}{g}}\int_0^{2\pi} du(1 + \tfrac{1}{2}K^2\sin^2 u + \cdots)$$

$$= \sqrt{\frac{l}{g}}\left(2\pi + \frac{2\pi}{4}K^2 + \cdots\right)$$

$$= 2\pi\sqrt{\frac{l}{g}}\left(1 + \frac{1}{4}\sin^2\frac{\phi_0}{2} + \cdots\right).$$

If $\phi_0 \ll 1$, then $\sin^2(\phi_0/2) \approx \phi_0^2/4$, and we have

$$T = 2\pi\sqrt{\frac{l}{g}}(1 + \tfrac{1}{16}\phi_0^2 + \cdots). \qquad 7$$

Problems

6.1 *a*. Show that if the total linear momentum of a system of particles is zero, the angular momentum of the system is the same about all origins.

b. Show that if the total force on a system of particles is zero, the torque on the system is the same about all origins.

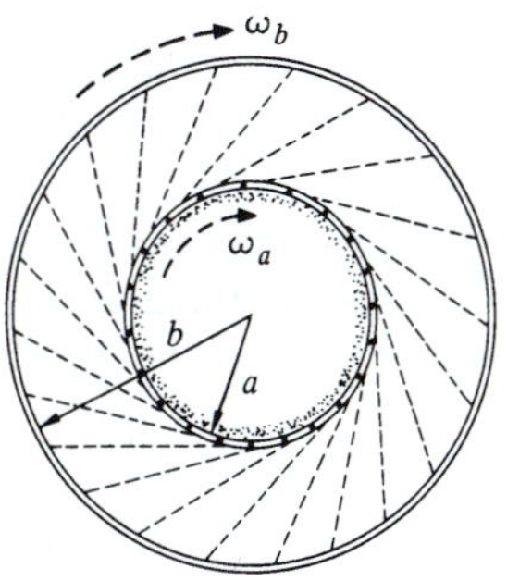

6.2 A drum of mass M_A and radius a rotates freely with initial angular velocity $\omega_A(0)$. A second drum with mass M_B and radius $b > a$ is mounted on the same axis and is at rest, although it is free to rotate. A thin layer of sand with mass M_s is distributed on the inner surface of the smaller drum. At $t = 0$, small perforations in the inner drum are opened. The sand starts to fly out at a constant rate λ and sticks to the outer drum. Find the subsequent angular velocities of the two drums ω_A and ω_B. Ignore the transit time of the sand.

Ans. clue. If $\lambda t = M_b$ and $b = 2a$, then $\omega_B = \omega_A(0)/8$

6.3 A ring of mass M and radius R lies on its side on a frictionless table. It is pivoted to the table at its rim. A bug of mass m walks around the ring with speed v, starting at the pivot. What is the rotational velocity of the ring when the bug is (*a*) halfway around and (*b*) back at the pivot.

Ans. clue. (*a*) If $m = M$, $\omega = v/3R$

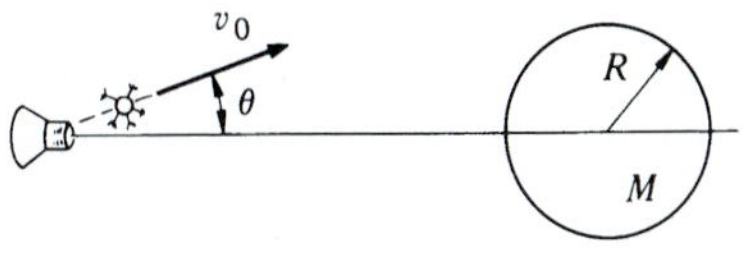

6.4 A spaceship is sent to investigate a planet of mass M and radius R. While hanging motionless in space at a distance $5R$ from the center of the planet, the ship fires an instrument package with speed v_0, as shown in the sketch. The package has mass m, which is much smaller than the mass of the spaceship. For what angle θ will the package just graze the surface of the planet?

6.5 A 3,000-lb car is parked on a 30° slope, facing uphill. The center of mass of the car is halfway between the front and rear wheels and is 2 ft above the ground. The wheels are 8 ft apart. Find the normal force exerted by the road on the front wheels and on the rear wheels.

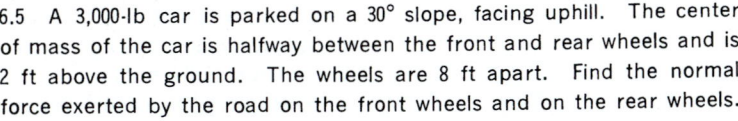

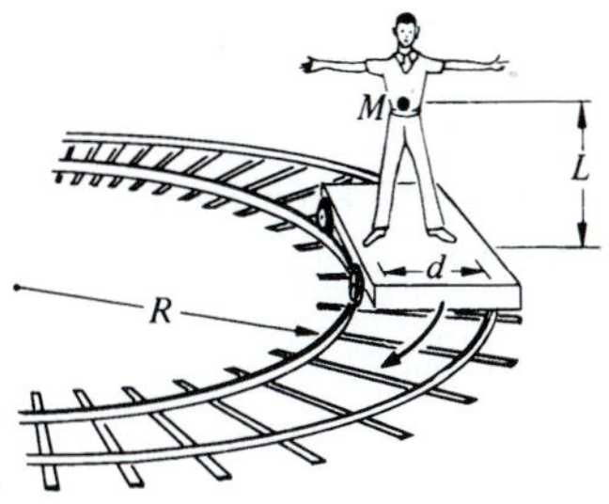

6.6 A man of mass M stands on a railroad car which is rounding an unbanked turn of radius R at speed v. His center of mass is height L above the car, and his feet are distance d apart. The man is facing the direction of motion. How much weight is on each of his feet?

6.7 Find the moment of inertia of a thin sheet of mass M in the shape of an equilateral triangle about an axis through a vertex, perpendicular to the sheet. The length of each side is L.

6.8 Find the moment of inertia of a uniform sphere of mass M and radius R about an axis through the center.

Ans. $I_0 = \frac{2}{5}MR^2$

6.9 A heavy uniform bar of mass M rests on top of two identical rollers which are continuously turned rapidly in opposite directions, as shown.

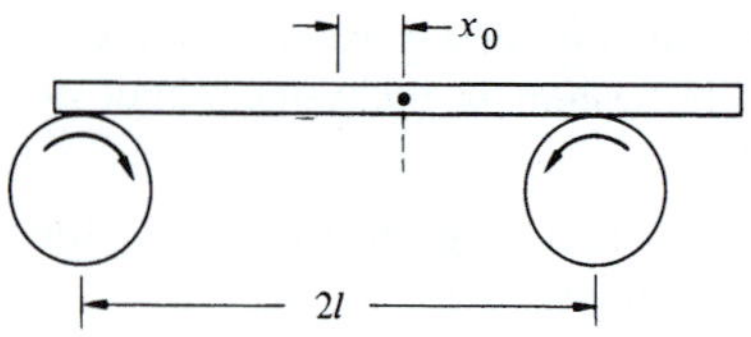

The centers of the rollers are a distance $2l$ apart. The coefficient of friction between the bar and the roller surfaces is μ, a constant independent of the relative speed of the two surfaces.

Initially the bar is held at rest with its center at distance x_0 from the midpoint of the rollers. At time $t = 0$ it is released. Find the subsequent motion of the bar.

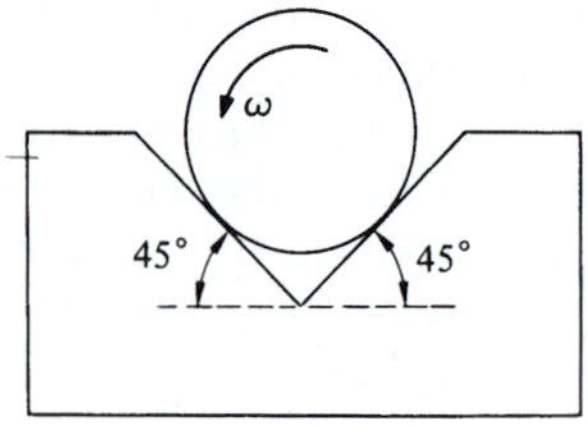

6.10 A cylinder of mass M and radius R is rotated in a uniform V groove with constant angular velocity ω. The coefficient of friction between the cylinder and each surface is μ. What torque must be applied to the cylinder to keep it rotating?

Ans. clue. If $\mu = 0.5$, $R = 0.1$ m, $W = 100$ N, then $\tau \approx 5.7$ N·m

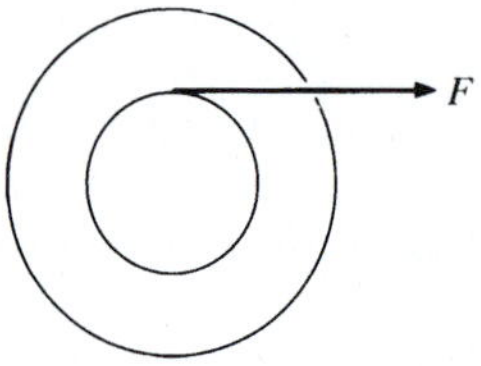

6.11 A wheel is attached to a fixed shaft, and the system is free to rotate without friction. To measure the moment of inertia of the wheel-shaft system, a tape of negligible mass wrapped around the shaft is pulled with a known constant force F. When a length L of tape has unwound, the system is rotating with angular speed ω_0. Find the moment of inertia of the system, I_0.

Ans. clue. If $F = 10$ N, $L = 5$ m, $\omega_0 = 0.5$ rad/s, then $I_0 = 400$ kg·m^2

6.12 A pivoted beam has a mass M_1 suspended from one end and an Atwood's machine suspended from the other (see sketch at left below). The frictionless pulley has negligible mass and dimension. Gravity is directed downward, and $M_2 > M_3$.

Find a relation between M_1, M_2, M_3, l_1, and l_2 which will ensure that the beam has no tendency to rotate just after the masses are released.

6.13 Mass m is attached to a post of radius R by a string (see right hand sketch below). Initially it is distance r from the center of the post and is moving tangentially with speed v_0. In case (*a*) the string passes through a hole in the center of the post at the top. The string is gradually shortened by drawing it through the hole. In case (*b*) the string wraps around the outside of the post.

What quantities are conserved in each case? Find the final speed of the mass when it hits the post for each case.

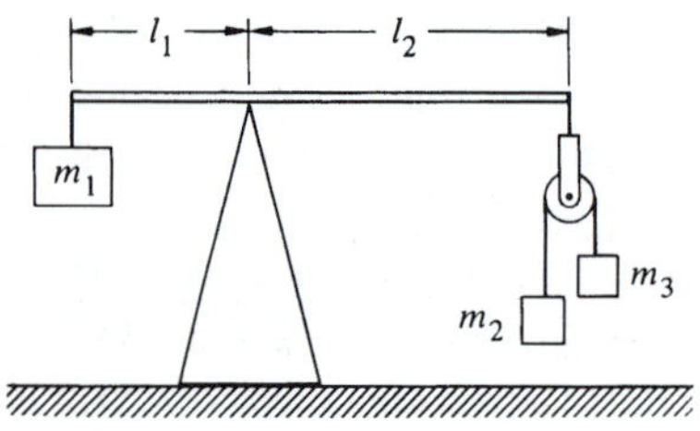

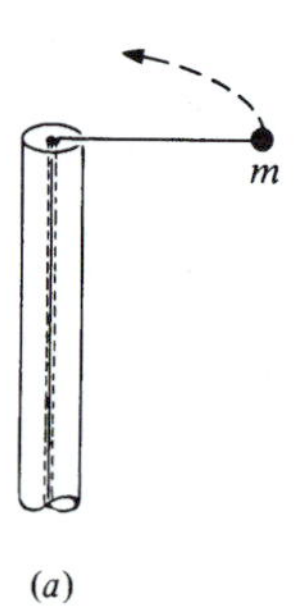

(*a*)

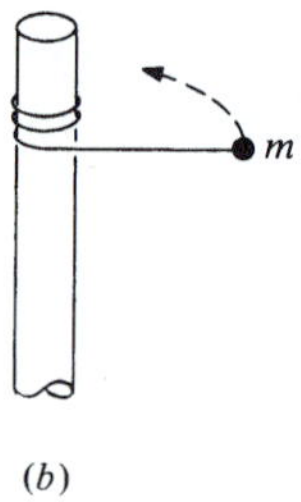

(*b*)

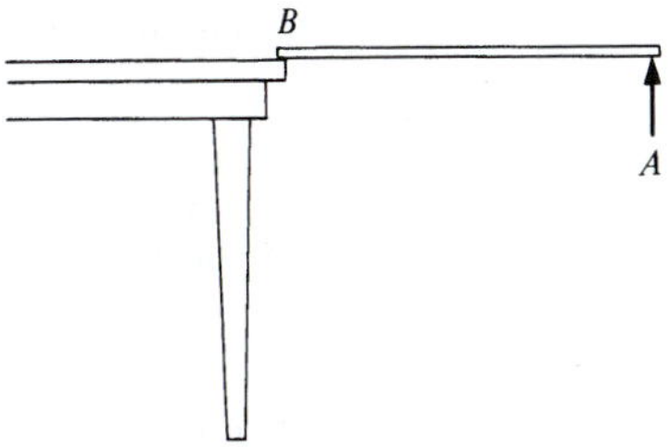

6.14 A uniform stick of mass M and length l is suspended horizontally with end B on the edge of a table, and the other end, A is held by hand. Point A is suddenly released. At the instant after release:

a. What is the torque about B?

b. What is the angular acceleration about B?

c. What is the vertical acceleration of the center of mass?

Ans. $3g/4$

d. From part c, find by inspection the vertical force at B.

Ans. $mg/4$

6.15 A pendulum is made of two disks each of mass M and radius R separated by a massless rod. One of the disks is pivoted through its center by a small pin. The disks hang in the same plane and their centers are a distance l apart. Find the period for small oscillations.

6.16 A physical pendulum is made of a uniform disk of mass M and radius R suspended from a rod of negligible mass. The distance from the pivot to the center of the disk is l. What value of l makes the period a minimum?

6.17 A rod of length l and mass m, pivoted at one end, is held by a spring at its midpoint and a spring at its far end, both pulling in opposite directions. The springs have spring constant k, and at equilibrium their pull is perpendicular to the rod. Find the frequency of small oscillations about the equilibrium position. See figure below left

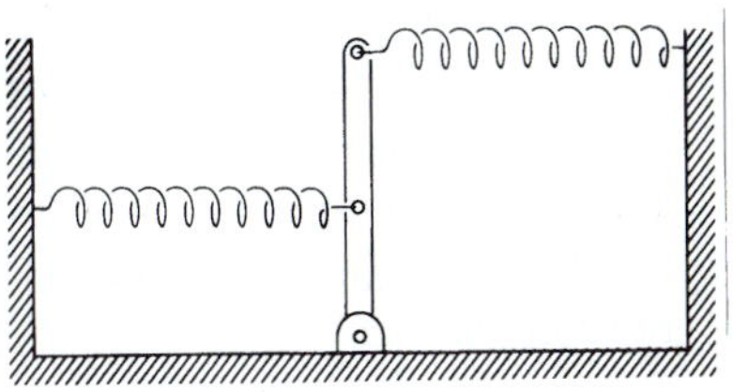

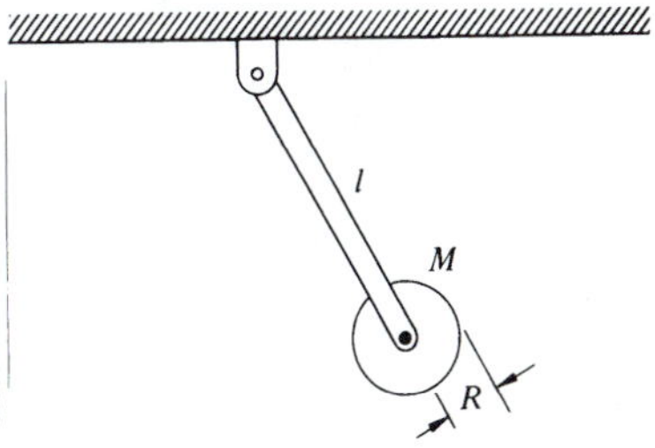

6.18 Find the period of a pendulum consisting of a disk of mass M and radius R fixed to the end of a rod of length l and mass m. How does the period change if the disk is mounted to the rod by a frictionless bearing so that it is perfectly free to spin? See figure above right

6.19 A solid disk of mass M and radius R is on a vertical shaft. The shaft is attached to a coil spring which exerts a linear restoring torque of magnitude $C\theta$, where θ is the angle measured from the static equilibrium position and C is a constant. Neglect the mass of the shaft and the spring, and assume the bearings to be frictionless.

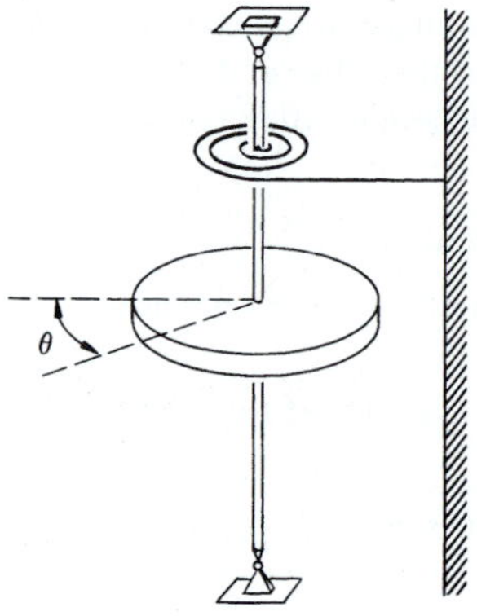

a. Show that the disk can undergo simple harmonic motion, and find the frequency of the motion.

b. Suppose that the disk is moving according to $\theta = \theta_0 \sin(\omega t)$, where ω is the frequency found in part a. At time $t_1 = \pi/\omega$, a ring of sticky putty of mass M and radius R is dropped concentrically on the disk. Find:

(1) The new frequency of the motion
(2) The new amplitude of the motion

6.20 A thin plank of mass M and length l is pivoted at one end (see figure below). The plank is released at 60° from the vertical. What is the magnitude and direction of the force on the pivot when the plank is horizontal?

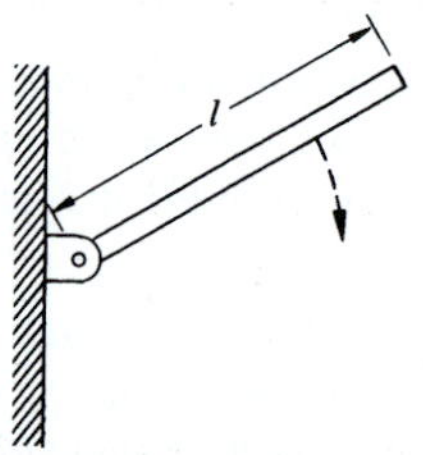

6.21 A cylinder of radius R and mass M rolls without slipping down a plane inclined at angle θ. The coefficient of friction is μ.

What is the maximum value of θ for the cylinder to roll without slipping?

Ans. $\theta = \arctan 3\mu$

6.22 A bead of mass m slides without friction on a rod that is made to rotate at a constant angular velocity ω. Neglect gravity.

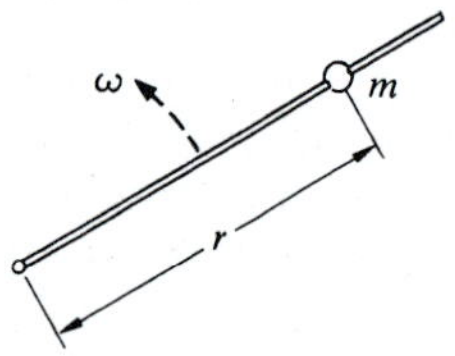

a. Show that $r = r_0 e^{\omega t}$ is a possible motion of the bead, where r_0 is the initial distance of the bead from the pivot.

b. For the motion described in part a, find the force exerted on the bead by the rod.

c. For the motion described above, find the power exerted by the agency which is turning the rod and show by direct calculation that this power equals the rate of change of kinetic energy of the bead.

6.23 A disk of mass M and radius R unwinds from a tape wrapped around it (see figure below at left). The tape passes over a frictionless pulley, and a mass m is suspended from the other end. Assume that the disk drops vertically.

a. Relate the accelerations of m and the disk, a and A, respectively, to the angular acceleration of the disk.

Ans. clue. If $A = 2a$, then $\alpha = \dfrac{3a}{R}$

b. Find a, A and α.

6.24 Drum A of mass M and radius R is suspended from a drum B also of mass M and radius R, which is free to rotate about its axis (see sketch below right). The suspension is in the form of a massless metal tape wound around the outside of each drum, and free to unwind, as shown. Gravity is directed downward. Both drums are initially at rest. Find the initial acceleration of drum A, assuming that it moves straight down.

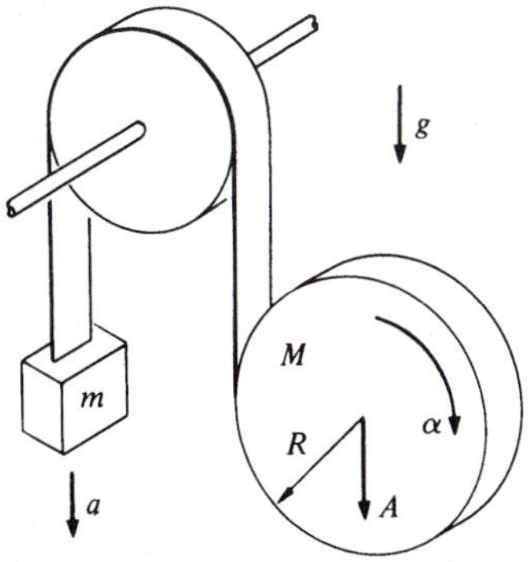

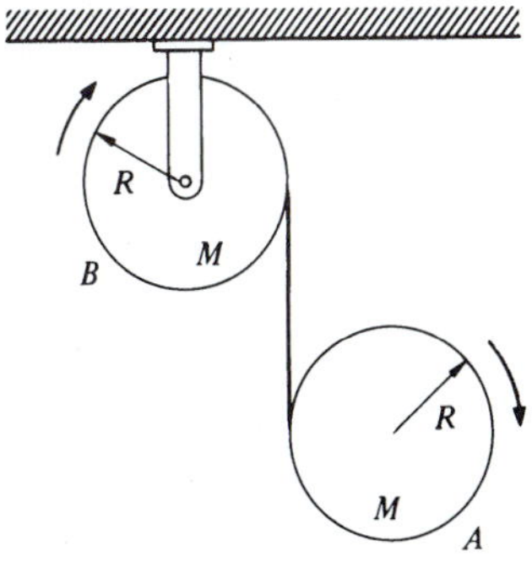

6.25 A marble of mass M and radius R is rolled up a plane of angle θ. If the initial velocity of the marble is v_0, what is the distance l it travels up the plane before it begins to roll back down?

Ans. clue. If $v_0 = 3$ m/s, $\theta = 30°$, then $l \approx 1.3$ m

6.26 A uniform sphere of mass M and radius R and a uniform cylinder of mass M and radius R are released simultaneously from rest at the top of an inclined plane. Which body reaches the bottom first if they both roll without slipping?

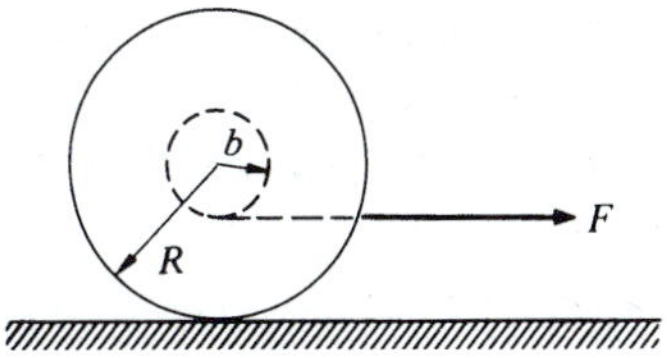

6.27 A Yo-Yo of mass M has an axle of radius b and a spool of radius R. Its moment of inertia can be taken to be $MR^2/2$. The Yo-Yo is placed upright on a table and the string is pulled with a horizontal force F as shown. The coefficient of friction between the Yo-Yo and the table is μ.

What is the maximum value of F for which the Yo-Yo will roll without slipping?

6.28 The Yo-Yo of the previous problem is pulled so that the string makes an angle θ with the horizontal. For what value of θ does the Yo-Yo have no tendency to rotate?

6.29 A Yo-Yo of mass M has an axle of radius b and a spool of radius R. Its moment of inertia can be taken to be $MR^2/2$ and the thickness of the string can be neglected. The Yo-Yo is released from rest.

a. What is the tension in the cord as the Yo-Yo descends and as it ascends?

b. The center of the Yo-Yo descends distance h before the string is fully unwound. Assuming that it reverses direction with uniform spin velocity, find the maximum force on the string while the Yo-Yo turns around.

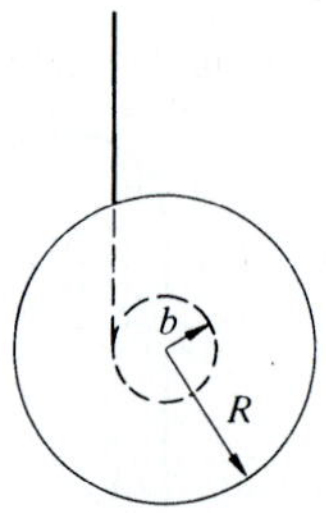

6.30 A bowling ball is thrown down the alley with speed v_0. Initially it slides without rolling, but due to friction it begins to roll. Show that its speed when it rolls without sliding is $\frac{5}{7}v_0$.

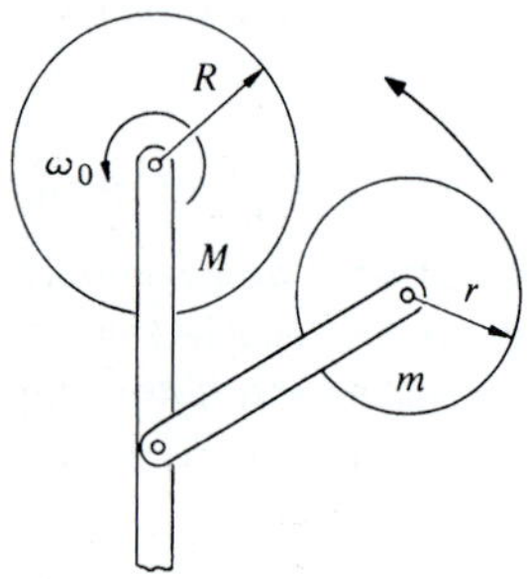

6.31 A cylinder of radius R spins with angular velocity ω_0. When the cylinder is gently laid on a plane, it skids for a short time and eventually rolls without slipping. What is the final angular velocity, ω_f?

Ans. clue. If $\omega_0 = 3$ rad/s, $\omega_f = 1$ rad/s

6.32 A solid rubber wheel of radius R and mass M rotates with angular velocity ω_0 about a frictionless pivot (see sketch at left). A second rubber wheel of radius r and mass m, also mounted on a frictionless pivot, is brought into contact with it. What is the final angular velocity of the first wheel?

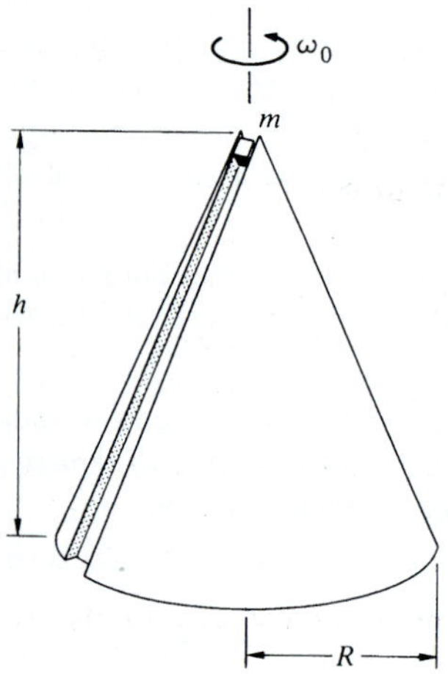

6.33 A cone of height h and base radius R is free to rotate about a fixed vertical axis. It has a thin groove cut in the surface. The cone is set rotating freely with angular speed ω_0, and a small block of mass m is released in the top of the frictionless groove and allowed to slide under gravity. Assume that the block stays in the groove. Take the moment of inertia of the cone about the vertical axis to be I_0.

a. What is the angular velocity of the cone when the block reaches the bottom?

b. Find the speed of the block in inertial space when it reaches the bottom.

6.34 A marble of radius b rolls back and forth in a shallow dish of radius R. Find the frequency of small oscillations. $R \gg b$.

Ans. $\omega = \sqrt{5g/7R}$

6.35 A cubical block of side L rests on a fixed cylindrical drum of radius R. Find the largest value of L for which the block is stable. See figure below left.

6.36 Two masses m_A and m_B are connected by a string of length l and lie on a frictionless table. The system is twirled and released with m_A instantaneously at rest and m_B moving with instantaneous velocity v_0 at right angles to the line of centers, as shown below right.

Find the subsequent motion of the system and the tension in the string.

Ans. clue. If $m_A = m_B = 2$ kg, $v_0 = 3$ m/s, $l = 0.5$ m, then $T = 18$ N

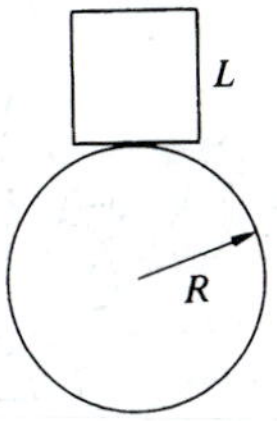

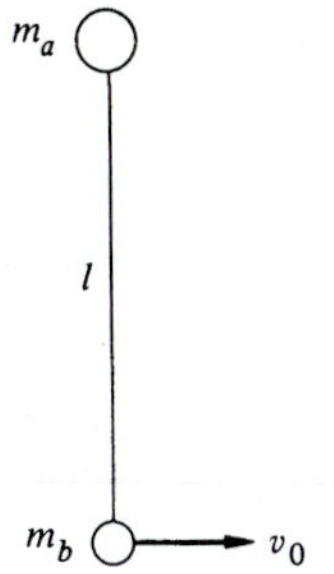

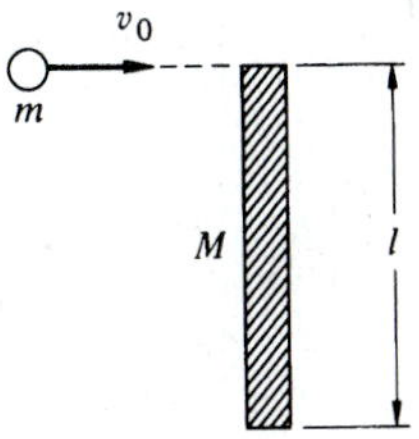

6.37 a. A plank of length $2l$ and mass M lies on a frictionless plane. A ball of mass m and speed v_0 strikes its end as shown. Find the final velocity of the ball, v_f, assuming that mechanical energy is conserved and that v_f is along the original line of motion.

b. Find v_f assuming that the stick is pivoted at the lower end.

Ans. clue. For $m = M$, (a) $v_f = 3v_0/5$; (b) $v_f = v_0/2$

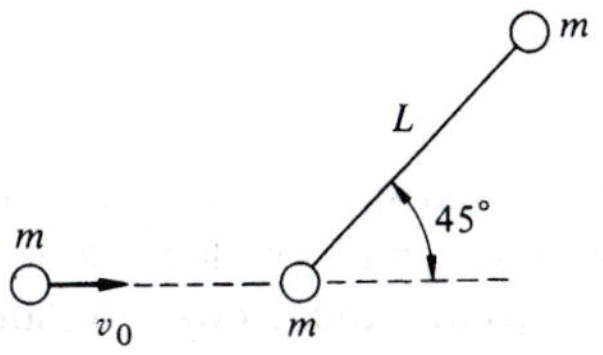

6.38 A rigid massless rod of length L joins two particles each of mass m. The rod lies on a frictionless table, and is struck by a particle of mass m and velocity v_0, moving as shown. After the collision, the projectile moves straight back.

Find the angular velocity of the rod about its center of mass after the collision, assuming that mechanical energy is conserved.

Ans. $\omega = (4\sqrt{2}/7)(v_0/L)$

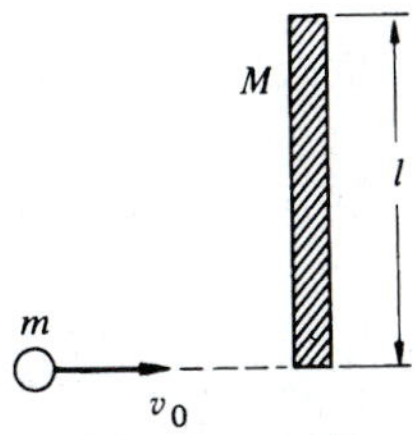

6.39 A boy of mass m runs on ice with velocity v_0 and steps on the end of a plank of length l and mass M which is perpendicular to his path.

a. Describe quantitatively the motion of the system after the boy is on the plank. Neglect friction with the ice.

b. One point on the plank is at rest immediately after the collision. Where is it?

Ans. $2l/3$ from the boy

6.40 A wheel with fine teeth is attached to the end of a spring with constant k and unstretched length l. For $x > l$, the wheel slips freely on

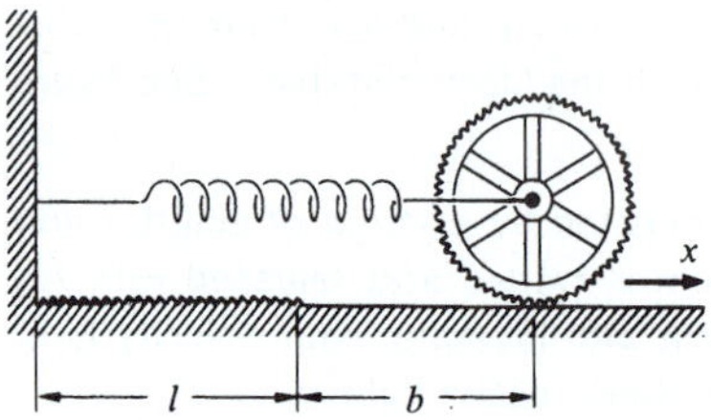

the surface, but for $x < l$ the teeth mesh with the teeth on the ground so that it cannot slip. Assume that all the mass of the wheel is in its rim.

a. The wheel is pulled to $x = l + b$ and released. How close will it come to the wall on its first trip?

b. How far out will it go as it leaves the wall?

c. What happens when the wheel next hits the gear track?

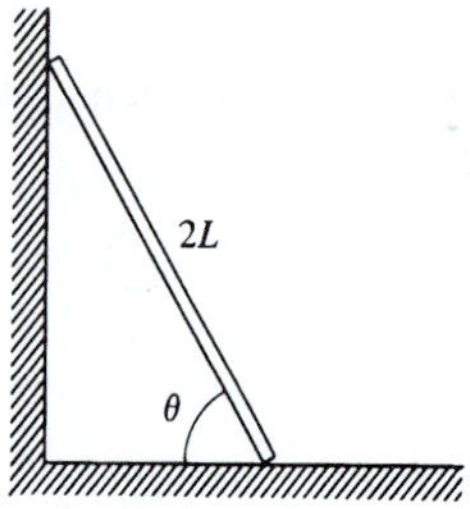

6.41 This problem utilizes most of the important laws introduced so far and it is worth a substantial effort. However, the problem is tricky (although not really complicated), so don't be alarmed if the solution eludes you.

A plank of length $2L$ leans against a wall. It starts to slip downward without friction. Show that the top of the plank loses contact with the wall when it is at two-thirds of its initial height.

Hint: Only a single variable is needed to describe the system. Note the motion of the center of mass.

7 RIGID BODY MOTION

7.1 Introduction

In the last chapter we analyzed the motion of rigid bodies undergoing fixed axis rotation. In this chapter we shall attack the more general problem of analyzing the motion of rigid bodies which can rotate about any axis. Rather than emphasize the formal mathematical details, we will try to gain insight into the basic principles. We will discuss the important features of the motion of gyroscopes and other devices which have large spin angular momentum, and we will also look at a variety of other systems. Our analysis is based on a very simple idea—that angular momentum is a vector. Although this is obvious from the definition, somehow its significance is often lost when one first encounters rigid body motion. Understanding the vector nature of angular momentum leads to a very simple and natural explanation for such a mysterious effect as the precession of a gyroscope.

A second topic which we shall treat in this chapter is the conservation of angular momentum. We touched on this in the last chapter but postponed any incisive discussion. Here the problem is physical subtlety rather than mathematical complexity.

7.2 The Vector Nature of Angular Velocity and Angular Momentum

In order to describe the rotational motion of a body we would like to introduce suitable coordinates. Recall that in the case of translational motion, our procedure was to choose some convenient coordinate system and to denote the position of the body by a vector **r**. The velocity and acceleration were then found by successively differentiating **r** with respect to time.

Suppose that we try to introduce angular coordinates θ_x, θ_y, and θ_z about the x, y, and z axes, respectively. Can we specify the angular orientation of the body by a vector?

$$\boldsymbol{\theta} \stackrel{?}{=} (\theta_x\hat{\mathbf{i}} + \theta_y\hat{\mathbf{j}} + \theta_z\hat{\mathbf{k}})$$

Unfortunately, this procedure can *not* be made to work; there is no way to construct a vector to represent an angular orientation.

The reason that $\theta_x\hat{\mathbf{i}}$ and $\theta_y\hat{\mathbf{j}}$ cannot be vectors is that the order in which we add them affects the final result: $\theta_x\hat{\mathbf{i}} + \theta_y\hat{\mathbf{j}} \neq \theta_y\hat{\mathbf{j}} + \theta_x\hat{\mathbf{i}}$, as we show explicitly in Example 7.1. For honest-to-goodness vectors like $x\hat{\mathbf{i}}$ and $y\hat{\mathbf{j}}$, $x\hat{\mathbf{i}} + y\hat{\mathbf{j}} = y\hat{\mathbf{j}} + x\hat{\mathbf{i}}$. Vector addition is commutative.

Example 7.1 **Rotations through Finite Angles**

Consider a can of maple syrup oriented as shown, and let us investigate what happens when we rotate it by an angle of $\pi/2$ around the x axis, and then by $\pi/2$ around the y axis, and compare the result with executing the same rotations but in reverse order.

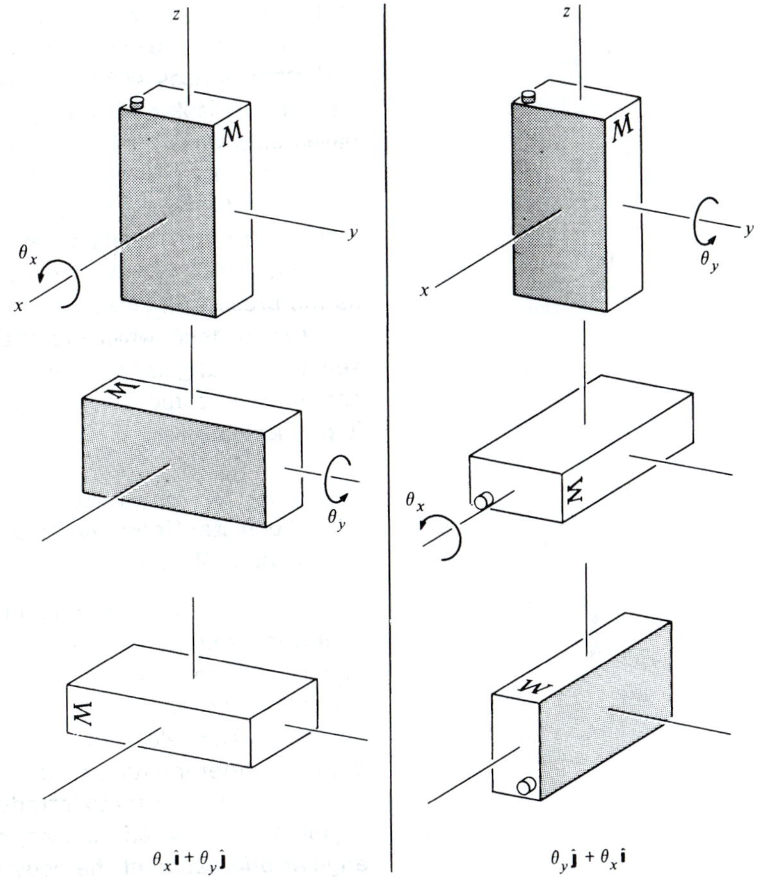

The diagram speaks for itself:

$\theta_x\hat{\mathbf{i}} + \theta_y\hat{\mathbf{j}} \neq \theta_y\hat{\mathbf{j}} + \theta_x\hat{\mathbf{i}}$.

Fortunately, all is not lost; although angular position cannot be represented by a vector, it turns out that angular velocity, the rate of change of angular position, is a perfectly good vector. We can define angular velocity by

$$\begin{aligned}\boldsymbol{\omega} &= \frac{d\theta_x}{dt}\hat{\mathbf{i}} + \frac{d\theta_y}{dt}\hat{\mathbf{j}} + \frac{d\theta_z}{dt}\hat{\mathbf{k}} \\ &= \omega_x\hat{\mathbf{i}} + \omega_y\hat{\mathbf{j}} + \omega_z\hat{\mathbf{k}}.\end{aligned}$$

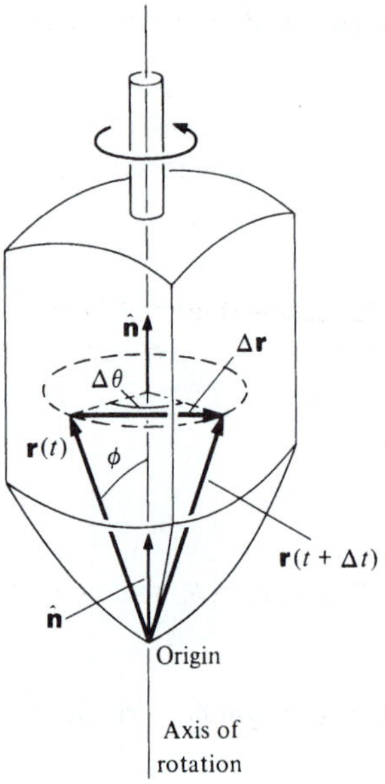

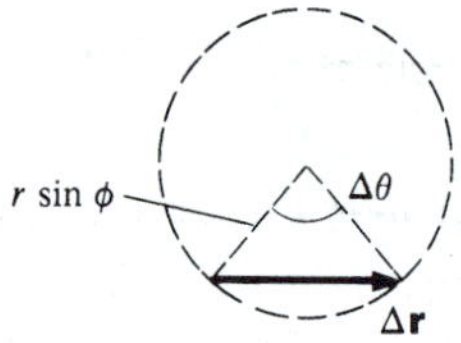

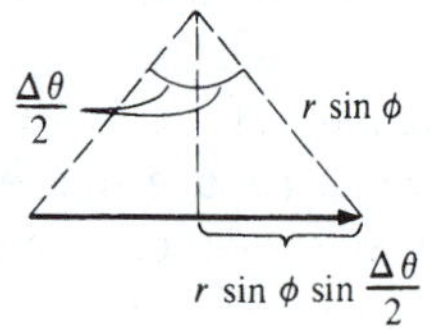

The important point is that although rotations through finite angles do not commute, infinitesimal rotations do commute, so that $\boldsymbol{\omega} = \lim_{\Delta t \to 0} (\Delta\boldsymbol{\theta}/\Delta t)$ represents a true vector. The reason for this is discussed in Note 7.1 at the end of the chapter. Assuming that angular velocity is indeed a vector, let us find how the velocity of any particle in a rotating rigid body is related to the angular velocity of the body.

Consider a rigid body rotating about some axis. We designate the instantaneous direction of the axis by $\hat{\mathbf{n}}$ and choose a coordinate system with its origin on the axis. The coordinate system is fixed in space and is inertial. As the body rotates, each of its particles describes a circle about the axis of rotation. A vector $\mathbf{r}$ from the origin to any particle tends to sweep out a cone. The drawing shows the result of rotation through angle $\Delta\theta$ about the axis along $\hat{\mathbf{n}}$. The angle ϕ between $\hat{\mathbf{n}}$ and $\mathbf{r}$ is constant, and the tip of $\mathbf{r}$ moves on a circle of radius $r \sin\phi$.

The magnitude of the displacement $|\Delta\mathbf{r}|$ is

$$|\Delta\mathbf{r}| = 2r \sin\phi \sin\frac{\Delta\theta}{2}.$$

For $\Delta\theta$ very small, we have

$$\sin\frac{\Delta\theta}{2} \approx \frac{\Delta\theta}{2} \qquad \text{and} \qquad |\Delta\mathbf{r}| \approx r \sin\phi\, \Delta\theta.$$

If $\Delta\theta$ occurs in time Δt, we have $|\Delta\mathbf{r}|/\Delta t \approx r \sin\phi\, (\Delta\theta/\Delta t)$. In the limit $\Delta t \to 0$,

$$\left|\frac{d\mathbf{r}}{dt}\right| = r \sin\phi \frac{d\theta}{dt}.$$

In the limit, $d\mathbf{r}/dt$ is tangential to the circle, as shown below. Recalling the definition of vector cross product (Sec. 1.2e), we see that the magnitude of $d\mathbf{r}/dt$, $|d\mathbf{r}/dt| = r \sin\phi\, d\theta/dt$, and its direction, perpendicular to the plane of $\mathbf{r}$ and $\hat{\mathbf{n}}$, are given cor-

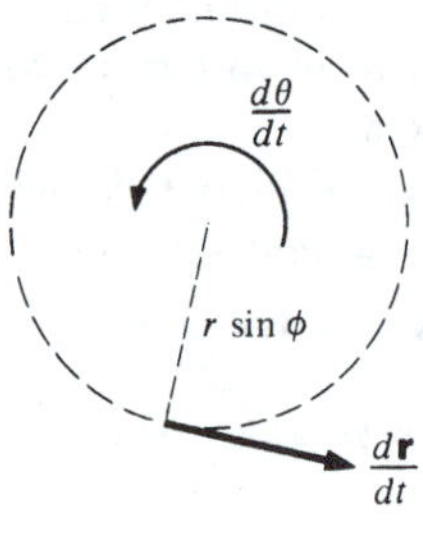

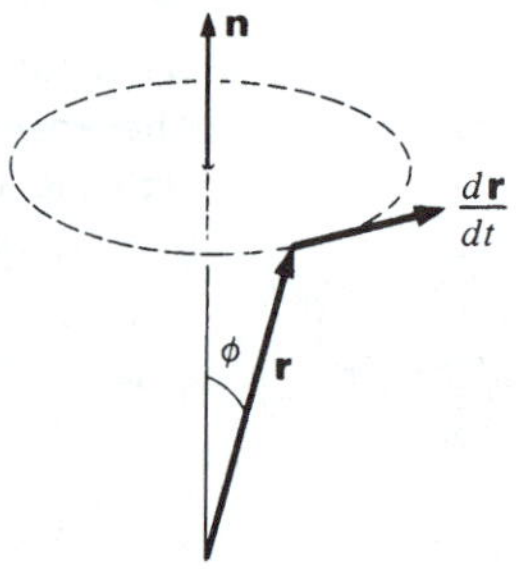

rectly by $d\mathbf{r}/dt = \hat{\mathbf{n}} \times \mathbf{r}\, d\theta/dt$. Since $d\mathbf{r}/dt = \mathbf{v}$ and $\hat{\mathbf{n}}\, d\theta/dt = \boldsymbol{\omega}$, we have

$$\frac{d\mathbf{r}}{dt} = \mathbf{v} = \boldsymbol{\omega} \times \mathbf{r}. \qquad 7.1$$

Example 7.2 **Rotation in the xy Plane**

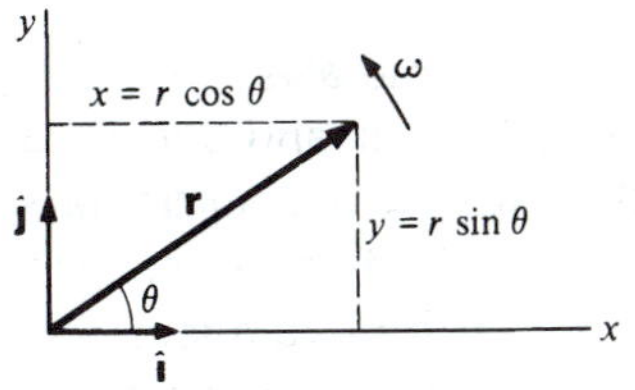

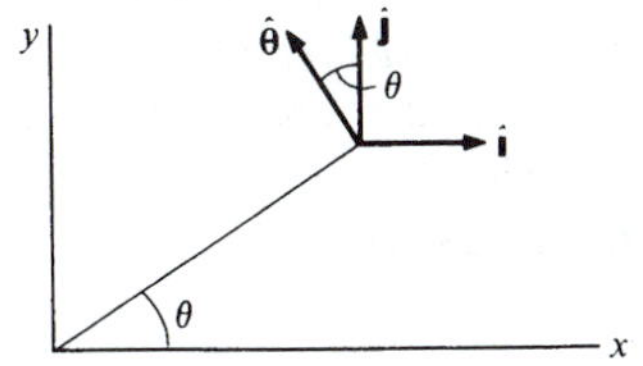

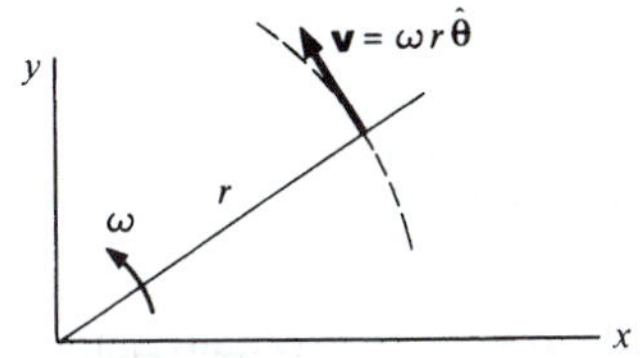

To connect Eq. (7.1) with a more familiar case—rotation in the xy plane—suppose that we evaluate $\mathbf{v}$ for the rotation of a particle about the z axis. We have $\boldsymbol{\omega} = \omega\hat{\mathbf{k}}$, and $\mathbf{r} = x\hat{\mathbf{i}} + y\hat{\mathbf{j}}$. Hence,

$$\begin{aligned}\mathbf{v} &= \boldsymbol{\omega} \times \mathbf{r}\\ &= \omega\hat{\mathbf{k}} \times (x\hat{\mathbf{i}} + y\hat{\mathbf{j}})\\ &= \omega(x\hat{\mathbf{j}} - y\hat{\mathbf{i}}).\end{aligned}$$

In plane polar coordinates $x = r\cos\theta$, $y = r\sin\theta$, and therefore

$$\mathbf{v} = \omega r(\hat{\mathbf{j}}\cos\theta - \hat{\mathbf{i}}\sin\theta).$$

But $\hat{\mathbf{j}}\cos\theta - \hat{\mathbf{i}}\sin\theta$ is a unit vector in the tangential direction $\hat{\boldsymbol{\theta}}$. Therefore,

$$\mathbf{v} = \omega r\hat{\boldsymbol{\theta}}.$$

This is the velocity of a particle moving in a circle of radius r at angular velocity ω.

It is sometimes difficult to appreciate at first the vector nature of angular velocity since we are used to visualizing rotation about a fixed axis, which involves only one component of angular velocity. We are generally much less familiar with simultaneous rotation about several axes.

We have seen that we can treat angular velocity as a vector in the relation $\mathbf{v} = \boldsymbol{\omega} \times \mathbf{r}$. It is important to assure ourselves that this relation remains valid if we resolve $\boldsymbol{\omega}$ into components like any other vector. In other words, if we write $\boldsymbol{\omega} = \boldsymbol{\omega}_1 + \boldsymbol{\omega}_2$, is it true that $\mathbf{v} = (\boldsymbol{\omega}_1 \times \mathbf{r}) + (\boldsymbol{\omega}_2 \times \mathbf{r})$? As the following example shows, the answer is yes.

Example 7.3 **Vector Nature of Angular Velocity**

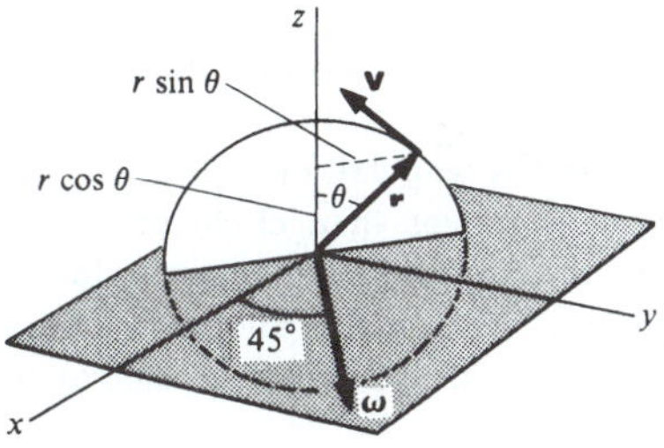

Consider a particle rotating in a vertical plane as shown in the sketch. The angular velocity $\boldsymbol{\omega}$ lies in the xy plane and makes an angle of 45° with the xy axes.

First we shall calculate $\mathbf{v}$ directly from the relation $\mathbf{v} = d\mathbf{r}/dt$. To find $\mathbf{r}$, note from the sketch at left that $z = r\cos\theta$, $x = -r\sin\theta/\sqrt{2}$ and $y = r\sin\theta/\sqrt{2}$. Hence,

$$\mathbf{r} = r\left(\frac{-1}{\sqrt{2}}\sin\theta\hat{\mathbf{i}} + \frac{1}{\sqrt{2}}\sin\theta\hat{\mathbf{j}} + \cos\theta\hat{\mathbf{k}}\right)$$

and differentiating, we have, since r = constant,

$$\begin{aligned}\frac{d\mathbf{r}}{dt} &= \mathbf{v}\\ &= r\left[\frac{-1}{\sqrt{2}}\cos\theta\hat{\mathbf{i}} + \frac{1}{\sqrt{2}}\cos\theta\hat{\mathbf{j}} - \sin\theta\hat{\mathbf{k}}\right]\frac{d\theta}{dt}\\ &= \omega r\left[\frac{-1}{\sqrt{2}}\cos\theta\hat{\mathbf{i}} + \frac{1}{\sqrt{2}}\cos\theta\hat{\mathbf{j}} - \sin\theta\hat{\mathbf{k}}\right],\end{aligned} \qquad 1$$

where we have used $d\theta/dt = \omega$.

Next we shall find the velocity from $\mathbf{v} = \boldsymbol{\omega} \times \mathbf{r}$. Assuming that $\boldsymbol{\omega}$ can be resolved into components,

$$\boldsymbol{\omega} = \frac{\omega}{\sqrt{2}}\hat{\mathbf{i}} + \frac{\omega}{\sqrt{2}}\hat{\mathbf{j}},$$

we have

$$\begin{aligned}\boldsymbol{\omega}\times\mathbf{r} &= \begin{vmatrix}\hat{\mathbf{i}} & \hat{\mathbf{j}} & \hat{\mathbf{k}}\\ \dfrac{\omega}{\sqrt{2}} & \dfrac{\omega}{\sqrt{2}} & 0\\ \dfrac{-r\sin\theta}{\sqrt{2}} & \dfrac{r\sin\theta}{\sqrt{2}} & r\cos\theta\end{vmatrix}\\ &= \omega r\left(\frac{-1}{\sqrt{2}}\cos\theta\hat{\mathbf{i}} + \frac{1}{\sqrt{2}}\cos\theta\hat{\mathbf{j}} - \sin\theta\hat{\mathbf{k}}\right)\end{aligned}$$

in agreement with Eq. (1).

As we expect, there is no problem in treating $\boldsymbol{\omega}$ like any other vector.

In the following example we shall see that a problem can be greatly simplified by resolving $\boldsymbol{\omega}$ into components along convenient axes. The example also demonstrates that angular momentum is not necessarily parallel to angular velocity.

Example 7.4 **Angular Momentum of a Rotating Skew Rod**

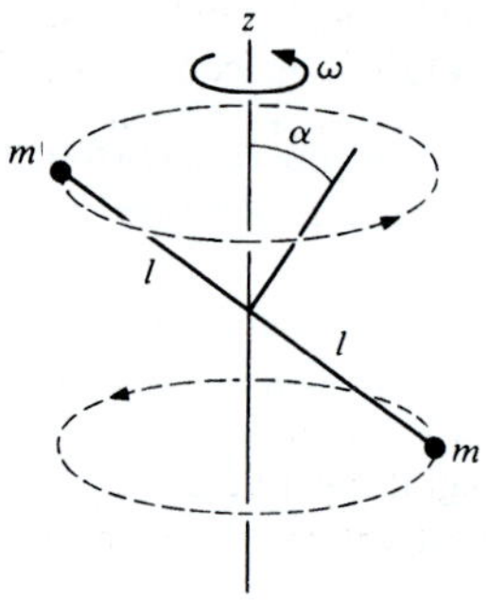

Consider a simple rigid body consisting of two particles of mass m separated by a massless rod of length $2l$. The midpoint of the rod is attached to a vertical axis which rotates at angular speed ω. The rod is skewed at angle α, as shown in the sketch. The problem is to find the angular momentum of the system.

The most direct method is to calculate the angular momentum from the definition $\mathbf{L} = \Sigma(\mathbf{r}_j \times \mathbf{p}_j)$. Each mass moves in a circle of radius $l\cos\alpha$ with angular speed ω. The momentum of each mass is $|\mathbf{p}| = m\omega l\cos\alpha$, tangential to the circular path. Taking the midpoint of the skew rod as origin, $|\mathbf{r}| = l$. $\mathbf{r}$ lies along the rod and is perpendicular to

p. Hence $|\mathbf{L}| = 2m\omega l^2 \cos\alpha$. **L** is perpendicular to the skew rod and lies in the plane of the rod and the z axis, as shown in the left hand drawing, below. **L** turns with the rod, and its tip traces out a circle about the z axis.

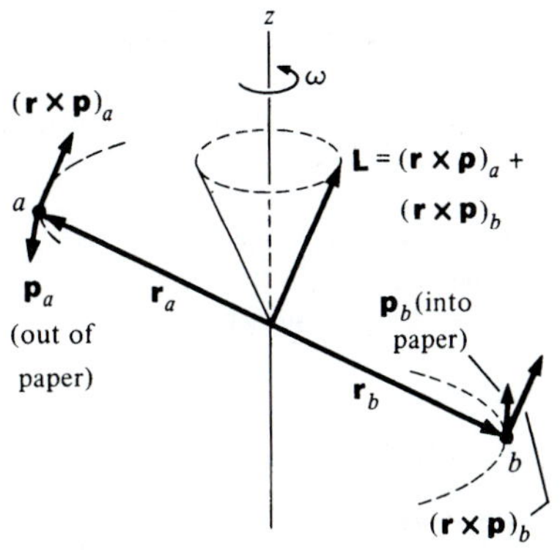

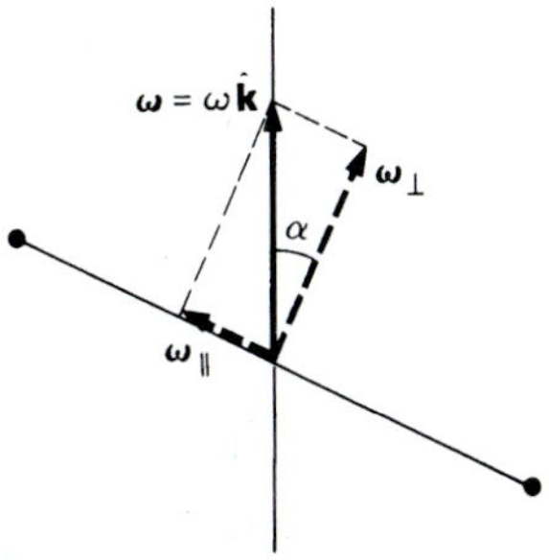

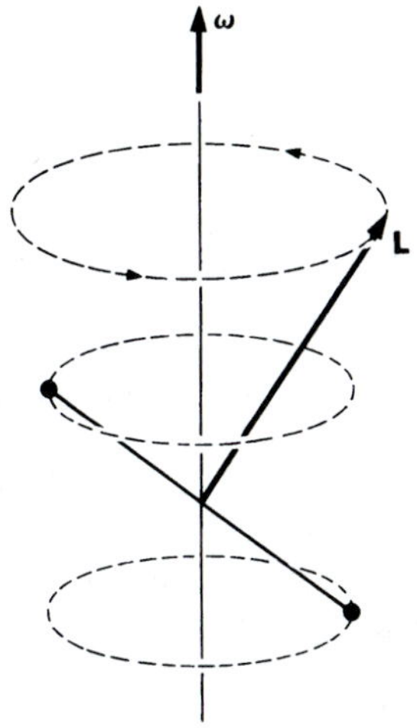

We now turn to a method for calculating **L** which emphasizes the vector nature of $\boldsymbol{\omega}$. First we resolve $\boldsymbol{\omega} = \omega\hat{\mathbf{k}}$ into components $\boldsymbol{\omega}_\perp$ and $\boldsymbol{\omega}_\parallel$, perpendicular and parallel to the skew rod. From the right hand drawing, above, we see that $\omega_\perp = \omega\cos\alpha$, and $\omega_\parallel = \omega\sin\alpha$.

Since the masses are point particles, $\boldsymbol{\omega}_\parallel$ produces no angular momentum. Hence, the angular momentum is due entirely to $\boldsymbol{\omega}_\perp$. The angular momentum is readily evaluated: the moment of inertia about the direction of $\boldsymbol{\omega}_\perp$ is $2ml^2$, and the magnitude of the angular momentum is

$$\begin{aligned} L &= I\omega_\perp \\ &= 2ml^2\omega_\perp \\ &= 2ml^2\omega\cos\alpha. \end{aligned}$$

L points along the direction of $\boldsymbol{\omega}_\perp$. Hence, **L** swings around with the rod; the tip of **L** traces out a circle about the z axis. (We encountered a similar situation in Example 6.2 with the conical pendulum.) Note that **L** is not parallel to $\boldsymbol{\omega}$. This is generally true for nonsymmetric bodies.

The dynamics of rigid body motion is governed by $\boldsymbol{\tau} = d\mathbf{L}/dt$. Before we attempt to apply this relation to complicated systems, let us gain some insight into its physical meaning by analyzing the torque on the rotating skew rod.

Example 7.5 **Torque on the Rotating Skew Rod**

In Example 7.4 we showed that the angular momentum of a uniformly rotating skew rod is constant in magnitude but changes in direction. **L** is fixed with respect to the rod and rotates in space with the rod.

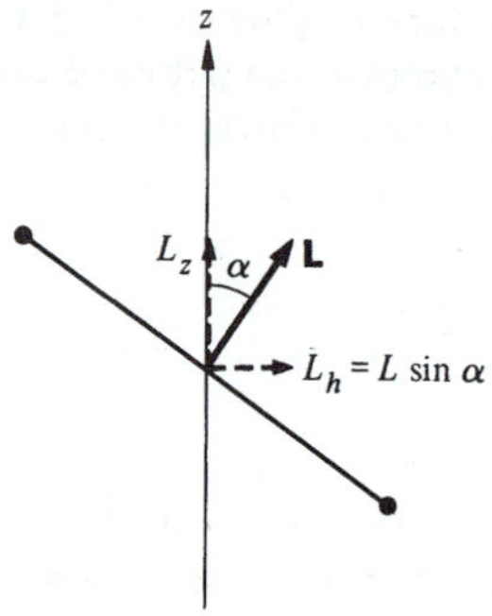

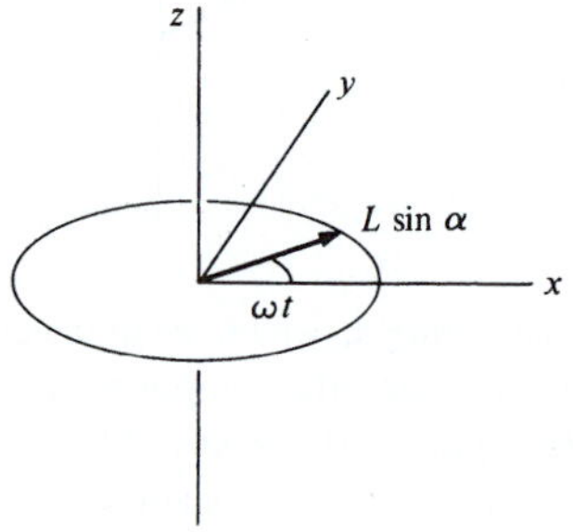

The torque on the rod is given by $\boldsymbol{\tau} = d\mathbf{L}/dt$. We can find $d\mathbf{L}/dt$ quite easily by decomposing $\mathbf{L}$ as shown in the sketch. (We followed a similar procedure in Example 6.6 for the conical pendulum.) The component L_z parallel to the z axis, $L\cos\alpha$, is constant. Hence, there is no torque in the z direction. The horizontal component of $\mathbf{L}$, $L_h = L\sin\alpha$, swings with the rod. If we choose xy axes so that L_h coincides with the x axis at $t = 0$, then at time t we have

$$\begin{aligned} L_x &= L_h \cos\omega t \\ &= L\sin\alpha\cos\omega t \\ L_y &= L_h \sin\omega t \\ &= L\sin\alpha\sin\omega t. \end{aligned}$$

Hence,

$$\mathbf{L} = L\sin\alpha(\hat{\mathbf{i}}\cos\omega t + \hat{\mathbf{j}}\sin\omega t) + L\cos\alpha\hat{\mathbf{k}}.$$

The torque is

$$\begin{aligned} \boldsymbol{\tau} &= \frac{d\mathbf{L}}{dt} \\ &= L\omega\sin\alpha(-\hat{\mathbf{i}}\sin\omega t + \hat{\mathbf{j}}\cos\omega t). \end{aligned}$$

Using $L = 2ml^2\omega\cos\alpha$, we obtain

$$\begin{aligned} \tau_x &= -2ml^2\omega^2\sin\alpha\cos\alpha\sin\omega t \\ \tau_y &= 2ml^2\omega^2\sin\alpha\cos\alpha\cos\omega t. \end{aligned}$$

Hence,

$$\begin{aligned} \tau &= \sqrt{\tau_x^2 + \tau_y^2} \\ &= 2ml^2\omega^2\sin\alpha\cos\alpha \\ &= \omega L\sin\alpha. \end{aligned}$$

Note that $\tau = 0$ for $\alpha = 0$ or $\alpha = \pi/2$. Do you see why? Also, can you see why the torque should be proportional to ω^2?

This analysis may seem roundabout, since the torque can be calculated directly by finding the force on each mass and using $\boldsymbol{\tau} = \Sigma\mathbf{r}_j \times \mathbf{f}_j$. However, the procedure used above is just as quick. Furthermore, it illustrates that angular velocity and angular momentum are *real* vectors which can be resolved into components along any axes we choose.

Example 7.6 Torque on the Rotating Skew Rod (Geometric Method)

In Example 7.5 we calculated the torque on the rotating skew rod by resolving $\mathbf{L}$ into components and using $\boldsymbol{\tau} = d\mathbf{L}/dt$. We repeat the calculation in this example using a geometric argument which emphasizes

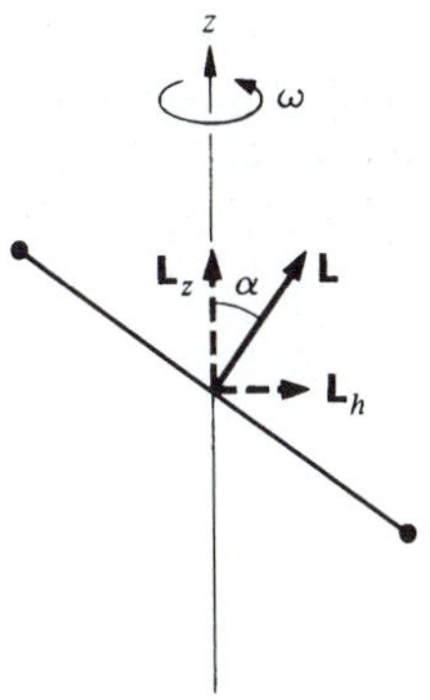

the connection between torque and the rate of change of **L**. This method illustrates a point of view that will be helpful in analyzing gyroscopic motion.

As in Example 7.5, we begin by resolving **L** into a vertical component $L_z = L \cos \alpha$ and a horizontal component $L_h = L \sin \alpha$ as shown in the sketch. Since $\mathbf{L}_z$ is constant, there is no torque about the z axis. $\mathbf{L}_h$ is constant in magnitude but is rotating with the rod. The time rate of change of **L** is due solely to this effect.

Once again we are dealing with a rotating vector. From Sec. 1.8 or Example 6.6, we know that $dL_h/dt = \omega L_h$. However, since it is so important to be able to visualize this result, we derive it once more. From the vector diagram we have

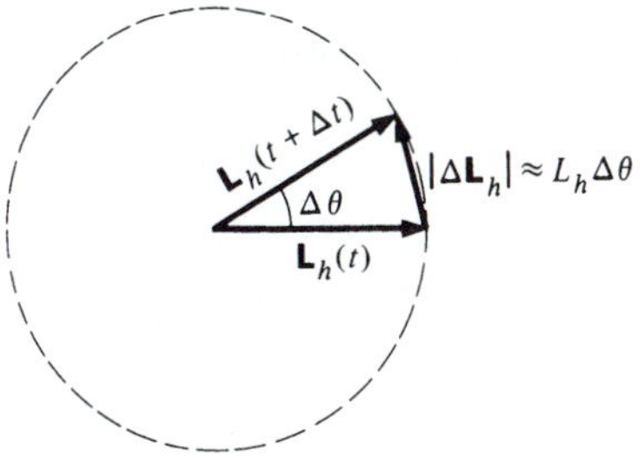

$$|\Delta \mathbf{L}_h| \approx |\mathbf{L}_h| \Delta\theta$$

$$\frac{dL_h}{dt} = L_h \frac{d\theta}{dt}$$

$$= L_h\omega.$$

The torque is given by

$$\tau = \frac{dL_h}{dt}$$

$$= L_h\omega$$

$$= \omega L \sin \alpha,$$

which is identical to the result of the last example. The torque $\boldsymbol{\tau}$ is parallel to $\Delta\mathbf{L}$ in the limit. For the skew rod, $\boldsymbol{\tau}$ is in the tangential direction in the horizontal plane and rotates with the rod.

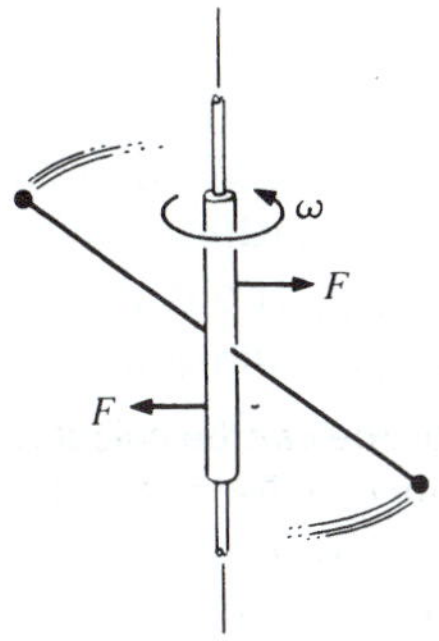

You may have thought that torque on a rotating system always causes the speed of rotation to change. In this problem the speed of rotation is constant, and the torque causes the direction of **L** to change. The torque is produced by the forces on the rotating bearing of the skew rod. For a real rod this would have to be an extended structure, something like a sleeve. The torque causes a time varying load on the sleeve which results in vibration and wear. Since there is no way for a uniform gravitational field to exert a torque on the skew rod, the rod is said to be *statically balanced*. However, there is a torque on the skew rod when it is rotating, which means that it is not *dynamically balanced*. Rotating machinery must be designed for dynamical balance if it is to run smoothly.

7.3 The Gyroscope

We now turn to some aspects of gyroscope motion which can be understood by using the basic concepts of angular momentum, torque, and the time derivative of a vector. We shall discuss each step carefully, since this is one area of physics where intuition may not be much help. Our treatment of the gyroscope in this section is by no means complete. Instead of finding the general motion of the gyroscope directly from the dynamical equations, we bypass this complicated mathematical problem and concentrate on uniform precession, a particularly simple and familiar type of gyroscope motion. Our aim is to show that uniform precession is consistent with $\boldsymbol{\tau} = d\mathbf{L}/dt$ and Newton's laws. While this approach cannot be completely satisfying, it does illuminate the physical principles involved.

The essentials of a gyroscope are a spinning flywheel and a suspension which allows the axle to assume any orientation. The familiar toy gyroscope shown in the drawing is quite adequate for our discussion. The end of the axle rests on a pylon, allowing the axis to take various orientations without constraint.

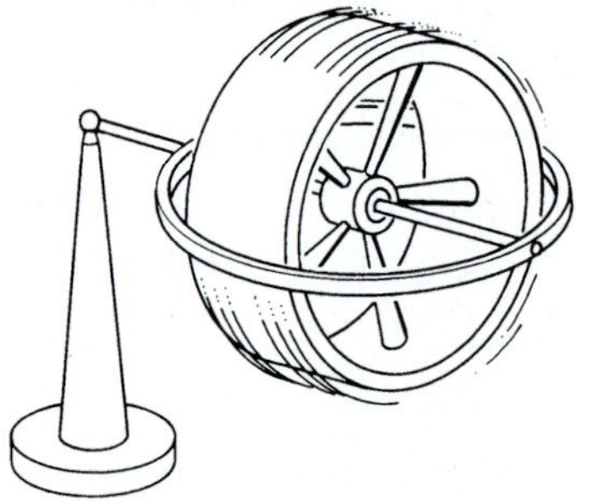

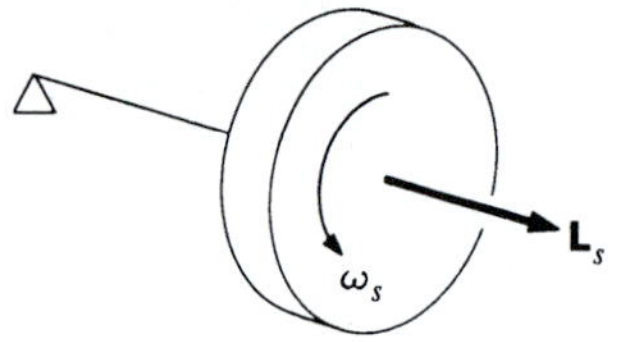

The right hand drawing above is a schematic representation of the gyroscope. The triangle represents the free pivot, and the flywheel spins in the direction shown.

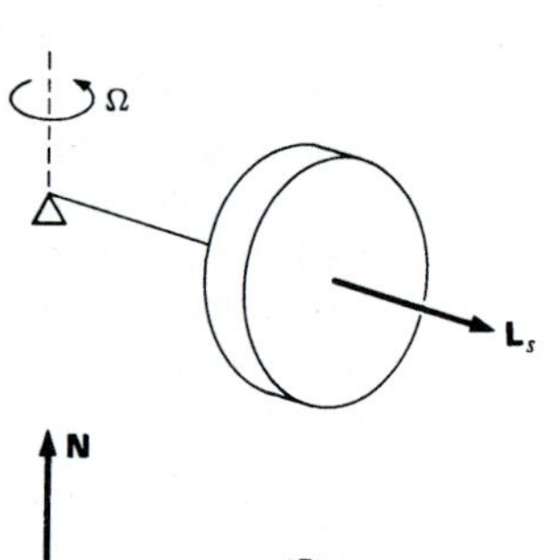

If the gyroscope is released horizontally with one end supported by the pivot, it wobbles off horizontally and then settles down to *uniform precession*, in which the axle slowly rotates about the vertical with constant angular velocity Ω. One's immediate impulse is to ask why the gyroscope does not fall. A possible answer is suggested by the force diagram. The total vertical force is $N - W$, where N is the vertical force exerted by the pivot and W is the weight. If $N = W$, the center of mass cannot fall.

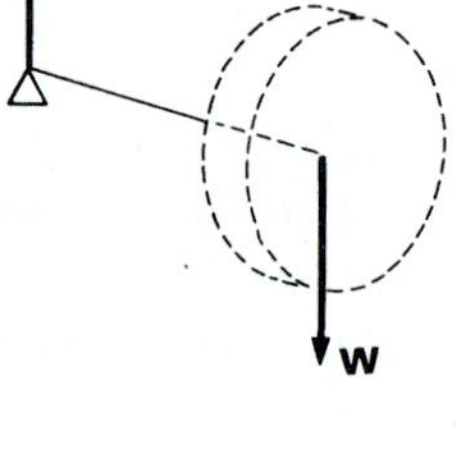

This explanation, which is quite correct, is not satisfactory. We have asked the wrong question. Instead of wondering why the gyroscope does not fall, we should ask why it does not swing about the pivot like a pendulum.

As a matter of fact, if the gyroscope is released with its flywheel stationary, it behaves exactly like a pendulum; instead of precessing horizontally, it swings vertically. The gyroscope precesses

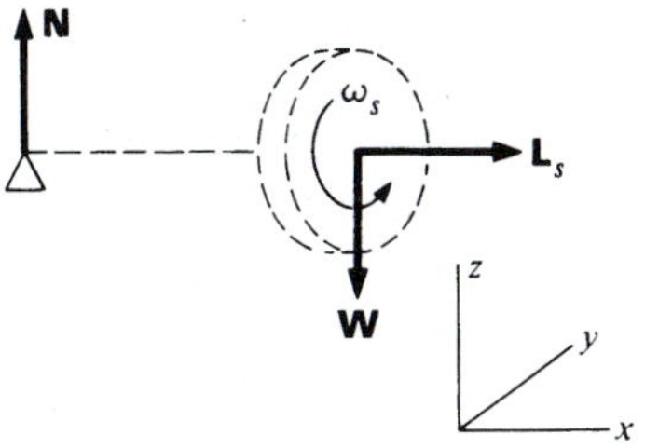

only if the flywheel is spinning rapidly. In this case, the large spin angular momentum of the flywheel dominates the dynamics of the system.

Nearly all of the gyroscope's angular momentum lies in $\mathbf{L}_s$, the spin angular momentum. $\mathbf{L}_s$ is directed along the axle and has magnitude $L_s = I_0\omega_s$, where I_0 is the moment of inertia of the flywheel about its axle. When the gyroscope precesses about the z axis, it has a small orbital angular momentum in the z direction. However, for uniform precession the orbital angular momentum is constant in magnitude and direction and plays no dynamical role. Consequently, we shall ignore it here.

$\mathbf{L}_s$ always points along the axle. As the gyroscope precesses, $\mathbf{L}_s$ rotates with it. (See figure a below.) We have encountered rotating vectors many times, most recently in Example 7.6. If the angular velocity of precession is Ω, the rate of change of $\mathbf{L}_s$ is given by

$$\left|\frac{d\mathbf{L}_s}{dt}\right| = \Omega L_s.$$

The direction of $d\mathbf{L}_s/dt$ is tangential to the horizontal circle swept out by $\mathbf{L}_s$. At the instant shown in figure b, $\mathbf{L}_s$ is in the x direction and $d\mathbf{L}_s/dt$ is in the y direction.

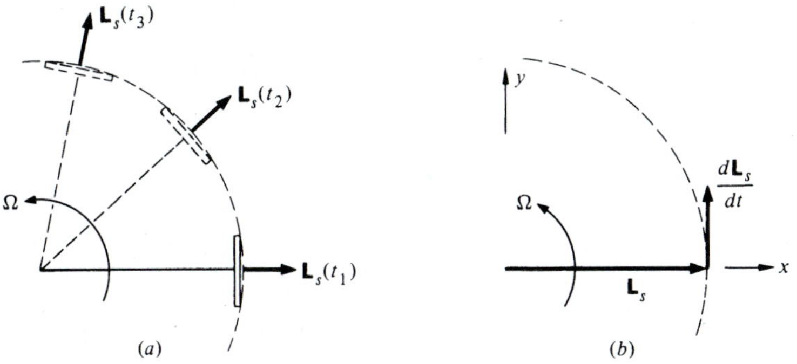

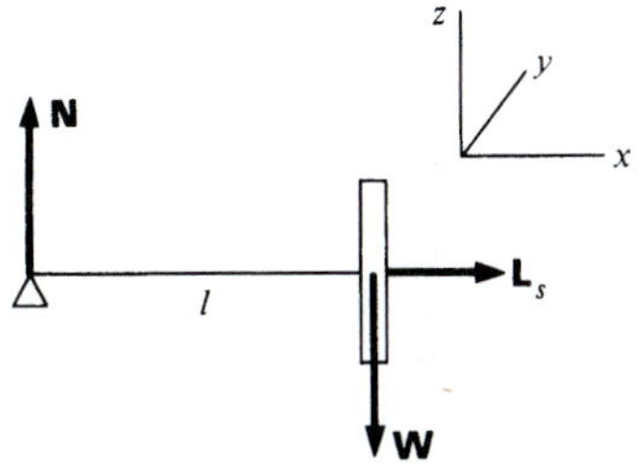

There must be a torque on the gyroscope to account for the change in $\mathbf{L}_s$. The source of the torque is apparent from the force diagram at left. If we take the pivot as the origin, the torque is due to the weight of the flywheel acting at the end of the axle. The magnitude of the torque is

$$\tau = lW.$$

$\boldsymbol{\tau}$ is in the y direction, parallel to $d\mathbf{L}_s/dt$, as we expect.

We can find the rate of precession Ω from the relation

$$\left|\frac{d\mathbf{L}_s}{dt}\right| = \tau.$$

Since $|d\mathbf{L}_s/dt| = \Omega L_s$ and $\tau = lW$, we have

$$\Omega L_s = lW.$$

or

$$\Omega = \frac{lW}{I_0\omega_s}. \qquad 7.2$$

Alternatively, we could have analyzed the motion about the center of mass. In this case the torque is $\tau_0 = Nl = Wl$ as before, since $N = W$.

Equation (7.2) indicates that Ω increases as the flywheel slows. This effect is easy to see with a toy gyroscope. Obviously Ω cannot increase indefinitely; eventually uniform precession gives way to a violent and erratic motion. This occurs when Ω becomes so large that we cannot neglect small changes in the angular momentum about the vertical axis due to frictional torque. However, as is shown in Note 7.2, uniform precession represents an exact solution to the dynamical equations governing the gyroscope.

Although we have assumed that the axle of the gyroscope is horizontal, the rate of uniform precession is independent of the angle of elevation, as the following example shows.

Example 7.7 **Gyroscope Precession**

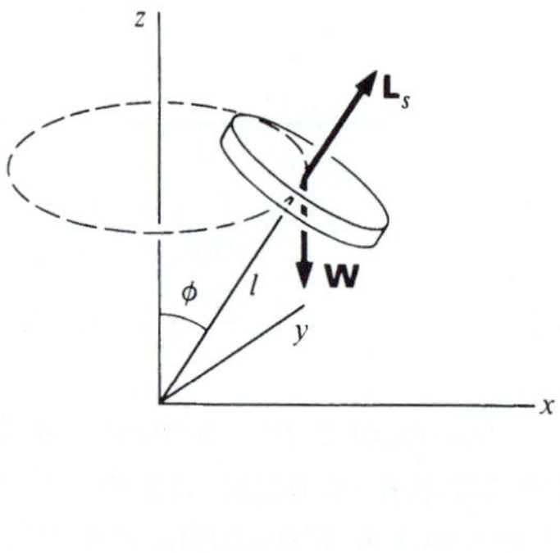

Consider a gyroscope in uniform precession with its axle at angle ϕ with the vertical. The component of $\mathbf{L}_s$ in the xy plane varies as the gyroscope precesses, while the component parallel to the z axis remains constant.

The horizontal component of $\mathbf{L}_s$ is $L_s \sin\phi$. Hence

$$|d\mathbf{L}_s/dt| = \Omega L_s \sin\phi.$$

The torque due to gravity is horizontal and has magnitude

$$\tau = l \sin\phi\, W.$$

We have

$$\Omega L_s \sin\phi = l \sin\phi\, W$$

$$\Omega = \frac{lW}{I_0\omega_s}.$$

The precessional velocity is independent of ϕ.

Our treatment shows that gyroscope precession is completely consistent with the dynamical equation $\boldsymbol{\tau} = d\mathbf{L}/dt$. The following example gives a more physical explanation of why a gyroscope precesses.

Example 7.8 **Why a Gyroscope Precesses**

Gyroscope precession is hard to understand because angular momentum is much less familiar to us than particle motion. However, the rotational dynamics of a simple rigid body can be understood directly in terms of Newton's laws. Rather than address ourselves specifically to the gyroscope, let us consider a rigid body consisting of two particles of mass m at either end of a rigid massless rod of length $2l$. Suppose that the rod is rotating in free space with its angular momentum $\mathbf{L}_s$ along the z direction. The speed of each mass is v_0. We shall show that an applied torque $\boldsymbol{\tau}$ causes $\mathbf{L}_s$ to precess with angular velocity $\Omega = \tau/L_s$.

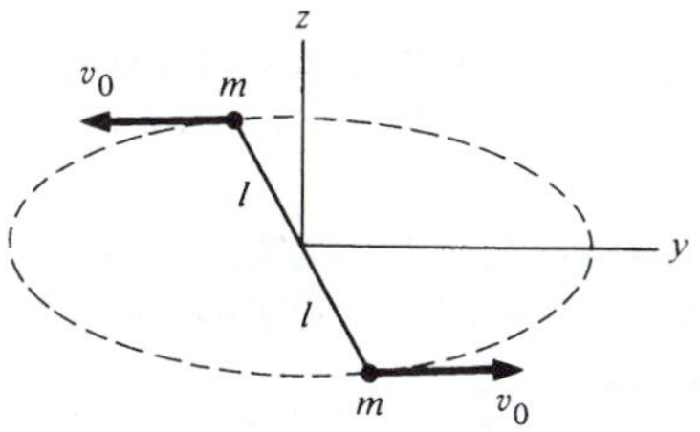

To simplify matters, suppose that the torque is applied only during a short time Δt while the rod is instantaneously oriented along the x axis. We assume that the torque is due to two equal and opposite forces F, as shown. (The total force is zero, and the center of mass remains at rest.) The momentum of each mass changes by

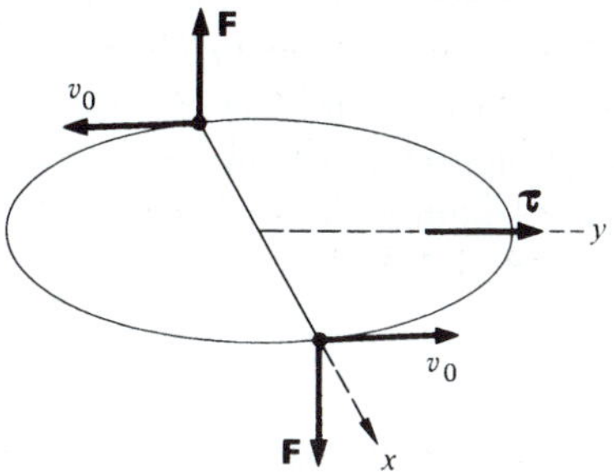

$$\Delta\mathbf{p} = m\,\Delta\mathbf{v} = \mathbf{F}\Delta t.$$

Since $\Delta\mathbf{v}$ is perpendicular to $\mathbf{v}_0$, the velocity of each mass changes direction, as shown at left below, and the rod rotates about a new direction.

The axis of rotation tilts by the angle

$$\Delta\phi \approx \frac{\Delta v}{v_0} = \frac{F\,\Delta t}{mv_0}.$$

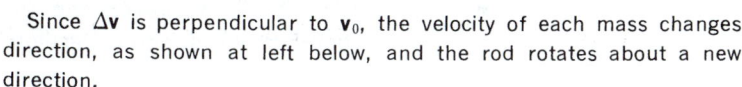

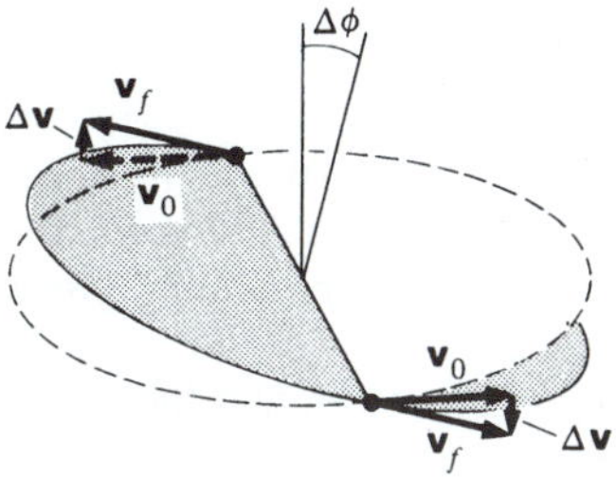

The torque on the system is $\tau = 2Fl$, and the angular momentum is $L_s = 2mv_0l$. Hence

$$\Delta\phi = \frac{F\,\Delta t}{mv_0} = \frac{2lF\,\Delta t}{2lmv_0} = \frac{\tau\,\Delta t}{L_s}.$$

The rate of precession while the torque is acting is therefore

$$\Omega = \frac{\Delta\phi}{\Delta t}$$

$$= \frac{\tau}{L_s},$$

which is identical to the result for gyroscope precession. Also, the change in the angular momentum, $\Delta\mathbf{L}_s$, is in the y direction parallel to the torque, as required.

This model gives some insight into why a torque causes a tilt in the axis of rotation of a spinning body. Although the argument can be elaborated to apply to an extended body like a gyroscope, the final result is equivalent to using $\boldsymbol{\tau} = d\mathbf{L}/dt$.

The discussion in this section applies to uniform precession, a very special case of gyroscope motion. We assumed at the beginning of our analysis that the gyroscope was executing this motion, but there are many other ways a gyroscope can move. For instance, if the free end of the axle is held at rest and suddenly released, the precessional velocity is instantaneously zero and the center of mass starts to fall. It is fascinating to see how this falling motion turns into uniform precession. We do this in Note 7.2 at the end of the chapter by a straightforward application of $\boldsymbol{\tau} = d\mathbf{L}/dt$. However, the treatment requires the general relation between $\mathbf{L}$ and $\boldsymbol{\omega}$ developed in Sec. 7.6.

7.4 Some Applications of Gyroscope Motion

In this section we present a few examples which show the application of angular momentum to rigid body motion.

Example 7.9 Precession of the Equinoxes

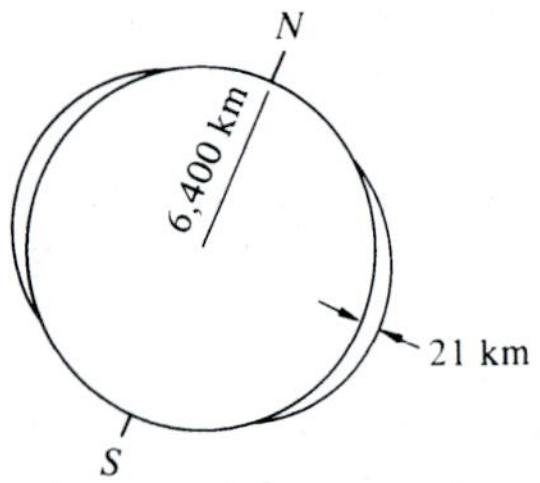

To a first approximation there are no torques on the earth and its angular momentum does not change in time. To this approximation, the earth's rotational speed is constant and its angular momentum always points in the same direction in space.

If we analyze the earth-sun system with more care, we find that there is a small torque on the earth. This causes the spin axis to slowly alter its direction, resulting in the phenomenon known as precession of the equinoxes.

The torque arises because of the interaction of the sun and moon with the nonspherical shape of the earth. The earth bulges slightly; its

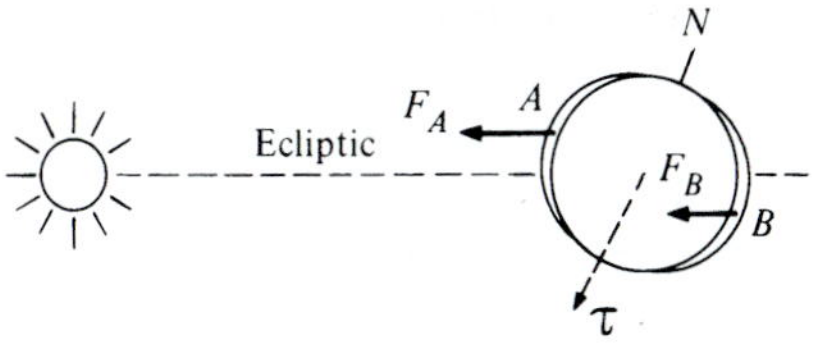

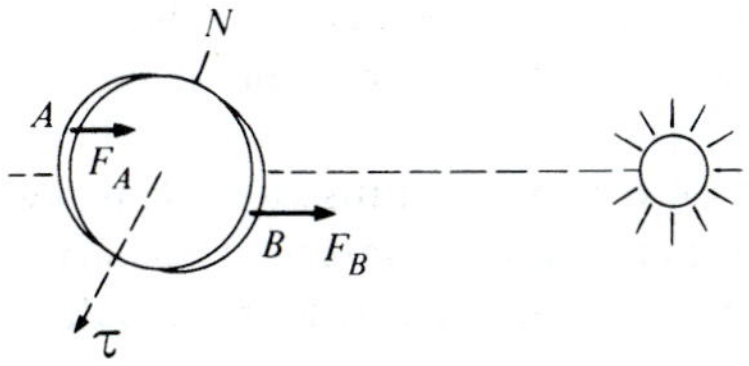

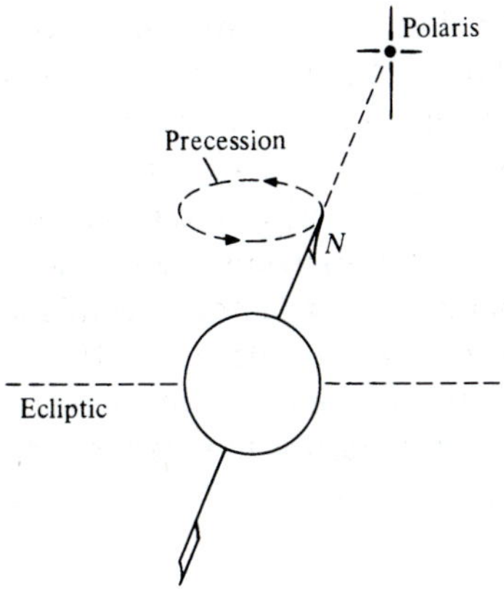

mean equatorial radius is 21 km greater than the polar radius. The gravitational force of the sun gives rise to a torque because the earth's axis of rotation is inclined with respect to the plane of the ecliptic (the orbital plane). During the winter, the part of the bulge above the ecliptic, A in the top sketch, is nearer the sun than the lower part B. The mass at A is therefore attracted more strongly by the sun than is the mass at B, as shown in the sketch. This results in a counterclockwise torque on the earth, out of the plane of the sketch. Six months later, when the earth is on the other side of the sun, B is attracted more strongly than A. However, the torque has the same direction in space as before. Midway between these extremes, the torque is zero. The average torque is perpendicular to the spin angular momentum and lies in the plane of the ecliptic. In a similar fashion, the moon exerts an average torque on the earth; this torque is about twice as great as that due to the sun.

The torque causes the spin axis to precess about a normal to the ecliptic. As the spin axis precesses, the torque remains perpendicular to it; the system acts like the gyroscope with tilted axis that we analyzed in Example 7.7.

The period of the precession is 26,000 years. 13,000 years from now, the polar axis will not point toward Polaris, the present north star; it will point $2 \times 23\frac{1}{2}° = 47°$ away. Orion and Sirius, those familiar winter guides, will then shine in the midsummer sky.

The spring equinox occurs at the instant the sun is directly over the equator in its apparent passage from south to north. Due to the precession of the earth's axis, the position of the sun at the equinox against the background of fixed stars shifts by 50 seconds of arc each year. This precession of the equinoxes was known to the ancients. It figures in the astrological scheme of cyclic history, which distinguishes twelve ages named by the constellation in which the sun lies at spring equinox. The present age is Pisces, and in 600 years it will be Aquarius.

Example 7.10 **The Gyrocompass Effect**

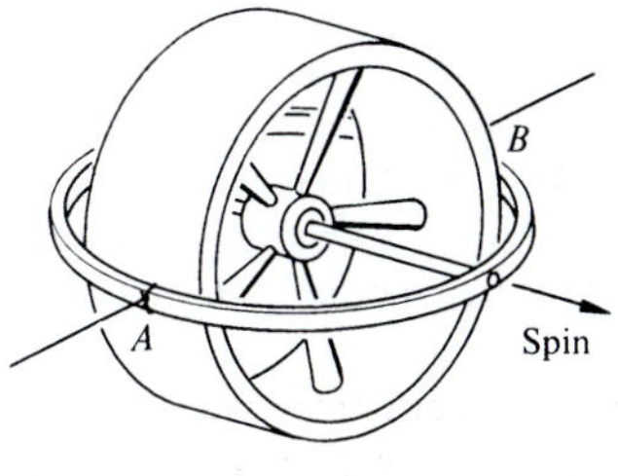

Try the following experiment with a toy gyroscope. Tie strings to the frame of the gyroscope at points A and B on opposite sides midway between the bearings of the spin axis. Hold the strings taut at arm's length with the spin axis horizontal. Now slowly pivot so that the spinning gyroscope moves in a circle with arm length radius. The gyroscope suddenly flips and comes to rest with its spin axis vertical, parallel to your axis of rotation. Rotation in the opposite direction causes the gyro to flip by 180°, making its spin axis again parallel to the rotation axis. (The spin axis tends to oscillate about the vertical, but friction in the horizontal axle quickly damps this motion.)

The gyrocompass is based on this effect. A flywheel free to rotate about two perpendicular axes tends to orient its spin axis parallel to the axis of rotation of the system. In the case of a gyrocompass, the "sys-

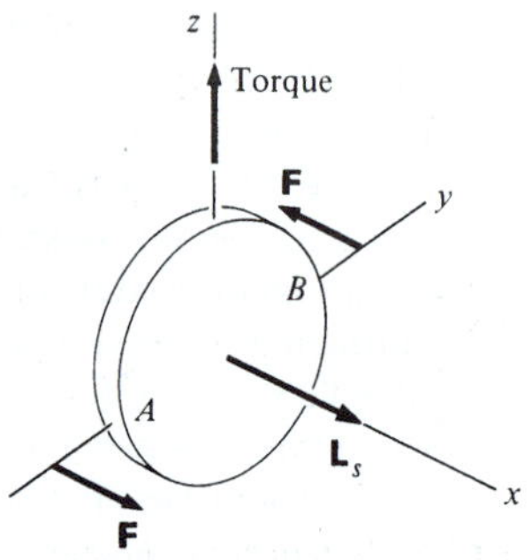

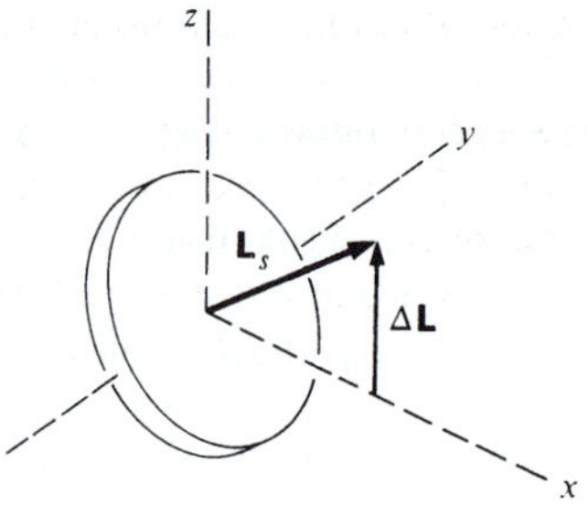

tem" is the earth; the compass comes to rest with its axis parallel to the polar axis.

We can understand the motion qualitatively by simple vector arguments. Assume that the axle is horizontal with $\mathbf{L}_s$ pointing along the x axis. Suppose that we attempt to turn the compass about the z axis. If we apply the forces shown, there is a torque along the z axis, τ_z, and the angular momentum along the z axis, L_z, starts to increase. If $\mathbf{L}_s$ were zero, L_z would be due entirely to rotation of the gyrocompass about the z axis: $L_z = I_z\omega_z$, where I_z is the moment of inertia about the z axis. However, when the flywheel is spinning, another way for L_z to change is for the gyrocompass to rotate around the AB axis, swinging $\mathbf{L}_s$ toward the z direction. Our experiment shows that if $\mathbf{L}_s$ is large, most of the torque goes into reorienting the spin angular momentum; only a small fraction goes toward rotating the gyrocompass about the z axis.

We can see why the effect is so pronounced by considering angular momentum along the y axis. The pivots at A and B allow the system to swing freely about the y axis, so there can be no torque along the y axis. Since L_y is initially zero, it must remain zero. As the gyrocompass starts to rotate about the z axis, L_s acquires a component in the y direction. At the same time, the gyrocompass and its frame begin to flip rapidly about the y axis. The angular momentum arising from this motion cancels the y component of $\mathbf{L}_s$. When $\mathbf{L}_s$ finally comes to rest parallel to the z axis, the motion of the frame no longer changes the direction of $\mathbf{L}_s$, and the spin axis remains stationary.

The earth is a rotating system, and a gyrocompass on the surface of the earth will line up with the polar axis, indicating true north. A practical gyrocompass is somewhat more complicated, however, since it must continue to indicate true north without responding to the motion of the ship or aircraft which it is guiding. In the next example we solve the dynamical equation for the gyrocompass and show how a gyrocompass fixed to the earth indicates true north.

Example 7.11 Gyrocompass Motion

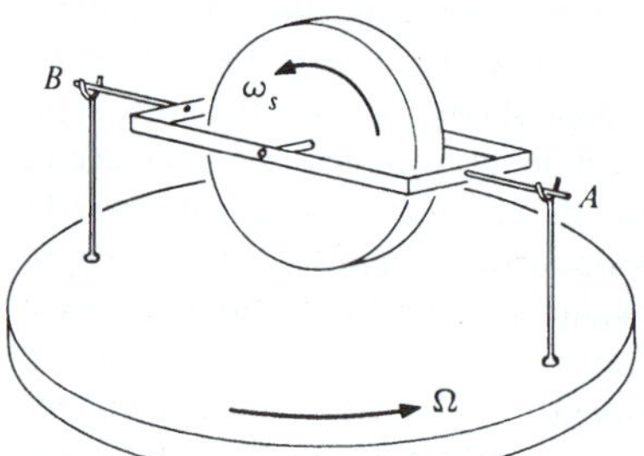

Consider a gyrocompass consisting of a balanced spinning disk held in a light frame supported by a horizontal axle. The assembly is on a turntable rotating at steady angular velocity Ω. The gyro has spin angular momentum $L_s = I_s\omega_s$ along the spin axis. In addition, it possesses angular momentum due to its bodily rotation about the vertical axis at rate Ω, and by virtue of rotation about the horizontal axle.

There cannot be any torque along the horizontal AB axis because that axle is pivoted. Hence, the angular momentum L_h along the AB direction is constant, and $dL_h/dt = 0$.

There are two contributions to dL_h/dt. If θ is the angle from the vertical to the spin axis, and $I_\perp$ is the moment of inertia about the AB axis, then $L_h = I_\perp\dot{\theta}$, and there is a contribution to dL_h/dt of $I_\perp\ddot{\theta}$.

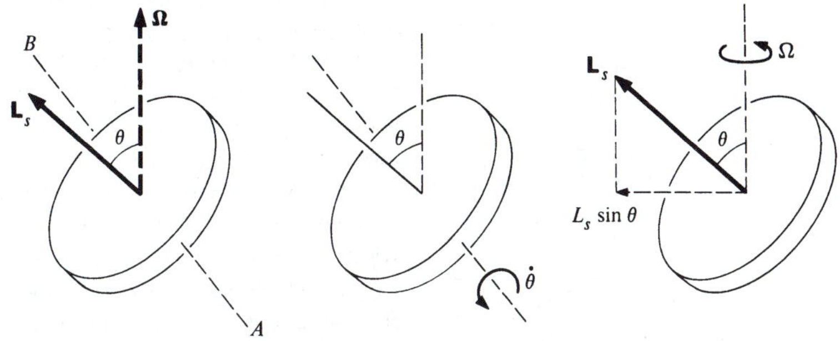

In addition, L_h can change because of a change in direction of $\mathbf{L}_s$, as we have learned from analyzing the precessing gyroscope. The horizontal component of $\mathbf{L}_s$ is $L_s \sin \theta$, and its rate of increase along the AB axis is $\Omega L_s \sin \theta$.

We have considered the two changes in L_h independently. It is plausible that the total change in L_h is the sum of the two changes; a rigorous justification can be given based on arguments presented in Sec. 7.7.

Adding the two contributions to dL_h/dt gives

$$\frac{dL_h}{dt} = I_\perp \ddot{\theta} + \Omega L_s \sin \theta.$$

Since $dL_h/dt = 0$, the equation of motion becomes

$$\ddot{\theta} + \left(\frac{L_s\Omega}{I_\perp}\right) \sin \theta = 0.$$

This is identical to the equation for a pendulum discussed in Sec. 6.6. When the spin axis is near the vertical, $\sin \theta \approx \theta$ and the gyro executes simple harmonic motion in θ:

$$\theta = \theta_0 \sin \beta t$$

where

$$\beta = \sqrt{\frac{L_s\Omega}{I_\perp}}$$
$$= \sqrt{\frac{\omega_s \Omega I_s}{I_\perp}}.$$

If there is a small amount of friction in the bearings at A and B, the amplitude of oscillation θ_0 will eventually become zero, and the spin axis comes to rest parallel to Ω.

To use the gyro as a compass, fix it to the earth with the AB axle vertical, and the frame free to turn. As the drawing on the next page shows, if λ is the latitude of the gyro, the component of the earth's angular velocity Ω_e perpendicular to the AB axle is the horizontal component $\Omega_e \cos \lambda$. The spin axis oscillates in the horizontal plane about the direction of the north pole, and eventually comes to rest pointing north.

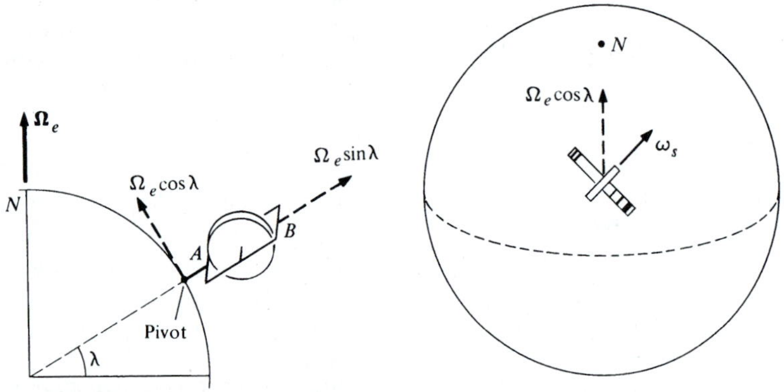

The period of small oscillations is $T = 2\pi/\beta = 2\pi \sqrt{I_\perp/(I_s\omega_s\Omega_e \cos \lambda)}$. For a thin disk $I_\perp/I_s = \frac{1}{2}$. $\Omega_e = 2\pi$ rad/day. With a gyro rotating at 20,000 rpm, the period at the equator is 11 s. Near the north pole the period becomes so long that the gyrocompass is not effective.

Example 7.12 The Stability of Rotating Objects

Angular momentum can make a freely moving object remarkably stable. For instance, spin angular momentum keeps a childs' rolling hoop upright even when it hits a bump; instead of falling, the hoop changes direction slightly and continues to roll. The effect of spin on a bullet provides another example. The spiral grooves, or rifling, in a gun's barrel give the bullet spin, which helps to stabilize it.

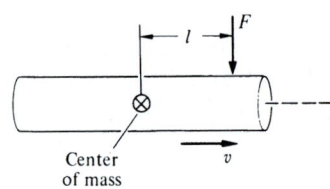

To analyze the effect of spin, consider a cylinder moving parallel to its axis. Suppose that a small perturbing force F acts on the cylinder for time Δt. F is perpendicular to the axis, and the point of application is a distance l from the center of mass.

We consider first the case where the cylinder has zero spin. The torque along the axis AA through the center of mass is $\tau = Fl$, and the "angular impulse" is $\tau\,\Delta t = Fl\,\Delta t$. The angular momentum acquired around the AA axis is

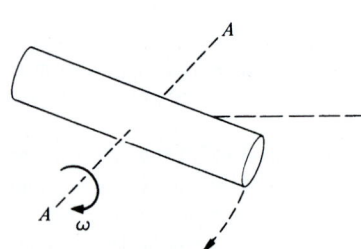

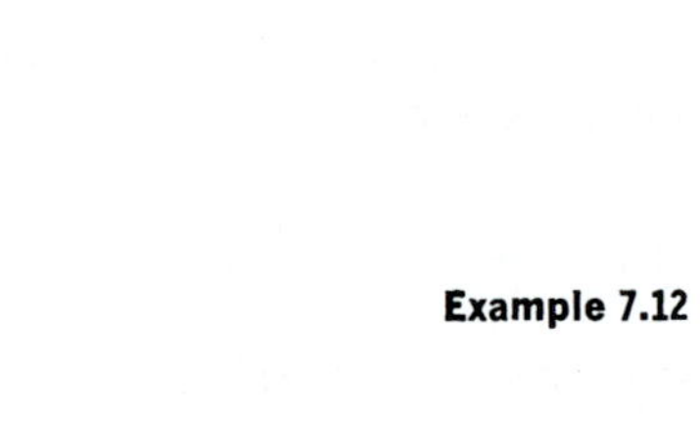

$$\Delta L_A = I_A(\omega - \omega_0) = Fl\,\Delta t.$$

Since ω_0, the initial angular velocity, is 0, the final angular velocity is given by

$$\omega = \frac{Fl\,\Delta t}{I_A}.$$

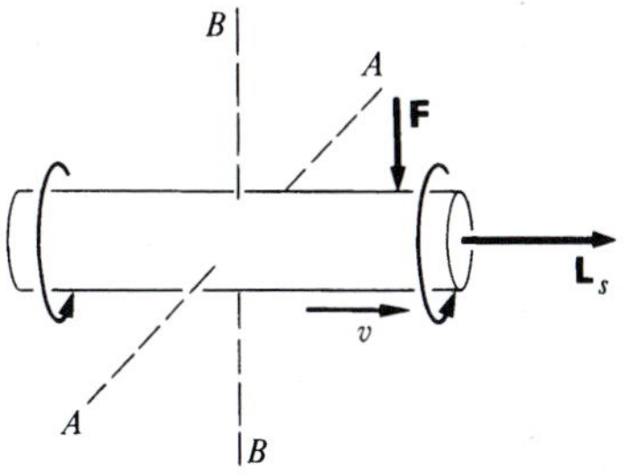

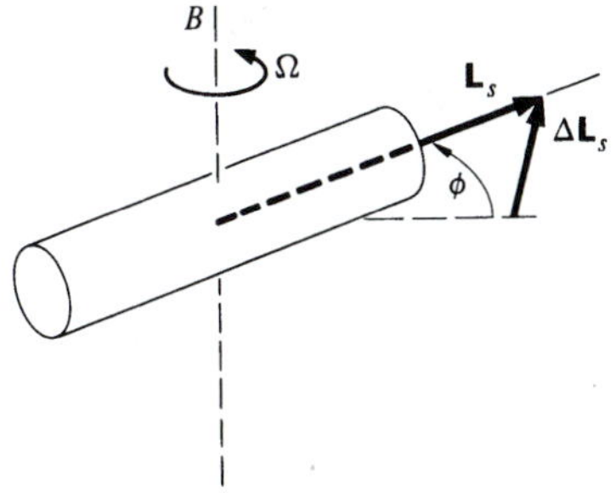

The effect of the blow is to give the cylinder angular velocity around the transverse axis; it starts to tumble.

Now consider the same situation, except that the cylinder is rapidly spinning with angular momentum $\mathbf{L}_s$. The situation is similar to that of the gyroscope: torque along the AA axis causes precession around the BB axis. The rate of precession while F acts is $dL_s/dt = \Delta L_s$, or

$$\Omega = \frac{Fl}{L_s}.$$

The angle through which the cylinder precesses is

$$\begin{aligned}\phi &= \Omega\,\Delta t\\ &= \frac{Fl\,\Delta t}{L_s}.\end{aligned}$$

Instead of starting to tumble, the cylinder slightly changes its orientation while the force is applied, and then stops precessing. The larger the spin, the smaller the angle and the less the effect of perturbations on the flight.

Note that spin has no effect on the center of mass motion. In both cases, the center of mass acquires velocity $\Delta\mathbf{v} = \mathbf{F}\,\Delta t/M$.

7.5 Conservation of Angular Momentum

Before tackling the general problem of rigid body motion, let us return to the question of whether or not the angular momentum of an isolated system is conserved. To start, we shall show that conservation of angular momentum does *not* follow from Newton's laws.

Consider a system of N particles with masses $m_1, m_2, \ldots, m_j, \ldots, m_N$. We assume that the system is isolated, so that the forces are due entirely to interactions between the particles. Let the force on particle j be

$$\mathbf{f}_j = \sum_{k=1}^{N} \mathbf{f}_{jk},$$

where $\mathbf{f}_{jk}$ is the force on particle j due to particle k. (In evaluating the sum, we can neglect the term with $k = j$, since $\mathbf{f}_{jj} = 0$, by Newton's third law.)

Let us choose an origin and calculate the torque $\boldsymbol{\tau}_j$ on particle j.

$$\begin{aligned}\boldsymbol{\tau}_j &= \mathbf{r}_j \times \mathbf{f}_j\\ &= \mathbf{r}_j \times \sum_k \mathbf{f}_{jk}.\end{aligned}$$

Let $\boldsymbol{\tau}_{jl}$ be the torque on j due to the particle l:

$$\boldsymbol{\tau}_{jl} = \mathbf{r}_j \times \mathbf{f}_{jl}.$$

Similarly, the torque on l due to j is

$$\boldsymbol{\tau}_{lj} = \mathbf{r}_l \times \mathbf{f}_{lj}.$$

The sum of these two torques is

$$\boldsymbol{\tau}_{jl} + \boldsymbol{\tau}_{lj} = \mathbf{r}_l \times \mathbf{f}_{lj} + \mathbf{r}_j \times \mathbf{f}_{jl}.$$

Since $\mathbf{f}_{jl} = -\mathbf{f}_{lj}$, we have

$$\begin{aligned}\boldsymbol{\tau}_{jl} + \boldsymbol{\tau}_{lj} &= (\mathbf{r}_l \times \mathbf{f}_{lj}) - (\mathbf{r}_j \times \mathbf{f}_{lj}) \\ &= (\mathbf{r}_l - \mathbf{r}_j) \times \mathbf{f}_{lj} \\ &= \mathbf{r}_{jl} \times \mathbf{f}_{lj},\end{aligned}$$

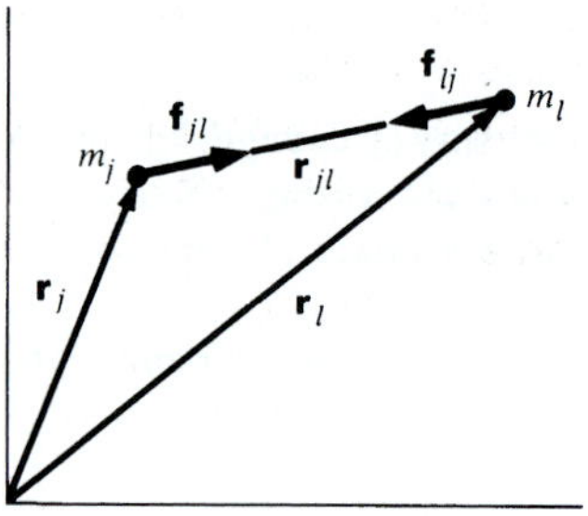

(a)

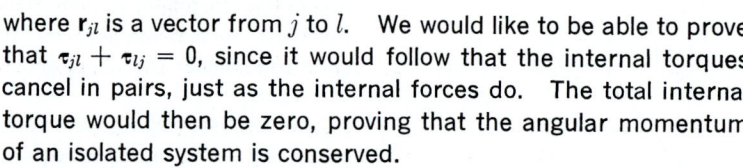

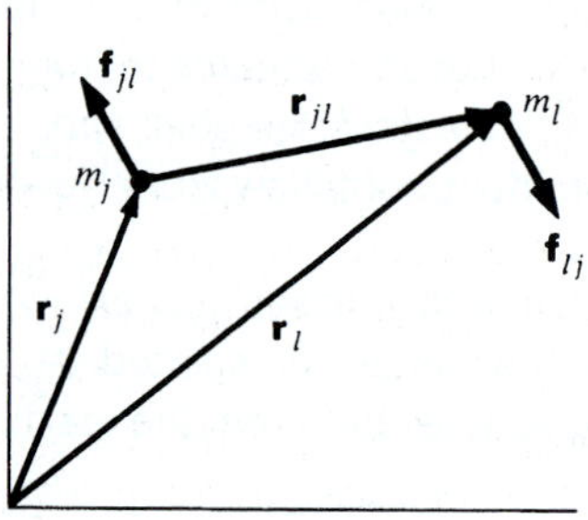

(b)

where $\mathbf{r}_{jl}$ is a vector from j to l. We would like to be able to prove that $\boldsymbol{\tau}_{jl} + \boldsymbol{\tau}_{lj} = 0$, since it would follow that the internal torques cancel in pairs, just as the internal forces do. The total internal torque would then be zero, proving that the angular momentum of an isolated system is conserved.

Since neither $\mathbf{r}_{jl}$ nor $\mathbf{f}_{lj}$ is zero, in order for the torque to vanish, $\mathbf{f}_{lj}$ must be parallel to $\mathbf{r}_{jl}$, as shown in figure (*a*). With respect to the situation in figure (*b*), however, the torque is not zero, and angular momentum is not conserved. Nevertheless, the forces are equal and opposite, and linear momentum is conserved.

The situation shown in figure (*a*) corresponds to the case of *central forces*, and we conclude that the conservation of angular momentum follows from Newton's laws in the case of central force motion. However, Newton's laws do not explicitly require forces to be central. We must conclude that Newton's laws have no direct bearing on whether or not the angular momentum of an isolated system is conserved, since these laws do not in themselves exclude the situation shown in figure (*b*).

It is possible to take exception to the argument above on the following grounds: although Newton's laws do not explicitly require forces to be central, they implicitly make this requirement because in their simplest form Newton's laws deal with particles. Particles are idealized masses which have no size and no structure. In this case, the force between isolated particles must be central, since the only vector defined in a two particle system is the vector $\mathbf{r}_{jl}$ from one particle to the other. For instance, suppose that we try to invent a force which lies at angle θ with respect to the interparticle axis, as shown in the diagram. There is no way to dis-

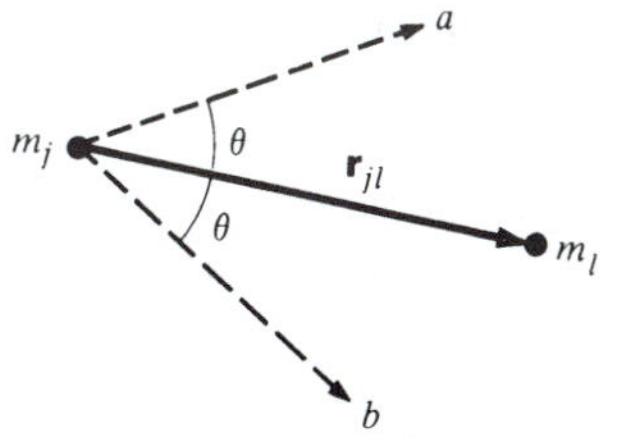

tinguish direction a from b, however; both are at angle θ with respect to $\mathbf{r}_{jl}$. An angle-dependent force cannot be defined using only the single vector $\mathbf{r}_{jl}$; the force between the two particles must be central.

The difficulty in discussing angular momentum in the context of newtonian ideas is that our understanding of nature now encompasses entities vastly different from simple particles. As an example, perhaps the electron comes closest to the newtonian idea of a particle. The electron has a well-defined mass and, as far as present knowledge goes, zero radius. In spite of this, the electron has something analogous to internal structure; it possesses spin angular momentum. It is paradoxical that an object with zero size should have angular momentum, but we must accept this paradox as one of the facts of nature.

Because the spin of an electron defines an additional direction in space, the force between two electrons need not be central. As an example, there might be a force

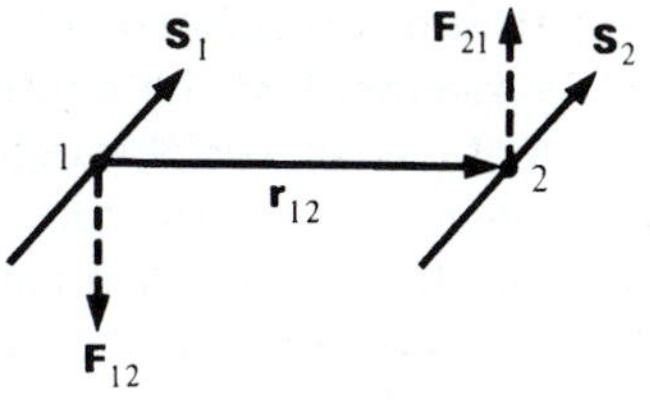

$$\mathbf{F}_{12} = C\mathbf{r}_{12} \times (\mathbf{S}_1 + \mathbf{S}_2)$$
$$\mathbf{F}_{21} = C\mathbf{r}_{21} \times (\mathbf{S}_1 + \mathbf{S}_2),$$

where C is some constant and $\mathbf{S}_i$ is a vector parallel to the angular momentum of the ith electron. The forces are equal and opposite but not central, and they produce a torque.

There are other possibilities for noncentral forces. Experimentally, the force between two charged particles moving with respect to each other is not central; the velocity provides a second axis on which the force depends. The angular momentum of the two particles actually changes. The apparent breakdown of conservation of angular momentum is due to neglect of an important part of the system, the electromagnetic field. Although the concept of a field is alien to particle mechanics, it turns out that fields have mechanical properties. They can possess energy, momentum, and angular momentum. When the angular momentum of the field is taken into account, the angular momentum of the entire particle-field system is conserved.

The situation, in brief, is that newtonian physics is incapable of predicting conservation of angular momentum, but no isolated system has yet been encountered experimentally for which angular momentum is not conserved. We conclude that conservation of angular momentum is an independent physical law, and until a contradiction is observed, our physical understanding must be guided by it.

7.6 Angular Momentum of a Rotating Rigid Body

Angular Momentum and the Tensor of Inertia

The governing equation for rigid body motion, $\boldsymbol{\tau} = d\mathbf{L}/dt$, bears a formal resemblance to the translational equation of motion $\mathbf{F} = d\mathbf{P}/dt$. However, there is an essential difference between them. Linear momentum and center of mass motion are simply related by $\mathbf{P} = M\mathbf{V}$, but the connection between $\mathbf{L}$ and $\boldsymbol{\omega}$ is not so direct. For fixed axis rotation, $L = I\omega$, and it is tempting to suppose that the general relation is $\mathbf{L} = I\boldsymbol{\omega}$, where I is a scalar, that is, a simple number. However, this cannot be correct, since we know from our study of the rotating skew rod, Example 7.4, that $\mathbf{L}$ and $\boldsymbol{\omega}$ are not necessarily parallel.

In this section, we shall develop the general relation between angular momentum and angular velocity, and in the next section we shall attack the problem of solving the equations of motion.

As we discussed in Chap. 6, an arbitrary displacement of a rigid body can be resolved into a displacement of the center of mass plus a rotation about some instantaneous axis through the center of mass. The translational motion is easily treated. We start from the general expressions for the angular momentum and torque of a rigid body, Eqs. (6.11) and (6.14):

$$\mathbf{L} = \mathbf{R} \times M\mathbf{V} + \Sigma \mathbf{r}'_j \times m_j \dot{\mathbf{r}}'_j \qquad 7.3$$

$$\boldsymbol{\tau} = \mathbf{R} \times \mathbf{F} + \Sigma \mathbf{r}'_j \times \mathbf{f}_j, \qquad 7.4$$

where $\mathbf{r}'_j$ is the position vector of m_j relative to the center of mass. Since $\boldsymbol{\tau} = d\mathbf{L}/dt$, we have

$$\mathbf{R} \times \mathbf{F} + \Sigma \mathbf{r}'_j \times \mathbf{f}_j = \frac{d}{dt}(\mathbf{R} \times M\mathbf{V}) + \frac{d}{dt}(\Sigma \mathbf{r}'_j \times m_j \dot{\mathbf{r}}'_j)$$

$$= \mathbf{R} \times M\mathbf{A} + \frac{d}{dt}(\Sigma \mathbf{r}'_j \times m_j \dot{\mathbf{r}}'_j).$$

Since $\mathbf{F} = M\mathbf{A}$, the terms involving $\mathbf{R}$ cancel, and we are left with

$$\Sigma \mathbf{r}'_j \times \mathbf{f}_j = \frac{d}{dt}(\Sigma \mathbf{r}'_j \times m_j \dot{\mathbf{r}}'_j). \qquad 7.5$$

The rotational motion can be found by taking torque and angular momentum about the center of mass, independent of the center of mass motion. The angular momentum $\mathbf{L}_0$ about the center of mass is

$$\mathbf{L}_0 = \Sigma \mathbf{r}'_j \times m_j \dot{\mathbf{r}}'_j. \qquad 7.6$$

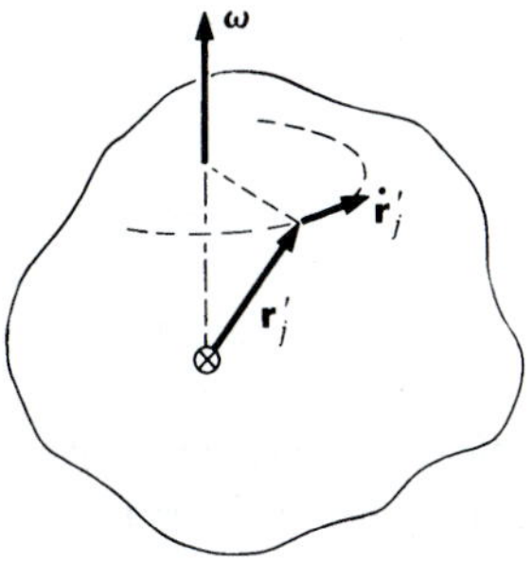

Our task is to express $\mathbf{L}_0$ in terms of the instantaneous angular velocity ω. Since $\mathbf{r}'_j$ is a rotating vector,

$$\dot{\mathbf{r}}'_j = \omega \times \mathbf{r}'_j.$$

Therefore,

$$\mathbf{L}_0 = \Sigma \mathbf{r}'_j \times m_j(\omega \times \mathbf{r}'_j).$$

To simplify the notation, we shall write $\mathbf{L}$ for $\mathbf{L}_0$ and $\mathbf{r}_j$ for $\mathbf{r}'_j$. Our result becomes

$$\mathbf{L} = \Sigma \mathbf{r}_j \times m_j(\omega \times \mathbf{r}_j). \qquad 7.7$$

This result looks complicated. As a matter of fact, it *is* complicated, but we can make it look simple. We will take the pedestrian approach of patiently evaluating the cross products in Eq. (7.7) using cartesian coordinates.[1]

Since $\omega = \omega_x\hat{\mathbf{i}} + \omega_y\hat{\mathbf{j}} + \omega_z\hat{\mathbf{k}}$, we have

$$\omega \times \mathbf{r} = (z\omega_y - y\omega_z)\hat{\mathbf{i}} + (x\omega_z - z\omega_x)\hat{\mathbf{j}} + (y\omega_x - x\omega_y)\hat{\mathbf{k}}. \qquad 7.8$$

Let us compute one component of $\mathbf{L}$, say L_x. Temporarily dropping the subscript j, we have

$$[\mathbf{r} \times (\omega \times \mathbf{r})]_x = y(\omega \times \mathbf{r})_z - z(\omega \times \mathbf{r})_y. \qquad 7.9$$

If we substitute the results of Eq. (7.8) into Eq. (7.9), the result is

$$\begin{aligned}[\mathbf{r} \times (\omega \times \mathbf{r})]_x &= y(y\omega_x - x\omega_y) - z(x\omega_z - z\omega_x) \\ &= (y^2 + z^2)\omega_x - xy\omega_y - xz\omega_z. \end{aligned} \qquad 7.10$$

Hence,

$$L_x = \Sigma m_j(y_j^2 + z_j^2)\omega_x - \Sigma m_j x_j y_j \omega_y - \Sigma m_j x_j z_j \omega_z. \qquad 7.11$$

Let us introduce the following symbols:

$$\begin{aligned} I_{xx} &= \Sigma m_j(y_j^2 + z_j^2) \\ I_{xy} &= -\Sigma m_j x_j y_j \\ I_{xz} &= -\Sigma m_j x_j z_j. \end{aligned} \qquad 7.12$$

I_{xx} is called a *moment of inertia*. It is identical to the moment of inertia introduced in the last chapter, $I = \Sigma m_j \rho_j^2$, provided that we take the axis in the x direction so that $\rho_j^2 = y_j^2 + z_j^2$. The quantities I_{xy} and I_{xz} are called *products of inertia*. They are symmetrical; for example, $I_{xy} = -\Sigma m_j x_j y_j = -\Sigma m_j y_j x_j = I_{yx}$.

To find L_y and L_z, we could repeat the derivation. However, a simpler method is to relabel the coordinates by letting $x \to y$,

[1] Another way is to use the vector identity $\mathbf{A} \times (\mathbf{B} \times \mathbf{C}) = (\mathbf{A} \cdot \mathbf{C})\mathbf{B} - (\mathbf{A} \cdot \mathbf{B})\mathbf{C}$.

$y \to z$, $z \to x$. If we make these substitutions in Eqs. (7.11) and (7.12), we obtain

$$L_x = I_{xx}\omega_x + I_{xy}\omega_y + I_{xz}\omega_z \qquad 7.13a$$

$$L_y = I_{yx}\omega_x + I_{yy}\omega_y + I_{yz}\omega_z \qquad 7.13b$$

$$L_z = I_{zx}\omega_x + I_{zy}\omega_y + I_{zz}\omega_z. \qquad 7.13c$$

This array of three equations is different from anything we have so far encountered. They include the results of the last chapter. For fixed axis rotation about the z direction, $\boldsymbol{\omega} = \omega\hat{\mathbf{k}}$ and Eq. (7.13c) reduces to

$$\begin{aligned} L_z &= I_{zz}\omega \\ &= \Sigma m_j(x_j^2 + y_j^2)\omega. \end{aligned}$$

However, Eq. (7.13) also shows that angular velocity in the z direction can produce angular momentum about *any* of the three coordinate axes. For example, if $\boldsymbol{\omega} = \omega\hat{\mathbf{k}}$, then $L_x = I_{xz}\omega$ and $L_y = I_{yz}\omega$. In fact, if we look at the set of equations for L_x, L_y, and L_z, we see that in each case the angular momentum about one axis depends on the angular velocity about *all three* axes. Both **L** and $\boldsymbol{\omega}$ are ordinary vectors, and **L** is proportional to $\boldsymbol{\omega}$ in the sense that doubling the components of $\boldsymbol{\omega}$ doubles the components of **L**. However, as we have already seen from the behavior of the rotating skew rod, Example 7.4, **L** does not necessarily point in the same direction as $\boldsymbol{\omega}$.

Example 7.13 Rotating Dumbbell

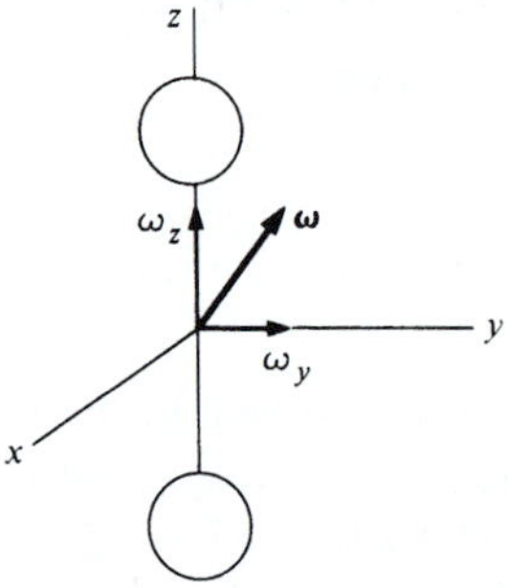

Consider a dumbbell made of two spheres of radius b and mass M separated by a thin rod. The distance between centers is $2l$. The body is rotating about some axis through its center of mass. At a certain instant the rod coincides with the z axis, and $\boldsymbol{\omega}$ lies in the yz plane, $\boldsymbol{\omega} = \omega_y\hat{\mathbf{j}} + \omega_z\hat{\mathbf{k}}$. What is **L**?

To find **L**, we need the moments and products of inertia. Fortunately, the products of inertia vanish for a symmetrical body lined up with the coordinate axes. For example, $I_{xy} = -\Sigma m_j x_j y_j = 0$, since for mass m_n located at (x_n, y_n) there is, in a symmetrical body, an equal mass located at $(x_n, -y_n)$; the contributions of these two masses to I_{xy} cancel. In this case Eq. (7.13) simplifies to

$$L_x = I_{xx}\omega_x$$

$$L_y = I_{yy}\omega_y$$

$$L_z = I_{zz}\omega_z.$$

The moment of inertia I_{zz} is just the moment of inertia of two spheres about their diameters.

$$I_{zz} = 2(\tfrac{2}{5}Mb^2) = \tfrac{4}{5}Mb^2.$$

In calculating I_{yy}, we can use the parallel axis theorem to find the moment of inertia of each sphere about the y axis.

$$\begin{aligned} I_{yy} &= 2(\tfrac{2}{5}Mb^2 + Ml^2) \\ &= \tfrac{4}{5}Mb^2 + 2Ml^2. \end{aligned}$$

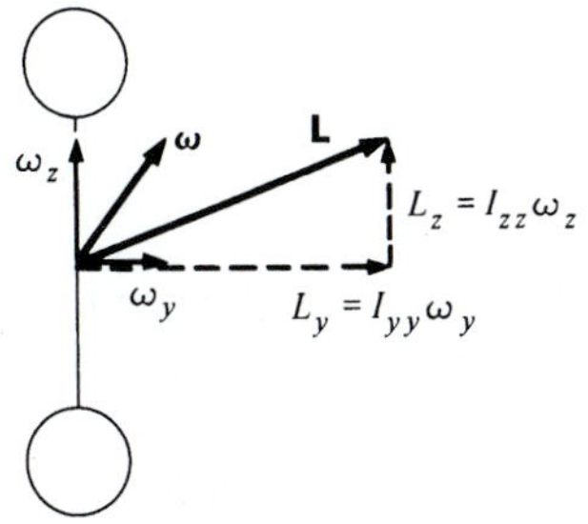

We have assumed that the rod has negligible mass.

Since $\boldsymbol{\omega} = \omega_y\hat{\mathbf{j}} + \omega_z\hat{\mathbf{k}}$,

$$\begin{aligned} L_x &= 0 \\ L_y &= I_{yy}\omega_y \\ L_z &= I_{zz}\omega_z. \end{aligned}$$

I_{yy} and I_{zz} are not equal; therefore $L_y/L_z \neq \omega_y/\omega_z$ and **L** is not parallel to $\boldsymbol{\omega}$, as the drawing shows.

Equations (7.13) are cumbersome, so that it is more convenient to write them in the following shorthand notation.

$$\mathbf{L} = \tilde{\mathbf{I}}\boldsymbol{\omega}. \qquad 7.14$$

This vector equation represents three equations, just as $\mathbf{F} = m\mathbf{a}$ represents three equations. The difference is that m is a simple scalar while $\tilde{\mathbf{I}}$ is a more complicated mathematical entity called a *tensor*. $\tilde{\mathbf{I}}$ is the *tensor of inertia*.

We are accustomed to displaying the components of some vector **A** in the form

$$\mathbf{A} = (A_x, A_y, A_z).$$

Similarly, the nine components of $\tilde{\mathbf{I}}$ can be tabulated in a 3×3 array:

$$\tilde{\mathbf{I}} = \begin{pmatrix} I_{xx} & I_{xy} & I_{xz} \\ I_{yx} & I_{yy} & I_{yz} \\ I_{zx} & I_{zy} & I_{zz} \end{pmatrix}. \qquad 7.15$$

Of the nine components, only six at most are different, since $I_{yx} = I_{xy}$, $I_{zx} = I_{xz}$, and $I_{yz} = I_{zy}$. The rule for multiplying $\boldsymbol{\omega}$ by $\tilde{\mathbf{I}}$ to find $\mathbf{L} = \tilde{\mathbf{I}}\boldsymbol{\omega}$ is defined by Eq. (7.13).

The following example illustrates the tensor of inertia.

Example 7.14 **The Tensor of Inertia for a Rotating Skew Rod**

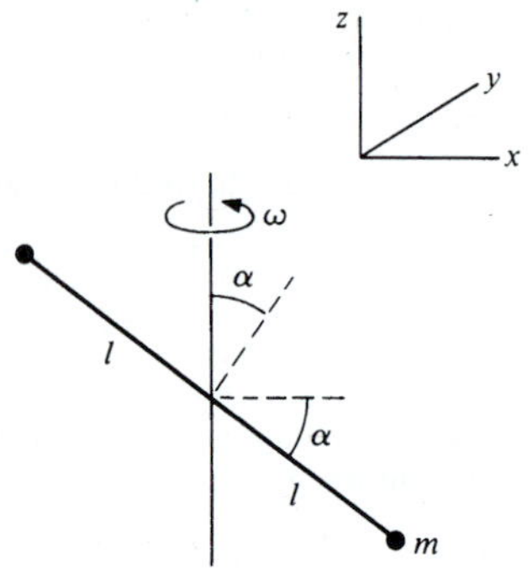

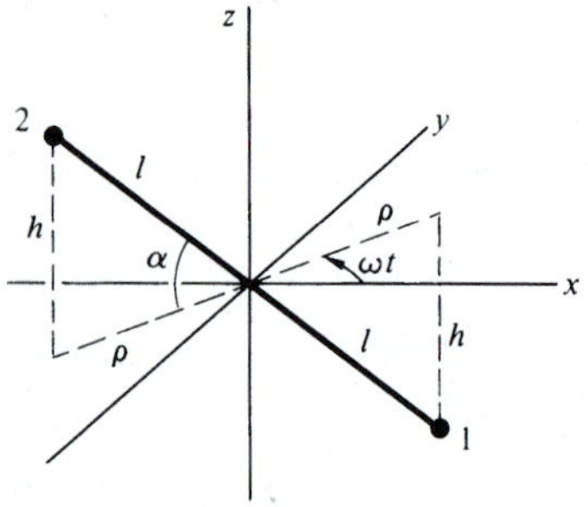

We found the angular momentum of a rotating skew rod from first principles in Example 7.3. Let us now find **L** for the same device by using $\mathbf{L} = \tilde{\mathbf{I}}\boldsymbol{\omega}$.

A massless rod of length $2l$ separates two equal masses m. The rod is skewed at angle α with the vertical, and rotates around the z axis with angular velocity ω. At $t = 0$ it lies in the xz plane. The coordinates of the particles at any other time are:

Particle 1	Particle 2
$x_1 = \rho \cos \omega t$	$x_2 = -\rho \cos \omega t$
$y_1 = \rho \sin \omega t$	$y_2 = -\rho \sin \omega t$
$z_1 = -h$	$z_2 = h,$

when $\rho = l \cos \alpha$ and $h = l \sin \alpha$.

The components of $\tilde{\mathbf{I}}$ can now be calculated from their definitions. For instance,

$$
\begin{aligned}
I_{xx} &= m_1(y_1{}^2 + z_1{}^2) + m_2(y_2{}^2 + z_2{}^2) \\
&= 2m(\rho^2 \sin^2 \omega t + h^2) \\
I_{zy} &= I_{yz} \\
&= -m_1 y_1 z_1 - m_2 y_2 z_2 \\
&= 2m\rho h \sin \omega t.
\end{aligned}
$$

The remaining terms are readily evaluated. We find:

$$
\mathbf{I} = 2m \begin{pmatrix} \rho^2 \sin^2 \omega t + h^2 & -\rho^2 \sin \omega t \cos \omega t & \rho h \cos \omega t \\ -\rho^2 \sin \omega t \cos \omega t & \rho^2 \cos^2 \omega t + h^2 & \rho h \sin \omega t \\ \rho h \cos \omega t & \rho h \sin \omega t & \rho^2 \end{pmatrix}.
$$

The common factor $2m$ multiplies each term.

Since $\boldsymbol{\omega} = (0,0,\omega)$, we have, from Eq. (7.13),

$$
\begin{aligned}
L_x &= 2m\rho h\omega \cos \omega t \\
L_y &= 2m\rho h\omega \sin \omega t \\
L_z &= 2m\rho^2\omega.
\end{aligned}
$$

We can differentiate **L** to find the applied torque:

$$
\begin{aligned}
\tau_x &= -2m\rho h\omega^2 \sin \omega t \\
\tau_y &= 2m\rho h\omega^2 \cos \omega t \\
\tau_z &= 0.
\end{aligned}
$$

The results are identical to those in Example 7.4, provided that we make the substitution $\rho h = l^2 \cos \alpha \sin \alpha$.

Principal Axes

If the symmetry axes of a uniform symmetric body coincide with the coordinate axes, the products of inertia are zero, as we saw in Example 7.13. In this case the tensor of inertia takes a simple diagonal form:

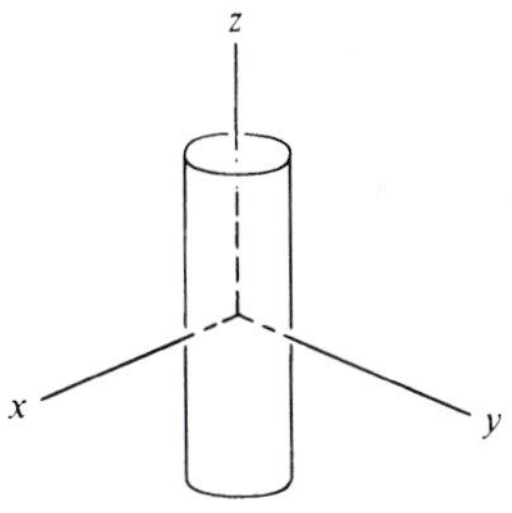

$$\tilde{\mathbf{I}} = \begin{pmatrix} I_{xx} & 0 & 0 \\ 0 & I_{yy} & 0 \\ 0 & 0 & I_{zz} \end{pmatrix}. \qquad 7.16$$

Remarkably enough, for a body of any shape and mass distribution, it is *always* possible to find a set of three orthogonal axes such that the products of inertia vanish. (The proof uses matrix algebra and is given in most texts on advanced dynamics.) Such axes are called *principal axes*. The tensor of inertia with respect to principal axes has a diagonal form.

For a uniform sphere, any perpendicular axes through the center are principal axes. For a body with cylindrical symmetry, the axis of revolution is a principal axis. The other two principal axes are mutually perpendicular and lie in a plane through the center of mass perpendicular to the axis of revolution.

Consider a rotating rigid body, and suppose that we introduce a coordinate system 1, 2, 3 which coincides instantaneously with the principal axes of the body. With respect to this coordinate system, the instantaneous angular velocity has components ω_1, ω_2, ω_3, and the components of **L** have the simple form

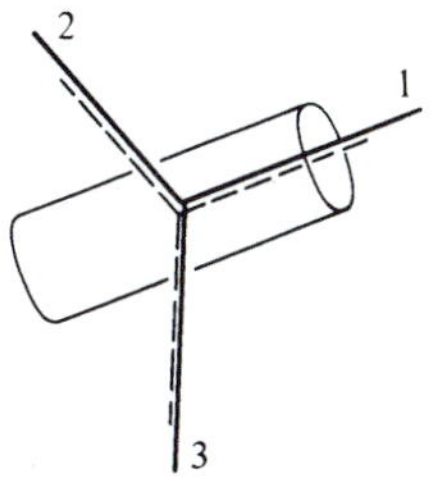

$$\begin{aligned} L_1 &= I_1\omega_1 \\ L_2 &= I_2\omega_2 \\ L_3 &= I_3\omega_3, \end{aligned} \qquad 7.17$$

where I_1, I_2, I_3 are the moments of inertia about the principal axes. In Sec. 7.7, we shall exploit Eq. (7.17) in our attack on the problem of rigid body dynamics.

Rotational Kinetic Energy

The kinetic energy of a rigid body is

$$K = \tfrac{1}{2}\Sigma m_j v_j^2.$$

To separate the translational and rotational contributions, we introduce center of mass coordinates:

$$\begin{aligned} \mathbf{r}_j &= \mathbf{R} + \mathbf{r}_j' \\ \mathbf{v}_j &= \mathbf{V} + \mathbf{v}_j'. \end{aligned}$$

We have

$$K = \tfrac{1}{2}\Sigma m_j(\mathbf{V} + \mathbf{v}_j')^2$$
$$= \tfrac{1}{2}MV^2 + \tfrac{1}{2}\Sigma m_j v_j'^2,$$

since the cross term $\mathbf{V} \cdot \Sigma m_j \mathbf{v}_j'$ is zero.

Using $\mathbf{v}_j' = \boldsymbol{\omega} \times \mathbf{r}_j'$, the kinetic energy of rotation becomes

$$K_{\text{rot}} = \tfrac{1}{2}\Sigma m_j \mathbf{v}_j'^2$$
$$= \tfrac{1}{2}\Sigma m_j(\boldsymbol{\omega} \times \mathbf{r}_j') \cdot (\boldsymbol{\omega} \times \mathbf{r}_j').$$

The right hand side can be simplified with the vector identity $(\mathbf{A} \times \mathbf{B}) \cdot \mathbf{C} = \mathbf{A} \cdot (\mathbf{B} \times \mathbf{C})$. Let $\mathbf{A} = \boldsymbol{\omega}$, $\mathbf{B} = \mathbf{r}_j'$, and $\mathbf{C} = \boldsymbol{\omega} \times \mathbf{r}_j'$. We obtain

$$K_{\text{rot}} = \tfrac{1}{2}\Sigma m_j \boldsymbol{\omega} \cdot [\mathbf{r}_j' \times (\boldsymbol{\omega} \times \mathbf{r}_j')]$$
$$= \tfrac{1}{2}\boldsymbol{\omega} \cdot \Sigma m_j \mathbf{r}_j' \times (\boldsymbol{\omega} \times \mathbf{r}_j').$$

The sum in the last term is the angular momentum $\mathbf{L}$ by Eq. (7.7). Therefore,

$$K_{\text{rot}} = \tfrac{1}{2}\boldsymbol{\omega} \cdot \mathbf{L}. \qquad 7.18$$

Rotational kinetic energy has a simple form when $\mathbf{L}$ and $\boldsymbol{\omega}$ are referred to principal axes. Using Eqs. (7.17) and (7.18) we have

$$K_{\text{rot}} = \tfrac{1}{2}\boldsymbol{\omega} \cdot \mathbf{L}$$
$$= \tfrac{1}{2}I_1\omega_1^2 + \tfrac{1}{2}I_2\omega_2^2 + \tfrac{1}{2}I_3\omega_3^2. \qquad 7.19$$

Alternatively,

$$K_{\text{rot}} = \frac{L_1^2}{2I_1} + \frac{L_2^2}{2I_2} + \frac{L_3^2}{2I_3}. \qquad 7.20$$

Example 7.15 Why Flying Saucers Make Better Spacecraft than Do Flying Cigars

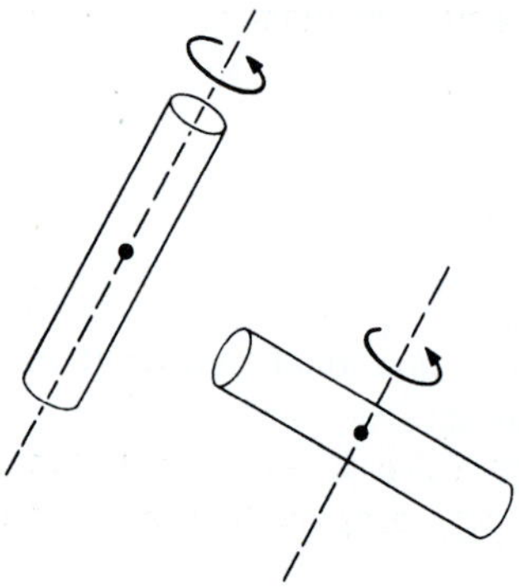

One of the early space satellites was cylindrical in shape and was put into orbit spinning around its long axis. To the designer's surprise, even though the spacecraft was torque-free, it began to wobble more and more, until finally it was spinning around a transverse axis.

The reason is that although $\mathbf{L}$ is strictly conserved for torque-free motion, kinetic energy of rotation can change if the body is not absolutely rigid. If the satellite is rotating slightly off the symmetry axis, each part of the body undergoes a time varying centripetal acceleration. The spacecraft warps and bends under the time varying force, and energy is dissipated by internal friction in the structure. The kinetic energy of rotation must therefore decrease. From Eq. (7.20), if the body is rotating about a single principal axis, $K_{\text{rot}} = L^2/2I$. K_{rot} is a minimum for the axis with greatest moment of inertia, and the motion is stable around that axis. For the cylindrical spacecraft, the initial axis of rotation had the minimum moment of inertia, and the motion was not stable.

A thin disk spinning about its cylindrical axis is inherently stable because the other two moments of inertia are only half as large. A cigar-shaped craft is unstable about its long axis and only neutrally stable about the transverse axes; there is no single axis of maximum moment of inertia.

Rotation about a Fixed Point

We showed at the beginning of this section that in analyzing the motion of a rotating and translating rigid body it is always correct to calculate torque and angular momentum about the center of mass. In some applications, however, one point of a body is fixed in space, like the pivot point of a gyroscope on a pylon. It is often convenient to analyze the motion using the fixed point as origin, since the center of mass motion need not be considered explicitly, and the constraint force at the pivot produces no torque.

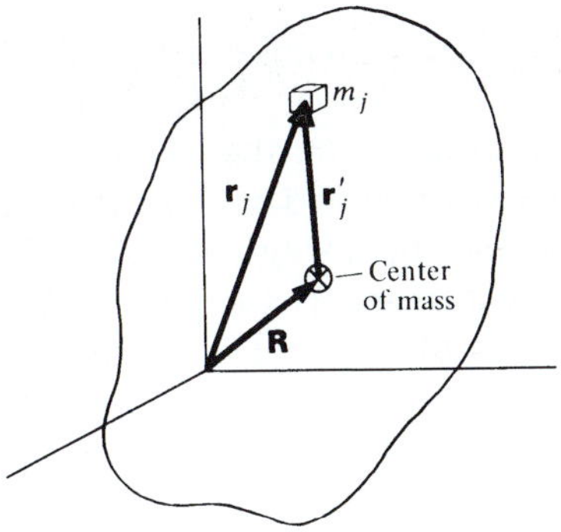

Taking the origin at the fixed point, let $\mathbf{r}_j$ be the position vector of particle m_j and let $\mathbf{R} = X\hat{\mathbf{i}} + Y\hat{\mathbf{j}} + Z\hat{\mathbf{k}}$ be the position vector of the center of mass. The torque about the origin is

$$\boldsymbol{\tau} = \Sigma \mathbf{r}_j \times \mathbf{f}_j,$$

where $\mathbf{f}_j$ is the force on m_j. If the angular velocity of the body is $\boldsymbol{\omega}$, the angular momentum about the origin is

$$\begin{aligned}\mathbf{L} &= \Sigma \mathbf{r}_j \times m_j \dot{\mathbf{r}}_j \\ &= \Sigma \mathbf{r}_j \times m_j(\boldsymbol{\omega} \times \mathbf{r}_j).\end{aligned}$$

This has the same form as Eq. (7.6), which we evaluated earlier in this section. Taking over the results wholesale, we have

$$\mathbf{L} = \tilde{\mathbf{I}}\boldsymbol{\omega}$$

where

$$\begin{aligned}I_{xx} &= \Sigma m_j(y_j^2 + z_j^2) \\ I_{xy} &= -\Sigma m_j x_j y_j\end{aligned}$$

etc.

Although this result is identical in form to Eq. (7.13), the components of $\tilde{\mathbf{I}}$ are now calculated with respect to the pivot point rather than the center of mass.

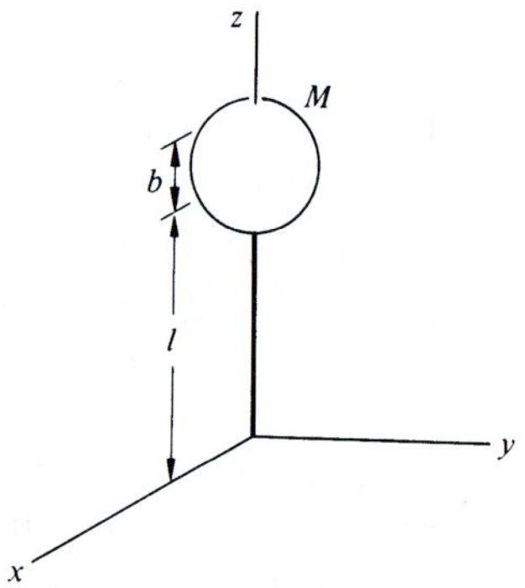

Once the tensor of inertia about the center of mass, $\tilde{\mathbf{I}}_0$, is known, $\tilde{\mathbf{I}}$ about any other origin can be found from a generalization of the parallel axis theorem of Example 6.9. Typical results, the proof of which we leave as a problem, are

$$I_{xx} = (I_0)_{xx} + M(Y^2 + Z^2)$$
$$I_{xy} = (I_0)_{xy} - MXY$$
etc. 7.21

Consider, for example, a sphere of mass M and radius b centered on the z axis a distance l from the origin. We have $I_{xx} = \frac{2}{5}Mb^2 + Ml^2$, $I_{yy} = \frac{2}{5}Mb^2 + Ml^2$, $I_{zz} = \frac{2}{5}Mb^2$.

7.7 Advanced Topics in the Dynamics of Rigid Body Rotation

Introduction

In this section we shall attack the general problem of rigid body rotation. However, none of the results will be needed in subsequent chapters, and the section can be skipped without loss of continuity.

The fundamental problem of rigid body dynamics is to find the orientation of a rotating body as a function of time, given the torque. The problem is difficult because of the complicated relation $\mathbf{L} = \tilde{\mathbf{I}}\boldsymbol{\omega}$ between angular momentum and angular velocity. We can make the problem look simpler by taking our coordinate system coincident with the principal axes of the body. With respect to principal axes, the tensor of inertia $\tilde{\mathbf{I}}$ is diagonal in form, and the components of $\mathbf{L}$ are

$$L_x = I_{xx}\omega_x$$
$$L_y = I_{yy}\omega_y$$
$$L_z = I_{zz}\omega_z.$$

However, the crux of the problem is that the principal axes are fixed to the body, whereas we need the components of $\mathbf{L}$ with respect to axes having a fixed orientation in space. As the body rotates, its principal axes move out of coincidence with the space-fixed system. The products of inertia are no longer zero in the space-fixed system and, worse yet, the components of $\tilde{\mathbf{I}}$ vary with time.

The situation appears hopelessly tangled, but if the principal axes do not stray far from the space-fixed system, we can find the motion using simple vector arguments. Leaving the general case for later, we illustrate this approach by finding the torque-free motion of a rigid body.

Torque-free Precession: Why the Earth Wobbles

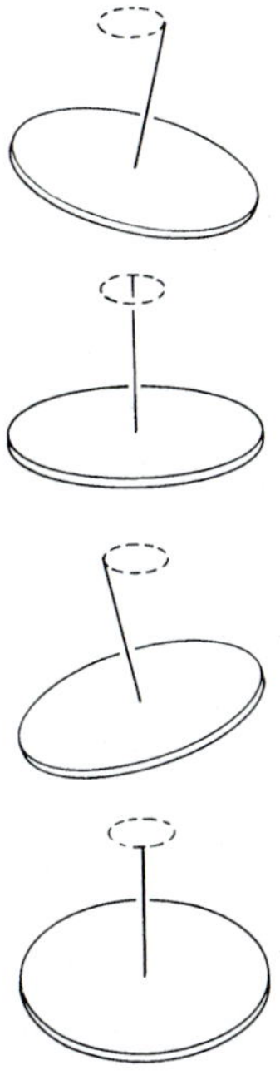

If you drop a spinning quarter with a slight flip, it will fall with a wobbling motion; the symmetry axis tends to rotate in space, as the sketch shows. Since there are no torques, the motion is known as torque-free precession.

Torque-free precession is a characteristic mode of rigid body motion. For example, the spin axis of the earth moves around the polar axis because of this effect. The physical explanation of the wobbling motion is related to our observation that **L** need not be parallel to $\boldsymbol{\omega}$. If there are no torques on the body, **L** is fixed in space, and $\boldsymbol{\omega}$ must move, as will be shown.

To avoid mathematical complexity, consider the special case of a cylindrically symmetric rigid body like a coin or an air suspension gyroscope. We shall assume that the precessional motion is small in amplitude, in order to apply small angle approximations.

Suppose that the body has a large spin angular momentum $L = I_s\omega_s$ along the main symmetry axis, where I_s is the moment of inertia and ω_s is the angular velocity about the symmetry axis. Let the body have small angular velocities about the other transverse axes.

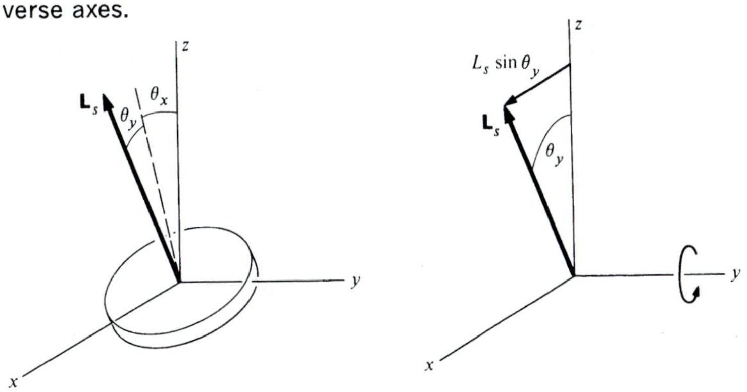

Suppose that $\mathbf{L}_s$ is always close to the z axis and makes angles $\theta_x \ll 1$ and $\theta_y \ll 1$ with the x and y axes. Note 7.1 on infinitesimal rotations shows that to first order, rotations about each axis can be considered separately. The contribution to L_x from rotation about the x axis is $L_x = d(I_{xx}\theta_x)/dt = I_{xx}\, d\theta_x/dt$. We have treated I_{xx} as a constant. The justification is that moments of inertia about principal axes are constant to first order for small angular displacements. Similarly, the products of inertia remain zero to first order. (The proofs are left as a problem.) Rotation about y also contributes to L_x by giving L_s a component $L_s \sin \theta_y$ in the x direction. Adding the two contributions, we have

$$L_x = I_{xx} \frac{d\theta_x}{dt} + L_s \sin \theta_y.$$

Similarly,

$$L_y = I_{yy} \frac{d\theta_y}{dt} - L_s \sin \theta_x.$$

By symmetry, $I_{xx} = I_{yy} \equiv I_\perp$. For small angles, $\sin \theta = \theta$ and $\cos \theta = 1$, to first order. Hence

$$L_x = I_\perp \frac{d\theta_x}{dt} + L_s\theta_y \qquad 7.22a$$

$$L_y = I_\perp \frac{d\theta_y}{dt} - L_s\theta_x. \qquad 7.22b$$

To the same order of approximation,

$$\begin{aligned} L_z &= L_s \\ &= I_s\omega_s. \end{aligned} \qquad 7.23$$

Since the torque is zero, $d\mathbf{L}/dt = 0$. Equation (7.23) then gives L_s = constant, ω_s = constant, and Eqs. (7.22) yield

$$I_\perp \frac{d^2\theta_x}{dt^2} + L_s \frac{d\theta_y}{dt} = 0 \qquad 7.24a$$

$$I_\perp \frac{d^2\theta_y}{dt^2} - L_s \frac{d\theta_x}{dt} = 0. \qquad 7.24b$$

If we let $\omega_x = d\theta_x/dt$, $\omega_y = d\theta_y/dt$, Eqs. (7.24) become

$$I_\perp \frac{d\omega_x}{dt} + L_s\omega_y = 0 \qquad 7.25a$$

$$I_\perp \frac{d\omega_y}{dt} - L_s\omega_x = 0. \qquad 7.25b$$

If we differentiate Eq. (7.25*a*) and substitute the value for $d\omega_y/dt$ in Eq. (7.25*b*), we obtain

$$\frac{I_\perp^2}{L_s} \frac{d^2\omega_x}{dt^2} + L_s\omega_x = 0$$

or

$$\frac{d^2\omega_x}{dt^2} + \gamma^2\omega_x = 0, \tag{7.26}$$

where

$$\gamma = \frac{L_s}{I_\perp}$$
$$= \omega_s \frac{I_s}{I_\perp}.$$

Equation (7.26) is the familiar equation for simple harmonic motion. The solution is

$$\omega_x = A \sin (\gamma t + \phi), \tag{7.27}$$

where A and ϕ are arbitrary constants. Substituting this in Eq. (7.25a) gives

$$\omega_y = -\frac{I_\perp}{L_s}\frac{d\omega_x}{dt}$$
$$= \frac{I_\perp}{I_s\omega_s} A\gamma \cos (\gamma t + \phi),$$

or

$$\omega_y = A \cos (\gamma t + \phi). \tag{7.28}$$

By integrating Eqs. (7.27) and (7.28) we obtain

$$\theta_x = \frac{A}{\gamma} \cos (\gamma t + \phi) + \theta_{x0}$$
$$\theta_y = -\frac{A}{\gamma} \sin (\gamma t + \phi) + \theta_{y0}, \tag{7.29}$$

where θ_{x0} and θ_{y0} are constants of integration. The first terms of Eq. (7.29) reveal that the axis rotates around a fixed direction in space. If we take that direction along the z axis, then $\theta_{x0} = \theta_{y0} = 0$. Assuming that at $t = 0$ $\theta_x = \theta_0$, $\theta_y = 0$, we have

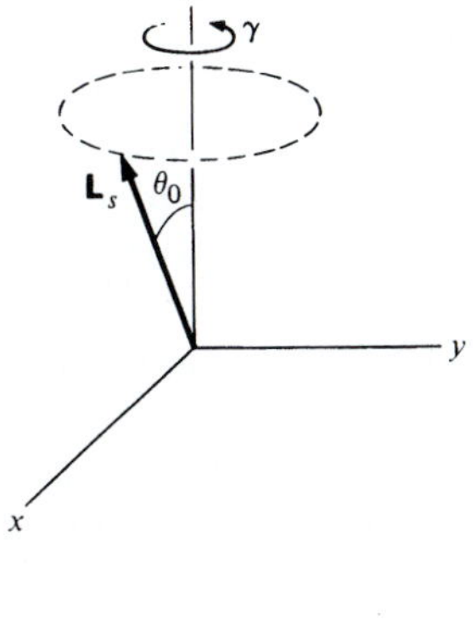

$$\theta_x = \theta_0 \cos \gamma t$$
$$\theta_y = \theta_0 \sin \gamma t, \tag{7.30}$$

where we have taken $A/\gamma = \theta_0$, $\phi = 0$.

Equation (7.30) describes torque-free precession. The frequency of the precessional motion is $\gamma = \omega_s I_s/I_\perp$. For a body flattened along the axis of symmetry, such as the oblate spheroid

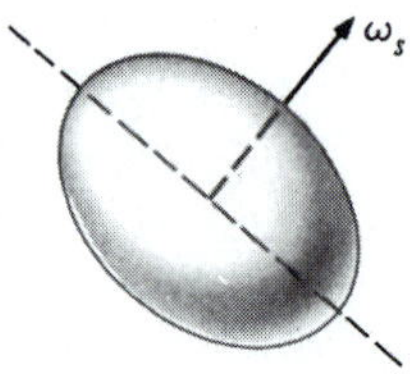

shown, $I_s > I_\perp$ and $\gamma > \omega_s$. For a thin coin, $I_s = 2I_\perp$ and $\gamma = 2\omega_s$. Thus, the falling quarter described earlier wobbles twice as fast as it spins.

The earth is an oblate spheroid and exhibits torque-free precession. The amplitude of the motion is small; the spin axis wanders about the polar axis by about 5 m at the North Pole. Since the earth itself is spinning, the apparent rate of precession to an earthbound observer is

$$\begin{aligned}\gamma' &= \gamma - \omega_s \\ &= \omega_s\left(\frac{I_s - I_\perp}{I_\perp}\right).\end{aligned} \qquad 7.31$$

For the earth, $(I_s - I_\perp)/I_\perp = \frac{1}{300}$, and the precessional motion should have a period of 300 days. However, the motion is quite irregular with an apparent period of about 430 days. The fluctuations arise from the elastic nature of the earth, which is significant for motions this small.

Note 7.2 on the nutating gyroscope illustrates another application of the small angle approximation that we have used.

Euler's Equations

We turn now to the task of deriving the exact equations of motion for a rigid body. In order to find $d\mathbf{L}/dt$, we shall calculate the change in the components of $\mathbf{L}$ in the time interval from t to $t + \Delta t$, using the small angle approximation. The results are correct only to first order, but they become exact when we take the limit $\Delta t \to 0$.

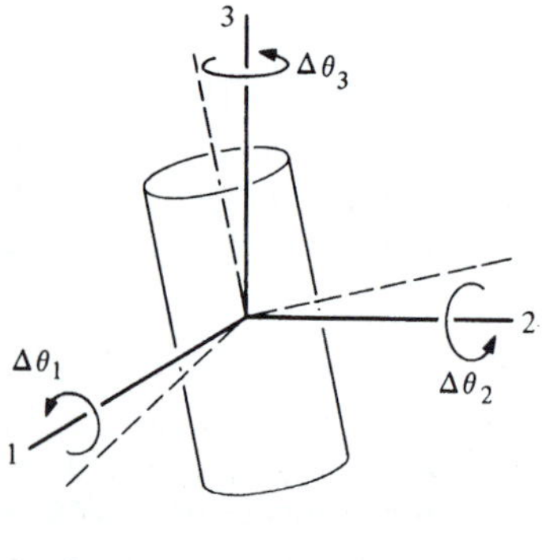

Let us introduce an inertial coordinate system which coincides with the instantaneous position of the body's principal axes at time t. We label the axes of the inertial system 1, 2, 3. Let the components of the angular velocity $\boldsymbol{\omega}$ at time t relative to the 1, 2, 3 system be ω_1, ω_2, ω_3. At the same instant, the components of $\mathbf{L}$ are $L_1 = I_1\omega_1$, $L_2 = I_2\omega_2$, $L_3 = I_3\omega_3$, where I_1, I_2, I_3 are the moments of inertia about the three principal axes.

In the time interval Δt, the principal axes rotate away from the 1, 2, 3 axes. To first order, the rotation angle about the 1 axis is $\Delta\theta_1 = \omega_1\,\Delta t$; similarly, $\Delta\theta_2 = \omega_2\,\Delta t$, $\Delta\theta_3 = \omega_3\,\Delta t$. The corresponding change $\Delta L_1 = L_1(t + \Delta t) - L_1(t)$ can be found to first order by treating the three rotations one by one, according to Note 7.1 on infinitesimal rotations. There are two ways L_1 can change. If ω_1 varies, $I_1\omega_1$ will change. In addition, rotations about the

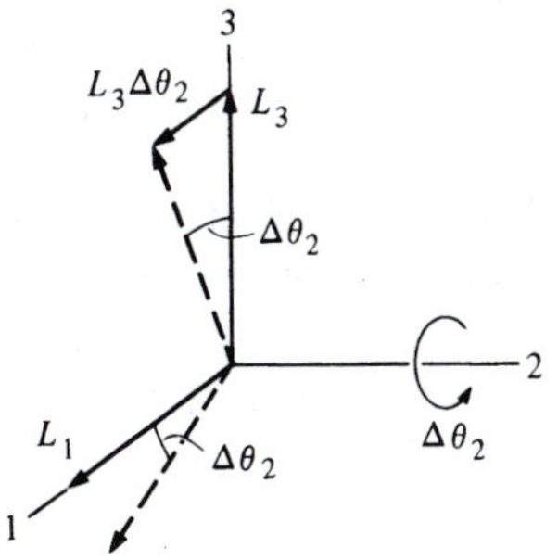

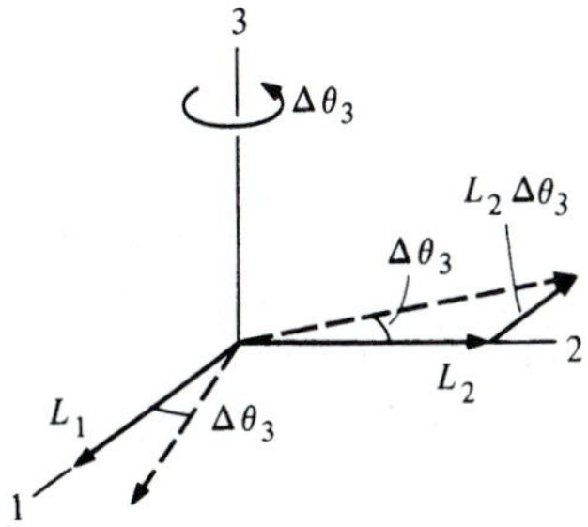

other two axes cause L_2 and L_3 to change direction, and this can contribute to angular momentum along the first axis.

The first contribution to ΔL_1 is from $\Delta(I_1\omega_1)$. Since the components of $\tilde{\mathbf{I}}$ are constant to first order for small angular displacements about the principal axes, $\Delta(I_1\omega_1) = I_1\,\Delta\omega_1$.

To find the remaining contributions to ΔL_1, consider first rotation about the 2 axis through angle $\Delta\theta_2$. This causes L_1 and L_3 to rotate as shown. The rotation of L_1 causes no change along the 1 axis to first order. However, the rotation of L_3 contributes $L_3\,\Delta\theta_2 = I_3\omega_3\,\Delta\theta_2$ along the 1 axis. Similarly, rotation about the 3 axis contributes $-L_2\,\Delta\theta_3 = -I_2\omega_2\,\Delta\theta_3$ to ΔL_1.

Adding all the contributions gives

$$\Delta L_1 = I_1\,\Delta\omega_1 + I_3\omega_3\,\Delta\theta_2 - I_2\omega_2\,\Delta\theta_3.$$

Dividing by Δt and taking the limit $\Delta t \to 0$ yields

$$\frac{dL_1}{dt} = I_1\frac{d\omega_1}{dt} + (I_3 - I_2)\omega_3\omega_2.$$

The other components can be treated in a similar fashion, or we can simply relabel the subscripts by $1 \to 2$, $2 \to 3$, $3 \to 1$. We find

$$\frac{dL_2}{dt} = I_2\frac{d\omega_2}{dt} + (I_1 - I_3)\omega_1\omega_3$$

$$\frac{dL_3}{dt} = I_3\frac{d\omega_3}{dt} + (I_2 - I_1)\omega_2\omega_1.$$

Since $\boldsymbol{\tau} = d\mathbf{L}/dt$,

$$\tau_1 = I_1\frac{d\omega_1}{dt} + (I_3 - I_2)\omega_3\omega_2$$

$$\tau_2 = I_2\frac{d\omega_2}{dt} + (I_1 - I_3)\omega_1\omega_3 \qquad 7.32$$

$$\tau_3 = I_3\frac{d\omega_3}{dt} + (I_2 - I_1)\omega_2\omega_1,$$

where τ_1, τ_2, τ_3 are the components of $\boldsymbol{\tau}$ along the axes of the inertial system 1, 2, 3. These equations were derived by Euler in the middle of the eighteenth century and are known as Euler's equations of rigid body motion.

Euler's equations are tricky to apply; thus, it is important to understand what they mean. At some time t we set up the 1,

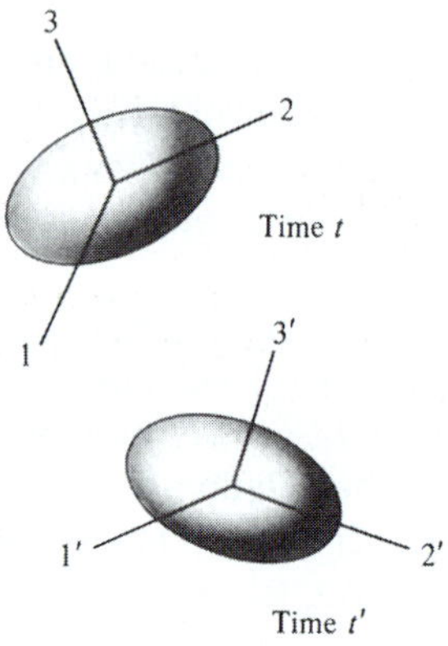

2, 3 inertial system to coincide with the instantaneous directions of the body's principal axes. τ_1, τ_2, τ_3 are the components of torque along the 1, 2, 3 axes at time t. Similarly, ω_1, ω_2, ω_3 are the components of $\boldsymbol{\omega}$ along the 1, 2, 3 axes at time t, and $d\omega_1/dt$, $d\omega_2/dt$, $d\omega_3/dt$ are the instantaneous rates of change of these components. Euler's equations relate these quantities at time t. To apply Euler's equations at another time t', we have to resolve $\boldsymbol{\tau}$ and $\boldsymbol{\omega}$ along the axes of a new inertial system 1′, 2′, 3′ which coincides with the principal axes at t'.

The difficulty is that Euler's equations do not show us how to find the orientation of these coordinate systems in space. Essentially, we have traded one problem for another; in the familiar x, y, z laboratory coordinate system, we know the disposition of the axes, but the components of the tensor of inertia vary in an unknown way. In the 1, 2, 3 system, the components of $\tilde{\mathbf{I}}$ are constant, but we do not know the orientation of the axes. Euler's equations cannot be integrated directly to give angles specifying the orientation of the body relative to the x, y, z laboratory system. Euler overcame this difficulty by expressing ω_1, ω_2, ω_3 in terms of a set of angles relating the principal axes to the axes of the x, y, z laboratory system.

In terms of these angles, Euler's equations are a set of coupled differential equations. The general equations are fairly complicated and are discussed in advanced texts. Fortunately, in many important applications we can find the motion from Euler's equations by using straightforward geometrical arguments. Here are a few examples.

Example 7.16 Stability of Rotational Motion

In principle, a pencil can be balanced on its point. In practice, the pencil falls almost immediately. Although a perfectly balanced pencil is in equilibrium, the equilibrium is not stable. If the pencil starts to tip because of some small perturbing force, the gravitational torque causes it to tip even further; the system continues to move away from equilibrium. A system is stable if displacement from equilibrium gives rise to forces which drive it back toward equilibrium. Similarly, a moving system is stable if it responds to a perturbing force by altering its motion only slightly. In contrast, an unstable system can have its motion drastically changed by a small perturbing force, possibly leading to catastrophic failure.

A rotating rigid body can exhibit either stable or unstable motion depending on the axis of rotation. The motion is stable for rotation about the axes of maximum or minimum moment of inertia but unstable for rotation about the axis with intermediate moment of inertia. The effect is easy to show: wrap a book with a rubber band and let it fall spinning about each of its principal axes in turn. I is maximum about axis

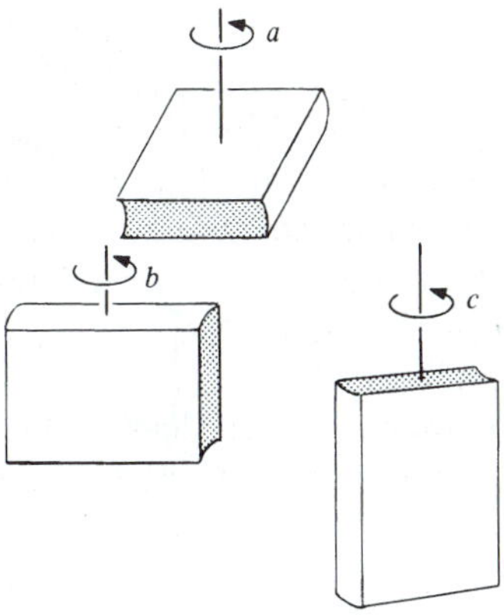

a and minimum about axis c; the motion is stable if the book is spun about either of these axes. However, if the book is spun about axis b, it tends to flop over as it spins, generally landing on its broad side.

To explain this behavior, we turn to Euler's equations. Suppose that the body is initially spinning with ω_1 = constant and $\omega_2 = 0$, $\omega_3 = 0$, and that immediately after a short perturbation, ω_2 and ω_3 are different from zero but very small compared with ω_1. Once the perturbation ends, the motion is torque-free and Euler's equations are:

$$I_1 \frac{d\omega_1}{dt} + (I_3 - I_2)\omega_2\omega_3 = 0 \qquad 1$$

$$I_2 \frac{d\omega_2}{dt} + (I_1 - I_3)\omega_1\omega_3 = 0 \qquad 2$$

$$I_3 \frac{d\omega_3}{dt} + (I_2 - I_1)\omega_1\omega_2 = 0. \qquad 3$$

Since ω_2 and ω_3 are very small at first, we can initially neglect the second term in Eq. (1). Therefore $I_1\, d\omega_1/dt = 0$, and ω_1 is constant.

If we differentiate Eq. (2) and substitute the value of $d\omega_3/dt$ from Eq. (3), we have

$$I_2 \frac{d^2\omega_2}{dt^2} - \frac{(I_1 - I_3)(I_2 - I_1)}{I_3}\omega_1{}^2\omega_2 = 0$$

or

$$\frac{d^2\omega_2}{dt^2} + A\omega_2 = 0 \qquad 4$$

where

$$A = \frac{(I_1 - I_2)(I_1 - I_3)}{I_2 I_3}\omega_1{}^2.$$

If I_1 is the largest or the smallest moment of inertia, $A > 0$ and Eq. (4) is the equation for simple harmonic motion. ω_2 oscillates at frequency $\sqrt{A}$ with bounded amplitude. It is easy to show that ω_3 also undergoes simple harmonic motion. Since ω_2 and ω_3 are bounded, the motion is stable. (It corresponds to the torque-free precession we calculated earlier.)

If I_1 is the intermediate moment of inertia, $A < 0$. In this case ω_2 and ω_3 tend to increase exponentially with time, and the motion is unstable.

Example 7.17 The Rotating Rod

Consider a uniform rod mounted on a horizontal frictionless axle through its center. The axle is carried on a turntable revolving with constant angular velocity Ω, with the center of the rod over the axis of the turntable. Let θ be the angle shown in the sketch. The problem is to find θ as a function of time.

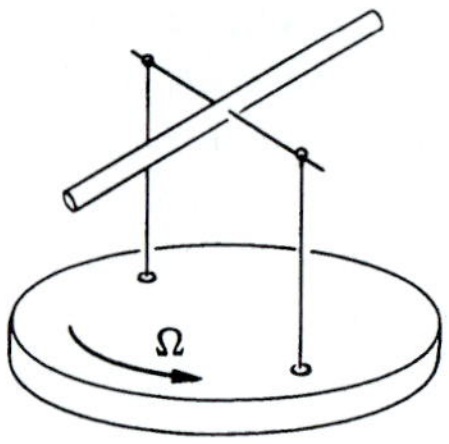

To apply Euler's equations, let principal axis 1 of the rod be along the axle, principal axis 2 be along the length of the rod, and principal axis 3 be in the vertical plane perpendicular to the rod. $\omega_1 = \dot{\theta}$, and by resolving Ω along the 2 and 3 directions we find $\omega_2 = \Omega \sin \theta$, $\omega_3 = \Omega \cos \theta$.

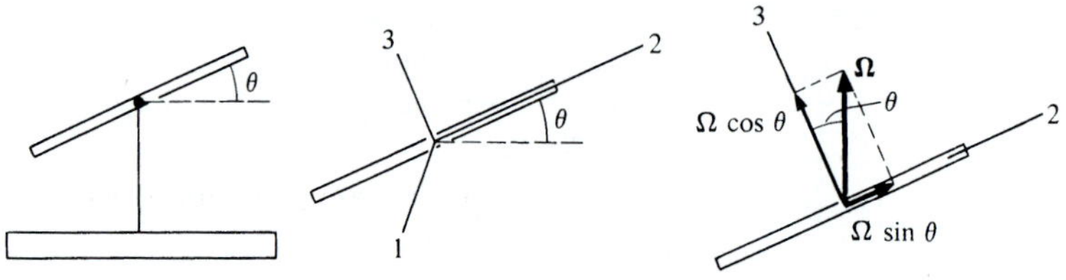

Since there is no torque about the 1 axis, the first of Euler's equations gives

$$I_1\ddot{\theta} + (I_3 - I_2)\Omega^2 \sin \theta \cos \theta = 0$$

or

$$2\ddot{\theta} + \left(\frac{I_3 - I_2}{I_1}\right)\Omega^2 \sin 2\theta = 0. \qquad 1$$

(We have used $\sin \theta \cos \theta = \frac{1}{2} \sin 2\theta$.)

Since $I_3 > I_2$, this is the equation for pendulum motion in the variable 2θ. For oscillations near the horizontal, $\sin 2\theta \approx 2\theta$ and Eq. (1) becomes

$$\ddot{\theta} + \left(\frac{I_3 - I_2}{I_1}\right)\Omega^2\theta = 0.$$

The motion is simple harmonic with angular frequency $\sqrt{(I_3 - I_2)/I_1}\,\Omega$.

Example 7.18 Euler's Equations and Torque-free Precession

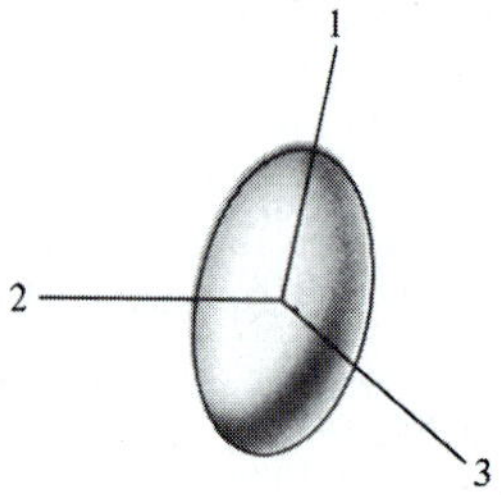

We discussed the torque-free motion of a cylindrically symmetric body earlier using the small angle approximation. In this example we shall obtain an exact solution by using Euler's equations.

Let the axis of cylindrical symmetry be principal axis 1 with moment of inertia I_1. The other two principal axes are perpendicular to the 1 axis, and $I_2 = I_3 = I_\perp$. From the first of Euler's equations

$$\tau_1 = I_1(d\omega_1/dt) + (I_3 - I_2)\omega_2\omega_3,$$

we have

$$0 = I_1\frac{d\omega_1}{dt},$$

which gives

$$\omega_1 = \text{constant} = \omega_s.$$

Principal axes 2 and 3 revolve at the constant angular velocity ω_s about the 1 axis.

The remaining Euler's equations are

$$0 = I_\perp \frac{d\omega_2}{dt} + (I_1 - I_\perp)\omega_s\omega_3 \qquad 1$$

$$0 = I_\perp \frac{d\omega_3}{dt} + (I_\perp - I_1)\omega_s\omega_2. \qquad 2$$

Differentiating the first equation and using the second to eliminate $d\omega_3/dt$ gives

$$\frac{d^2\omega_2}{dt^2} + \left(\frac{I_1 - I_\perp}{I_\perp}\right)^2 \omega_s^2\omega_2 = 0.$$

The angular velocity component ω_2 executes simple harmonic motion with angular frequency

$$\Gamma = \left|\frac{I_1 - I_\perp}{I_\perp}\right| \omega_s.$$

Thus, ω_2 is given by $\omega_2 = \omega_\perp \cos \Gamma t$ where the amplitude $\omega_\perp$ is determined by initial conditions. Then, if $I_1 > I_\perp$, Eq. (1) gives

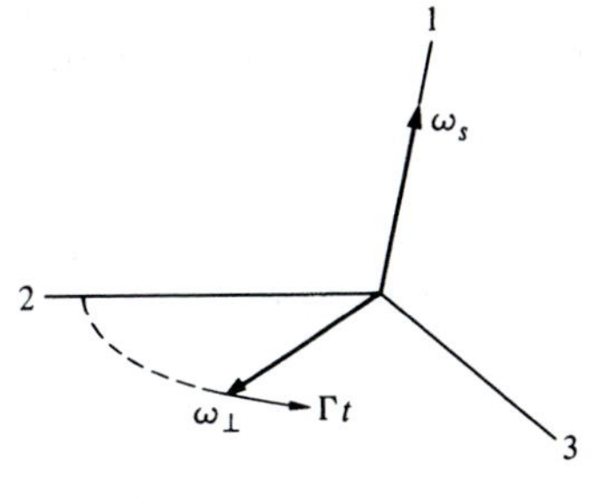

$$\omega_3 = -\frac{1}{\Gamma}\frac{d\omega_2}{dt}$$
$$= \omega_\perp \sin \Gamma t.$$

As the drawing shows, ω_2 and ω_3 are the components of a vector $\boldsymbol{\omega}_\perp$ which rotates in the 2-3 plane at rate Γ. Thus, an observer fixed to the body would see $\boldsymbol{\omega}$ rotate relative to the body about the 1 axis at angular frequency Γ. Since the 1, 2, 3 axes are fixed to the body and the body is rotating about the 1 axis at rate ω_s, the rotational speed of $\boldsymbol{\omega}$ to an observer fixed in space is

$$\Gamma + \omega_s = \frac{I_1}{I_\perp}\omega_s.$$

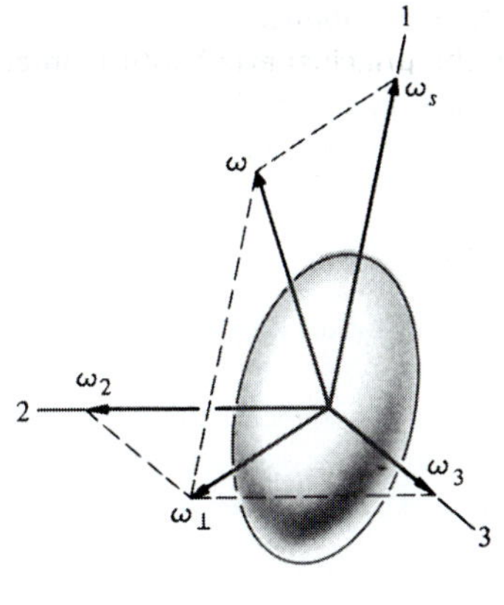

Euler's equations have told us how the angular velocity moves relative to the body, but we have yet to find the actual motion of the body in space. Here we must use our ingenuity. We know the motion of $\boldsymbol{\omega}$ relative to the body, and we also know that for torque-free motion, **L** is constant. As we shall show, this is enough to find the actual motion of the body.

The diagram at the top of the next page shows $\boldsymbol{\omega}$ and **L** at some instant of time. Since $L \cos \alpha = I_1\omega_s$, and ω_s and L are constant, α must be constant as well. Hence, the relative position of all the vectors in the diagram never changes. The only possible motion is for the

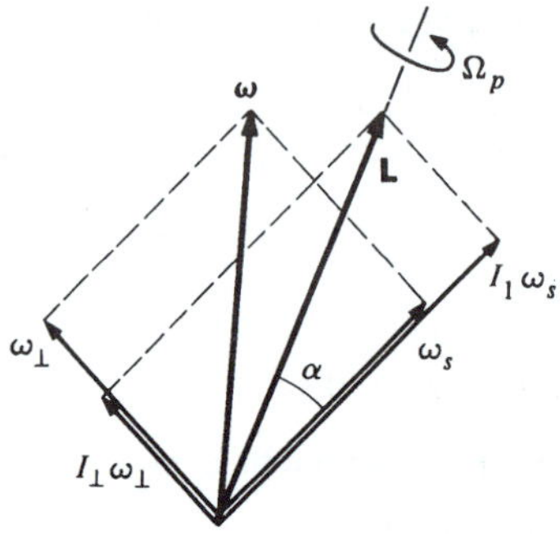

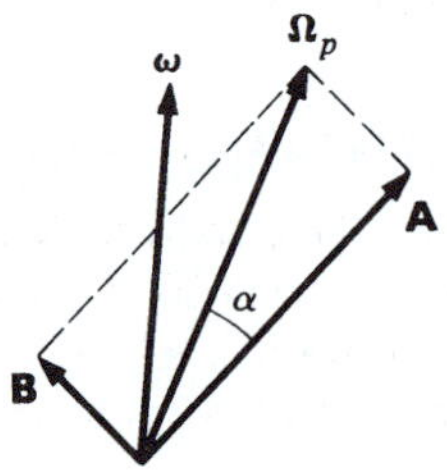

diagram to rotate about **L** with some "precessional" angular velocity Ω_p. (Bear in mind that the diagram is moving relative to the body; Ω_p is greater than ω_s.)

The remaining problem is to find Ω_p. We have shown that $\boldsymbol{\omega}$ precesses about $\boldsymbol{\omega}_s$ in space at rate $\Gamma + \omega_s$. To relate this to $\boldsymbol{\Omega}_p$, resolve $\boldsymbol{\Omega}_p$ into a vector **A** along $\boldsymbol{\omega}_s$ and a vector **B** perpendicular to $\boldsymbol{\omega}_s$. The magnitudes are $A = \Omega_p \cos\alpha$, $B = \Omega_p \sin\alpha$. The rotation **A** turns $\boldsymbol{\omega}$ about $\boldsymbol{\omega}_s$, but the rotation **B** does not. Hence the rate at which $\boldsymbol{\omega}$ precesses about $\boldsymbol{\omega}_s$ is $\Omega_p \cos\alpha$. Equating this to $\Gamma + \omega_s$,

$$\Omega_p \cos\alpha = \Gamma + \omega_s$$
$$= \frac{I_1}{I_\perp}\omega_s$$

or

$$\Omega_p = \frac{I_1\omega_s}{I_\perp \cos\alpha}.$$

The precessional angular velocity Ω_p represents the rate at which the symmetry axis rotates about the fixed direction **L**. It is the frequency of wobble we observe when we flip a spinning coin. Earlier in this section we found that the rate at which the symmetry axis rotates about a space-fixed direction is $I_1\omega_s/I$ in the small angle approximation. The result agrees with Ω_p in the limit $\alpha \to 0$.

Note 7.1 Finite and Infinitesimal Rotations

In this note we shall demonstrate that finite rotations do not commute, but that infinitesimal rotations do. By an infinitesimal rotation we mean one for which all powers of the rotation angle beyond the first can be neglected.

Consider rotation of an object through angle α about an axis $\hat{\mathbf{n}}_\alpha$ followed by a rotation through β about axis $\hat{\mathbf{n}}_\beta$. It is not possible to specify the orientation of the body by a vector because if the rotations are performed in opposite order, we do not obtain the same final orientation. To show this, we shall consider the effect of successive rotations on a vector **r**. Let $\mathbf{r}_\alpha$ be the result of rotating **r** through α about $\hat{\mathbf{n}}_\alpha$, and $\mathbf{r}_{\alpha\beta}$ be the result of rotating $\mathbf{r}_\alpha$ through β about $\hat{\mathbf{n}}_\beta$. We shall show that

$$\mathbf{r}_{\alpha\beta} \neq \mathbf{r}_{\beta\alpha}.$$

However, we shall find that for $\alpha \ll 1$, $\beta \ll 1$, $\mathbf{r}_{\alpha\beta} = \mathbf{r}_{\beta\alpha}$ to first order, and there is therefore no ambiguity in the orientation angle vector for infinitesimal rotations.

Consider the effect of successive rotation on a vector initially along the x axis, $\mathbf{r} = r\hat{\mathbf{i}}$, first through angle α about the z axis and then through angle β about the y axis. Although this is a special case, it illustrates the important features of a general proof.

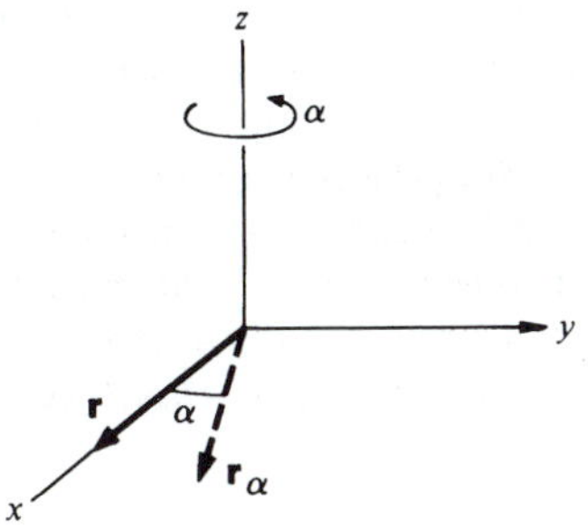

First rotation: through angle α about z axis.

$$\mathbf{r} = r\hat{\mathbf{i}}$$

$$\mathbf{r}_\alpha = r\cos\alpha\hat{\mathbf{i}} + r\sin\alpha\hat{\mathbf{j}},$$

since $|\mathbf{r}_\alpha| = |\mathbf{r}| = r$.

Second rotation: through angle β about y axis.
The component $r\sin\alpha\hat{\mathbf{j}}$ is unchanged by this rotation.

$$\mathbf{r}_{\alpha\beta} = r\cos\alpha\,(\cos\beta\hat{\mathbf{i}} - \sin\beta\hat{\mathbf{k}}) + r\sin\alpha\hat{\mathbf{j}}$$

$$= r\cos\alpha\cos\beta\hat{\mathbf{i}} + r\sin\alpha\hat{\mathbf{j}} - r\cos\alpha\sin\beta\hat{\mathbf{k}} \quad 1$$

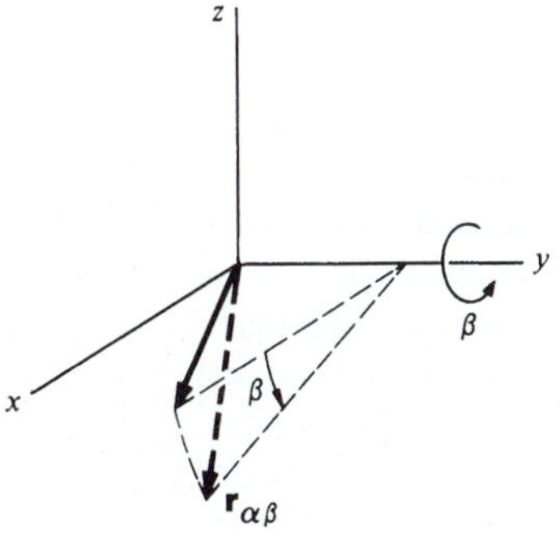

To find $\mathbf{r}_{\beta\alpha}$, we go through the same argument in reverse order. The result is

$$\mathbf{r}_{\beta\alpha} = r\cos\alpha\cos\beta\hat{\mathbf{i}} + r\cos\beta\sin\alpha\hat{\mathbf{j}} - r\sin\beta\hat{\mathbf{k}}. \quad 2$$

From Eqs. (1) and (2), $\mathbf{r}_{\alpha\beta}$ and $\mathbf{r}_{\beta\alpha}$ differ in the y and z components. Suppose that we represent the angles by $\Delta\alpha$ and $\Delta\beta$, as in the lower two drawings, and take $\Delta\alpha \ll 1$, $\Delta\beta \ll 1$. If we neglect all terms of second order and higher, so that $\sin\Delta\theta \approx \Delta\theta$, $\cos\Delta\theta \approx 1$, Eq. (1) becomes

$$\mathbf{r}_{\alpha\beta} = r\hat{\mathbf{i}} + r\,\Delta\alpha\hat{\mathbf{j}} - r\,\Delta\beta\hat{\mathbf{k}}. \quad 3$$

Equation (3) becomes

$$\mathbf{r}_{\beta\alpha} = r\hat{\mathbf{i}} + r\,\Delta\alpha\hat{\mathbf{j}} - r\,\Delta\beta\hat{\mathbf{k}}. \quad 4$$

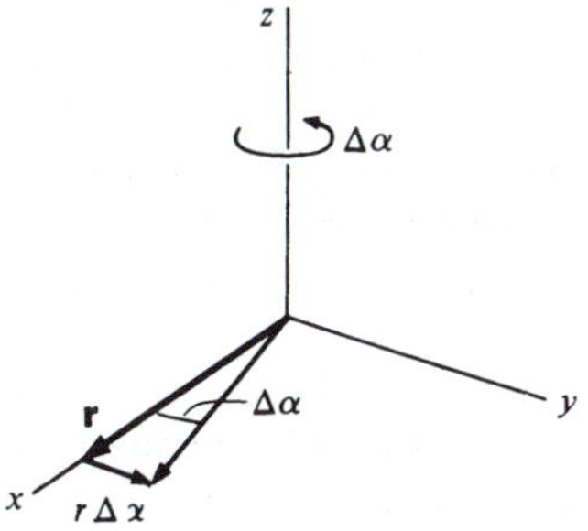

Hence $\mathbf{r}_{\alpha\beta} = \mathbf{r}_{\beta\alpha}$ to first order for small rotations, and the vector

$$\Delta\boldsymbol{\theta} = \Delta\beta\hat{\mathbf{j}} + \Delta\alpha\hat{\mathbf{k}}$$

is well defined. In particular, the displacement of $\mathbf{r}$ is

$$\Delta\mathbf{r} = \mathbf{r}_{\text{final}} - \mathbf{r}_{\text{initial}}$$

$$= \mathbf{r}_{\alpha\beta} - r\hat{\mathbf{i}}$$

$$= r\,\Delta\alpha\hat{\mathbf{j}} - r\,\Delta\beta\hat{\mathbf{k}} = \Delta\boldsymbol{\theta} \times \mathbf{r}.$$

If the displacement occurs in time Δt, the velocity is

$$\mathbf{v} = \frac{d\mathbf{r}}{dt}$$

$$= \lim_{\Delta t\to 0} \frac{\Delta\boldsymbol{\theta} \times \mathbf{r}}{\Delta t}$$

$$= \boldsymbol{\omega} \times \mathbf{r},$$

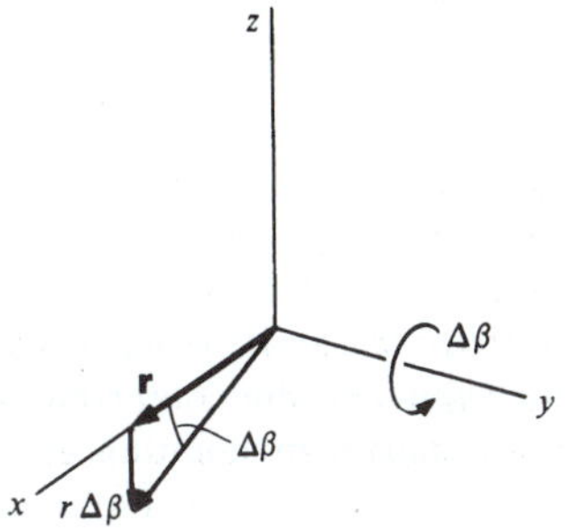

where

$$\boldsymbol{\omega} = \lim_{\Delta t\to 0} \frac{\Delta\boldsymbol{\theta}}{\Delta t}.$$

In our example, $\boldsymbol{\omega} = (d\beta/dt)\hat{\mathbf{j}} + (d\alpha/dt)\hat{\mathbf{k}}$.

Our results in Eq. (3) or (4) indicate that the effect of infinitesimal rotations can be found by considering the rotations independently one at a time. To first order, the effect of rotating $\mathbf{r} = r\hat{\mathbf{i}}$ through $\Delta\alpha$ about z is to generate a y component $r\,\Delta\alpha\hat{\mathbf{j}}$. The effect of rotating $\mathbf{r}$ through $\Delta\beta$ about y is to generate a z component, $-r\,\Delta\beta\hat{\mathbf{k}}$. The total change in $\mathbf{r}$ to first order is the sum of the two effects,

$$\Delta\mathbf{r} = r\,\Delta\alpha\hat{\mathbf{j}} - r\,\Delta\beta\hat{\mathbf{k}},$$

in agreement with Eq. (3) or (4).

Note 7.2 More about Gyroscopes

In Sec. 7.3 we used simple vector arguments to discuss the uniform precession of a gyroscope. However, uniform precession is not the most general form of gyroscope motion. For instance, a gyroscope released with its axle at rest horizontally does not instantaneously start to precess. Instead, the center of mass begins to fall. The falling motion is rapidly converted to an undulatory motion called *nutation.* If the undulations are damped out by friction in the bearings, the gyroscope eventually settles into uniform precession. The purpose of this note is to show how nutation occurs, using a small angle approximation. (The same method is used in Sec. 7.7 to explain torque-free precession.)

Consider a gyroscope consisting of a flywheel on a shaft of length l whose other end is attached to a universal pivot. The flywheel is set spinning rapidly and the axle is released from the horizontal. What is the motion?

Since it is natural to consider the motion in terms of rotation about the fixed pivot point, we introduce a coordinate system with its origin at the pivot.

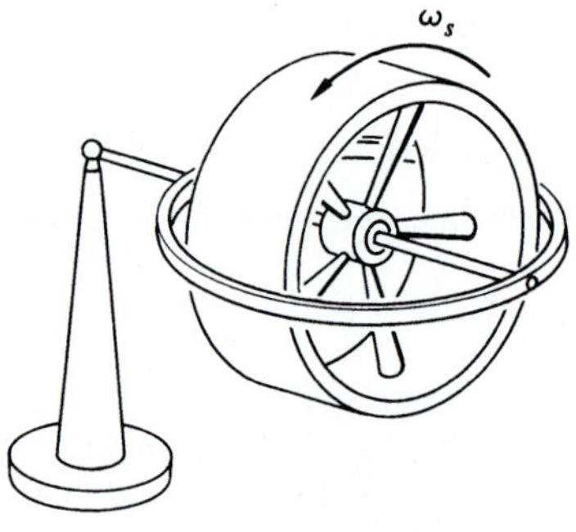

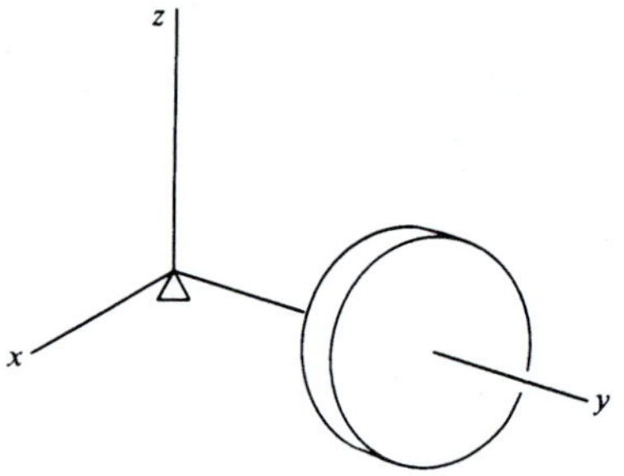

Assume for the moment that the gyroscope is not spinning but that the axle is rotating about the pivot. In order to calculate the angular momentum about the origin, we shall need a generalization of the parallel axis theorem of Example 6.9. Consider the angular momentum due to rotation of the axle about the z axis at rate ω_z. If the moment of inertia

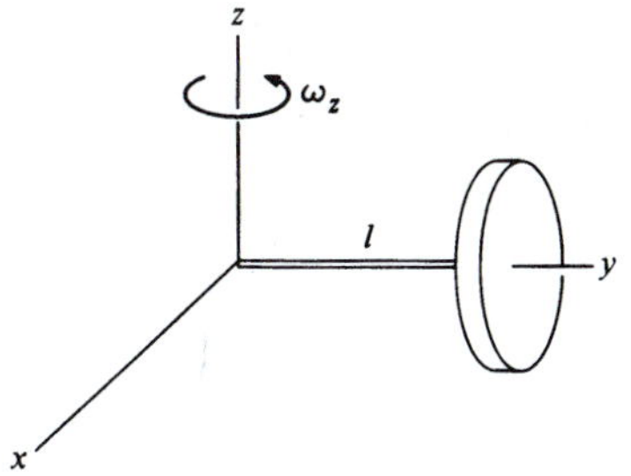

of the disk around a vertical axis through the center of mass is I_{zz}, then the moment of inertia about the z axis through the pivot is $I_{zz} + Ml^2$. The proof of this is straightforward, and we leave it as a problem. If we let $I_{zz} + Ml^2 = I_p$, then $L_z = \omega_z I_p$. By symmetry, the moment of inertia about the x axis is $I_{xx} + Ml^2 = I_p$, so that $L_x = \omega_x I_p$.

The results above are exact when the gyroscope lies along the y axis, as in the drawing, and they are true to first order in angle for small angles of tilt around the y axis.

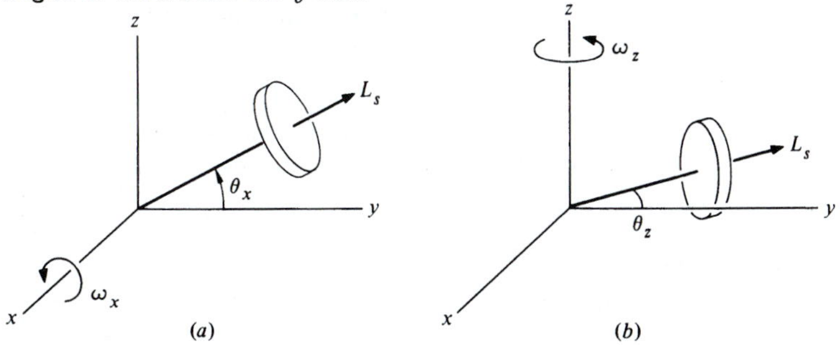

(a) (b)

Now suppose that the flywheel is set spinning at rate ω_s. If the moment of inertia along the axle is I_s, then the spin angular momentum is $L_s = I_s\omega_s$.

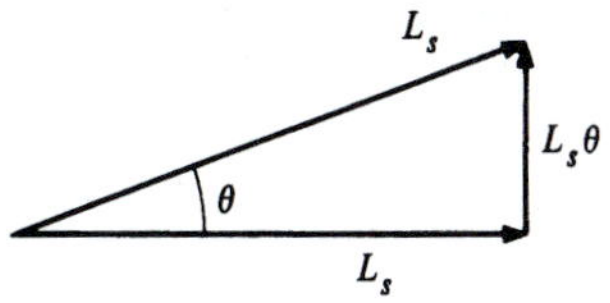

There are two kinds of contributions to the angular momentum associated with small angular displacements from the y axis. From rotation of the system as a whole with angular velocity ω, we have angular momentum contributions of the form $I_p\omega$. In addition, as the gyroscope moves away from the y axis, components of $\mathbf{L}_s$ can be generated in the x and z directions. For small angular displacements θ, such components will be of the form $L_s\theta$.

For small angular displacements, $\theta_x \ll 1$ about the x axis and $\theta_z \ll 1$ about the z axis, the rotations can be considered independently and their effects added.

a. *Rotation about the x Axis* (fig. a)
Suppose that the axle has rotated about the x axis through angle $\theta_x \ll 1$, and has instantaneous angular velocity ω_x. Then

$$
\begin{aligned}
L_x &= I_p\omega_x \\
L_y &= L_s \cos\theta_x \approx L_s \\
L_z &= L_s \sin\theta_x \approx L_s\theta_x.
\end{aligned} \tag{1}
$$

b. *Rotation about the z Axis* (fig. b)
For a rotation by $\theta_z \ll 1$ about the z axis, a similar argument gives

$$
\begin{aligned}
L_x &= -L_s \sin\theta_z \approx -L_s\theta_z \\
L_y &= L_s \cos\theta_z \approx L_s \\
L_z &= I_p\omega_z.
\end{aligned} \tag{2}
$$

Equations (1) and (2) show that the rotations θ_x and θ_z leave L_y unchanged to first order. However, the rotations give rise to first order contributions to L_x and L_z. From Eqs. (1) and (2) we find

$$\begin{aligned} L_x &= I_p\omega_x - L_s\theta_z \\ L_y &= L_s \\ L_z &= I_p\omega_z + L_s\theta_x. \end{aligned} \quad 3$$

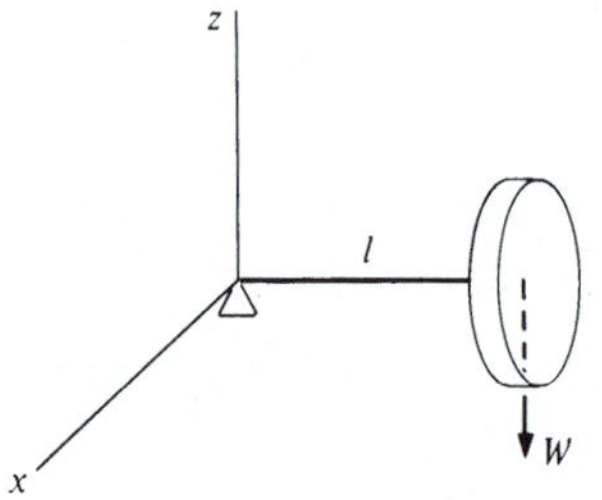

The instantaneous torque about the origin is

$$\tau_x = -lW, \quad 4$$

where l is the length of the axle and W is the weight of the gyro. Since $\boldsymbol{\tau} = d\mathbf{L}/dt$, Eqs. (3) and (4) give

$$I_p\dot{\omega}_x - L_s\omega_z = -lW \quad 5a$$

$$\dot{L}_s = 0 \quad 5b$$

$$I_p\dot{\omega}_z + L_s\omega_x = 0, \quad 5c$$

where we have used $\dot{\theta}_z = \omega_z$, $\dot{\theta}_x = \omega_x$.

Equation (5b) assures us that the spin is constant, as we expect for a flywheel with good bearings. If we differentiate Eq. (5a), we obtain

$$I_p\ddot{\omega}_x - L_s\dot{\omega}_z = 0.$$

Substituting the result $\dot{\omega}_z = -L_s\omega_x/I_p$ from Eq. (5c) gives

$$\ddot{\omega}_x + \frac{L_s^2}{I_p^2}\omega_x = 0.$$

If we let $\gamma = L_s/I_p = \omega_s I_s/I_p$, this becomes

$$\ddot{\omega}_x + \gamma^2\omega_x = 0.$$

We have the familiar equation for simple harmonic motion. The solution is

$$\omega_x = A\cos(\gamma t + \phi), \quad 6$$

where A and φ are arbitrary constants.

We can use Eq. (5a) to find ω_z:

$$\omega_z = \frac{lW}{L_s} + \frac{I_p}{L_s}\dot{\omega}_x.$$

Substituting the result $\dot{\omega}_x = -A\gamma\sin(\gamma t + \phi)$ from Eq. (6) gives

$$\begin{aligned} \omega_z &= \frac{lW}{L_s} - \frac{I_p}{L_s}A\gamma\sin(\gamma t + \phi) \\ &= \frac{lW}{L_s} - A\sin(\gamma t + \phi). \end{aligned} \quad 7$$

We can integrate Eqs. (6) and (7) to obtain

$$\theta_x = B \sin(\gamma t + \phi) + C \qquad 8a$$

$$\theta_z = \frac{lW}{L_s} t + B \cos(\gamma t + \phi) + D, \qquad 8b$$

where $B = A/\gamma$, and C, D are constants of integration.

The motion of the gyroscope depends on the constants B, ϕ, C, and D in Eq. (8), and these depend on the initial conditions. We consider three separate cases.

CASE 1. UNIFORM PRECESSION

If we take $B = 0$, and $C = D = 0$, Eq. (8) gives

$$\theta_x = 0$$

$$\theta_z = lW \frac{t}{L_s}. \qquad 9$$

This corresponds to the case of uniform precession we treated in Sec. 7.3. The rate of precession is $d\theta_z/dt = lW/L_s$, as in Eq. (7.2). If the gyroscope is moving in uniform precession at $t = 0$, it will continue to do so.

CASE 2. TORQUE-FREE PRECESSION

If we "turn off" gravity so that W is zero, then Eq. (8) gives, with $C = D = 0$,

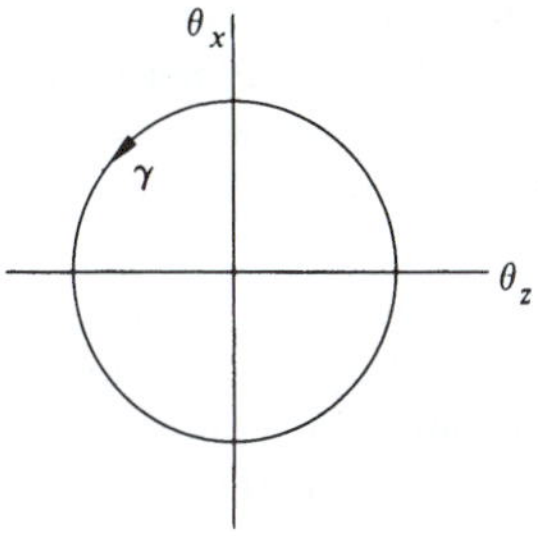

$$\theta_x = B \sin(\gamma t + \phi) \qquad 10$$

$$\theta_z = B \cos(\gamma t + \phi).$$

The tip of the axle moves in a circle about the y axis. The amplitude of the motion depends on the initial conditions. This is identical to the torque-free precession discussed in Sec. 7.7.

CASE 3. NUTATION

Suppose that the axle is released from rest along the y axis at $t = 0$. The initial conditions at $t = 0$ on the x motion are $(\theta_x)_0 = (d\theta_x/dt)_0 = 0$. From Eq. (8$a$) we obtain

$$B \sin \phi + C = 0$$

$$B\gamma \cos \phi = 0.$$

Assuming for the moment that B is not zero, we have $\phi = \pi/2$, $C = -B$. Equation (8b) then becomes

$$\theta_z = \frac{lW}{L_s} t - B \sin \gamma t + D.$$

From the initial conditions on the z motion, $(\theta_z)_0 = (d\theta_z/dt)_0 = 0$, we obtain

$$D = 0$$

$$-B\gamma + \frac{lW}{L_s} = 0$$

or

$$B = \frac{lW}{\gamma L_s}.$$

Inserting these results in Eq. (8) gives

$$\theta_x = \frac{lW}{\gamma L_s}(\cos \gamma t - 1)$$
$$\theta_z = \frac{lW}{\gamma L_s}(\gamma t - \sin \gamma t). \qquad 11$$

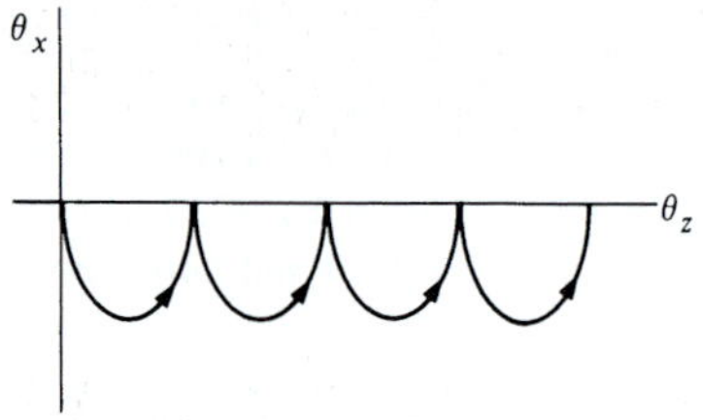

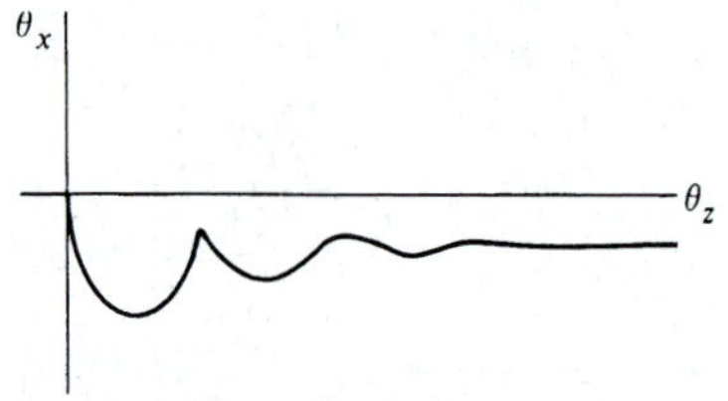

Damped nutation

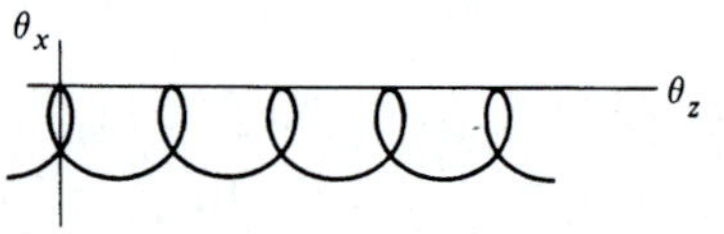

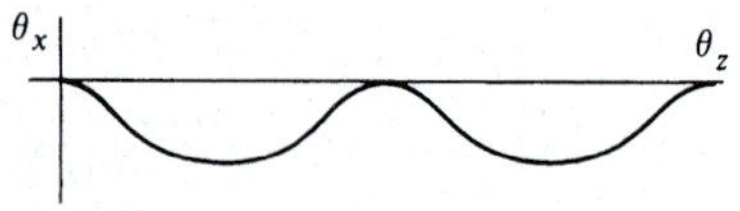

The motion described by Eq. (11) is illustrated in the sketch. As time increases, the tip of the axle traces out a cycloidal path. The dipping motion of the axle is called *nutation*. The motion is easy to see with a well-made gyroscope. Note that the initial motion of the axle is vertically down; the gyro starts to fall when it is released. Eventually the nutation dies out due to friction in the pivot, and the motion turns into uniform precession, as shown in the second sketch. The axle is left with a slight dip after the nutation is damped; this keeps the total angular momentum about the z axis zero. The rotational energy of precession comes from the fall of the center of mass. Other nutational motions are also possible, depending on the initial conditions; the lower two sketches show two possible cases. These can all be described by Eq. (8) by suitable choices of the constants.

We made the approximation that $\theta_x \ll 1$, $\theta_z \ll 1$, but because of precession, θ_z increases linearly with time, so that the approximation inevitably breaks down. This is not a problem if we examine the motion for one period of nutation. The nutational motion repeats itself whenever $\gamma t = 2\pi$. The period of the nutation is $T = 2\pi/\gamma$. If θ_z is small during one period, then we can mentally start the problem over at the end of the period with a new coordinate system having its y axis again along the direction of the axle. The restriction on θ_z is then that $\Omega T \ll 1$, or

$$\frac{2\pi\Omega}{\gamma} \ll 1.$$

Our solution breaks down if the rate of precession becomes comparable to the rate of nutation. More vividly, we require the gyroscope to nutate many times as it precesses through a full turn.

In a toy gyroscope, friction is so large that it is practically impossible to observe nutation. However, in the air suspension gyroscope, friction is so small that nutation is easy to observe. The rotor of this gyroscope

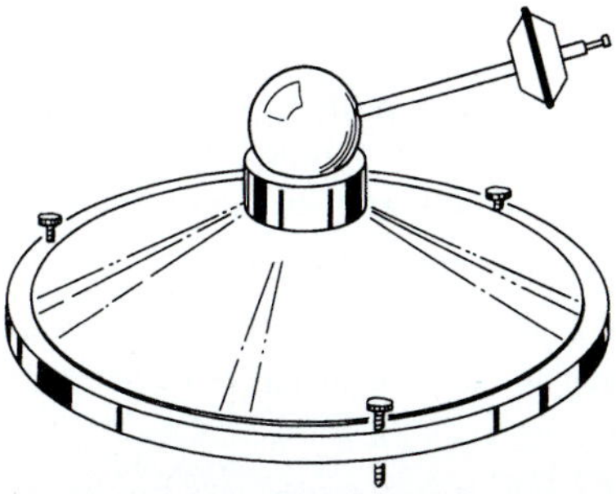

is a massive metal sphere which rests in a close fitting cup. The sphere is suspended on a film of air which flows from an orifice at the bottom of the cup. Torque is applied by the weight of a small mass on a rod protruding radially from the sphere. The pictures below are photographs of a stroboscopic light source reflected from a small bead on the end of the rod. The three modes of precession are apparent; by studying the distance between the dots you can discern the variation in speed of the rod through the precession cycle.

Problems 7.1 A thin hoop of mass M and radius R rolls without slipping about the z axis. It is supported by an axle of length R through its center, as shown. The hoop circles around the z axis with angular speed Ω.

a. What is the instantaneous angular velocity $\boldsymbol{\omega}$ of the hoop?

b. What is the angular momentum $\mathbf{L}$ of the hoop? Is $\mathbf{L}$ parallel to $\boldsymbol{\omega}$? (*Note:* the moment of inertia of a hoop for an axis along its diameter is $\frac{1}{2}MR^2$.)

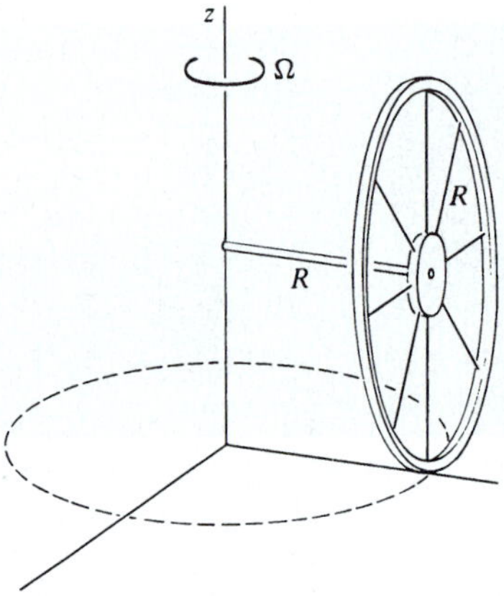

7.2 A flywheel of moment of inertia I_0 rotates with angular velocity ω_0 at the middle of an axle of length $2l$. Each end of the axle is attached to a support by a spring which is stretched to length l and provides tension T. You may assume that T remains constant for small displacements of the axle. The supports are fixed to a table which rotates at constant angular velocity, Ω, where $\Omega \ll \omega_0$. The center of mass of the flywheel is directly over the center of rotation of the table. Neglect gravity and assume that the motion is completely uniform so that nutational effects are absent. The problem is to find the direction of the axle with respect to a straight line between the supports.

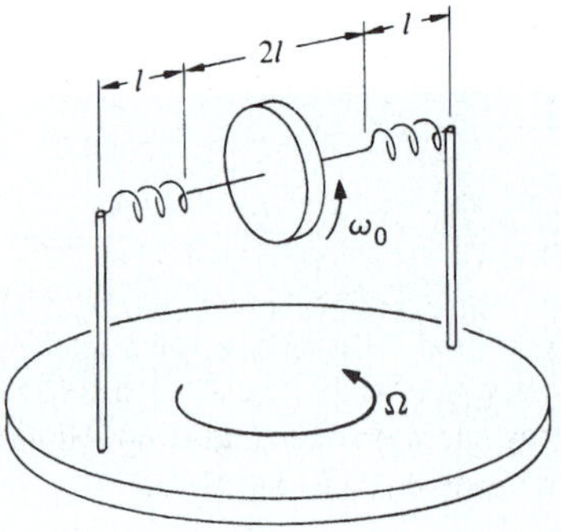

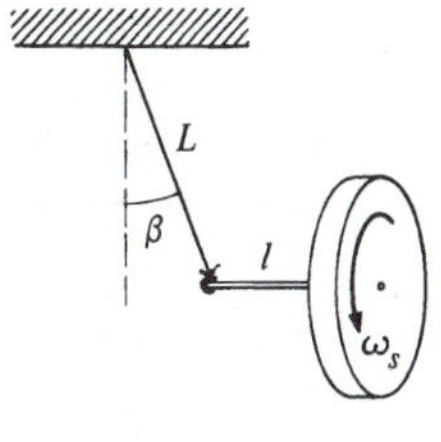

7.3 A gyroscope wheel is at one end of an axle of length l. The other end of the axle is suspended from a string of length L. The wheel is set into motion so that it executes uniform precession in the horizontal plane. The wheel has mass M and moment of inertia about its center of mass I_0. Its spin angular velocity is ω_s. Neglect the mass of the shaft and of the string.

Find the angle β that the string makes with the vertical. Assume that β is so small that approximations like $\sin\beta \approx \beta$ are justified.

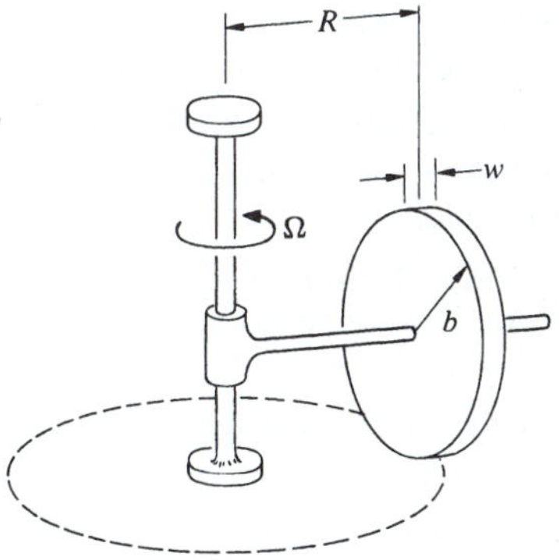

7.4 In an old-fashioned rolling mill, grain is ground by a disk-shaped millstone which rolls in a circle on a flat surface driven by a vertical shaft. Because of the stone's angular momentum, the contact force with the surface can be considerably greater than the weight of the wheel.

Assume that the millstone is a uniform disk of mass M, radius b, and width w, and that it rolls without slipping in a circle of radius R with angular velocity Ω. Find the contact force. Assume that the millstone is closely fitted to the axle so that it cannot tip, and that $w \ll R$.

Ans. clue. If $\Omega^2 b = 2g$, the force is twice the weight

7.5 When an automobile rounds a curve at high speed, the loading (weight distribution) on the wheels is markedly changed. For sufficiently high speeds the loading on the inside wheels goes to zero, at which point the car starts to roll over. This tendency can be avoided by mounting a large spinning flywheel on the car.

a. In what direction should the flywheel be mounted, and what should be the sense of rotation, to help equalize the loading? (Be sure that your method works for the car turning in either direction.)

b. Show that for a disk-shaped flywheel of mass m and radius R, the requirement for equal loading is that the angular velocity of the flywheel, ω, is related to the velocity of the car v by

$$\omega = 2v\frac{ML}{mR^2},$$

where M is the total mass of the car and flywheel, and L is the height of the center of mass of the car (including the flywheel) above the road. Assume that the road is unbanked.

7.6 If you start a coin rolling on a table with care, you can make it roll in a circle. The coin "leans" inward, with its axis tilted. The radius of the coin is b. The radius of the circle traced by the coin's center of mass is R, and the velocity of its center of mass is v. The coin rolls without slipping. Find the angle ϕ that the coin's axis makes with the horizontal. You may use the small angle approximations $\sin\phi = \phi$, $\cos\phi = 1$, and terms of order ϕ^2 are negligible.

Ans. $\phi = 3v^2/2gR$

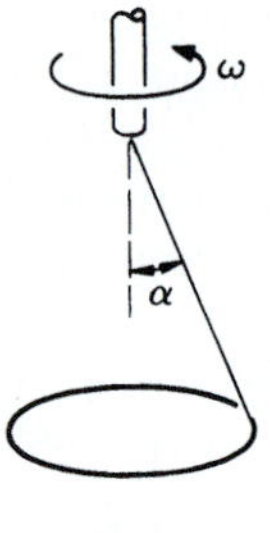

7.7 A thin hoop of mass M and radius R is suspended from a string through a point on the rim of the hoop. If the support is turned with high angular velocity ω, the hoop will spin as shown, with its plane nearly horizontal and its center nearly on the axis of the support. The string makes angle α with the vertical.

a. Find, approximately, the small angle β between the plane of the hoop and the horizontal.

b. Find, approximately, the radius of the small circle traced out by the center of mass about the vertical axis. (With skill you can demonstrate this motion with a rope. It is a favorite cowboy lariat trick.)

7.8 A child's hoop of mass M and radius b rolls in a straight line with velocity v. Its top is given a light tap with a stick at right angles to the direction of motion. The impulse of the blow is I.

a. Show that this results in a deflection of the line of rolling by angle $\phi = I/Mv$, assuming that the gyroscope approximation holds and neglecting friction with the ground.

b. Show that the gyroscope approximation is valid provided $F \ll \dfrac{2Mv^2}{b}$, where F is the peak applied force.

7.9 This problem involves investigating the effect of the angular momentum of a bicycle's wheels on the stability of the bicycle and rider. Assume that the center of mass of the bike and rider is height $2l$ above the ground. Each wheel has mass m, radius l, and moment of inertia ml^2. The bicycle moves with velocity V in a circular path of radius R. Show that it leans through an angle given by

$$\tan\phi = \frac{V^2}{Rg}\left(1 + \frac{m}{M}\right),$$

where M is the total mass.

The last term in parentheses would be absent if angular momentum were neglected. Do you think that it is important? How important is it for a bike without a rider?

7.10 Latitude can be measured with a gyro by mounting the gyro with its axle horizontal and lying along the east-west axis.

a. Show that the gyro can remain stationary when its spin axis is parallel to the polar axis and is at the latitude angle λ with the horizontal.

b. If the gyro is released with the spin axis at a small angle to the polar axis show that the gyro spin axis will oscillate about the polar axis with a frequency $\omega_{osc} = \sqrt{I_1\omega_s\Omega_e/I_\perp}$, where I_1 is the moment of inertia of the gyro about its spin axis, $I_\perp$ is its moment of inertia about the fixed horizontal axis, and $\mathbf{\Omega}_e$ is the earth's rotational angular velocity.

What value of ω_{osc} is expected for a gyro rotating at 40,000 rpm, assuming that it is a thin disk and that the mounting frame makes no contribution to the moment of inertia?

7.11 A particle of mass m is located at $x = 2$, $y = 0$, $z = 3$.

a. Find its moments and products of inertia relative to the origin.

b. The particle undergoes pure rotation about the z axis through a small angle α. Show that its moments and products of inertia are unchanged to first order in α if $\alpha \ll 1$.

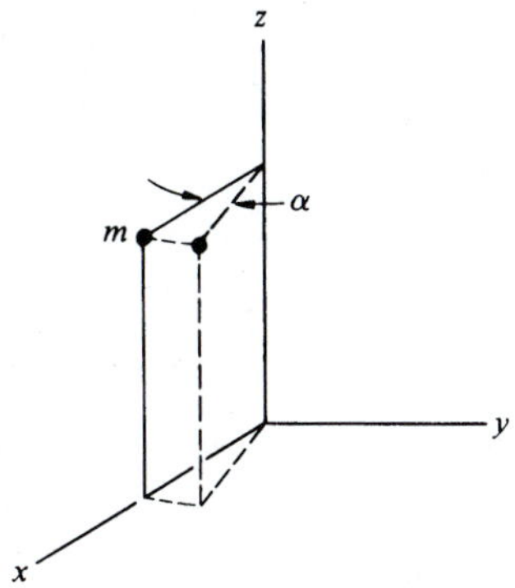

8 NONINERTIAL SYSTEMS AND FICTITIOUS FORCES

8.1 Introduction

In discussing the principles of dynamics in Chap. 2, we stressed that Newton's second law $\mathbf{F} = m\mathbf{a}$ holds true only in inertial coordinate systems. We have so far avoided noninertial systems in order not to obscure our goal of understanding the physical nature of forces and accelerations. Since that goal has largely been realized, in this chapter we turn to the use of noninertial systems. Our purpose is twofold. By introducing noninertial systems we can simplify many problems; from this point of view, the use of noninertial systems represents one more computational tool. However, consideration of noninertial systems enables us to explore some of the conceptual difficulties of classical mechanics, and the second goal of this chapter is to gain deeper insight into Newton's laws, the properties of space, and the meaning of inertia.

We start by developing a formal procedure for relating observations in different inertial systems.

8.2 The Galilean Transformations

In this section we shall show that any coordinate system moving uniformly with respect to an inertial system is also inertial. This result is so transparent that it hardly warrants formal proof. However, the argument will be helpful in the next section when we analyze noninertial systems.

Suppose that two physicists, α and β, set out to observe a series of events such as the position of a body of mass m as a function of time. Each has his own set of measuring instruments and each works in his own laboratory. α has confirmed by separate experiments that Newton's laws hold accurately in his laboratory. His reference frame is therefore inertial. How can he predict whether or not β's system is also inertial?

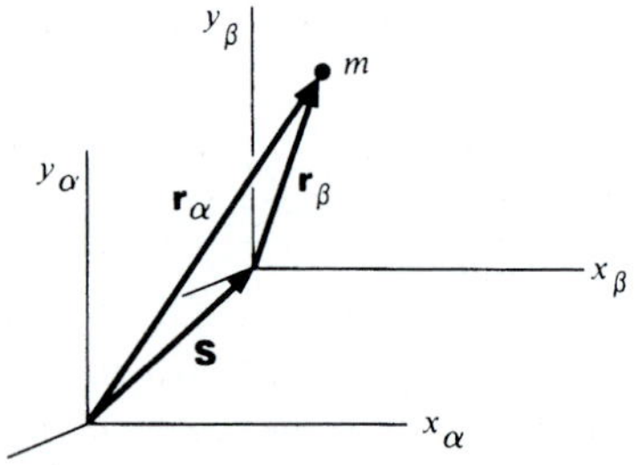

For simplicity, α and β agree to use cartesian coordinate systems with identical scale units. In general, their coordinate systems do not coincide. Leaving rotations for later, we suppose for the time being that the systems are in relative motion but that corresponding axes are parallel. Let the position of mass m be given by $\mathbf{r}_\alpha$ in α's system, and $\mathbf{r}_\beta$ in β's system. If the origins of the two systems are displaced by $\mathbf{S}$, as shown in the sketch, then

$$\mathbf{r}_\beta = \mathbf{r}_\alpha - \mathbf{S}. \qquad 8.1$$

If physicist α sees the mass accelerating at rate $\mathbf{a}_\alpha = \ddot{\mathbf{r}}_\alpha$, he concludes from Newton's second law that there is a force on m given by

$$\mathbf{F}_\alpha = m\mathbf{a}_\alpha.$$

Physicist β observes m to be accelerating at rate $\mathbf{a}_\beta$, as if it were acted on by a force

$$\mathbf{F}_\beta = m\mathbf{a}_\beta.$$

What is the relation between $\mathbf{F}_\beta$ and the true force $\mathbf{F}_\alpha$ measured in an inertial system?

It is a simple matter to relate the accelerations in the two systems. Successive differentiation with respect to time of Eq. (8.1) yields

$$\mathbf{v}_\beta = \mathbf{v}_\alpha - \mathbf{V}$$

$$\mathbf{a}_\beta = \mathbf{a}_\alpha - \mathbf{A}. \qquad 8.2$$

If $\mathbf{V} = \dot{\mathbf{S}}$ is constant, the relative motion is uniform and $\mathbf{A} = 0$. In this case $\mathbf{a}_\beta = \mathbf{a}_\beta$, and

$$\mathbf{F}_\beta = m\mathbf{a}_\beta = m\mathbf{a}_\alpha$$

$$= \mathbf{F}_\alpha.$$

The force is the same in both systems. The equations of motion in a system moving uniformly with respect to an inertial system are identical to those in the inertial system. It follows that all systems translating uniformly relative to an inertial system are inertial. This simple result leads to something of an enigma. Although it would be appealing to single out a coordinate system absolutely at rest, there is no dynamical way to distinguish one inertial system from another. Nature provides no clue to absolute rest.

We have tacitly made a number of plausible assumptions in the above argument. In the first place, we have assumed that both observers use the same scale for measuring distance. To assure this, α and β must calibrate their scales with the same standard of length. If α determines that the length of a certain rod at rest in his system is L_α, we expect that β will measure the same length. This is indeed the case if there is no motion between the two systems. However, it is not generally true. If β moves parallel to the rod with uniform velocity v, he will measure a length $L_\beta = L_\alpha(1 - v^2/c^2)^{\frac{1}{2}}$, where c is the velocity of light. This result follows from the theory of special relativity. The contraction of the moving rod, known as the Lorentz contraction, is discussed in Sec. 12.3.

A second assumption we have made is that time is the same in both systems. That is, if α determines that the time between two events is T_α, then we assumed that β will observe the same interval. Here again the assumption breaks down at high velocities. As discussed in Sec. 13.3, β finds that the interval he measures is $T_\beta = T_\alpha/(1 - v^2/c^2)^{\frac{1}{2}}$. Once again nature provides an unexpected result.

The reason these results are so unexpected is that our notions of space and time come chiefly from immediate contact with the world around us, and this never involves velocities remotely near the velocity of light. If we normally moved with speeds approaching the velocity of light, we would take these results for granted. As it is, even the highest "everyday" velocities are low compared with the velocity of light. For instance, the velocity of an artificial satellite around the earth is about 8 km/s. In this case $v^2/c^2 \approx 10^{-9}$, and length and time are altered by only one part in a billion.

A third assumption is that the observers agree on the value of the mass. However, mass is defined by experiments which involve both time and distance, and so this assumption must also be examined. As mentioned in our discussion of momentum, if an object at rest has mass m_0, the most useful quantity corresponding to mass for an observer moving with velocity v is $m = m_0/(1 - v^2/c^2)^{\frac{1}{2}}$.

Now that we are aware of some of the complexities, let us defer consideration of special relativity until Chaps. 11 to 14 and for the time being limit our discussion to situations where $v \ll c$. In this case the classical ideas of space, time, and mass are valid to high accuracy. The following equations then relate measurements made by α and β, provided that their coordinate systems move with uniform relative velocity $\mathbf{V}$. We choose the origins of the coordinate systems to coincide at $t = 0$ so that $\mathbf{S} = \mathbf{V}t$. Then from Eq. (8.1) we have

$$\mathbf{r}_\beta = \mathbf{r}_\alpha - \mathbf{V}t \qquad 8.3$$

$$t_\beta = t_\alpha.$$

The time relation is generally assumed implicitly.

This set of relations, called *transformations*, gives the prescription for transforming coordinates of an event from one coordinate system to another. Equations (8.3) transform coordinates between inertial systems and are known as the *Galilean transformations*. Since force is unchanged by the Galilean transformations, observers in different inertial systems obtain the same dynamical equations. It follows that the forms of the laws of physics are the same in all inertial systems. Otherwise, different observers would make different predictions; for instance, if one observer predicts the collision of two particles, another observer might not. The assertion that the forms of the laws of physics are the same in all inertial systems is known as the *principle of relativity*. Although the principle of relativity played only a minor role in the development of classical mechanics, its role in Einstein's theory of relativity is crucial. This is discussed further in Chap. 11, where it is also shown that the Galilean transformations are not universally valid but must be replaced by a more general transformation law, the Lorentz transformation. However, the Galilean transformations are accurate for $v \ll c$, and we shall take them to be exact in this chapter.

8.3 Uniformly Accelerating Systems

Next we turn our attention to the appearance of physical laws to an observer in a system accelerating at rate $\mathbf{A}$ with respect to an inertial system. To simplify notation we shall drop the subscripts α and β and label quantities in noninertial systems by primes. Thus, Eq. (8.2), $\mathbf{a}_\beta = \mathbf{a}_\alpha - \mathbf{A}$, becomes

$$\mathbf{a}' = \mathbf{a} - \mathbf{A},$$

where $\mathbf{A}$ is the acceleration of the primed system as measured in the inertial system.

In the accelerating system the apparent force is

$$\begin{aligned}\mathbf{F}' &= m\mathbf{a}' \\ &= m\mathbf{a} - m\mathbf{A}.\end{aligned}$$

$m\mathbf{a}$ is the true force $\mathbf{F}$ due to physical interactions. Hence,

$$\mathbf{F}' = \mathbf{F} - m\mathbf{A}.$$

We can write this as

$$\mathbf{F}' = \mathbf{F} + \mathbf{F}_{\text{fict}},$$

where

$$\mathbf{F}_{\text{fict}} \equiv -m\mathbf{A}.$$

$\mathbf{F}_{fict}$ is called a *fictitious force*.[1] The fictitious force experienced in a uniformly accelerating system is uniform and proportional to the mass, like a gravitational force. However, fictitious forces originate in the acceleration of the coordinate system, not in interaction between bodies.

Here are two examples illustrating the use of fictitious forces.

Example 8.1 **The Apparent Force of Gravity**

A small weight of mass m hangs from a string in an automobile which accelerates at rate A. What is the static angle of the string from the vertical, and what is its tension?

We shall analyze the problem both in an inertial frame and in a frame accelerating with the car.

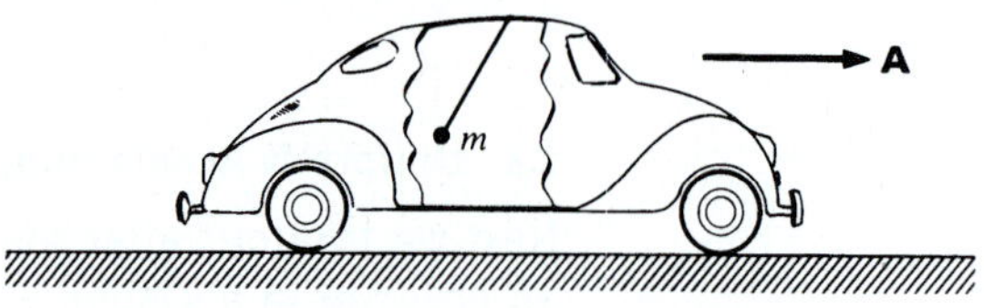

Inertial system

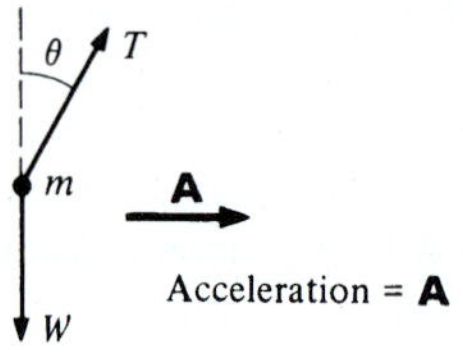

System accelerating with auto

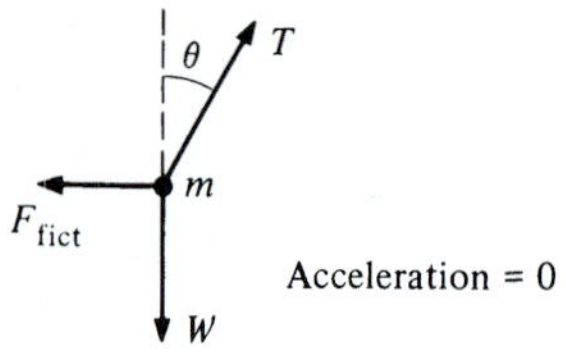

$$T\cos\theta - W = 0$$
$$T\sin\theta = MA$$
$$\tan\theta = \frac{MA}{W} = \frac{A}{g}$$
$$T = M(g^2 + A^2)^{1/2}$$

$$T\cos\theta - W = 0$$
$$T\sin\theta - F_{\text{fict}} = 0$$
$$F_{\text{fict}} = -MA$$
$$\tan\theta = \frac{A}{g}$$
$$T = M(g^2 + A^2)^{1/2}$$

From the point of view of a passenger in the accelerating car, the fictitious force acts like a horizontal gravitational force. The effective gravitational force is the vector sum of the real and fictitious forces. How would a helium-filled balloon held on a string in the accelerating car behave?

[1] Sometimes $\mathbf{F}_{fict}$ is called an *inertial force*. However, the term fictitious force more clearly emphasizes that $\mathbf{F}_{fict}$ does not arise from physical interactions.

The fictitious force in a uniformly accelerating system behaves exactly like a constant gravitational force; the fictitious force is constant and is proportional to the mass. The fictitious force on an extended body therefore acts at the center of mass.

Example 8.2 **Cylinder on an Accelerating Plank**

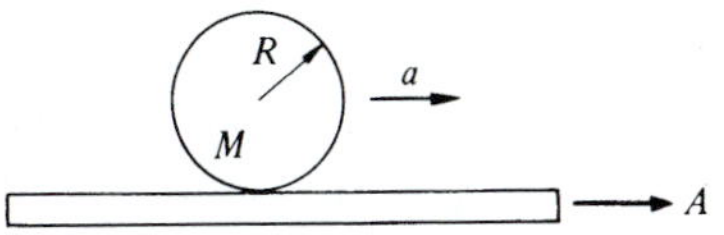

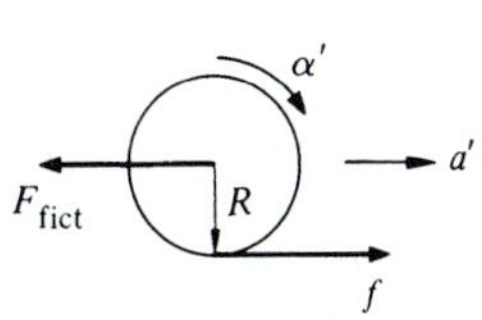

A cylinder of mass M and radius R rolls without slipping on a plank which is accelerated at the rate A. Find the acceleration of the cylinder.

The force diagram for the horizontal force on the cylinder as viewed in a system accelerating with the plank is shown in the sketch. a' is the acceleration of the cylinder as observed in a system fixed to the plank. f is the friction force, and $F_{\text{fict}} = MA$ with the direction shown.

The equations of motion in the system fixed to the accelerating plank are

$$f - F_{\text{fict}} = Ma'$$
$$Rf = -I_0\alpha'.$$

The cylinder rolls on the plank without slipping, so

$$\alpha' R = a'.$$

These yield

$$Ma' = -I_0\frac{a'}{R^2} - F_{\text{fict}}$$

$$a' = -\frac{F_{\text{fict}}}{M + I_0/R^2}.$$

Since $I_0 = MR^2/2$, and $F_{\text{fict}} = MA$, we have

$$a' = -\tfrac{2}{3}A.$$

The acceleration of the cylinder in an inertial system is

$$a = A + a'$$
$$= \tfrac{1}{3}A.$$

Example 8.1 and 8.2 can be worked with about the same ease in either an inertial or an accelerating system. Here is a problem which is rather complicated to solve in an inertial system (try it), but which is almost trivial in an accelerating system.

Example 8.3 **Pendulum in an Accelerating Car**

Consider again the car and weight on a string of Example 8.1, but now assume that the car is at rest with the weight hanging vertically. The car suddenly accelerates at rate A. The problem is to find the maximum angle ϕ through which the weight swings. ϕ is larger than the equilibrium position due to the sudden acceleration.

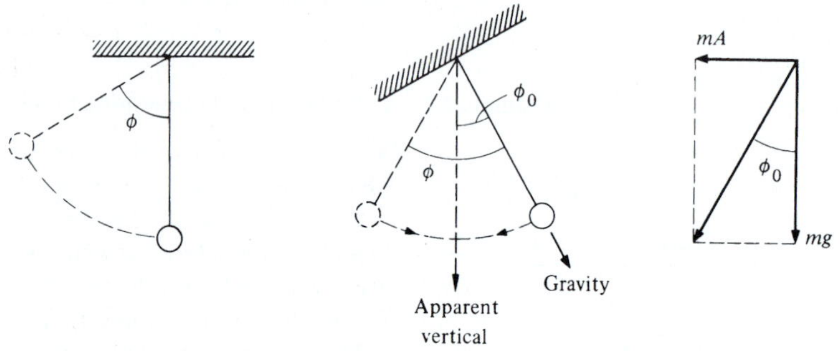

In a system accelerating with the car, the bob behaves like a pendulum in a gravitational field in which "down" is at an angle ϕ_0 from the true vertical. From Example 8.1, $\phi_0 = \arctan(A/g)$. The pendulum is initially at rest, so that it swings back and forth with amplitude ϕ_0 about the apparent vertical direction. Hence, $\phi = 2\phi_0 = 2\arctan(A/g)$.

8.4 The Principle of Equivalence

The laws of physics in a uniformly accelerating system are identical to those in an inertial system provided that we introduce a fictitious force on each particle, $\mathbf{F}_{\text{fict}} = -m\mathbf{A}$. $\mathbf{F}_{\text{fict}}$ is indistinguishable from the force due to a uniform gravitational field $\mathbf{g} = -\mathbf{A}$; both the gravitational force and the fictitious force are constant forces proportional to the mass. In a local gravitational field $\mathbf{g}$, a free particle of mass m experiences a force $\mathbf{F} = m\mathbf{g}$. Consider the same particle in a noninertial system uniformly accelerating at rate $\mathbf{A} = -\mathbf{g}$, with no gravitational field nor any other interaction. The apparent force is $\mathbf{F}_{\text{fict}} = -m\mathbf{A} = m\mathbf{g}$, as before. Is there any way to distinguish physically between these different situations?

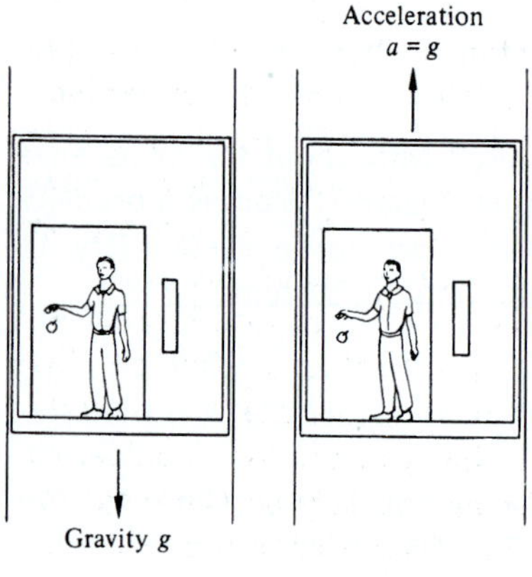

The significance of this question was first pointed out by Einstein, who illustrated the problem with the following "gedanken" experiment. (A gedanken, or thought, experiment is meant to be thought about rather than carried out.)

A man is holding an apple in an elevator at rest in a gravitational field g. He lets go of the apple, and it falls with a downward acceleration $a = g$. Now consider the same man in the same elevator, but let the elevator be in free space accelerating upward at rate $a = g$. The man again lets go of the apple, and it again appears to him to accelerate down at rate g. From his point of view the two situations are identical. He cannot distinguish between acceleration of the elevator and a gravitational field.

The point becomes even more apparent in the case of the elevator freely falling in the gravitational field. The elevator and all its contents accelerate downward at rate g. If the man releases the apple, it will float as if the elevator were motionless in free space. Einstein pointed out that the downward acceleration of the elevator exactly cancels the local gravitational field. From the point of view of an observer in the elevator, there is no way to determine whether the elevator is in free space or whether it is falling in a gravitational field.

This apparently simple idea, known as the *principle of equivalence*, underlies Einstein's general theory of relativity, and all other theories of gravitation. We summarize the principle of equivalence as follows: there is no way to distinguish locally between a uniform gravitational acceleration $\mathbf{g}$ and an acceleration of the coordinate system $\mathbf{A} = -\mathbf{g}$. By saying that there is no way to distinguish *locally*, we mean that there is no way to distinguish from within a sufficiently confined system. The reason that Einstein put his observer in an elevator was to define such an enclosed system. For instance, if you are in an elevator and observe that free objects accelerate toward the floor at rate a, there are two possible explanations:

1. There is a gravitational field down, $g = a$, and the elevator is at rest (or moving uniformly) in the field.
2. There is no gravitational field, but the elevator is accelerating up at rate a.

To distinguish between these alternatives, you must look out of the elevator. Suppose, for instance, that you see an apple suddenly drop from a nearby tree and fall down with acceleration a. The most likely explanation is that you and the tree are at rest in a downward gravitational field of magnitude $g = a$. However, it is conceivable that your elevator and the tree are both at rest on a giant elevator which is accelerating up at rate a.

To choose between these alternatives you must look farther off. If you see that you have an upward acceleration a relative to the fixed stars, that is, if the stars appear to accelerate down at rate a, the only possible explanation is that you are in a noninertial system; your elevator and the tree are actually accelerating up. The alternative is the impossible conclusion that you are at rest in a gravitational field which extends uniformly *through all of space*. But such fields do not exist; real forces arise from interactions between real bodies, and for sufficiently large separations the forces always decrease. Hence it is most unphysical to invoke a uniform gravitational field extending throughout space.

This, then, is the difference between a gravitational field and an accelerating coordinate system. Real fields are local; at large distances they decrease. An accelerating coordinate system is nonlocal; the acceleration extends uniformly throughout space. Only for small systems are the two indistinguishable.

Although these ideas may sound somewhat abstract, the next two examples show that they have direct physical consequences.

Example 8.4 The Driving Force of the Tides

The earth is in free fall toward the sun, and according to the principle of equivalence it should be impossible to observe the sun's gravitational force in an earthbound system. However, the equivalence principle applies only to local systems. The earth is so large that appreciable nonlocal effects like the tides can be observed. In this example we shall discuss the origin of the tides to see what is meant by a nonlocal effect.

The tides arise because of variations in the apparent gravitational field of the sun and the moon at different points on the earth's surface. Although the moon's effect is larger than the sun's, we shall consider only the sun for purposes of illustration.

The gravitational field of the sun at the center of the earth is

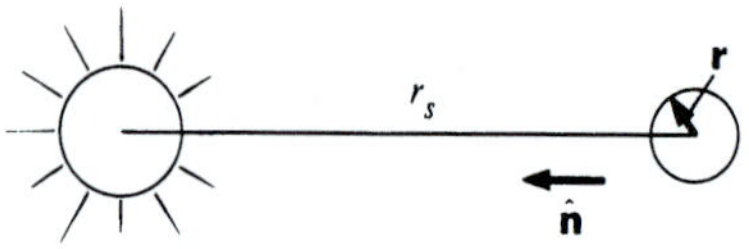

$$\mathbf{G}_0 = GM_s \frac{\hat{\mathbf{n}}}{r_s^2},$$

where M_s is the sun's mass, r_s is the distance between the center of the sun and the center of the earth, and $\hat{\mathbf{n}}$ is the unit vector from the earth toward the sun. The earth accelerates toward the sun at rate $\mathbf{A} = \mathbf{G}_0$.

If $\mathbf{G}(\mathbf{r})$ is the gravitational field of the sun at some point $\mathbf{r}$ on the earth, where the origin of $\mathbf{r}$ is the center of the earth, then the force on mass m at $\mathbf{r}$ is

$$\mathbf{F} = m\mathbf{G}(\mathbf{r}).$$

The apparent force to an earthbound observer is

$$\mathbf{F}' = \mathbf{F} - m\mathbf{A} = m[\mathbf{G}(\mathbf{r}) - \mathbf{G}_0].$$

The apparent field is

$$\mathbf{G}'(\mathbf{r}) = \frac{\mathbf{F}'}{m}$$
$$= \mathbf{G}(\mathbf{r}) - \mathbf{G}_0.$$

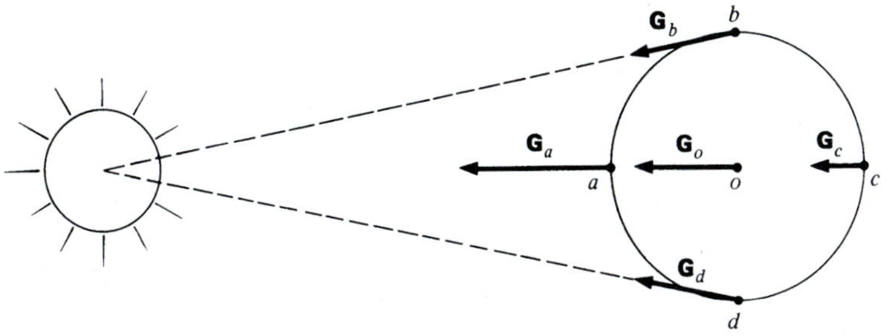

The drawing above shows the true field **G(r)** at different points on the earth's surface. (The variations are exaggerated.) G_a is larger than G_0 since a is closer to the sun than the center of the earth. Similarly, G_c is less than G_0. The magnitudes of $\mathbf{G}_b$ and $\mathbf{G}_c$ are approximately the same as the magnitude of $\mathbf{G}_0$, but their directions are slightly different.

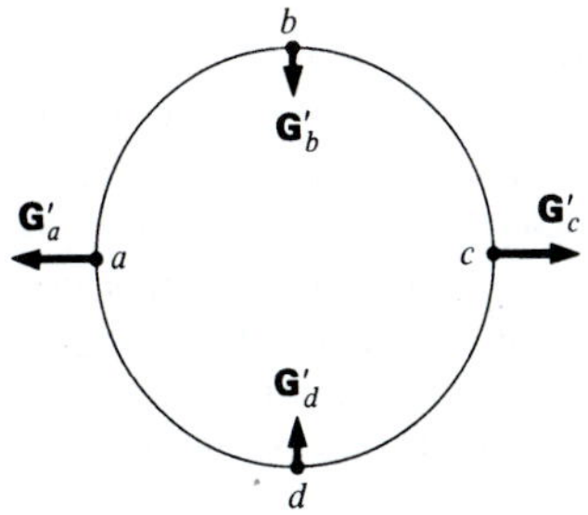

The apparent field $\mathbf{G}' = \mathbf{G} - \mathbf{G}_0$ is shown in the drawing at left. We now evaluate $\mathbf{G}'$ at each of the points indicated.

1. $\mathbf{G}'_a$ AND $\mathbf{G}'_c$

The distance from a to the center of the sun is $r_s - R_e$ where R_e is the earth's radius. The magnitude of the sun's field at a is

$$G_a = \frac{GM_s}{(r_s - R_e)^2}.$$

$\mathbf{G}_a$ is parallel to $\mathbf{G}_0$. The magnitude of the apparent field at a is

$$\begin{aligned}
G'_a &= G_a - G_0 \\
&= \frac{GM_s}{(r_s - R_e)^2} - \frac{GM_s}{r_s^2} \\
&= \frac{GM_s}{r_s^2}\left[\frac{1}{[1 - (R_e/r_s)]^2} - 1\right].
\end{aligned}$$

Since $R_e/r_s = 6.4 \times 10^3 \text{ km}/1.5 \times 10^8 \text{ km} = 4.3 \times 10^{-5} \ll 1$, we have

$$\begin{aligned}
G'_a &= G_0\left[\left(1 - \frac{R_e}{r_s}\right)^{-2} - 1\right] \\
&= G_0\left[1 + 2\frac{R_e}{r_s} + \cdots - 1\right] \\
&= 2G_0\frac{R_e}{r_s},
\end{aligned}$$

where we have neglected terms of order $(R_e/r_s)^2$ and higher.

The analysis at c is similar, except that the distance to the sun is $r_s + R_e$ instead of $r_s - R_e$. We obtain

$$G_c' = -2G_0 \frac{R_e}{r_s}.$$

Note that $\mathbf{G}_a'$ and $\mathbf{G}_c'$ point radially out from the earth.

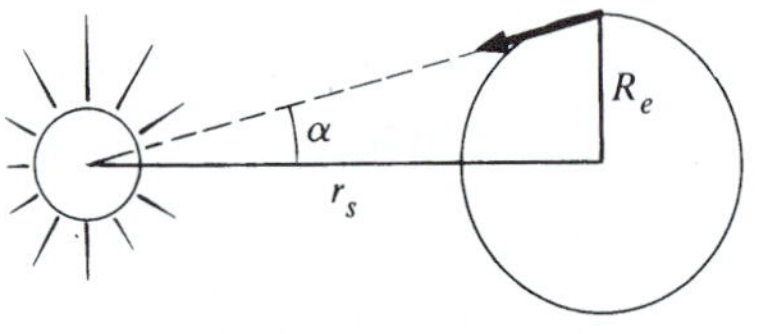

2. $\mathbf{G}_b'$ AND $\mathbf{G}_d'$

Points b and d are, to excellent approximation, the same distance from the sun as the center of the earth. However, $\mathbf{G}_b$ is not parallel to $\mathbf{G}_0$; the angle between them is $\alpha \approx R_e/r_s = 4.3 \times 10^{-5}$. To this approximation

$$G_b' = G_0\alpha$$
$$= G_0 \frac{R_e}{r_s}.$$

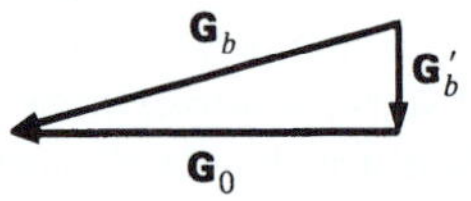

By symmetry, $\mathbf{G}_d'$ is equal and opposite to $\mathbf{G}_b'$. Both $\mathbf{G}_b'$ and $\mathbf{G}_d'$ point toward the center of the earth.

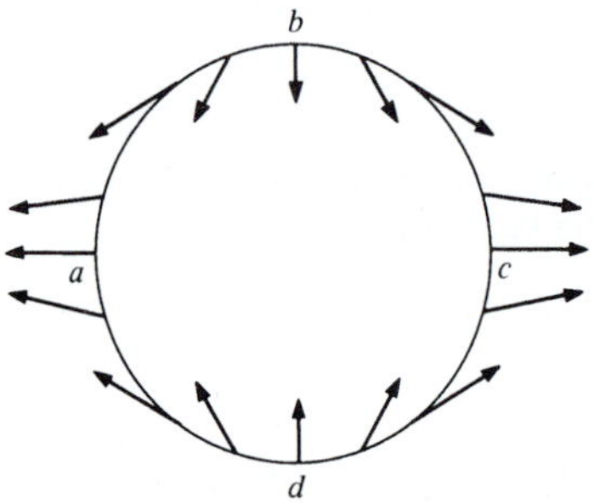

The sketch shows $\mathbf{G}'(\mathbf{r})$ at various points on the earth's surface. This diagram is the starting point for analyzing the tides. The forces at a and c tend to lift the oceans, and the forces at b and d tend to depress them. If the earth were uniformly covered with water, the tangential force components would cause the two tidal bulges to sweep around the globe with the sun. This picture explains the twice daily ebb and flood of the tides, but the actual motions depend in a complicated way on the response of the oceans as the earth rotates, and on features of local topography.

We can estimate the magnitude of tidal effects quite easily, as the next example shows.

Example 8.5 **Equilibrium Height of the Tide**

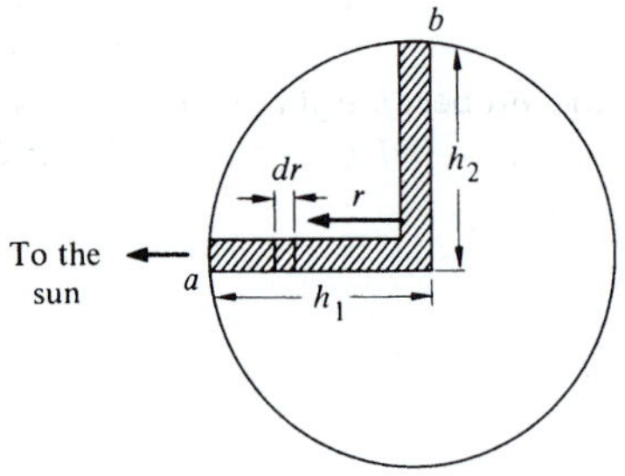

The following argument is based on a model devised by Newton. Pretend that two wells full of water run from the surface of the earth to the center, where they join. One is along the earth-sun axis and the other is perpendicular. For equilibrium, the pressures at the bottom of the wells must be identical.

The pressure due to a short column of water of height dr is $\rho g(r)dr$, where ρ is the density and $g(r)$ is the effective gravitational field at r. The condition for equilibrium is

$$\int_0^{h_1} \rho g_1(r)\,dr = \int_0^{h_2} \rho g_2(r)\,dr.$$

h_1 and h_2 are the distances from the center of the earth to the surface of the respective water columns. If we assume that the water is incompressible, so that ρ is constant, then the equilibrium condition becomes

$$\int_0^{h_1} g_1(r)\,dr = \int_0^{h_2} g_2(r)\,dr.$$

The problem is to calculate the difference $h_1 - h_2 = \Delta h_s$, the height of the tide due to the sun. We shall assume that the earth is spherical and neglect effects due to its rotation.

The effective field toward the center of the earth along column 1 is $g_1(r) = g(r) - G_1'(r)$, where $g(r)$ is the gravitational field of the earth and $G_1'(r)$ is the effective field of the sun along column 1. (The negative sign indicates that $G_1'(r)$ is directed radially out.) In the last example we evaluated $G_1'(R_e) = G_a' = 2GM_sR_e/r_s^3$. The effective field along column 1 is obtained by substituting r for R_e. Hence,

$$G_1'(r) = \frac{2GM_sr}{r_s^3}$$
$$= 2Cr,$$

where $C = GM_s/r_s^3$.

Putting these together, we obtain

$$g_1(r) = G(r) - 2Cr.$$

By the same reasoning we obtain

$$g_2(r) = g(r) + G_2'(r)$$
$$= g(r) + Cr.$$

The condition for equilibrium is

$$\int_0^{h_1} [g(r) - 2Cr]\,dr = \int_0^{h_2} [g(r) + Cr]\,dr,$$

or, rearranging,

$$\int_0^{h_1} g(r)\,dr - \int_0^{h_2} g(r)\,dr = \int_0^{h_1} 2Cr\,dr + \int_0^{h_2} Cr\,dr.$$

We can combine the integrals on the left hand side to give $\int_{h_1}^{h_2} g(r)\,dr$. Since h_1 and h_2 are close to the earth's radius, $g(r)$ can be taken as constant in the integral. $g(r) = g(R_e) = g$, the acceleration due to gravity at the earth's surface. The integrals on the left become $g(h_1 - h_2) = g\,\Delta h_s$. The integrals on the right can be combined by taking $h_1 \approx h_2 \approx R_e$, and they yield $\int_0^{R_e} 3Cr\,dr = \frac{3}{2}CR_e^2$. The final result is

$$g\,\Delta h_s = \tfrac{3}{2}CR_e^2.$$

By using $g = GM_e/R_e^2$, $C = GM_s/r_s^3$, we find

$$\Delta h_s = \frac{3}{2}\frac{M_s}{M_e}\left(\frac{R_e}{r_s}\right)^3 R_e.$$

From the numerical values

$M_s = 1.99 \times 10^{33}$ g $\quad r_s = 1.49 \times 10^{13}$ cm

$M_e = 5.98 \times 10^{27}$ g $\quad R_e = 6.37 \times 10^8$ cm, we obtain

$\Delta h_s = 24.0$ cm.

The identical argument for the moon gives

$$\Delta h_m = \frac{3}{2}\frac{M_m}{M_e}\left(\frac{R_e}{r_m}\right)^3 R_e.$$

Inserting $M_m = 7.34 \times 10^{25}$ g, $r_m = 3.84 \times 10^{10}$ cm, we obtain $\Delta h_m =$ 53.5 cm. We see that the moon's effect is about twice as large as the sun's, even though the sun's gravitational field at the earth is about 200 times stronger than the moon's. The reason is that the tidal force depends on the gradient of the gravitational field. The moon is so close that its field varies considerably across the earth, whereas the field of the distant sun is more nearly constant.

The strongest tides, called the spring tides, occur at the new and full moon when the moon and sun act together. Midway between, at the quarters of the moon, occur the weak neap tides. The ratio of the driving forces in these two cases is

$$\frac{\Delta h_{\text{spring}}}{\Delta h_{\text{neap}}} = \frac{\Delta h_m + \Delta h_s}{\Delta h_m - \Delta h_s} \approx 3.$$

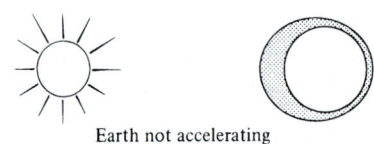

Earth not accelerating

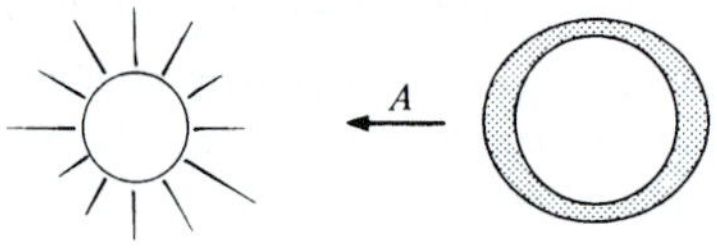

Earth in free fall

The tides offer convincing evidence that the earth is in free fall toward the sun. If the earth were attracted by the sun but not in free fall, there would be only a single tide, whereas free fall results in two tides a day, as the sketches illustrate. The fact that we can sense the sun's gravitational field from a body in free fall does not contradict the principle of equivalence. The height of the tide depends on the ratio of the earth's radius to the sun's distance, R_e/r_s. However, for a system to be local with respect to a gravitational field, the variation of the field must be negligible over the dimensions of the system. The earth would be a local system if R_e were negligible compared with r_s, but then there would be no tides. Hence, the tides demonstrate that the earth is too large to constitute a local system in the sun's field.

There have been a number of experimental investigations of the principle of equivalence, since in spite of its apparent simplicity, far-reaching conclusions follow from it. For example, the principle of equivalence demands that gravitational force be strictly proportional to inertial mass. An alternative statement is that the ratio of gravitational mass to inertial mass must be the same for all matter, where the gravitational mass is the mass which enters the gravitational force equation and the inertial mass is the mass which appears in Newton's second law. Hence, if an object with gravitational mass M_{gr} and inertial mass M_{in} interacts with an object of gravitational mass M_0, we have

$$\mathbf{F} = -\frac{GM_0M_{gr}\hat{\mathbf{r}}}{r^2}.$$

Since the acceleration is $\mathbf{F}/M_{in}$,

$$\mathbf{a} = -\frac{GM_0}{r^2}\left(\frac{M_{gr}}{M_{in}}\right)\hat{\mathbf{r}}. \qquad 8.4$$

The equivalence principle requires M_{gr}/M_{in} to be the same for all objects, since otherwise it would be possible to distinguish locally between a gravitational field and an acceleration. For instance, suppose that for object A, M_{gr}/M_{in} is twice as large as for object B. If we release both objects in an Einstein elevator and they fall with the same acceleration, the only possible conclusion is that the elevator is actually accelerating up. On the other hand, if A falls with twice the acceleration of B, we know that the acceleration must be due to a gravitational field. The upward acceleration of the elevator would be distinguishable from a downward gravitational field, in defiance of the principle of equivalence.

The ratio M_{gr}/M_{in} is taken to be 1 in Newton's law of gravitation. Any other choice for the ratio would be reflected in a different value for G, since experimentally the only requirement is that $G(M_{gr}/M_{in}) = 6.67 \times 10^{-11}$ N·m^2/kg^2.

Newton investigated the equivalence of inertial and gravitational mass by studying the period of a pendulum with interchangeable bobs. The equation of motion for the bob in the small angle approximation is

$$M_{in}l\ddot{\theta} + M_{gr}g\theta = 0.$$

The period of the pendulum is

$$T = \frac{2\pi}{\omega}$$

$$= 2\pi\sqrt{\frac{l}{g}}\sqrt{\frac{M_{in}}{M_{gr}}}.$$

Newton's experiment consisted of looking for a variation in T using bobs of different composition. He found no such change and, from an estimate of the sensitivity of the method, concluded that M_{gr}/M_{in} is constant to better than one part in a thousand for common materials.

The most compelling evidence for the principle of equivalence comes from an experiment devised by the Hungarian physicist Baron Roland von Eötvös at the turn of the century. (The experiments were completed in 1908 but the results were not published until 1922, three years after von Eötvös' death.) The method and technique of von Eötvös' experiment were refined by R. H. Dicke and his collaborators at Princeton University, and it is this work, completed in 1963, which we shall now outline.[1]

Consider a torsion balance consisting of two masses A and B of different composition at each end of a bar which hangs from a thin fiber so that it can rotate only about the vertical axis. The masses are attracted by the earth and also by the sun. The gravitational force due to the earth is vertical and causes no rotation of the balance, but as we now show, the sun's attraction will cause a rotation if the principle of equivalence is violated.

Assume that the sun is on the horizon, as shown in the sketch, and that the horizontal bar is perpendicular to the sun-earth axis. According to Eq. (8.4) the accelerations of the masses due to the sun are

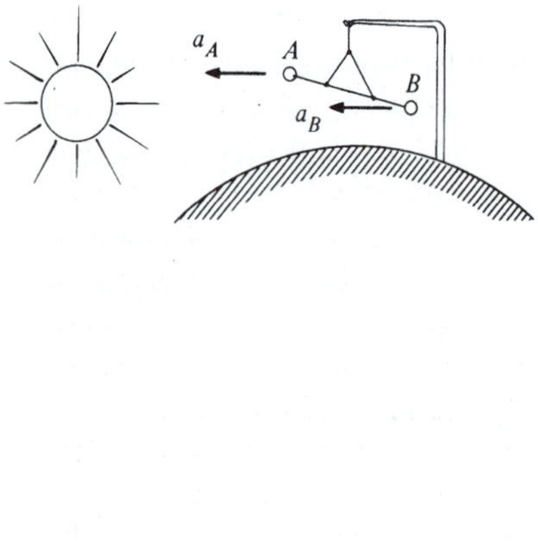

$$a_A = \frac{GM_s}{r_s^2}\left[\frac{M_{gr}(A)}{M_{in}(A)}\right]$$

$$a_B = \frac{GM_s}{r_s^2}\left[\frac{M_{gr}(B)}{M_{in}(B)}\right],$$

where M_s is the gravitational mass of the sun, and r_s is the distance between sun and earth. The acceleration of the masses in a coordinate system fixed to the earth are

$$a'_A = a_A - a_0$$

$$a'_B = a_B - a_0,$$

where a_0 is the acceleration of the earth toward the sun. (Acceleration due to the rotation of the earth plays no role and we neglect it.)

If the principle of equivalence is obeyed, $a'_A = a'_B$ and the bar has no tendency to rotate about the fiber. However, if the two masses A and B have different ratios of gravitational to inertial mass, then one will accelerate more than the other. The balance

[1] An account of the experiment is given in an article by R. H. Dicke in *Scientific American*, vol. 205, no. 84, December, 1961.

will rotate until the restoring torque of the suspension fiber brings it to rest. As the earth rotates, the apparent direction of the sun changes; the equilibrium position of the balance moves with a 24-h period.

Dicke's apparatus was capable of detecting the deflection caused by a variation of 1 part in 10^{11} in the ratio of gravitational to inertial mass, but no effect was found to this accuracy.

The principle of equivalence is generally regarded as a fundamental law of physics. We have used it to discuss the ratio of gravitational to inertial mass. Surprisingly enough, it can also be used to show that clocks run at different rates in different gravitational fields. A simple argument showing how the principle of equivalence forces us to give up the classical notion of time is presented in Note 8.1.

8.5 Physics in a Rotating Coordinate System

The transformation from an inertial coordinate system to a rotating system is fundamentally different from the transformation to a translating system. A coordinate system translating uniformly relative to an inertial system is also inertial; the transformation leaves the laws of motion unaffected. In contrast, a uniformly rotating system is intrinsically noninertial. Rotational motion is accelerating motion, and the laws of physics always involve fictitious forces when referred to a rotating reference frame. The fictitious forces do not have the simple form of a uniform gravitational field, as in the case of a uniformly accelerating system, but involve several terms, including one which is velocity dependent. However, in spite of these complications, rotating coordinate systems can be very helpful. In certain cases the fictitious forces actually simplify the form of the equations of motion. In other cases it is more natural to introduce the fictitious forces than to describe the motion with inertial coordinates. A good example is the physics of airflow over the surface of the earth. It is easier to explain the rotational motion of weather systems in terms of fictitious forces than to use inertial coordinates which must then be related to coordinates on the rotating earth.

If a particle of mass m is accelerating at rate $\mathbf{a}$ with respect to inertial coordinates and at rate $\mathbf{a}_{\text{rot}}$ with respect to a rotating coordinate system, then the equation of motion in the inertial system is

$$\mathbf{F} = m\mathbf{a}.$$

We would like to write the equation of motion in the rotating system as

$$\mathbf{F}_{\text{rot}} = m\mathbf{a}_{\text{rot}}.$$

If the accelerations of m in the two systems are related by

$$\mathbf{a} = \mathbf{a}_{\text{rot}} + \mathbf{A},$$

where $\mathbf{A}$ is the relative acceleration, then

$$\mathbf{F}_{\text{rot}} = m(\mathbf{a} - \mathbf{A})$$
$$= \mathbf{F} + \mathbf{F}_{\text{fict}},$$

where $\mathbf{F}_{\text{fict}} = -m\mathbf{A}$. So far the argument is identical to that in Sec. 8.3. Our task now is to find $\mathbf{A}$ for a rotating system.

One way of evaluating $\mathbf{A}$ is to find the transformation connecting the inertial and rotating coordinates and then to differentiate. However, there is a much simpler and more general method, which consists of finding a transformation rule relating the time derivatives of any vector in inertial and rotating coordinates. In order to motivate the derivation, we proceed by first finding the relation between the velocity of a particle measured in an inertial system, $\mathbf{v}_{\text{in}}$, and the velocity measured in a rotating system, $\mathbf{v}_{\text{rot}}$.

Time Derivatives and Rotating Coordinates

We are interested in pure rotation without translation, and so we consider a rotating system x', y', z' whose origin coincides with the origin of an inertial system x, y, z. Suppose, for the sake of the argument, that the x', y', z' system is rotating so that the z and z' axes always coincide. Thus, the angular velocity of the rotating system, $\mathbf{\Omega}$, lies along the z axis. Furthermore, let the x and x' axes coincide instantaneously at time t. Imagine now that a particle has position vector $\mathbf{r}(t)$ in the xz plane (and $x'z'$ plane) at time t.

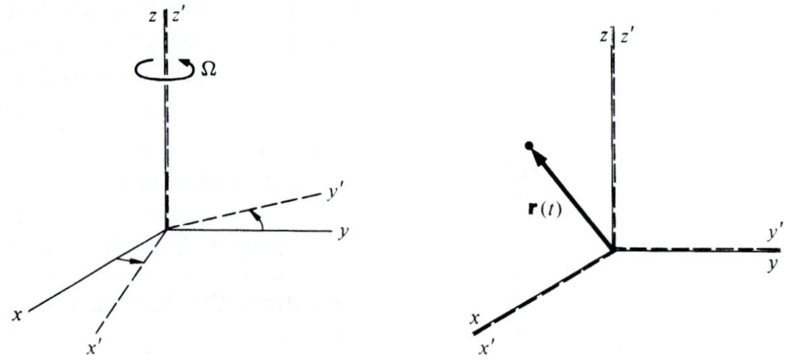

At time $t + \Delta t$, the position vector is $\mathbf{r}(t + \Delta t)$, and, from the figure at left below the displacement of the particle in the inertial system is

$$\Delta\mathbf{r} = \mathbf{r}(t + \Delta t) - \mathbf{r}(t).$$

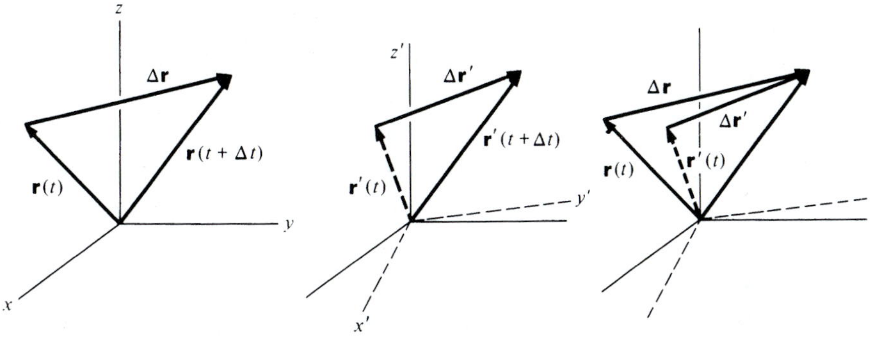

The situation is different for an observer in the rotating coordinate system. He also notes the same final position vector $\mathbf{r}(t + \Delta t)$, but in calculating the displacement he remembers that the initial position vector in his coordinate system $\mathbf{r}'(t)$ was in the $x'z'$ plane. The displacement he measures relative to his coordinates is $\Delta\mathbf{r}' = \mathbf{r}(t + \Delta t) - \mathbf{r}'(t)$, as in the figure at right above however, the $x'z'$ plane is now rotated away from its earlier position and, as we see from the drawing at left, $\Delta\mathbf{r}$ and $\Delta\mathbf{r}'$ are not the same

$$\Delta\mathbf{r} = \Delta\mathbf{r}' + \mathbf{r}'(t) - \mathbf{r}(t).$$

Consequently, the velocity is different in the two frames.

Since $\mathbf{r}'(t)$ and $\mathbf{r}(t)$ differ only by a pure rotation, we can use the result of Sec. 7.2 to write

$$\mathbf{r}'(t) - \mathbf{r}(t) = (\boldsymbol{\Omega} \times \mathbf{r})\,\Delta t.$$

Hence,

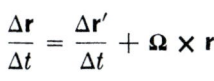

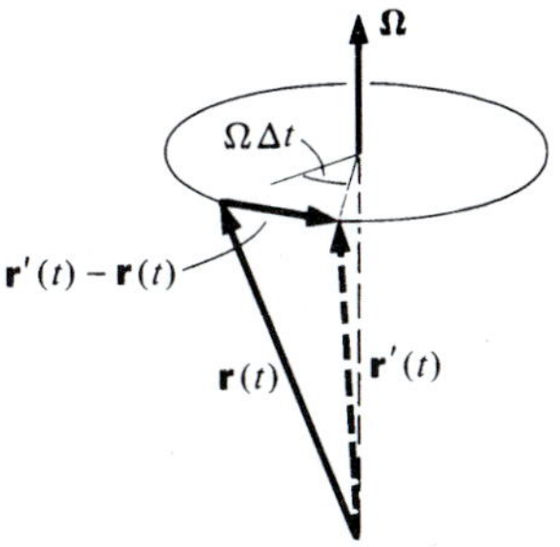

$$\frac{\Delta\mathbf{r}}{\Delta t} = \frac{\Delta\mathbf{r}'}{\Delta t} + \boldsymbol{\Omega} \times \mathbf{r}.$$

Taking the limit $\Delta t \to 0$ yields

$$\mathbf{v}_{\text{in}} = \mathbf{v}_{\text{rot}} + \boldsymbol{\Omega} \times \mathbf{r}. \qquad 8.5$$

It is important to realize that Eq. (8.5) is a general vector relation; the proof did not employ the special arrangement of axes we used to illustrate the derivation.

An alternative way to write Eq. (8.5) is

$$\left(\frac{d\mathbf{r}}{dt}\right)_{\text{in}} = \left(\frac{d\mathbf{r}}{dt}\right)_{\text{rot}} + \mathbf{\Omega} \times \mathbf{r}. \qquad 8.6$$

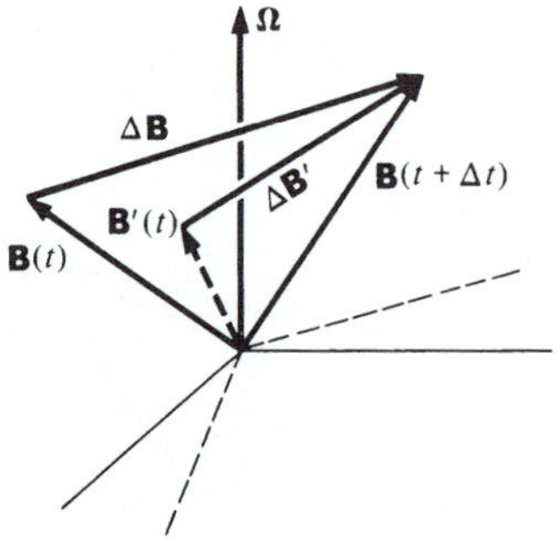

Since our proof used only the geometric properties of $\mathbf{r}$, Eq. (8.6) can immediately be generalized for any vector $\mathbf{B}$, as the sketch indicates.

$$\left(\frac{d\mathbf{B}}{dt}\right)_{\text{in}} = \left(\frac{d\mathbf{B}}{dt}\right)_{\text{rot}} + \mathbf{\Omega} \times \mathbf{B}. \qquad 8.7$$

When applying Eq. (8.7), keep in mind that $\mathbf{B}$ is instantaneously the same in both systems; it is only the time rates of change which differ. Note 8.2 presents an alternative derivation of Eq. (8.7).

Acceleration Relative to Rotating Coordinates

We can use Eq. (8.7) to relate the acceleration observed in a rotating system, $\mathbf{a}_{\text{rot}} = (d\mathbf{v}_{\text{rot}}/dt)_{\text{rot}}$, to the acceleration in an inertial system, $\mathbf{a}_{\text{in}} = (d\mathbf{v}_{\text{in}}/dt)_{\text{in}}$. Applying Eq. (8.7) to $\mathbf{v}_{\text{in}}$ gives

$$\mathbf{a}_{\text{in}} = \left(\frac{d\mathbf{v}_{\text{in}}}{dt}\right)_{\text{in}} = \left(\frac{d\mathbf{v}_{\text{in}}}{dt}\right)_{\text{rot}} + \mathbf{\Omega} \times \mathbf{v}_{\text{in}}.$$

Using

$$\mathbf{v}_{\text{in}} = \mathbf{v}_{\text{rot}} + \mathbf{\Omega} \times \mathbf{r}$$

we have

$$\mathbf{a}_{\text{in}} = \left[\frac{d}{dt}(\mathbf{v}_{\text{rot}} + \mathbf{\Omega} \times \mathbf{r})\right]_{\text{rot}} + \mathbf{\Omega} \times \mathbf{v}_{\text{rot}} + \mathbf{\Omega} \times (\mathbf{\Omega} \times \mathbf{r}).$$

We shall assume that $\mathbf{\Omega}$ is constant, since this is the case generally needed in practice. Hence

$$\mathbf{a}_{\text{in}} = \mathbf{a}_{\text{rot}} + \mathbf{\Omega} \times \left(\frac{d\mathbf{r}}{dt}\right)_{\text{rot}} + \mathbf{\Omega} \times \mathbf{v}_{\text{rot}} + \mathbf{\Omega} \times (\mathbf{\Omega} \times \mathbf{r}),$$

or

$$\mathbf{a}_{\text{in}} = \mathbf{a}_{\text{rot}} + 2\mathbf{\Omega} \times \mathbf{v}_{\text{rot}} + \mathbf{\Omega} \times (\mathbf{\Omega} \times \mathbf{r}). \qquad 8.8$$

Let us examine the various contributions to $\mathbf{a}_{\text{in}}$ in Eq. (8.8). The term $\mathbf{a}_{\text{rot}}$ is simply the acceleration measured in the rotating coordinate system; there is nothing mysterious here. For example, if we measure the acceleration of a car or plane in a coordinate system fixed to the rotating earth, we are measuring $\mathbf{a}_{\text{rot}}$.

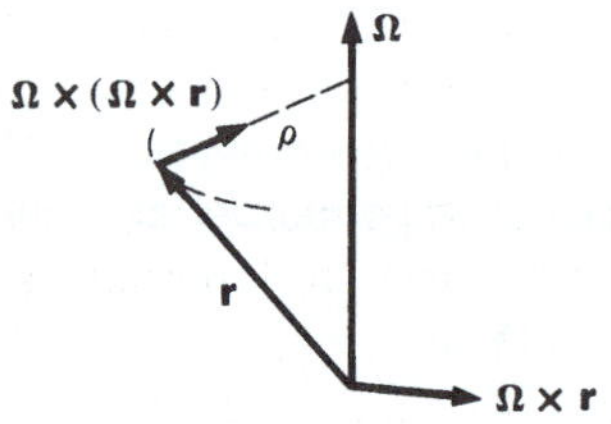

To see the origin of the term $\mathbf{\Omega} \times (\mathbf{\Omega} \times \mathbf{r})$, note first that $\mathbf{\Omega} \times \mathbf{r}$ is perpendicular to the plane of $\mathbf{\Omega}$ and $\mathbf{r}$ and has magnitude $\Omega\rho$, where ρ is the perpendicular distance from the axis of rotation

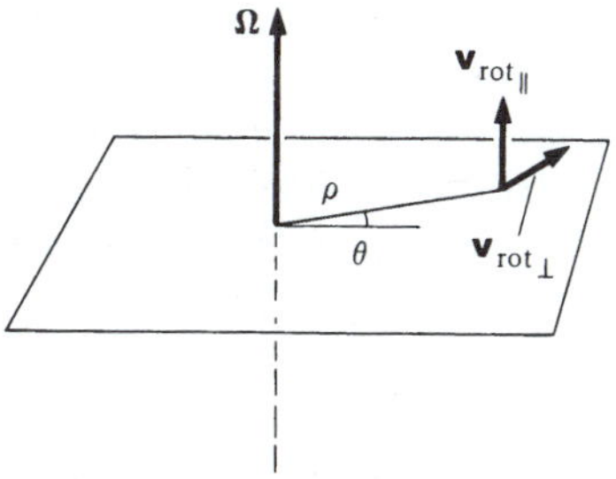

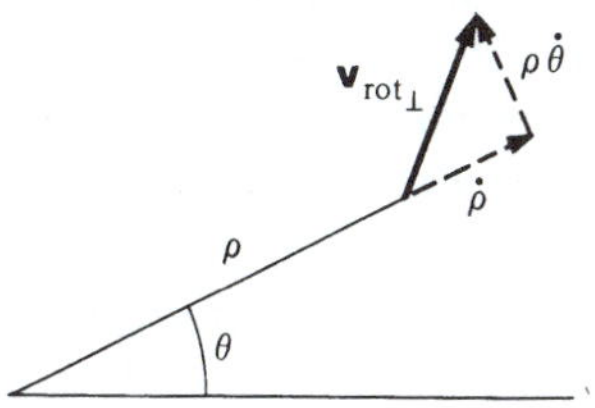

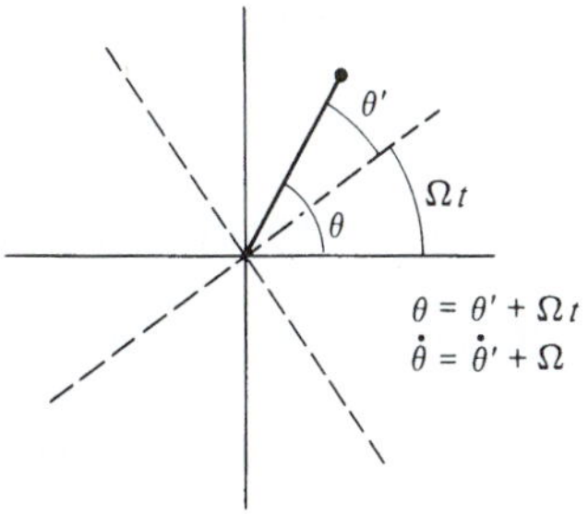

to the tip of $\mathbf{r}$. Hence $\mathbf{\Omega} \times (\mathbf{\Omega} \times \mathbf{r})$ is directed radially inward toward the axis of rotation and has magnitude $\Omega^2\rho$. It is a centripetal acceleration, arising because every point at rest in the rotating system is actually moving in a circular path in inertial space.

The term $2\mathbf{\Omega} \times \mathbf{v}_{rot}$ is the general vector expression for the Coriolis acceleration in three dimensions. If $\mathbf{v}_{rot}$ is resolved into components $\mathbf{v}_{rot_\parallel}$ and $\mathbf{v}_{rot_\perp}$, parallel and perpendicular to $\mathbf{\Omega}$, respectively, only $\mathbf{v}_{rot_\perp}$ contributes to $2\mathbf{\Omega} \times \mathbf{v}_{rot}$. Hence, the coriolis acceleration is perpendicular to $\mathbf{\Omega}$. Here is how it arises:

The radial component $\dot\rho$ of $\mathbf{v}_{rot_\perp}$ contributes $2\Omega\dot\rho$ in the tangential direction to $\mathbf{a}_{in}$. This is simply the Coriolis term we found in Sec. 1.9 for motion in inertial space with angular velocity Ω and radial velocity $\dot\rho$. The tangential component $\rho\dot\theta'$ of $\mathbf{v}_{rot_\perp}$ contributes $2\Omega\rho\dot\theta'$ toward the rotation axis. To see the origin of this term, note that in inertial space the instantaneous angular velocity is $\dot\theta = \dot\theta' + \Omega$ and the centripetal acceleration term in $\mathbf{a}_{in}$ is

$$\begin{aligned}\rho\dot\theta^2 &= \rho(\dot\theta' + \Omega)^2 \\ &= \rho\dot\theta'^2 + 2\Omega\rho\dot\theta' + \rho\Omega^2.\end{aligned}$$

The three terms on the right correspond to the three terms on the right of Eq. (8.8). $\rho\dot\theta'^2$ is part of $\mathbf{a}_{rot}$, $2\Omega\rho\dot\theta'$ follows from $2\mathbf{\Omega} \times \mathbf{v}_{rot}$ as we have shown, and $\rho\Omega^2$ comes from $\mathbf{\Omega} \times (\mathbf{\Omega} \times \mathbf{r})$.

The Apparent Force in a Rotating Coordinate System

From Eq. (8.8) we have

$$\mathbf{a}_{rot} = \mathbf{a}_{in} - 2\mathbf{\Omega} \times \mathbf{v}_{rot} - \mathbf{\Omega} \times (\mathbf{\Omega} \times \mathbf{r}).$$

The force observed in the rotating system is

$$\begin{aligned}\mathbf{F}_{rot} &= m\mathbf{a}_{rot} = m\mathbf{a}_{in} - m[2\mathbf{\Omega} \times \mathbf{v}_{rot} + \mathbf{\Omega} \times (\mathbf{\Omega} \times \mathbf{r})] \\ &= \mathbf{F} + \mathbf{F}_{fict},\end{aligned}$$

where the fictitious force is

$$\mathbf{F}_{fict} = -2m\mathbf{\Omega} \times \mathbf{v}_{rot} - m\mathbf{\Omega} \times (\mathbf{\Omega} \times \mathbf{r}).$$

The first term on the right is called the *Coriolis force*, and the second term, which points outward from the rotation axis, is called the *centrifugal force*.

The Coriolis and centrifugal forces are nonphysical; they arise from kinematics and are not due to physical interactions. For instance, the centrifugal force actually increases with ρ, whereas real forces always decrease with distance. Nevertheless, the

Coriolis and centrifugal forces seem quite real to an observer in a rotating frame. When we drive a çar too fast around a curve, it skids outward as if pushed by the centrifugal force. From the standpoint of an observer in an inertial frame, however, what has happened is that the sideward force exerted by the road on the tires is not adequate to keep the car turning with the road.

There is a natural human tendency to describe rotational motion with a rotating system. For instance, if we whirl a rock on a string, we instinctively say that centrifugal force is pulling the rock outward. In a coordinate system rotating with the rock, this is correct; the rock is stationary and the centrifugal force is in balance with the tension in the string. In an inertial system there is no centrifugal force; the rock is accelerating radially due to the force exerted by the string. Either system is valid for analyzing the problem. However, it is essential not to confuse the systems by trying to use fictitious forces in inertial frames.

Here are some examples to illustrate the use of rotating coordinates.

Example 8.6 Surface of a Rotating Liquid

A bucket of water spins with angular speed ω. What shape does the water's surface assume?

In a coordinate system rotating with the bucket, the problem is purely static. Consider the force on a small volume of water of mass m at the surface of the liquid. For equilibrium, the total force on m must be zero. The forces are the contact force $\mathbf{F}_0$, the weight $\mathbf{W}$, and the fictitious force $\mathbf{F}_{\text{fict}}$, which is radial.

$$F_0 \cos\phi - W = 0$$

$$-F_0 \sin\phi + F_{\text{fict}} = 0,$$

where $F_{\text{fict}} = m\Omega^2 r = m\omega^2 r$, since $\Omega = \omega$ for a coordinate system rotating with the bucket.

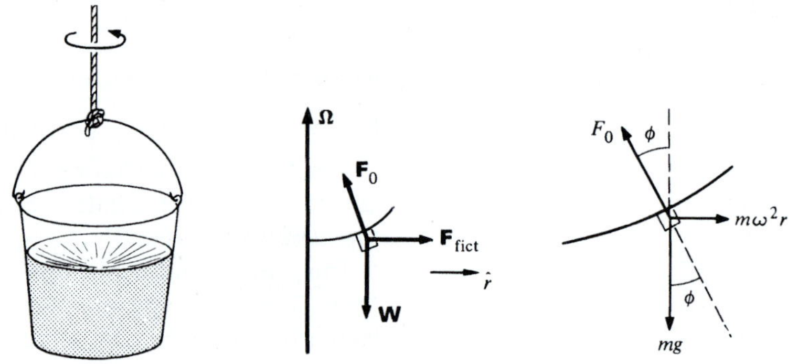

Solving these equations for ϕ yields

$$\phi = \arctan \frac{\omega^2 r}{g}.$$

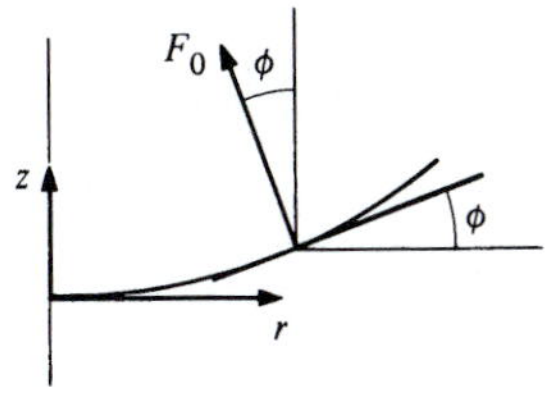

Unlike solids, liquids cannot exert a static force tangential to the surface. Hence $\mathbf{F}_0$, the force on m due to the neighboring liquid, must be perpendicular to the surface. The slope of the surface at any point is therefore

$$\frac{dz}{dr} = \tan \phi$$

$$= \frac{\omega^2 r}{g}.$$

We can integrate this relation to find the equation of the surface $z = f(r)$.

We have

$$\int dz = \frac{\omega^2}{g} \int r\, dr$$

$$z = \frac{1}{2} \frac{\omega^2}{g} r^2,$$

where we have taken $z = 0$ on the axis at the surface of the liquid. The surface is a paraboloid of revolution.

Example 8.7 The Coriolis Force

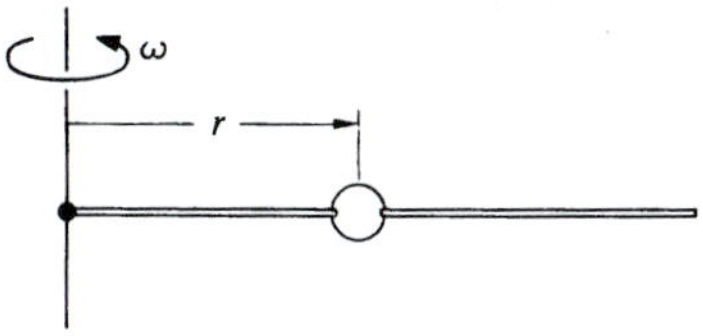

A bead slides without friction on a rigid wire rotating at constant angular speed ω. The problem is to find the force exerted by the wire on the bead.

In a coordinate system rotating with the wire the motion is purely radial. The sketch shows the force diagram in the rotating system. F_{cent} is the centrifugal force and F_{Cor} is the Coriolis force. Since the wire is frictionless, the contact force N is normal to the wire. (We neglect gravity.) In the rotating system the equations of motion are

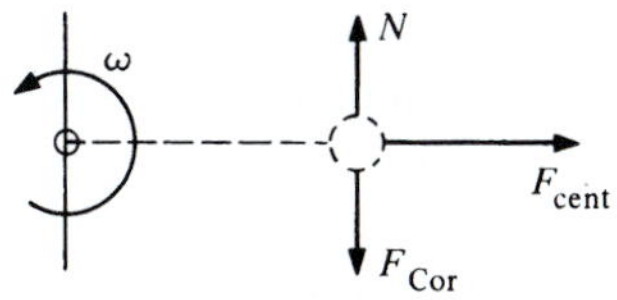

$$F_{\text{cent}} = m\ddot{r}$$

$$N - F_{\text{Cor}} = 0.$$

Using $F_{\text{cent}} = m\omega^2 r$, the first equation gives

$$m\ddot{r} - m\omega^2 r = 0,$$

which has the solution

$$r = Ae^{\omega t} + Be^{-\omega t},$$

where A and B are constants depending on the initial conditions.

The tangential equation of motion, which expresses the fact that there is no tangential acceleration in the rotating system, gives

$$N = F_{\text{Cor}} = 2m\dot{r}\omega$$
$$= 2m\omega^2(Ae^{\omega t} - Be^{-\omega t}).$$

To complete the problem, we must be given the initial conditions which specify A and B.

Example 8.8 Deflection of a Falling Mass

Because of the Coriolis force, falling objects on the earth are deflected horizontally. For instance, a mass dropped from a tower lands to the east of a plumb line from the release point. In this example we shall calculate the deflection for a mass m dropped from a tower of height h at the equator.

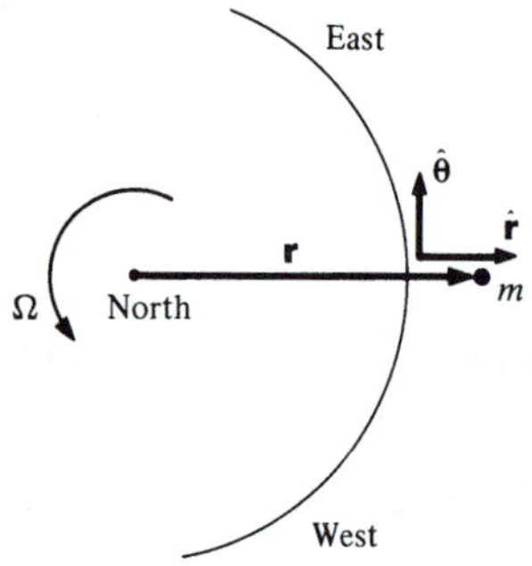

In the coordinate system r, θ fixed to the earth (with the tangential direction toward the east) the apparent force on m is

$$\mathbf{F} = -mg\hat{\mathbf{r}} - 2m\mathbf{\Omega} \times \mathbf{v}_{\text{rot}} - m\mathbf{\Omega} \times (\mathbf{\Omega} \times \mathbf{r}).$$

The gravitational and centrifugal forces are radial, and if m is dropped from rest, the Coriolis force is in the equatorial plane. Thus the motion of m is confined to the equatorial plane, and we have

$$\mathbf{v}_{\text{rot}} = \dot{r}\hat{\mathbf{r}} + r\dot{\theta}\hat{\mathbf{\theta}}.$$

Using $\mathbf{\Omega} \times \mathbf{v}_{\text{rot}} = \Omega\dot{r}\hat{\mathbf{\theta}} - r\Omega\dot{\theta}\hat{\mathbf{r}}$, and $\mathbf{\Omega} \times (\mathbf{\Omega} \times \mathbf{r}) = -\Omega^2 r\hat{\mathbf{r}}$, we obtain

$$F_r = -mg + 2m\Omega\dot{\theta}r + m\Omega^2 r,$$

$$F_\theta = -2m\dot{r}\Omega.$$

The radial equation of motion is

$$m\ddot{r} - mr\dot{\theta}^2 = -mg + 2m\Omega\dot{\theta}r + m\Omega^2 r.$$

To an excellent approximation, m falls vertically and $\dot{\theta} \ll \Omega$. We can therefore omit the terms $mr\dot{\theta}^2$ and $2m\Omega\dot{\theta}r$ in comparison with $m\Omega^2 r$. Thus

$$\ddot{r} = -g + \Omega^2 r. \qquad 1$$

The tangential equation of motion is

$$mr\ddot{\theta} + 2m\dot{r}\dot{\theta} = -2m\dot{r}\Omega.$$

To the same approximation $\dot{\theta} \ll \Omega$ we have

$$r\ddot{\theta} = -2\dot{r}\Omega. \qquad 2$$

During the fall, r changes only slightly, from $R_e + h$ to R_e, where R_e is the radius of the earth, and we can take g to be constant and $r \approx R_e$. Equation (1) becomes

$$\begin{aligned}\ddot{r} &= -g + \Omega^2 R_e \\ &= -g',\end{aligned}$$

where $g' = g - \Omega^2 R_e$ is the acceleration due to the gravitational force minus a centrifugal term. g' is the apparent acceleration due to gravity, and since this is customarily denoted by g, we shall henceforth drop the prime. The solution of the radial equation of motion $\ddot{r} = -g$ is

$$\begin{aligned}\dot{r} &= -gt \\ r &= r_0 - \tfrac{1}{2}gt^2. \end{aligned} \qquad 3$$

If we insert $\dot{r} = -gt$ in the tangential equation of motion, Eq. (2), we have

$$r\ddot{\theta} = 2gt\Omega$$

or

$$\ddot{\theta} = \frac{2g\Omega}{R_e}t,$$

where we have used $r \approx R_e$. Hence

$$\dot{\theta} = \frac{g\Omega}{R_e}t^2$$

and

$$\theta = \frac{1}{3}\frac{g\Omega}{R_e}t^3. \qquad 4$$

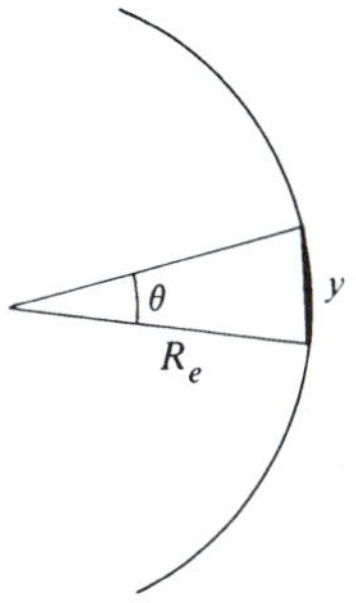

The horizontal deflection of m is $y \approx R_e\theta$ or

$$y = \tfrac{1}{3}g\Omega t^3.$$

The time T to fall distance h is given by

$$\begin{aligned}r - r_0 &= -h \\ &= -\tfrac{1}{2}gT^2\end{aligned}$$

so that

$$T = \sqrt{\frac{2h}{g}} \qquad \text{and} \qquad y = \frac{1}{3}g\Omega\left(\frac{2h}{g}\right)^{\frac{3}{2}}.$$

For a tower 50 m high,

$$y \approx 0.77 \text{ cm}.$$

θ is positive, and the deflection is toward the east.

Example 8.9 Motion on the Rotating Earth

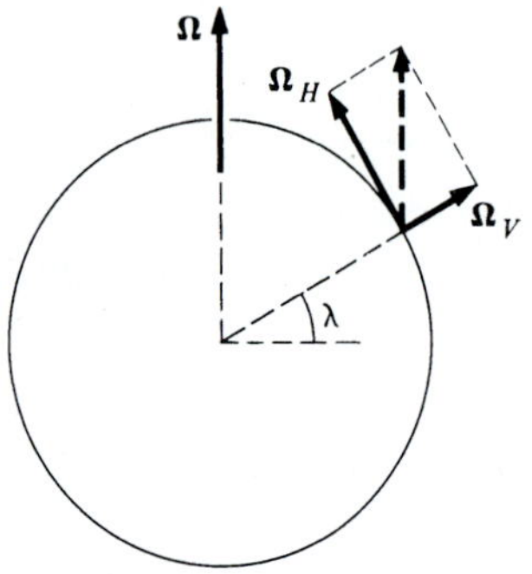

A surprising effect of the Coriolis force is that it turns straight line motion on a rotating sphere into circular motion. As we shall show in this example, for a velocity **v** tangential to the sphere (like the velocity of a wind over the earth's surface) the horizontal component of the Coriolis force is perpendicular to **v** and its magnitude is independent of the direction of **v**.

Consider a particle of mass m moving with velocity **v** at latitude λ on the surface of a sphere. The sphere is rotating with angular velocity $\mathbf{\Omega}$. If we decompose $\mathbf{\Omega}$ into a vertical part $\mathbf{\Omega}_V$ and a horizontal part $\mathbf{\Omega}_H$, the Coriolis force is

$$\mathbf{F} = -2m\mathbf{\Omega} \times \mathbf{v}$$
$$= -2m(\mathbf{\Omega}_V \times \mathbf{v} + \mathbf{\Omega}_H \times \mathbf{v}).$$

$\mathbf{\Omega}_H$ and **v** are horizontal, so that $\mathbf{\Omega}_H \times \mathbf{v}$ is vertical. Thus the horizontal Coriolis force, $\mathbf{F}_H$, arises solely from the term $\mathbf{\Omega}_V \times \mathbf{v}$. $\mathbf{\Omega}_V$ is perpendicular to **v** and $\mathbf{\Omega}_V \times \mathbf{v}$ has magnitude $v\Omega_V$, independent of the direction of **v**, as we wished to prove.

We can write the result in a more explicit form. If $\hat{\mathbf{r}}$ is a unit vector perpendicular to the surface at latitude λ, $\mathbf{\Omega}_V = \Omega \sin \lambda \hat{\mathbf{r}}$ and

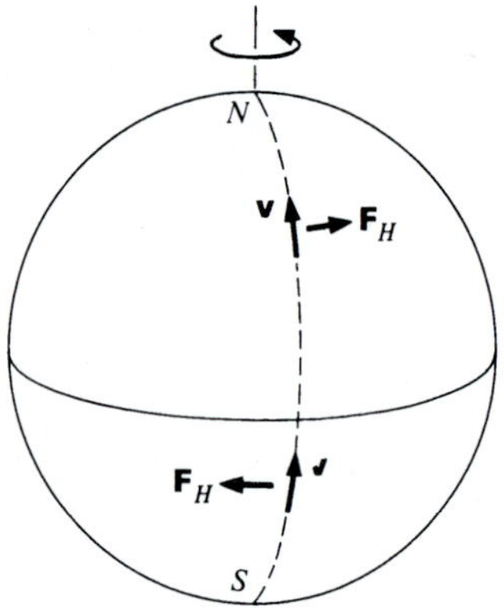

$$\mathbf{F}_H = -2m\Omega \sin \lambda\, \hat{\mathbf{r}} \times \mathbf{v}.$$

The magnitude of $\mathbf{F}_H$ is

$$F_H = 2mv\Omega \sin \lambda.$$

$\mathbf{F}_H$ is always perpendicular to **v**, and in the absence of other horizontal forces it would produce circular motion, clockwise in the northern hemisphere and counterclockwise in the southern. Air flow on the earth is strongly influenced by the Coriolis force and without it stable circular weather patterns could not form. However, to understand the dynamics of weather systems we must also include other forces, as the next example discusses.

Example 8.10 Weather Systems

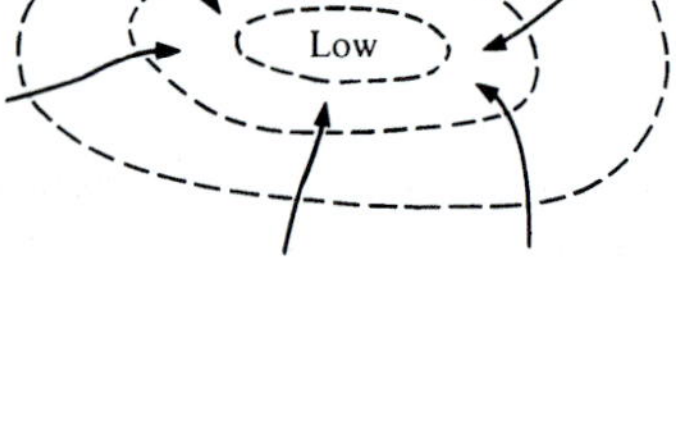

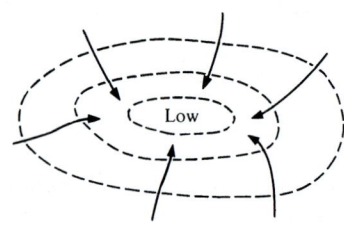

Imagine that a region of low pressure occurs in the atmosphere, perhaps because of differential heating of the air. The closed curves in the sketch represent lines of constant pressure, or *isobars*. There is a force on each element of air due to the pressure gradient, and in the absence of other forces winds would blow inward, quickly equalizing the pressure difference.

However, the wind pattern is markedly altered by the Coriolis force. As the air begins to flow inward, it is deflected sideways by the Coriolis

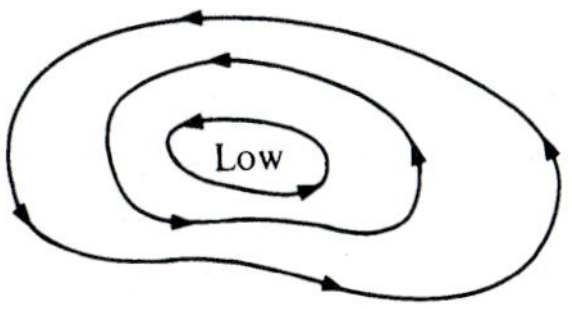

force, as shown in figure *a*. (The drawing is for the northern hemisphere.) The result is that the wind circulates counterclockwise about the low along the isobars, as in the sketch at left. Similarly, wind circulates clockwise about regions of high pressure in the southern hemisphere. Since the Coriolis force is essentially zero near the equator, circular weather systems cannot form there and the weather tends to be uniform.

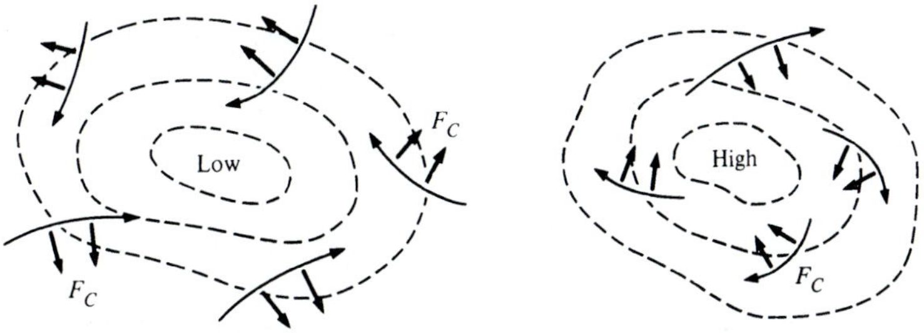

(*a*) (*b*)

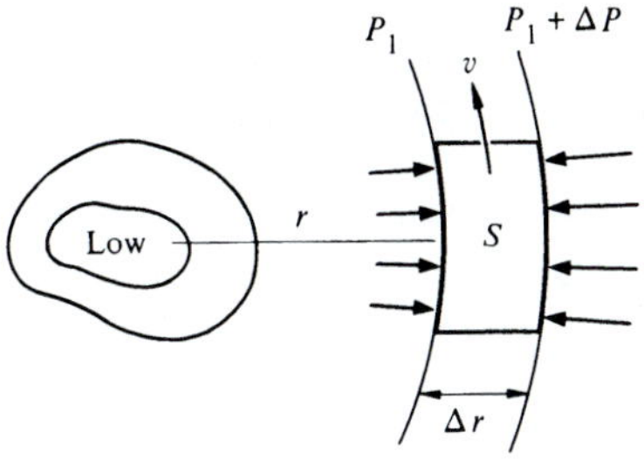

In order to analyze the motion, consider the forces on a parcel of air which is rotating about a low. The pressure force on the face along the isobar P_1 is P_1S, where S is the area of the inner face, as shown in the sketch. The force on the outer face is $(P_1 + \Delta P)S$, and the net pressure force is $(\Delta P)S$ inward. The Coriolis force is $2mv\Omega \sin \lambda$, where m is the mass of the parcel and v its velocity. The air is rotating counterclockwise about the low, so that the Coriolis force is outward. Hence, the radial equation of motion for steady circular flow is

$$\frac{mv^2}{r} = (\Delta P)S - 2mv\Omega \sin \lambda.$$

The volume of the parcel is $\Delta r\, S$, where Δr is the distance between the isobars, and the mass is $w\, \Delta r\, S$, where w is the density of air, assumed constant. Inserting this in the equation of motion and taking the limit $\Delta r \to 0$ yields

$$\frac{v^2}{r} = \frac{1}{w}\frac{dP}{dr} - 2v\Omega \sin \lambda. \qquad 1$$

Air masses do not rotate as rigid bodies. Near the center of the low, where the pressure gradient dP/dr is large, wind velocities are highest. Far from the center, v^2/r is small and can be neglected. Equation (1) predicts that far from the center the wind speed is

$$v = \frac{1}{2\Omega \sin \lambda}\frac{1}{w}\frac{dP}{dr}. \qquad 2$$

The density of air at sea level is 1.3 kg/m^3 and atmospheric pressure is $P_{at} = 10^5$ N/m^2. dP/dr can be estimated by looking at a weather map.

Far from a high or low, a typical gradient is 3 millibars over 100 km $\approx$ 3×10^{-3} N/m^3, and at latitude 45° Eq. (2) gives

$$\begin{aligned} v &= 22 \text{ m/s} \\ &= 50 \text{ mi/h.} \end{aligned}$$

Near the ground this speed is reduced by friction with the land, but at higher altitudes Eq. (2) can be applied with good accuracy.

A hurricane is an intense compact low in which the pressure gradient can be as high as 30×10^{-3} N/m^3. Hurricane winds are so strong that the v^2/r term in Eq. (1) cannot be neglected. Solving Eq. (1) for v we find

$$v = \sqrt{(r\Omega \sin \lambda)^2 + \frac{r}{w}\frac{dP}{dr}} - r\Omega \sin \lambda. \qquad 3$$

At a distance 100 km from the eye of a hurricane at latitude 20°, Eq. (3) predicts a wind speed of 45 m/s $\approx$ 100 mi/h for a pressure gradient of 30×10^{-3} N/m^3. This is in reasonable agreement with weather observations. At larger radii, the wind speed drops because of a decrease in the pressure gradient.

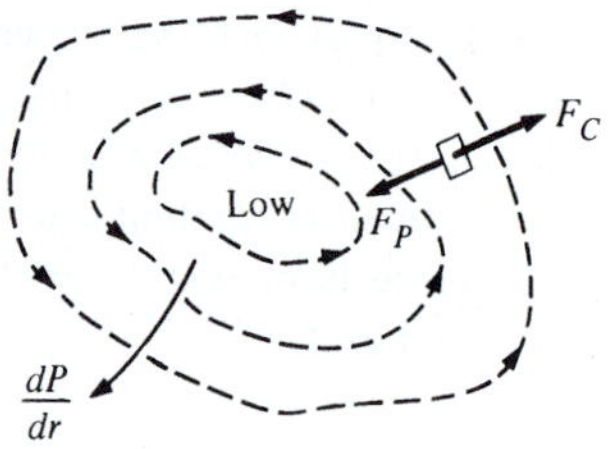

There is an interesting difference between lows and highs. In a low, the pressure force is inward and the Coriolis force is outward, whereas in a high, the directions of the forces are reversed. The radial equation of motion for air circulating around a high is

$$\frac{v^2}{r} = 2v\Omega \sin \lambda - \frac{1}{w}\left|\frac{dP}{dr}\right|. \qquad 4$$

Solving Eq. (4) for v yields

$$v = r\Omega \sin \lambda - \sqrt{(r\Omega \sin \lambda)^2 - \frac{r}{w}\left|\frac{dP}{dr}\right|}. \qquad 5$$

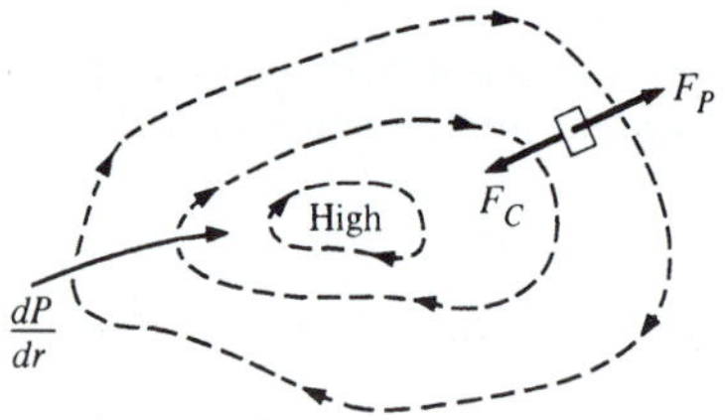

We see from Eq. (5) that if $1/w|dP/dr| > r(\Omega \sin \lambda)^2$, the high cannot form; the Coriolis force is too weak to supply the needed centripetal acceleration against the large outward pressure force. For this reason, storms like hurricanes are always low pressure systems; the strong inward pressure force helps hold a low together.

The Foucault pendulum provides one of the most dramatic demonstrations that the earth is a noninertial system. The pendulum is simply a heavy bob hanging from a long wire mounted to swing freely in any direction. As the pendulum swings back and forth, the plane of motion precesses slowly about the vertical, taking about a day and a half for a complete rotation in the midlatitudes. The precession is a result of the earth's rotation.

The plane of motion tends to stay fixed in inertial space while the earth rotates beneath it.

In the 1850s Foucault hung a pendulum 67 m long from the dome of the Pantheon in Paris. The bob precessed almost a centimeter on each swing, and it presented the first direct evidence that the earth is indeed rotating. The pendulum became the rage of Paris.

The next example uses our analysis of the Coriolis force to calculate the motion of the Foucault pendulum in a simple way.

Example 8.11 **The Foucault Pendulum**

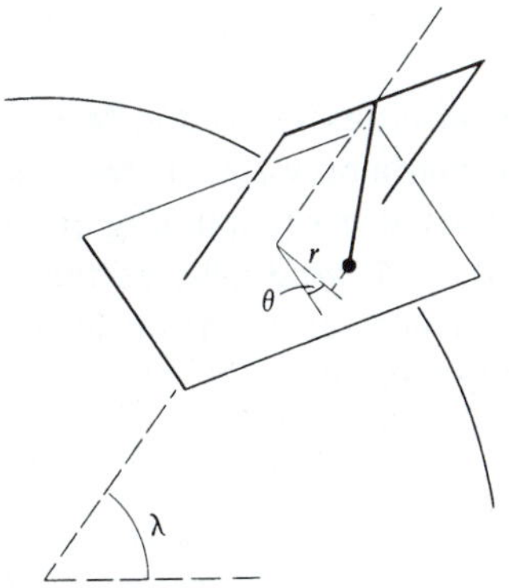

Consider a pendulum of mass m which is swinging with frequency $\gamma = \sqrt{g/l}$, where l is the length of the pendulum. If we describe the position of the pendulum's bob in the horizontal plane by coordinates r, θ, then

$$r = r_0 \sin \gamma t,$$

where r_0 is the amplitude of the motion. In the absence of the Coriolis force, there are no tangential forces and θ is constant.

The horizontal Coriolis force $\mathbf{F}_{\text{CH}}$ is

$$\mathbf{F}_{\text{CH}} = -2m\Omega \sin \lambda \dot{r} \hat{\boldsymbol{\theta}}.$$

Hence, the tangential equation of motion, $ma_\theta = F_{\text{CH}}$, becomes

$$m(r\ddot{\theta} + 2\dot{r}\dot{\theta}) = -2m\Omega \sin \lambda\, \dot{r}$$

or

$$r\ddot{\theta} + 2\dot{r}\dot{\theta} = -2\Omega \sin \lambda\, \dot{r}.$$

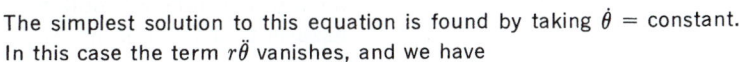

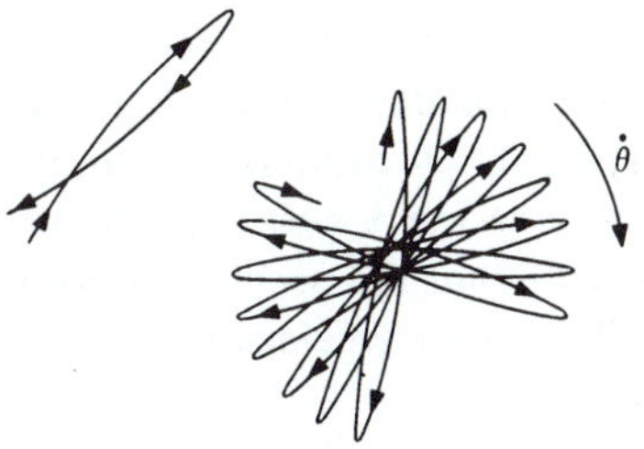

The simplest solution to this equation is found by taking $\dot{\theta}$ = constant. In this case the term $r\ddot{\theta}$ vanishes, and we have

$$\dot{\theta} = -\Omega \sin \lambda.$$

The pendulum precesses uniformly in a clockwise direction. The time for the plane of oscillation to rotate once is

$$\begin{aligned} T &= \frac{2\pi}{\dot{\theta}} \\ &= \frac{2\pi}{\Omega \sin \lambda} \\ &= \frac{24\text{ h}}{\sin \lambda}. \end{aligned}$$

Thus, at a latitude of 45°, the Foucault pendulum rotates once in 34 h.

At the North Pole the period of precession is 24 h; the pendulum rotates clockwise with respect to the earth at the same rate as the earth rotates counterclockwise. With respect to inertial space the plane of motion remains fixed.

In addition to its dramatic display of the earth's rotation, the Foucault pendulum embodies a profound mystery. Consider, for instance, a Foucault pendulum at the North Pole. The precession is obviously an artifact; the plane of motion stays fixed while the earth rotates beneath it. The plane of the pendulum remains fixed relative to the fixed stars. Why should this be? How does the pendulum "know" that it must swing in a plane which is stationary relative to the fixed stars instead of, say, in a plane which rotates at some uniform rate?

This question puzzled Newton, who described it in terms of the following experiment: if a bucket contains water at rest, the surface of the water is flat. If the bucket is set spinning at a steady rate, the water at first lags behind, but gradually, as the water's rotational speed increases, the surface takes on the form of the parabola of revolution discussed in Example 8.6. If the bucket is suddenly stopped, the concavity of the water's surface persists for some time. It is evidently not motion relative to the bucket that is important in determining the shape of the liquid surface. So long as the water rotates, the surface is depressed. Newton concluded that rotational motion is absolute, since by observing the water's surface it is possible to detect rotation without reference to outside objects.

From one point of view there is really no paradox to the absolute nature of rotational motion. The principle of galilean invariance asserts that there is no way to detect locally the uniform translational motion of a system. However, this does not limit our ability to detect the *acceleration* of a system. A rotating system accelerates in a most nonuniform way. At every point the acceleration is directed toward the axis of rotation; the acceleration points out the axis. Our ability to detect such an acceleration in no way contradicts galilean invariance.

Nevertheless, there is an engima. Both the rotating bucket and the Foucault pendulum maintain their motion *relative to the fixed stars.* How do the fixed stars determine an inertial system? What prevents the plane of the pendulum from rotating with respect to the fixed stars? Why is the surface of the water in the rotating bucket flat only when the bucket is at rest with respect to the fixed stars? Ernst Mach, who in 1883 wrote the first incisive critique of newtonian physics, put the matter this way. Suppose that we keep a bucket of water fixed and rotate all the stars. Physically there is no way to distinguish this from the original case where the bucket is rotated, and we expect the surface of the water to again assume a parabolic shape. Apparently the motion of the water in the bucket depends on the motion of matter far off in the universe. To put it more dramatically, suppose that we eliminate the stars, one by one, until only our bucket remains. What will happen now if we rotate the bucket? There is no way for us to predict the motion of the water in the bucket—the inertial properties of space might be totally different. We have a most peculiar situation. The local properties of space depend on far-off matter, yet when we rotate the water, the surface *immediately* starts to deflect. There is no time for signals to travel to the distant stars and return. How does the water in the bucket "know" what the rest of the universe is doing?

The principle that the inertial properties of space depend on the existence of far-off matter is known as Mach's principle. The principle is accepted by many physicists, but it can lead to strange conclusions. For instance, there is no reason to believe that matter in the universe is uniformly distributed around the earth; the solar system is located well out in the limb of our galaxy, and matter in our galaxy is concentrated predominantly in a very thin plane. If inertia is due to far-off matter, then we might well expect it to be different in different directions so that the value of mass would depend on the direction of acceleration. No such effects have ever been observed. Inertia remains a mystery.

Note 8.1 The Equivalence Principle and the Gravitational Red Shift

Radiating atoms emit light at only certain characteristic wavelengths. If light from atoms in the strong gravitational field of dense stars is analyzed spectroscopically, the characteristic wavelengths are observed to be slightly increased, shifted toward the red. We can visualize atoms as clocks which "tick" at characteristic frequencies. The shift toward longer wavelengths, known as the gravitational red shift, corresponds to a slowing of the clocks. The gravitational red shift implies that clocks in a gravitational field appear to run slow when viewed from outside the field. As we shall show, the origin of the effect lies in the nature of space, time, and gravity, not in the trivial effect of gravity on mechanical clocks.

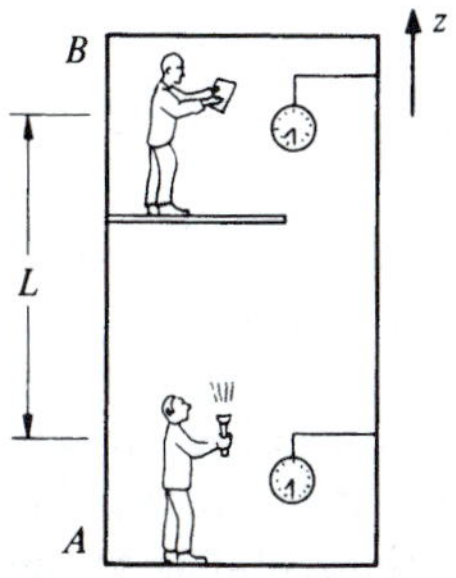

(*a*)

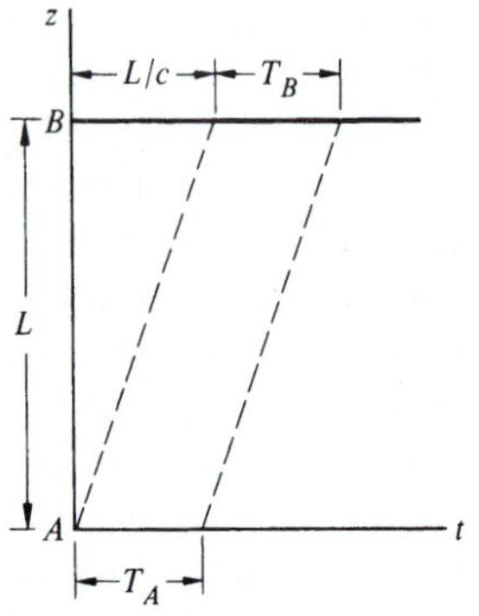

(b)

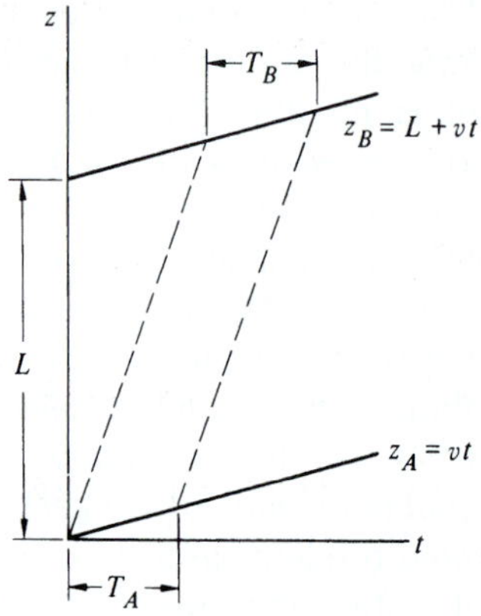

(c)

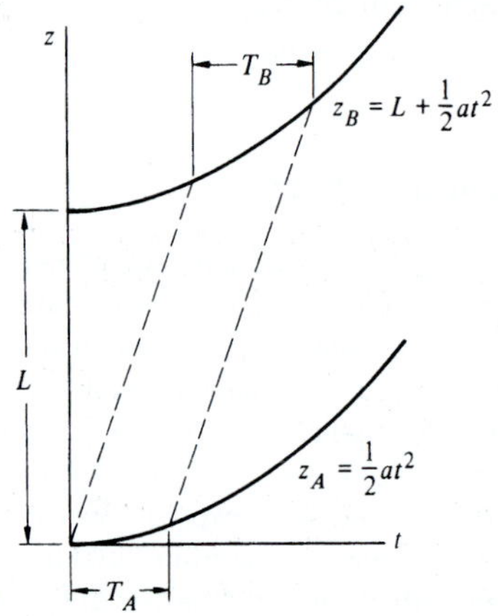

(d)

It is rather startling to see how the equivalence principle, which is so simple and nonmathematical, leads directly to a connection between space, time, and gravity. To show the connection we must use an elementary result from the theory of relativity; it is impossible to transmit information faster than the velocity of light, $c = 3 \times 10^8$ m/s. However, this is the only relativistic idea needed; aside from this, our argument is completely classical.

Consider two scientists, A and B, separated by distance L as shown in sketch (a). A has a clock and a light which he flashes at intervals separated by time T_A. The signals are received by B, who notes the interval between pulses, T_B, with his own clock. A plot of vertical distance versus time is shown for two light pulses in (b). The pulses are delayed by the transit time, L/c, but the interval T_B is the same as T_A. Hence, if A transmits the pulses at, say, 1-s intervals, so that $T_A = 1$ s, then B's clock will read 1 s between the arrival of successive pulses.

Now consider the situation if both observers move upward uniformly with speed v, as shown in sketch (c). Although both scientists move during the time interval, they move equally, and we still have $T_B = T_A$.

The situation is entirely different if both observers are accelerating upward at uniform rate a as shown in sketch (d). A and B start from rest, and the graph of distance versus time is a parabola. Since A and B have the same acceleration, the curves are parallel, separated by distance L at each instant. It is apparent from the sketch that $T_B > T_A$, since the second pulse travels farther than the first and has a longer transit time. The effect is purely kinematical.

Now, by the principle of equivalence, A and B cannot distinguish between their upward accelerating system and a system at rest in a downward gravitational field with magnitude $g = a$. Thus, if the experiment is repeated in a system at rest in a gravitational field, the equivalence principle requires that $T_B > T_A$, as before. If $T_A = 1$ s, B will observe an interval greater than 1 s between successive pulses. B will conclude that A's clock is running slow. This is the origin of the gravitational red shift.

By applying the argument quantitatively, the following approximate result is readily obtained:

$$\frac{\Delta T}{T} = \frac{T_B - T_A}{T_A} = \frac{gL}{c^2},$$

where it is assumed that $\Delta T/T \ll 1$.

On earth the gravitational red shift is $\Delta T/T = 10^{-16} L$, where L is in meters. In spite of its small size, the effect has been measured and confirmed to an accuracy of 1 percent. The experiment was done by Pound, Rebka, and Snyder at Harvard University. The "clock" was the frequency of a gamma ray, and by using a technique known as Mössbauer absorption they were able to measure accurately the gravitational red shift due to a vertical displacement of 25 m.

Note 8.2 Rotating Coordinate Transformation

In this note we present an analytical derivation of Eq. (8.7) relating the time derivative of any vector **B** as observed in a rotating coordinate system to the time derivative observed in an inertial system. If the system x', y', z' rotates at rate $\mathbf{\Omega}$ with respect to the inertial system x, y, z, we shall prove that the time derivatives in the two systems of any vector **B** are related by

$$\left(\frac{d\mathbf{B}}{dt}\right)_{\text{in}} = \left(\frac{d\mathbf{B}}{dt}\right)_{\text{rot}} + \mathbf{\Omega} \times \mathbf{B}. \qquad 1$$

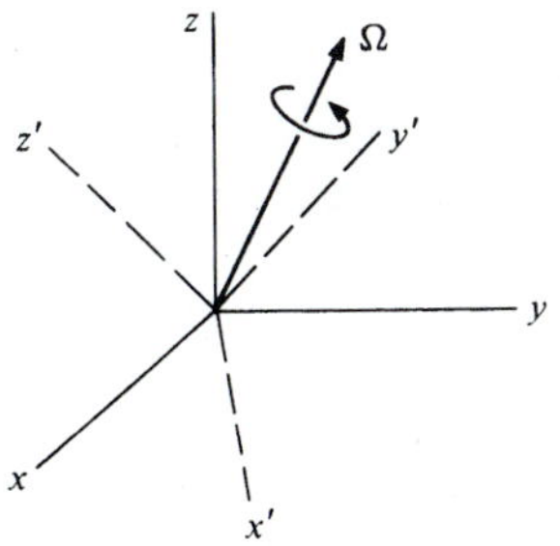

Consider an inertial coordinate system x, y, z and a coordinate system x', y', z' which rotates with respect to the inertial system at angular velocity $\mathbf{\Omega}$. The origins coincide. We can describe an arbitrary vector **B** by components along base vectors of either coordinate system. Thus, we have

$$\mathbf{B} = B_x\hat{\mathbf{i}} + B_y\hat{\mathbf{j}} + B_z\hat{\mathbf{k}} \qquad 2$$

or, alternatively,

$$\mathbf{B} = B_x'\hat{\mathbf{i}}' + B_y'\hat{\mathbf{j}}' + B_z'\hat{\mathbf{k}}', \qquad 3$$

where $\hat{\mathbf{i}}$, $\hat{\mathbf{j}}$, $\hat{\mathbf{k}}$ are the base vectors along the inertial axes and $\hat{\mathbf{i}}'$, $\hat{\mathbf{j}}'$, $\hat{\mathbf{k}}'$ are the base vectors along the rotating axes.

We now find an expression for the time derivative of **B** in each coordinate system. By differentiating Eq. (2) we have

$$\left(\frac{d\mathbf{B}}{dt}\right) = \frac{d}{dt}(B_x\hat{\mathbf{i}} + B_y\hat{\mathbf{j}} + B_z\hat{\mathbf{k}}).$$

The x, y, z system is inertial so that $\hat{\mathbf{i}}$, $\hat{\mathbf{j}}$, and $\hat{\mathbf{k}}$ are fixed in space. We have

$$\frac{d\mathbf{B}}{dt} = \frac{dB_x}{dt}\hat{\mathbf{i}} + \frac{dB_y}{dt}\hat{\mathbf{j}} + \frac{dB_z}{dt}\hat{\mathbf{k}}, \qquad 4$$

which is the familiar expression for the time derivative of a vector in cartesian coordinates. We designate this expression by $(d\mathbf{B}/dt)_{\text{in}}$.

If we differentiate Eq. (3) we obtain

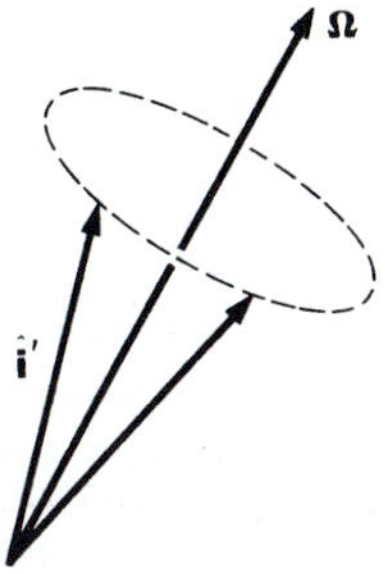

$$\left(\frac{d\mathbf{B}}{dt}\right) = \left(\frac{dB_x'}{dt}\hat{\mathbf{i}}' + \frac{dB_y'}{dt}\hat{\mathbf{j}}' + \frac{dB_z'}{dt}\hat{\mathbf{k}}'\right) + \left(B_x'\frac{d\hat{\mathbf{i}}'}{dt} + B_y'\frac{d\hat{\mathbf{j}}'}{dt} + B_z'\frac{d\hat{\mathbf{k}}'}{dt}\right). \qquad 5$$

The first term is the time derivative of **B** with respect to the $x'y'z'$ axes; this is the rate of change of **B** which would be measured by an observer in the rotating system, $(d\mathbf{B}/dt)_{\text{rot}}$. To evaluate the second term, note that since $\hat{\mathbf{i}}'$ is a unit vector, it can change only in direction, not in magnitude; thus $\hat{\mathbf{i}}'$ undergoes pure rotation. In Sec. 7.2 we found that the time derivative of a vector **r** of constant magnitude rotating with angular velocity $\boldsymbol{\omega}$ is $d\mathbf{r}/dt = \boldsymbol{\omega} \times \mathbf{r}$. We can use this result to evaluate $d\hat{\mathbf{i}}'/dt$. Let $\mathbf{r}$ lie along the x' axis and have unit magnitude: $\mathbf{r} = \hat{\mathbf{i}}'$. Hence

$$\frac{d\hat{\mathbf{i}}'}{dt} = \boldsymbol{\Omega} \times \hat{\mathbf{i}}'.$$

Similarly,

$$\frac{d\hat{\mathbf{j}}'}{dt} = \boldsymbol{\Omega} \times \hat{\mathbf{j}}' \quad \text{and} \quad \frac{d\hat{\mathbf{k}}'}{dt} = \boldsymbol{\Omega} \times \hat{\mathbf{k}}'.$$

The second term in Eq. (5) becomes

$$B'_x(\boldsymbol{\Omega} \times \hat{\mathbf{i}}') + B'_y(\boldsymbol{\Omega} \times \hat{\mathbf{j}}') + B'_z(\boldsymbol{\Omega} \times \hat{\mathbf{k}}') = \boldsymbol{\Omega} \times (B'_x\hat{\mathbf{i}}' + B'_y\hat{\mathbf{j}}' + B'_z\hat{\mathbf{k}}')$$
$$= \boldsymbol{\Omega} \times \mathbf{B}.$$

Equation (5) becomes

$$\left(\frac{d\mathbf{B}}{dt}\right)_{\text{in}} = \left(\frac{d\mathbf{B}}{dt}\right)_{\text{rot}} + \boldsymbol{\Omega} \times \mathbf{B}, \qquad 6$$

which is the desired result.

Since $\mathbf{B}$ is an arbitrary vector, this result is quite general; it can be applied to any vector we choose. It is important to be clear on the meaning of Eq. (6). The vector $\mathbf{B}$ itself is the same in both the inertial and the rotating coordinate systems. (For this reason there is no subscript to $\mathbf{B}$ in the term $\boldsymbol{\Omega} \times \mathbf{B}$.) It is only the time derivative of $\mathbf{B}$ which depends on the coordinate system. For instance, a vector which is constant in one system will change with time in the other.

Problems

8.1 A uniform thin rod of length L and mass M is pivoted at one end. The pivot is attached to the top of a car accelerating at rate A, as shown.

a. What is the equilibrium value of the angle θ between the rod and the top of the car?

b. Suppose that the rod is displaced a small angle ϕ from equilibrium. What is its motion for small ϕ?

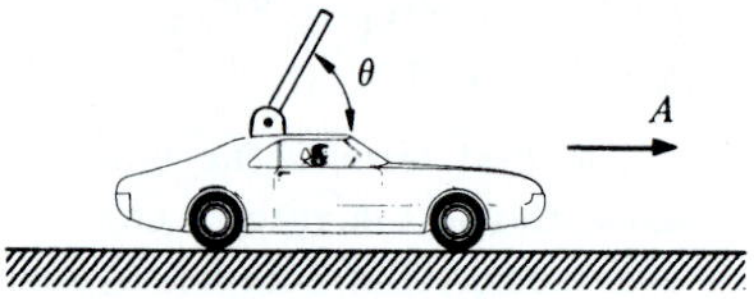

8.2 A truck at rest has one door fully open, as shown. The truck accelerates forward at constant rate A, and the door begins to swing shut.

The door is uniform and solid, has total mass M, height h, and width w. Neglect air resistance.

a. Find the instantaneous angular velocity of the door about its hinges when it has swung through 90°.

b. Find the horizontal force on the door when it has swung through 90°.

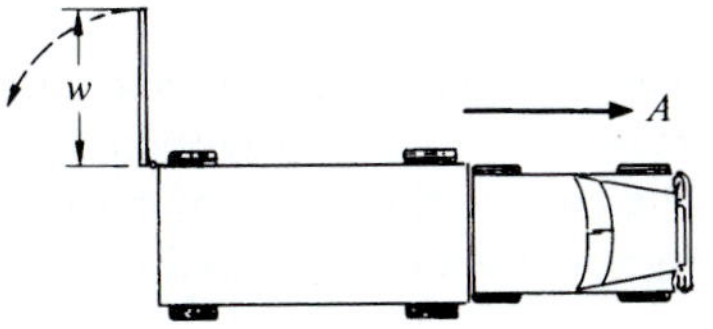

8.3 A pendulum is at rest with its bob pointing toward the center of the earth. The support of the pendulum is moved horizontally with uniform acceleration a, and the pendulum starts to swing. Neglect rotation of the earth. Consider the motion of the pendulum as the pivot moves over a small distance d subtending an angle $\theta_0 \approx d/R_e \ll 1$ at the center of the earth. Show that if the period of the pendulum is $2\pi\sqrt{R_e/g}$, the pendulum will continue to point toward the center of the earth, if effects of order $\theta_0{}^2$ and higher are neglected.

8.4 The center of mass of a 3,200-lb car is midway between the wheels and 2 ft above the ground. The wheels are 8 ft apart.

a. What is the minimum acceleration A of the car so that the front wheels just begin to lift off the ground?

b. If the car decelerates at rate g, what is the normal force on the front wheels and on the rear wheels?

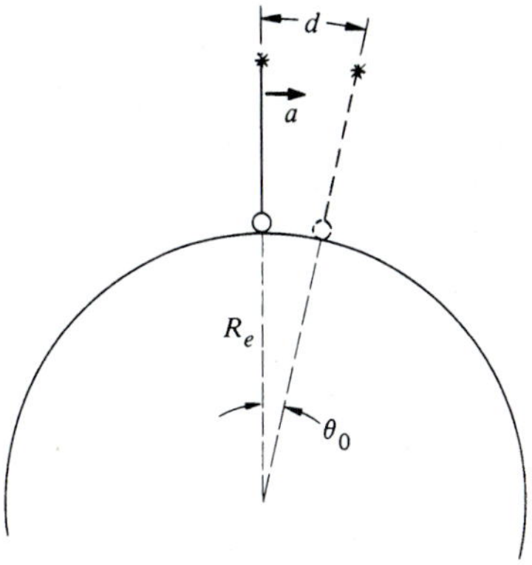

8.5 Many applications for gyroscopes have been found in navigational systems. For instance, gyroscopes can be used to measure acceleration. Consider a gyroscope spinning at high speed ω_s. The gyroscope is attached to a vehicle by a universal pivot P. If the vehicle accelerates in the direction perpendicular to the spin axis at rate a, then the gyroscope will precess about the acceleration axis, as shown in the sketch. The total angle of precession, θ, is measured. Show that if the system starts from rest, the final velocity of the vehicle is given by

$$v = \frac{I_s\omega_s}{Ml}\,\theta,$$

where $I_s\omega_s$ is the gyroscope's spin angular momentum, M is the total mass of the pivoted portion of the gyroscope, and l is the distance from the pivot to the center of mass. (Such a system is called an integrating gyro, since it automatically integrates the acceleration to give the velocity.)

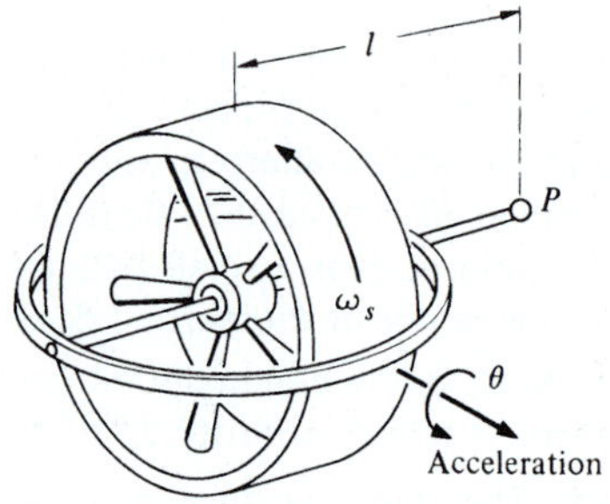

8.6 A top of mass M spins with angular speed ω_S about its axis, as shown. The moment of inertia of the top about the spin axis is I_0, and the center of mass of the top is a distance l from the point. The axis is inclined at angle ϕ with respect to the vertical, and the top is undergoing uniform precession. Gravity is directed downward. The top is in an elevator, with its tip held to the elevator floor by a frictionless pivot. Find the rate of precession, Ω, clearly indicating its direction, in each of the following cases:

a. The elevator at rest

b. The elevator accelerating down at rate $2g$

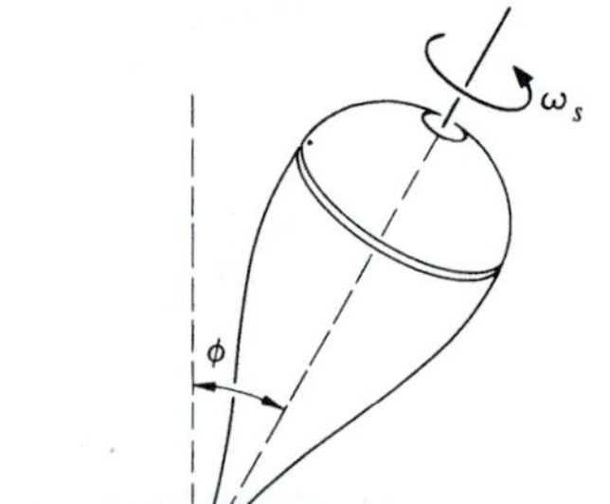

8.7 Find the difference in the apparent force of gravity at the equator and the poles, assuming that the earth is spherical.

8.8 Derive the familiar expression for velocity in plane polar coordinates, $\mathbf{v} = \dot{r}\hat{\mathbf{r}} + r\dot{\theta}\hat{\boldsymbol{\theta}}$, by examining the motion of a particle in a rotating coordinate system in which the velocity is instantaneously radial.

8.9 A 400-ton train runs south at a speed of 60 mi/h at a latitude of 60° north.

a. What is the horizontal force on the tracks?

b. What is the direction of the force?

Ans. (*a*) Approximately 300 lb

8.10 The acceleration due to gravity measured in an earthbound coordinate system is denoted by g. However, because of the earth's rotation, g differs from the true acceleration due to gravity, g_0. Assuming that the earth is perfectly round, with radius R_e and angular velocity Ω_e, find g as a function of latitude λ. (Assuming the earth to be round is actually not justified—the contributions to the variation of g with latitude due to the polar flattening is comparable to the effect calculated here.)

Ans. $g = g_0[1 - (2x - x^2)\cos^2\lambda]^{\frac{1}{2}}$, where $x = R_e\Omega_e^2/g_0$

8.11 A high speed hydrofoil races across the ocean at the equator at a speed of 200 mi/h. Let the acceleration of gravity for an observer at rest on the earth be g. Find the fractional change in gravity, $\Delta g/g$, measured by a passenger on the hydrofoil when the hydrofoil heads in the following directions:

a. East

b. West

c. South

d. North

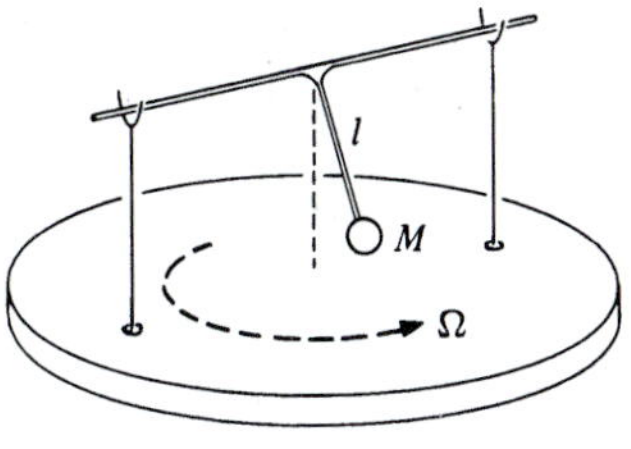

8.12 A pendulum is rigidly fixed to an axle held by two supports so that it can swing only in a plane perpendicular to the axle. The pendulum consists of a mass M attached to a massless rod of length l. The supports are mounted on a platform which rotates with constant angular velocity Ω. Find the pendulum's frequency assuming that the amplitude is small.

9 CENTRAL FORCE MOTION

9.1 Introduction

It was Newton's fascination with planetary motion that led him to formulate his laws of motion and the law of universal gravitation. His success in explaining Kepler's empirical laws of planetary motion was an overwhelming argument in favor of the new mechanics and marked the beginning of modern mathematical physics. Planetary motion and the more general problem of motion under a central force continue to play an important role in most branches of physics and turn up in such topics as particle scattering, atomic structure, and space navigation.

In this chapter we apply newtonian physics to the general problem of central force motion. We shall start by looking at some of the general features of a system of two particles interacting with a central force $f(r)\hat{\mathbf{r}}$, where $f(r)$ is any function of the distance r between the particles and $\hat{\mathbf{r}}$ is a unit vector along the line of centers. After making a simple change of coordinates, we shall show how to find a complete solution by using the conservation laws of angular momentum and energy. Finally, we shall apply these results to the case of planetary motion, $f(r) \propto 1/r^2$, and show how they predict Kepler's empirical laws.

9.2 Central Force Motion as a One Body Problem

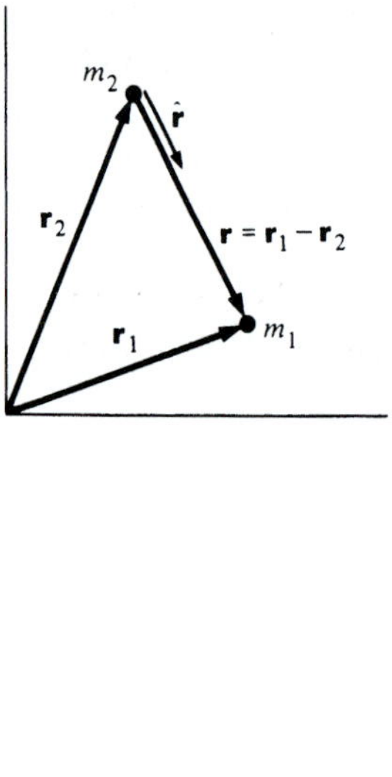

Consider an isolated system consisting of two particles interacting under a central force $f(r)$. The masses of the particles are m_1 and m_2 and their position vectors are $\mathbf{r}_1$ and $\mathbf{r}_2$. We have

$$\begin{aligned} \mathbf{r} &= \mathbf{r}_1 - \mathbf{r}_2 \\ r &= |\mathbf{r}| \\ &= |\mathbf{r}_1 - \mathbf{r}_2|. \end{aligned} \qquad 9.1$$

The equations of motion are

$$m_1\ddot{\mathbf{r}}_1 = f(r)\hat{\mathbf{r}} \qquad 9.2a$$

$$m_2\ddot{\mathbf{r}}_2 = -f(r)\hat{\mathbf{r}}. \qquad 9.2b$$

The force is attractive for $f(r) < 0$ and repulsive for $f(r) > 0$. Equations (9.2a and b) are coupled together by $\mathbf{r}$; the behavior of $\mathbf{r}_1$ and $\mathbf{r}_2$ depends on $\mathbf{r} = \mathbf{r}_1 - \mathbf{r}_2$. We shall show that the problem is easier to handle if we replace $\mathbf{r}_1$ and $\mathbf{r}_2$ by $\mathbf{r} = \mathbf{r}_1 - \mathbf{r}_2$ and the center of mass vector $\mathbf{R} = (m_1\mathbf{r}_1 + m_2\mathbf{r}_2)/(m_1 + m_2)$. The equation of motion for $\mathbf{R}$ is trivial since there are no external forces. The equation for $\mathbf{r}$ turns out to be like the equation of motion of a single particle and has a straightforward solution.

The equation of motion for **R** is

$$\ddot{\mathbf{R}} = 0,$$

which has the simple solution

$$\mathbf{R} = \mathbf{R}_0 + \mathbf{V}t. \qquad 9.3$$

The constant vectors $\mathbf{R}_0$ and $\mathbf{V}$ depend on the choice of coordinate system and the initial conditions. If we are clever enough to take the origin at the center of mass, $\mathbf{R}_0 = 0$ and $\mathbf{V} = 0$.

To find the equation of motion for $\mathbf{r}$ we divide Eq. (9.2a) by m_1 and Eq. (9.2b) by m_2 and subtract. This gives

$$\ddot{\mathbf{r}}_1 - \ddot{\mathbf{r}}_2 = \left(\frac{1}{m_1} + \frac{1}{m_2}\right) f(r)\hat{\mathbf{r}}$$

or

$$\left(\frac{m_1 m_2}{m_1 + m_2}\right)(\ddot{\mathbf{r}}_1 - \ddot{\mathbf{r}}_2) = f(r)\hat{\mathbf{r}}.$$

Denoting $m_1 m_2/(m_1 + m_2)$ by μ, the *reduced mass*, and using $\ddot{\mathbf{r}}_1 - \ddot{\mathbf{r}}_2 = \ddot{\mathbf{r}}$, we have

$$\mu\ddot{\mathbf{r}} = f(r)\hat{\mathbf{r}}. \qquad 9.4$$

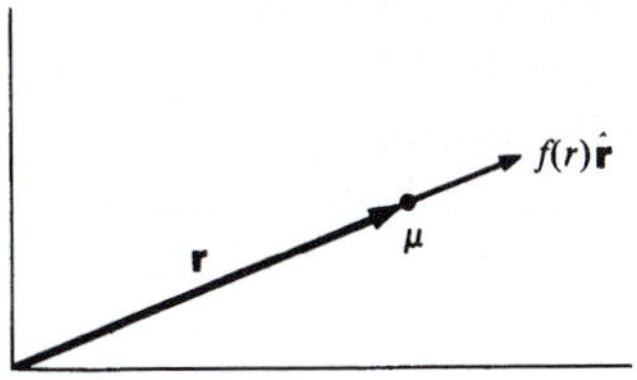

Equation (9.4) is identical to the equation of motion for a particle of mass μ acted on by a force $f(r)\hat{\mathbf{r}}$; no trace of the two particle problem remains. The two particle problem has been transformed to a one particle problem. (Unfortunately, the method cannot be generalized. There is no way to reduce the equations of motion for three or more particles to equivalent one body equations, and for this reason the exact solution of the three body problem is unknown.)

The problem now is to find $\mathbf{r}$ as a function of time from Eq. (9.4). Once we know $\mathbf{r}$, we can easily find $\mathbf{r}_1$ and $\mathbf{r}_2$ by using the relations

$$\mathbf{r} = \mathbf{r}_1 - \mathbf{r}_2 \qquad 9.5a$$

$$\mathbf{R} = \frac{m_1\mathbf{r}_1 + m_2\mathbf{r}_2}{m_1 + m_2}. \qquad 9.5b$$

Solving for $\mathbf{r}_1$ and $\mathbf{r}_2$ gives

$$\mathbf{r}_1 = \mathbf{R} + \left(\frac{m_2}{m_1 + m_2}\right)\mathbf{r} \qquad 9.6a$$

$$\mathbf{r}_2 = \mathbf{R} - \left(\frac{m_1}{m_1 + m_2}\right)\mathbf{r}. \qquad 9.6b$$

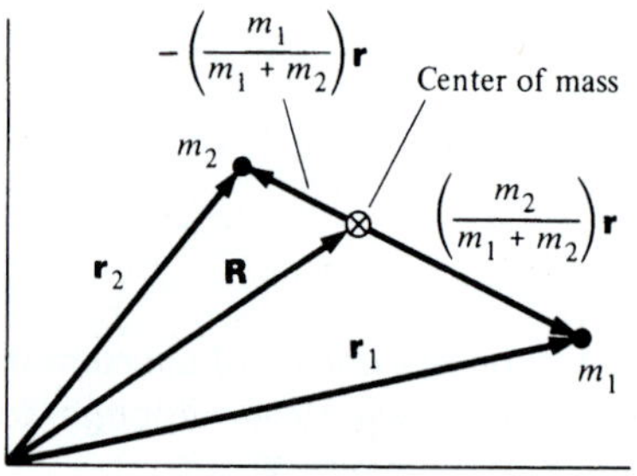

$m_2\mathbf{r}/(m_1 + m_2)$ and $-m_1\mathbf{r}/(m_1 + m_2)$ are the position vectors of m_1 and m_2 relative to the center of mass, as the sketch shows.

The complete solution of $\mu\ddot{\mathbf{r}} = f(r)\,\hat{\mathbf{r}}$ depends on the particular form of $f(r)$. However, a number of the properties of central force motion hold true in general regardless of the form of $f(r)$, and we turn next to investigate these.

9.3 General Properties of Central Force Motion

The equation $\mu\ddot{\mathbf{r}} = f(r)\,\hat{\mathbf{r}}$ is a vector equation, and although only a single particle is involved, there are three components to be considered. In this section we shall see how to use the conservation laws to find some general properties of the solution and to reduce the equation to an equation in a single scalar variable.

The Motion Is Confined to a Plane

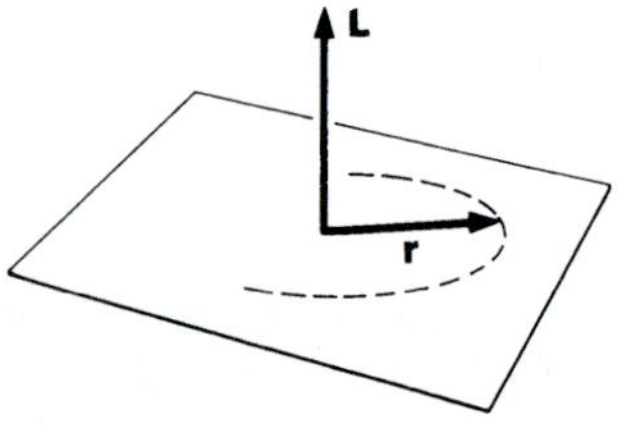

The central force $f(r)\,\hat{\mathbf{r}}$ is along $\mathbf{r}$ and can exert no torque on the reduced mass μ. Hence, the angular momentum $\mathbf{L}$ of μ is constant. It is easy to show that this implies that the motion of μ is confined to a plane. Since $\mathbf{L} = \mathbf{r} \times \mu\mathbf{v}$, where $\mathbf{v} = \dot{\mathbf{r}}$, $\mathbf{r}$ is always perpendicular to $\mathbf{L}$ by the properties of the cross product. However, $\mathbf{L}$ is fixed in space, and it follows that $\mathbf{r}$ can only move in the plane perpendicular to $\mathbf{L}$ through the origin.

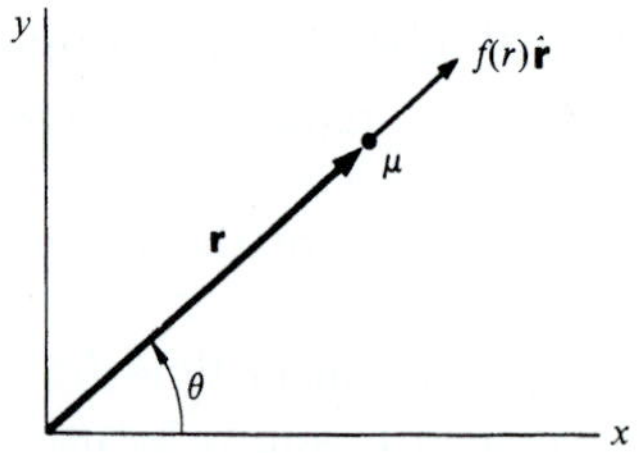

Since the motion is confined to a plane, we can, without loss of generality, choose our coordinate system so that the motion is in the xy plane. Introducing polar coordinates, the equation of motion $\mu\ddot{\mathbf{r}} = f(r)\,\hat{\mathbf{r}}$ becomes

$$\mu(\ddot{r} - r\dot{\theta}^2) = f(r) \qquad 9.7a$$

$$\mu(r\ddot{\theta} + 2\dot{r}\dot{\theta}) = 0. \qquad 9.7b$$

The Energy and Angular Momentum Are Constants of the Motion

We have reduced the problem to two dimensions by using the fact that the direction of $\mathbf{L}$ is constant. There are two other important constants of central force motion: the magnitude of the angular momentum $|\mathbf{L}| \equiv l$, and the total energy E. Using l and E, we can solve the problem of central force motion more easily and with greater physical insight than by working with Eqs. (9.7a and b).

The angular momentum of μ has magnitude

$$l = \mu r v_\theta = \mu r^2\dot{\theta}. \qquad 9.8a$$

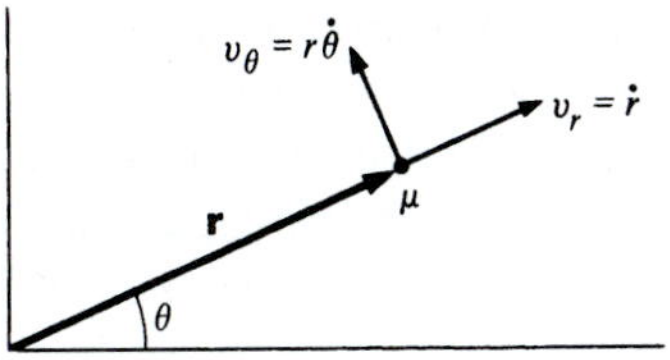

The total energy of μ is

$$E = \tfrac{1}{2}\mu v^2 + U(r)$$
$$= \tfrac{1}{2}\mu(\dot{r}^2 + r^2\dot{\theta}^2) + U(r), \qquad 9.8b$$

where the potential energy $U(r)$ is given by

$$U(r) - U(r_a) = -\int_{r_a}^{r} f(r)\,dr.$$

The constant $U(r_a)$ is not physically significant and so we can leave r_a unspecified; adding a constant to the energy has no effect on the motion.

We can eliminate $\dot{\theta}$ from Eq. (9.8b) by using Eq. (9.8a). The result is

$$E = \frac{1}{2}\mu\dot{r}^2 + \frac{1}{2}\frac{l^2}{\mu r^2} + U(r). \qquad 9.9$$

This looks like the equation of motion of a particle moving in one dimension; all reference to θ is gone. We can press the parallel further by introducing

$$U_{\text{eff}}(r) = \frac{1}{2}\frac{l^2}{\mu r^2} + U(r), \qquad 9.10$$

so that

$$E = \tfrac{1}{2}\mu\dot{r}^2 + U_{\text{eff}}(r). \qquad 9.11$$

U_{eff} is called the *effective potential energy*. Often it is referred to simply as the *effective potential*. U_{eff} differs from the true potential $U(r)$ by the term $l^2/2\mu r^2$, called the *centrifugal potential*.

The formal solution of Eq. (9.11) is

$$\frac{dr}{dt} = \sqrt{\frac{2}{\mu}(E - U_{\text{eff}})} \qquad 9.12$$

or

$$\int_{r_0}^{r} \frac{dr}{\sqrt{(2/\mu)(E - U_{\text{eff}})}} = t - t_0. \qquad 9.13$$

Equation (9.13) gives us r as a function of t, although the integral may have to be done numerically in some cases. To find θ as a function of t, we can use the solution for r in Eq. (9.8a):

$$\frac{d\theta}{dt} = \frac{l}{\mu r^2}. \qquad 9.14$$

Since r is known as a function of t from Eq. (9.13), it is possible to integrate to find θ:

$$\theta - \theta_0 = \int_{t_0}^{t} \frac{l}{\mu r^2}\,dt. \qquad 9.15$$

Often we are interested in the path of the particle, which means knowing r as a function of θ rather than as a function of time. We call $r(\theta)$ the *orbit* of the particle. (The term is used even if the trajectory does not close on itself.) Dividing Eq. (9.14) by Eq. (9.12) gives

$$\frac{d\theta}{dr} = \frac{l}{\mu r^2} \frac{1}{\sqrt{(2/\mu)(E - U_{\text{eff}})}}. \qquad 9.16$$

This completes the formal solution of the central force problem. We can obtain $r(t)$, $\theta(t)$, or $r(\theta)$ as we please; all we need to do is evaluate the appropriate integrals.

You may have noticed that we found the solution without using the equations of motion, Eqs. (9.7a and b). Actually, we did use them, but in a disguised form. For instance, differentiating $l = \mu r^2\dot{\theta}$ with respect to time gives $0 = \mu r^2\ddot{\theta} + 2r\dot{r}\dot{\theta}$ or

$$\mu(r\ddot{\theta} + 2\dot{r}\dot{\theta}) = 0,$$

which is identical to the tangential equation of motion, Eq. (9.7b). Similarly, differentiation of the energy equation with respect to time gives the radial equation of motion, Eq. (9.7a).

The Law of Equal Areas

We have already shown in Example 6.3 that for any central force, **r** sweeps out equal areas in equal times. This general property of central force motion is a direct consequence of the fact that the angular momentum is constant.

9.4 Finding the Motion in Real Problems

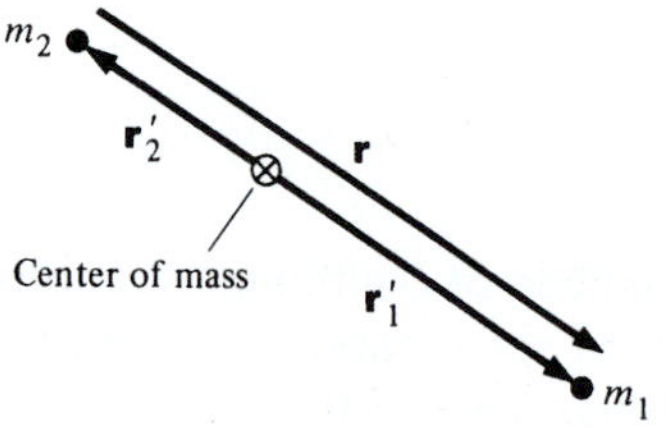

In order to apply the solution for the motion which we found in the last section, we need to relate the position vectors of m_1 and m_2 to **r** and evaluate l and E.

From Eqs. (9.6a and b) the position vectors of m_1 and m_2 relative to the center of mass are

$$\mathbf{r}_1' = \frac{m_2}{m_1 + m_2}\mathbf{r} \qquad 9.17a$$

$$\mathbf{r}_2' = -\frac{m_1}{m_1 + m_2}\mathbf{r}. \qquad 9.17b$$

$\mathbf{r}_1'$ and $\mathbf{r}_2'$ lie along **r**. They remain back to back in the plane of motion. Hence, m_1 and m_2 move about their center of mass in the fixed plane, separated by distance r.

In many problems, like the motion of a planet around the sun, the masses of the two particles are very different. If $m_2 \gg m_1$, Eqs. (9.17a and b) become

$$\mathbf{r}_1' \approx \mathbf{r}$$
$$\mathbf{r}_2' \approx 0.$$

The reduced mass μ is approximately m_1, and the center of mass lies at m_2. In this case the more massive particle is essentially fixed at the origin, and there is no important difference between the actual two particle problem and the equivalent one particle problem.

In the one particle problem the angular momentum is

$$\mathbf{L} = \mu\mathbf{r} \times \mathbf{v}.$$

It is easy to show that $\mathbf{L}$ is simply the angular momentum of m_1 and m_2 about the center of mass, $\mathbf{L}_c$.

$$\mathbf{L}_c = m_1\mathbf{r}_1' \times \mathbf{v}_1' + m_2\mathbf{r}_2' \times \mathbf{v}_2',$$

where $\mathbf{v}_1' = \dot{\mathbf{r}}_1'$ and $\mathbf{v}_2' = \dot{\mathbf{r}}_2'$. Using Eqs. (9.17$a$ and b) we have

$$\begin{aligned}\mathbf{L}_c &= \frac{m_1m_2}{m_1+m_2}\mathbf{r} \times \mathbf{v}_1' - \frac{m_1m_2}{m_1+m_2}\mathbf{r} \times \mathbf{v}_2' \\ &= \mu\mathbf{r} \times (\mathbf{v}_1' - \mathbf{v}_2') \\ &= \mu\mathbf{r} \times \mathbf{v} \\ &= \mathbf{L}.\end{aligned}$$

Similarly, the total energy E is the energy of m_1 and m_2 relative to their center of mass, E_c.

$$E_c = \tfrac{1}{2}m_1(\mathbf{v}_1' \cdot \mathbf{v}_1') + \tfrac{1}{2}m_2(\mathbf{v}_2' \cdot \mathbf{v}_2') + U(r).$$

From Eqs. (9.17a and b), we have $m_1\mathbf{v}_1' = \mu\mathbf{v}$ and $m_2\mathbf{v}_2' = -\mu\mathbf{v}$. Hence,

$$\begin{aligned}E_c &= \tfrac{1}{2}\mu\mathbf{v} \cdot (\mathbf{v}_1' - \mathbf{v}_2') + U(r) \\ &= \tfrac{1}{2}\mu(\mathbf{v} \cdot \mathbf{v}) + U(r) \\ &= E.\end{aligned}$$

9.5 The Energy Equation and Energy Diagrams

In Sec. 9.3 we found two equivalent ways of writing E, the total energy in the center of mass system. According to Eq. (9.8b),

$$E = \tfrac{1}{2}\mu v^2 + U(r),$$

and according to Eq. (9.11),

$$E = \tfrac{1}{2}\mu\dot{r}^2 + U_{\text{eff}}(r).$$

We generally need to use both these forms in analyzing central force motion. The first form, $\frac{1}{2}\mu v^2 + U(r)$, is handy for evaluating E; all we need to know is the relative speed and position at some instant. However, $v^2 = \dot{r}^2 + (r\dot{\theta})^2$, and this dependence on two coordinates, r and θ, makes it difficult to visualize the motion. In contrast, the second form, $\frac{1}{2}\mu\dot{r}^2 + U_{\text{eff}}(r)$ depends on the single coordinate r. In fact, it is identical to the equation for the energy of a particle of mass μ constrained to move along a straight line with kinetic energy $\frac{1}{2}\mu\dot{r}^2$ and potential energy $U_{\text{eff}}(r)$. The coordinate θ is completely suppressed—the kinetic energy associated with the tangential motion, $\frac{1}{2}\mu(r\dot{\theta})^2$, is accounted for in the effective potential by the relations

$$\frac{1}{2}\mu(r\dot{\theta})^2 = \frac{l^2}{2\mu r^2}$$

$$U_{\text{eff}}(r) = \frac{l^2}{2\mu r^2} + U(r).$$

The equation

$$E = \frac{1}{2}\mu\dot{r}^2 + U_{\text{eff}}(r)$$

involves only the radial motion. Consequently, we can use the energy diagram technique developed in Chap. 4 to find the qualitative features of the radial motion.

To see how the method works, let's start by looking at a very simple system, two noninteracting particles.

Example 9.1 Noninteracting Particles

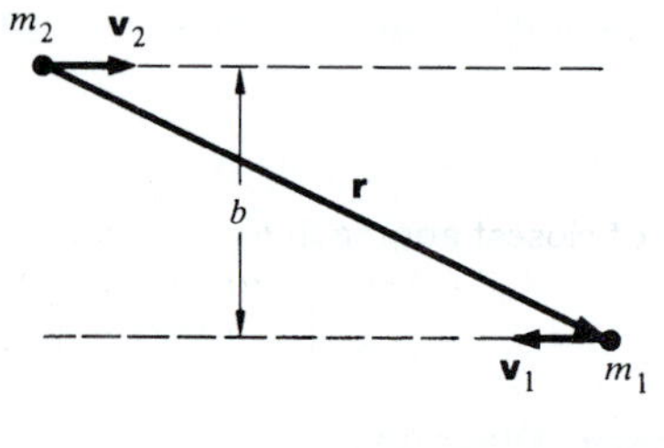

Two noninteracting particles m_1 and m_2 move toward each other with velocities $\mathbf{v}_1$ and $\mathbf{v}_2$. Their paths are offset by distance b, as shown in the sketch. Let us investigate the equivalent one body description of this system.

The relative velocity is

$$\begin{aligned}\mathbf{v}_0 &= \dot{\mathbf{r}} \\ &= \dot{\mathbf{r}}_1 - \dot{\mathbf{r}}_2 \\ &= \mathbf{v}_1 - \mathbf{v}_2.\end{aligned}$$

$\mathbf{v}_0$ is constant since $\mathbf{v}_1$ and $\mathbf{v}_2$ are constant. The energy of the system relative to the center of mass is

$$E = \frac{1}{2}\mu v_0^2 + U(r) = \frac{1}{2}\mu v_0^2,$$

since $U(r) = 0$ for noninteracting particles.

In order to draw the energy diagram we need to find the effective potential

$$U_{\text{eff}} = \frac{l^2}{2\mu r^2} + U(r) = \frac{l^2}{2\mu r^2}.$$

We could evaluate l by direct computation, but it is simpler to use the relation

$$E = \tfrac{1}{2}\mu\dot{r}^2 + \frac{l^2}{2\mu r^2}$$
$$= \tfrac{1}{2}\mu v_0^2.$$

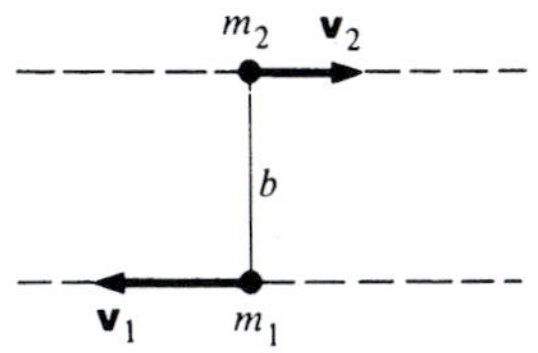

When m_1 and m_2 pass each other, $r = b$ and $\dot{r} = 0$. Hence

$$\frac{l^2}{2\mu b^2} = \tfrac{1}{2}\mu v_0^2,$$

$$l = \mu b v_0,$$

and

$$U_{\text{eff}} = \tfrac{1}{2}\mu v_0^2 \frac{b^2}{r^2}.$$

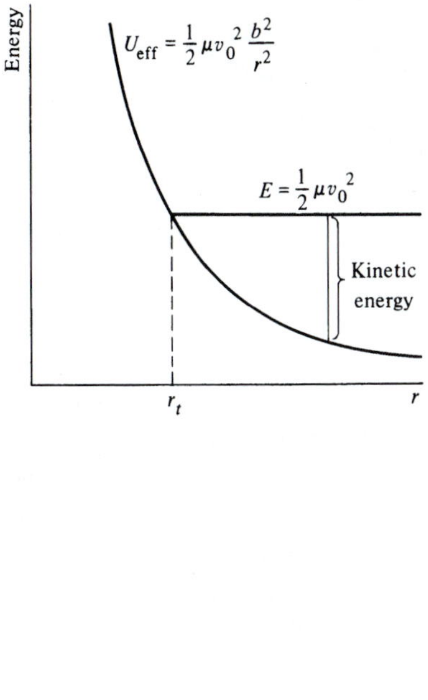

The energy diagram is shown in the sketch. The kinetic energy associated with radial motion is

$$K = \tfrac{1}{2}\mu\dot{r}^2$$
$$= E - U_{\text{eff}}.$$

K is never negative so that the motion is restricted to regions where $E - U_{\text{eff}} \geq 0$. Initially r is very large. As the particles approach, the kinetic energy decreases, vanishing at the *turning point* r_t, where the radial velocity is zero and the motion is purely tangential. At the turning point $E = U_{\text{eff}}(r_t)$, which gives

$$\tfrac{1}{2}\mu v_0^2 = \tfrac{1}{2}\mu v_0^2 \frac{b^2}{r_t^2}$$

or

$$r_t = b$$

as we expect, since r_t is the distance of closest approach of the particles. Once the turning point is passed, r increases and the particles separate. In our one dimensional picture, the particle μ "bounces off" the barrier of the effective potential.

Now let us apply energy diagrams to the meatier problem of planetary motion. For the attractive gravitational force,

$$f(r) = -\frac{Gm_1m_2}{r^2}$$

$$U(r) = -\frac{Gm_1m_2}{r}.$$

(By the usual convention, we take $U(\infty) = 0$.) The effective potential energy is

$$U_{\text{eff}} = -\frac{Gm_1m_2}{r} + \frac{l^2}{2\mu r^2}.$$

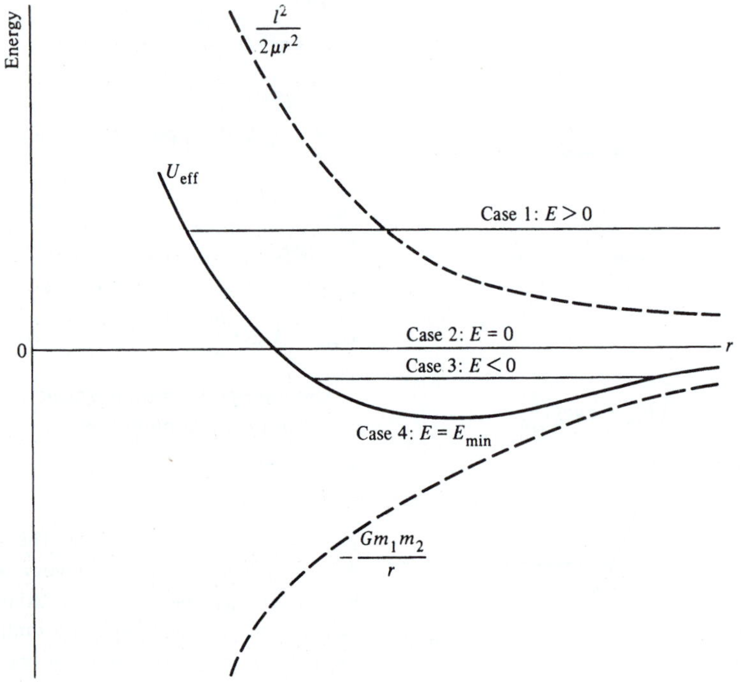

If $l \neq 0$, the repulsive centrifugal potential $l^2/(2\mu r^2)$ dominates at small r, whereas the attractive gravitational potential $-Gm_1m_2/r$ dominates at large r. The drawing shows the energy diagram with various values of the total energy. The kinetic energy of radial motion is $K = E - U_{\text{eff}}$, and the motion is restricted to regions where $K \geq 0$. The nature of the motion is determined by the total energy. Here are the various possibilities:

1. $E > 0$: r is unbounded for large values but must exceed a certain minimum if $l \neq 0$. The particles are kept apart by the "centrifugal barrier."

2. $E = 0$: This is qualitatively similar to case 1 but on the boundary between unbounded and bounded motion.

3. $E < 0$: The motion is bounded for both large and small r. The two particles form a bound system.

4. $E = E_{\min}$: r is restricted to one value. The particles stay a constant distance from one another.

In the next section we shall find that case 1 corresponds to motion in a hyperbola; case 2, to a parabola; case 3, to an ellipse; and case 4, to a circle.

There is one other possibility, $l = 0$. In this case the particles move along a straight line on a collision course, since when l is zero there is no centrifugal barrier to hold them apart.

Example 9.2 **The Capture of Comets**

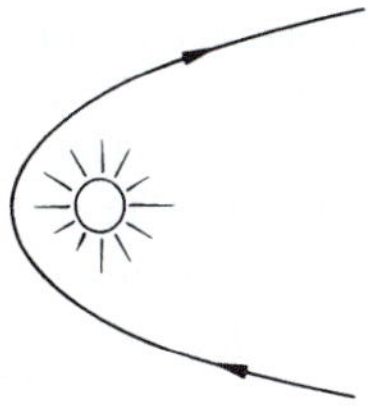

Suppose that a comet with $E > 0$ drifts into the solar system. From our discussion of the energy diagram for motion under a gravitational force, the comet will approach the sun and then swing away, never to return. In order for the comet to become a member of the solar system, its energy would have to be reduced to a negative value. However, the gravitational force is conservative and the comet's total energy cannot change.

The situation is quite different if more than two bodies are involved. For instance, if the comet is deflected by a massive planet like Jupiter, it can transfer energy to the planet and so become trapped in the solar system.

Suppose that a comet is heading outward from the sun toward the orbit of Jupiter, as shown in the sketch. Let the velocity of the comet before it starts to interact appreciably with Jupiter be $\mathbf{v}_i$, and let Jupiter's velocity be $\mathbf{V}$. For simplicity we shall assume that the orbits are not appreciably deflected by the sun during the time of interaction.

In the comet-Jupiter center of mass system Jupiter is essentially at rest, and the center of mass velocity of the comet is $\mathbf{v}_{ic} = \mathbf{v}_i - \mathbf{V}$, as shown in figure a.

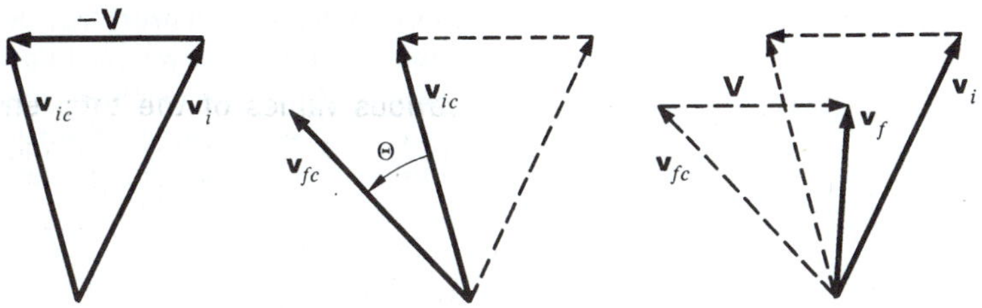

In the center of mass system the path of the comet is deflected, but the final speed is equal to the initial speed v_{ic}. Hence, the interaction merely rotates $\mathbf{v}_{ic}$ through some angle Θ to a new direction $\mathbf{v}_{fc}$, as shown in Fig. b. The final velocity in the space fixed system is

$$\mathbf{v}_f = \mathbf{v}_{fc} + \mathbf{V}.$$

Figure c shows $\mathbf{v}_f$ and, for comparison, $\mathbf{v}_i$. For the deflection shown, $v_f < v_i$, and the comet's energy has decreased. Conversely, if the deflection is in the opposite direction, interaction with Jupiter would increase the energy, possibly freeing a bound comet from the solar system. A large proportion of known comets have energies close to zero, so close that it is often difficult to determine from observations whether the orbit is elliptic ($E < 0$) or hyperbolic ($E > 0$). The interaction of a comet with Jupiter is therefore often sufficient to change the orbit from unbound to bound, or vice versa.

This mechanism for picking up energy from a planet can be used to accelerate an interplanetary spacecraft. By picking the orbit cleverly, the spacecraft can "hop" from planet to planet with a great saving in fuel.

The process we have described may seem to contradict the idea that the gravitational force is strictly conservative. Only gravity acts on the comet and yet its total energy can change. The reason is that the comet experiences a time-dependent gravitational force, and time-dependent forces are intrinsically nonconservative. Nevertheless, the total energy of the entire system is conserved, as we expect.

Example 9.3 Perturbed Circular Orbit

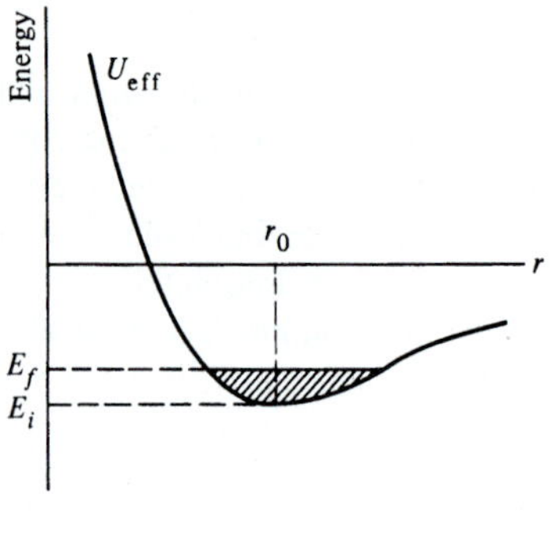

A satellite of mass m orbits the earth in a circle of radius r_0. One of its engines is fired briefly toward the center of the earth, changing the energy of the satellite but not its angular momentum. The problem is to find the new orbit.

The energy diagram shows the initial energy E_i and the final energy E_f. Note that firing the engine radially does not change the effective potential because l is not altered. Since the earth's mass M_e is much greater than m, the reduced mass is nearly m and the earth is effectively fixed.

If E_f is not much greater than E_i, the energy diagram shows that r never differs much from r_0. Rather than solve the planetary motion problem exactly, as we shall do in the next section, we instead approximate $U_{\text{eff}}(r)$ in the neighborhood of r_0 by a parabolic potential. As we know from our analysis of small oscillations of a particle about equilibrium, Sec. 4.10, the resulting radial motion of the satellite will be simple harmonic motion about r_0 to good accuracy.

The effective potential is, with $C \equiv GmM_e$,

$$U_{\text{eff}}(r) = -\frac{C}{r} + \frac{l^2}{2mr^2}.$$

The minimum of U_{eff} is at $r = r_0$. Since the slope is zero there, we have

$$\left.\frac{dU_{\text{eff}}}{dr}\right|_{r_0} = 0$$

$$= \frac{C}{r_0^2} - \frac{l^2}{mr_0^3},$$

which gives

$$l = \sqrt{mCr_0}. \tag{1}$$

(This result can also be found by applying Newton's second law to circular motion.) As we recall from Sec. 4.10, the frequency of oscillation of the system, which we shall denote by β, is

$$\beta = \sqrt{\frac{k}{m}},$$

where

$$k = \left.\frac{d^2U_{\text{eff}}}{dr^2}\right|_{r_0}. \tag{2}$$

This is readily evaluated to yield

$$\beta = \sqrt{\frac{C}{mr_0^3}} = \frac{l}{mr_0^2}. \tag{3}$$

Hence, the radial position is given by

$$r = r_0 + A \sin \beta t. \tag{4}$$

We have omitted the term $B \cos \beta t$ in order to satisfy the initial condition $r(0) = r_0$. Although we could calculate the amplitude A in terms of E_f, we shall not bother with the algebra here except to note that $A \ll r_0$ for E_f nearly equal to E_i.

To find the new orbit, we must eliminate t and express r as a function of θ. For the circular orbit,

$$\dot{\theta} = \frac{l}{mr_0^2}, \qquad \text{or} \tag{5}$$

$$\theta = \left(\frac{l}{mr_0^2}\right)t. \tag{6}$$

Equation (5) is accurate enough for our purposes, even though the radius oscillates slightly after the engine is fired; t occurs only in a small correction term to r in Eq. (4), and we are neglecting terms of order A and higher.

From Eqs. (1) and (5) we see that the frequency of rotation of the satellite around the earth is

$$\frac{l}{mr_0^2} = \frac{\sqrt{mCr_0}}{mr_0^2} = \sqrt{\frac{C}{mr_0^3}}$$

and

$$\theta = \frac{l}{mr_0^2}t = \beta t. \tag{7}$$

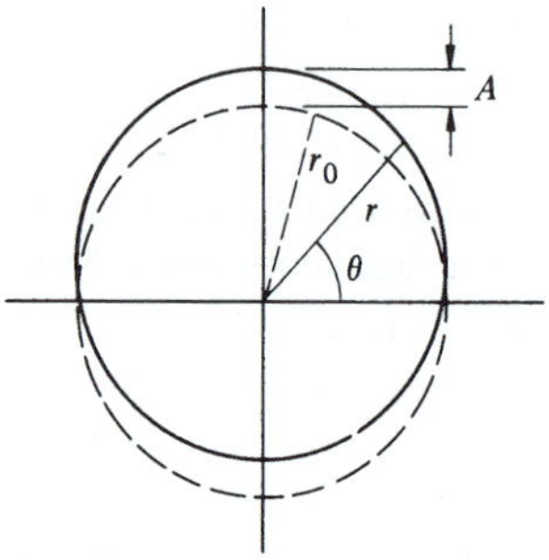

Surprisingly, the frequency of rotation is identical to the frequency of radial oscillation. If we substitute Eq. (7) in Eq. (4), we obtain

$$r = r_0 + A \sin \theta. \tag{8}$$

The new orbit is shown as the solid line in the sketch. The orbit looks almost circular, but it is no longer centered on the earth.

As we shall show in Sec. 9.6, the exact orbit for $E = E_f$ is an ellipse with the equation

$$r = \frac{r_0}{1 - (A/r_0) \sin \theta}.$$

If $A/r_0 \ll 1$,

$$\begin{aligned} r &= \frac{r_0}{1 - (A/r_0) \sin \theta} \\ &\approx r_0 \left(1 + \frac{A}{r_0} \sin \theta\right) \\ &= r_0 + A \sin \theta. \end{aligned}$$

To first order in A, Eq. (8) is the equation of an ellipse. However, the exact calculation is harder to derive (and to digest) than is the approximate result we found by using the energy diagram.

9.6 Planetary Motion

Let us now solve the main problem of the chapter—finding the orbit for the gravitational interaction

$$U(r) = -G\frac{Mm}{r} \equiv -\frac{C}{r},$$

where M is the mass of the sun and m is the mass of a planet. Alternatively, M could be the mass of a planet and m the mass of a satellite. Before proceeding with the calculation, it might be useful to consider whether or not this is a realistic description of the interaction of the sun and a planet. If both bodies were homogeneous spheres, they would interact like point particles as we saw in Note 2.1, and our formula would be exact. However, most of the members of the solar system are neither perfectly homogeneous nor perfectly spherical. For example, satellites around the moon are perturbed by mass concentrations ("mascons") in the moon, and the planet Mercury may be slightly perturbed by an equatorial bulge of the sun. Furthermore, the solar system is by no means a two body system. Each planet is attracted by all the other planets as well as by the sun.

Fortunately, none of these effects is particularly large. Most of the mass of the solar system is in the sun, so that the attraction of the planets for each other is quite feeble. The largest interaction is between Jupiter and Saturn. The effect of this perturbation is chiefly to change the speed of each planet, so that the law of equal areas no longer holds exactly. However, the perturbation never shifts Jupiter by more than a few minutes of arc from its expected position (one minute of arc is approximately equal to one-thirtieth the moon's diameter as seen from the earth). In practice, one first calculates planetary orbits neglecting the other planets and then calculates small corrections to the orbits due to their presence. Such a procedure is called a perturbation method. (The transuranic planets were actually discovered by their small perturbing effects on the orbits of the known outer planets.) Furthermore, if a body is not quite homogeneous or spherically symmetric, its gravitational field can be shown to have terms depending on $1/r^3$, $1/r^4$, etc., in addition to the main $1/r^2$ term. The coefficients depend on the size of the body compared with r; over the span of the solar system the higher order terms become negligible, although they may be important for a nearby satellite.

Returning to our idealized planetary motion problem $U(r) = -C/r$, we find that the equation for the orbit Eq. (9.16) becomes, using indefinite integrals,

$$\theta - \theta_0 = l \int \frac{dr}{r(2\mu E r^2 + 2\mu C r - l^2)^{\frac{1}{2}}},$$

where θ_0 is a constant of integration. The integral over r is listed in tables of integrals. The result is

$$\theta - \theta_0 = \arcsin\left(\frac{\mu C r - l^2}{r\sqrt{\mu^2 C^2 + 2\mu E l^2}}\right)$$

or

$$\mu C r - l^2 = r\sqrt{\mu^2 C^2 + 2\mu E l^2}\sin(\theta - \theta_0).$$

Solving for r,

$$r = \frac{(l^2/\mu C)}{1 - \sqrt{1 + (2El^2/\mu C^2)}\sin(\theta - \theta_0)}. \qquad 9.18$$

The usual convention is to take $\theta_0 = -\pi/2$ and to introduce the parameters

$$r_0 \equiv \frac{l^2}{\mu C} \qquad 9.19$$

$$\epsilon \equiv \sqrt{1 + \frac{2El^2}{\mu C^2}}. \qquad 9.20$$

Physically, r_0 is the radius of the circular orbit corresponding to the given values of l, μ, and C. The dimensionless parameter ϵ, called the *eccentricity*, characterizes the shape of the orbit, as we shall see. With these replacements, Eq. (9.18) becomes

$$r = \frac{r_0}{1 - \epsilon \cos \theta}. \qquad 9.21$$

Equation (9.21) looks more familiar in cartesian coordinates $x = r \cos \theta$, $y = r \sin \theta$. Rewriting it in the form $r - \epsilon r \cos \theta = r_0$, we have

$$\sqrt{x^2 + y^2} - \epsilon x = r_0$$

or

$$(1 - \epsilon^2)x^2 - 2r_0\epsilon x + y^2 = r_0^2. \qquad 9.22$$

Here are the possibilities:

1. $\epsilon > 1$: The coefficients of x^2 and y^2 are unequal and opposite in sign; the equation has the form $y^2 - Ax^2 - Bx =$ constant, which is the equation of a *hyperbola*. From Eq. (9.20), $\epsilon > 1$ whenever $E > 0$.

2. $\epsilon = 1$: Eq. (9.22) becomes

$$x = \frac{y^2}{2r_0} - \frac{r_0}{2}.$$

This is the equation of a *parabola*. $\epsilon = 1$ when $E = 0$.

3. $0 \leq \epsilon < 1$: The coefficients of x^2 and y^2 are unequal but of the same sign; the equation has the form $y^2 + Ax^2 - Bx =$ constant, which is the equation of an *ellipse*. The term linear in x means that the geometric center of the ellipse is not at the origin of coordinates. As proved in Note 9.1, one focus of the ellipse is at the origin. For $\epsilon < 1$, the allowed values of E are

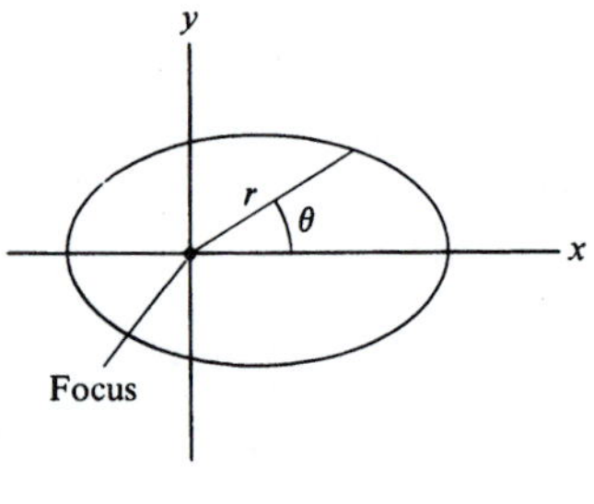

$$-\frac{\mu C^2}{2l^2} \leq E < 0.$$

When $E = -\mu C^2/2l^2$, $\epsilon = 0$ and the equation of the orbit becomes $x^2 + y^2 = r_0^2$; the ellipse degenerates to a *circle*.

Example 9.4 **Hyperbolic Orbits**

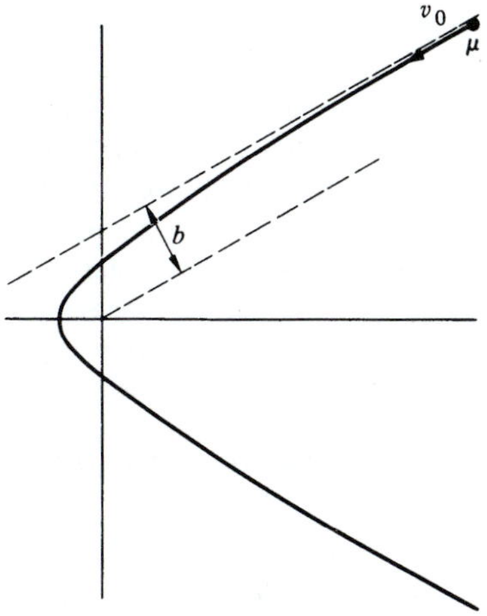

In order to use the orbit equation we must be able to express the orbit in terms of experimentally accessible parameters. For example, if the orbit is unbound, we might know the energy and the initial trajectory.

In this example we shall show how to relate some experimental parameters to the trajectory for the case of a hyperbolic orbit. The results could apply to the motion of a comet about the sun, or to the trajectory of a charged particle scattering off an atomic nucleus.

Let the speed of μ be v_0 when μ is far from the origin, and let the initial path pass the origin at distance b, as shown. b is commonly called the *impact parameter.* The angular momentum l and energy E are

$$l = \mu v_0 b$$
$$E = \tfrac{1}{2}\mu v_0^2.$$

For an inverse square force, $U(r) = -C/r$ and the equation of the orbit is

$$r = \frac{r_0}{1 - \epsilon\cos\theta},$$

where

$$r_0 = \frac{l^2}{\mu C} = \frac{\mu v_0^2 b^2}{C} = \frac{2Eb^2}{C}.$$

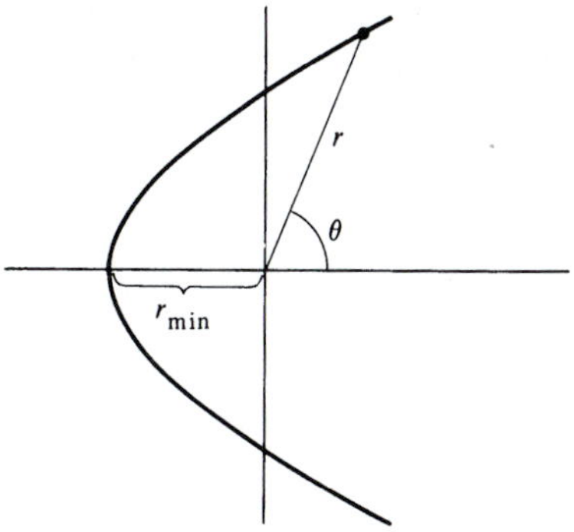

and

$$\epsilon = \sqrt{1 + \frac{2El^2}{\mu C^2}} = \sqrt{1 + \left(\frac{2Eb}{C}\right)^2}.$$

When $\theta = \pi$, $r = r_{\min}$,

$$r_{\min} = \frac{r_0}{1+\epsilon} = \frac{2Eb^2/C}{1 + \sqrt{1 + (2Eb/C)^2}}.$$

For $E \to \infty$, $r_{\min} \to b$. Hence $0 < r_{\min} < b$.

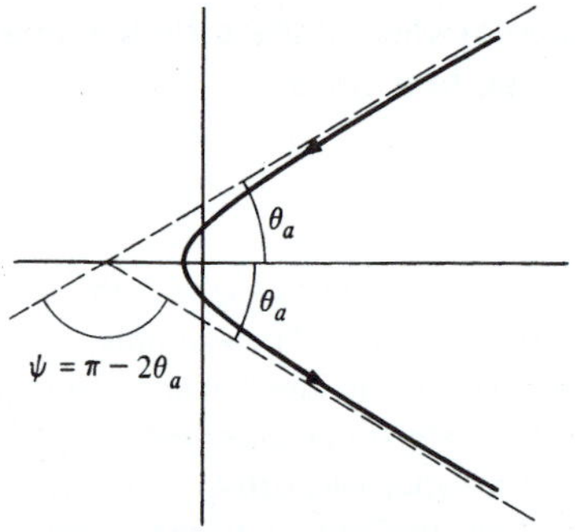

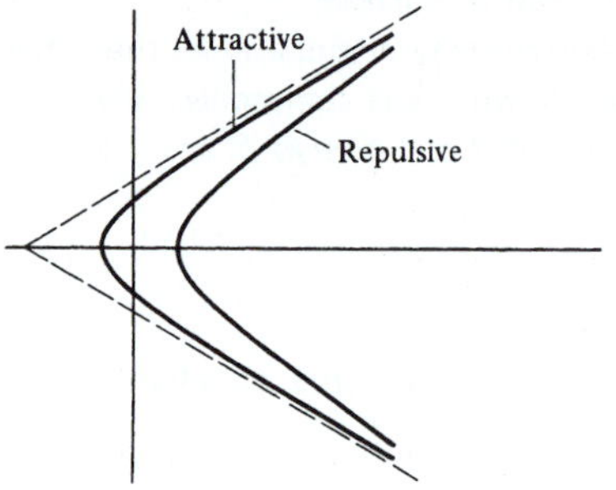

The angle of the asymptotes θ_a can be found from the orbit equation by letting $r \to \infty$. We find

$$\cos\,\theta_a = \frac{1}{\epsilon}.$$

In the interaction, μ is deflected through the angle $\psi = \pi - 2\theta_a$. The deflection angle ψ approaches 180° if $(2Eb/C)^2 \ll 1$.

Rutherford's classic experiment that established the nuclear model of the atom showed that fast alpha particles (doubly charged helium nuclei) interact with single atoms in thin gold foils according to the Coulomb potential $U(r) = -C'/r$. He found that the alpha particles followed hyperbolic orbits even when r_{min} was much less than the radius of the atom, proving that the charge of an atom must be concentrated in a small volume, the nucleus. Surprisingly, Rutherford was unable to determine whether the gold nuclei attracted ($C' > 0$) or repelled ($C' < 0$) alpha particles. The eccentricity, hence the scattering angle, depends on $(2Eb/C')^2$, making it impossible to determine the algebraic sign of the strength parameter C'.

Elliptical orbits ($E < 0$, $0 \leq \epsilon < 1$) are so important it is worth looking at their properties in more detail. From the orbit equation, Eq. (9.21),

$$r = \frac{r_0}{1 - \epsilon\cos\theta},$$

The maximum value of r occurs at $\theta = 0$:

$$r_{\text{max}} = \frac{r_0}{1 - \epsilon}. \qquad 9.23$$

the minimum value of r occurs at $\theta = \pi$:

$$r_{\text{min}} = \frac{r_0}{1 + \epsilon}. \qquad 9.24$$

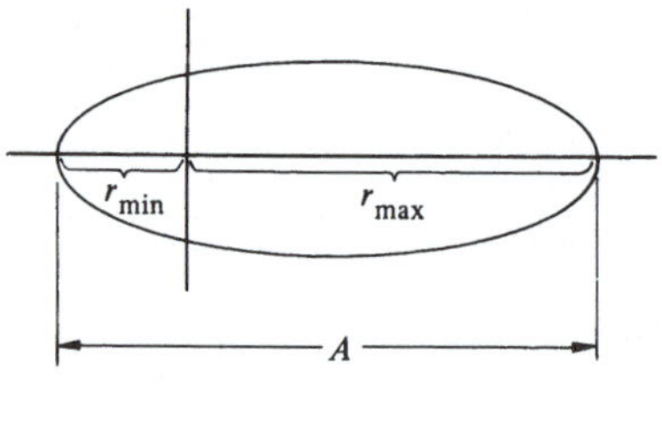

The length of the major axis is

$$\begin{aligned} A &= r_{\text{min}} + r_{\text{max}} \\ &= r_0\left(\frac{1}{1+\epsilon} + \frac{1}{1-\epsilon}\right) \\ &= \frac{2r_0}{1-\epsilon^2}. \qquad 9.25 \end{aligned}$$

Expressing r_0 and ϵ in terms of E, l, μ, C by Eq. (9.19) and (9.20) gives

$$A = \frac{2r_0}{1-\epsilon^2}$$
$$= \frac{2l^2/(\mu C)}{1-[1+2El^2/(\mu C^2)]}$$
$$= \frac{C}{(-E)}. \qquad 9.26$$

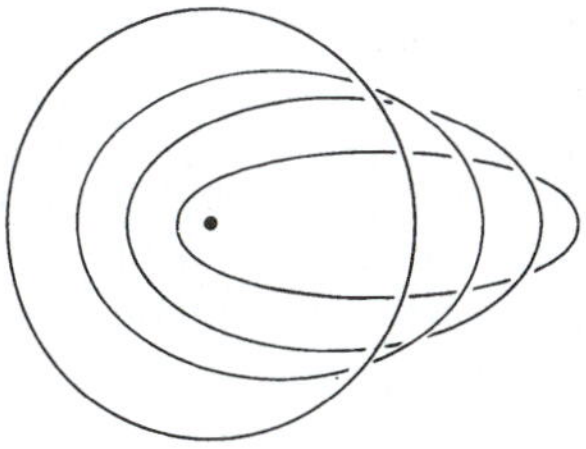

The length of the major axis is independent of l; orbits with the same major axis have the same energy. For instance, all the orbits in the sketch correspond to the same value of E.

The ratio r_{max}/r_{min} is

$$\frac{r_{max}}{r_{min}} = \frac{r_0/(1-\epsilon)}{r_0/(1+\epsilon)}$$
$$= \frac{1+\epsilon}{1-\epsilon}.$$

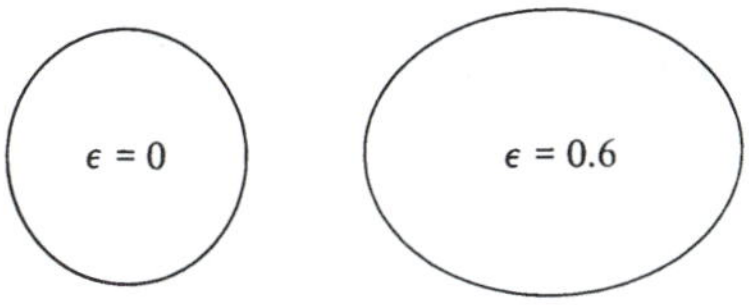

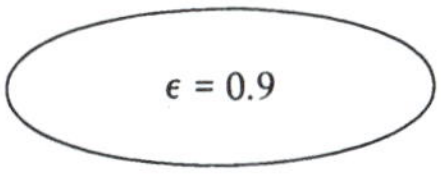

When ϵ is near zero, $r_{max}/r_{min} \approx 1$ and the ellipse is nearly circular. When ϵ is near 1, the ellipse is very elongated. The shape of the ellipse is determined entirely by ϵ; r_0 only supplies the scale.

Table 9.1 gives the eccentricities of the orbits of the planets and Halley's comet. The table reveals why the Ptolemaic theory of circles moving on circles was reasonably successful in dealing with early observations. All the planetary orbits, except those of Mercury and Pluto, have eccentricities near zero and are nearly circular. Mercury is never far from the sun and is hard to observe, and Pluto was not discovered until 1930, so that neither of these

TABLE 9.1

PLANET	ECCENTRICITY
Mercury	0.206
Venus	0.007
Earth	0.017
Mars	0.093
Jupiter	0.048
Saturn	0.055
Uranus	0.051
Neptune	0.007
Pluto	0.252
Halley's Comet	0.967

planets was an impediment to the Ptolemaists. Mars has the most eccentric orbit of the easily observable planets, and its motion was a stumbling block to the Ptolemaic theory. Kepler discovered his laws of planetary motion by trying to fit his calculations to Brahe's accurate observations of Mars' orbit.

Note 9.1 derives the geometric properties of elliptical orbits. We turn now to some examples.

Example 9.5 **Satellite Orbit**

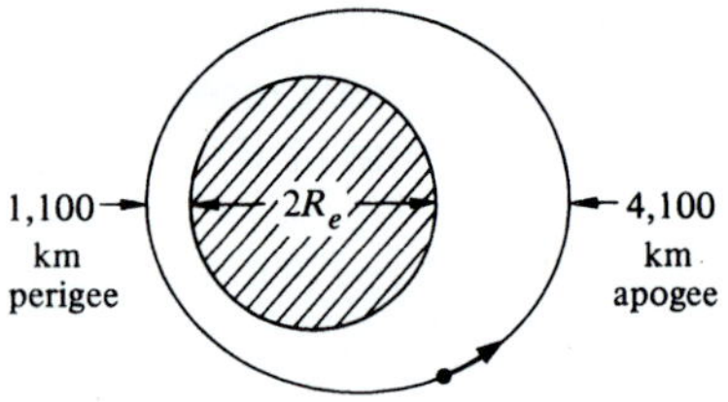

A satellite of mass $m = 2{,}000$ kg is in elliptic orbit about the earth. At perigee (closest approach to the earth) it has an altitude of 1,100 km and at apogee (farthest distance from the earth) its altitude is 4,100 km. What are the satellite's energy E and angular momentum l? How fast is it traveling at perigee and at apogee?

Since $m \ll M_e$, we can take $\mu \approx m$ and assume that the earth is fixed. The radius of the earth is $R_e = 6{,}400$ km, and the major axis of the orbit is therefore

$$\begin{aligned} A &= [1{,}100 + 4{,}100 + 2(6{,}400)]\text{km} \\ &= 1.8 \times 10^7 \text{ m.} \end{aligned}$$

Knowing A, we can find E from Eq. (9.26):

$$A = \frac{C}{(-E)} \qquad \text{or} \qquad E = -\frac{C}{A}.$$

$C = GmM_e = mgR_e^2$, since $g = GM_e/R_e^2$. Numerically,

$$C = (2 \times 10^3)(9.8)(6.4 \times 10^6)^2 = 8.0 \times 10^{17} \text{ J·m.}$$

$$\begin{aligned} E &= -\frac{C}{A} \\ &= -4.5 \times 10^{10} \text{ J.} \end{aligned}$$

The initial energy of the satellite before launch was

$$\begin{aligned} E_i &= -\frac{GmM_e}{R_e} \\ &= -\frac{C}{R_e} \\ &= -12.5 \times 10^{10} \text{ J.} \end{aligned}$$

The energy needed to put the satellite into orbit, neglecting losses due to friction, is $E - E_i = 8 \times 10^{10}$ J.

We can find the angular momentum from the eccentricity. Since

$$r_{\min} = \frac{r_0}{1 + \epsilon} \qquad \text{and} \qquad r_{\max} = \frac{r_0}{1 - \epsilon},$$

we have

$$(1 + \epsilon)r_{\min} = (1 - \epsilon)r_{\max}$$

and

$$\begin{aligned}\epsilon &= \frac{r_{\max} - r_{\min}}{r_{\max} + r_{\min}} \\ &= \frac{r_{\max} - r_{\min}}{A} \\ &= \frac{3 \times 10^3}{1.8 \times 10^4} \\ &= \frac{1}{6}.\end{aligned}$$

From the definition of ϵ, Eq. (9.20),

$$\epsilon^2 = 1 + \frac{2El^2}{mC^2}$$

which yields

$$l = 1.2 \times 10^{14} \text{ kg·m}^2/\text{s}.$$

We can find the speed v of the satellite at any r from the energy equation

$$E = \frac{1}{2}mv^2 - \frac{C}{r}.$$

At perigee, $r = (1{,}100 + 6{,}400)$ km $= 7.5 \times 10^6$ m, and the speed at perigee is

$$v_p = 7{,}900 \text{ m/s}.$$

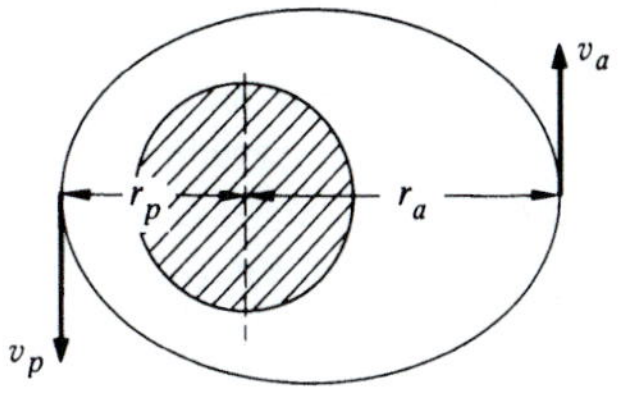

To find the speed at apogee, v_a, most simply, note that at apogee and perigee the velocity of the satellite is purely tangential. Hence, by conservation of angular momentum,

$$\mu v_p r_p = \mu v_a r_a,$$

and we find that

$$\begin{aligned}v_a &= \frac{v_p r_p}{r_a} \\ &= 5{,}600 \text{ m/s}.\end{aligned}$$

Suppose that a body is projected from the surface of the earth with initial velocity v_0. If v_0 is less than the escape velocity, 1.12×10^4 m/s, the total energy of the body is negative, and it travels in an elliptic orbit with one focus at the center of earth. As the drawing on the left shows, the body inevitably returns to earth.

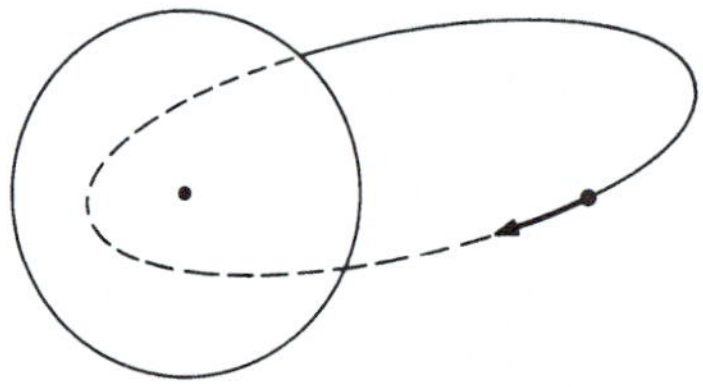

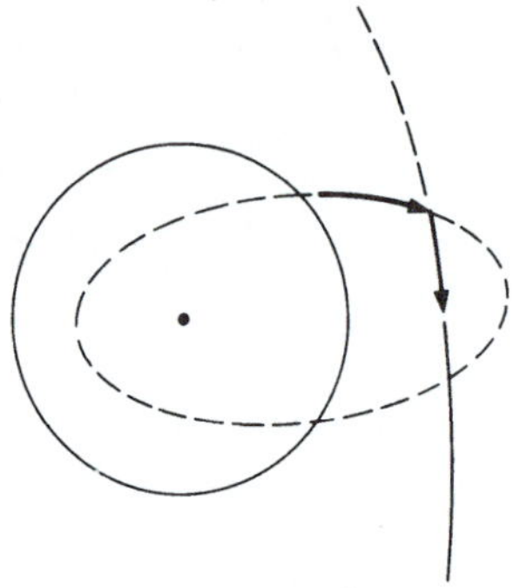

In order to put a spacecraft into orbit around the earth, the magnitude and direction of its velocity must be altered at a point where the old and new orbits intersect. Orbit transfer maneuvers are frequently needed in astronautics. For example, on an Apollo moon flight the vehicle is first put into near earth orbit and is then transferred to a trajectory toward the moon. The next example illustrates the physical principles of orbit transfer.

Example 9.6 **Satellite Maneuver**

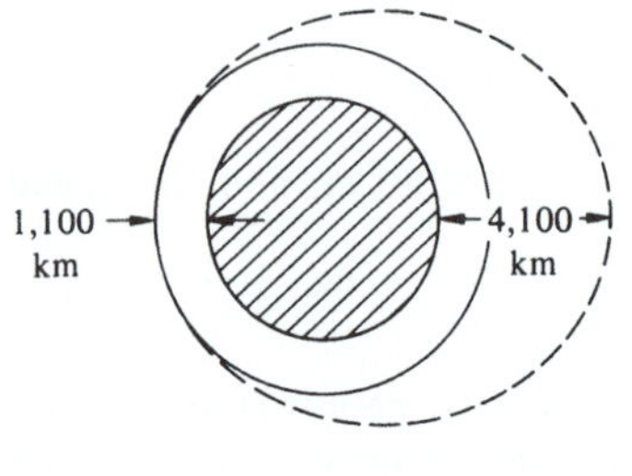

One of the commonest orbit maneuvers is the transfer between an elliptical and a circular orbit. This maneuver is used to inject spacecrafts into high orbits around the earth, or to put a planetary exploration satellite into a low orbit for surface inspection.

Suppose, for instance, that we want to transfer the satellite of Example 9.5 into a circular orbit at perigee, as shown in the sketch. Let E and l be the initial energy and angular momentum of the satellite and let E', l' be the parameters for the new orbit.

We start our analysis by finding E, l, E', l'. For simplicity, we shall assume that the amount of fuel burned by the satellite's rockets at transfer is negligible compared with the satellite's mass $m = 2{,}000$ kg.

From Eq. (9.26), $E = -C/A$. Since $A/r_p = 18 \times 10^6/(7.5 \times 10^6) = \frac{12}{5}$, we have

$$E = -\frac{5}{12}\frac{C}{r_p}. \qquad 1$$

r_p is the radius at perigee, hence the radius of the desired circular orbit.

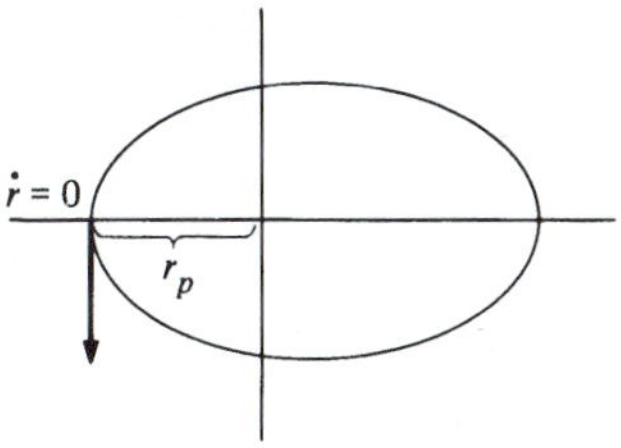

An easy way to find l is to use the one dimensional energy equation, Eq. (9.9):

$$E = \frac{1}{2} m\dot{r}^2 + \frac{l^2}{2mr^2} - \frac{C}{r}. \qquad 2$$

At perigee, $\dot{r} = 0$ and $r = r_p$, and we find

$$l^2 = \tfrac{7}{6} mCr_p. \qquad 3$$

For the circular orbit, the major axis is $2r_p$ and therefore

$$E' = -\frac{C}{2r_p}. \qquad 4$$

$\dot{r} = 0$ for the circular orbit, and from the one dimensional energy equation,

$$E' = \frac{l'^2}{2mr_p^2} - \frac{C}{r_p},$$

which yields

$$l'^2 = mCr_p. \qquad 5$$

How can we switch from E, l to E', l'? Since $E' < E$ and $l' < l$, we want to apply a braking thrust in order to reduce both the energy and the angular momentum. Thrust in the radial direction at perigee changes the energy but not the angular momentum, whereas tangential thrust changes both parameters. The old and new orbits are tangential where they intersect, and we might suspect that tangential thrust alone would be sufficient. We now show that this is correct.

At perigee, **v** is purely tangential, and tangential thrust changes the speed from v to v'. From the energy equation,

$$E = \frac{1}{2} mv^2 - \frac{C}{r},$$

and at perigee

$$v^2 = \frac{2}{m}\left(E + \frac{C}{r_p}\right)$$
$$= \frac{7}{6}\frac{C}{mr_p},$$

using Eq. (1). Similarly,

$$v'^2 = \frac{2}{m}\left(E' + \frac{C}{r_p}\right)$$
$$= \frac{C}{mr_p},$$

using Eq. (4).

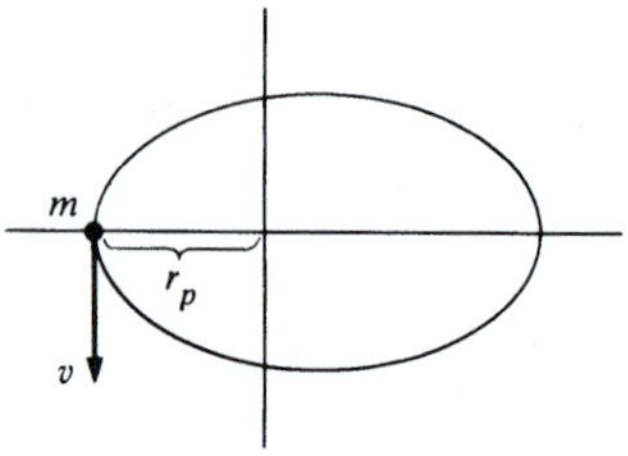

We now check to see if the angular momentum has its required value. At perigee, **v** is perpendicular to **r** and

$$l = mr_pv$$
$$= mr_p\sqrt{\frac{7}{6}\frac{C}{mr_p}}$$
$$= \sqrt{\frac{7}{6}mr_pC},$$

as we have already found, Eq. (3). Similarly,

$$l' = mr_pv'$$
$$= mr_p\sqrt{\frac{C}{mr_p}}$$
$$= \sqrt{mr_pC},$$

which is the required value according to Eq. (5).

The maneuver can be executed by applying a braking thrust tangential to the orbit at perigee to reduce the speed of the satellite from $v = \sqrt{7C/(6mr_p)}$ = 7,900 m/s to $v' = \sqrt{C/(mr_p)}$ = 7,300 m/s.

Practical orbit maneuvers are generally planned to economize on the fuel. According to our discussion of rockets in Sec. 3.5, if the mass of the spacecraft changes from M_i to $M_i - \Delta M$ during the rocket burn, its velocity changes by

$$\Delta\mathbf{v} = -\mathbf{u}\ln\left(\frac{M_i}{M_i - \Delta M}\right).$$

Therefore, the smaller the change in speed required by a maneuver, the more economical of fuel it is.

The maneuver described in this example reaches the maximum efficiency. At transfer,

$$E - E' = \tfrac{1}{2}mv^2 - \tfrac{1}{2}mv'^2$$
$$= \tfrac{1}{2}mv^2 - \tfrac{1}{2}m(\mathbf{v} - \Delta\mathbf{v})^2$$
$$\approx m\mathbf{v}\cdot\Delta\mathbf{v}.$$

$|\mathbf{v}|$ is greatest at perigee, and since $\Delta\mathbf{v}$ is parallel to **v**, $|\Delta\mathbf{v}|$ is least there to obtain the needed value of $E - E'$.

9.7 Kepler's Laws

Johannes Kepler was the assistant of the sixteenth century Danish astronomer Tycho Brahe. They had a remarkable combination of talents. Brahe made planetary measurements of unprecedented accuracy, and Kepler had the mathematical genius and fortitude to show that Brahe's data could be fitted into three simple empirical laws. The task was formidable. It took Kepler 18 years of laborious calculation to obtain the following three laws:

1. Each planet moves in an ellipse with the sun at one focus.

2. The radius vector from the sun to a planet sweeps out equal areas in equal times.

3. The period of revolution T of a planet about the sun is related to the major axis of the ellipse A by

$$T^2 = kA^3,$$

where k is the same for all the planets.

Kepler's first law follows from the results of the last section; elliptic orbits are characteristic of the inverse square law force. The second law is a general feature of central force motion as we demonstrated in Example 6.3.

Kepler's third law is easily proved by the following trick: We start with the definition of angular momentum, Eq. (9.8a),

$$l = \mu r^2 \frac{d\theta}{dt},$$

which can be written

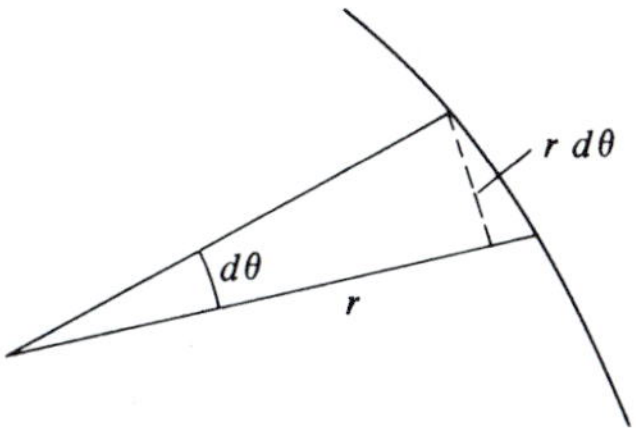

$$\frac{l}{2\mu} dt = \tfrac{1}{2} r^2\, d\theta. \qquad 9.27$$

But $\frac{1}{2} r^2\, d\theta$ is a differential element of area in polar coordinates. Over one complete period, the whole area of the ellipse is swept out, and integration of Eq. (9.27) yields

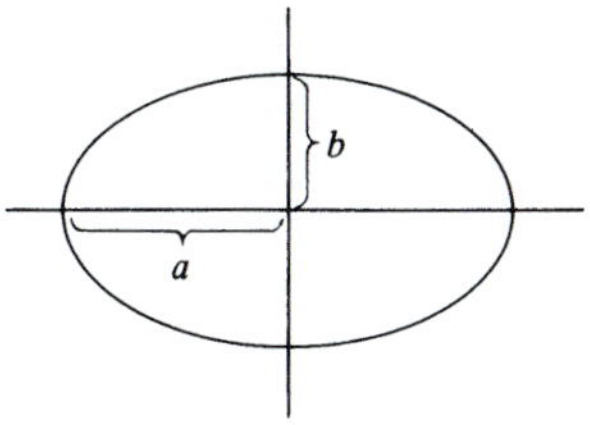

$$\frac{l}{2\mu} T = \text{area of ellipse} = \pi ab, \qquad 9.28$$

where $a = A/2$ is the semimajor axis and b is the semiminor axis. From Eq. (9.26),

$$a = \frac{C}{(-2E)},$$

and from Note 9.1,

$$b = \frac{l}{\sqrt{-2\mu E}},$$

Equation (9.28) becomes

$$T^2 = \frac{4\mu^2}{l^2}\pi^2 a^2 b^2$$

$$= \frac{\pi^2 \mu C^2}{(-2E^3)}$$

$$= \frac{\pi^2 \mu}{2C} A^3, \qquad 9.29$$

using $A = C/(-E)$. Since $C = GMm$ and $\mu = Mm/(M + m)$, we obtain finally

$$T^2 = \frac{\pi^2}{2(M + m)G} A^3. \qquad 9.30$$

This result shows that Kepler's third law is not exact; T^2/A^3 depends slightly on the planet's mass. However, even for Jupiter, the largest planet, m/M is only 1/1,000, so that Kepler's third law holds to good accuracy in the solar system.

Kepler's laws also apply to the motion of satellites around a planet. In Table 9.2 we show how his third law, the law of periods, holds for a number of artificial earth satellites. The ratio A^3/T^2 is constant to a fraction of a percent, although the periods vary by nearly a factor of 100. A more refined check would take into account the nonspherical shape of the earth and perturbations due to the moon.

TABLE 9.2*

SATELLITE	ϵ	A, km	T, min	A^3/T^2
Cosmos 358	0.002	13,823	95.2	2.91×10^8
Explorer 17	0.047	13,928	96.39	2.91×10^8
Cosmos 374	0.104	15,446	112.3	2.92×10^8
Cosmos 382	0.260	18,117	143	2.91×10^8
ATS 2	0.455	24,123	219.7	2.91×10^8
15th Molniya I	0.738	52,537	706	2.91×10^8
Ers 13	0.887	117,390	2,352	2.92×10^8
Ogo 3	0.901	135,270	2,917	2.91×10^8
Explorer 34	0.940	224,150	6,225	2.91×10^8
Explorer 28	0.952	273,740	8,400	2.91×10^8

* Data taken from the data catalogs of the National Space Science Data Center and the World Data Center A. The catalogs give satellite altitudes relative to the surface of the earth; we assumed the diameter of the earth to be 12,757 km in calculating *A*.

Example 9.7 **The Law of Periods**

Here is a more general way of deriving the law of periods. Starting from Eq. (9.13) we have, with $U(r) = -C/r$,

$$\int_{t_a}^{t_b} dt = \mu \int_{r_a}^{r_b} \frac{r\,dr}{(2\mu Er^2 + 2\mu Cr - l^2)^{\frac{1}{2}}}.$$

The integral is listed in standard tables. For the case of interest, $E < 0$, we find

$$t_b - t_a = \left.\frac{\sqrt{2\mu Er^2 + 2\mu Cr - l^2}}{2E}\right|_{r_a}^{r_b} - \left(\frac{\mu C}{2E}\right)\frac{1}{\sqrt{-2\mu E}}\arcsin\left.\left(\frac{-2\mu Er - \mu C}{\sqrt{\mu^2C^2 + 2\mu El^2}}\right)\right|_{r_a}^{r_b}$$

Fortunately this result can be greatly simplified. For a complete period, $t_b - t_a = T$, and $r_b = r_a$. The first term on the right hand side vanishes, and in the second term, the arcsine changes by 2π. The result is

$$T = \frac{\pi\mu C}{(-E)}\frac{1}{\sqrt{-2\mu E}}$$

or

$$T^2 = \frac{\pi^2\mu C^2}{(-2E^3)}$$

$$= \frac{\pi^2\mu}{2C}A^3,$$

as we found earlier, Eq. (9.29).

Note 9.1 **Properties of the Ellipse**

The equation of any conic section is, in polar coordinates,

$$r = \frac{r_0}{1 - \epsilon\cos\theta}. \quad 1$$

Converting to cartesian coordinates $r = \sqrt{x^2 + y^2}$, $x = r\cos\theta$, Eq. (1) becomes

$$(1 - \epsilon^2)x^2 - 2r_0x + y^2 = r_0^2. \quad 2$$

The ellipse corresponds to the case $0 \leq \epsilon < 1$. The ellipse described by Eqs. (1) and (2) is symmetrical about the x axis, but its center does not lie at the origin.

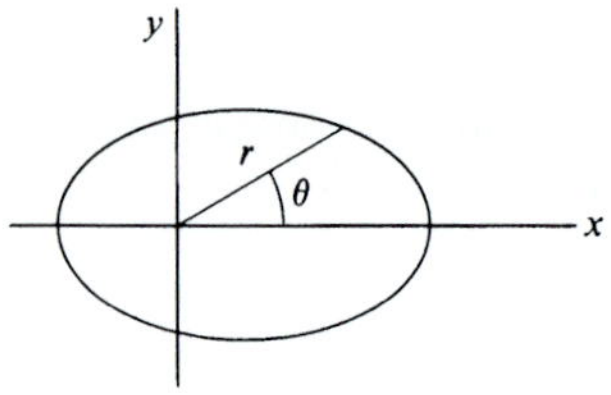

We can use Eq. (1) to determine the important dimensions of the ellipse. The maximum value of r, which occurs at $\theta = 0$, is

$$r_{\max} = \frac{r_0}{1 - \epsilon}.$$

The minimum value of r, which occurs at $\theta = \pi$, is

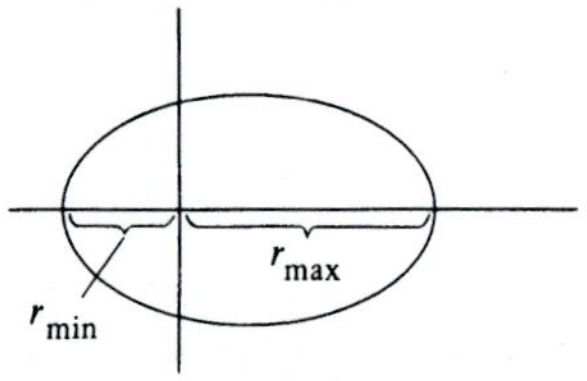

$$r_{\min} = \frac{r_0}{1 + \epsilon}.$$

The major axis is

$$\begin{aligned} A &= r_{\max} + r_{\min} \\ &= r_0\left(\frac{1}{1-\epsilon} + \frac{1}{1+\epsilon}\right) \\ &= \frac{2r_0}{1-\epsilon^2}. \end{aligned} \qquad 3$$

The semimajor axis is

$$\begin{aligned} a &= \frac{A}{2} \\ &= \frac{r_0}{1-\epsilon^2}. \end{aligned}$$

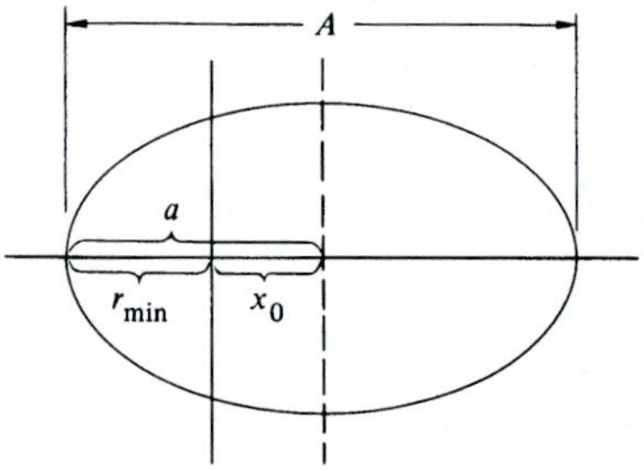

The distance from the origin to the center of the ellipse is

$$\begin{aligned} x_0 &= a - r_{\min} \\ &= r_0\left(\frac{1}{1-\epsilon^2} - \frac{1}{1+\epsilon}\right) \\ &= \frac{r_0\epsilon}{1-\epsilon^2}. \end{aligned} \qquad 4$$

We see that the eccentricity is equal to the ratio x_0/a.

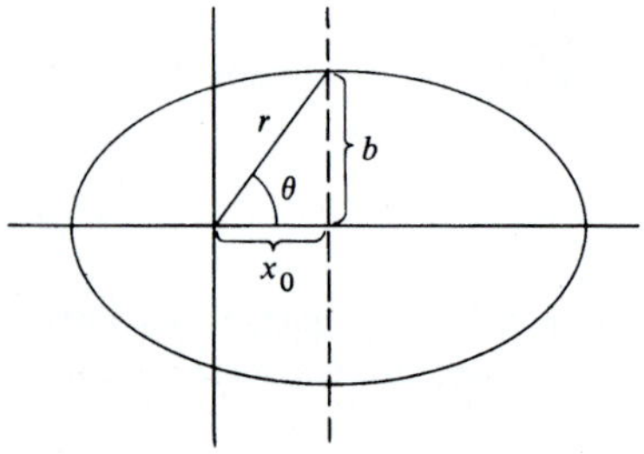

To find the length of the semiminor axis $b = \sqrt{r^2 - x_0^2}$, note that the tip of the semiminor axis has angular coordinates given by $\cos\theta = x_0/r$. We have

$$\begin{aligned} r &= \frac{r_0}{1 - \epsilon\cos\theta} \\ &= \frac{r_0}{1 - \epsilon x_0/r} \end{aligned}$$

or

$$r = r_0 + \epsilon x_0 = r_0\left(1 + \frac{\epsilon^2}{1 - \epsilon^2}\right)$$

$$= \frac{r_0}{1 - \epsilon^2}.$$

Hence,

$$b = \sqrt{r^2 - x_0{}^2} = \left(\frac{r_0}{1 - \epsilon^2}\right)\sqrt{1 - \epsilon^2}$$

$$= \frac{r_0}{\sqrt{1 - \epsilon^2}}.$$

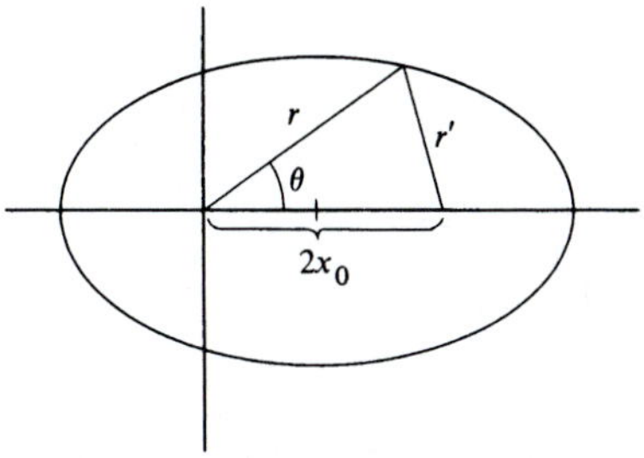

Finally, we shall prove that the origin lies at a focus of the ellipse. According to the definition of an ellipse, the sum of the distances from the foci to a point on the ellipse is a constant. Hence, for the ellipse shown in the sketch we need to prove $r + r' =$ constant. By the law of cosines,

$$r'^2 = r^2 + 4x_0{}^2 - 4rx_0 \cos\theta. \qquad 5$$

From Eq. (1) we find that

$$r\cos\theta = \frac{r - r_0}{\epsilon}.$$

Equation (5) becomes

$$r'^2 = r^2 - \frac{4x_0}{\epsilon}r + 4x_0{}^2 + \frac{4r_0x_0}{\epsilon}.$$

Using the relation $x_0 = r_0\epsilon/(1 - \epsilon^2)$ from Eq. (4) gives

$$r'^2 = r^2 - \left(\frac{4r_0}{1 - \epsilon^2}\right)r + \frac{4r_0{}^2\epsilon^2}{(1 - \epsilon^2)^2} + \frac{4r_0{}^2}{(1 - \epsilon^2)}$$

$$= r^2 - \left(\frac{4r_0}{1 - \epsilon^2}\right)r + \frac{4r_0{}^2}{(1 - \epsilon^2)^2}.$$

The right hand side is a perfect square.

$$r' = \pm\left(r - \frac{2r_0}{1 - \epsilon^2}\right)$$

$$= \pm(r - A).$$

Since $A > r$, we must choose the negative sign to keep $r' > 0$. Therefore,

$$r' + r = A$$

$$= \text{constant}.$$

To conclude, we list a few of our results in terms of E, l, μ, C for the inverse square force problem $U(r) = -C/r$. When using these formulas, E must be taken to be a negative number. From Eqs. (9.19) and (9.20),

$$r_0 = \frac{l^2}{\mu C}$$

and

$$\epsilon = \sqrt{1 + 2El^2/(\mu C^2)}.$$

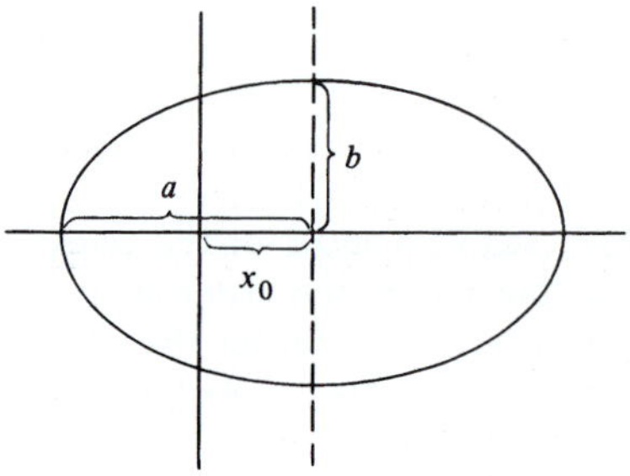

Hence,

$$\text{semimajor axis } a = \frac{r_0}{1-\epsilon^2} = \frac{C}{-2E}$$

$$\text{semiminor axis } b = \frac{r_0}{\sqrt{1-\epsilon^2}} = \frac{l}{\sqrt{-2\mu E}}$$

$$x_0 = \frac{r_0\epsilon}{1-\epsilon^2} = \left(\frac{C}{-2E}\right)\sqrt{1 + \frac{2El^2}{\mu C^2}}.$$

Problems

9.1 Obtain Eqs. (9.7a and b) by differentiating Eqs. (9.8a and b) with respect to time.

9.2 A particle of mass 50 g moves under an attractive central force of magnitude $4r^3$ dynes. The angular momentum is equal to 1,000 g·cm²/s.

a. Find the effective potential energy.

b. Indicate on a sketch of the effective potential the total energy for circular motion.

c. The radius of the particle's orbit varies between r_0 and $2r_0$. Find r_0.

Ans. (c) $r_0 \approx 2.8$ cm

9.3 A particle moves in a circle under the influence of an inverse cube law force. Show that the particle can also move with uniform radial velocity, either in or out. (This is an example of unstable motion. Any slight perturbation to the circular orbit will start the particle moving radially, and it will continue to do so.) Find θ as a function of r for motion with uniform radial velocity v.

9.4 For what values of n are circular orbits stable with the potential energy $U(r) = -A/r^n$, where $A > 0$?

9.5 A 2-kg mass on a frictionless table is attached to one end of a massless spring. The other end of the spring is held by a frictionless pivot. The spring produces a force of magnitude $3r$ newtons on the mass, where

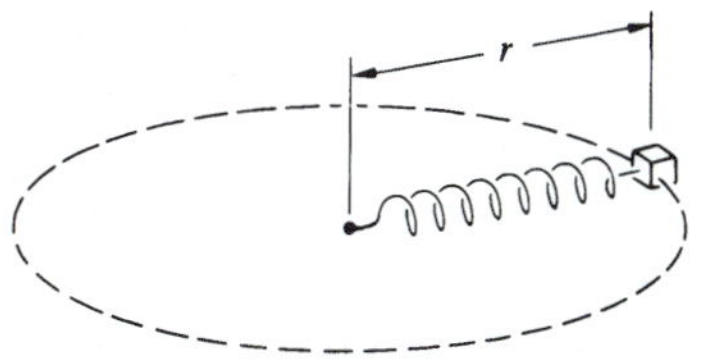

r is the distance in meters from the pivot to the mass. The mass moves in a circle and has a total energy of 12 J.

a. Find the radius of the orbit and the velocity of the mass.

b. The mass is struck by a sudden sharp blow, giving it instantaneous velocity of 1 m/s radially outward. Show the state of the system before and after the blow on a sketch of the energy diagram.

c. For the new orbit, find the maximum and minimum values of r.

9.6 A particle of mass m moves under an attractive central force Kr^4 with angular momentum l. For what energy will the motion be circular, and what is the radius of the circle? Find the frequency of radial oscillations if the particle is given a small radial impulse.

9.7 A rocket is in elliptic orbit around the earth. To put it into an escape orbit, its engine is fired briefly, changing the rocket's velocity by $\Delta\mathbf{V}$. Where in the orbit, and in what direction, should the firing occur to attain escape with a minimum value of ΔV?

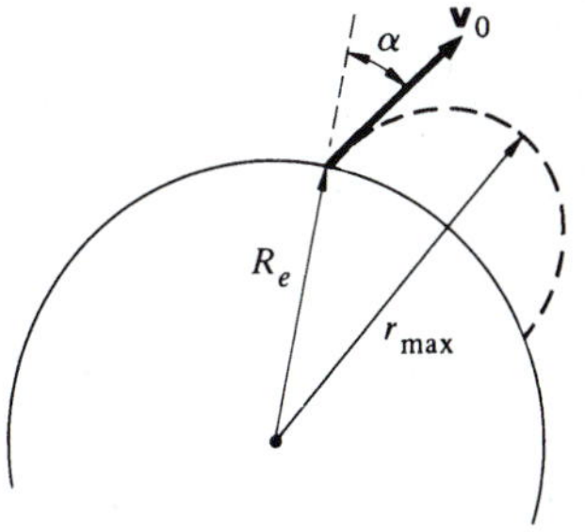

9.8 A projectile of mass m is fired from the surface of the earth at an angle α from the vertical. The initial speed v_0 is equal to $\sqrt{GM_e/R_e}$. How high does the projectile rise? Neglect air resistance and the earth's rotation. (*Hint*: It is probably easier to apply the conservation laws directly instead of using the orbit equations.)

Ans. clue. If $\alpha = 60°$, then $r_{max} = 3R_e/2$

9.9 Halley's comet is in an elliptic orbit about the sun. The eccentricity of the orbit is 0.967 and the period is 76 years. The mass of the sun is 2×10^{30} kg, and $G = 6.67 \times 10^{-11}$ N·m²/kg².

a. Using these data, determine the distance of Halley's comet from the sun at perihelion and at aphelion.

b. What is the speed of Halley's comet when it is closest to the sun?

9.10 *a.* A satellite of mass m is in circular orbit about the earth. The radius of the orbit is r_0 and the mass of the earth is M_e. Find the total mechanical energy of the satellite.

b. Now suppose that the satellite moves in the extreme upper atmosphere of the earth where it is retarded by a constant feeble friction force f. The satellite will slowly spiral toward the earth. Since the friction force is weak, the change in radius will be very slow. We can therefore assume that at any instant the satellite is effectively in a circular orbit of average radius r. Find the approximate change in radius per revolution of the satellite, Δr.

c. Find the approximate change in kinetic energy of the satellite per revolution, ΔK.

Ans. (c) $\Delta K = +2\pi rf$ (note the sign!)

9.11 Before landing men on the moon, the Apollo 11 space vehicle was put into orbit about the moon. The mass of the vehicle was 9,979 kg and the period of the orbit was 88 min. The maximum and minimum distances from the center of the moon were 1,861 km and 1,838 km. Assuming the moon to be a uniform spherical body, what is the mass of the moon according to these data? $G = 6.67 \times 10^{-11}$ N·m²/kg².

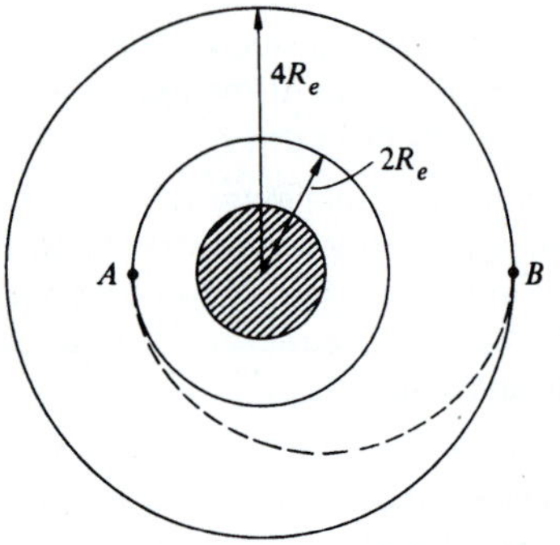

9.12 A space vehicle is in circular orbit about the earth. The mass of the vehicle is 3,000 kg and the radius of the orbit is $2R_e = 12{,}800$ km. It is desired to transfer the vehicle to a circular orbit of radius $4R_e$.

a. What is the minimum energy expenditure required for the transfer?

b. An efficient way to accomplish the transfer is to use a semielliptical orbit (known as a Hohmann transfer orbit), as shown. What velocity changes are required at the points of intersection, A and B?

10 THE HARMONIC OSCILLATOR

10.1 Introduction and Review

The motion of a mass on a spring, better known as a harmonic oscillator, is familiar to us from Chaps. 2 and 4 and from numerous problems. However, so far we have considered only the idealized case in which friction is absent and there are no external forces. In this chapter we shall investigate the effect of friction on the oscillator, and then study the motion when the mass is subjected to a driving force which is a periodic function of time. Finally, we shall use the harmonic oscillator to illustrate a remarkable result—the possibility of predicting how a mechanical system will respond to an applied driving force of any given frequency merely by studying what the system does when it is put into motion and allowed to move freely.

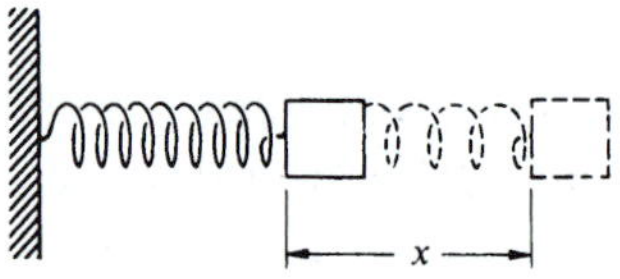

We begin by reviewing the properties of the frictionless harmonic oscillator which we discussed at the end of Chap. 2. The prototype oscillator is a mass m acted on by a spring force $F_{\text{spring}} = -kx$, where x is the displacement from equilibrium. The equation of motion is $m\ddot{x} = -kx$, or

$$m\ddot{x} + kx = 0. \qquad 10.1$$

The solution is

$$x = B \sin \omega_0 t + C \cos \omega_0 t, \qquad 10.2$$

where

$$\omega_0 = \sqrt{\frac{k}{m}}. \qquad 10.3$$

We shall use ω_0 rather than ω, as in previous chapters, to represent the natural frequency of the oscillator. B and C are arbitrary constants which can be evaluated from a set of given initial conditions, such as the position and the velocity at a particular time.

Standard Form of the Solution

We can rewrite Eq. (10.2) in the following more convenient form:

$$x = A \cos (\omega_0 t + \phi), \qquad 10.4$$

where A and ϕ are constants. To show the correspondence between Eqs. (10.2) and (10.4) we make use of the trigonometric identity

$$\cos (\alpha + \beta) = \cos \alpha \cos \beta - \sin \alpha \sin \beta.$$

By applying this to Eq. (10.4) and equating Eqs. (10.2) and (10.4), we obtain

$$A \cos \omega_0 t \cos \phi - A \sin \omega_0 t \sin \phi = B \sin \omega_0 t + C \cos \omega_0 t.$$

For this to be true at all times, the coefficients of the terms in $\sin \omega_0 t$ and $\cos \omega_0 t$ must be separately equal. Hence, we have

$$A \cos \phi = C$$
$$A \sin \phi = -B, \qquad 10.5a$$

which are readily solved to yield

$$A = (B^2 + C^2)^{\frac{1}{2}}$$
$$\tan \phi = -\frac{B}{C}. \qquad 10.5b$$

This result shows that the two expressions Eqs. (10.2) and (10.4) for the general motion of the harmonic oscillator are equivalent. We shall generally use Eq. (10.4) as the standard form for the motion of a frictionless harmonic oscillator.

Nomenclature

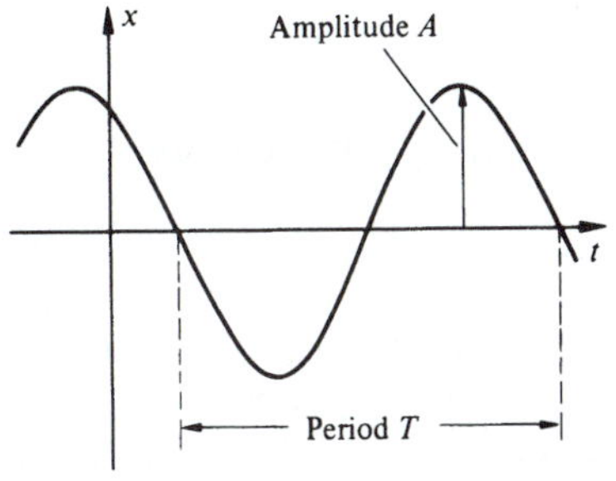

There are a number of definitions with which we should be familiar. Consider the expression

$$x = A \cos (\omega_0 t + \phi).$$

x is the instantaneous displacement of the particle at time t.

A is the amplitude of the motion, measured from zero displacement to a maximum.

ω_0 is the frequency (or angular frequency) of motion. $\omega_0 = \sqrt{k/m}$ rad/s. The circular frequency $\nu = \omega_0/2\pi$ Hz (1 Hz = 1 cycle per second).

ϕ is the phase factor or phase angle.

T is the period of the motion, the time required to execute one complete cycle. $T = 2\pi/\omega_0$.

Example 10.1 **Initial Conditions and the Frictionless Harmonic Oscillator**

Suppose that at time $t = 0$ the position of the mass is $x(0)$ and its velocity $v(0)$. If we express the motion in the form of Eq. (10.2) we have

$$x = B \sin \omega_0 t + C \cos \omega_0 t$$
$$v = \dot{x}$$
$$= \omega_0 B \cos \omega_0 t - \omega_0 C \sin \omega_0 t.$$

Evaluating these at $t = 0$ gives

$$C = x(0)$$

$$B = \frac{\dot{x}(0)}{\omega_0}.$$

If we begin with the standard form $x = A\cos(\omega_0 t + \phi)$, the displacement and velocity are

$$x = A\cos(\omega_0 t + \phi)$$

$$v = -\omega_0 A\sin(\omega_0 t + \phi).$$

For $t = 0$,

$$x(0) = A\cos\phi$$

$$v(0) = -\omega_0 A\sin\phi,$$

from which we find

$$A = \sqrt{x(0)^2 + \left[\frac{v(0)}{\omega_0}\right]^2}$$

$$\tan\phi = \frac{-v(0)}{\omega_0 x(0)}.$$

Energy Considerations

If we take the potential energy to be 0 at $x = 0$, we have

$$U = \tfrac{1}{2}kx^2$$

$$= \tfrac{1}{2}kA^2\cos^2(\omega_0 t + \phi). \qquad 10.6a$$

The kinetic energy is

$$K = \tfrac{1}{2}mv^2$$

$$= \tfrac{1}{2}m\omega_0^2 A^2\sin^2(\omega_0 t + \phi), \qquad 10.6b$$

where we have used

$$v = \dot{x} = -\omega_0 A\sin(\omega_0 t + \phi).$$

Since $\omega_0^2 = k/m$, Eq. (10.6b) becomes $K = \tfrac{1}{2}kA^2\sin^2(\omega_0 t + \phi)$. The total energy is

$$E = K + U = \tfrac{1}{2}kA^2[\cos^2(\omega_0 t + \phi) + \sin^2(\omega_0 t + \phi)]$$

$$E = \tfrac{1}{2}kA^2. \qquad 10.7$$

Hence, the total energy is constant, a familiar feature of motion when only conservative forces act.

Time Average Values

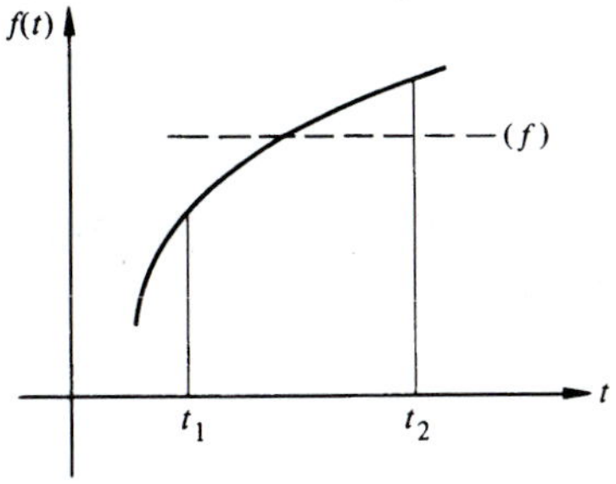

In the following sections we need the concept of the *time average value* $\langle f \rangle$ of a function $f(t)$. Consider $f(t)$, some function of time, and an interval $t_1 \leq t \leq t_2$ as shown in the sketch. $\langle f \rangle$, the time average value of $f(t)$, is defined so that the rectangular area shown in the sketch, $(t_2 - t_1)\langle f \rangle$, equals the actual area under the curve between t_1 and t_2:

$$(t_2 - t_1)\langle f \rangle = \int_{t_1}^{t_2} f(t)\, dt$$

or

$$\langle f \rangle = \frac{1}{(t_2 - t_1)} \int_{t_1}^{t_2} f(t)\, dt.$$

To make this idea more concrete, suppose that $f(t)$ represents the rate of flow of water into a bucket in liters per second. Then the volume of water passing into the bucket in a short interval dt is $f(t)\, dt$, and the total volume passing into the bucket in the interval $t_2 - t_1$ is $\int_{t_1}^{t_2} f(t)\, dt$. If the flow were steady, the rate would have to be $\langle f \rangle$ for the same volume of water to accumulate in the time interval $t_2 - t_1$.

For our work with the harmonic oscillator we shall need the time averages of $\sin(\omega t)$ and $\sin^2(\omega t)$ over one cycle of oscillation. Here is a graphical device for calculating these averages. The first sketch shows $\sin\theta$ for the interval $0 \leq \theta \leq 2\pi$, where $\theta = \omega t$. It is apparent that the area above the axis equals the area below the axis, so that $\langle \sin\theta \rangle = 0$. In the second sketch, we show $\sin^2\theta$. This varies between 0 and 1, and by symmetry we see that its average value is $\frac{1}{2}$. Thus $\langle \sin^2\theta \rangle = \frac{1}{2}$. By identical arguments $\langle \cos\theta \rangle = 0$, $\langle \cos^2\theta \rangle = \frac{1}{2}$, and you can also show graphically that these results hold as long as the average is taken over a whole period of oscillation, irrespective of the starting point. These results can also be proven analytically; we leave this task for a problem.

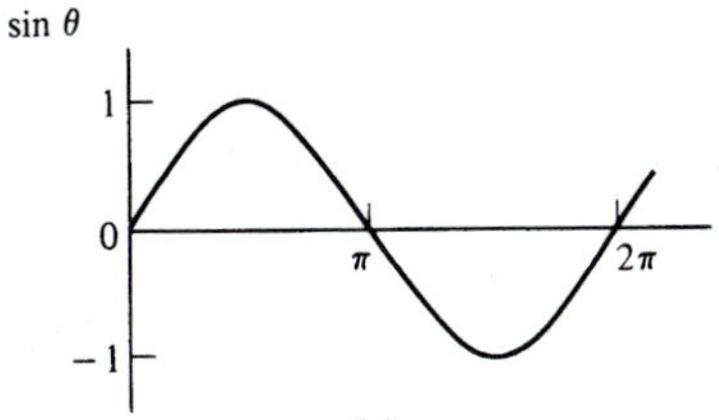

(*a*)

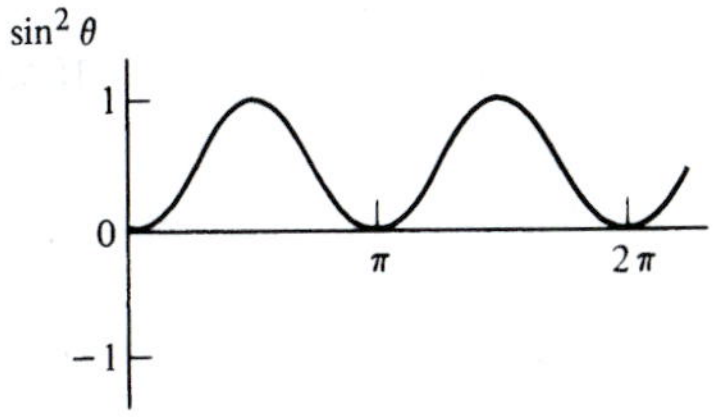

(*b*)

Average Energy

Returning to the frictionless harmonic oscillator, we can now evaluate the time average values of the potential and kinetic energies over one period of oscillation $0 \leq t \leq T$. From Eq. (10.6a),

$$U = \tfrac{1}{2}kA^2 \cos^2 (\omega_0 t + \phi)$$
$$\langle U \rangle = \tfrac{1}{2}kA^2 \langle \cos^2 (\omega_0 t + \phi) \rangle$$
$$= \tfrac{1}{4}kA^2.$$

(We have used $\langle \cos^2 \theta \rangle = \frac{1}{2}$ for an average over one period.) Similarly, from Eq. (10.6b),

$$\langle K \rangle = \tfrac{1}{2}m\omega_0{}^2A^2 \langle \sin^2 (\omega_0 t + \phi) \rangle$$
$$= \tfrac{1}{4}m\omega_0{}^2A^2.$$

Since $\omega_0{}^2 = k/m$, we have

$$\langle K \rangle = \tfrac{1}{4}kA^2$$
$$= \langle U \rangle.$$

The time average kinetic and potential energies are equal. When friction is present, this is no longer exactly true.

10.2 The Damped Harmonic Oscillator

Our next step is to consider the effect of friction on the harmonic oscillator. We are going to restrict our discussion to a very special form of friction force, the viscous force. Such a force arises when an object moves through a fluid, either liquid or gas, at speeds which are not so large as to cause turbulence. In this case the friction force f is of the form

$$f = -bv,$$

where b is a constant of proportionality that depends on the shape of the mass and the medium through which it moves, and where v is the instantaneous velocity. Although this is a special friction force, we should emphasize that it is the type most often encountered and that our analysis has wide applicability. Although the discussion here is devoted to a mechanical oscillator, equations of identical form describe many other oscillating systems. For example, electric current can oscillate in certain electric circuits; the electrical resistance of the circuit plays a role exactly analogous to a viscous retarding force.

The total force acting on the mass m is

$$F = F_{\text{spring}} + f$$
$$= -kx - bv.$$

The equation of motion is

$$m\ddot{x} = -kx - b\dot{x},$$

which can be rewritten as

$$\ddot{x} + \gamma\dot{x} + \omega_0{}^2 x = 0. \qquad 10.8$$

Here γ stands for b/m and, as before, $\omega_0{}^2 = k/m$. The units of γ are second^{-1}.

Equation (10.8) is a more complicated differential equation than any we have yet encountered. We leave the details of the solution for Note 10.1 and merely state the result here:

$$x = Ae^{-(\gamma/2)t} \cos (\omega_1 t + \phi). \qquad 10.9$$

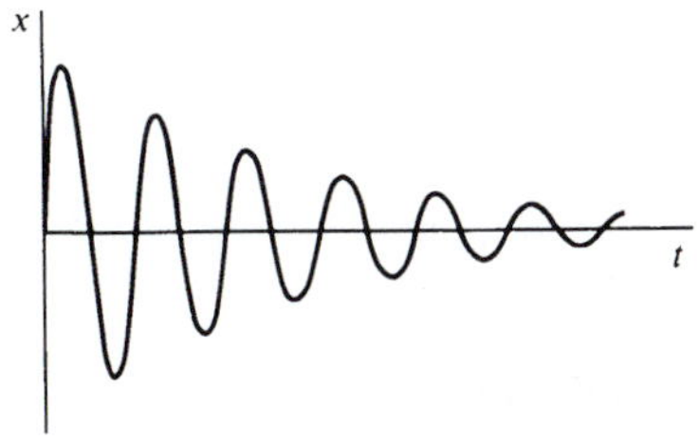

A and ϕ again stand for arbitrary constants and

$$\omega_1 = \sqrt{\omega_0{}^2 - \frac{\gamma^2}{4}}. \qquad 10.10$$

This solution is valid when $\omega_0{}^2 - \gamma^2/4 > 0$, or, equivalently, $\gamma < 2\omega_0$. (Other cases are discussed in Note 10.1). Substituting Eq. 10.9 into Eq. (10.8) to verify the solution makes a good exercise.

The motion described by Eq. (10.9) is known as *damped harmonic motion*. A typical case is shown in the top sketch. The motion is reminiscent of the undamped harmonic motion described in the last section. In fact, we can rewrite Eq. (10.9) as

$$x = A(t) \cos (\omega_1 t + \phi),$$

where

$$A(t) = Ae^{-(\gamma/2)t}. \qquad 10.11$$

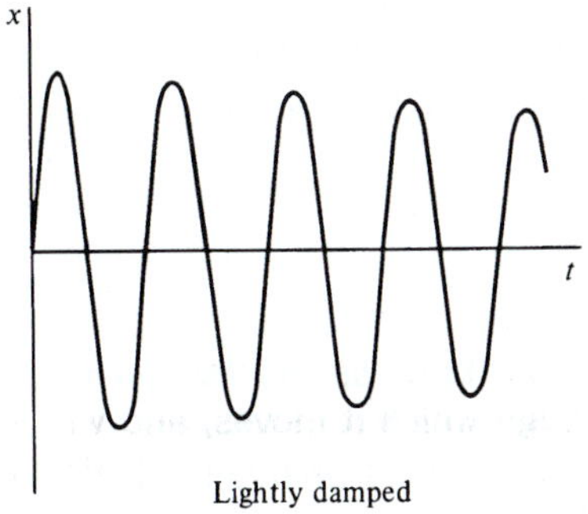

Lightly damped

The motion is similar to the undamped case except that the amplitude decreases exponentially in time and the frequency of oscillation ω_1 is less than the undamped frequency ω_0. Incidentally, although the concept of a definite frequency can be strictly applied only to a pure sine or cosine function, ω_1 is commonly called the frequency of oscillation. The zero crossings of the function $Ae^{-(\gamma/2)t} \cos (\omega_1 t + \phi)$ are separated by equal time intervals $T = 2\pi/\omega_1$, but the peaks do not lie halfway between them.

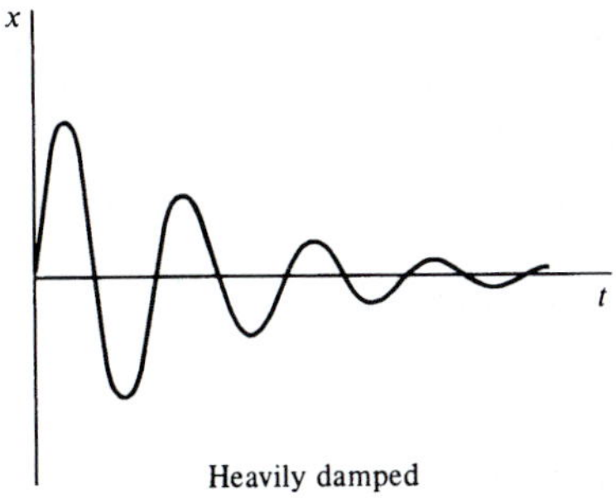

Heavily damped

Before we investigate damped harmonic motion quantitatively, it will be helpful to look at it qualitatively. The essential features of the motion depend on the ratio γ/ω_1. If $\gamma/\omega_1 \ll 1$, $A(t)$ decreases very little during the time the cosine makes many zero crossings; in this regime, the motion is called lightly damped.

If γ/ω_1 is comparatively large, $A(t)$ tends rapidly to zero while the cosine makes only a few oscillations. This motion is called heavily damped. For light damping, $\omega_1 \approx \omega_0$, but for heavy damping ω_1 can be significantly smaller than ω_0.

Energy

By considering the energy of the system we can see why the amplitude must decrease with time. From the work-energy theorem of Chap. 4,

$$E(t) = E(0) + W_{\text{frict}},$$

where

$$E(t) = \tfrac{1}{2}mv^2 + \tfrac{1}{2}kx^2 = K(t) + U(t)$$

and

$$W_{\text{frict}} = \text{work done by friction from time 0 to time } t.$$

The dissipative friction force, $f = -bv$, opposes the motion. Hence,

$$\begin{aligned} W_{\text{frict}} &= \int_{x(0)}^{x(t)} f\,dx \\ &= \int_0^t fv\,dt \\ &= -\int_0^t bv^2\,dt < 0. \end{aligned} \qquad 10.12$$

Physically, $E(t)$ decreases with time because the friction force continually dissipates energy. We can find how $E(t)$ depends on time by calculating the kinetic and potential energies $K(t)$ and $U(t)$.

To evaluate $K(t) = \tfrac{1}{2}mv^2$ we need the velocity v. The time derivative of Eq. (10.9) gives

$$\begin{aligned} v &= -Ae^{-(\gamma/2)t}\left[\omega_1 \sin(\omega_1 t + \phi) + \frac{\gamma}{2}\cos(\omega_1 t + \phi)\right] \\ &= -\omega_1 A e^{-(\gamma/2)t}\left[\sin(\omega_1 t + \phi) + \frac{1}{2}\left(\frac{\gamma}{\omega_1}\right)\cos(\omega_1 t + \phi)\right]. \end{aligned} \qquad 10.13$$

If the motion is only lightly damped, $\gamma/\omega_1 \ll 1$, and the coefficient of the second term in the bracket is small. Let us assume that the damping is so small that we can neglect the second term entirely. In this case we have

$$v = -\omega_1 A e^{-(\gamma/2)t} \sin(\omega_1 t + \phi),$$

and

$$\begin{aligned} K(t) &= \tfrac{1}{2}mv^2 \\ &= \tfrac{1}{2}m\omega_1^2 A^2 e^{-\gamma t} \sin^2(\omega_1 t + \phi). \end{aligned} \qquad 10.14a$$

The potential energy is

$$\begin{aligned} U(t) &= \tfrac{1}{2}kx^2 \\ &= \tfrac{1}{2}kA^2 e^{-\gamma t} \cos^2(\omega_1 t + \phi) \end{aligned} \qquad 10.14b$$

and the total energy is

$$\begin{aligned} E(t) &= K(t) + U(t) \\ &= \tfrac{1}{2}A^2 e^{-\gamma t}[m\omega_1^2 \sin^2(\omega_1 t + \phi) + k\cos^2(\omega_1 t + \phi)]. \end{aligned}$$

Since the damping is assumed to be small, we can simplify the term in brackets by replacing ω_1^2 by ω_0^2† and using the relation $\omega_0^2 = k/m$.

$$\begin{aligned} E(t) &= \tfrac{1}{2}A^2 e^{-\gamma t}[k\cos^2(\omega_1 t + \phi) + k\sin^2(\omega_1 t + \phi)] \\ &= \tfrac{1}{2}kA^2 e^{-\gamma t} \end{aligned} \qquad 10.15$$

At $t = 0$ the energy of the system is

$$E_0 = \tfrac{1}{2}kA^2$$

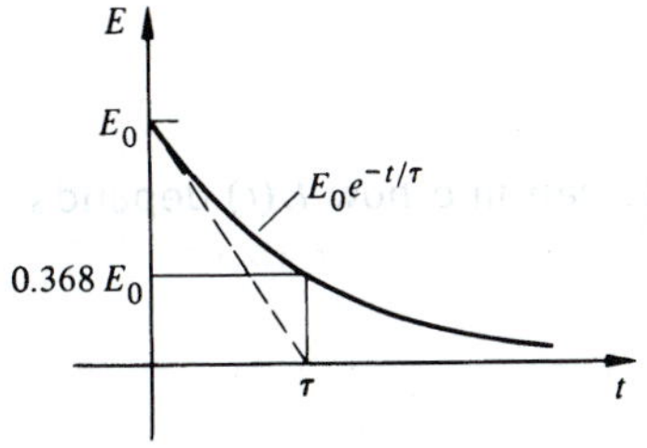

and we can rewrite Eq. (10.15) as

$$E(t) = E_0 e^{-\gamma t}. \qquad 10.16$$

This is a remarkably simple result. The energy decreases exponentially in time.

The decay can be characterized by the time τ required for the energy to drop to $e^{-1} = 0.368$ of its initial value.

$$\begin{aligned} E(\tau) &= E_0 e^{-\gamma\tau} \\ &= e^{-1}E_0. \end{aligned}$$

† This approximation can be justified for $\gamma/\omega_1 \ll 1$ as follows:

$$\begin{aligned} \omega_0^2 &= \omega_1^2 + \frac{\gamma^2}{4} \\ &= \omega_1^2\left[1 + \frac{1}{4}\left(\frac{\gamma}{\omega_1}\right)^2\right] \\ &\approx \omega_1^2. \end{aligned}$$

Hence, $\gamma\tau = 1$.

$$\tau = \frac{1}{\gamma} = \frac{m}{b}. \qquad 10.17$$

τ is often called the *damping time* (or, alternatively, the *time constant* or characteristic time) of the system. In the limit of light damping, $\gamma \to 0$ and $\tau \to \infty$; E is effectively constant and the system behaves like an undamped oscillator.

The Q of an Oscillator

The degree of damping of an oscillator is often specified by a dimensionless parameter Q, the *quality factor*, defined by

$$Q = \frac{\text{energy stored in the oscillator}}{\text{energy dissipated per radian}}. \qquad 10.18$$

By energy dissipated per radian we mean the energy lost during the time it takes the system to oscillate through one radian. In the period $T = 2\pi/\omega_1$, the system oscillates through 2π radians. Thus the time to oscillate through one radian is $T/2\pi = 1/\omega_1$.

Q is easily calculated for the lightly damped case. The rate of change of energy is, from Eq. (10.16),

$$\begin{aligned}\frac{dE}{dt} &= -\gamma E_0 e^{-\gamma t}\\ &= -\gamma E.\end{aligned}$$

The energy dissipated in a short time Δt is the positive quantity

$$\begin{aligned}\Delta E &\approx \left|\frac{dE}{dt}\right| \Delta t\\ &= \gamma E\,\Delta t.\end{aligned}$$

One radian of oscillation requires time $\Delta t = 1/\omega_1$, and the energy dissipated is $\gamma E/\omega_1$. Hence, the quality factor is

$$Q = \frac{E}{\gamma E/\omega_1} = \frac{\omega_1}{\gamma} \approx \frac{\omega_0}{\gamma}. \qquad 10.19$$

A lightly damped oscillator has $Q \gg 1$. A heavily damped system loses its energy rapidly and its Q is low. A tuning fork has a Q of a thousand or so, whereas a superconducting microwave cavity can have a Q in excess of 10^7. An undamped oscillator has infinite Q.

Example 10.2 **The Q of Two Simple Oscillators**

A musician's tuning fork rings at A above middle C, 440 Hz. A sound level meter indicates that the sound intensity decreases by a factor of 5 in 4 s. What is the Q of the tuning fork?

The sound intensity from the tuning fork is proportional to the energy of oscillation. Since the energy of a damped oscillator decreases as $e^{-\gamma t}$, we can find γ by taking the ratio of the energy at $t = 0$ to that at $t = 4$ s.

$$5 = \frac{E(0)e^{(0)}}{E(0)e^{-4\gamma}} = e^{4\gamma}$$

Hence

$$4\gamma = \ln 5 = 1.6$$
$$\gamma = 0.4 \text{ s}^{-1},$$

and

$$Q = \frac{\omega_1}{\gamma} = \frac{2\pi(440)}{0.4}$$
$$\approx 7000.$$

The energy loss is due primarily to the heating of the metal as it bends. Air friction and energy loss to the mounting point also contribute. (The symmetrical design of a tuning fork minimizes loss to the mount.) Incidentally, if you try this experiment, bear in mind that the ear is a poor sound level meter because it does not respond linearly to sound intensity; its response is more nearly logarithmic.

A rubber band exhibits a much lower Q than a tuning fork primarily because of the internal friction generated by the coiling of the long chain molecules. In one experiment, a paperweight suspended from a hefty rubber band had a period of 1.2 s and the amplitude of oscillation decreased by a factor of 2 after three periods. What is the estimated Q of this system?

From Eq. (10.11) the amplitude is given by $Ae^{(-\gamma/2)t}$. The ratio of the amplitude at $t = 0$ to that at $t = 3(1.2) = 3.6$ s is

$$2 = \frac{Ae^{(0)}}{Ae^{-3.6\gamma/2}}.$$

Hence

$$1.8\gamma = \ln 2 = 0.69$$

or

$$\gamma = 0.39 \text{ s}^{-1}.$$

Therefore

$$Q \approx \frac{\omega_1}{\gamma}$$
$$= \frac{2\pi/T}{0.39}$$
$$= \frac{2\pi/1.2}{0.39}$$
$$= 13.$$

You may wonder whether it is justifiable to use the light damping result, $Q = \omega_1/\gamma$, when Q is so low. The approximations involved introduce errors of order $(\gamma/\omega_1)^2 = (1/Q)^2$. For $Q > 10$ the error is less than 1 percent.

It is interesting to note that the damping constants for the tuning fork and for the rubber band are very nearly the same. The tuning fork has a much higher Q, however, because it goes through many more cycles of oscillation in one damping time and loses correspondingly less of its energy per cycle.

Example 10.3 **Graphical Analysis of a Damped Oscillator**

The illustration is drawn from a photograph of an oscilloscope trace of the displacement of an oscillating system versus time. We immediately recognize that the system is a damped harmonic oscillator. The frequency ω_1 and quality factor Q can be found from the photograph.

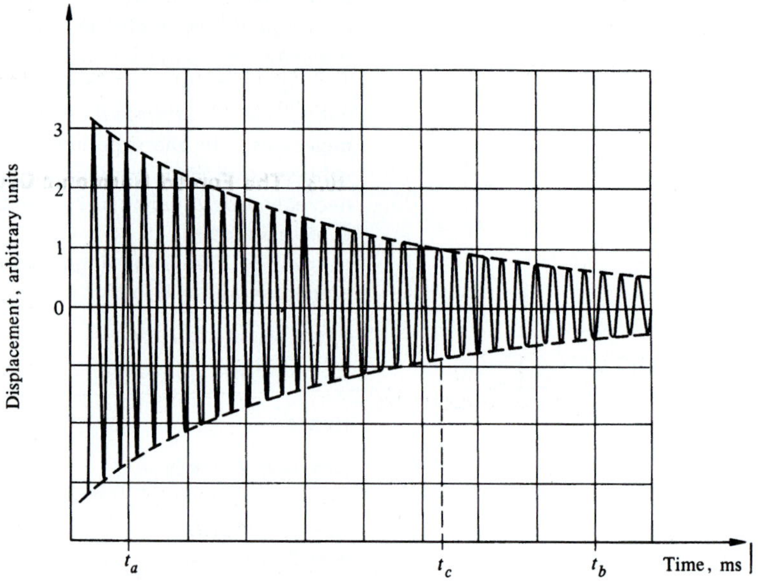

The time interval from t_a to t_b is 8 ms. There are 28.5 cycles (i.e., complete periods) in this interval. (Check this for yourself from the data.) The period of oscillation is $T = 8 \times 10^{-3}$ s/28.5 $= 2.81 \times 10^{-4}$ s. The angular frequency is $\omega_1 = 2\pi/T = 22{,}400$ rad/s. The corresponding circular frequency is $\nu = \omega_1/2\pi = 3{,}560$ Hz.

In order to obtain the quality factor $Q = \omega_1/\gamma$, the damping constant must be known. From Eq. (10.11) the amplitude is $Ae^{-(\gamma/2)t}$. This function describes the *envelope* of the displacement curve. The envelope has been drawn with a dashed curve on the photograph. At time t_a the envelope has magnitude $A_a = 2.75$ units. When the envelope decays by a factor $e^{-1} = 0.368$, its magnitude is 1.01 units. From the photograph this occurs at $t_c = 5.35$ ms, measured from t_a. Hence, $e^{-(\gamma/2)t_c} = e^{-1}$, or $\gamma = 2/t_c = 374\ \text{s}^{-1}$. The quality factor is $Q = \omega_1/\gamma = 60$.

Now for a word about the system. This is not a mechanical oscillator, nor even an electrical oscillator. The signal is produced by radiating atomic electrons in a small volume of hydrogen gas. The signal was greatly amplified for oscilloscope display. Furthermore, the atoms were actually radiating at 9.2×10^9 Hz. Since this is much too high for the oscilloscope to follow, the frequency was translated to a lower value by electronic means. This did not affect the shape of the envelope, and our measured value of γ is correct. If we use the true value of the frequency of the atomic system, we find that the actual Q is

$$Q = \frac{2\pi\nu}{\gamma} = \frac{2\pi \times 9.2 \times 10^9}{374} = 1.6 \times 10^8.$$

Such a high Q is virtually unattainable for mechanical systems, although it is not unusual in an atomic system.

10.3 The Forced Harmonic Oscillator

The Undamped Forced Oscillator

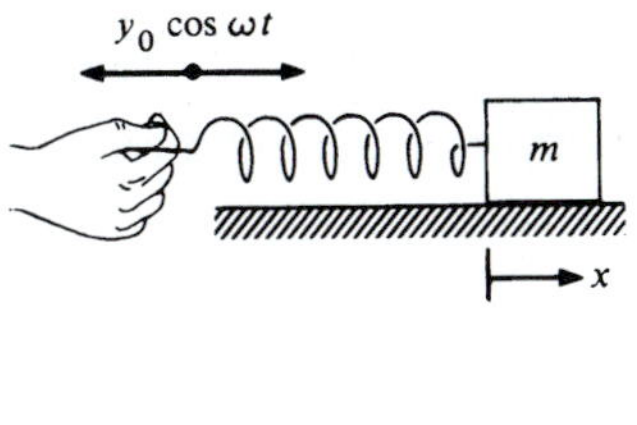

We next investigate the effect of an applied time varying force $F(t)$ on a frictionless harmonic oscillator. In the case of a mass on a spring, the force can be applied by jiggling the end of the spring. To be concrete, suppose that the end of the spring moves according to $y = y_0 \cos \omega t$, as shown in the sketch. The change in length of the spring from its equilibrium length is $x - y$, where x is the position of the mass. The equation of motion, neglecting damping, is $m\ddot{x} = -k(x - y)$, or, since $y = y_0 \cos \omega t$,

$$m\ddot{x} + kx = F_0 \cos \omega t \qquad 10.20$$

where $F_0 = ky_0$. $F_0 \cos \omega t$ is called the *driving force*. F_0 is the amplitude of the driving force (note that F_0 has the dimensions of force) and ω is the driving frequency, a quantity we are free to vary.

It is apparent that we have chosen a very special form for the driving force in Eq. (10.21). Nevertheless, the solution is of quite general interest. It turns out that any periodic function of time can be represented as a sum of sinusoidal terms (this is the basis of Fourier's theorem), so that understanding the response of the harmonic oscillator to the force $F_0 \cos \omega t$ lays the groundwork for finding the response to any periodic force. Furthermore, many important cases involve the simple sinusoidal force we assume here; two examples are the response of a bound electron to an electromagnetic field (a problem which arises in the classical theory of the scattering of light) and the tidal response of a lake to the periodic force of the moon or sun. So, without further apology we turn to the solution of Eq. (10.20).

A general procedure for solving Eq. (10.20) is given in Note 10.2, but in fact this equation is so simple that we can guess the correct solution by the following argument: the right hand side of the equation varies in time as $\cos \omega t$. It seems plausible that the left hand side involves the same time dependence. We try the solution

$$x = A \cos \omega t.$$

Substituting this in Eq. (10.20) yields

$$(-m\omega^2 + k)A \cos \omega t = F_0 \cos \omega t,$$

which is valid provided that we choose

$$A = \frac{F_0}{k - m\omega^2} = \frac{F_0}{m} \frac{1}{\omega_0^2 - \omega^2}, \qquad 10.21$$

where $\omega_0^2 = k/m$, as in the last section. Our solution becomes

$$x = \frac{F_0}{m} \frac{1}{\omega_0^2 - \omega^2} \cos \omega t. \qquad 10.22$$

The solution we found in Eq. (10.22) is quite different in nature from the solution of Eq. (10.4) or (10.9). There are no arbitrary constants in Eq. (10.22); the motion is fully determined. Physically, this is surprising, since we should be able to specify the initial position and velocity of any particle obeying Newton's laws. The difficulty is that although the solution in Eq. (10.22) is correct, it is not complete. The complete solution is[1]

$$x = \frac{F_0}{m} \frac{1}{\omega_0^2 - \omega^2} \cos \omega t + B \cos (\omega_0 t + \phi), \tag{10.23}$$

where B and ϕ are arbitrary. As we have seen in Sec. 10.1, the term $B \cos (\omega_0 t + \phi)$ is the general solution for the motion of the free undamped oscillator, $m\ddot{x} + kx = 0$. For a damped system, the amplitude B would decrease exponentially in time and eventually we would be left with the steady-state solution

$$x = \frac{F_0}{m} \frac{1}{\omega_0^2 - \omega^2} \cos \omega t.$$

The effects of the initial conditions die out given enough time. In the remainder of this chapter we shall concentrate on the steady-state solution.

Resonance

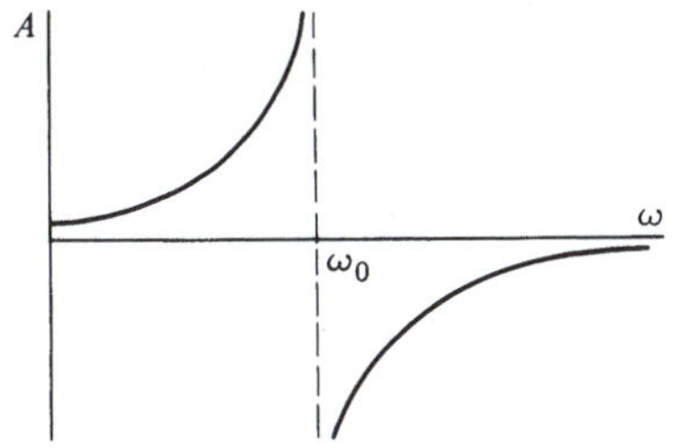

The amplitude of oscillation, Eq. (10.21), is shown in the sketch as a function of the driving frequency ω. A approaches zero as $\omega \to \infty$ and has a finite value at $\omega = 0$, but it increases without limit at $\omega = \omega_0$, when the oscillator is driven at its natural frequency. This great increase of the amplitude when a system is driven at a certain frequency is known as *resonance*. ω_0 is often called the resonance, or natural, frequency of this system. Equation (10.21) predicts that $A \to \infty$ as $\omega \to \omega_0$, but since no physical system can have infinite amplitude, it is apparent that our solution is inadequate at resonance. The difficulty is due to our neglect of friction; when we take friction into account, we shall see that although the amplitude may be large at resonance, it remains finite.

Equation (10.21) asserts that A is positive for $\omega < \omega_0$ and negative for $\omega > \omega_0$. Negative amplitude means that if the force varies as $\cos \omega t$, the displacement varies as $-\cos \omega t$. Since $-\cos \omega t = \cos (\omega t + \pi)$, the negative sign is equivalent to a phase shift of π radians (i.e., 180°) between the driving force and the

[1] This solution can be verified by direct substitution. In the language of differential equations, the first term on the right in Eq. (10.23) is a particular solution and the second term, $B \cos (\omega t + \phi)$, is the general solution of the homogeneous equation $m\ddot{x} + kx = 0$. These two terms represent the complete solution.

displacement. For $\omega < \omega_0$, the displacement is in phase with the driving force. This phase change through resonance of 180°, which is characteristic of all oscillating systems, is easily demonstrated.

Example 10.4 Forced Harmonic Oscillator Demonstration

Break a long rubber band and suspend something like a heavy pocket knife from one end, holding the other end in your hand. The resonant frequency ω_0 is easily determined by observing the free motion. Now slowly jiggle your hand at a frequency $\omega < \omega_0$: the weight will move in phase with your hand. If you jiggle the system with $\omega > \omega_0$, you will find that the weight always moves in the opposite direction to your hand. For a given amplitude of motion of your hand, the weight moves with decreasing amplitude as ω is increased above ω_0. If you try to jiggle the system at resonance $\omega = \omega_0$, the amplitude increases so much that the weight either flies up in the air or hits your hand. In either case the system no longer behaves like a simple oscillator.

The phenomenon of resonance has both positive and negative aspects in practice. By operating at the resonance frequency of a system we can obtain a response of large amplitude for a very small driving force. Organ pipes utilize this principle effectively, and resonant electric circuits enable us to tune our radios to the desired frequency. On the negative side, we do not want motions of large amplitude in the springs of an automobile or in the crankshaft of its engine. To reduce response at resonance a dissipative friction force is needed. We turn now to the analysis of the forced damped oscillator.

The Forced Damped Harmonic Oscillator

If the motion of the oscillating mass is opposed by a viscous retarding force $-bv$, the total force is

$$\begin{aligned} F &= F_{\text{spring}} + F_{\text{viscous}} + F_{\text{driving}} \\ &= -kx - bv + F_0 \cos \omega t \end{aligned}$$

and the equation of motion can be written

$$m\ddot{x} + b\dot{x} + kx = F_0 \cos \omega t.$$

Dividing by m and using $\gamma = b/m$, $\omega_0^2 = k/m$, we have the standard form

$$\ddot{x} + \gamma\dot{x} + \omega_0^2 x = \frac{F_0}{m} \cos \omega t. \qquad 10.24$$

To find the steady-state solution we could again try the trick of taking $x = A \cos \omega t$. However, the term $\gamma \dot{x}$ introduces a term in $\sin \omega t$ which does not appear on the right hand side, so that this trial solution is not adequate. This suggests that we try $x = B \cos \omega t + C \sin \omega t = A \cos (\omega t + \phi)$. If this is substituted into Eq. (10.23), you will find that the solution indeed fits provided that A and ϕ have the values

$$A = \frac{F_0}{m} \frac{1}{[(\omega_0{}^2 - \omega^2)^2 + (\omega\gamma)^2]^{\frac{1}{2}}},$$

$$\phi = \arctan\left(\frac{\gamma\omega}{\omega_0{}^2 - \omega^2}\right). \qquad 10.25$$

A somewhat more formal method for obtaining this solution is presented in Note 10.2.

The behavior of A and ϕ as functions of ω depends markedly on the ratio γ/ω_0 as the sketches show. For light damping, A is maximum for $\omega = \omega_0$, and the amplitude at resonance is

$$A(\omega_0) = \frac{F_0}{m\omega_0\gamma}.$$

As $\gamma \to 0$, $A(\omega_0) \to \infty$, as we expect for an undamped oscillator. Note also that as $\gamma \to 0$, the phase change occurs more and more abruptly. In the limit $\gamma = 0$, the phase changes from 0 to $-\pi$ when $\omega = \omega_0$.

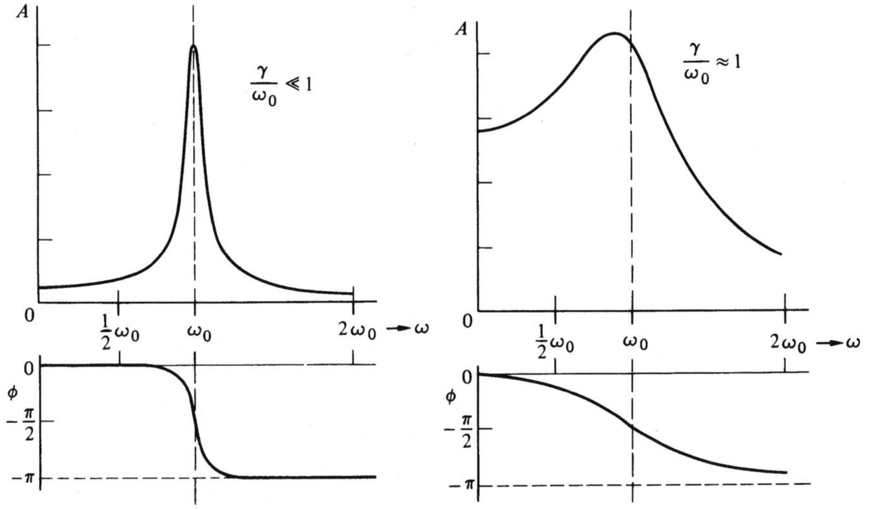

Resonance in a Lightly Damped System: The Quality Factor Q

Energy considerations simplified our discussion of the undriven damped oscillator in Sec. 10.2, and, similarly, they will be useful to us in the driven case. For the steady-state motion, the amplitude is constant in time. Using

$$x = A\cos(\omega t + \phi) \qquad \text{and} \qquad v = -\omega A \sin(\omega t + \phi),$$

we have

$$K(t) = \tfrac{1}{2}mv^2 = \tfrac{1}{2}m\omega^2 A^2 \sin^2(\omega t + \phi)$$
$$U(t) = \tfrac{1}{2}kx^2 = \tfrac{1}{2}kA^2 \cos^2(\omega t + \phi)$$

and

$$\begin{aligned} E(t) &= K(t) + U(t) \\ &= \tfrac{1}{2}A^2[m\omega^2 \sin^2(\omega t + \phi) + k\cos^2(\omega t + \phi)]. \end{aligned}$$

The energy is time-dependent and our analysis is simplified if we focus on time average values, as we did in Sec. 10.1. Since $\langle \cos^2(\omega t + \phi)\rangle = \langle \sin^2(\omega t + \phi)\rangle = \frac{1}{2}$, for an average over one period, we have

$$\begin{aligned} \langle K\rangle &= \tfrac{1}{4}m\omega^2 A^2 \\ \langle U\rangle &= \tfrac{1}{4}kA^2 \\ \langle E\rangle &= \tfrac{1}{4}A^2(m\omega^2 + k) \\ &= \tfrac{1}{4}mA^2(\omega^2 + \omega_0^2). \end{aligned} \qquad 10.26$$

Let us consider how $\langle E\rangle$ varies as a function of ω. Using Eq. (10.25) for A,

$$\langle E(\omega)\rangle = \frac{1}{4}\frac{F_0^2}{m}\frac{(\omega^2 + \omega_0^2)}{[(\omega_0^2 - \omega^2)^2 + (\omega\gamma)^2]}. \qquad 10.27$$

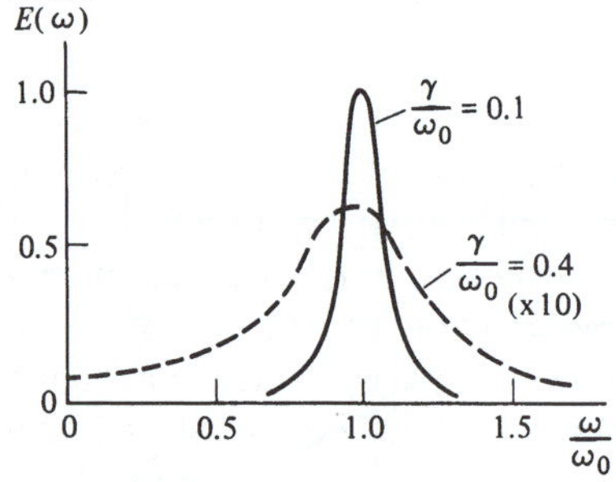

This expression is exact but awkward. It can be written in a much simpler approximate form for the case of light damping, where $\gamma \ll \omega_0$. To see this, consider the sketch of $\langle E(\omega)\rangle$ for $\gamma/\omega_0 = 0.1$ and $\gamma/\omega_0 = 0.4$. For γ sufficiently small, $\langle E(\omega)\rangle$ is effectively zero except near resonance. Hence, there is not much error introduced by replacing ω by ω_0 everywhere in Eq. (10.27) except in the term $(\omega_0^2 - \omega^2)^2$ in the denominator, since this term varies rapidly near resonance. Even this term can be simplified as

$$(\omega_0^2 - \omega^2) = [(\omega_0 + \omega)(\omega_0 - \omega)] \approx 2\omega_0(\omega_0 - \omega).$$

With this approximation, $\langle E(\omega)\rangle$ takes the simple form

$$\langle E(\omega)\rangle = \frac{1}{4}\frac{F_0^2}{m}\frac{2\omega_0^2}{4\omega_0^2(\omega-\omega_0)^2+\omega_0^2\gamma^2}$$

$$= \frac{1}{8}\frac{F_0^2}{m}\frac{1}{(\omega-\omega_0)^2+(\gamma/2)^2}. \qquad 10.28$$

The plot of the function $[(\omega-\omega_0)^2+(\gamma/2)^2]^{-1}$, which contains the entire frequency dependence of $\langle E(\omega)\rangle$, is called a *resonance curve*, or *lorentzian*. Resonance curves for several values of γ are plotted below. For concreteness, we have taken $\omega_0 = 8$ rad/s. γ is given in units of s^{-1}.

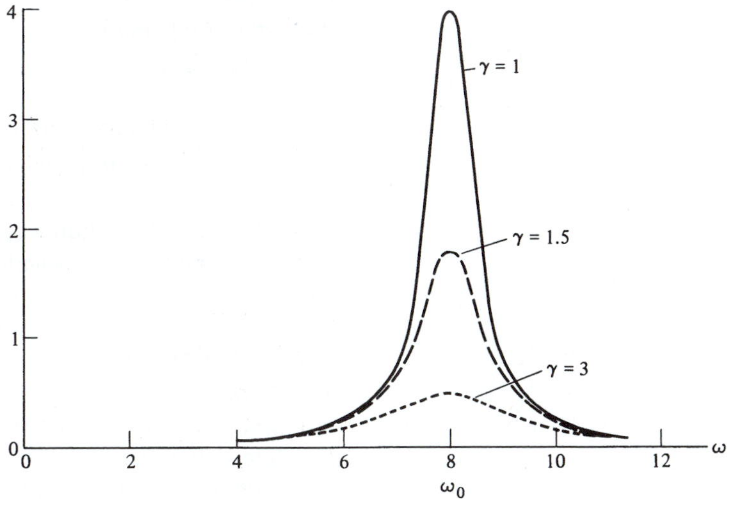

Let us look more closely at the resonance curve. Its maximum height is $4/\gamma^2$. It falls to one-half maximum when

$$(\omega-\omega_0)^2 = (\gamma/2)^2$$

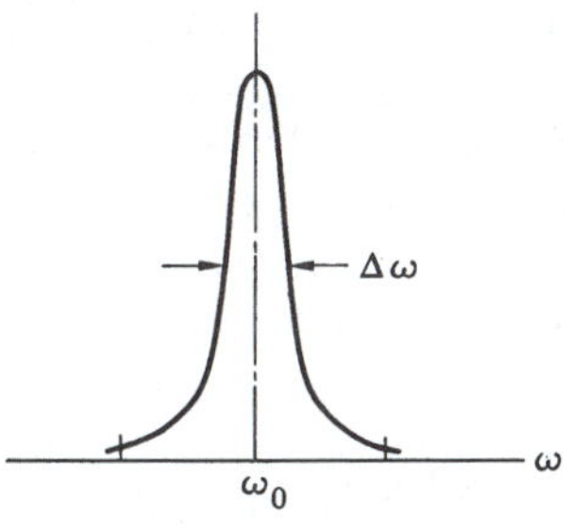

or when $\omega - \omega_0 = \pm\gamma/2$. The full width of the curve at half maximum value is often called the *resonance width*. If the resonance curve drops to half its maximum value at ω_+ on the high frequency side, and at ω_- on the low frequency side, then the resonance width is $\omega_+ - \omega_- = 2(\gamma/2) = \gamma$. The resonance width is denoted by $\Delta\omega$ in the sketch at left. In general, we have

$$\Delta\omega = \gamma. \qquad 10.29$$

As γ decreases the curve becomes higher and narrower, the range of frequency over which the system responds becomes smaller, and the oscillator becomes increasingly selective in frequency.

The frequency-selective property of an oscillator can be characterized in a simple fashion by Q, the quality factor introduced in Sec. 10.2. Recall that Q is defined as the ratio of energy stored in the oscillator to energy lost per radian of oscillation. For a lightly damped system oscillating freely, Q has the value

$$Q = \frac{\omega_0}{\gamma},$$

as we showed in Eq. (10.19). The same oscillator, when driven, has a resonance curve with frequency width $\Delta\omega = \gamma$. Hence, the ratio of resonance frequency to the width of the resonance curve, $\omega_0/\Delta\omega$, is $\omega_0/\gamma = Q$. In fact, Q is often defined by

$$Q = \frac{\text{resonance frequency}}{\text{frequency width of resonance curve}}. \tag{10.30}$$

Incidentally, if we had applied the definition of Q in terms of energy to the driven oscillator, the result would have been the same, $Q = \omega_0/\gamma$. The proof of this is left for a problem.

Although Q is fundamentally defined in terms of energy, its chief use in practice is to characterize the frequency response of a system. The drawing shows two resonance curves with different Q's. The heights at resonance have been made equal to facilitate comparison of the widths. It is apparent that the system with $Q = 10$ is considerably more selective than that with $Q = 5$. As pointed out in Example 10.3, certain atomic systems can have a Q greater than 10^8. The sharpness of the resonance curve means that the system will not respond unless driven very near its resonance frequency. Since the resonance frequency is determined by atomic constants, the frequency of oscillation is essentially independent of external influences. Frequencies from such "atomic clocks" are so accurate that they have superseded astronomical time standards.

Example 10.5 Vibration Eliminator

Occasionally one needs to reduce the effect of floor vibrations on a delicate apparatus such as a sensitive balance or a precision optical system. This can be accomplished by mounting the apparatus on an "air

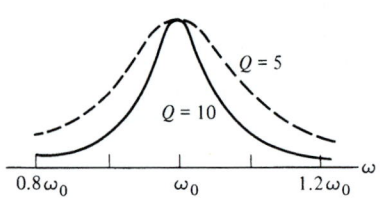

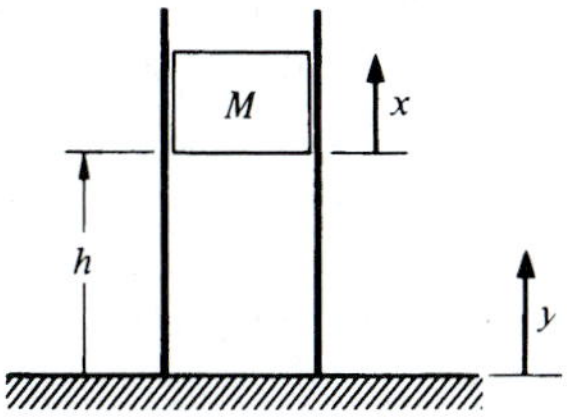

table" whose legs are hollow tubes with pistons supported by air pressure. One such leg is shown schematically in the drawing. The area of the column is A, and the mass it supports is M.

The static forces on M are related by the equilibrium condition

$$P_0A = Mg + P_{\text{at}}A,$$

where P_0 is the pressure of gas in the cylinder at equilibrium and P_{at} is the atmospheric pressure on the upper face of M. For some air tables, the weight Mg is much greater than the atmospheric force, and we shall neglect the term $P_{\text{at}}A$ in the following. Hence,

$$P_0A = Mg.$$

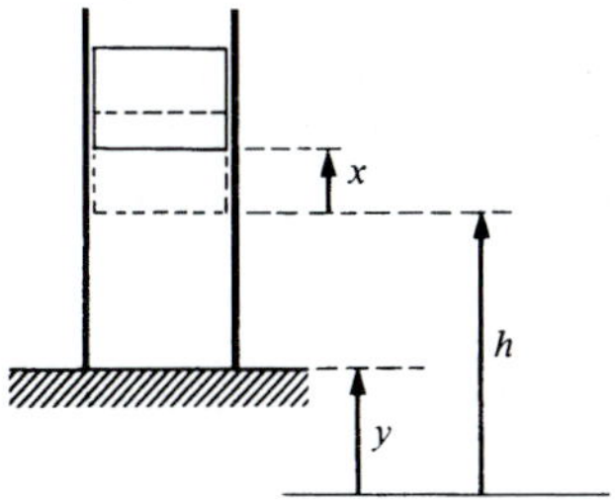

Inertial frame

The equilibrium height of M is h. Let x be the displacement of M from equilibrium relative to an inertial frame. The smaller the value of x, the more nearly motionless the table top will be in inertial space. Floor vibrations cause the lower end of the table leg to move vertically a distance y. When M moves relative to the floor, the volume and the pressure of the trapped gas change. If P is the instantaneous pressure in the cylinder, the equation of motion of M is

$$M\ddot{x} = PA - Mg.$$

According to Boyle's law, the pressure in the cylinder varies inversely with volume for a gas at constant temperature. Therefore

$$\begin{aligned} PV &= \text{constant} \\ &= P_0V_0 \\ &= P_0Ah. \end{aligned}$$

The volume V is

$$V = A(h + x - y).$$

Therefore

$$\begin{aligned} P &= \frac{P_0V_0}{V} = P_0\frac{h}{h + x - y} \\ &\approx P_0\left(1 - \frac{x}{h} + \frac{y}{h}\right). \end{aligned}$$

In the last step we have assumed that the displacements x and y are small compared with h, the height of the table leg.

The equation of motion becomes

$$M\ddot{x} = P_0A\left(1 - \frac{x}{h} + \frac{y}{h}\right) - Mg.$$

Since we are neglecting the atmospheric force, $P_0A = Mg$, and the equation of motion is simply

$$M\ddot{x} = \frac{Mg}{h}(-x + y)$$

$$\ddot{x} + \frac{g}{h}x = \frac{g}{h}y.$$

If the floor vibration is $y = y_0 \cos \omega t$, M moves like an undamped driven oscillator. Using Eq. (10.22) we see that the solution of the equation is

$$x = x_0 \cos \omega t,$$

where

$$x_0 = y_0 \frac{\omega_0^2}{\omega_0^2 - \omega^2}$$

and

$$\omega_0 = \sqrt{\frac{g}{h}}.$$

The object of the air suspension is to make the ratio

$$\frac{x_0}{y_0} = \frac{\omega_0^2}{(\omega_0^2 - \omega^2)}$$

as small as possible. For $\omega \ll \omega_0$, $x_0 = y_0$ and the vibration is transmitted without reduction. For $\omega \gg \omega_0$, $x_0/y_0 = -\omega_0^2/\omega^2$, and the amplitude of vibration is reduced. Thus, for the vibration eliminator to be successful, the resonance frequency must be low compared with the driving frequency. Since $\omega_0 = \sqrt{g/h}$, this requires as long a leg as possible. (Note that the resonance frequency is independent of the mass, a surprising aspect of this type of support.)

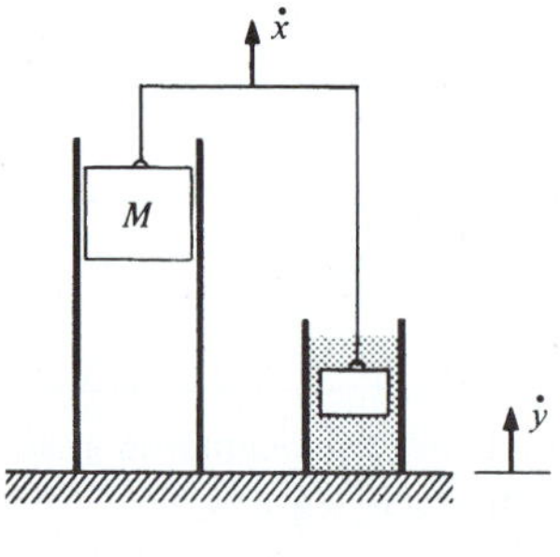

The system suffers from one fatal flaw; if vibration occurs near the resonant frequency, the vibration eliminator becomes a vibration amplifier. To avoid this, some damping mechanism must be provided. Often this is accomplished with a device called a *dashpot*, which consists of a piston in a cylinder of oil. The dashpot provides a viscous retarding force $-bv$, where v is the relative velocity of its ends.

$$v = \dot{x} - \dot{y}.$$

The equation of motion is

$$M\ddot{x} = \frac{Mg}{h}(-x + y) - b(\dot{x} - \dot{y})$$

$$\ddot{x} + \gamma\dot{x} + \omega_0^2 x = \omega_0^2 y + \gamma\dot{y},$$

where

$$\gamma = \frac{b}{M} \qquad \text{and} \qquad \omega_0^2 = \frac{g}{h}.$$

With $y = y_0 \cos \omega t$, this is the equation of a driven damped oscillator. However, the motion of the floor has introduced an additional driving term $\gamma \dot{y} = -\gamma\omega y_0 \sin \omega t$. The steady-state amplitude x_0 can be found by substituting $x = x_0 \cos(\omega t + \phi)$ in the equation. A simpler method is to use complex variables, as outlined in Notes 10.1 and 10.2. Let

$$y = y_0 e^{i\omega t}$$
$$x = x_0 e^{i\omega t}.$$

y_0 and x_0 are now complex numbers. Substituting in the equation of motion gives

$$(-\omega^2 + i\omega\gamma + \omega_0^2)x_0 e^{i\omega t} = (\omega_0^2 + i\omega\gamma)y_0 e^{i\omega t}$$

$$x_0 = \left[\frac{\omega_0^2 + i\omega\gamma}{(\omega_0^2 - \omega^2) + i\omega\gamma}\right] y_0.$$

We are interested in the ratio of the magnitudes, $|x_0|/|y_0|$.

$$\frac{|x_0|}{|y_0|} = \sqrt{\frac{x_0 x_0^*}{y_0 y_0^*}}$$
$$= \left[\frac{\omega_0^4 + (\omega\gamma)^2}{(\omega_0^2 - \omega^2)^2 + (\omega\gamma)^2}\right]^{\frac{1}{2}}.$$

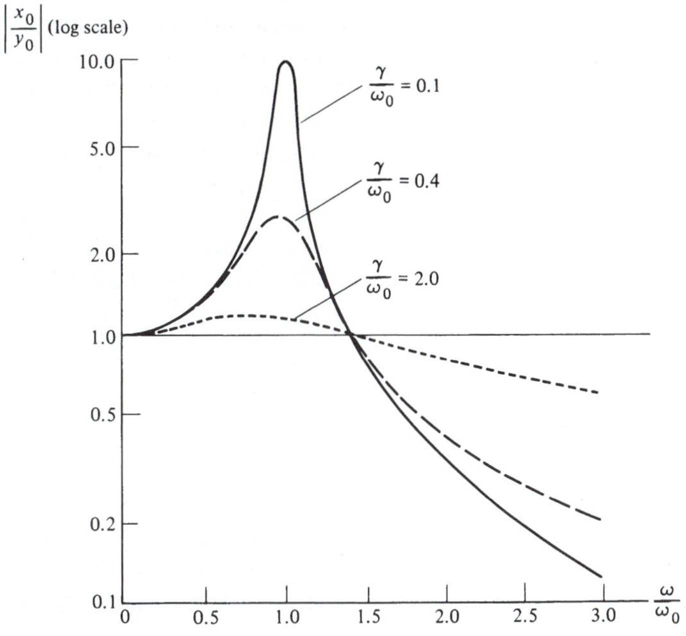

The graph shows $|x_0|/|y_0|$ versus ω/ω_0 for various values of γ/ω_0. For ω/ω_0 less than about 1.5, $|x_0|/|y_0| > 1$. The vibration is actually enhanced, showing that even with damping it is essential to reduce the resonance frequency below the driving frequency. When ω/ω_0 is greater than 1.5, $|x_0|/|y_0| < 1$. For these higher frequencies, the vibration isolation is more effective the smaller the damping. However, small damping increases the danger from vibrations near resonance. Practical air tables have resonance frequencies of 1 Hz or less.

Many vibration elimination systems use springs instead of an air suspension. However, this does not change the form of the equation of motion. Often coil springs are used in automobiles to isolate the chassis from road vibrations. Damping is provided by shock absorbers, a type of dashpot. The resonance frequency is $\omega_0 = \sqrt{k/M}$, where k is the spring constant and M is the mass. If a smooth turnpike ride is the chief consideration, one wants a massive car with weak damping and soft springs. Such a car is difficult to control on a bumpy road where resonance can be excited. The best suspensions are heavily damped and feel rather stiff. The danger in driving with defective shock absorbers is that the car may be thrown out of control if it is excited at resonance by bumps in the road.

10.4 Response in Time Versus Response in Frequency

The smaller the damping of a free oscillator, the more slowly its energy is dissipated. The same oscillator, when driven, becomes increasingly more frequency selective as the damping is decreased. As we shall now show, the time dependence of the free oscillator and the frequency dependence of the driven oscillator are intimately related.

Recall from Eq. (10.16) that the energy of a free oscillator is

$$E(t) = E_0 e^{-\gamma t}.$$

The damping time is $\tau = 1/\gamma$.

Next, consider the response in frequency of the same oscillator when it is driven by a force $F_0 \cos \omega t$. From Eq. (10.29) the width of the resonance curve is

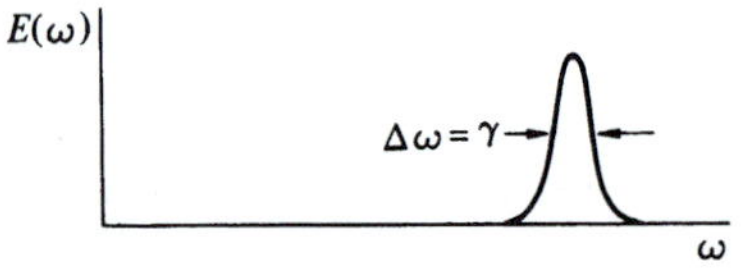

Forced oscillator

$$\Delta\omega = \gamma.$$

The damping time τ and the resonance curve width $\Delta\omega$ obey

$$\tau\,\Delta\omega = 1. \qquad 10.31$$

According to this result it is impossible to design an oscillator with arbitrary damping time and resonance width; if we choose one, the other is automatically fixed by Eq. (10.31).

Equation (10.31) has many implications for the design of mechanical and electrical systems. Any element which is highly frequency selective will oscillate for a long time if it is accidentally perturbed. Furthermore, such an element will take a long time to reach the steady state when a driving force is applied because the effects of the initial conditions die out only slowly. More generally, Eq. (10.31) plays a fundamental role in quantum mechanics; it is closely related to one form of the Heisenberg uncertainty principle.

Note 10.1 Solution of the Equation of Motion for the Undriven Damped Oscillator

THE USE OF COMPLEX VARIABLES

All the equations of motion in this chapter can be solved simply by using complex variables.[1] Here is a summary of the algebra of complex numbers.

1. Every complex number z can be written in the cartesian form $x + iy$, where $i^2 = -1$. x is the *real* part of z, and y is the *imaginary* part. The sum of two complex numbers $z_1 = x_1 + iy_1$ and $z_2 = x_2 + iy_2$ is the complex number $z_1 + z_2 = (x_1 + x_2) + i(y_1 + y_2)$. The product of z_1 and z_2 is

$$\begin{aligned} z_1z_2 &= (x_1 + iy_1)(x_2 + iy_2) \\ &= x_1x_2 + ix_1y_2 + iy_1x_2 + i^2y_1y_2 \\ &= (x_1x_2 - y_1y_2) + i(x_1y_2 + y_1x_2). \end{aligned}$$

If two complex numbers are equal, the real parts and the imaginary parts are respectively equal.

$$x_1 + iy_1 = x_2 + iy_2$$

implies that

$$\begin{aligned} x_1 &= x_2 \\ y_1 &= y_2. \end{aligned}$$

2. $z^* \equiv x - iy$ is the *complex conjugate* of $z = x + iy$. The quantity $|z| = \sqrt{zz^*}$ is the *magnitude* of z.

$$\begin{aligned} |z| &= \sqrt{zz^*} \\ &= [(x + iy)(x - iy)]^{\frac{1}{2}} \\ &= \sqrt{x^2 + y^2}. \end{aligned}$$

[1] A simple treatment of the algebra of complex numbers may be found in most of the calculus texts listed at the end of Chap. 1.

3. Every complex number z can be written in the polar form $re^{i\theta}$. r is a real number, the *modulus*, and θ is the *argument*. To go from cartesian to polar form we use De Moivre's theorem

$$e^{i\theta} = \cos\theta + i\sin\theta.$$

Hence,

$$\begin{aligned} re^{i\theta} &= r\cos\theta + ir\sin\theta \\ &= x + iy, \end{aligned}$$

from which it follows that

$$\begin{aligned} x &= r\cos\theta \\ y &= r\sin\theta \end{aligned}$$

and

$$\begin{aligned} r &= \sqrt{x^2 + y^2} \\ \theta &= \arctan\frac{y}{x}. \end{aligned}$$

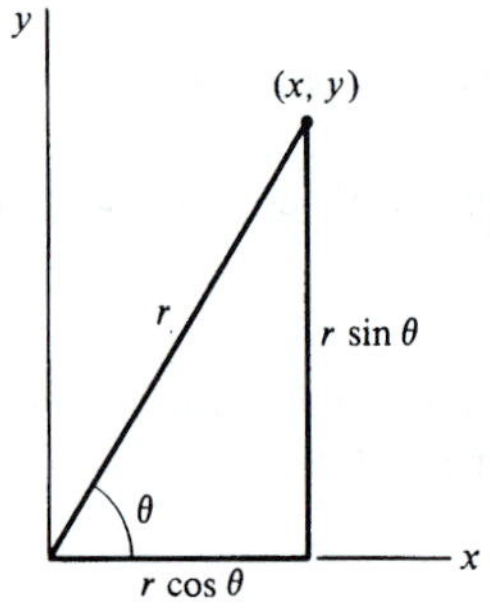

We see that $r = |z|$.

Complex numbers can be represented graphically. Let the horizontal axis be the real (x) axis, and the vertical axis be the imaginary (y) axis. The complex number $x + iy$ is represented by the point (x,y). As the sketch shows, introduction of the polar form is analogous to the use of plane polar coordinates.

Here are some examples:

1. Express $z = (3 + 4i)/(2 + i)$ in cartesian form. The method is to multiply numerator and denominator by the complex conjugate of the denominator.

$$\begin{aligned} z &= \frac{3 + 4i}{2 + i} \\ &= \frac{3 + 4i}{2 + i}\cdot\frac{2 - i}{2 - i} \\ &= \frac{6 + 8i - 3i - 4i^2}{4 + 2i - 2i - i^2} \\ &= \frac{10 + 5i}{5} \\ &= 2 + i \end{aligned}$$

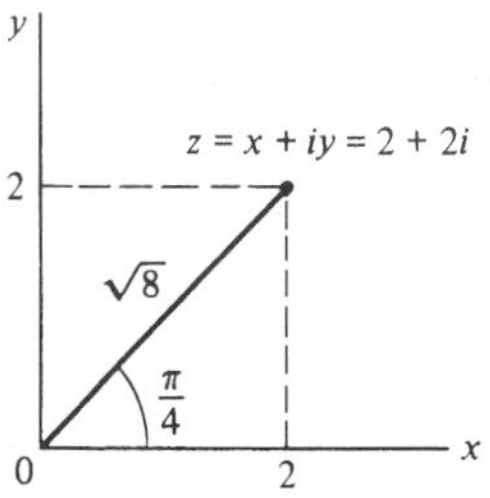

2. Express $z = 2 + 2i$ in polar form.

$$r = |z| = \sqrt{2^2 + 2^2} = \sqrt{8}$$

$$\theta = \arctan\frac{2}{2} = \frac{\pi}{4}$$

$$z = \sqrt{8}e^{i\pi/4}$$

THE DAMPED OSCILLATOR

We turn now to the equation for the damped oscillator.

$$\ddot{x} + \gamma\dot{x} + \omega_0^2 x = 0 \qquad 1$$

To cast this into complex form we introduce the companion equation

$$\ddot{y} + \gamma\dot{y} + \omega_0^2 y = 0. \qquad 2$$

Multiplying Eq. (2) by i and adding it to Eq. (1) yields

$$\ddot{z} + \gamma\dot{z} + \omega_0^2 z = 0. \qquad 3$$

Note that either the real or imaginary part of z is an acceptable solution for the equation of motion.

Since the coefficients of the derivatives of z are all constants, a natural choice for the solution of Eq. (3) is

$$z = z_0 e^{\alpha t}, \qquad 4$$

where z_0 and α are independent of t. With this trial solution Eq. (3) yields

$$\alpha^2 z_0 e^{\alpha t} + \alpha\gamma z_0 e^{\alpha t} + \omega_0^2 z_0 e^{\alpha t} = 0.$$

Dividing out the common factor $z_0 e^{\alpha t}$, we have

$$\alpha^2 + \alpha\gamma + \omega_0^2 = 0, \qquad 5$$

which has the solution

$$\alpha = -\frac{\gamma}{2} \pm \sqrt{\frac{\gamma^2}{4} - \omega_0^2}. \qquad 6$$

Let us call the two roots α_1 and α_2. We see that our solution can be written as

$$z = z_A e^{\alpha_1 t} + z_B e^{\alpha_2 t},$$

where z_A and z_B are constants.

There are three possible forms of the solution, depending on whether α is real or complex. We consider these solutions in turn.

Case 1 Light Damping: $\gamma^2 \ll 4\omega_0^2$

In this case $\sqrt{\gamma^2/4 - \omega_0^2}$ is imaginary and we can write

$$\alpha = -\frac{\gamma}{2} \pm i\sqrt{\omega_0^2 - \frac{\gamma^2}{4}} \qquad 7$$

$$= -\frac{\gamma}{2} \pm i\omega_1,$$

where

$$\omega_1 = \sqrt{\omega_0^2 - \frac{\gamma^2}{4}}.$$

The solution is

$$z = e^{-(\gamma/2)t}\,(z_1 e^{i\omega_1 t} + z_2 e^{-i\omega_1 t}),$$

where z_1 and z_2 are complex constants. In order to find the real part of z we write the complex numbers in cartesian form.

$$x + iy = e^{-(\gamma/2)t}[(x_1 + iy_1)(\cos\omega_1 t + i\sin\omega_1 t) + (x_2 + iy_2)(\cos\omega_1 t - i\sin\omega_1 t)]$$

The real part x is

$$x = e^{-(\gamma/2)t}(B\cos\omega_1 t + C\sin\omega_1 t)$$

or

$$x = Ae^{-(\gamma/2)t}\cos(\omega_1 t + \phi), \qquad 8$$

where A and ϕ are new arbitrary constants. This is the result quoted in Eq. (10.9). Incidentally, the imaginary part of z, which is also an acceptable solution, has exactly the same form.

Case 2 Heavy Damping: $\gamma^2/4 > \omega_0^2$

In this case, $\sqrt{\gamma^2/4 - \omega_0^2}$ is real and Eq. (5) has the solution

$$\alpha = -\frac{\gamma}{2} \pm \frac{\gamma}{2}\sqrt{1 - \frac{4\omega_0^2}{\gamma^2}}.$$

Both roots are negative, and we have

$$z = z_1 e^{-|\alpha_1|t} + z_2 e^{-|\alpha_2|t}. \qquad 9$$

The exponentials are real. The real part of z is

$$x = Ae^{-|\alpha_1|t} + Be^{-|\alpha_2|t}. \qquad 10$$

This solution has no oscillatory behavior; the motion is known as *overdamped.*

Case 3 Critical Damping: $\gamma^2/4 = \omega_0^2$

If $\gamma^2/4 = \omega_0^2$ we have only the single root

$$\alpha = -\frac{\gamma}{2}.$$

The corresponding solution is

$$x = Ae^{-(\gamma/2)t}. \tag{11}$$

However, this solution is incomplete. Mathematically, the solution of a second order linear differential equation always involves two arbitrary constants. Physically, the solution must have two constants to allow us to specify the initial position and initial velocity of the oscillator. As described in texts on differential equations, the second solution can be found by using a "variation of parameters" trial solution.

$$x = u(t)e^{(-\gamma/2)(t)}.$$

Substituting in Eq. (1) and recalling that $\gamma = 2\omega_0$ for this case, we find that $u(t)$ must satisfy the equation

$$\ddot{u} = 0.$$

Hence,

$$u = a + bt$$

and the general solution is

$$x = Ae^{-(\gamma/2)t} + Bte^{-(\gamma/2)t}. \tag{12}$$

Note 10.2 Solution of the Equation of Motion for the Forced Oscillator

We wish to solve

$$\ddot{x} + \gamma\dot{x} + \omega_0^2 x = \frac{F_0}{m}\cos\omega t. \tag{1}$$

Consider the companion equation

$$\ddot{y} + \gamma\dot{y} + \omega_0^2 y = \frac{F_0}{m}\sin\omega t. \tag{2}$$

Multiplying Eq. (2) by i and adding to Eq. (1) yields

$$\ddot{z} + \gamma\dot{z} + \omega_0^2 z = \frac{F_0}{m}e^{i\omega t}. \tag{3}$$

z must vary as $e^{i\omega t}$, so we try

$$z = z_0 e^{i\omega t}.$$

Inserting this in Eq. (3) gives

$$(-\omega^2 + i\omega\gamma + \omega_0{}^2)z_0 e^{i\omega t} = \frac{F_0}{m} e^{i\omega t}$$

or

$$z_0 = \frac{F_0}{m} \frac{1}{\omega_0{}^2 - \omega^2 + i\omega\gamma}.$$

We can put z_0 into cartesian form by multiplying numerator and denominator by the complex conjugate of the denominator.

$$\begin{aligned} z_0 &= \frac{F_0}{m} \frac{1}{(\omega_0{}^2 - \omega^2) + i\omega\gamma} \frac{(\omega_0{}^2 - \omega^2) - i\omega\gamma}{(\omega_0{}^2 - \omega^2) - i\omega\gamma} \\ &= \frac{F_0}{m} \frac{(\omega_0{}^2 - \omega^2) - i\omega\gamma}{(\omega_0{}^2 - \omega^2)^2 + (\omega\gamma)^2} \end{aligned}$$

In polar form, $z_0 = Re^{i\phi}$, where

$$\begin{aligned} R &= \sqrt{z_0 z_0^*} \\ &= \frac{F_0}{m} \left[\frac{1}{(\omega_0{}^2 - \omega^2)^2 + (\omega\gamma)^2}\right]^{\frac{1}{2}} \end{aligned} \tag{4}$$

and

$$\phi = \arctan\left(\frac{\omega\gamma}{\omega^2 - \omega_0{}^2}\right). \tag{5}$$

The complete solution is

$$z = Re^{i\phi}e^{i\omega t},$$

which has the real part

$$x = R \cos(\omega t + \phi).$$

The steady-state motion is completely specified by the amplitude R and the phase angle ϕ. Both R and ϕ are contained implicitly in the single complex number z_0.

Problems

10.1 Show by direct calculation that $\langle\overline{\sin^2(\omega t)}\rangle = \frac{1}{2}$, where the time average is taken over any complete period $t_1 \le t \le t_1 + 2\pi/\omega$.

Show also that $\langle\overline{\sin(\omega t)\cos(\omega t)}\rangle = 0$ when the average is over a complete period.

10.2 A 0.3-kg mass is attached to a spring and oscillates at 2 Hz with a Q of 60. Find the spring constant and damping constant.

10.3 In an undamped free harmonic oscillator the motion is given by $x = A \sin \omega_0 t$. The displacement is maximum exactly midway between the zero crossings.

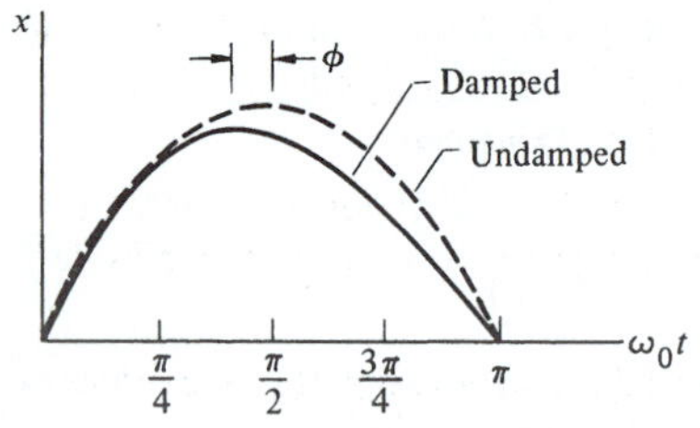

In a damped oscillator the motion is no longer sinusoidal, and the maximum is advanced before the midpoint of the zero crossings. Show that the maximum is advanced by a phase angle ϕ given approximately by

$$\phi = \frac{1}{2Q},$$

where we assume that Q is large.

10.4 The *logarithmic decrement* δ is defined to be the natural logarithm of the ratio of successive maximum displacements (in the same direction) of a free damped oscillator. Show that $\delta = \pi/Q$.

Find the spring constant k and damping constant b of a damped oscillator having a mass of 5 kg, frequency of oscillation 0.5 Hz, and logarithmic decrement 0.02.

10.5 If the damping constant of a free oscillator is given by $\gamma = 2\omega_0$, the system is said to be critically damped. Show by direct substitution that in this case the motion is given by

$$x = (A + Bt)e^{-(\gamma/2)t},$$

where A and B are constants.

A critically damped oscillator is at rest at equilibrium. At $t = 0$ it is given a blow of total impulse I. Sketch the motion, and find the time at which the velocity starts to decrease.

10.6 *a.* A mass of 10 kg falls 50 cm onto the platform of a spring scale, and sticks. The platform eventually comes to rest 10 cm below its initial position. The mass of the platform is 2 kg. Find the spring constant.

b. It is desired to put in a damping system so that the scale comes to rest in minimum time without overshoot. This means that the scale must be critically damped (see Note 10.1). Find the necessary damping constant and the equation for the motion of the platform after the mass hits.

10.7 Find the driving frequency for which the velocity of a forced damped oscillator is exactly in phase with the driving force.

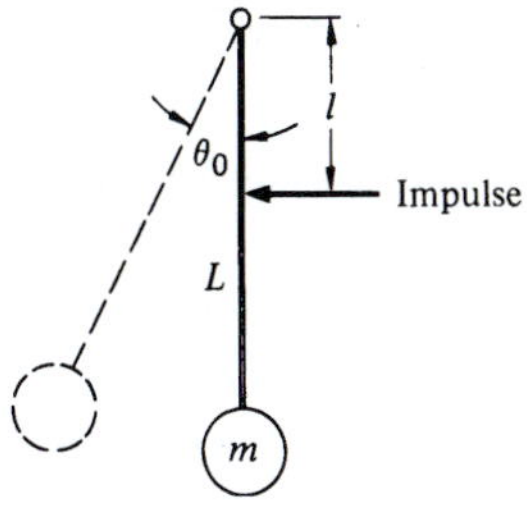

10.8 The pendulum of a grandfather's clock activates an escapement mechanism every time it passes through the vertical. The escapement is under tension (provided by a hanging weight) and gives the pendulum a small impulse a distance l from the pivot. The energy transferred by this impulse compensates for the energy dissipated by friction, so that the pendulum swings with a constant amplitude.

a. What is the impulse needed to sustain the motion of a pendulum of length L and mass m, with an amplitude of swing θ_0 and quality factor Q?

b. Why is it desirable for the pendulum to engage the escapement as it passes vertical rather than at some other point of the cycle?

10.9 Show that for a lightly damped forced oscillator

$$\frac{\text{average energy stored in the oscillator}}{\text{average energy dissipated per radian}} \approx \frac{\omega_0}{\gamma} = Q.$$

10.10 A small cuckoo clock has a pendulum 25 cm long with a mass of 10 g and a period of 1 s. The clock is powered by a 200-g weight which falls 2 m between the daily windings. The amplitude of the swing is 0.2 rad. What is the Q of the clock? How long would the clock run if it were powered by a battery with 1 J capacity?

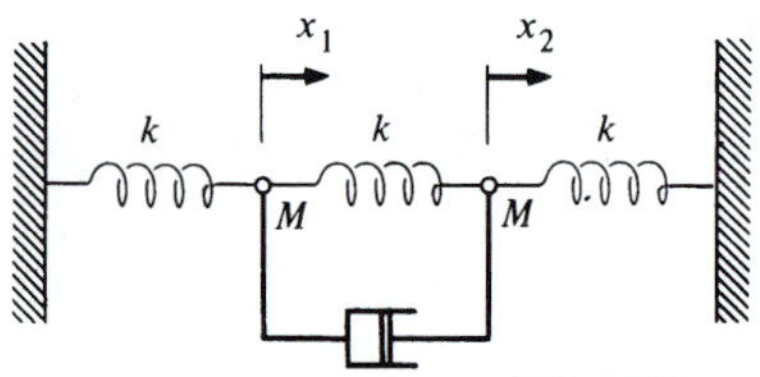

10.11 Two particles, each of mass M, are hung between three identical springs. Each spring is massless and has spring constant k. Neglect gravity. The masses are connected as shown to a dashpot of negligible mass.

The dashpot exerts a force bv, where v is the relative velocity of its two ends. The force opposes the motion. Let x_1 and x_2 be the displacements of the two masses from equilibrium.

a. Find the equation of motion for each mass.

b. Show that the equations of motion can be solved in terms of the new dependent variables $y_1 = x_1 + x_2$ and $y_2 = x_1 - x_2$.

c. Show that if the masses are initially at rest and mass 1 is given an initial velocity v_0, the motion of the masses after a sufficiently long time is

$$x_1 = x_2$$

$$= \frac{v_0}{2\omega} \sin \omega t.$$

Evaluate ω.

10.12 The motion of a damped oscillator driven by an applied force $F_0 \cos \omega t$ is given by $x_a(t) = A \cos (\omega t + \phi)$, where A and ϕ are given by Eq. (10.25). Consider an oscillator which is released from rest at $t = 0$. Its motion must satisfy $x(0) = 0$, $v(0) = 0$, but after a very long time, we expect that $x(t) = x_a(t)$. To satisfy these conditions we can take as the solution

$$x(t) = x_a(t) + x_b(t),$$

where $x_b(t)$ is the solution to the equation motion of the free damped oscillator, Eq. (10.8).

a. Show that if $x_a(t)$ satisfies the equation of motion for the forced damped oscillator, then so does $x(t) = x_a(t) + x_b(t)$, where $x_b(t)$ satisfies the equation of motion of the free damped oscillator, Eq. (10.25).

b. Choose the arbitrary constants in $x_b(t)$ so that $x(t)$ satisfies the initial conditions. [$x_b(t)$ is given by Eq. (10.9). Note that A and ϕ here are arbitrary.]

c. Sketch the resulting motion for the case where the oscillator is driven at resonance.

11 THE SPECIAL THEORY OF RELATIVITY

11.1 The Need for a New Mode of Thought

In some ways the structure of physics resembles a mansion whose outward form is apparent to the casual visitor but whose inner life —the customs and rituals which give a special outlook and kinship to its occupants—require time and effort to comprehend. Indeed, initiation into this special knowledge is the goal of our present endeavor. In the first ten chapters we introduced and applied the fundamental laws of classical mechanics; hopefully you now feel familiar with these laws and have come to appreciate their beauty, their essential simplicity, and their power.

Unfortunately, in order to present dynamics in a concise and tidy form, we have generally sidestepped discussion of how physics actually grew. In Chaps. 11 through 14 we are going to discuss one of the great achievements of modern physics, the special theory of relativity. Rather than present the theory as a completed structure—a simple set of postulates with the rules for their application—we shall depart from our previous style and look into the background of the theory and its rationale.

If the structure of physics is a mansion, it is a mansion of ancient origin. It is founded on the remains of prehistoric hovels where man first kept track of the moon and tried to understand the simple patterns of nature. Traces of antiquity lie hidden in the site: Phoenician and Egyptian, Babylonian, and, of course, Greek. Compass and straightedge lie scattered among lodestone and amber, artifacts of astrologer and alchemist. The mansion is built on the debris of false starts and painful struggles to understand nature honestly. None of this is visible, however, and we take the present structure much for granted. The outer shell was built in the seventeenth century by Kepler, Galileo, Newton, and others, such as Huygens, Hooke, Leibniz, Bernoulli, and Boyle. The major architects have one characteristic in common: while extending the external dimensions of the mansion by applying physics to new areas, they also deepened its foundations by advancing our knowledge of the fundamental laws. The greatest of these figures is Newton, who revealed the laws of dynamics and of gravity, cornerstones of modern science. At the same time he vigorously applied physics to the natural world. Newton executed meticulous experiments in heat flow, optics, and the motion of bodies under viscous forces; he investigated the shape of the moon, the tides along the coast of England, and how to build bridges.

The momentum generated by Newton's discoveries gave physics

an impetus which is still very much with us. The eighteenth and nineteenth centuries saw a flowering of science as physicists such as Euler, Lagrange, Laplace, Faraday, and Maxwell extended our knowledge of the physical world. However, their efforts were directed at upward extension of the mansion; Newton's account of the fundamental laws of physics was so overwhelming, and so successful, that not until the last quarter of the nineteenth century was there a serious attempt to investigate the foundations.

It was the German physicist Ernst Mach who first successfully challenged newtonian thought. Although Mach's work left newtonian physics more or less intact, his thinking was crucial in the revolution shortly to come. In 1883 Mach published his text "The Science of Mechanics," which incorporated a critique of newtonian physics, the first incisive criticism of Newton's theory of dynamics. In addition to presenting a lucid account of newtonian mechanics, the text incorporates several significant contributions to the fundamentals of mechanics. Mach clarified newtonian dynamics by carefully analyzing Newton's explanation of the dynamical laws, taking care to distinguish between definitions, derived results, and statements of physical law. Mach's approach is now widely accepted; our discussion of Newton's laws in Chap. 2 is very much in Mach's spirit.

"The Science of Mechanics" raised the question of the distinction between absolute and relative motion. Mach pointed out Newton's ambivalence on this subject, although he went on to show that the question was irrelevant to the application of newtonian dynamics. In the process he dwelt on the problem of inertia and enunciated the principle that now bears his name: inertia is not an intrinsic property of matter or space but depends on the existence of all matter in the universe. We encountered Mach's principle in our discussion of fictitious forces in Chap. 8, but we shall not dwell on it here for it turns out that the problem of inertia was not the crucial difficulty with newtonian mechanics.

The fundamental weakness in newtonian dynamics, as Mach pointed out, centers on Newton's conception of space and time. In a preface to his dynamical theory, Newton avowed that he would forgo abstract speculation and deal only with observable facts. Although such a point of view is now commonplace, at the time it represented a tremendous intellectual leap. Before Newton, the business of natural philosophy was to explain the reasons for things, to find a rational account for the workings of nature, rather than to describe natural phenomena quantitatively. Newton essentially reversed the priorities. Against the criticism that his theory of universal gravitation merely described gravity without accounting for its origin, Newton replied "I do not make hypotheses."

Unfortunately, Newton was not completely faithful to his resolve to avoid abstract speculation and to deal only with demonstrable facts. In particular, consider the following description of time that appears in the "Principia." (The excerpt is condensed.)

> Absolute, true and mathematical time, of itself and by its own true nature, flows uniformly on, without regard to anything external.
>
> Relative, apparent and common time is some sensible and external measure of absolute time estimated by the motions of bodies, whether accurate or inequable, and is commonly employed in place of true time; as an hour, a day, a month, a year.

Mach comments that "it would appear as though Newton in the remarks cited here still stood under the influence of medieval philosophy, as though he had grown unfaithful to his resolve to investigate only actual facts." Mach goes on to point out that since time is necessarily measured by the repetitive motion of some physical system, for instance the pendulum of a clock or the revolution of the earth about the sun, then the properties of time must be connected with the laws which describe the motions of physical systems. Simply put, Newton's idea of time without clocks is metaphysical; to understand the properties of time we must observe the properties of clocks. A simple idea? Yes, indeed, were it not for the fact that the idea of absolute time is so natural that the eventual consequences of Mach's position, the relativistic description of time, still come as something of a shock to the student of science.

There are similar weaknesses in the newtonian view of space. Mach argued that since position in space is determined with measuring rods, the properties of space can be understood only by investigating the properties of meter sticks. We must look to nature to understand space, not to platonic ideals.

Mach's special contribution was to examine the most elemental aspects of newtonian thought, to look critically at matters which seem too simple to discuss, and to insist that we turn to experience to understand the properties of nature rather than to rely on abstractions of the mind. From this point of view, Newton's assumptions about space and time must be regarded merely as postulates. Classical mechanics follows from these postulates, but other assumptions are possible and from them different laws of dynamics could follow.

Mach's critique had little immediate effect, but its influence was eventually profound. In particular, the youthful Einstein, while a student at the Polytechnic Institute in Zurich in the period 1897–1900, was much attracted by Mach's ideas on the foundations of newtonian physics and by Mach's insistence that physical concepts be defined in terms of observables. However, the immediate cause for the overthrow of newtonian physics was not Mach's criticisms of newtonian thought. The difficulties lay with Maxwell's electromagnetic theory, the crowning achievement of classical physics. Traditionally, the problem is presented in terms of a single crucial experiment that decisively condemned classical physics, the Michelson-Morley experiment, and most treatments of special relativity take this experiment as the point of departure. We shall follow this tradition, but we should point out that history is not that simple. In the first place, Albert A. Michelson, who conceived and executed the experiment, never regarded it as crucial. Michelson viewed the experiment as a flop for not giving the expected result, a view he maintained long after its full significance became known. Furthermore, it now appears that the Michelson-Morley experiment played little, if any, role in Einstein's thinking. In fact, there is good reason to believe that Einstein knew nothing of the experiment until after he had published his theory of relativity in 1906. Nevertheless, the Michelson-Morley experiment so clearly dramatizes the essential dilemma of electromagnetic theory that we shall bow to tradition and take it as our starting point.

11.2 The Michelson-Morley Experiment

The problem to which Michelson devoted himself was that of determining the effect of the earth's motion on the velocity of light. Briefly, Maxwell's electromagnetic theory (1861) predicted that electromagnetic disturbances in empty space would propagate at 3×10^8 m/s—the speed of light. The simplest disturbance is a periodic wave, and the evidence was overwhelming that light consisted of electromagnetic waves. However, there were conceptual difficulties.

The only waves previously known to physics were mechanical waves propagating in solids, liquids, and gases. A sound wave in air, for example, consists of alternate regions of higher and lower pressure propagating with a speed of 330 m/s, somewhat less than the speed of molecular motion. The speed of mechanical waves in metals is higher, typically 5,000 m/s, and increases with the strength of the "spring forces" between neighboring atoms.

Electromagnetic wave propagation seemed to be a very different sort of thing. The ether, the medium which supposedly supported the electromagnetic disturbance, had to be immensely rigid to give a speed of 3×10^8 m/s. At the same time it had to be insubstantial enough not to interfere with the motion of the planets. Maxwell's theory itself made no essential reference to the ether, but Maxwell and his contemporaries were unable to accept the idea of waves propagating in empty space.

The speed of a sound wave v_s depends on the properties of the medium. If we observe a sound wave from a coordinate system moving relative to the medium, the speed of sound will appear to be greater or less than v_s, depending on whether we move in the direction of propagation or against it. Similarly, Maxwell pointed out that the speed of the earth as it circled the sun, 3×10^4 m/s, should change the apparent speed of light.

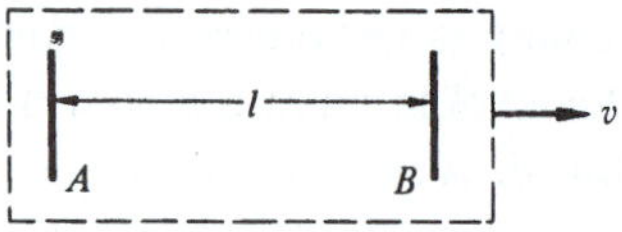

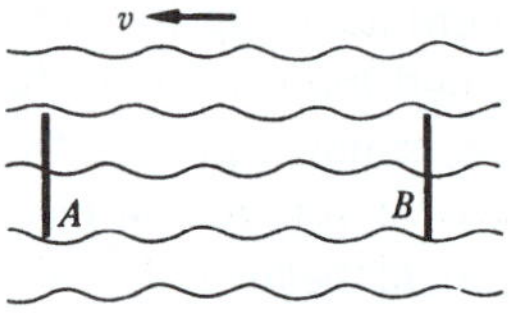

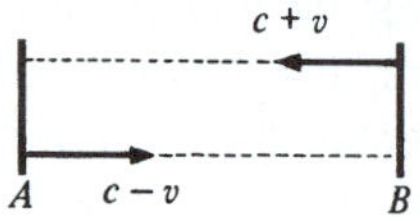

Suppose that light makes a round trip ABA between two points A and B separated by distance l. The apparatus is moving through the ether to the right, as shown in the upper drawing. Relative to the apparatus, the ether is moving to the left, as shown in the second drawing. The velocity of light relative to the apparatus is $c + v$ to the left, and $c - v$ to the right.

The transit time from A to B is $t_1 = l/(c - v)$, and from B to A it is $t_2 = l/(c + v)$. If the apparatus were at rest, t_1 and t_2 would have the value $t_0 = l/c$. The effect of the earth's motion is to delay the return of the light signal by

$$\begin{aligned}\Delta t &= t_1 + t_2 - 2t_0 \\ &= \frac{l}{c - v} + \frac{l}{c + v} - 2\frac{l}{c} \\ &= \frac{l}{c}\left(\frac{1}{1 - v/c} + \frac{1}{1 + v/c} - 2\right) \\ &= 2\frac{l}{c}\left(\frac{1}{1 - v^2/c^2} - 1\right) \\ &\approx 2\frac{l}{c}\frac{v^2}{c^2}.\end{aligned}$$

For the earth in orbit $v/c = 10^{-4}$, and if we take l to be typical of a laboratory apparatus, $l = 1$ m, then $\Delta t = 2 \times 1/(3 \times 10^8) \times$

$10^{-8} \approx 7 \times 10^{-17}$ s, an interval much too small to be measured directly. Fortunately, Michelson was not discouraged. In 1881 he came up with the following solution.

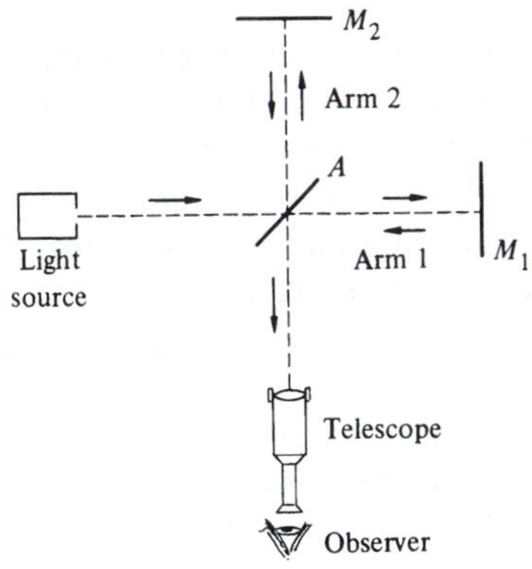

Rather than measure the time of transit of one light beam, Michelson observed the *difference* between the transit times of two beams. His device is sketched at the left. The light from the source is split into two beams by a thinly silvered mirror, A. Half the light passes through mirror A to mirror M_1, where it is reflected back to mirror A and then to the observer. The other half of the light from the source is diverted up the second arm and strikes mirror M_2, which reflects it to the observer. If the two arms are identical, the light waves recombine at mirror A just as if they had never separated: the observer sees an illuminated field of view. The situation is drastically altered if either beam suffers a delay. Suppose, for instance, that beam 1 is delayed by exactly one-half cycle of oscillation. The waves arrive in opposite phase and exactly cancel each other: the observer's field is dark.

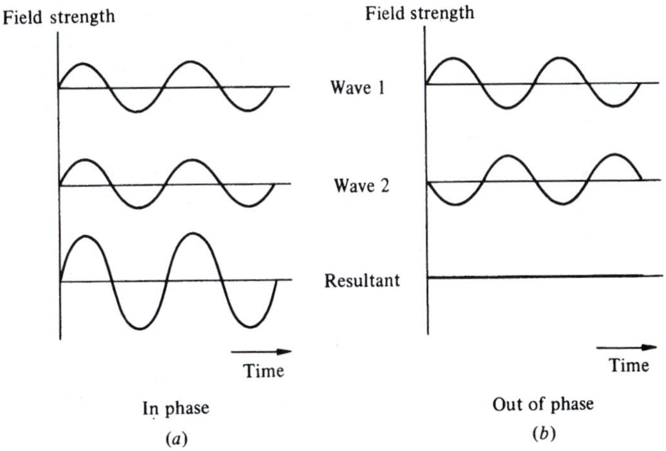

The two cases are shown in the sketches above. The vertical displacement corresponds to the strength of the electric field of light at the observer's eye. The fields of the two beams add vectorially. For visible light the period of the wave is typically 10^{-15} s, too fast for our eyes to follow. Rather, our eyes respond to the average power of the wave which is proportional to the square of the resultant field. Thus, beams in phase, sketch (a), give steady bright illumination, and beams out of phase, sketch (b), give darkness.

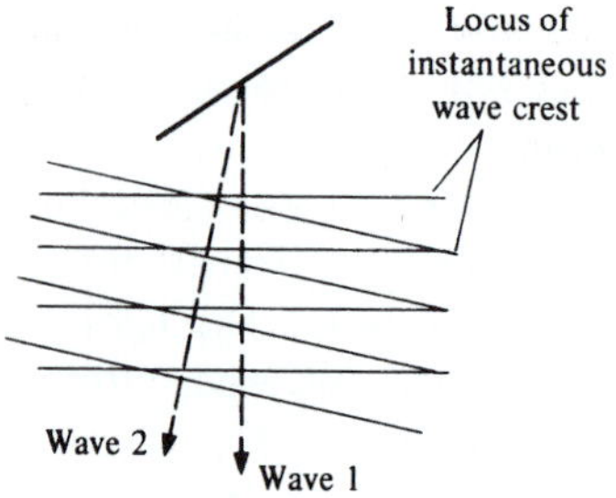

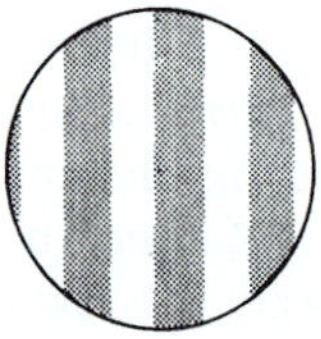

Usually one of the mirrors is slightly tilted. This produces a gradual time delay across the returning wavefront, as shown in the first sketch, and the two interfering waves go in and out of phase across the field of view. The observer sees alternate light and dark bands, as in the second sketch. If the length of either arm is changed, the fringe pattern shifts; a change in path of one wavelength shifts the pattern by one fringe. Since the light traverses each arm twice, once in each direction, a change in the length of either arm by one-half wavelength produces a shift of one fringe. With care it is possible to measure a small fraction of a fringe shift; one can readily observe a path change of one-hundredth wavelength, approximately 10^{-8} m. (Michelson also used his interferometer to measure the length of the standard meter bar; he essentially created the field of high precision measurement.)

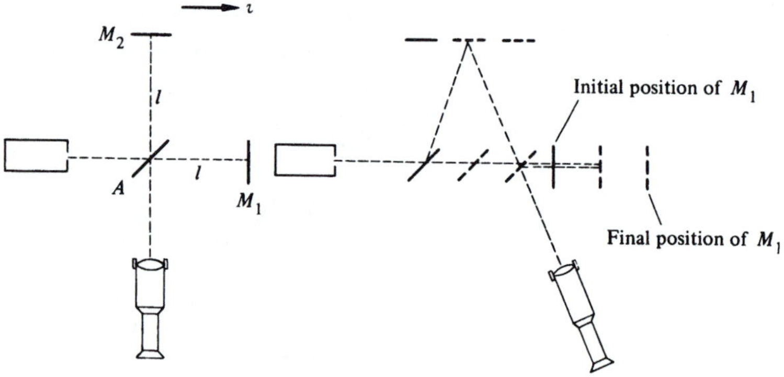

Suppose that the interferometer is oriented so that one axis lies along the direction of motion of the earth, as shown. The time for the wave to travel from mirror A to mirror M_1 and back is

$$\begin{aligned} T_1 &= \frac{l}{c-v} + \frac{l}{c+v} \\ &= \frac{l}{c}\left(\frac{1}{1-v/c} + \frac{1}{1+v/c}\right) \\ &= \frac{2l}{c}\left(\frac{1}{1-v^2/c^2}\right) \approx \frac{2l}{c}\left(1+\frac{v^2}{c^2}\right), \end{aligned}$$

where l is the length of the arm. There is also a time delay along arm 2, but this is a trifle more subtle to calculate. (Michelson overlooked it in the first report of his experiment in 1882.) For

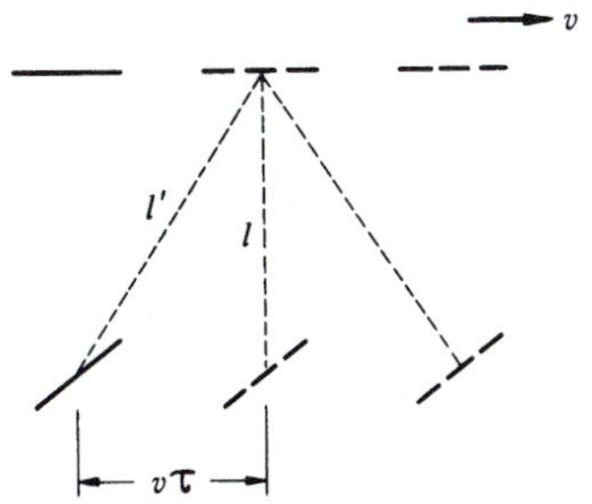

the beam to return to its initial point on the thinly silvered mirror, it must traverse the angular path shown at left. Let τ be the time it takes the wavefront to go from mirror A to mirror M_2. The distance actually traversed is $l' = (l^2 + v^2\tau^2)^{\frac{1}{2}}$ and, since $l' = c\tau$ we have

$$\tau = \frac{(l^2 + v^2\tau^2)^{\frac{1}{2}}}{c}$$

or

$$\tau^2 = \frac{l^2}{c^2} + \frac{v^2}{c^2}\tau^2.$$

It follows that

$$\tau = \frac{l}{c}\frac{1}{\sqrt{1 - v^2/c^2}}.$$

The time for the wave to travel from mirror A to mirror M_2 and back is

$$\begin{aligned} T_2 &= 2\tau \\ &= 2\frac{l}{c}\frac{1}{\sqrt{1 - v^2/c^2}} \\ &\approx 2\frac{l}{c}\left(1 + \frac{1}{2}\frac{v^2}{c^2}\right). \end{aligned}$$

The difference between the travel times of the beams is

$$\begin{aligned} \Delta T &= T_1 - T_2 \\ &= \frac{l}{c}\frac{v^2}{c^2}. \end{aligned}$$

The delay ΔT shifts the fringe pattern from where it would be if the earth were at rest. However, there is a major problem: the fringe scale has no "zero," since the arms cannot be made identical in length to the needed accuracy. Michelson hit upon the idea of watching the fringes as the apparatus is rotated by 90°. The rotation effectively interchanges arms 1 and 2. The change in the delay between the two positions is $2\Delta T$, and the corresponding fringe shift is readily calculated. If λ is the wavelength of the illuminating light, a time delay of λ/c will shift the pattern by one fringe. Thus, the time delay $2\Delta T$ will shift the pattern N fringes, where

$$N = \frac{2\Delta T}{(\lambda/c)} = \frac{2l}{\lambda}\frac{v^2}{c^2}.$$

If the arms have unequal lengths, l_1 and l_2, this result still holds, provided that we replace $2l$ by $l_1 + l_2$.

In Michelson's first apparatus, the arm length was 1.2 m, or, as he put it, 2×10^6 wavelengths of yellow (sodium) light. Since $v/c = 10^{-4}$, we expect

$$N = 2(2 \times 10^6)(10^{-4})^2 = 0.04.$$

Although this is not a large shift, Michelson had adequate resolution to see it. To his disappointment, he found no measurable shift in the fringe pattern. A much more refined experiment, executed with E. W. Morley, in 1887, used multiple reflections to increase the expected shift to 0.4 fringe. Although a shift as small as 0.01 fringe could have been detected, no effect was seen. The experiment has been repeated many times since, but always with negative results. It appears that we are unable to detect our motion through the ether.

11.3 The Postulates of Special Relativity

The elusive nature of the ether presented physics with a troublesome enigma. Maxwell attempted to devise a mechanical model of the ether, but as he continued to develop his theory of light, the ether played a less and less important role, until finally it was altogether absent. The ether vanished like the Cheshire Cat, leaving only a smile behind. After the Michelson-Morley experiment, even the smile had vanished. Numerous attempts to explain the null results of the Michelson-Morley experiment introduced such complexity as to threaten the foundations of electromagnetic theory. The most successful attempt was the hypothesis suggested independently by FitzGerald and by Lorentz that motion of the earth through the ether caused a shortening of one arm of the Michelson interferometer by exactly the amount required to eliminate the fringe shift. However, their speculations were based on an assumed model of atomic forces, and even though they arrived at some of the formulas shortly to be obtained by Einstein, their reasoning was far less general. Other theories which involved such artifacts as drag of the ether by the earth were even less productive.

The Universal Velocity

It is an indication of Einstein's genius that the troublesome problem of the ether pointed the way not to complexity and elaboration but to a simplification that unified the basic concepts of physics. Einstein viewed the difficulty with the ether not as stemming from a fault of electromagnetic theory but as arising from an error in basic dynamical principles. He argued that since the velocity of light predicted by electromagnetic theory, c, involves no reference to a medium, c must be a universal constant, the same for all observers. Thus, if we measure the speed of light from a source, the result will always be c, independent of our motion. This is in marked contrast to the case of sound waves, for example, where the observed speed depends on motion of the observer with respect to the medium. The ideas of a universal velocity was indeed a bold hypothesis, contrary to all previous experience and, for many of Einstein's contemporaries, defying common sense. But common sense is often a poor guide. Einstein once remarked that common sense consists of all the prejudices one learns before the age of eighteen.

The Principle of Relativity

The special theory of relativity involves one additional postulate—the assertion that the laws of physics have the same form with respect to all inertial systems. This principle, known as the principle of relativity, was not novel; Galileo is credited with first pointing out that there is no way to determine whether one is moving uniformly or is at rest, and Newton, although troubled by this point, gave it a rigorous expression in his dynamical laws in which acceleration, not velocity, is paramount. The principle of relativity played only a minor role in the development of classical mechanics; Einstein elevated it to a keystone of dynamics. He extended the principle to include not only the laws of mechanics but also the laws of electromagnetic interaction and, by supposition, all the laws of physics. Furthermore, in his hands the principle of relativity became an important working principle in discovering the correct form of physical laws. We can only surmise the sources of his inspiration, but they must have included the following consideration. If the velocity of light were not a universal constant, that is, if the ether could be detected, then the principle of relativity would fail; a special inertial frame would be singled out, the one at rest in the ether. However, the form of Maxwell's equations, as well as the failure of any experiment to detect motion through the ether, suggests that the speed of light is constant, independent of the motion of the source. Our inability to detect absolute motion, either with light or with newtonian forces, implies that absolute motion has no role in physics.

Whereas most physicists regarded the absence of the ether as a paradox, Einstein saw that its absence preserved the simplicity of the principle of relativity. His view was essentially conservative; he insisted on preserving the principle of relativity which the ether would destroy. Apparently the urge toward simplicity was fundamental to his personality.[1] The special theory of relativity was the simplest way to preserve the unity of classical physics. In fact, as we shall see in the closing chapter, special relativity actually simplifies newtonian thought by combining space and time in a natural fashion from which the various conservation laws follow as a single entity.

The Postulates of Special Relativity

To summarize, the postulates of special relativity are:

The laws of physics have the same form in all inertial systems.

The velocity of light in empty space is a universal constant, the same for all observers.

The mathematical expression of the special theory of relativity is embodied in the Lorentz transformations—a simple prescription for relating events in different inertial systems. Contrary to the mystique, the mathematics of relativity is quite simple: elementary algebra will suffice. The reasoning is also simple, but it has a deceptive simplicity. We start by looking once more at the Galilean transformations.

[1] Einstein had much in common with Newton. In the second book of his "Principia," Newton states his rules of scientific reasoning. Rule 1 is: "We are to admit no more causes of natural things than such as are both true and sufficient to explain their appearances. . . . Nature is pleased with simplicity. . ."

11.4 The Galilean Transformations

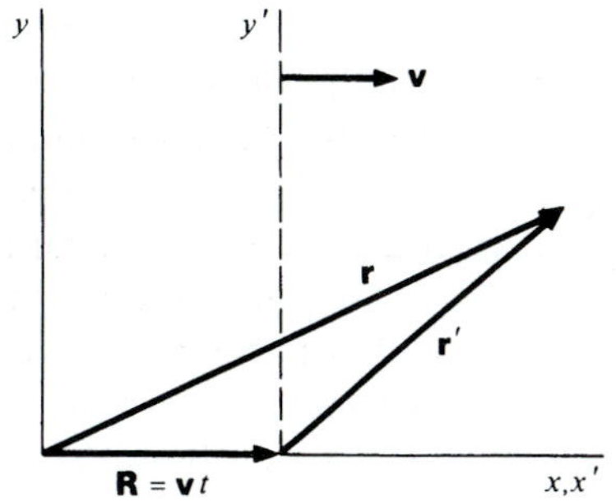

Let us review for a moment the newtonian way of viewing an event in different coordinate systems. Consider an inertial system x, y, z, in which we are at rest, and a second inertial system x', y', z', which is translating uniformly in the $+x$ direction with velocity v. For convenience, we take the origins to coincide at $t = 0$, and take the axes to be parallel.

If a particular point in space has coordinates $\mathbf{r} = (x,y,z)$ in our "rest" system, the corresponding coordinates in the moving system are $\mathbf{r}' = (x',y',z')$. These are related by

$$\mathbf{r}' = \mathbf{r} - \mathbf{R},$$

where

$$\mathbf{R} = \mathbf{v}t.$$

Since v is in the x direction, we have

$$\begin{aligned} x' &= x - vt \\ y' &= y \\ z' &= z \\ t' &= t. \end{aligned} \qquad 11.1$$

The last equation is listed merely for completeness. It follows from the newtonian idea of an "absolute" time, and it is so taken for granted that it is generally omitted in discussions of classical physics.

Equations (11.1) are known as the *Galilean transformations.* Since the laws of newtonian mechanics hold in all inertial systems, they are unaffected by these transformations. The classical principle of relativity asserts that the laws of mechanics are unchanged by the Galilean transformations. The following example illustrates the meaning of this statement.

Example 11.1 **The Galilean Transformations**

Consider how we might discover the law of force between two isolated bodies from observations of their motion. For example, the problem might be to discover the law of gravitation from data on the elliptical orbit of one of Jupiter's moons. If m_1 and m_2 are the masses of the moon and of Jupiter, respectively, and $\mathbf{r}_1$ and $\mathbf{r}_2$ are their positions relative to an astronomer on the earth, we have

$$m_1\ddot{\mathbf{r}}_1 = \mathbf{F}(r)$$

$$m_2\ddot{\mathbf{r}}_2 = -\mathbf{F}(r),$$

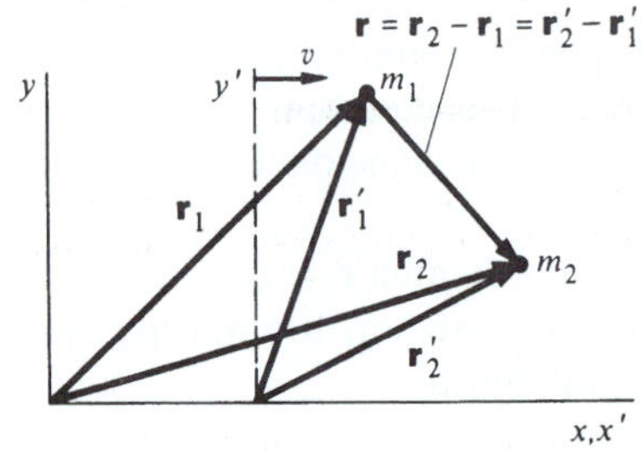

where we assume that **F**, the force between the bodies, depends only on their separation $r = |\mathbf{r}_1 - \mathbf{r}_2|$. (Including the effect of the sun makes the analysis more cumbersome without changing the conclusions.)

From our data on $\mathbf{r}_1(t)$ we can evaluate $\ddot{\mathbf{r}}_1$, which yields the value of **F**, (or $\mathbf{F}/m_1$, to be more precise). In principle, this is the procedure Newton followed in discovering the law of universal gravitation. Suppose that the data show $\mathbf{F}(r) = -Gm_1m_2\hat{\mathbf{r}}/r^2$.

Now let us consider the problem from the point of view of an astronomer in a spacecraft observatory which is flying by the earth. According to the principle of relativity he must obtain the same force law. The situation is represented in the drawing. x, y is the earthbound system, x', y' is the spacecraft system, and v is the relative velocity.

In the x', y' system the astronomer concludes that the force on m_1 is

$$\mathbf{F}'(r') = m_1\ddot{\mathbf{r}}_1'.$$

However,

$$\mathbf{r}_1 = \mathbf{r}_1' + \mathbf{v}t$$
$$\dot{\mathbf{r}}_1 = \dot{\mathbf{r}}_1' + \mathbf{v}$$
$$\ddot{\mathbf{r}}_1 = \ddot{\mathbf{r}}_1'.$$

Hence,

$$\begin{aligned}\mathbf{F}'(r') &= m_1\ddot{\mathbf{r}}_1' \\ &= m_1\ddot{\mathbf{r}}_1 \\ &= \mathbf{F}(r).\end{aligned}$$

Since $r' = r$, $\mathbf{F}'(r') = \mathbf{F}'(r)$. But we have just shown that $\mathbf{F}'(r') = \mathbf{F}(r)$. Hence,

$$\begin{aligned}\mathbf{F}'(r) &= \mathbf{F}(r) \\ &= -\frac{Gm_1m_2}{r^2}\hat{\mathbf{r}}.\end{aligned}$$

The law of force is identical to the one found on earth. This is what we mean when we say that the two inertial systems are equivalent. If the form of the law, or the value of G, were different in the two systems, we could make a judgment about the speed of a coordinate system by investigating the law of gravitation in that system. The systems would not be equivalent.

Example 11.1 is almost trivial, since the force depends on the separation of the two particles, a quantity which is unchanged (invariant) under the Galilean transformations. In newtonian physics, all forces are due to interactions between particles, interactions which depend on the *relative* coordinates of the particles. Consequently they are invariant under the Galilean transformations.

What happens to the equation for a light signal under the Galilean transformations? The following example shows the difficulty that arises.

Example 11.2 **A Light Pulse as Described by the Galilean Transformations**

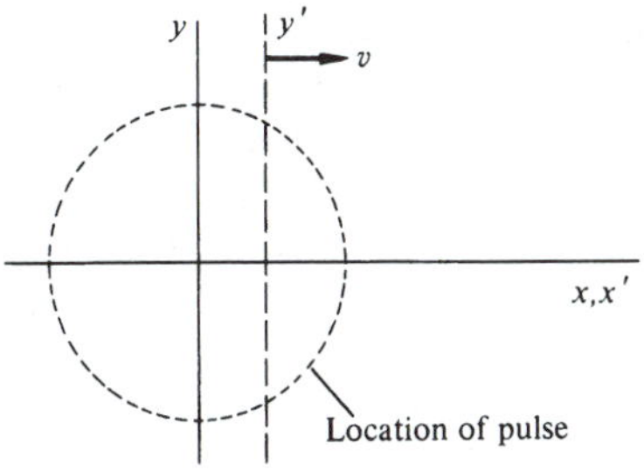

At $t = 0$ a pulse of light is emitted isotropically in the x, y system. It travels outward with velocity c. The equation for the wavefront along the x axis is

$$x = ct.$$

In the x', y' system, the equation for the wavefront along the x' axis is

$$\begin{aligned} x' &= x - vt \\ &= (c - v)t, \end{aligned}$$

where v is the relative velocity of the two systems.

The x' velocity of the pulse in the x', y' system is

$$\frac{dx'}{dt} = c - v.$$

But this is contrary to the postulate that the speed of light is a universal constant c for all observers. Clearly, the Galilean transformations are inadequate.

11.5 The Lorentz Transformations

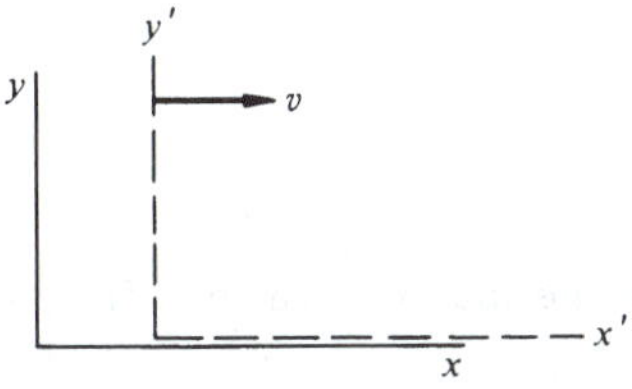

Since the Galilean transformations do not satisfy the postulate that the speed of light is a universal constant, Einstein proposed an alternate prescription for describing the same event in different inertial systems. Let us refer once more to our standard systems, the rest system, x, y, z, t and the system x', y', z', t' which moves with velocity v along the positive x axis. The origins coincide at $t = t' = 0$. We take the most general transformation relating the coordinates of a given event in the two systems to be of the form

$$x' = Ax + Bt \qquad 11.2a$$

$$y' = y \qquad 11.2b$$

$$z' = z \qquad 11.2c$$

$$t' = Cx + Dt. \qquad 11.2d$$

The transformations are linear, for otherwise there would not be a simple one-to-one relation between events in the different systems. For instance, a nonlinear transformation would predict acceleration in one system even if the velocity were constant in

TABLE 11.1

EVENT	COOR-DINATES (x,y,t)	COOR-DINATES (x',y',t')	TRANSFORMATION LAW		RESULT
Observer in (x,y) sees origin of (x',y') move along x axis with velocity v.	$x = vt$	$x' = 0$	$x' = Ax + Bt$ $0 = Avt + Bt$	11.2a	$B = -Av$
Observer in (x',y') sees origin of (x,y) move along x' axis with velocity $-v$.	$x = 0$	$x' = -vt'$	$x' = A(x - vt)$ $t' = Cx + Dt$ $A(0 - vt) = -v(0 + Dt)$	11.2a 11.2d	$D = A$
A light pulse is sent out from origin along x axis at $t = 0$. Its location is given by:	$x = ct$	$x' = ct'$	$x' = A(x - vt)$ $t' = Cx + At$ $A(ct - vt) = c(Cct + At)$	11.2a 11.2d	$C = -\frac{Av}{c^2}$
A light pulse is emitted along the y axis in (x,y) at $t = 0$. In (x',y') the pulse has components along the x' and y' axes. The velocity of the pulse is c in both systems. Its coordinates are given by:	$x = 0$ $y = ct$	$x'^2 + y'^2 = c^2t'^2$	$x' = A(x - vt)$ $y' = y$ $t' = A(-vx/c^2 + t)$ $A^2(0 - vt)^2 + (ct)^2 = c^2A^2[-(v/c^2)0 + t]^2$	11.2a 11.2b 11.2d	†$A = \frac{1}{\sqrt{1 - v^2/c^2}}$

† In general, $A = \pm 1/\sqrt{1 - v^2/c^2}$. We choose the positive root; otherwise, in the limit $v = 0$ we would find $x' = -x$ rather than $x' = x$ as we require.

the other, clearly an unacceptable property for a transformation between inertial systems. We have assumed that the y' and z' axes are left unchanged by the transformation for reasons of symmetry, which we shall discuss later.

Equations (11.2) contain four unknown constants. To evaluate these we consider four cases in which we know *a priori* how an event appears in the two systems. This is carried out in Table 11.1.

Inserting the results of Table 11.1 into Eq. (11.2) gives

$$\begin{aligned} x' &= \frac{1}{\sqrt{1 - v^2/c^2}}(x - vt) \\ y' &= y \\ z' &= z \\ t' &= \frac{1}{\sqrt{1 - v^2/c^2}}\left(t - \frac{vx}{c^2}\right) \end{aligned} \qquad 11.3$$

It is a straightforward matter to solve these equations algebraically for x, y, z, t in terms of x', y', z', t'. Alternatively, we can simply interchange the labels and reverse the sign of v, because the only difference between the systems is the direction of the relative velocity. The result is

$$\begin{aligned} x &= \frac{1}{\sqrt{1 - v^2/c^2}}(x' + vt') \\ y &= y' \\ z &= z' \\ t &= \frac{1}{\sqrt{1 - v^2/c^2}}\left(t' + \frac{vx'}{c^2}\right) \end{aligned} \qquad 11.4$$

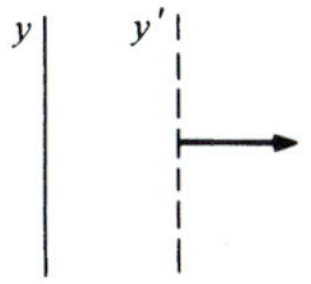

(a)

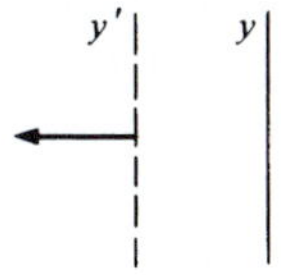

(b)

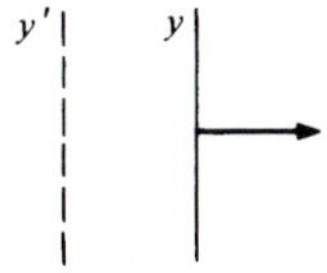

(c)

Equations (11.3) and (11.4) are the *Lorentz transformations*, the prescription for relating the coordinates of an event in different inertial systems so as to satisfy the postulates of special relativity. In the following chapters we shall explore their consequences. We conclude the present discussion by explaining the argument for assuming $y = y'$, $z = z'$.

Consider a section of the y and y' axes as shown in figure (a). The y' axis is moving to the right with velocity v.

If we look at the systems from behind the paper, the situation appears as in sketch (b).

Since only relative motion is important, Figure (b) is equivalent to (c). However, (c) is identical to (a) except that y' and y are interchanged. We conclude that the y and y' axes are indistinguishable and $y = y'$. By a similar argument $z = z'$.

Problems

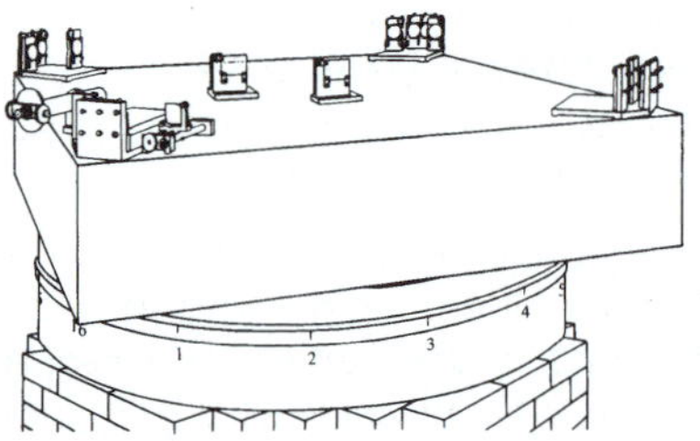

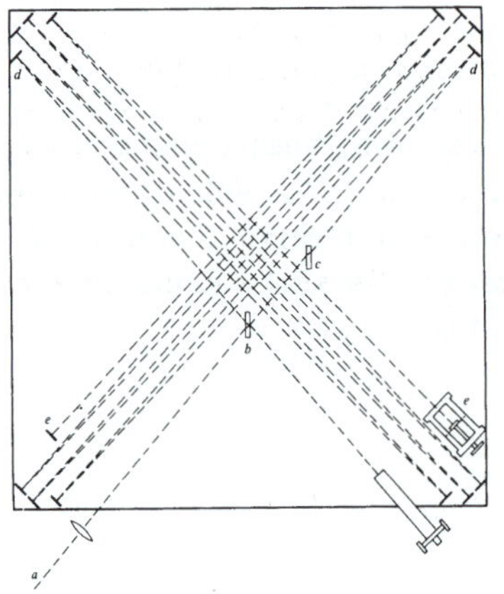

11.1 The Michelson-Morley experiment was carried out at the Case School of Applied Science (now Case-Western Reserve University) in 1887. The apparatus was a refined version of the interferometer used by Michelson in his initial search in Berlin during 1881. The interferometer was mounted on a granite slab 5 ft square and 14 in thick resting on a float riding in a mercury-filled trough. The effective length of the interferometer arms was lengthened to 11 m by the use of mirrors. The light source was the yellow line of sodium, $\lambda = 590 \times 10^{-9}$ m. Michelson and Morley found no systematic shift of fringe with direction, although they could have detected a shift as small as one-hundredth fringe.

How does the upper limit to the earth's velocity through the ether set by this experiment compare with the earth's orbital velocity around the sun, 30 km/s? See drawing on page 458.

11.2 If the two arms of a Michelson interferometer have lengths l_1 and l_2, show that the fringe shift when the interferometer is rotated by 90° with respect to the velocity v through the ether is

$$N = \frac{l_1 + l_2}{\lambda} \frac{v^2}{c^2},$$

where λ is the wavelength of the light.

11.3 The Irish physicist G. F. FitzGerald and the Dutch physicist H. A. Lorentz independently tried to explain the null result of the Michelson-Morley experiment by the following hypothesis: motion of a body through the ether sets up a strain which causes the body to contract along the line of motion by the factor $1 - \frac{1}{2}v^2/c^2$. Show that this hypothesis accounts for the absence of a fringe shift in the Michelson-Morley experiment. (The hypothesis was disproved in 1932 by R. J. Kennedy and E. M. Thorndike, who repeated the Michelson-Morley experiment with an interferometer having arms of different lengths.)

11.4 The Michelson-Morley experiment is known as a second order experiment because the observed effect depends on $(v/c)^2$. Consider the following first order experiment.

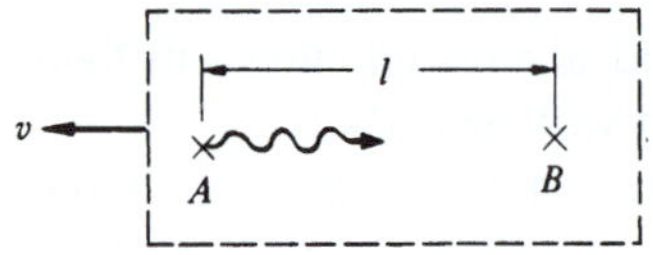

At time $t = 0$, observer A sends a signal to observer B a distance l away. B records the arrival time. Assume that the system is moving through the ether with speed v in the direction shown. Suppose that the laboratory is then rotated 180° with respect to the velocity, reversing the positions of A and B. At time $t = T$, A sends a second signal to B.

a. Show that the interval B observes between the arrival of the signals is $T + \Delta T$, where

$$\Delta T = \frac{2l}{c}\frac{v}{c}$$

to order $(v/c)^3$.

b. Assume that the experiment is done between a clock on the ground and one in a satellite overhead. For an orbit with a 24-h period, $l = 5.6R_e$, where R_e is the earth's radius. Present atomic clocks approach a stability of 1 part in 10^{14}. What is the smallest value of v that this experiment could detect using such clocks?

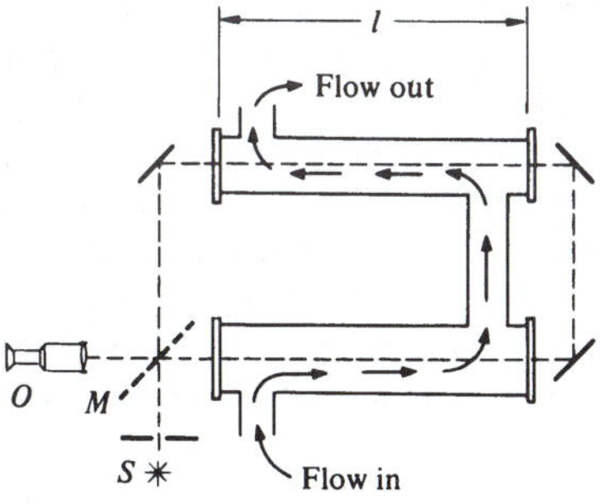

11.5 In 1851 H. L. Fizeau investigated the velocity of light through a moving medium using the interferometer shown. Light of wavelength λ from a source S is split into two beams by the mirror M. The beams travel around the interferometer in opposite directions and are combined at the telescope of the observer, O, who sees a fringe pattern. Two arms of the interferometer pass through water-filled tubes of length l with flat glass end plates. The water runs through the tubes, so that one of the light beams travels downstream while the other goes upstream. The velocity of light in water at rest is c/n, where n is the refractive index of water. If we assume that the velocity of the water is added to the velocity of light in the downstream direction, and subtracted in the upstream direction, show that the fringe shift which occurs when the water flows with velocity v is

$$N = 4n^2 \frac{l}{\lambda c} v.$$

(The actual fringe shift measured by Fizeau was

$$N = \frac{4n^2 l}{\lambda c} fv,$$

where $f = 1 - 1/n^2$. f, known as the Fresnel drag coefficient, was postulated in 1818, but it was not satisfactorily explained until the advent of relativity. It is derived in the next chapter.)

12 RELATIVISTIC KINEMATICS

12.1 Introduction

The special theory of relativity demands that we examine and modify the familiar results of newtonian physics. We must start by reconsidering kinematics, the most elementary aspect of mechanics, a topic apparently so simple that we gave little thought to its foundations in our earlier discussion. As we pointed out in the last chapter, classical kinematics obeys the Galilean transformations. We must now develop the kinematics appropriate to the Lorentz transformations.

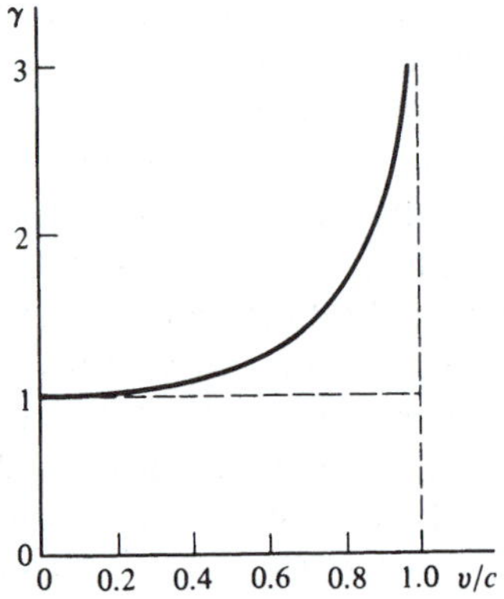

The Lorentz transformations are simplified by introducing

$$\gamma = \frac{1}{\sqrt{1 - v^2/c^2}}.$$

Since $(v/c)^2 \leq 1$, γ is greater than or equal to one. The Lorentz transformations, Eqs. (11.3) and (11.4), then take the form

$$\begin{aligned} x' &= \gamma(x - vt) \qquad & x &= \gamma(x' + vt') \\ y' &= y & y &= y' \\ z' &= z & z &= z' \\ t' &= \gamma\left(t - \frac{xv}{c^2}\right) & t &= \gamma\left(t' + \frac{x'v}{c^2}\right). \end{aligned} \tag{12.1}$$

It is important to understand clearly the function of the Lorentz transformations, for the lore of relativity is filled with so-called paradoxes (generally simple mistakes) in which the Lorentz transformations are misapplied and lead to contradictory results. The Lorentz transformations relate the coordinates of a *single event* in one inertial system to the coordinates of the *same event* in a second inertial system. Examples of single events are:

A light pulse leaves the point $x = 3$ m, $y = 7$ m, $z = -4$ m at $t = 5$ s.

The origin of the x', y', z' system passes the origin of the x, y, z system at time t.

One end of a stick lies at the point x', y', z', at time t'.

A bearer of evil tidings bursts into the king's chamber at midnight.

Single events are characterized by a set of definite values for the coordinates x, y, z, t. More complicated events can be described by a collection of single events. For example, consider a stick lying along the y axis. The location of the stick is defined by *two* single events—the coordinates of its end points at a particular time.

Before setting out to apply the Lorentz transformations, we should consider carefully how to determine the coordinates of an event. Often we speak of "an observer"; for instance, "an observer in the x', y' system sees a flash of light at $x' = 1$, $y' = 3$, $t' = 0$." This is a handy way to describe observations, but there are conceptual difficulties with the idea of a single observer. Consider an observer who notes that a pulse of light leaves the origin at $t = 0$, and finds that at time t_A the pulse is at $x_A = ct_A$. To make such an observation he would have to move to position x_A before the light arrived there—he would have to move faster than light. As we shall see, this is impossible. However, it is nevertheless possible to record the coordinates of any series of events we please by assuming that we have many observers stationed throughout space. Each one has his own clock, and each is assigned to a specific location, x, y, z. Every time an event occurs at a particular location, the local observer notes the time. Later, all the observers send reports to a central office which prepares a complete record of the times and locations of all events in the system. When we talk of "an observer," we mean someone who has, at least in principle, a copy of this record.

In order for the procedure to work it is essential that all the clocks run at the same rate and that they be synchronized. There is a subtle point here, for synchronized clocks will not appear to agree unless they are at the same location. For example, suppose that we use a powerful telescope to look at a clock on the moon. Since it takes light approximately 1 s to travel from the moon to the earth, a moon clock should indicate 1 s before noon when an earth clock indicates noon, provided that the two clocks are properly synchronized. Similarly, the earth clock should appear to be 1 s late to an observer on the moon. By extension, this procedure can be used to synchronize all the clocks in a particular inertial system.

12.2 Simultaneity and the Order of Events

We have an intuitive idea of what is meant when we say that two events are simultaneous. With respect to a given coordinate system, two events are simultaneous if their time coordinates have the same value. However, as the following example shows, events which are simultaneous in one coordinate system are not necessarily simultaneous in a second coordinate system.

Example 12.1 Simultaneity

Consider a railwayman standing at the middle of a freight car of length $2L$. He flicks on his lantern and a light pulse travels out in all directions

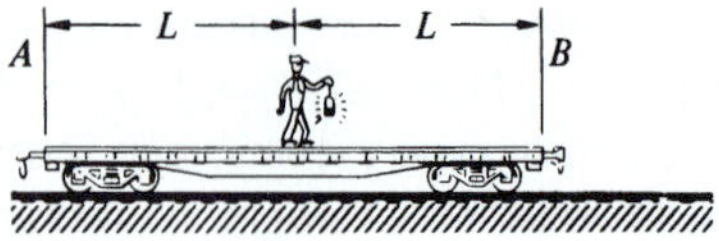

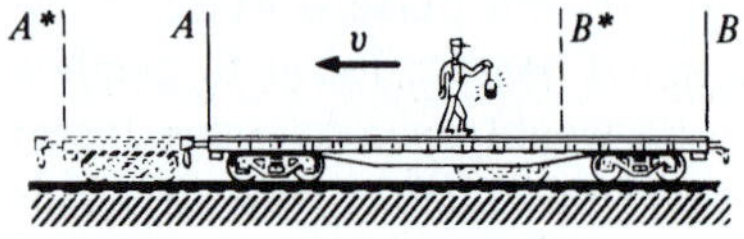

with the velocity c. Light arrives at the two ends of the car after a time interval L/c. In this system, the freight car's rest system, the light arrives simultaneously at A and B.

Now let us observe the same situation from a different frame, one moving to the right with velocity v. In this frame the freight car moves to the left with velocity v. As observed in this frame the light still has velocity c, according to the second postulate of special relativity. However, during the transit time, A moves to A^* and B moves to B^*. It is apparent that the pulse arrives at B^* before A^*; the events are not simultaneous in this frame.

Example 12.1 shows that once we accept the postulates of relativity we are forced to abandon the intuitive idea of simultaneity. The Lorentz transformations, which embody the postulates of relativity, allow us to calculate the times of events in two different systems.

Example 12.2 An Application of the Lorentz Transformations

How do we find the time of arrival of the light pulse at each end of the freight car in the last example? The problem is trivial in the rest frame. Take the origin of coordinates at the center of the car, and take $t = 0$ at the instant the lantern flashes. The two events are

Event 1:

$$\text{Pulse arrives at end } A \begin{cases} x_1 = -L \\ t_1 = \dfrac{L}{c} = T \end{cases}$$

Event 2:

$$\text{Pulse arrives at end } B \begin{cases} x_2 = L \\ t_2 = \dfrac{L}{c} = T \end{cases}$$

To find the time of the events in the moving system we apply the Lorentz transformations for the time coordinates.

Event 1:

$$\begin{aligned} t_1' &= \gamma\left(t_1 - \frac{vx_1}{c^2}\right) \\ &= \gamma\left(T + \frac{vL}{c^2}\right) \\ &= \frac{1}{\sqrt{1 - v^2/c^2}}\left(T + \frac{v}{c}T\right) \\ &= T\sqrt{\frac{1 + v/c}{1 - v/c}}. \end{aligned}$$

Event 2:

$$t_2' = \gamma\left(t_2 - \frac{vx_2}{c^2}\right)$$

$$= T\sqrt{\frac{1 - v/c}{1 + v/c}}.$$

In the moving system, the pulse arrives at B (event 2) earlier than it arrives at A, as we anticipated.

As we saw in the last two examples, simultaneity is not a particularly fundamental property of events; it depends on the coordinate system. Is it possible to find a coordinate system in which any two events are simultaneous? As the following example shows, there are two classes of events. For two given events, we can either find a coordinate system in which the events are simultaneous or one in which the events occur at the same point in space.

Example 12.3 **The Order of Events: Timelike and Spacelike Intervals**

Two events A and B have the following coordinates in the x, y system.

Event A:

x_A, t_A.

Event B:

x_B, t_B.

(For both events, $y = 0$.)

The distance L and time T separating the events in the x, y system are

$$L = x_B - x_A$$

$$T = t_B - t_A.$$

For concreteness, we take L and T to be positive. To find the coordinates in the x', y' system we use the Lorentz transformations, Eq. (12.1):

$$x_A' = \gamma(x_A - vt_A)$$

$$t_A' = \gamma\left(t_A - \frac{vx_A}{c^2}\right)$$

$$x_B' = \gamma(x_B - vt_B)$$

$$t_B' = \gamma\left(t_B - \frac{vx_B}{c^2}\right).$$

The distance L' between the events in the x', y' system is

$$L' = x'_B - x'_A$$
$$= \gamma[x_B - x_A - v(t_B - t_A)]$$
$$L' = \gamma(L - vT).$$

Similarly,

$$T' = \gamma\left(T - \frac{vL}{c^2}\right).$$

Assuming that v is always less than c, it follows that if $L > cT$, L' is always positive, while T' can be positive, negative, or zero. Such an interval is called *spacelike,* since it is possible to choose a system in which the events are simultaneous, namely, a system moving with $v = c^2T/L$. On the other hand, if $L < cT$, T' is always positive, whereas L' can be positive, negative, or zero. The interval is then known as *timelike,* since it is possible to find a coordinate system in which the events occur at the same point.

12.3 The Lorentz Contraction and Time Dilation

Two dramatic results of the special theory of relativity are that a meter stick is shorter when moving than when it is at rest, and that a moving clock runs slow. These results are quite real: the experimental evidence for relativity is so overwhelming that physicists now regard such kinematic effects as commonplace.

The Lorentz Contraction

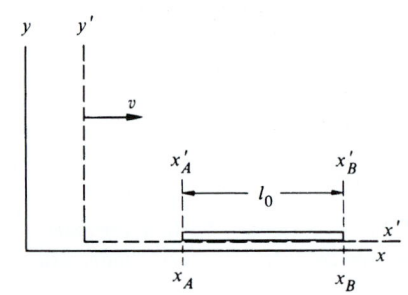

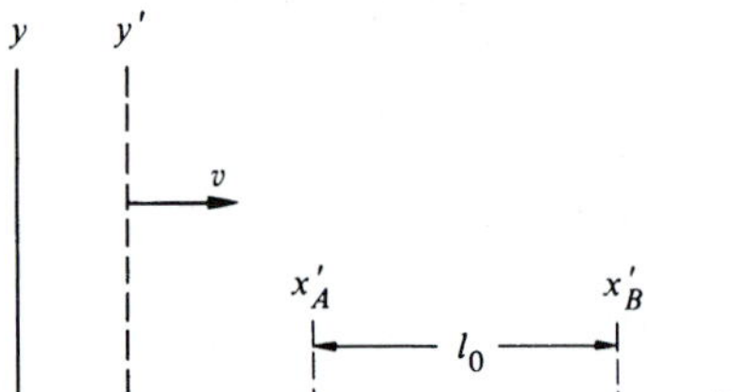

Consider a stick at rest in the x', y' system, lying along the x' axis with its ends at x'_A and x'_B. The length of the stick is $l_0 = x'_B - x'_A$. l_0 is called the "rest," or "proper," length of the stick: it is what we normally mean when we talk of length. The system x', y' is called the rest, or proper, system of the stick.

Now let us determine the length of the stick l in the system in which the observer is at rest. This system, known as the "laboratory" system, has coordinates x, y. In the laboratory system the stick moves to the right with velocity v.

The length of a stick is the distance between its ends at the same instant of time. The end points must be determined simultaneously in the *lab* system; we must find the correspondence between x' and x at some value of t. This is readily accomplished by applying the Lorentz transformation $x' = \gamma(x - vt)$. We have

$$x'_B = \gamma(x_B - vt)$$
$$x'_A = \gamma(x_A - vt).$$

Subtracting, we obtain $l_0 = \gamma l$, or

$$l = \frac{l_0}{\gamma} = l_0\sqrt{1 - \frac{v^2}{c^2}}.$$

l is shorter than l_0: the meter stick is contracted. As $v \rightarrow c$, $l \rightarrow 0$. This shortening, known as the Lorentz contraction, occurs only along the direction of motion: if the stick lay along the y axis, we would use the transformation $y' = y$ to find $l_0 = l$.

A word of caution. The following argument is fallacious—but it is easy to get trapped by it. "In the rest system, the end of the stick has coordinates x'_A and x'_B at some time $t' = 0$. To find the length in the lab system we use $x = \gamma(x' + vt')$, and obtain $l = \gamma l_0$. Hence, the moving stick looks long." The error is that the end points must be measured simultaneously in the *lab* system. These measurements will not be simultaneous in the rest system, but this is of no consequence.

Example 12.4 The Orientation of a Moving Rod

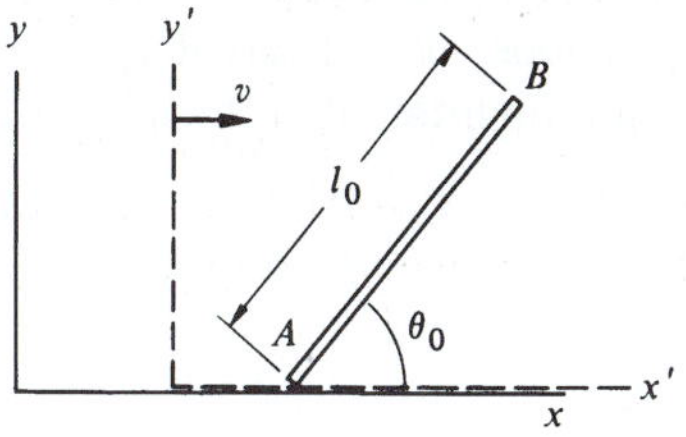

A rod of length l_0 lies in the $x'y'$ plane of its rest system and makes an angle θ_0 with the x' axis. What is the length and orientation of the rod in the lab system x, y in which the rod moves to the right with velocity v?

Designate the ends of the rod A and B. In the rest system these points have coordinates

$$A: \quad x'_A = 0 \qquad y'_A = 0$$
$$B: \quad x'_B = l_0 \cos\theta_0 \qquad y'_B = l_0 \sin\theta_0.$$

We require the coordinates of A and B in the lab system at a time t. We use $x' = \gamma(x - vt)$, $y' = y$ to obtain:

$$A: \quad x'_A = 0 = \gamma(x_A - vt) \qquad y'_A = 0 = y_A$$
$$B: \quad x'_B = l_0 \cos\theta_0 = \gamma(x_B - vt) \qquad y'_B = l_0 \sin\theta_0 = y_B.$$

Hence,

$$x_B - x_A = \frac{l_0 \cos\theta_0}{\gamma}$$
$$y_B - y_A = l_0 \sin\theta_0.$$

The length is

$$\begin{aligned} l &= [(x_B - x_A)^2 + (y_B - y_A)^2]^{\frac{1}{2}} \\ &= l_0\left[\left(1 - \frac{v^2}{c^2}\right)\cos^2\theta_0 + \sin^2\theta_0\right]^{\frac{1}{2}} \\ &= l_0\left[1 - \frac{v^2}{c^2}\cos^2\theta_0\right]^{\frac{1}{2}}. \end{aligned}$$

The angle that the rod makes with the x axis is

$$\theta = \arctan \frac{y_B - y_A}{x_B - x_A}$$
$$= \arctan\left(\gamma \frac{\sin \theta_0}{\cos \theta_0}\right)$$
$$= \arctan (\gamma \tan \theta_0).$$

The moving rod is both contracted and rotated.

Time Dilation

Next we investigate the effect of motion on time. Consider a clock at rest in the x', y' system and consider two events A and B, both occurring at the same point x'_0:

$$A: \quad x'_0 \quad t'_A$$
$$B: \quad x'_0 \quad t'_B.$$

The interval $\tau = t'_B - t'_A$ is the time interval between the events in the rest system. It is called the *proper time* interval.

In order to find the corresponding time interval in the laboratory system we use $t = \gamma(t' + x'v/c^2)$.

$$t_A = \gamma\left(t'_A + \frac{vx'_0}{c^2}\right)$$
$$t_B = \gamma\left(t'_B + \frac{vx'_0}{c^2}\right).$$

Subtracting to obtain $T = t_B - t_A$, we find

$$T = \gamma(t'_B - t'_A)$$
$$= \gamma\tau$$
$$= \frac{\tau}{\sqrt{1 - v^2/c^2}}.$$

The time interval in the laboratory system is greater than that in the rest system; the moving clock runs slow. This effect, known as *time dilation*, has important practical consequences.

Example 12.5 Time Dilation and Meson Decay

The lifetime of cosmic ray μ mesons (muons) has become a classic demonstration of time dilation. The effect was first observed by B. Rossi and

D. B. Hall[1] and is the subject of an excellent film by D. H. Frisch and J. H. Smith.[2]

The experiment hinges on the fact that the muon is an unstable particle which spontaneously decays into an electron and two neutrinos. The meson carries either a positive or negative charge and decays into either a positive electron (positron, e^+) or ordinary electron (e^-).

Symbolically, we can write

$$\mu^{\pm} \rightarrow e^{\pm} + \nu + \bar{\nu}.$$

ν stands for neutrino and $\bar{\nu}$ for antineutrino. The decay of the μ meson is typical of radioactive decay processes: if there are $N(0)$ muons at $t = 0$, the number at time t is

$$N(t) = N(0)e^{-t/\tau},$$

where τ, the mean lifetime, is 2.15×10^{-6} s. Muons can be observed by stopping them in dense absorbers and detecting the decay electron, which comes off with an energy of about 40 MeV (1 MeV = 1 million electronvolts = 1.6×10^{-13} J).

μ mesons are formed in abundance when high energy cosmic ray protons enter the earth's upper atmosphere. The protons lose energy rapidly, and at the altitude of a typical mountaintop, 2,000 m, there are few left. However, the muons penetrate far through the earth's atmosphere and many reach the ground.

The muons descend through the earth's atmosphere with a velocity close to c. The minimum time to descend 2,000 m is then

$$T = \frac{2 \times 10^3 \text{ m}}{3 \times 10^8 \text{ m/s}}$$
$$= 7 \times 10^{-6} \text{ s}.$$

This is more than three times the lifetime; $T/\tau \approx 3$.

The experiment consists of comparing the flux of μ mesons at the top of a mountain with the flux at sea level. We can safely neglect the formation of new mesons in the lower atmosphere or the loss of mesons due to absorption in air. One might expect

$$\frac{\text{flux at sea level}}{\text{flux at mountaintop}} = e^{-T/\tau}$$
$$= 0.045.$$

[1] B. Rossi and D. B. Hall, *Physical Review,* vol. 59, p. 223, 1941.
[2] An account of the experiment demonstrated in the film is given by D. H. Frisch and J. H. Smith, *American Journal of Physics,* vol. 31, p. 342, 1963.

However, the experimental result disagrees sharply: the ratio is 0.7, corresponding to $T/\tau = 0.3$, which is smaller than the predicted ratio by a factor of 10.

The resolution of the disagreement is that we have neglected time dilation. The lifetime τ refers to the decay of a meson at rest. The μ mesons in the atmosphere are moving at high speed with respect to the laboratories on the mountaintop and at its base. When the muon moves rapidly, the lifetime τ' we observe is increased by time dilation. The observed lifetime is

$$\tau' = \gamma\tau = \frac{\tau}{\sqrt{1 - v^2/c^2}}.$$

To account for the observed muon decay rate, we require $\gamma = 10$. This was found to be the case: by measuring the energy of the mesons, γ was determined, and within experimental error it agreed with the prediction from relativity.

Example 12.6 The Role of Time Dilation in an Atomic Clock

Possibly you have looked through a spectroscope at the light from an atomic discharge lamp. Each line of the spectrum is composed of the light emitted when an atom makes a transition between two of its internal energy states. The lines have different colors because the frequency ν of the light is proportional to the energy change ΔE in the transition. (Atomic spectra are discussed in more detail in Sec. 6.8.) If ΔE is of the order of electron volts, the emitted light is in the optical region ($\nu \approx 10^{15}$ Hz). There are some transitions, however, for which the energy change is so small that the emitted radiation is in the microwave region ($\nu \approx 10^{10}$ Hz). These microwave signals can be detected and amplified electronically. Since the oscillation frequency depends almost entirely on the internal structure of the atom, the signals can servê as a frequency reference to govern the rate of an atomic clock. Atomic clocks are highly stable and relatively immune to external influences.

Each atom radiating at its natural frequency serves as a miniature clock. The atoms are frequently in a gas and move randomly with thermal velocities. Because of their thermal motion, the clocks are not at rest with respect to the laboratory and the observed frequency is shifted by time dilation.

Consider an atom which is radiating its characteristic frequency ν_0 in the rest frame. We can think of the atom's internal harmonic motion as akin to the swinging motion of the pendulum of a grandfather's clock: each cycle corresponds to a complete swing of the pendulum. If the period of the swing is τ_0 seconds in the rest frame, the period in the laboratory is $\tau = \gamma\tau_0$. The observed frequency in the laboratory system is

$$\nu = \frac{1}{\tau} = \frac{1}{\gamma\tau_0} = \frac{\nu_0}{\gamma}$$

$$= \nu_0\sqrt{1 - \frac{v^2}{c^2}}.$$

The shift in the frequency is $\delta\nu = \nu - \nu_0$. If $v^2/c^2 \ll 1$, $\gamma \approx 1 - \frac{1}{2}v^2/c^2$, and the fractional change in frequency is

$$\frac{\delta\nu}{\nu_0} = \frac{\nu - \nu_0}{\nu_0} = -\frac{1}{2}\frac{v^2}{c^2}. \qquad 1$$

A handy way to evaluate the term on the right is to multiply numerator and denominator by M, the mass of the atom:

$$\frac{\delta\nu}{\nu_0} = -\frac{\frac{1}{2}Mv^2}{Mc^2}$$

$\frac{1}{2}Mv^2$ is the kinetic energy due to thermal motion of the atom. This energy increases with the temperature of the gas, and according to an elementary result of statistical mechanics,

$$\tfrac{1}{2}M\overline{v^2} = \tfrac{3}{2}kT,$$

where $\overline{v^2}$ is the average squared velocity, $k = 1.38 \times 10^{-23}$ J/deg is Boltzmann's constant, and T is the absolute temperature.

In the atomic clock known as the hydrogen maser, the reference frequency arises from a transition in atomic hydrogen. M is close to the mass of a proton, 1.67×10^{-27} kg, and using $c = 3 \times 10^8$ m/s, we obtain from Eq. (1),

$$\frac{\delta\nu}{\nu} = -\frac{\frac{3}{2} \times 1.38 \times 10^{-23}}{1.67 \times 10^{-27} \times 9 \times 10^{16}}T$$

$$= 1.4 \times 10^{-13}\,T.$$

At room temperature, $T = 300$ K (300 degrees on the absolute temperature scale or 27°C), we have

$$\frac{\delta\nu}{\nu} = -4.2 \times 10^{-11}.$$

This, believe it or not, is a sizable effect. In order to correct for time dilation to an accuracy of 1 part in 10^{13}, it is necessary to know the temperature of the radiating atoms to an accuracy of one degree. However, if one wishes to compare frequencies to parts in 10^{15}, the absolute temperature must be known to millidegrees, a much harder task.

12.4 The Relativistic Transformation of Velocity

The starship Enterprise silently glides to the east with speed $0.9c$. At the same time, the starship Fleagle glides to the west with speed $0.9c$. Classically, the relative speed of the ships is $1.8c$, and the Fleagle's crew would see the Enterprise moving away with a speed faster than light. According to special relativity the picture is quite different. To show this we need the relativistic law for the addition of velocities.

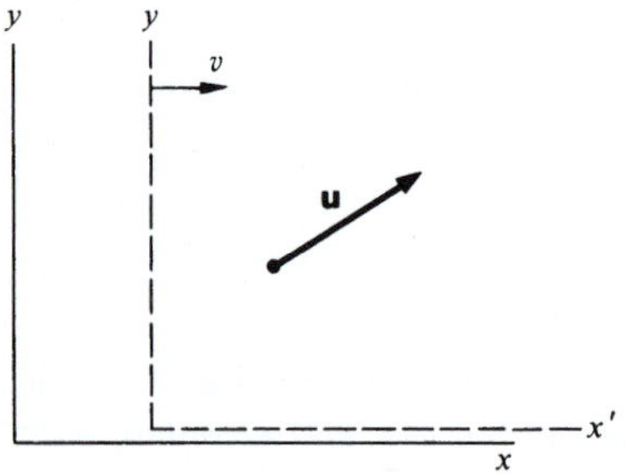

Consider a particle with instantaneous velocity $\mathbf{u} = (u_x, u_y)$ in the x, y, z, t system. Our task is to find the corresponding components u'_x, u'_y in the x', y', z', t' system, which moves with speed v along the positive x axis.

From the definition of velocity, we have, in the unprimed system,

$$u_x = \lim_{\Delta t \to 0} \frac{\Delta x}{\Delta t} \qquad u_y = \lim_{\Delta t \to 0} \frac{\Delta y}{\Delta t}.$$

The corresponding components in the primed system are

$$u'_x = \lim_{\Delta t' \to 0} \frac{\Delta x'}{\Delta t'} \qquad u'_y = \lim_{\Delta t' \to 0} \frac{\Delta y'}{\Delta t'}.$$

The problem is to relate displacements and time intervals in the primed system to those in the unprimed system. Using the procedure of Example 12.2 (or simply writing the Lorentz transformations for differentials), we find

$$\Delta x' = \gamma(\Delta x - v\,\Delta t)$$
$$\Delta y' = \Delta y$$
$$\Delta t' = \gamma\left(\Delta t - \frac{v}{c^2}\Delta x\right).$$

Hence,

$$\frac{\Delta x'}{\Delta t'} = \frac{\gamma(\Delta x - v\Delta t)}{\gamma[\Delta t - (v/c^2)\Delta x]} = \frac{\Delta x/\Delta t - v}{1 - (v/c^2)(\Delta x/\Delta t)}.$$

Next we take the limit $\Delta t \to 0$. Since $\Delta x = u_x \Delta t$, $\Delta x \to 0$ when $\Delta t \to 0$. The Lorentz transformations show that $\Delta x'$ and $\Delta t'$ also approach zero. Using $u_x' = \lim_{\Delta t' \to 0} (\Delta x'/\Delta t')$, we obtain

$$u_x' = \frac{u_x - v}{1 - vu_x/c^2}. \tag{12.2a}$$

Similarly,

$$u_y' = \frac{u_y}{\gamma[1 - vu_x/c^2]}. \tag{12.2b}$$

By symmetry, u_z' behaves like u_y':

$$u_z' = \frac{u_z}{\gamma[1 - vu_x/c^2]}. \tag{12.2c}$$

These transformations can be inverted by changing the sign of v:

$$u_x = \frac{u_x' + v}{1 + vu_x'/c^2} \tag{12.3a}$$

$$u_y = \frac{u_y'}{\gamma[1 + vu_x'/c^2]} \tag{12.3b}$$

$$u_z = \frac{u_z'}{\gamma[1 + vu_x'/c^2]}. \tag{12.3c}$$

In these formulas, $\gamma = 1/\sqrt{1 - v^2/c^2}$ as before.

Equation (12.2a) or (12.3a) is the relativistic law for the addition of velocities. For $v \ll c$, we obtain the Galilean result $u_x' = u_x - v$.

Returning to the problem of the two starships, let $u_x = 0.9c$ be the speed of the Enterprise relative to the ground, and $v = -0.9c$ be the speed of the Fleagle relative to the ground. The velocity of the Enterprise relative to the Fleagle is, from Eq. (12.2a),

$$\begin{aligned} u_x' &= \frac{0.9c - (0.9c)}{1 - [(-0.9c)(0.9c)]} \\ &= \frac{1.8c}{1.81} \\ &= 0.99c. \end{aligned}$$

The relative speed is less than c. The relativistic transformation of velocities assures that we cannot exceed the velocity of light by changing reference frames.

The limiting case is $u_x = c$. The velocity in the moving system is then

$$u_x' = \frac{c - v}{1 - vc/c^2}$$
$$= c,$$

independent of v. This agrees with the postulate we originally built into the Lorentz transformations: the velocity of light is the same for all observers. Furthermore, it suggests that the velocity of light plays the role of an ultimate speed in the theory of relativity.

Example 12.7 **The Speed of Light in a Moving Medium**

As an exercise in the relativistic addition of velocities, let us find how the motion of a medium, such as water, influences the speed of light.

The velocity of light in matter is less than c. The index of refraction, n, is used to specify the speed in a medium:

$$n = \frac{c}{\text{velocity of light in the medium}}$$

$n = 1$ corresponds to empty space; in matter $n > 1$. The slowing can be appreciable: for water $n = 1.3$.

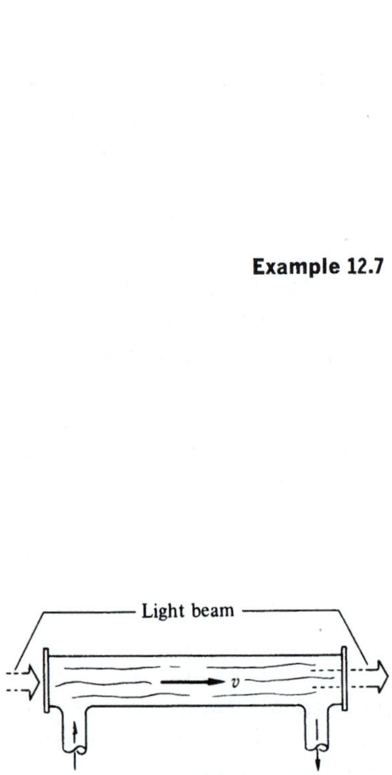

The problem is to find the speed of light through a moving liquid. For instance, consider a tube filled with water. If the water is at rest, the velocity of light in the water with respect to the laboratory is $u = c/n$. What is the speed of light when the water is flowing with speed v?

Consider the speed of light in water as observed in a coordinate system x', y' moving with the water. The speed is

$$u' = \frac{c}{n}.$$

The speed in the laboratory is, by Eq. (12.3a),

$$u = \frac{u' + v}{1 + u'v/c^2} = \frac{c/n + v}{1 + v/nc} = \frac{c}{n}\left(\frac{1 + nv/c}{1 + v/nc}\right).$$

If we expand the last term and neglect terms of order $(v/c)^2$ and smaller, we obtain

$$u = \frac{c}{n}\left(1 + \frac{nv}{c} - \frac{v}{nc}\right)$$
$$= \frac{c}{n} + v\left(1 - \frac{1}{n^2}\right).$$

The light appears to be dragged by the fluid, but not completely. Only the fraction $f = 1 - 1/n^2$ of the fluid velocity is added to the speed of light c/n. This effect was observed experimentally in 1851 by Fizeau, although it was not explained satisfactorily until the advent of relativity.

12.5 The Doppler Effect

The roar of a car or motorcycle zooming past is characterized by a rapid drop in pitch as the vehicle goes by. The effect is quite noticeable if you listen for it at the side of a road. (It is the sound most people make when trying to mimic a near miss by a speeding car.) The decrease in frequency of all the sounds from the car as it goes by is due to the Doppler effect. In general, the Doppler effect is a shift in frequency due to the motion of a source or an observer. The Doppler shift occurs for light as well as sound. Our knowledge of the motion of distant receding galaxies comes from studies of the Doppler shift of their spectral lines. More prosaic applications of the Doppler effect include satellite tracking and radar speed traps.

We shall start by examining the Doppler shift in sound—a situation we can treat classically.

The Doppler Shift in Sound

Sound travels through a medium, such as air, with a speed w determined by the properties of the medium, independent of the motion of the source.

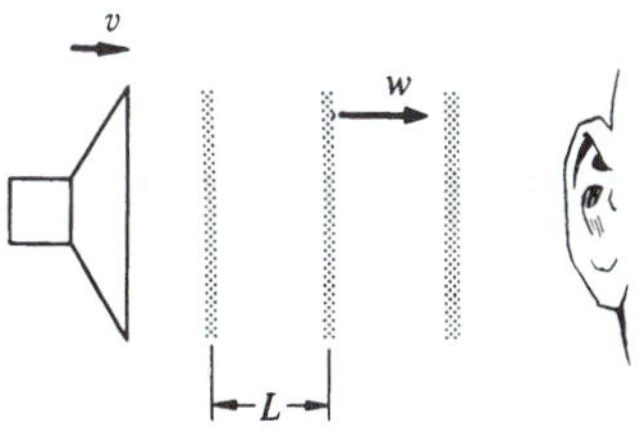

Consider a source of sound which is moving with velocity v through the medium toward an observer at rest. To simplify the geometry we shall restrict ourselves for the present to the case where the observer is along the line of motion. We can regard the sound as a regular series of pulses separated by time $\tau_0 = 1/\nu_0$, where ν_0 is the number of pulses per second generated by the source. (ν_0 corresponds to the frequency of sound from the source.) The situation is shown in the sketch.

In time T the sound travels a distance wT, and if the pulses are separated by distance L, the number reaching the observer is wT/L. The rate at which the pulses arrive is w/L, and this is the frequency of sound ν_D heard by the observer:

$$\nu_D = \frac{w}{L}.$$

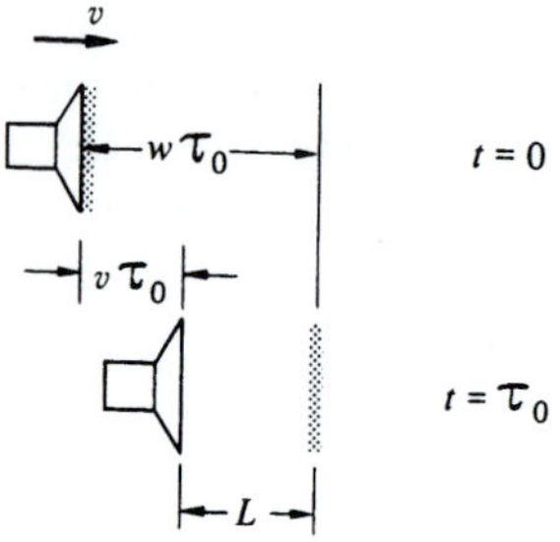

To determine L, consider a pulse emitted at $t = 0$ and the next pulse emitted at $t = \tau_0$. During the interval τ_0 the first pulse travels distance $w\tau_0$ in the medium, and the source travels distance $v\tau_0$. The distance between the pulses is therefore

$$L = w\tau_0 - v\tau_0$$
$$= (w - v)\frac{1}{\nu_0}.$$

Hence,

$$\nu_D = \frac{w}{L}$$
$$= \nu_0 \frac{w}{w - v}$$

or

$$\nu_D = \nu_0 \frac{1}{1 - (v/w)}. \qquad \text{(Moving source.)} \qquad 12.4$$

For an approaching source, v is positive and $\nu_D > \nu_0$. For a receding source, v is negative and $\nu_D < \nu_0$. Qualitatively, this accounts for the drop in pitch of the sound of a car as it goes by.

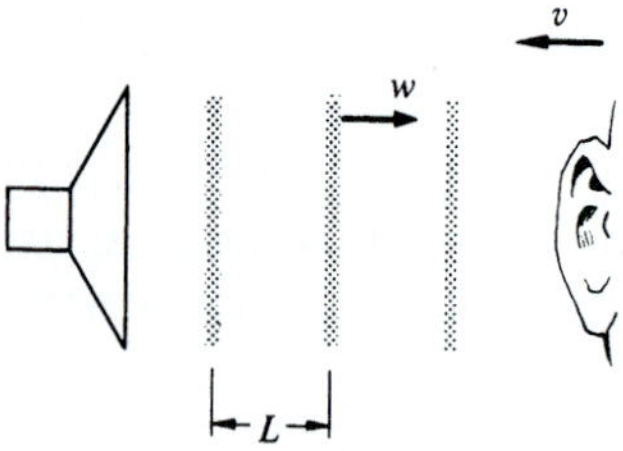

The situation is somewhat different if the source is at rest in the medium and the observer is moving with speed v toward the source. The situation is shown in the sketch. The speed of the pulses relative to the observer is $w + v$. The rate at which pulses arrive is

$$\nu_D = \frac{w + v}{L}.$$

Since the source is at rest, $L = w\tau_0 = w/\nu_0$, and

$$\nu_D = \nu_0 \frac{w + v}{w} = \nu_0 \left(1 + \frac{v}{w}\right). \qquad \text{(Moving observer.)} \qquad 12.5$$

This differs from the result for a moving source, Eq. (12.4), although the results agree to order v/w. The situation is not symmetric; if ν_0, v, and w are known, we can tell whether it is the observer or the source which is moving by measuring ν_D carefully. The reason is that in the case of sound there is a medium, the air, to which motion can be referred.

If it were possible to apply these results to light waves in space, we would be able to distinguish which of two inertial systems was at rest. This would contradict the principle of special relativity that only the relative motion of inertial systems is observable. To resolve this difficulty, we turn now to a relativistic derivation of the Doppler effect.

Relativistic Doppler Effect

A light source flashes with period $\tau_0 = 1/\nu_0$ in its rest frame. The source is moving toward an observer with velocity v. Due to time dilation, the period in the observer's rest frame is

$$\tau = \gamma\tau_0.$$

Since the speed of light is a universal constant, the pulses arrive at the observer with speed c. It is for this reason that the relative velocity alone plays a role in the Doppler effect for light. In the classical case, the pulses arrive with a speed dependent on the state of motion of the observer relative to the medium.

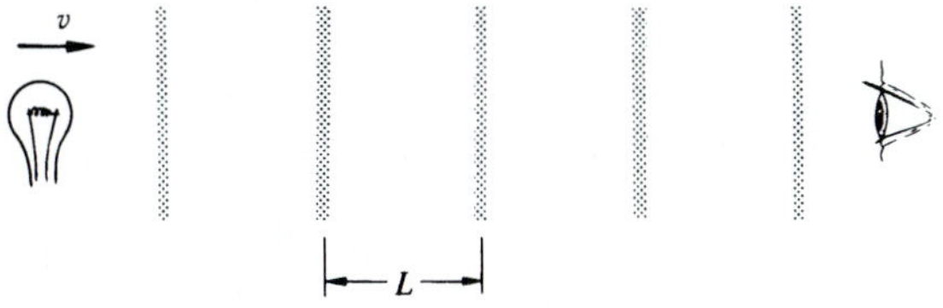

The frequency of the pulses is $\nu_D = c/L$, where L is the separation in the observer's frame. Since the source is moving toward the observer,

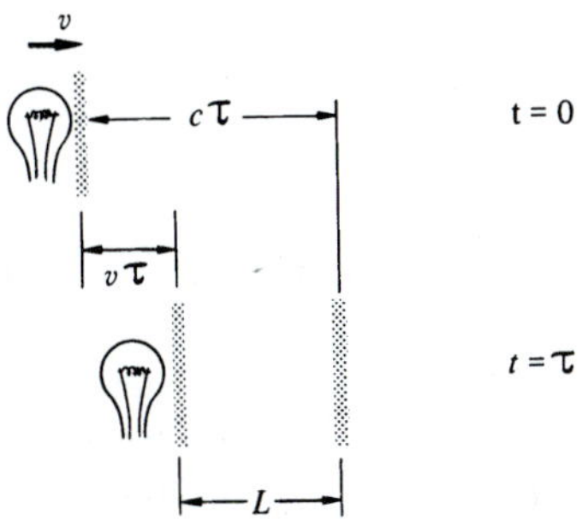

$$L = c\tau - v\tau = (c - v)\tau$$

and

$$\nu_D = \frac{c}{(c - v)\tau}$$

$$= \frac{1}{1 - v/c}\frac{1}{\gamma\tau_0}$$

or

$$\nu_D = \nu_0\frac{\sqrt{1 - v^2/c^2}}{1 - v/c}.$$

This reduces to

$$\nu_D = \nu_0\sqrt{\frac{1 + v/c}{1 - v/c}}. \qquad 12.6$$

ν_D is the frequency in the observer's rest frame and v is the relative speed of source and observer. As we expect, there is no mention of motion relative to a medium. The relativistic result plays no favorites with the classical results; it disagrees with both and, in fact, turns out to be their geometric mean.

The Doppler Effect for an Observer off the Line of Motion

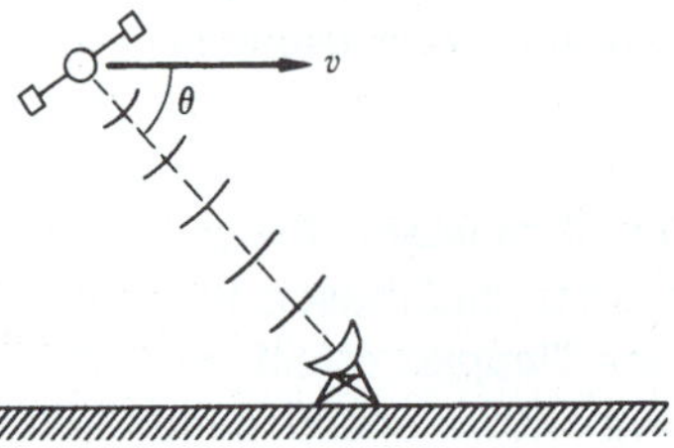

So far we have restricted ourselves to the Doppler effect for a source and observer along the line of motion. However, consider a satellite broadcasting a radio beacon signal to a ground tracking station which monitors the Doppler shifted frequency. Although our earlier results do not apply to such a case, we can readily generalize the method to find the Doppler effect when the observer is at angle θ from the line of motion. We shall again visualize the source as a flashing light. The period of the flashes in the observer's rest frame is $\tau = \gamma\tau_0$, as before. The frequency seen by the observer is c/L. Since the source moves distance $v\tau$ between flashes, it is apparent from the lower sketch that

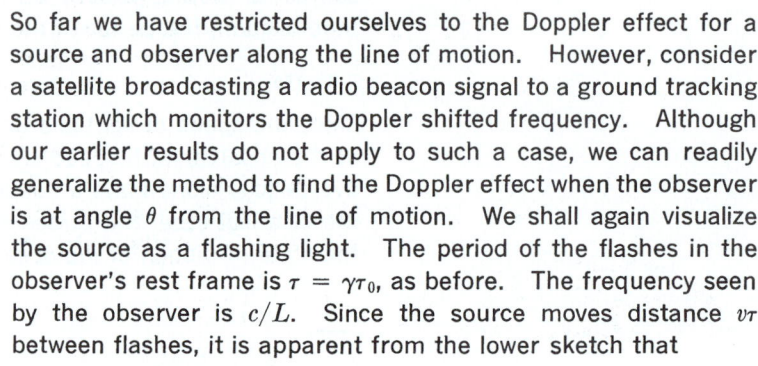

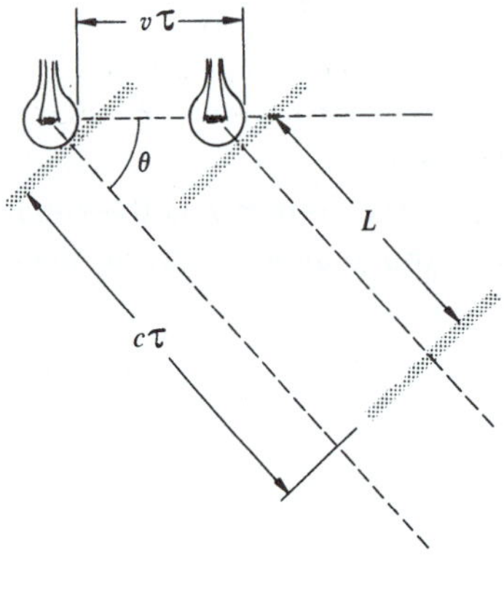

$$\begin{aligned} L &= c\tau - v\tau \cos\theta \\ &= (c - v\cos\theta)\tau. \end{aligned}$$

(We assume that the source and observer are so far apart that θ is effectively constant between pulses.) Hence

$$\begin{aligned} \nu_D &= \frac{c}{L} \\ &= \frac{c}{(c - v\cos\theta)\tau_0\gamma} \end{aligned}$$

or

$$\nu_D = \nu_0 \frac{\sqrt{1 - v^2/c^2}}{1 - (v/c)\cos\theta}. \qquad 12.7$$

In this result, θ is the angle measured in the rest frame of the observer. Along the line of motion, $\theta = 0$ and we recover our previous result for that case, Eq. (12.6). At $\theta = \pi/2$ the relative velocity between source and observer is zero. However, even in this case there is a shift in frequency; ν_D differs from ν_0 by the factor $\sqrt{1 - v^2/c^2}$. This "transverse" Doppler effect is due to time dilation. The flashing lamp is effectively a moving clock.

The relativistic Doppler effect agrees with the classical result to order v/c, so that any experiment to differentiate between them must be sensitive to effects of order $(v/c)^2$, a difficult task.

The relativistic expression was confirmed by Ives and Stilwell in 1938 by observations on the spectral light from fast moving atoms.

One of the more interesting practical applications of the Doppler effect is in navigational systems, as the following example explains.

Example 12.8 **Doppler Navigation**

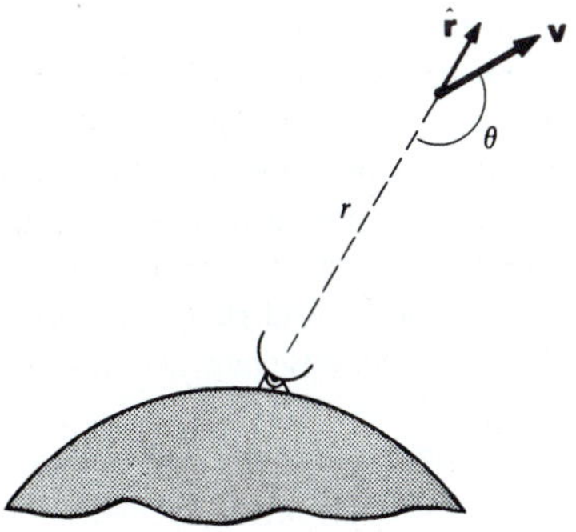

The Doppler effect can be used to track a moving body, such as a satellite, from a reference point on the earth. The method is remarkably accurate; changes in the position of a satellite 10^8 m away can be determined to a fraction of a centimeter.

Consider a satellite moving with velocity $\mathbf{v}$ at some distance r from a ground station. An oscillator on the satellite broadcasts a signal with proper frequency ν_0. Since $v \ll c$ for satellites, we can approximate Eq. (12.7) by retaining only terms of order v/c. Then the frequency ν_D received by the ground station can be written

$$\nu_D \approx \frac{\nu_0}{1 - (v/c)\cos\theta}$$

$$\approx \nu_0\left(1 + \frac{v}{c}\cos\theta\right).$$

There is an oscillator in the ground station identical to the one in the satellite, and by simple electronic methods the difference frequency ("beat" frequency) $\nu_D - \nu_0$ can be measured:

$$\nu_D - \nu_0 = \nu_0 \frac{v}{c}\cos\theta.$$

The radial velocity of the satellite is

$$\frac{dr}{dt} = \hat{\mathbf{r}} \cdot \mathbf{v}$$

$$= -v\cos\theta.$$

Hence

$$\frac{dr}{dt} = -\frac{c}{\nu_0}(\nu_D - \nu_0)$$

$$= -\lambda_0(\nu_D - \nu_0),$$

where $\lambda_0 = c/\nu_0$ is the wavelength of the radiation.

ν_D varies in time as the satellite's velocity and direction change. To find the total radial distance traveled between times T_a and T_b, we integrate the above expression with respect to time:

$$\int_{T_a}^{T_b}\left(\frac{dr}{dt}\right)dt = -\lambda_0\int_{T_a}^{T_b}(\nu_D - \nu_0)\,dt$$

$$r_b - r_a = -\lambda_0\int_{T_a}^{T_b}(\nu_D - \nu_0)\,dt.$$

The integral is the number of cycles N_{ba} of the beat frequency which occur in the interval T_a to T_b. (One cycle occurs in a time $\tau = 1/(\nu_D - \nu_0)$, so that $\int dt/\tau$ is the total number of cycles.) Hence

$$r_b - r_a = -\lambda_0 N_{ba}.$$

This result has a simple interpretation: whenever the radial distance increases by one wavelength, the phase of the beat signal decreases one cycle. Similarly, when the radial distance decreases one wavelength, the phase of the beat signal increases by one cycle.

Satellite communication systems operate at a typical wavelength of 10 cm, and since the beat signal can be measured to a fraction of a cycle, satellites can be tracked to about 1 cm. If the satellite and ground-based oscillators do not each stay tuned to the same frequency, ν_0, there will be an error in the beat frequency. To avoid this problem a two-way Doppler tracking system can be used in which a signal from the ground is broadcast to the satellite which then amplifies it and relays it back to the ground. This has the added advantage of doubling the Doppler shift, increasing the resolution by a factor of 2.

We sketched the principles of Doppler navigation for the classical case $v \ll c$. For certain tracking applications the precision is so high that relativistic effects must be taken into account.

As we have already shown, a Doppler tracking system also gives the instantaneous radial velocity of the satellite $v_r = -c(\nu_D - \nu_0)/\nu_0$. This is particularly handy, since both velocity and position are needed to check satellite trajectories. A more prosaic use of this result is in police radar speed monitors: a microwave signal is reflected from an oncoming car and the beat frequency of the reflected signal reveals the car's speed.

12.6 The Twin Paradox

The kinematical effects we have analyzed in this chapter depend on the *relative* velocity of two systems; such phenomena as Lorentz contraction, time dilation, and the Doppler shift give no clue as to which of two systems is at rest and which is moving, nor can they do so within the framework of relativity, which postulates that all inertial systems are equivalent. There is no such equivalence between *non*inertial systems. Indeed, there is little difficulty in deciding whether or not an isolated system is accelerating.

Failure to appreciate this point was responsible for a vociferous controversy over the so-called "twin paradox." The problem is of interest because it affords a good illustration of the physical difference between inertial and noninertial systems.

The paradox is as follows: two identical twins, Castor and Pollux, A and B for short, have identical clocks. B sets out on a long space voyage while A remains home. A constantly observes B's clock and sees that it is running slow due to time dilation. Eventually B returns home. Since B's clock has run slow throughout the trip, A concludes that B is younger than A at the end of the journey. But suppose we look at the situation from B's point of view. Since time dilation depends only on relative motion, during the trip B sees A's clock running slow, and when the trip is finished B concludes that A is younger than B. Obviously both twins can't be right. Is either twin really younger?

The explanation lies in the fact that the situation is *not* equivalent from the point of view of each twin. A's system is inertial throughout, but B must change his velocity at some time in order to return to the starting point. While the velocity is changing, B's system is not inertial. There is no doubt as to which twin is really accelerating. If each were carrying an accelerometer, such as a mass on a spring, A's would remain at zero while B's would show a large deflection at the turning point. It is apparent that the systems are not equivalent.

We cannot apply special relativity to determine the coordinates of events in noninertial frames. Fortunately, it is possible to determine what B will observe during turnaround by introducing the idea of the Doppler shift.

To make the argument quantitative, suppose that the relative velocity is v. A observes that B travels away a distance L in time $T = L/v$. B then rapidly reverses his motion and returns with the same velocity. The time for the return trip is also T. We shall neglect the time it takes B to reverse his motion since if T is sufficiently long, the turnaround time is negligible. (Nothing anomalous happens to B's clock during turnaround; A simply observes a varying dilation factor while the velocity is changing.)

Neglecting this small turnaround correction, A observes a total elapsed time T'_B on B's moving clock which is related to the time on A's own clock $T_A = 2T$ by

$$\begin{aligned} T'_B &= \frac{T_A}{\gamma} \\ &= T_A\sqrt{1 - \frac{v^2}{c^2}}. \end{aligned} \tag{12.8}$$

A concludes that B is younger.

$$\frac{\text{aging of } B}{\text{aging of } A} = \frac{T'_B}{T_A} = \sqrt{1 - \frac{v^2}{c^2}} \qquad \text{(As viewed by } A.) \tag{12.9}$$

Now let us look at the situation from B's point of view. Except for the turnaround time, B's observations are similar to A's. B sees A go away for distance L with velocity $-v$ and return. This takes time $T_B = 2T$ on B's clock, and if B sees time T'_A elapse on A's clock, then

$$T'_A = \frac{T_B}{\gamma} = T_B\sqrt{1 - \frac{v^2}{c^2}}. \qquad 12.10$$

B seems to conclude that A is younger.

$$\frac{\text{aging of } B}{\text{aging of } A} \stackrel{?}{=} \frac{T_B}{T'_A} = \frac{1}{\sqrt{1 - v^2/c^2}}. \qquad \text{(As viewed by } B.\text{)} \qquad 12.11$$

This is the paradox: A thinks that B is younger and B thinks that A is younger.

Now consider what happens to B during turnaround. He experiences an acceleration as if he were in a gravitational field. According to the discussion of the principle of equivalence in Chap. 8, clocks run at different rates in a gravitational field—this is the origin of the gravitational red shift. For this reason, B sees A's clock run fast during turnaround and, as we shall show, this puts A's clock ahead. However, instead of involving the gravitational red shift, we shall derive the result from simple kinematics.

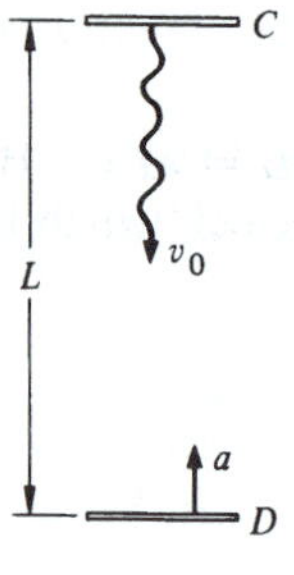

Consider a clock C which has period τ_0 in its rest frame and which emits signals at frequency $\nu_0 = 1/\tau_0$. An observer D is at rest a distance L away and starts accelerating toward C at rate a when the signal of frequency ν_0 leaves C. The signal arrives at time $t_0 \approx L/c$. (We assume that D has not moved appreciably in time t_0, and that his velocity is so low that relativistic effects are negligible.) When the signal arrives, D is moving toward C at velocity $v = at_0 = aL/c$ and the observed frequency, ν', is Doppler shifted. From Eq. (12.6) we have

$$\nu' = \nu_0\sqrt{\frac{1 + v/c}{1 - v/c}} \approx \nu_0\left(1 + \frac{v}{c}\right) = \nu_0\left(1 + \frac{aL}{c^2}\right),$$

where we have neglected terms of order $(v/c)^2$. Since $\nu' > \nu_0$, C's clock appears to run faster than if there were no acceleration. If D's clock records a time interval

$$T_D = 1/\nu',$$

then C's clock marks off an interval

$$T_C = 1/\nu_0.$$

Hence,

$$\begin{aligned} T_C &= T_D \frac{\nu'}{\nu_0} \\ &= T_D\left(1 + \frac{aL}{c^2}\right). \end{aligned}$$

Applying this to the twins, suppose that B accelerates uniformly at rate a toward A during turnaround. B notes on his own clock that the turnaround time is T_t. He notes that A's clock marks off an interval

$$T_t' = T_t\left(1 + \frac{aL}{c^2}\right).$$

Since the velocity changes by $2v$ during turnaround, $T_t = 2v/a$. Therefore,

$$\begin{aligned} T_t' &= T_t + \frac{2v}{a}\frac{aL}{c^2} \\ &= T_t + \frac{2vL}{c^2}. \end{aligned}$$

The total length of the trip is $2L = vT_B$. Hence, the total time that B observes on A's clock during turnaround is

$$T_t' = T_t + \frac{v^2}{c^2}T_B.$$

The total time that B observes on A's clock during the entire trip is

$$\begin{aligned} (T_A')_{\text{total}} &= T_A' + T_t + \frac{v^2}{c^2}T_B \\ &= T_B\sqrt{1 - \frac{v^2}{c^2}} + T_t + \frac{v^2}{c^2}T_B, \end{aligned}$$

where we have used $T'_A = T_B/\gamma$, Eq. (12.10). We shall again neglect the turnaround time. The Doppler shift correction during turnaround is valid to order v^2/c^2 and to this approximation,

$$(T'_A)_{\text{total}} = T_B\left(1 - \frac{1}{2}\frac{v^2}{c^2} + \frac{v^2}{c^2}\right)$$
$$= T_B\left(1 + \frac{1}{2}\frac{v^2}{c^2}\right).$$

The result of this argument is that from B's point of view,

$$\frac{\text{aging of } B}{\text{aging of } A} = \frac{T_B}{(T'_A)_{\text{total}}} = \frac{1}{1 + \frac{1}{2}v^2/c^2} \approx 1 - \frac{1}{2}\frac{v^2}{c^2}.$$

We have already shown, Eq. (12.9), that from A's point of view

$$\frac{\text{aging of } B}{\text{aging of } A} = \frac{T'_B}{T_A} = \sqrt{1 - \frac{v^2}{c^2}} \approx 1 - \frac{1}{2}\frac{v^2}{c^2}.$$

The formerly identical twins are in agreement; A has aged more than B. The paradox is resolved.

Our analysis is valid only to order v^2/c^2. To this order, the special theory of relativity led to no contradictions as long as we treated the accelerated reference frame separately. An exact calculation appears to require the general theory of relativity.

Problems

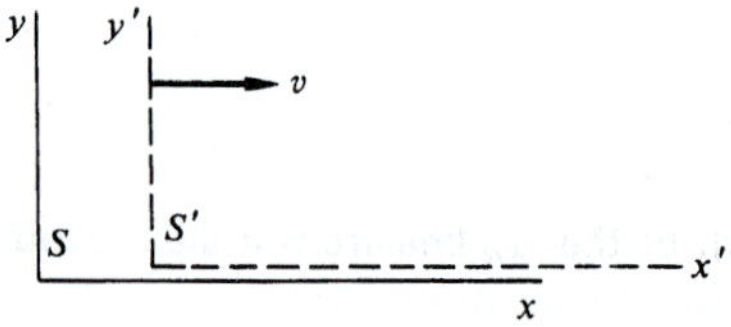

In these problems S refers to an inertial system x, y, z, t and S' refers to an inertial system x', y', z', t', moving along the x axis with speed v relative to S. The origins coincide at $t = t' = 0$. Take $c = 3 \times 10^8$ m/s.

12.1 Assume that $v = 0.6c$. Find the coordinates in S' of the following events.

a. $x = 4$ m, $t = 0$ s.

b. $x = 4$ m, $t = 1$ s.

c. $x = 1.8 \times 10^8$ m, $t = 1$ s.

d. $x = 10^9$ m, $t = 2$ s.

12.2 An event occurs in S at $x = 6 \times 10^8$ m, and in S' at $x' = 6 \times 10^8$ m, $t' = 4$ s. Find the relative velocity of the systems.

12.3 The clock in the sketch on the opposite page can provide an intuitive explanation of the time dilation formula. The clock consists of a flash tube, mirror, and phototube. The flash tube emits a pulse of light which

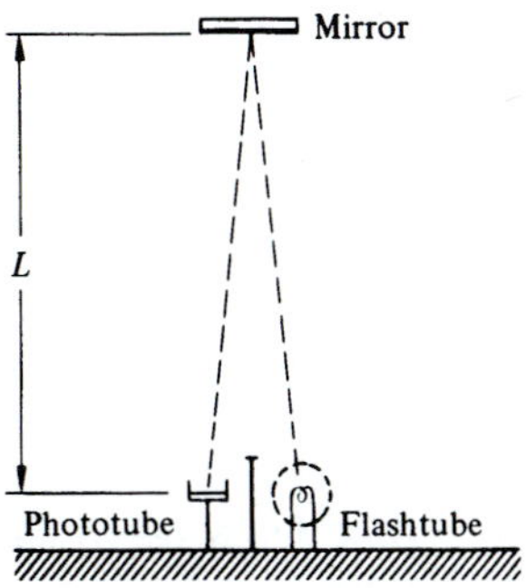

travels distance L to the mirror and is reflected to the phototube. Every time a pulse hits the phototube it triggers the flash tube. Neglecting time delay in the triggering circuits, the period of the clock is $\tau_0 = 2L/c$.

Now examine the clock in a coordinate system moving to the left with uniform velocity v. In this system the clock appears to move to the right with velocity v. Find the period of the clock in the moving system by direct calculation, using only the assumptions that c is a universal constant, and that distance perpendicular to the line of motion is unaffected by the motion. The result should be identical to that given by the Lorentz transformations: $\tau = \tau_0/\sqrt{1 - v^2/c^2}$.

12.4 A light beam is emitted at angle θ_0 with respect to the x' axis in S'.

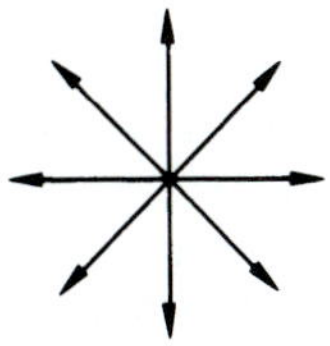

Rest frame

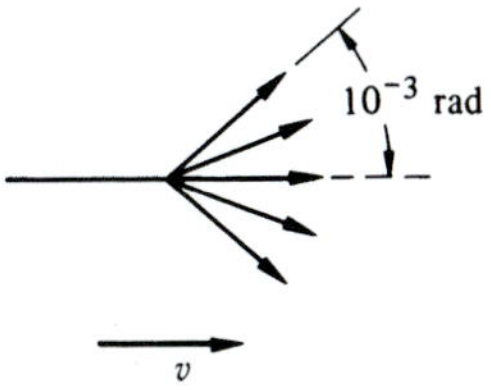

a. Find the angle θ the beam makes with respect to the x axis in S.

Ans. $\cos\theta = (\cos\theta_0 + v/c)/(1 + v/c\cos\theta_0)$

b. A source which radiates light uniformly in all directions in its rest frame radiates strongly in the forward direction in a frame in which it is moving with speed v close to c. This is called the headlight effect; it is very pronounced in synchrotrons in which electrons moving at relativistic speeds emit light in a narrow cone in the forward direction. Using the result of part a, find the speed of a source for which half the radiation is emitted in a cone subtending 10^{-3} rad.

Ans. $v = c(1 - 5 \times 10^{-7})$

12.5 An observer sees two spaceships flying apart with speed $0.99c$. What is the speed of one spaceship as viewed by the other?

Ans. $0.99995c$

12.6 A rod of proper length l_0 oriented parallel to the x axis moves with speed u along the x axis in S. What is the length measured by an observer in S'?

Ans. $l = l_0[(c^2 - v^2)(c^2 - u^2)]^{\frac{1}{2}}/(c^2 - uv)$

12.7 One of the most prominent spectral lines of hydrogen is the H_α line, a bright red line with a wavelength of 656.1×10^{-9} m.

a. What is the expected wavelength of the H_α line from a star receding with a speed of 3,000 km/s?

Ans. 662.7×10^{-9} m

b. The H_α line measured on earth from opposite ends of the sun's equator differ in wavelength by 9×10^{-12} m. Assuming that the effect is caused by rotation of the sun, find the period of rotation. The diameter of the sun is 1.4×10^6 km.

Ans. 25 d

12.8 The frequency of light reflected from a moving mirror undergoes a Doppler shift because of the motion of the image. Find the Doppler shift of light reflected directly back from a mirror which is approaching the observer with speed v, and show that it is the same as if the image were moving toward the observer at speed $2v/(1 + v^2/c^2)$.

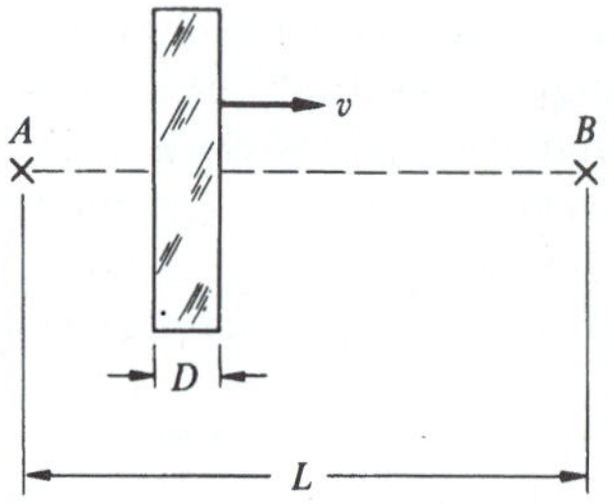

12.9 A slab of glass moves to the right with speed v. A flash of light is emitted from A and passes through the glass to arrive at B, a distance L away. The glass has thickness D in its rest frame, and the speed of light in the glass is c/n. How long does it take the light to go from A to B?

Ans. clue. If $v = 0$, $T = [L + (n - 1)D]/c$; if $v = c$, $T = L/c$

12.10 Here is the pole-vaulter paradox. A pole-vaulter and a farmer have the following bet: the pole-vaulter has a pole of length l_0, and the farmer has a barn $\frac{3}{4}l_0$ long. The farmer bets that he can shut the door of the barn with the pole completely inside. The bet being made, the farmer asks the pole-vaulter to run into the barn with a speed of $v = c\sqrt{3}/2$. In this case the farmer observes the pole to be Lorentz contracted to $l = l_0/2$, and the pole fits into the barn with ease. He slams the door the instant the pole is inside, and claims the bet. The pole-vaulter disagrees: he sees the barn contracted by a factor of 2, and so the pole can't possibly fit inside. How would you settle the disagreement? Is the Lorentz contraction "real" in this problem? (*Hint*: Consider events at the ends of the pole from the point of view of each observer.)

12.11 The relativistic transformation of acceleration from S' to S can be found by extending the procedure of Sec. 12.4. The most useful transformation is for the case in which the particle is instantaneously at rest in S' but is accelerating at rate a_0 in S', parallel to the x' axis.

Show that for this case the x acceleration in S is given by $a_x = a_0/\gamma^3$.

12.12 The relativistic transformation for acceleration derived in the last problem shows the impossibility of accelerating a system to a velocity greater than c. Consider a rocketship which accelerates at constant rate a_0 as measured by an accelerometer carried aboard, for instance a a mass stretching a spring.

a. Find the speed after time t for an observer in the system in which the rocketship was originally at rest.

Ans. $v = a_0t/\gamma$, or $v = a_0t/\sqrt{1 + (a_0t/c)^2}$

b. The speed predicted classically is $v_0 = a_0t$. What is the actual speed for the following cases: $v_0 = 10^{-3}c$, c, 10^3c.

Ans. $v = v_0(1 - 5 \times 10^{-7})$, $c/\sqrt{2}$, $c(1 - 5 \times 10^{-7})$

12.13 A young man voyages to the nearest star, α Centauri, 4.3 light-years away. He travels in a spaceship at a velocity of $c/5$. When he returns to earth, how much younger is he than his twin brother who stayed home?

12.14 Any quantity which is left unchanged by the Lorentz transformations is called a ***Lorentz invariant***. Show that Δs is a Lorentz invariant, where

$$\Delta s^2 = (c\,\Delta t)^2 - (\Delta x^2 + \Delta y^2 + \Delta z^2).$$

Here Δt is the interval between two events and $(\Delta x^2 + \Delta y^2 + \Delta x^2)^{\frac{1}{2}}$ is the distance between them in the same inertial system.

13 RELATIVISTIC MOMENTUM AND ENERGY

13.1 Momentum

In the last chapter we saw how the postulates of special relativity lead in a natural way to kinematical relations which agree with newtonian relations at low velocity but depart markedly for velocities approaching c. We turn now to the problem of investigating the implications of special relativity for dynamics. One approach would be to develop a formal procedure for writing the laws of physics in a form which satisfies the postulates of special relativity. Such a procedure is actually possible; it involves the concepts of four-vectors and relativistic invariance, and we shall pursue it in the next chapter. However, here we shall take another approach, one which is not as powerful or as economical as the method of four-vectors, but which has the advantage of using physical arguments to show the relation between the familiar concepts of classical mechanics and their relativistic counterparts.

First we shall focus on conservation of momentum and find what modifications are needed to preserve this principle in relativistic mechanics. This is a technique often used in extending the frontiers of physics: by reformulating conservation laws so that they are preserved in new situations, we are quite naturally led to generalizations of familiar concepts. In particular, as the following argument shows, we must modify our idea of mass to preserve conservation of momentum under relativistic transformations.

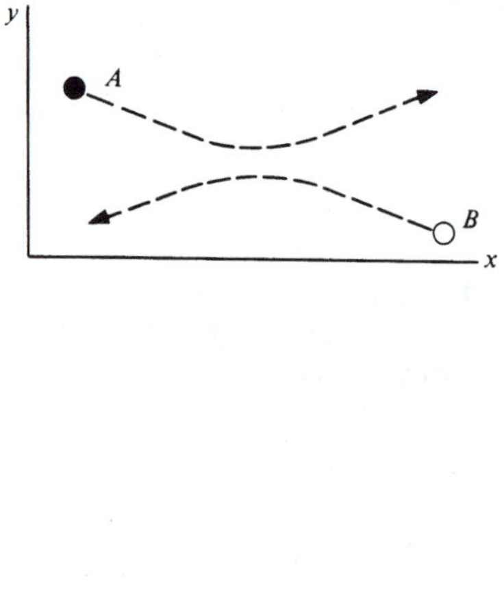

Consider a glancing elastic collision between two identical particles, A and B. We are going to view the collision in two special frames: A's frame, the frame moving along the x axis with A, and B's frame, the frame moving along the x axis with B. We

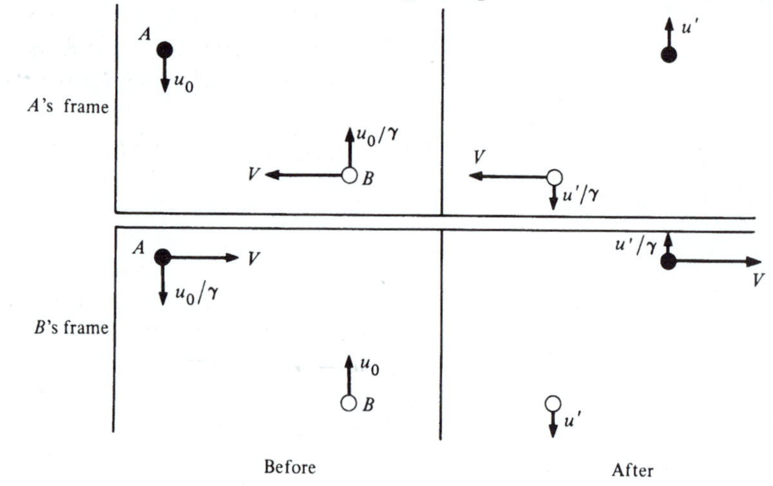

take the collisions to be completely symmetrical. Each particle has the same y speed u_0 in its own frame before the collision, as shown in the sketches. The effect of the collision is to alter the y velocities but leave the x velocities unchanged.

The relative x velocity of the frames is V and by the law of transformation of velocities, Eq. (12.2), the y velocity of the opposite particle in each frame is $u_0/\gamma = u_0\sqrt{1 - V^2/c^2}$.

After the collisions the y velocities have reversed their directions as shown in the sketches. The situation remains symmetrical. If the y speed of A and B in their own frames is u', the y speed of the other particle is u'/γ.

Our task is to find a conserved quantity analogous to classical momentum. We suppose that the momentum of a particle moving with velocity **w** is

$$\mathbf{p} = m(w)\mathbf{w},$$

where $m(w)$ is a scalar quantity, yet to be determined, analogous to newtonian mass, but one which may depend on the speed w.

The x momentum in A's frame is due entirely to particle B. Before the collision B's speed is $w = (V^2 + u_0^2/\gamma^2)^{\frac{1}{2}}$, and after the collision it is $w' = (V^2 + u'^2/\gamma^2)^{\frac{1}{2}}$. Imposing conservation of momentum in the x direction yields

$$m(w)V = m(w')V.$$

It follows that $w = w'$, so that

$$u' = u_0.$$

Next we write the statement of the conservation of momentum in the y direction, as evaluated in A's frame. Equating the y momentum before and after the collision gives

$$-m(u_0)u_0 + m(w)\frac{u_0}{\gamma} = m(u_0)u_0 - m(w)\frac{u_0}{\gamma}$$

or

$$m(w) = \gamma m(u_0).$$

In the limit $u_0 \rightarrow 0$, $m(u_0) \rightarrow m(0)$, which we take to be the newtonian mass, or "rest mass" m_0 of the particle. In this limit, $w = V$. Hence

$$m(V) = \gamma m(0) = \frac{m_0}{\sqrt{1 - V^2/c^2}}. \qquad 13.1$$

We have found the dependence of m on speed. In general, therefore,

$$\mathbf{p} = \frac{m_0\mathbf{u}}{\sqrt{1 - u^2/c^2}} = m\mathbf{u}$$

for a particle moving with arbitrary velocity $\mathbf{u}$, where

$$m = \frac{m_0}{\sqrt{1 - u^2/c^2}}. \qquad 13.2$$

Example 13.1 Velocity Dependence of the Electron's Mass

At the beginning of the twentieth century there were several speculative theories which predicted that the mass of an electron varies with its speed. These theories were based on various models of the structure of the electron. The principal theories were those of Abraham (1902), which predicted $m = m_0[1 + \frac{2}{5}(v^2/c^2)]$ for $v \ll c$,† and of Lorentz (1904), which gave $m = m_0/\sqrt{1 - v^2/c^2} \approx m_0[1 + \frac{1}{2}(v^2/c^2)]$. The Abraham theory, which retained the idea of the ether drift and absolute motion, predicted no time dilation effect. Lorentz' result, while identical in form to that published by Einstein in 1905, was derived using the ad hoc Lorentz contraction and did not possess the generality of Einstein's theory.

Experimental work on the effect of velocity on the electron's mass was initiated by Kaufmann in Göttingen in 1902. His data favored the theory of Abraham, and in a 1906 paper he rejected the Lorentz-Einstein results. However, further work by Bestelmeyer (1907) in Göttingen and Bucherer (1909) in Bonn revealed errors in Kaufmann's work and confirmed the Lorentz-Einstein formula.

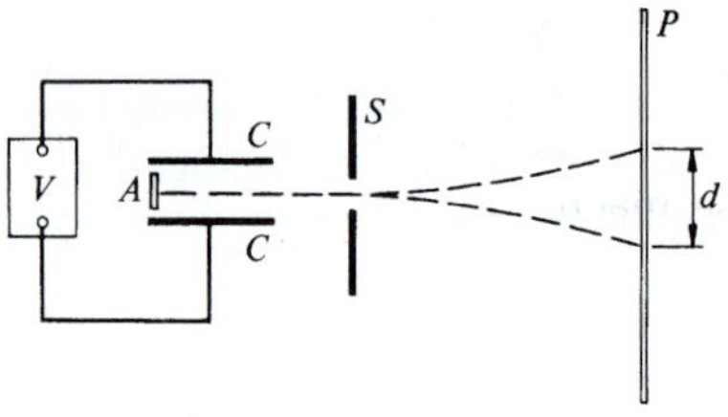

Physicists were in agreement that the force on a moving electron in an applied electric field $\mathbf{E}$ and magnetic field $\mathbf{B}$ is $q(\mathbf{E} + \mathbf{v} \times \mathbf{B})$ (the units are SI), where q is the electron's charge and $\mathbf{v}$ its velocity. Bucherer employed this force law in the apparatus shown at left. The apparatus is evacuated and immersed in an external magnetic field $\mathbf{B}$ perpendicular to the plane of the sketch. The source of the electrons A is a button of radioactive material, generally radium salts. The emitted electrons ("beta rays") have a broad energy spectrum extending to 1 MeV or so. To select a single speed, the electrons are passed through a "velocity filter" composed of a transverse electric field $\mathbf{E}$ (produced between two parallel metal plates C by the battery V) together with the magnetic field $\mathbf{B}$. $\mathbf{E}$, $\mathbf{B}$, and $\mathbf{v}$ are mutually perpendicular. The transverse force is

† Abraham's full result was

$$m = m_0 \frac{3}{4}\frac{1}{\beta^2}\left[\left(\frac{1 + \beta^2}{2\beta}\right) \ln\left(\frac{1 + \beta}{1 - \beta}\right) - 1\right],$$

where $\beta = v/c$.

zero when $qE = qvB$, so that electrons with $v = E/B$ are undeflected and are able to pass through the slit S.

Beyond S only the magnetic field acts. The electrons move with constant speed v and are bent into a circular path by the magnetic force $q\mathbf{v} \times \mathbf{B}$. The radius of curvature R is given by $mv^2/R = qvB$, or $R = mv/qB = (m/q)(E/B^2)$.

The electrons eventually strike the photographic plate P, leaving a trace. By reversing **E** and **B**, the sense of deflection is reversed. R is found from a measurement of the total deflection d and the known geometry of the apparatus. E and B are found by standard techniques. By finding R for different velocities, the velocity dependence of m/q can be studied. We believe that charge does not vary with velocity (otherwise an atom would not stay strictly neutral in spite of how the energy of its electrons varied), so that the variation of m/q can be attributed to variation in m alone.

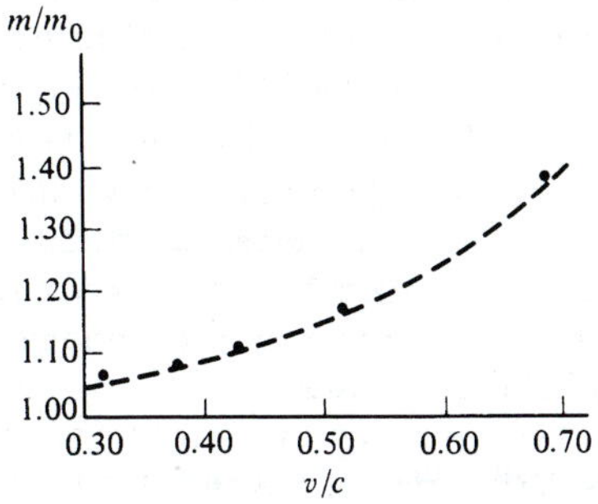

The graph shows Bucherer's data together with a dashed line corresponding to the Einstein prediction $m = m_0/\sqrt{1 - v^2/c^2}$. The agreement is striking.

The velocity filter with crossed **E** and **B** fields was used by Bestelmeyer and by Bucherer. (Bucherer attributes the design to J. J. Thomson, discoverer of the electron.) Kaufmann, on the other hand, used transverse **E** and **B** fields which were parallel to one another, and this probably caused his erroneous results. His configuration did not select velocities; instead, all the electrons were spread into a two dimensional trace on the photographic plate. Electrons of different speeds followed different deflected paths between the plates C, and nonuniformity of the **E** field gave rise to substantial errors.

In recent years the relativistic equations of motion have been used to design high energy electron and proton accelerators. For protons, accelerators have been operated with m/m_0 up to 200, while for electrons the ratio $m/m_0 = 40{,}000$ has been reached. The successful operation of these machines leaves no doubt that the relativistic results are accurate.

13.2 Energy

By generalizing the classical concept of energy, we can find a corresponding relativistic quantity which is also conserved. From the discussion in Chap. 4 we can write the kinetic energy of a particle, K, as

$$K_b - K_a = \int_a^b \frac{d\mathbf{p}}{dt} \cdot d\mathbf{r}.$$

For a classical particle moving with velocity $\mathbf{u}$, $\mathbf{p} = m\mathbf{u}$, where m is constant. Then

$$K_b - K_a = \int_a^b \frac{d}{dt}(m\mathbf{u}) \cdot d\mathbf{r}$$
$$= \int_a^b m \frac{d\mathbf{u}}{dt} \cdot \mathbf{u}\, dt$$
$$= \int_a^b m\mathbf{u} \cdot d\mathbf{u}.$$

Using the identity $\mathbf{u} \cdot d\mathbf{u} = \frac{1}{2}d(\mathbf{u} \cdot \mathbf{u}) = \frac{1}{2}d(u^2) = u\, du$, we obtain $K_b - K_a = \frac{1}{2}mu_b^2 - \frac{1}{2}mu_a^2$.

It is natural to try the same procedure starting with the relativistic expression for momentum $\mathbf{p} = m_0\mathbf{u}/\sqrt{1 - u^2/c^2}$.

$$K_b - K_a = \int_a^b \frac{d\mathbf{p}}{dt} \cdot d\mathbf{r}$$
$$= \int_a^b \frac{d}{dt}\left[\frac{m_0\mathbf{u}}{\sqrt{1 - u^2/c^2}}\right] \cdot \mathbf{u}\, dt$$
$$= \int_a^b \mathbf{u} \cdot d\left[\frac{m_0\mathbf{u}}{\sqrt{1 - u^2/c^2}}\right]$$

The integrand is $\mathbf{u} \cdot d\mathbf{p} = d(\mathbf{u} \cdot \mathbf{p}) - \mathbf{p} \cdot d\mathbf{u}$. Therefore

$$K_b - K_a = (\mathbf{u} \cdot \mathbf{p})\Big|_a^b - \int_a^b \mathbf{p} \cdot d\mathbf{u}$$
$$= \frac{m_0u^2}{\sqrt{1 - u^2/c^2}}\Bigg|_a^b - \int_a^b \frac{m_0u\, du}{\sqrt{1 - u^2/c^2}},$$

where we have used the earlier identity $\mathbf{u} \cdot d\mathbf{u} = u\, du$. The integral is elementary, and we find

$$K_b - K_a = \frac{m_0u^2}{\sqrt{1 - u^2/c^2}}\Bigg|_a^b + m_0c^2\sqrt{1 - \frac{u^2}{c^2}}\Bigg|_a^b.$$

Take point b as arbitrary, and let the particle be at rest at point a, $u_a = 0$.

$$K = \frac{m_0u^2}{\sqrt{1 - u^2/c^2}} + m_0c^2\sqrt{1 - \frac{u^2}{c^2}} - m_0c^2$$
$$= \frac{m_0[u^2 + c^2(1 - u^2/c^2)]}{\sqrt{1 - u^2/c^2}} - m_0c^2$$
$$= \frac{m_0c^2}{\sqrt{1 - u^2/c^2}} - m_0c^2$$

or

$$K = mc^2 - m_0c^2, \tag{13.3}$$

where $m = m_0/\sqrt{1 - u^2/c^2}$.

This expression for kinetic energy bears little resemblance to its classical counterpart. However, in the limit $u \ll c$, the relativistic result should approach the classical expression $K = \frac{1}{2}mu^2$. This is indeed the case, as we see by making the approximation $1/\sqrt{1 - u^2/c^2} \approx 1 + \frac{1}{2}u^2/c^2$. Then

$$\begin{aligned} K &= \frac{m_0c^2}{\sqrt{1 - u^2/c^2}} - m_0c^2 \\ &\approx m_0c^2\left(1 + \frac{1}{2}\frac{u^2}{c^2} - 1\right) \\ &= \tfrac{1}{2}m_0u^2. \end{aligned}$$

The kinetic energy arises from the work done on the particle to bring it from rest to speed u. Suppose that we rewrite Eq. (13.3) as

$$\begin{aligned} mc^2 &= K + m_0c^2 \\ &= \text{work done on particle} + m_0c^2. \end{aligned} \tag{13.4}$$

Einstein proposed the following bold interpretation of this result: mc^2 is the *total* energy E of the particle. The first term arises from external work; the second term, m_0c^2, represents the "rest" energy the particle possesses by virtue of its mass. In summary

$$E = mc^2. \tag{13.5}$$

It is important to realize that Einstein's generalization goes far beyond the classical conservation law for mechanical energy. Thus, if energy ΔE is added to a body, its mass will change by $\Delta m = \Delta E/c^2$, irrespective of the form of energy. ΔE could represent mechanical work, heat energy, the absorption of light, or any other form of energy. In relativity the classical distinction between mechanical energy and other forms of energy disappears. Relativity treats all forms of energy on an equal footing, in contrast to classical physics where each form of energy must be treated as a special case. The conservation of total energy $E = mc^2$ is a consequence of the very structure of relativity. In the next chapter we shall show that the conservation laws for energy and momentum are different aspects of a single, more general, conservation law.

The following example illustrates the relativistic concept of energy and the validity of the conservation laws in different inertial frames.

Example 13.2 **Relativistic Energy and Momentum in an Inelastic Collision**

Suppose that two identical particles collide with equal and opposite velocities and stick together. Classically, the initial kinetic energy is $2(\frac{1}{2}MV^2) = MV^2$, where M is the newtonian mass. By conservation of momentum the mass $2M$ is at rest and has zero kinetic energy. In the language of Chap. 4 we say that mechanical energy MV^2 was lost as heat. As we shall see, this distinction between forms of energy does not occur in relativity.

Now consider the same collision relativistically, as seen in the original frame x, y, and in a frame x', y' moving with one of the particles. By the relativistic transformation of velocities, Eq. (12.2),

$$U = \frac{2V}{1 + V^2/c^2} \qquad 1$$

in the direction shown.

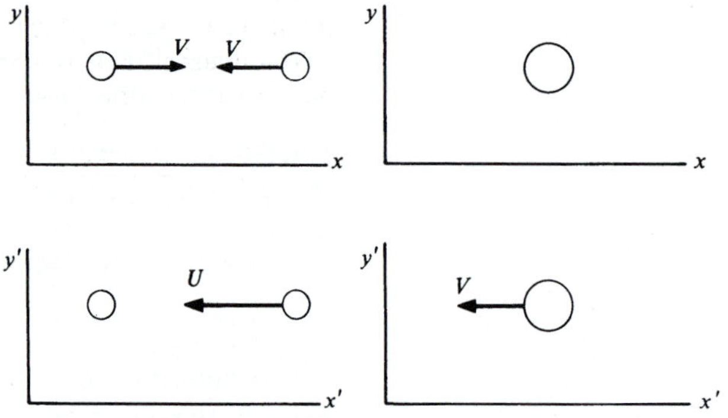

Before After

Let the rest mass of each particle be M_{0i} before the collision and M_{0f} after the collision. In the x, y frame, momentum is obviously conserved. The total energy before the collision is $2M_{0i}c^2/\sqrt{1 - V^2/c^2}$, and after the collision the energy is $2M_{0f}c^2$. No external work was done on the particles, and the total energy is unchanged. Therefore,

$$\frac{2M_{0i}c^2}{\sqrt{1 - V^2/c^2}} = 2M_{0f}c^2$$

or

$$M_{0f} = \frac{M_{0i}}{\sqrt{1 - V^2/c^2}}. \qquad 2$$

The final rest mass is greater than the initial rest mass because the particles are warmer. To see this, we take the low velocity approximation

$$M_{0f} \approx M_{0i}\left(1 + \frac{1}{2}\frac{V^2}{c^2}\right)$$

The increase in rest energy for the two particles is $2(M_{0f} - M_{0i})c^2 \approx 2(\frac{1}{2}M_{0i}V^2)$, which corresponds to the loss of classical kinetic energy. Now, however, the kinetic energy is not "lost"—it is present as a mass increase.

By the postulate that all inertial frames are equivalent, the conservation laws must hold in the x', y' frame as well. If our assumed conservation laws possess this necessary property, we have in the x', y' frame

$$\frac{M_{0i}U}{\sqrt{1 - U^2/c^2}} = \frac{2M_{0f}V}{\sqrt{1 - V^2/c^2}} \qquad 3$$

by conservation of momentum and

$$M_{0i}c^2 + \frac{M_{0i}c^2}{\sqrt{1 - U^2/c^2}} = \frac{2M_{0f}c^2}{\sqrt{1 - V^2/c^2}} \qquad 4$$

by the conservation of energy.

The question now is whether Eqs. (3) and (4) are consistent with our earlier results, Eqs. (1) and (2). To check Eq. (3), we use Eq. (1) to write

$$\begin{aligned} 1 - \frac{U^2}{c^2} &= 1 - \frac{4V^2/c^2}{(1 + V^2/c^2)^2} \\ &= \frac{(1 - V^2/c^2)^2}{(1 + V^2/c^2)^2}. \end{aligned} \qquad 5$$

From Eqs. (1) and (5),

$$\begin{aligned} \frac{U}{\sqrt{1 - U^2/c^2}} &= \frac{2V}{(1 + V^2/c^2)}\,\frac{(1 + V^2/c^2)}{(1 - V^2/c^2)} \\ &= \frac{2V}{1 - V^2/c^2} \end{aligned}$$

and the left hand side of Eq. (3) becomes

$$\frac{M_{0i}U}{\sqrt{1 - U^2/c^2}} = \frac{2M_{0i}V}{1 - V^2/c^2}. \qquad 6$$

From Eq. (2), $M_{0i} = M_{0f}\sqrt{1 - V^2/c^2}$, and Eq. (6) reduces to

$$\frac{M_{0i}U}{\sqrt{1 - U^2/c^2}} = \frac{2M_{0f}V}{\sqrt{1 - V^2/c^2}},$$

which is identical to Eq. (3). Similarly, it is not hard to show that Eq. (4) is also consistent.

We see from Eq. (6) that if we had assumed that rest mass was unchanged in the collision, $M_{0i} = M_{0f}$, the conservation law for momentum (or for energy) would not be correct in the second inertial frame. The relativistic description of energy plays an essential part in maintaining the validity of the conservation laws in all inertial frames.

Example 13.3 The Equivalence of Mass and Energy

In 1932 Cockcroft and Walton, two young British physicists, successfully operated the first high energy proton accelerator and succeeded in causing a nuclear disintegration. Their experiment provided one of the earliest confirmations of the relativistic mass-energy relation.

Briefly, their accelerator consisted of a power supply that could reach 600 kV and a source of protons (hydrogen nuclei). The power supply used an ingenious arrangement of capacitors and rectifiers to quadruple the voltage of a 150-kV supply. The protons were supplied by an electrical discharge in hydrogen and were accelerated in vacuum by the applied high voltage.

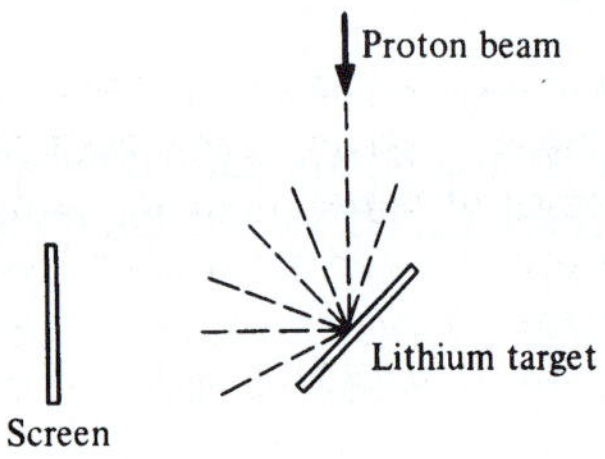

Cockcroft and Walton studied the effect of the protons on a target of ^{7}Li (lithium, having atomic mass 7). A zinc sulfide fluorescent screen, located nearby, emitted occasional flashes, or scintillations. By various tests they determined that the scintillations were due to alpha particles, the nuclei of helium, ^{4}He. Their interpretation was that the ^{7}Li captures a proton and that the resulting nucleus of mass 8 immediately disintegrates into two alpha particles. We can write the reaction as

$$^1\text{H} + {}^7\text{Li} \rightarrow {}^4\text{He} + {}^4\text{He}.$$

The mass energy equation for the reaction is

$$K(^1\text{H}) + [M(^1\text{H}) + M(^7\text{Li})]c^2 = 2K(^4\text{He}) + 2M(^4\text{He})c^2$$

where $K(^1\text{H})$ is the kinetic energy of the incident proton, $K(^4\text{He})$ is the kinetic energy of each of the emitted alpha particles, and $M(^1\text{H})$ is the proton rest mass, etc. (The initial momentum of the proton is negligible, and the two alpha particles are emitted back to back with equal energy by conservation of momentum.)

We can rewrite the mass-energy equation as

$$K = \Delta Mc^2,$$

where $K = 2K(^4\text{He}) - K(^1\text{H})$, and ΔM is the initial rest mass minus the final rest mass.

The energy of the alpha particles was determined by measuring their range. Cockcroft and Walton obtained the value $K = 17.2$ MeV (1 MeV $= 10^6$ eV $= 1.6 \times 10^{-13}$ J).

The relative masses of the nuclei were known from mass spectrometer measurements. In atomic mass units, amu, defined so that $M(^{16}\text{O}) = 16$, the values available to Cockcroft and Walton were

$$M(^1\text{H}) = 1.0072$$
$$M(^7\text{Li}) = 7.0104 \pm 0.0030$$
$$M(^4\text{He}) = 4.0011.$$

These yield

$$\begin{aligned}\Delta M &= (1.0072 + 7.0104) - 2(4.0011) \\ &= (0.0154 \pm 0.0030)\ \text{amu}.\end{aligned}$$

The rest energy of 1 amu is 931 MeV and therefore

$$\Delta Mc^2 = (14.3 \pm 2.7)\ \text{MeV}.$$

The difference between K and ΔMc^2 is $(17.2 - 14.3)$ MeV $= 2.9$ MeV, slightly larger than the experimental uncertainty of 2.7 MeV. However, the experimental uncertainty always represents an estimate, not a precise limit, and the result can be taken as consistent with the relation $K = \Delta Mc^2$.

It is clear that the masses must be known to high accuracy for studying the energy balance in nuclear reactions. Modern techniques of mass spectrometry have achieved an accuracy of better than 10^{-5} amu, and the mass-energy equivalence has been amply confirmed. According to a modern table of masses, the decrease in rest mass in the reaction studied by Cockcroft and Walton is $\Delta Mc^2 = (17.3468 \pm 0.0012)$ MeV.

Often it is useful to express the total energy of a free particle in terms of its momentum. Classically the relation is

$$E = \tfrac{1}{2}mv^2 = \frac{p^2}{2m}.$$

To find the equivalent relativistic expression we must combine the relativistic momentum

$$\mathbf{p} = m\mathbf{u} = \frac{m_0\mathbf{u}}{\sqrt{1 - u^2/c^2}} = m_0\mathbf{u}\gamma \qquad 13.6$$

with the energy

$$E = mc^2 = m_0c^2\gamma. \qquad 13.7$$

Squaring Eq. (13.6) gives

$$p^2 = \frac{m_0{}^2u^2}{1 - u^2/c^2},$$

which we solve for γ as follows:

$$\frac{u^2}{c^2} = \frac{p^2}{p^2 + m_0{}^2c^2}$$

$$\gamma = \frac{1}{\sqrt{1 - u^2/c^2}}$$

$$= \sqrt{1 + \frac{p^2}{m_0{}^2c^2}}.$$

Inserting this in Eq. (13.7), we have

$$E = m_0c^2 \sqrt{1 + \frac{p^2}{m_0{}^2c^2}}.$$

The square of this equation is algebraically somewhat simpler and is the form usually employed.

$$E^2 = (pc)^2 + (m_0c^2)^2 \qquad 13.8$$

We have derived the relativistic expressions for momentum and energy by invoking conservation laws. However, we have not dealt with the role of force in relativity. It is possible to attack this problem by considering the form of the equations of motion in various coordinate systems. We shall develop a systematic way of doing this in the next chapter, and so we defer the problem of force for the present.

For convenience, here is a summary of the important dynamical formulas we have developed so far.

$$\mathbf{p} = m\mathbf{u} = m_0\mathbf{u}\gamma \qquad 13.9$$

$$K = mc^2 - m_0c^2 = m_0c^2(\gamma - 1) \qquad 13.10$$

$$E = mc^2 = m_0c^2\gamma \qquad 13.11$$

$$E^2 = (pc)^2 + (m_0c^2)^2 \qquad 13.12$$

13.3 Massless Particles

A surprising consequence of the relativistic energy-momentum relation is the possibility of "massless" particles—particles which possess momentum and energy but no rest mass. If we take $m_0 = 0$ in the relation

$$E^2 = (pc)^2 + (m_0c^2)^2,$$

the result is

$$E = pc. \qquad 13.13$$

We take the positive root on the plausible assumption that particles whose energy decreases with increasing momentum would be unstable.

In order to have nonzero momentum we must have a finite value for

$$\mathbf{p} = m_0\mathbf{u} \Big/ \sqrt{1 - \frac{u^2}{c^2}}.$$

in the limit $m_0 \to 0$. This is only possible if $u \to c$ as $m_0 \to 0$; massless particles must travel at the speed of light.

The principal massless particle known to physics is the *photon*, the particle of light. Photons interact electromagnetically with electrons and other charged particles and are easy to detect with photographic films, phototubes, or the eye. The *neutrino*, which is associated with the weak forces of radioactive beta decay, is believed to be massless, but it interacts so weakly with matter that its direct detection is extremely difficult. (The sun is a copious source of neutrinos, but most of the solar neutrinos which reach the earth pass through it without interacting.) Experiments have shown that the neutrino rest mass is no larger than 1/2,000 the rest mass of the electron, and it could well be zero. There are theoretical reasons for believing in the existence of the graviton, a massless particle associated with the gravitational force. The graviton's interaction with matter is so weak that it has not yet been detected.

We owe the concept of the photon to Einstein, who introduced it in his pioneering paper on the photoelectric effect published a few months before his work on relativity.[1] Briefly, Einstein proposed that the energy of a light wave can only be transmitted to matter in discrete amounts, or quanta, of value $h\nu$, where h is Planck's constant 6.63×10^{-34} J/Hz, and ν is the frequency of the light wave in hertz. The arguments for this proposal grew out of Einstein's concern with problems in classical electromagnetic theory and considerations of Planck's quantum hypothesis,

[1] Within a period of one year Einstein wrote four papers, each of which became a classic, on the photoelectric effect, relativity, brownian motion, and the quantum theory of the heat capacity of solids. It was for his work on the photoelectric effect, not relativity, that Einstein received the Nobel Prize for Physics in 1921. Relativity was so encumbered with philosophical and political implications that the Nobel committee refused to acknowledge it. This regrettable incident was unique in the history of the prize.

a theory constructed by Planck in 1900 to overcome difficulties in classical statistical mechanics. Although we cannot develop here the background necessary to justify Einstein's theory of the photon, perhaps the following experimental evidence will help make the photon seem plausible.

Example 13.4 The Photoelectric Effect

In 1887 Heinrich Hertz discovered that metals can give off electrons when illuminated by ultraviolet light. This process, the *photoelectric effect*, represents the direct conversion of light into mechanical energy (here, the kinetic energy of the electron). Einstein predicted that the energy an electron absorbs from a beam of light at frequency ν is exactly $h\nu$. If the electron loses a certain amount of energy W in leaving the metal, then the kinetic energy of the emitted electron is

$$K = h\nu - W.$$

W is known as the *work function* of the metal. The work function is typically a few electron volts, but unfortunately it depends on the chemical state of the metal surface, making the photoelectric effect a difficult matter to investigate. Millikan overcame this problem in 1916 by working with metal surfaces prepared in a high vacuum system. The kinetic energy was determined by measuring the photocurrent collected on a plate near the metal and applying an electric potential between the plate and photosurface just adequate to stop the current. If the potential is $-V$, then the energy lost by the electrons as they travel to the plate is $(-e)(-V)$. At cutoff we have $V = V_c$ and

$$eV_c = h\nu - W.$$

Millikan observed the cutoff voltage as a function of frequency for several alkali metals. In accord with Einstein's formula, he found that V_c was a linear function of ν, with slope h/e, and that V_c was independent of the intensity of the light.

If the energy of light were absorbed by the electron according to the classical picture, the electrons would have a wide energy distribution depending on the intensity of the light, in sharp disagreement with Millikan's results. The fact that light can interfere with itself, as in the Michelson interferometer, is compelling evidence that light has wave properties. Nevertheless, the photoelectric effect illustrates that light also has particle properties. Einstein's energy relation, $E = h\nu$, provides the link between these apparently conflicting descriptions of light by relating the energy of the particle to the frequency of the wave.

Example 13.5 Radiation Pressure of Light

A consequence of Maxwell's electromagnetic theory is that a light wave carries momentum which it will transfer to a surface when it is reflected or absorbed. The result, as we know from our study of momentum in Chap. 3, is a pressure on the surface. The calculation of radiation pressure is complicated using the wave theory of light, but with the photon picture it is simple.

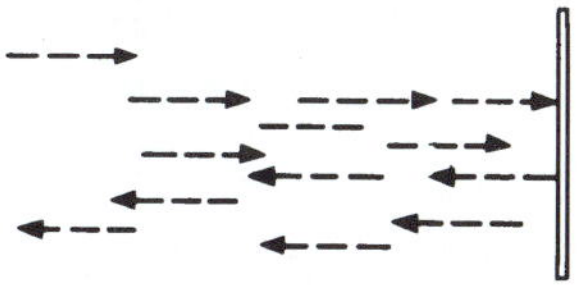

Consider a stream of photons striking a perfectly reflecting mirror at normal incidence. The initial momentum of each photon is $p = E/c$, and the total change in momentum in the reflection is $2p = 2E/c$. If there are n photons incident per unit area per second, the total momentum change per second is $2nE/c$, and this is equal to the force per unit area exerted on the mirror by the light. Hence the radiation pressure P is

$$P = \frac{2nE}{c} = \frac{2I}{c},$$

where $I = nE$ is the intensity of the light, the power per unit area. Similarly, the radiation pressure on a perfect absorber is I/c.

The average intensity of sunlight falling on the earth's surface at normal incidence, known as the *solar constant*, is $I \approx 1{,}000\ \text{W/m}^2$. The radiation pressure on a mirror due to sunlight is therefore $P = 2I/c = 7 \times 10^{-6}\ \text{N/m}^2$, a very small pressure. (Atmospheric pressure is $10^5\ \text{N/m}^2$.) On the cosmic scale, however, radiation pressure is large; it helps keep stars from collapsing under their own gravitational forces.

Since the photon is a completely relativistic particle, newtonian physics provides little insight into its properties. For instance, unlike classical particles, photons can be created and destroyed; the absorption of light by matter corresponds to the destruction of photons, while the process of radiation corresponds to the creation of photons. Nevertheless, the familiar laws of conservation of momentum and energy, as generalized in the theory of relativity, are sufficiently powerful to let us draw conclusions about processes involving photons without a detailed knowledge of the interaction, as the following examples illustrate.

Example 13.6 The Compton Effect

The special theory of relativity was not widely accepted by the 1920s partly because of the radical nature of its concepts, but also because there was little experimental evidence. In 1922 Arthur Compton performed a refined experiment on the scattering of x-rays from matter which left little doubt that relativistic dynamics was valid.

A photon of visible light has energy in the range of 1 to 2 eV, but photons of much higher energy can be obtained from x-ray machines, radioactive sources, or particle accelerators. X-ray photons have energies typically in the range 10 to 100 keV, and their wavelengths can be measured with high accuracy by the technique of crystal diffraction.

When a photon collides with a free electron, the conservation laws require that the photon lose a portion of its energy. The outgoing photon therefore has a longer wavelength than the primary photon, and this shift in wavelength, first observed by Compton, is known as the *Compton effect*.

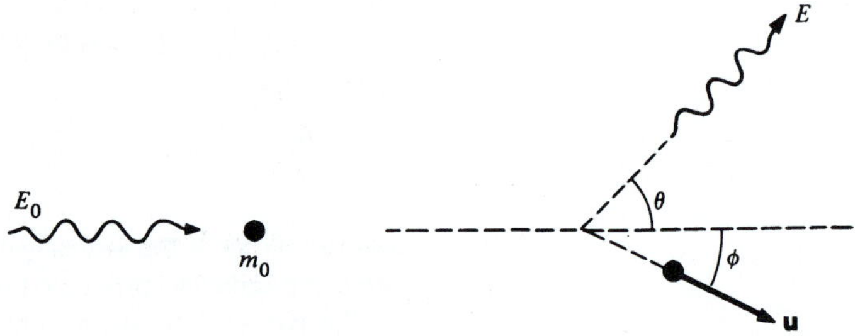

Let the photon have initial energy E_0 and momentum E_0/c, and suppose that the electron is initially at rest. After the collision, the electron is scattered at angle ϕ with velocity **u** and the photon is scattered at angle θ with energy E. Let $E_e = m_0c^2/\sqrt{1 - u^2/c^2}$ be the final electron energy and $\mathbf{p} = m\mathbf{u}$ the momentum. Then, by conservation of energy,

$$E_0 + m_0c^2 = E + E_e. \qquad 1$$

By conservation of momentum,

$$\frac{E_0}{c} = \frac{E}{c}\cos\theta + p\cos\phi \qquad 2$$

$$0 = \frac{E}{c}\sin\theta - p\sin\phi. \qquad 3$$

Our object is to eliminate reference to the electron and find E as a function of θ, since Compton detected only the outgoing photon in his experiments. Equations (2) and (3) can be written

$$(E_0 - E\cos\theta)^2 = (pc)^2\cos^2\phi$$

$$(E\sin\theta)^2 = (pc)^2\sin^2\phi.$$

Adding,

$$E_0^2 - 2E_0E\cos\theta + E^2 = (pc)^2 \qquad 4$$

$$= E_e^2 - (m_0c^2)^2,$$

where we have used the energy-momentum relation, Eq. (13.12). Using Eq. (1) to eliminate E_e, Eq. (4) becomes

$$E_0^2 - 2E_0E\cos\theta + E^2 = (E_0 + m_0c^2 - E)^2 - (m_0c^2)^2,$$

which reduces to

$$E = \frac{E_0}{1 + (E_0/m_0c^2)(1 - \cos\theta)}. \qquad 5$$

Note that E is always greater than zero, which means that a free electron cannot absorb a photon.

Compton measured wavelengths rather than energies in his experiment. From the Einstein frequency condition, $E_0 = h\nu_0 = hc/\lambda_0$ and $E = hc/\lambda$, where λ_0 and λ are the wavelengths of the incoming and outgoing photons, respectively. In terms of wavelength, Eq. (5) takes the simple form

$$\lambda = \lambda_0 + \frac{h}{m_0c}(1 - \cos\theta).$$

The quantity h/m_0c is known as the *Compton wavelength* of the electron and has the value

$$\frac{h}{m_0c} = 2.426 \times 10^{-12}\ \text{m}$$
$$= 0.02426\ \text{Å},$$

where 1 Å $= 10^{-10}$ m.

The shift in wavelength at a given angle is independent of the initial photon energy:

$$\lambda - \lambda_0 = \frac{h}{m_0c}(1 - \cos\theta).$$

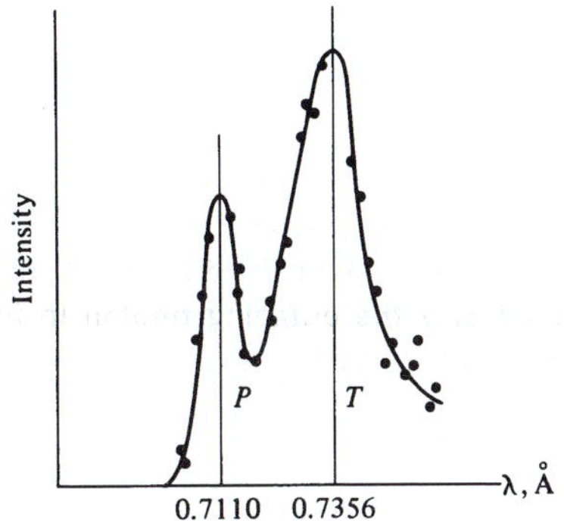

The figure shows one of Compton's results for $\lambda_0 = 0.711$ Å and $\theta = 90°$, where peak P is due to primary photons and peak T to the Compton scattered photons from a block of graphite. The measured wavelength shift is approximately 0.0246 Å and the calculated value is 0.02426 Å. The difference is less than the estimated uncertainty due to the limited resolution of the spectrometer and other experimental limitations.

In our analysis we assumed that the electron was free and at rest. For sufficiently high proton energies, this is a good approximation for electrons in the outer shells of light atoms. If the motion of the electrons is taken into account, the Compton peak is broadened. (The broadening of peak T in the figure compared with P shows this effect.)

If the binding energy of the electron is comparable to the photon energy, momentum and energy can be transferred to the atom as a whole, and the photon can be completely absorbed.

Example 13.7 **Pair Production**

We have already seen two ways by which a photon can lose energy in matter, photoelectric absorption and Compton scattering. If a photon's energy is sufficiently high, it can also lose energy in matter by the mechanism of *pair production*. The rest mass of an electron is $m_0c^2 =$ 0.511 MeV. Can a photon of this energy create an electron? The answer is no, since this would require the creation of a single electric charge. As far as we know, electric charge is conserved in all physical processes. However, if equal amounts of positive and negative charge are created, the total charge remains zero and charge is conserved. Hence, it is possible to create an electron-positron pair (e^--e^+), two particles having the same mass but opposite charge.

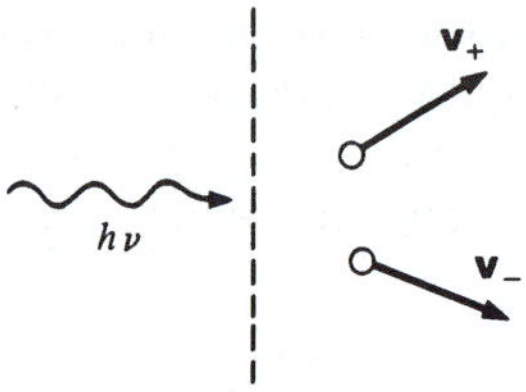

A single photon of energy $2m_0c^2$ or greater has enough energy to form an e^--e^+ pair, but the process cannot occur in free space because it would not conserve momentum. If we imagine that the process occurs, conservation of energy gives

$$h\nu = m_+c^2 + m_-c^2 = (\gamma_+ + \gamma_-)m_0c^2,$$

or

$$\frac{h\nu}{c} = (\gamma_+ + \gamma_-)m_0c,$$

while conservation of momentum gives

$$h\nu/c = |\gamma_+\mathbf{v}_+ + \gamma_-\mathbf{v}_-|m_0.$$

These equations cannot be satisfied simultaneously because

$$(\gamma_+ + \gamma_-)c > |\gamma_+\mathbf{v}_+ + \gamma_-\mathbf{v}_-|.$$

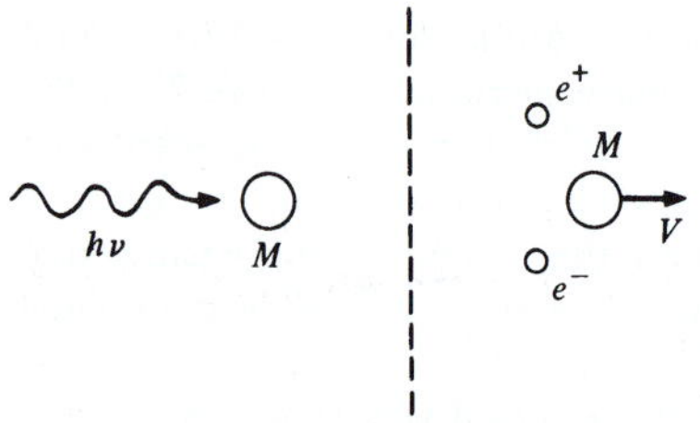

Pair production is possible if a third particle is available for carrying off the excess momentum. For instance, suppose that the photon hits a nucleus of rest mass M and creates an e^--e^+ pair at rest. We have

$$h\nu + Mc^2 = 2m_0c^2 + Mc^2\gamma.$$

Since nuclei are much more massive than electrons, let us assume that $h\nu \ll Mc^2$. (For hydrogen, the lightest atom, this means that $h\nu \ll$ 940 MeV.) In this case the atom will not attain relativistic speeds and we can make the classical approximation

$$\begin{aligned} h\nu &= 2m_0c^2 + Mc^2(\gamma - 1) \\ &\approx 2m_0c^2 + \tfrac{1}{2}MV^2. \end{aligned}$$

To the same approximation, conservation of momentum yields

$$\frac{h\nu}{c} = MV.$$

Substituting this in the energy expression gives

$$h\nu = 2m_0c^2 + \frac{1}{2}\frac{(h\nu)^2}{Mc^2} \approx 2m_0c^2,$$

since we have already assumed $h\nu \ll Mc^2$. The threshold for pair production in matter is therefore $2m_0c^2 = 1.02$ MeV. The nucleus plays an essentially passive role, but by providing for momentum conservation it allows an otherwise forbidden process to occur.

Example 13.8 The Photon Picture of the Doppler Effect

In Chap. 12 we discussed the Doppler effect from the standpoint of wave theory, but we can also treat it using the photon picture. Consider first an atom with rest mass M_0, held stationary. If the atom emits a photon of energy $h\nu_0$, its new rest mass is given by $M_0'c^2 = M_0c^2 - h\nu_0$.

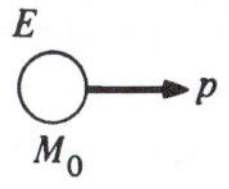

Suppose now that the atom moves freely with velocity **u** before emitting the photon. The atom's energy is $E = M_0c^2/\sqrt{1 - u^2/c^2}$ and its momentum is $p = M_0u/\sqrt{1 - u^2/c^2}$. After the emission of a photon of energy $h\nu$ the atom has velocity **u**′, rest mass M_0', energy E', and momentum p'. For simplicity, we consider the photon to be emitted along the line of motion. By conservation of energy and momentum we have

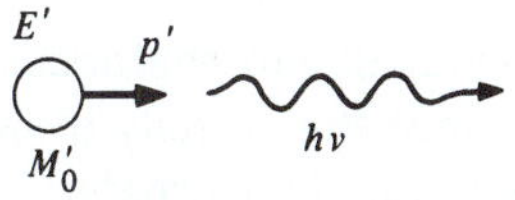

$$E = E' + h\nu \qquad 1$$

$$p = p' + \frac{h\nu}{c}. \qquad 2$$

Therefore,

$$(E - h\nu)^2 = E'^2$$

$$(pc - h\nu)^2 = (p'c)^2$$

and

$$(E - h\nu)^2 - (pc - h\nu)^2 = E'^2 - (p'c)^2 = (M_0'c^2)^2 \qquad 3$$

by the energy-momentum relation. Expanding the left hand side and using $E^2 - (pc)^2 = (M_0c^2)^2$, we obtain

$$(M_0c^2)^2 - 2Eh\nu + 2pch\nu = (M_0'c^2)^2$$
$$= (M_0c^2 - h\nu_0)^2.$$

Simplifying, we find that

$$\nu = \nu_0 \frac{(2M_0c^2 - h\nu_0)}{2(E - pc)}.$$

However,

$$E - pc = \frac{M_0c^2}{\sqrt{1 - u^2/c^2}}\left(1 - \frac{u}{c}\right)$$
$$= M_0c^2\sqrt{\frac{1 - u/c}{1 + u/c}}.$$

Hence,

$$\nu = \nu_0 \left(1 - \frac{h\nu_0}{2M_0c^2}\right) \sqrt{\frac{1 + u/c}{1 - u/c}}.$$

The term $h\nu_0/2M_0c^2$ represents a decrease in the photon energy due to the recoil energy of the atom. For a massive source, this term is negligible and

$$\nu = \nu_0 \sqrt{\frac{1 + u/c}{1 - u/c}},$$

in agreement with the result of the last chapter, Eq. (12.6).

Although it is always satisfying to derive a result by different arguments, perhaps the chief interest in this exercise is to show how two completely different views of light, wave and particle, lead to exactly the same prediction for the shift in frequency of radiation from a moving source.

13.4 Does Light Travel at the Velocity of Light?

Although the title of this section may sound rhetorical, the question is not trivial. It is apparent that the velocity of light plays a special role in relativity. In fact, Einstein created the special theory of relativity primarily from considerations of Maxwell's electromagnetic theory, the theory of light. However, it is important to realize that the real significance of the velocity of light is that it exemplifies a *universal* velocity, a velocity whose value is the same for an observer in any inertial system. There can be only *one* such universal velocity in the theory of relativity, as the following argument shows.

Suppose that there is a second universal velocity c^* representing the velocity of some phenomenon other than light—perhaps the speed of gravitons or neutrinos. Let us call the phenomenon Γ. Consider a light pulse and a Γ pulse emitted along the x axis from the origin of the x, y system at $t = 0$. The pulses travel according to:

Light: $x_l = ct$

Γ: $x_\Gamma = c^*t.$

The relative velocity of the two pulses is

$$u = \frac{d}{dt}(x_\Gamma - x_l)$$

$$= c^* - c.$$

Now consider the same pulses in the x', y' system which is moving along the x axis with velocity V. Since c^* and c are universal velocities, the loci of the pulses must be given by

$$x'_l = ct'$$
$$x'_{\mathrm{r}} = c^*t'.$$

The relative velocities of the two pulses is

$$\begin{aligned} u' &= \frac{d}{dt'}[x'_{\mathrm{r}} - x'_l] \\ &= c^* - c, \end{aligned}$$

as before. But the relativistic transformation of velocities gives

$$c' = \frac{c - V}{1 - cV/c^2} = c$$
$$(c^*)' = \frac{c^* - V}{1 - c^*V/c^2}.$$

Thus, the Lorentz transformations predict that

$$\begin{aligned} u' &= (c^*)' - c \\ &= \frac{c^* - V}{1 - c^*V/c^2} - c. \end{aligned}$$

This disagrees with the result above, $u' = c^* - c$, unless $c^* = c$, in which case $u = 0$ and $u' = 0$. We conclude that there can be only one universal velocity.

If this argument seems rather formal, perhaps the following explanation will help. The theory of relativity satisfies the postulate of relativity: all inertial coordinate systems are equivalent. It also satisfies the postulate that the velocity of light is a universal constant: all observers in inertial systems will obtain the same result for the velocity of a particular light signal. However, the theory of relativity cannot accommodate more than one such universal velocity; if we try to introduce a second universal velocity, the whole edifice of relativity collapses. In particular, we can no longer obtain a consistent recipe for relating coordinates of events in different systems.

With this background, perhaps we can rephrase the title of this section more meaningfully as "does light travel with the universal velocity?" The question is actually quite interesting and a matter of current investigation.

Example 13.9 The Rest Mass of the Photon

If the photon had a nonzero rest mass, the velocity of light would differ from c. If we let m_p represent the rest mass of a photon, we would have

$$E = \gamma m_p c^2.$$

If we assume that the photon energy-frequency relation $E = h\nu$ remains valid, then squaring the equation above gives

$$(h\nu)^2 = (m_p c^2)^2 \frac{1}{1 - v^2/c^2},$$

or, after rearranging,

$$\frac{v^2}{c^2} = 1 - \frac{\nu_0^2}{\nu^2},$$

where $h\nu_0 = m_p c^2$. ν_0 plays the role of a characteristic frequency of the photon: $h\nu_0$ is the rest energy of the photon. If $\nu_0 = 0$, we have $v = c$. Otherwise, the velocity of light depends on frequency. Behavior such as this is well known when light passes through a refractive medium such as glass or water; it is known as *dispersion.* The question for experiment to decide is whether or not empty space exhibits dispersion.

There have been a number of recent attempts to set a limit on the rest mass of the photon (or, better still, to measure it, although at present there is no compelling reason to believe that the rest mass is not zero).

Example 13.10 Light from a Pulsar

Pulsars are stars that emit regular bursts of energy at repetition frequencies from 30 to 0.1 Hz. They were discovered in 1968 and their unexpected properties have been a source of much excitement among astronomers and astrophysicists. Perhaps the most interesting pulsar is the one in the Crab nebula. It has the highest frequency, 30 Hz, and is the only one so far observed which pulses in the optical and x-ray regions, as well as at radio frequencies. The pulses are quite sharp, and their arrival time can be measured to an accuracy of microseconds. It is known that light from the pulsar at different optical wavelengths arrives simultaneously within the experimental resolving time. We can use these facts to set a limit on the rest mass of the photon.

It takes light 5,000 years to reach us from the Crab nebula. Suppose that signals at two different frequencies travel with a small difference in velocity, Δv, and arrive at slightly different times, T and $T + \Delta T$. Since $T = L/v$, where L is the distance from the Crab nebula, we have

$$\Delta v = -\frac{L}{T^2}\Delta T$$

or

$$\frac{\Delta v}{v} = -\frac{\Delta T}{T}.$$

No such velocity difference has been observed, but by estimating the sensitivity of the experiment we can set an upper limit to Δv. ΔT can be measured to an accuracy of about 2×10^{-3} s, and using $T = 5 \times 10^3$ years $= 1.5 \times 10^{11}$ s, we have

$$\left|\frac{\Delta v}{c}\right| = \left|\frac{\Delta T}{T}\right| < \frac{2 \times 10^{-3}}{1.5 \times 10^{11}} \approx 10^{-14},$$

where we have taken $v \approx c$.

Now let us translate this limit on Δv into a limit on the possible rest mass of a photon. From the result of the last example,

$$\frac{v^2}{c^2} = 1 - \frac{\nu_0^2}{\nu^2}.$$

Consider signals at two different frequencies, ν_1 and ν_2. We have

$$\frac{v_1^2 - v_2^2}{c^2} = \nu_0^2\left(\frac{1}{\nu_2^2} - \frac{1}{\nu_1^2}\right).$$

The left hand side can be written

$$\frac{(v_1 - v_2)(v_1 + v_2)}{c^2} \approx 2\frac{\Delta v}{c},$$

where we have taken $(v_1 - v_2) = \Delta v$, and $v_1 + v_2 \approx 2c$. For observations made in the optical region we can take $\nu_1 = 8 \times 10^{14}$ Hz (blue) and $\nu_2 = 5 \times 10^{14}$ Hz (red). Then, using the limit $\Delta v/c < 2 \times 10^{-16}$, we have

$$2 \times 2 \times 10^{-16} > \frac{\nu_0^2}{10^{28}}\left(\frac{1}{5^2} - \frac{1}{8^2}\right) = 2.4 \times 10^{-30}\nu_0^2$$

or

$$\nu_0 < 10^7 \text{ Hz}.$$

This gives an upper limit to the photon rest mass of

$$m_p = \frac{h\nu_0}{c^2} < 10^{-40} \text{ kg}.$$

An even lower limit to the photon rest mass can be found by observing the arrival time of radio pulses from the Crab nebula. The analysis is somewhat more complicated because of the effect of free electrons in interstellar space. The result is that the rest mass of the photon has an upper limit of 10^{-47} kg.

Problems

13.1 It is estimated that a cosmic ray primary proton can have energy up to 10^{13} MeV (almost 10^8 greater than the highest energy achieved with an accelerator). Our galaxy has a diameter of about 10^5 light-years. How long does it take the proton to traverse the galaxy, in its own rest frame? (1 eV $= 1.6 \times 10^{-19}$ J, $M_p = 1.67 \times 10^{-27}$ kg.)

13.2 When working with particles it is important to know when relativistic effects have to be considered.

A particle of rest mass m_0 is moving with speed v. Its classical kinetic energy is $K_{\text{cl}} = m_0v^2/2$. Let K_{rel} be the relativistic expression for its kinetic energy.

a. By expanding $K_{\text{rel}}/K_{\text{cl}}$ in powers of v^2/c^2, estimate the value of v^2/c^2 for which K_{rel} differs from K_{cl} by 10 percent.

b. For this value of v^2/c^2, what is the kinetic energy in MeV of
(1) An electron ($m_0c^2 = 0.51$ MeV)
(2) A proton ($m_0c^2 = 930$ MeV)

13.3 In newtonian mechanics, the kinetic energy of a mass m moving with velocity $\mathbf{v}$ is $K = mv^2/2 = p^2/(2m)$ where $\mathbf{p} = m\mathbf{v}$. Hence, the change in kinetic energy due to a small change in momentum is $dK = \mathbf{p} \cdot d\mathbf{p}/m = \mathbf{v} \cdot d\mathbf{p}$.

Show that the relation $dK = \mathbf{v} \cdot d\mathbf{p}$ also holds in relativistic mechanics.

13.4 Two particles of rest mass m_0 approach each other with equal and opposite velocity v, in the laboratory frame. What is the total energy of one particle as measured in the rest frame of the other?

Ans. clue. If $v^2/c^2 = \frac{1}{2}$, $E = 3m_0c^2$

13.5 A particle of rest mass m and speed v collides and sticks to a stationary particle of mass M. What is the final speed of the composite particle?

Ans. $v_f = \gamma vm/(\gamma m + M)$, where $\gamma = (1 - v^2/c^2)^{-\frac{1}{2}}$

13.6 A particle of rest mass m_0 and kinetic energy xm_0c^2, where x is some number, strikes and sticks to an identical particle at rest. What is the rest mass of the resultant particle?

Ans. clue. If $x = 6$, $m = 4m_0$

13.7 In the laboratory frame a particle of rest mass m_0 and speed v is moving toward a particle of mass m_0 at rest.

What is the speed of the inertial frame in which the total momentum of the system is zero?

Ans. clue. If $v^2/c^2 = \frac{3}{4}$, the speed is $2v/3$

13.8 A photon of energy E_0 and wavelength λ_0 collides head on with a free electron of rest mass m_0 and speed v, as shown. The photon is scattered at 90°.

a. Find the energy E of the scattered photon.

Ans. $E = [E_0(1 + v/c)]/(1 + E_0/E_i)$, where $E_i = m_0c^2/\sqrt{1 - v^2/c^2}$

b. The outer electrons in a carbon atom move with speed $v/c \approx 6 \times 10^{-3}$. Using the result of part a, estimate the broadening in wavelength of the Compton scattered peak from graphite for $\lambda_0 = 0.711 \times 10^{-10}$ m and 90° scattering. The rest mass of an electron is 0.51 MeV and $h/(m_0c) = 2.426 \times 10^{-12}$ m. Neglect the binding of the electrons. Compare your result with Compton's data shown in Example 13.6.

13.9 The solar constant, the average energy per unit area falling on the earth, is 1.4×10^3 W/m². How does the total force of sunlight compare with the sun's gravitational force on the earth?

Sufficiently small particles can be ejected from the solar system by the radiation pressure of sunlight. Assuming a specific gravity of 5, what is the radius of the largest particle which can be ejected?

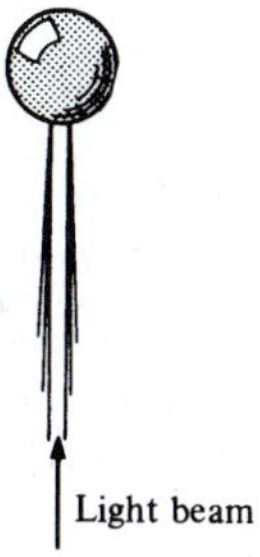

13.10 A 1-kW light beam from a laser is used to levitate a solid aluminum sphere by focusing it on the sphere from below. What is the diameter of the sphere, assuming that it floats freely in the light beam? The density of aluminum is 2.7 g/cm³.

13.11 A photon of energy E_0 collides with a free particle of mass m_0 at rest. If the scattered photon flies off at angle θ, what is the scattering angle of the particle, ϕ?

Ans. $\cot \phi = (1 + E_0/m_0c^2) \tan (\theta/2)$

14 FOUR-VECTORS AND RELATIVISTIC INVARIANCE

14.1 Introduction

When a major advance in physics is made, old concepts inevitably lose importance and points of view which previously were of minor interest move to the center. Thus, with the advent of relativity the concept of the ether vanished, taking with it the problem of absolute motion. At the same time, the transformation properties of physical laws, previously of little interest, took on central importance. As we shall see in this chapter, transformation theory provides a powerful tool for generalizing nonrelativistic concepts and for testing the relativistic correctness of physical laws. Furthermore, it is a useful guide in the search for new laws. By using transformation theory we shall derive in a natural way the important results of relativity that we found by ad hoc arguments in the preceding chapters. This approach emphasizes the mathematical structure of physics and the nature of symmetry; it illustrates a characteristic mode of thought in contemporary physics.

To introduce the methods of transformation theory, we defer relativity for the moment and turn first to the transformation properties of ordinary vectors in three dimensions.

14.2 Vectors and Transformations

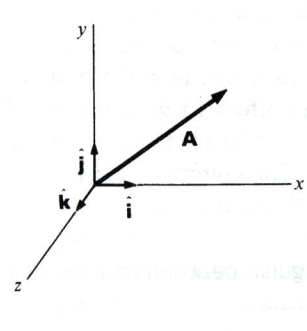

In Chap. 1 we defined vectors as "directed line segments"; with the help of transformation theory we can develop a more fundamental definition.

To motivate the argument and to illustrate the ideas of transformation theory we shall rely at first on our intuitive concept of vectors. Consider vector **A**, which represents some physical quantity such as force or velocity. To describe **A** in component form we introduce an orthogonal coordinate system x, y, z with unit base vectors $\hat{\mathbf{i}}$, $\hat{\mathbf{j}}$, $\hat{\mathbf{k}}$. **A** can then be written

$$\mathbf{A} = A_x\hat{\mathbf{i}} + A_y\hat{\mathbf{j}} + A_z\hat{\mathbf{k}}.$$

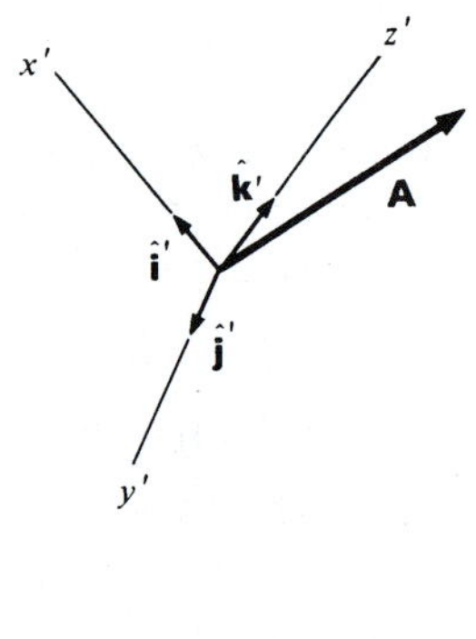

The coordinate system is not an essential part of the physics; it is a construct we introduce for convenience. We are perfectly free to use some other orthogonal coordinate system x', y', z' with base vectors $\hat{\mathbf{i}}'$, $\hat{\mathbf{j}}'$, $\hat{\mathbf{k}}'$. Let the x', y', z' system have the same origin as the x, y, z system, in which case the two systems are related by a rotation. In the primed system,

$$\mathbf{A} = A'_x\hat{\mathbf{i}}' + A'_y\hat{\mathbf{j}}' + A'_z\hat{\mathbf{k}}'.$$

For a general coordinate rotation, the components A'_x, A'_y, A'_z have a definite relation to the components A_x, A_y, A_z. Equating the two expressions for **A** gives

$$A'_x\hat{\mathbf{i}}' + A'_y\hat{\mathbf{j}}' + A'_z\hat{\mathbf{k}}' = A_x\hat{\mathbf{i}} + A_y\hat{\mathbf{j}} + A_z\hat{\mathbf{k}}.$$

If we take the dot product of both sides with $\hat{\mathbf{i}}'$ we obtain

$$A'_x = A_x(\hat{\mathbf{i}}' \cdot \hat{\mathbf{i}}) + A_y(\hat{\mathbf{i}}' \cdot \hat{\mathbf{j}}) + A_z(\hat{\mathbf{i}}' \cdot \hat{\mathbf{k}}) \qquad 14.1a$$

Similarly,

$$A'_y = A_x(\hat{\mathbf{j}}' \cdot \hat{\mathbf{i}}) + A_y(\hat{\mathbf{j}}' \cdot \hat{\mathbf{j}}) + A_z(\hat{\mathbf{j}}' \cdot \hat{\mathbf{k}}) \qquad 14.1b$$

$$A'_z = A_x(\hat{\mathbf{k}}' \cdot \hat{\mathbf{i}}) + A_y(\hat{\mathbf{k}}' \cdot \hat{\mathbf{j}}) + A_z(\hat{\mathbf{k}}' \cdot \hat{\mathbf{k}}). \qquad 14.1c$$

The coefficients $(\hat{\mathbf{i}}' \cdot \hat{\mathbf{i}})$, $(\hat{\mathbf{i}}' \cdot \hat{\mathbf{j}})$, etc., are numbers which are determined by the given rotation; they do not depend on **A**.

We derived Eq. (14.1) from our concept of vectors as directed line segments, but now we shall reverse the order and use Eq. (14.1) to define vectors. A vector in three dimensions is a set of three numbers which transform under a rotation of the coordinate system according to Eq. (14.1). It is easy to show that the vector algebra developed in Chap. 1 is consistent with our new definition of a vector. For example, the sum of two vectors is a vector, and the time derivative of a vector is also a vector.

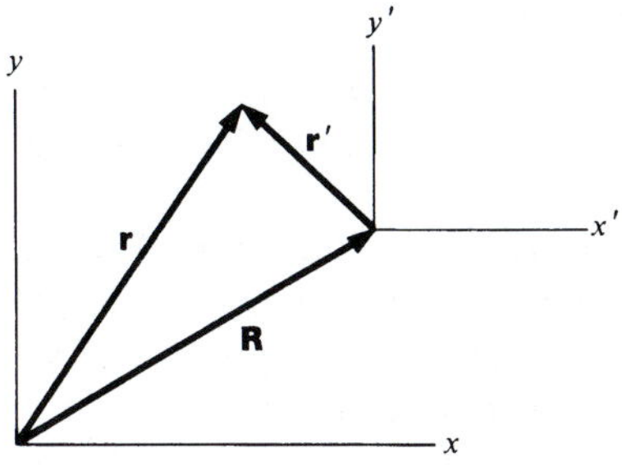

$\mathbf{r}' = \mathbf{r} - \mathbf{R}$

We should point out that the general displacement of a coordinate system is composed of a translation as well as a rotation. The reason that we referred only to rotations in the definition of a vector is that translations have no effect on the components of a vector. The sole exception is the position vector **r**, which is defined with respect to a specific origin. The components of **r** transform under rotations according to Eq. (14.1), but **r** can be distinguished from true vectors such as **F** and **v** by its transformation properties under translation. We can distinguish between true vectors, position vectors, and other mathematical entities by investigating how they behave under all possible transformations.

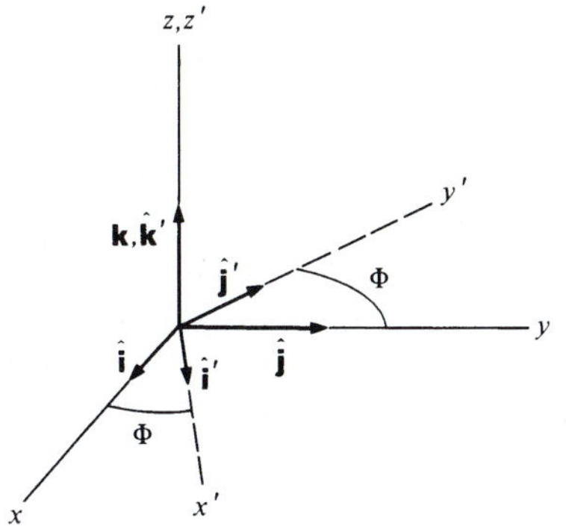

Rotation about the z axis

Equation (14.1) is completely general, but usually it is convenient to work with a special case such as a rotation of coordinates through angle Φ around the z axis, as shown in the sketch. We have

$(\hat{\mathbf{i}}' \cdot \hat{\mathbf{i}}) = \cos\Phi$	$(\hat{\mathbf{j}}' \cdot \hat{\mathbf{i}}) = -\sin\Phi$	$(\hat{\mathbf{k}}' \cdot \hat{\mathbf{i}}) = 0$
$(\hat{\mathbf{i}}' \cdot \hat{\mathbf{j}}) = \sin\Phi$	$(\hat{\mathbf{j}}' \cdot \hat{\mathbf{j}}) = \cos\Phi$	$(\hat{\mathbf{k}}' \cdot \hat{\mathbf{j}}) = 0$
$(\hat{\mathbf{i}}' \cdot \hat{\mathbf{k}}) = 0$	$(\hat{\mathbf{j}}' \cdot \hat{\mathbf{k}}) = 0$	$(\hat{\mathbf{k}}' \cdot \hat{\mathbf{k}}) = 1.$

Hence the components of any vector $\mathbf{A} = (A_x,A_y,A_z)$ must transform according to the relations

$$A'_x = A_x \cos\Phi + A_y \sin\Phi$$
$$A'_y = -A_x \sin\Phi + A_y \cos\Phi \qquad 14.2$$
$$A'_z = A_z.$$

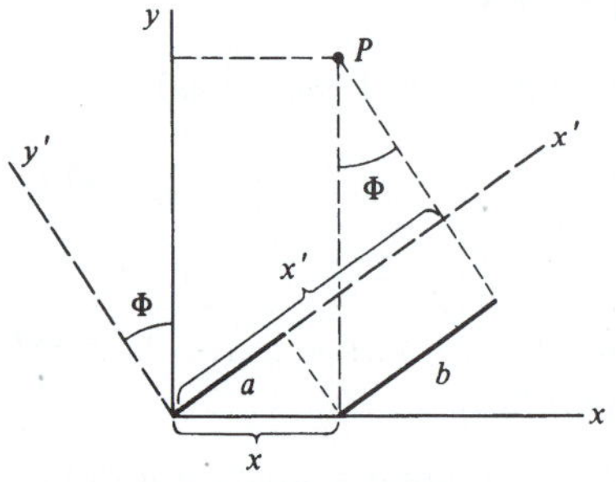

$x' = a + b = x \cos\Phi + y \sin\Phi$

For example, let $\mathbf{A} = \mathbf{r} = (x,y,z)$. Then

$$x' = x\cos\Phi + y\sin\Phi$$
$$y' = -x\sin\Phi + y\cos\Phi$$
$$z' = z.$$

These relations can be independently verified from the geometry. The drawing shows how the x' coordinate of point P is related to the coordinates (x,y).

Example 14.1 **Transformation Properties of the Vector Product**

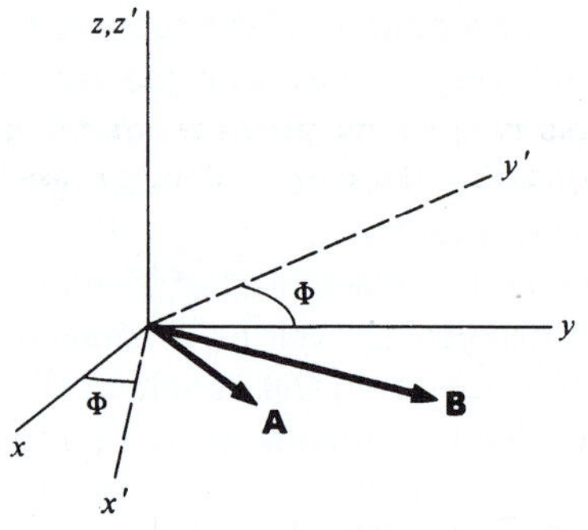

In Chap. 1 we gave an essentially geometrical definition of the vector product. To demonstrate our new definition of a vector we shall prove that the components of the vector product transform as the components of a vector. For simplicity, we consider two coordinate systems, x, y, z and x', y', z', which differ by a rotation through angle Φ around the z axis, and two vectors **A** and **B** in the x, y plane. From the definition of vector product we have

$$\mathbf{C} = \mathbf{A}\times\mathbf{B} = \begin{vmatrix} \hat{\mathbf{i}} & \hat{\mathbf{j}} & \hat{\mathbf{k}} \\ A_x & A_y & 0 \\ B_x & B_y & 0 \end{vmatrix} = \begin{vmatrix} \hat{\mathbf{i}}' & \hat{\mathbf{j}}' & \hat{\mathbf{k}}' \\ A'_x & A'_y & 0 \\ B'_x & B'_y & 0 \end{vmatrix}.$$

In the x, y, z system the components of **C** are

$$C_x = 0 \qquad 1a$$
$$C_y = 0 \qquad 1b$$
$$C_z = A_xB_y - A_yB_x \qquad 1c$$

and in the x', y', z' system they are

$$C'_x = 0 \qquad 2a$$
$$C'_y = 0 \qquad 2b$$
$$C'_z = A'_xB'_y - A'_yB'_x. \qquad 2c$$

If **C** is a vector, its components must obey the transformation law, Eq. (14.2):

$$C'_x = C_x\cos\Phi + C_y\sin\Phi \qquad 3a$$
$$C'_y = -C_x\sin\Phi + C_y\cos\Phi \qquad 3b$$
$$C'_z = C_z. \qquad 3c$$

Equations ($3a$) and ($3b$) are identically satisfied by Eqs. (1) and (2). To prove Eq. ($3c$), we need to show that $A'_xB'_y - A'_yB'_x = A_xB_y - A_yB_x$. From Eq. (14.2) we have

$$A'_x = A_x \cos\Phi + A_y \sin\Phi$$
$$A'_y = -A_x \sin\Phi + A_y \cos\Phi$$
$$B'_x = B_x \cos\Phi + B_y \sin\Phi$$
$$B'_y = -B_x \sin\Phi + B_y \cos\Phi.$$

Hence,

$$\begin{aligned} A'_xB'_y - A'_yB'_x &= (A_x \cos\Phi + A_y \sin\Phi)(-B_x \sin\Phi + B_y \cos\Phi) \\ &\quad - (-A_x \sin\Phi + A_y \cos\Phi)(B_x \cos\Phi + B_y \sin\Phi) \\ &= A_xB_y - A_yB_x \\ &= C_z. \end{aligned}$$

This proves that all three components of the vector product transform like the components of a vector so that the vector product is, in fact, a vector.

Example 14.2 **A Nonvector**

To give a counterexample to the cross product, suppose that we try to introduce a new type of vector multiplication, the vector "double cross" product **C** = **A** ×× **B** defined by

$$C_x = A_yB_z + A_zB_y$$
$$C_y = A_zB_x + A_xB_z$$
$$C_z = A_xB_y + A_yB_x.$$

Is **C** actually a vector?

If we again take the case **A** $= (A_x,A_y,0)$, **B** $= (B_x,B_y,0)$, we have

$$C_x = 0$$
$$C_y = 0$$
$$C_z = A_xB_y + A_yB_x.$$

In the x', y', z' system the components are

$$C'_x = A'_yB'_z + A'_zB'_y = 0$$
$$C'_y = A'_zB'_z + A'_zB'_x = 0$$
$$C'_z = A'_xB'_y + A'_yB'_x.$$

The first two equations obey the transformation rule, Eqs. ($3a$) and ($3b$) of Example 14.1. However, when we evaluate the last equation we find that

$$\begin{aligned} C'_z &= (A_x \cos\Phi + A_y \sin\Phi)(-B_x \sin\Phi + B_y \cos\Phi) \\ &\quad + (-A_x \sin\Phi + A_y \cos\Phi)(B_x \cos\Phi + B_y \sin\Phi) \\ &= (A_xB_y + A_yB_x)(\cos^2\Phi - \sin^2\Phi) - 2(A_xB_x - A_yB_y)\cos\Phi \sin\Phi. \end{aligned}$$

It is apparent that $C'_z \neq C_z$, so that Eq. (3c) of Example 14.1 is not satisfied. The elements generated by the double cross product are not the components of a vector, and the double cross product is a useless operation.

Invariants of a Transformation

Any quantity which is unchanged by a general coordinate transformation is called an *invariant* of the transformation. Invariants play an important role in physics. They are the only entities suitable for the construction of physical laws, since the principle of relativity requires that the results of physical theories be independent of the choice of coordinate system (provided, of course, that the system is inertial).

We have so far encountered two classes of invariants—scalars and vectors. Scalars are single numbers and are unaffected by the choice of coordinates. Vectors are invariant under rotations of the coordinates by construction; we designed the transformation rule, Eq. (14.1), to assure this.

Any mathematical entity which is invariant under a rotation of coordinates is called a *tensor.* A scalar is a tensor of zeroth rank, and a vector is a tensor of the first rank. Tensors of higher rank also exist; the moment of inertia introduced in Chap. 7 is a tensor of the second rank.

The Transformation Properties of Physical Laws

We have used vector notation wherever possible because of its simplicity; one vector equation is easier to handle than three scalar equations. However, from the point of transformation theory, vectors have a deeper significance. Since we must be able to use any coordinate system we choose for describing physical events, it is essential that we be able to write physical laws in a form independent of coordinate systems. Thus, if an equation represents a statement of a physical law, both sides of the equation must transform the same way under a change of coordinates. For example, consider the equation for motion along some axis j: $F_j = ma_j$. Assuming m is a scalar, ma_j must be a component of a vector, since acceleration is a vector. Thus, F_j is a component of a vector along the same axis, and the general form of the equation must be $\mathbf{F} = m\mathbf{a}$. Once the law is in vector form, we can easily find the motion along any set of axes we choose. From this point of view, the vector nature of force, including the rule for superposition of forces, is a mathematical consequence of the requirement that the laws of motion be valid in all inertial systems.

The question arises as to whether the law of superposition of forces is a physical law or simply a mathematical result. It is, in fact, both. Experimentally, we find that the translation of a body can be described by only three independent equations, one for each coordinate axis; this implies that force has three independent components. According to transformation theory, the only three component entity suitable for describing physical laws is a vector, and vectors obey the law of superposition.

Scalar Invariants

We can use the dot product to combine two vectors to form a scalar. Since scalars are independent of the coordinate system, the dot product of two vectors is called a *scalar invariant.*

Let us show explicitly that the dot product **A · B** is a scalar invariant under rotations. Considering a rotation about the z axis for simplicity, we use Eq. (15.2) to obtain

$$\begin{aligned} A'_xB'_x + A'_yB'_y + A'_zB'_z = &\\ &(A_x \cos \Phi + A_y \sin \Phi)(B_x \cos \Phi + B_y \sin \Phi)\\ &+ (-A_x \sin \Phi + A_y \cos \Phi)\\ &\quad(-B_x \sin \Phi + B_y \cos \Phi) + (A_zB_z)\\ &= A_xB_x + A_yB_y + A_zB_z. \end{aligned}$$

In particular, the dot product of a vector with itself, called the *norm* of the vector, is a scalar invariant:

$$A'^2_x + A'^2_y + A'^2_z = A_x{}^2 + A_y{}^2 + A_z{}^2.$$

The norm of the position vector **r** changes under a translation of coordinates but is invariant under pure rotations. We can use this to define a rotation of coordinates: it is any transformation which leaves $r^2 = x^2 + y^2 + z^2$ invariant.

14.3 Minkowski Space and Four-vectors

As we have discussed, it must be possible to express the laws of classical physics using entities like scalars and vectors, which are invariant under rotations of the coordinates x, y, z. From the mathematical point of view the Lorentz transformations have much in common with a spatial rotation: they are both linear transformations from one set of coordinates to another.

CHANGE OF COORDINATES UNDER A ROTATION	CHANGE OF COORDINATES UNDER LORENTZ TRANSFORMATION
$x' = x \cos \Phi + y \sin \Phi$	$x' = \gamma x - \gamma v t$
$y' = -x \sin \Phi + y \cos \Phi$	$y' = y$
$z' = z$	$z' = z$
$(t' = t)$	$t' = -(\gamma v/c^2)x + \gamma t$

Our object in this section is to find a way to write physical laws so that they are invariant under the Lorentz transformations. This assures that the laws will have the same form for observers in all inertial frames as required by the first postulate of relativity.

We shall start from the observation made in 1908 by the mathematician Minkowski that, with a slight change of notation, the Lorentz transformations represent a rotation in a four dimensional space. To introduce his line of reasoning, we return to the second postulate of relativity: the speed of light is the same for observers in all inertial frames. Consider two inertial systems x, y, z, t and x', y', z', t' moving with relative speed v in the x direction. If their origins coincide at $t = 0$ and a short light pulse is sent out from the origin at that instant, the locus of the pulse in the x, y, z, t system is $r = ct$, or

$$x^2 + y^2 + z^2 = (ct)^2,$$

while in the x', y', z', t' system it is

$$x'^2 + y'^2 + z'^2 = (ct')^2.$$

Comparing, we see that the quantity $x^2 + y^2 + z^2 - (ct)^2$ is equal to zero in each coordinate system; it appears to be a scalar invariant under the Lorentz transformations. We can show this directly by employing the Lorentz transformations, Eq. (11.3):

$$\begin{aligned} x'^2 + y'^2 + z'^2 - (ct')^2 &= \gamma^2(x - vt)^2 + y^2 + z^2 - \gamma^2 c^2\left(t - \frac{vx}{c^2}\right)^2 \\ &= \frac{1}{1 - v^2/c^2}\left[x^2\left(1 - \frac{v^2}{c^2}\right)\right. \\ &\qquad \left. - c^2t^2\left(1 - \frac{v^2}{c^2}\right)\right] + y^2 + z^2 \\ &= x^2 + y^2 + z^2 - (ct)^2. \end{aligned} \qquad 14.3$$

In ordinary three dimensional space, the only transformation that leaves $x^2 + y^2 + z^2$ unchanged is a rotation. Minkowski considered a four dimensional space with coordinates x_1, x_2, x_3, x_4, where $x_1 = x$, $x_2 = y$, $x_3 = z$ and $x_4 = ict\,(i^2 = -1)$. With these coordinates,

$$x^2 + y^2 + z^2 - (ct)^2 = x_1{}^2 + x_2{}^2 + x_3{}^2 + x_4{}^2$$

and Eq. (14.3) can be written

$$x_1'^2 + x_2'^2 + x_3'^2 + x_4'^2 = x_1{}^2 + x_2{}^2 + x_3{}^2 + x_4{}^2.$$

It is apparent that $x_1{}^2 + x_2{}^2 + x_3{}^2 + x_4{}^2$ is invariant under Lorentz transformations; by analogy with the three dimensional case, the Lorentz transformations represent a rotation of coordinates. The analogy also suggests that x_1, x_2, x_3, x_4 are the components of a true four dimensional vector.

The transformation rules for $(x_1,x_2,x_3,x_4) = (x,y,z,ict)$ are readily obtained from the Lorentz transformations.

$$\begin{aligned} x_1' &= \gamma(x_1 + i\beta x_4) \\ x_2' &= x_2 \\ x_3' &= x_3 \\ x_4' &= \gamma(x_4 - i\beta x_1), \end{aligned}$$

where $\beta = v/c$. (As usual, to simplify the algebra we restrict ourselves to systems whose relative motion is in the x direction.) It follows that any true four dimensional vector must transform in the same fashion. Such vectors are known as *four-vectors*. Thus the transformation rule for a four-vector $\underset{\rightarrow}{\mathbf{A}} = (A_1,A_2,A_3,A_4)$ is

$$\begin{aligned} A_1' &= \gamma(A_1 + i\beta A_4) \\ A_2' &= A_2 \\ A_3' &= A_3 \\ A_4' &= \gamma(A_4 - i\beta A_1). \end{aligned} \qquad 14.4$$

As we expect, the norm of $\underset{\rightarrow}{\mathbf{A}}$ is a Lorentz invariant.

$$A_1'^2 + A_2'^2 + A_3'^2 + A_4'^2 = A_1{}^2 + A_2{}^2 + A_3{}^2 + A_4{}^2.$$

The factor of c gives A_4 the same dimensions as the other components. From Eq. (14.4), we see that if A_1 is a real number, A_4 must be imaginary, as in the four-vector $\underset{\rightarrow}{\mathbf{s}} = (x,y,z,ict)$. The fact that the fourth component is imaginary arises from the essential difference between space and time.

In Minkowski's formulation of relativity, an event specified by x, y, z, t is viewed as a point x_1, x_2, x_3, x_4 in space-time. Minkowski called the four dimensional space-time manifold "world," although it has come to be called *Minkowski space*. A point in Minkowski space is called a *world point*. As a particle moves in space and time its successive world points trace out a *world line*.

The location of a world point is specified by its position four-vector

$$\underset{\rightarrow}{\mathbf{s}} = (x_1, x_2, x_3, x_4).$$

The Lorentz transformations, which relate an event in different coordinate systems, represent a transformation of the components of $\underset{\rightarrow}{\mathbf{s}}$ from one coordinate system to another.

The displacement between two world points is

$$\Delta\underset{\rightarrow}{\mathbf{s}} = (\Delta x, \Delta y, \Delta z, ic\,\Delta t)$$

or, in differential form,

$$d\underset{\rightarrow}{\mathbf{s}} = (dx, dy, dz, ic\,dt).$$

Since $d\underset{\rightarrow}{\mathbf{s}}$ is a four-vector, its norm is a Lorentz invariant. The norm is

$$ds^2 = dx^2 + dy^2 + dz^2 - c^2\,dt^2.$$

A related Lorentz invariant that will be useful to us is $d\tau^2 = -ds^2/c^2$.

$$d\tau^2 = dt^2 - \frac{1}{c^2}(dx^2 + dy^2 + dz^2)$$

$d\tau$ has a simple interpretation. Consider a displacement $d\underset{\rightarrow}{\mathbf{s}}$ between two world points of a moving particle. In the rest frame of the particle, the space coordinates are constant, and therefore $dx = dy = dz = 0$. Thus $d\tau = dt$ in the rest frame; the world points are separated only in time. $d\tau$ is the time interval measured in the rest frame, and for this reason τ is known as the *proper time*.

Example 14.3 Time Dilation

We rederive the Einstein time dilation formula to show the power of four-vectors.

Consider an observer at rest in the x', y', z', t' system. In this system, the proper time interval between two world points is $d\tau = dt'$. In the x, y, z, t system moving with velocity $\underset{\rightarrow}{\mathbf{v}}$ relative to the first frame, the interval between the same points is given by

$$dt^2 - \frac{1}{c^2}(dx^2 + dy^2 + dz^2).$$

Since $d\tau^2$ is a Lorentz invariant, its value for the same world points is the same in all frames. Hence, we can equate its value in the rest frame to its value in the second frame.

$$dt'^2 = dt^2 - \frac{1}{c^2}(dx^2 + dy^2 + dz^2)$$

or

$$\left(\frac{dt'}{dt}\right)^2 = 1 - \frac{1}{c^2}\left[\left(\frac{dx}{dt}\right)^2 + \left(\frac{dy}{dt}\right)^2 + \left(\frac{dz}{dt}\right)^2\right].$$

Since $(dx/dt)^2 + (dy/dt)^2 + (dz/dt)^2 = v^2$, we have

$$\left(\frac{dt'}{dt}\right)^2 = 1 - \frac{v^2}{c^2}$$

or

$$dt = \frac{dt'}{\sqrt{1 - v^2/c^2}} = \gamma\, d\tau.$$

In contrast to the derivation of Sec. 13.3, this treatment avoids hypothetical experiments and discussions of simultaneity.

Example 14.4 Construction of a Four-vector: The Four-velocity

In ordinary three dimensional space, dividing a vector by a scalar (a rotational invariant) yields another vector. Similarly, dividing a four-vector by a Lorentz invariant yields another four-vector.

Consider the displacement four-vector

$$d\underset{\rightarrow}{\mathbf{s}} = (dx, dy, dz, ic\, dt).$$

Dividing by the Lorentz invariant $d\tau$, we obtain a new four-vector

$$\frac{d\underset{\rightarrow}{\mathbf{s}}}{d\tau} = \left(\frac{dx}{d\tau}, \frac{dy}{d\tau}, \frac{dz}{d\tau}, ic\frac{dt}{d\tau}\right). \qquad 1$$

By analogy with the three dimensional case, we call $d\underset{\rightarrow}{\mathbf{s}}/d\tau$ the *four-velocity* $\underset{\rightarrow}{\mathbf{u}}$.

In the rest frame of the particle, $dx = dy = dz = 0$, and $d\tau = dt$. For a particle at rest

$$\underset{\rightarrow}{\mathbf{u}} = (0, 0, 0, ic). \qquad 2$$

The norm of $\underset{\rightarrow}{\mathbf{u}}$ is $(\underset{\rightarrow}{\mathbf{u}})^2 = -c^2$ and it has the same value in all frames. Obviously the four-velocity $\underset{\rightarrow}{\mathbf{u}}$ is very different physically from $\mathbf{u}$, the familiar three dimensional velocity.

We now wish to find an expression for the four-velocity of a moving particle. Let the x, y, z, t system move with velocity $-\mathbf{u}$ relative to the rest frame of the particle. Using the time dilation formula of Example 14.3, we can write

$$dt = \gamma \, d\tau,$$

where dt is now the time interval in the moving frame. Using this in Eq. (1),

$$\begin{aligned} \underset{\to}{\mathbf{u}} &= \gamma \left(\frac{dx}{dt}, \frac{dy}{dt}, \frac{dz}{dt}, ic \right) \\ &= \gamma(\mathbf{u}, ic), \end{aligned} \qquad 3$$

where $\gamma = 1/\sqrt{1 - u^2/c^2}$.

We shall use $\underset{\to}{\mathbf{u}}$ in the next section to derive the momentum-energy four-vector. However, we shall first demonstrate how to transform a four-vector from one frame to another.

Example 14.5 **The Relativistic Addition of Velocities**

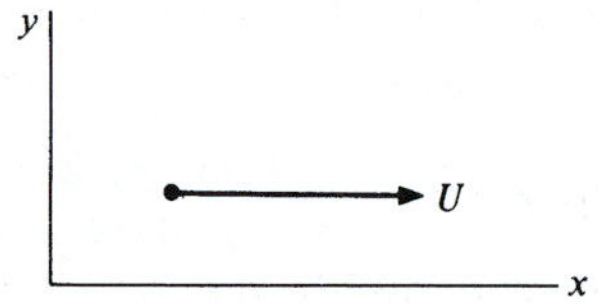

We can easily derive the formula for the relativistic addition of velocities by transforming the four-velocity $\underset{\to}{\mathbf{u}} = \gamma(\mathbf{u}, ic)$ into successive frames with the aid of Eq. (14.4).

Consider a particle moving along the x direction of the x, y, z, t system with speed U. In this frame,

$$\underset{\to}{\mathbf{u}} = (u_1, u_2, u_3, u_4) = \Gamma(U, 0, 0, ic),$$

where $\Gamma = 1/\sqrt{1 - U^2/c^2}$. Consider a second frame x', y', z', t' moving along the x direction with speed v relative to the first frame. In this frame, the four-velocity of the particle is

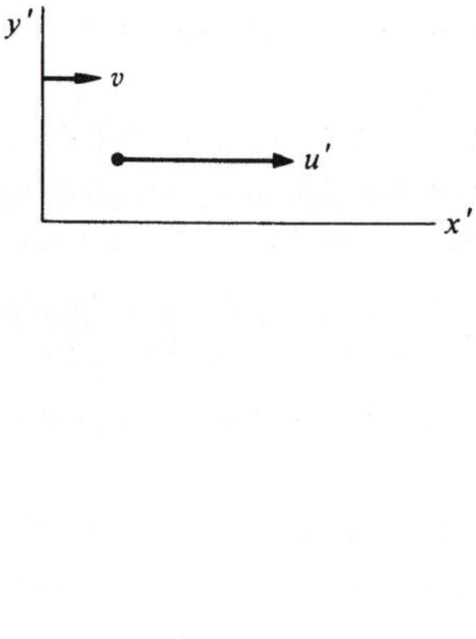

$$\begin{aligned} \underset{\to}{\mathbf{u}}' &= (u_1', u_2', u_3', u_4') \\ &= \gamma'(\mathbf{u}', ic), \end{aligned}$$

where $\gamma' = 1/\sqrt{1 - u'^2/c^2}$. u' is the speed of the particle in the x', y', z', t' frame.

From the transformation rule, Eq. (14.4), and using $u_1 = \Gamma U$, $u_2 = 0$, $u_3 = 0$, $u_4 = i\Gamma c$,

$$\begin{aligned} u_1' &= \gamma(u_1 + i\beta u_4) = \gamma\Gamma(U - v) \\ u_2' &= u_2 = 0 \\ u_3' &= u_3 = 0 \\ u_4' &= \gamma(u_4 - i\beta u_1) = i\gamma\Gamma\left(c - \frac{vU}{c}\right) = ic\gamma\Gamma\left(1 - \frac{vU}{c^2}\right), \end{aligned}$$

where $\gamma = 1/\sqrt{1 - v^2/c^2}$ and $\beta = v/c$. Hence,

$$\underset{\rightarrow}{\mathbf{u}}' = \gamma'(\mathbf{u}', ic)$$
$$= \gamma\Gamma\left[U - v, 0, 0, ic\left(1 - \frac{vU}{c^2}\right)\right].$$

Equating components,

$$\gamma' u' = \gamma\Gamma(U - v)$$

and

$$\gamma' = \gamma\Gamma\left(1 - \frac{vU}{c^2}\right).$$

Therefore,

$$u' = (\gamma\Gamma/\gamma')(U - v)$$
$$= \frac{U - v}{1 - vU/c^2},$$

which is Einstein's velocity addition formula for the case we are considering. The same procedure can be used to add nonparallel velocities.

14.4 The Momentum-energy Four-vector

In the last chapter we obtained expressions for the relativistic momentum and energy by rather labored arguments based on a hypothetical two body collision. In this section we shall obtain the same results in a much more direct manner by simply constructing a momentum-energy four-vector. We shall also obtain the relativistic expression for force, a difficult quantity to derive by the methods of the last chapter.

Our starting point is the observation that the classical momentum $m_0\mathbf{u}$ is not relativistically invariant since the classical velocity is not a four-vector. However, we found the form of the four-velocity $\underset{\rightarrow}{\mathbf{u}}$ in Example 14.4. Since the rest mass m_0 is a Lorentz invariant, the product $m_0\underset{\rightarrow}{\mathbf{u}}$ is a four-vector. It is natural to identify this with the relativistic momentum, and we therefore define the four-momentum

$$\underset{\rightarrow}{\mathbf{p}} = m_0\underset{\rightarrow}{\mathbf{u}}$$
$$= \gamma(m_0\mathbf{u}, im_0c)$$
$$= (m\mathbf{u}, imc)$$

or

$$\underset{\rightarrow}{\mathbf{p}} = (\mathbf{p}, imc). \tag{14.5}$$

Does the four-momentum obey a conservation law? Classically, the rate of change of momentum is equal to the applied force, so that the momentum of an isolated system is conserved. However, it is not obvious whether the four-momentum is similarly conserved since we have not developed a relativistic expression for force. Recall that we obtained the four-velocity by dividing $d\underset{\rightarrow}{\mathbf{s}}$ by the Lorentz invariant $d\tau$. Let us apply the same method to obtain the "time derivative" of $\underset{\rightarrow}{\mathbf{p}}$, and then *define* this equal to the four-force.

$$\underset{\rightarrow}{\mathbf{F}} = \frac{d\underset{\rightarrow}{\mathbf{p}}}{d\tau} = \left(\frac{d\mathbf{p}}{d\tau}, i\frac{d}{d\tau}mc\right) \tag{14.6}$$

$\underset{\rightarrow}{\mathbf{F}}$ is known as the *Minkowski force.*

If dt is the time interval in the observer's frame corresponding to the interval of proper time $d\tau$, then $dt = \gamma\, d\tau$ and we have

$$\underset{\rightarrow}{\mathbf{F}} = \gamma\left(\frac{d\mathbf{p}}{dt}, i\frac{d}{dt}mc\right).$$

In the classical limit, $d\mathbf{p}/dt = \mathbf{F}$. In order to conserve the momentum of an isolated system, we retain the identification of force with rate of change of momentum in all inertial systems. The Minkowski force becomes

$$\underset{\rightarrow}{\mathbf{F}} = \gamma\left(\mathbf{F}, i\frac{d}{dt}mc\right). \tag{14.7}$$

We have constructed $\underset{\rightarrow}{\mathbf{F}}$ so that four-momentum is conserved when the four-force is zero. Like all four-vectors, $\underset{\rightarrow}{\mathbf{F}}$ is relativistically invariant; if it is zero in one frame, it is zero in every frame. This assures us that if four-momentum is conserved in one inertial frame, it must be conserved in all inertial frames.

To interpret the fourth, or timelike component of $\underset{\rightarrow}{\mathbf{p}} = (\mathbf{p}, imc)$, we recall that classically $\mathbf{F} \cdot \mathbf{u}$ represents the rate at which work is done on a particle. By the work-energy theorem, $\mathbf{F} \cdot \mathbf{u} = dE/dt$, where E is the total energy. With this inspiration, let us examine $\underset{\rightarrow}{\mathbf{F}} \cdot \underset{\rightarrow}{\mathbf{u}}$ for a particle moving with velocity $\mathbf{u}$. Since $\underset{\rightarrow}{\mathbf{u}} = \gamma(\mathbf{u}, ic)$,

$$\underset{\rightarrow}{\mathbf{F}} \cdot \underset{\rightarrow}{\mathbf{u}} = \gamma^2\left(\mathbf{F} \cdot \mathbf{u} - \frac{d}{dt}mc^2\right).$$

Since the scalar product of two four-vectors is a Lorentz invariant, we are free to evaluate it in any frame we please. Let us evaluate $\mathbf{F} \cdot \mathbf{u}$ in the rest frame of the particle. In this frame, $(d\mathbf{p}/dt) \cdot \mathbf{u} = 0$ since $\mathbf{u} = 0$. We also have

$$\left.\frac{d}{dt} mc^2\right|_{u=0} = \left.\frac{d}{dt}\left(\frac{m_0c^2}{\sqrt{1 - u^2/c^2}}\right)\right|_{u=0}$$

$$= \left.\frac{m_0u\,(du/dt)}{(1 - u^2/c^2)^{\frac{3}{2}}}\right|_{u=0} = 0.$$

Hence $\underset{\rightarrow}{\mathbf{F}} \cdot \underset{\rightarrow}{\mathbf{u}} = 0.$

$$\mathbf{F} \cdot \mathbf{u} - \frac{d}{dt} mc^2 = 0$$

or

$$\mathbf{F} \cdot \mathbf{u} = \frac{d}{dt} mc^2.$$

This relativistic result bears a close resemblance to the classical relation $\mathbf{F} \cdot \mathbf{u} = dE/dt$. We conclude that the relativistic equivalent of total energy is

$$E = mc^2.$$

The four-momentum becomes

$$\underset{\rightarrow}{\mathbf{p}} = (\mathbf{p}, imc) = \left(\mathbf{p}, \frac{iE}{c}\right). \qquad 14.8$$

$\underset{\rightarrow}{\mathbf{p}}$ is often called the *momentum-energy four-vector.*

We can generate a Lorentz invariant by taking the norm of $\underset{\rightarrow}{\mathbf{p}}$.

$$\underset{\rightarrow}{\mathbf{p}} \cdot \underset{\rightarrow}{\mathbf{p}} = p^2 - \frac{E^2}{c^2} = m_0{}^2\gamma^2(u^2 - c^2) = -m_0{}^2c^2.$$

Hence,

$$E^2 = p^2c^2 + (m_0c^2)^2,$$

a familiar result.

The Minkowski approach of generating four-vectors has led us in a natural way to relativistically correct expressions for momentum and energy. With this approach the conservation laws for energy and momentum appear as a single law: the conservation of four-momentum. In relativity, momentum and energy are different aspects of a single entity; this represents a significant simplification over classical physics, where the concepts are essentially unrelated.

We conclude this section with a few applications of the momentum-energy four-vector.

Example 14.6 **The Doppler Effect, Once More**

We have derived the relativistic expression for the Doppler effect by two different approaches: from a geometrical argument in Section 12.5 and by a dynamical argument in Example 13.8. In this example we obtain the same result by a third, much simpler, approach—four-vector invariance.

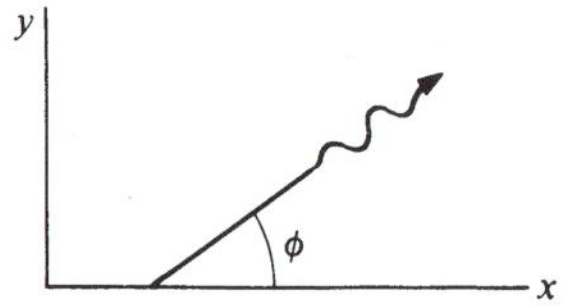

Consider a photon with energy $E = h\nu$ and momentum $h\nu/c$ traveling in the xy plane at angle ϕ with the x axis. The momentum in the x, y system is $\mathbf{p} = (h\nu/c)(\cos\phi, \sin\phi, 0)$. The momentum-energy four-vector is

$$\underset{\rightarrow}{\mathbf{p}} = \left(\mathbf{p}, \frac{iE}{c}\right)$$

$$= \frac{h\nu}{c}(\cos\phi, \sin\phi, 0, i).$$

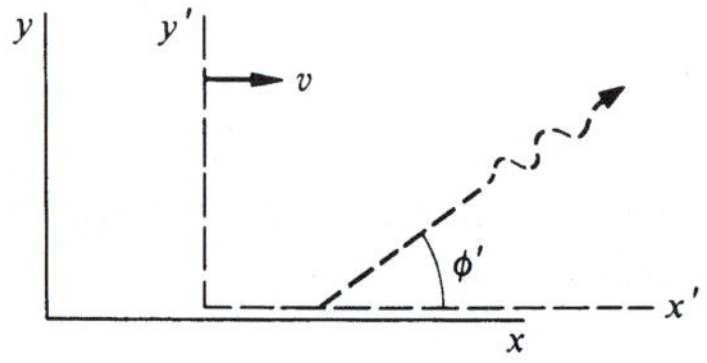

In the x', y' system shown in the sketch, the four-momentum can be written

$$\underset{\rightarrow}{\mathbf{p}} = \frac{h\nu'}{c}(\cos\phi', \sin\phi', 0, i).$$

From Eq. (14.4) we have $p_4' = \gamma[p_4 - i(v/c)p_1]$. Hence,

$$i\frac{h\nu'}{c} = \gamma\left(i\frac{h\nu}{c} - i\frac{v}{c}\frac{h\nu}{c}\cos\phi\right)$$

$$\nu' = \gamma\nu\left(1 - \frac{v}{c}\cos\phi\right)$$

or

$$\nu = \frac{\nu'}{\gamma}\frac{1}{1 - (v/c)\cos\phi}$$

$$= \nu'\frac{\sqrt{1 - v^2/c^2}}{1 - (v/c)\cos\phi}$$

identical to our earlier result, Eq. (12.7).

Example 14.7 Relativistic Center of Mass Systems

The center of mass system we used in Chap. 4 to analyze collision problems is the coordinate system in which the spatial momentum is zero. In this example, we shall find the relativistic transformation from the laboratory system to the zero momentum frame.

Consider a collision between two particles with rest masses M_1 and M_2. Let particle 1 be moving with velocity **u** in the laboratory system and particle 2 be at rest. The momentum-energy four-vector of each particle is

$$\underset{\rightarrow}{\mathbf{p}}_1 = \left(p_1, 0, 0, \frac{iE_1}{c}\right)$$

$$\underset{\rightarrow}{\mathbf{p}}_2 = \left(0, 0, 0, \frac{iE_2}{c}\right).$$

The total momentum-energy is

$$\underset{\rightarrow}{\mathbf{P}} = \underset{\rightarrow}{\mathbf{p}}_1 + \underset{\rightarrow}{\mathbf{p}}_2 = \left(p_1, 0, 0, i\frac{E_1 + E_2}{c}\right). \qquad 1$$

In a frame moving along the x axis with speed V the spatial components of $\underset{\rightarrow}{\mathbf{P}}$ are, by Eq. (14.4),

$$P_1' = \Gamma\left(p_1 - V\frac{E_1 + E_2}{c^2}\right)$$

$$P_2' = 0 \qquad 2$$

$$P_3' = 0,$$

where $\Gamma = 1/\sqrt{1 - V^2/c^2}$.

In the center of mass system, $\mathbf{P}' = 0$. From Eq. (2) we see that the speed of this frame with respect to the laboratory frame is

$$V = \frac{p_1c^2}{E_1 + E_2}. \qquad 3$$

The energy available for physical processes such as the production of new particles or other inelastic events is the total energy in the center of mass system E'. In the center of mass frame, the momentum-energy four-vector is

$$\left(0, 0, 0, \frac{iE'}{c}\right). \qquad 4$$

We can find E' by using the invariance of the norm of $\underset{\rightarrow}{\mathbf{P}}$. From Eqs. (1) and (4),

$$-\frac{E'^2}{c^2} = p_1^2 - \frac{(E_1 + E_2)^2}{c^2}$$

or

$$E'^2 = (M_1c^2)^2 + 2E_1E_2 + E_2^2,$$

where we have used $p_1^2c^2 = E_1^2 - (M_1c^2)^2$. For our problem, $E_1 = \gamma M_1c^2$ and $E_2 = M_2c^2$, where $\gamma = 1/\sqrt{1 - u^2/c^2}$. Hence,

$$E' = (M_1^2 + M_2^2 + 2\gamma M_1M_2)^{\frac{1}{2}}c^2. \tag{5}$$

The total energy in the laboratory system is

$$E = (\gamma M_1 + M_2)c^2, \tag{6}$$

and the fraction of the initial energy available for physical processes is

$$\frac{E'}{E} = \frac{(M_1^2 + M_2^2 + 2\gamma M_1M_2)^{\frac{1}{2}}}{\gamma M_1 + M_2}. \tag{7}$$

An important practical case is that of equal masses $M_1 = M_2$. Equation (7) becomes

$$\frac{E'}{E} = \frac{\sqrt{2}\sqrt{1+\gamma}}{1+\gamma}$$

$$= \frac{\sqrt{2}}{\sqrt{1+\gamma}}. \tag{8}$$

In the low velocity limit, $\gamma = 1$ and $E'/E = 1$. At low speeds, most of the energy is in rest mass energy and kinetic energy is relatively unimportant. To discuss the high-speed limit, it is convenient to write Eq. (8) in terms of the projectile energy $E_1 = \gamma Mc^2$.

$$\frac{E'}{E} = \frac{\sqrt{2}}{\sqrt{1 + E_1/Mc^2}}.$$

For $E_1 \gg Mc^2$, we have

$$\frac{E'}{E} \approx \frac{\sqrt{2Mc^2}}{\sqrt{E_1}}.$$

The useful fraction of energy decreases as $E_1^{-\frac{1}{2}}$. For example, the proton synchrotron at the National Accelerator Laboratory in Batavia, Illinois, can accelerate protons to an energy of 300 GeV (1 GeV $= 10^9$ eV). Since the rest mass of the proton is about 1 GeV, we see that for protons colliding with a hydrogen target, $E'/E \approx \sqrt{3}/\sqrt{200} \approx 0.1$. Only 30 GeV is available for interesting experiments.

By using identical beams colliding head on, the laboratory frame becomes the center of mass frame, and the total energy is available for inelastic events. This technique of colliding beams has been used extensively in electron accelerators and has proved feasible in proton machines as well.

Example 14.8 **Pair Production in Electron-electron Collisions**

In Example 13.7 we analyzed pair production, the process by which a photon collides with an electron to create an electron-positron pair. The threshold energy for the process was found to be $E = 2m_0c^2 = 1.02$ MeV, where $m_0c^2 = 0.51$ MeV is the rest energy of the electron or positron.

A related process is the production of an electron-positron pair by the collision of two electrons:

$$e^- + e^- \rightarrow e^- + e^- + (e^- + e^+).$$

The reaction evidently satisfies conservation of charge. The problem is to find the threshold energy for the process.

To describe the dynamics of the problem we introduce the following four-momenta:

$\underset{\rightarrow}{\mathbf{p}}_1$: electron 1 before the collision

$\underset{\rightarrow}{\mathbf{p}}_2$: electron 2 before the collision

$\underset{\rightarrow}{\mathbf{p}}_3$: electron 1 after the collision

$\underset{\rightarrow}{\mathbf{p}}_4$: electron 2 after the collision

$\underset{\rightarrow}{\mathbf{p}}_5$: electron created in e^--e^+ pair

$\underset{\rightarrow}{\mathbf{p}}_6$: positron created in e^--e^+ pair

Then conservation of four-momentum gives

$$\underset{\rightarrow}{\mathbf{p}}_1 + \underset{\rightarrow}{\mathbf{p}}_2 = \underset{\rightarrow}{\mathbf{p}}_3 + \underset{\rightarrow}{\mathbf{p}}_4 + \underset{\rightarrow}{\mathbf{p}}_5 + \underset{\rightarrow}{\mathbf{p}}_6.$$

Squaring, we have

$$(\underset{\rightarrow}{\mathbf{p}}_1 + \underset{\rightarrow}{\mathbf{p}}_2)^2 = (\underset{\rightarrow}{\mathbf{p}}_3 + \underset{\rightarrow}{\mathbf{p}}_4 + \underset{\rightarrow}{\mathbf{p}}_5 + \underset{\rightarrow}{\mathbf{p}}_6)^2. \qquad 1$$

Since each side of the equation is Lorentz invariant, we can compute the terms in whatever reference frame is most convenient.

Let us compute the left hand side of Eq. (1) in the laboratory frame. Taking particle 2 to be initially at rest, we have

$$\underset{\rightarrow}{\mathbf{p}}_1 = \left(\mathbf{p}_1, i\frac{E_1}{c}\right) \qquad \underset{\rightarrow}{\mathbf{p}}_2 = (0, im_0c)$$

and

$$\begin{aligned}(\underset{\rightarrow}{\mathbf{p}}_1 + \underset{\rightarrow}{\mathbf{p}}_2)^2 &= \underset{\rightarrow}{\mathbf{p}}_1^2 + \underset{\rightarrow}{\mathbf{p}}_2^2 + 2\underset{\rightarrow}{\mathbf{p}}_1 \cdot \underset{\rightarrow}{\mathbf{p}}_2 \\ &= -2(m_0c)^2 - 2m_0E_1, \qquad 2\end{aligned}$$

where we have used $\underset{\rightarrow}{\mathbf{p}}^2 = p^2 - E^2/c^2 = -m_0{}^2c^2$, valid for any particle.

The right hand side of Eq. (1) is most conveniently calculated in the center of mass frame. At threshold, all four particles are at rest. (This minimizes the energy and is consistent with the requirement that the total spatial momentum be zero in the center of mass frame.) Hence $\underset{\to}{\mathbf{p}}_3$, $\underset{\to}{\mathbf{p}}_4$, $\underset{\to}{\mathbf{p}}_5$, $\underset{\to}{\mathbf{p}}_6$ all have the form $(0,0,0,im_0c)$, and the right hand side of Eq. (1) becomes

$$(0,0,0,4im_0c)^2 = -16(m_0c)^2. \qquad 3$$

Substituting Eqs. (2) and (3) in Eq. (1) gives

$$-2(m_0c)^2 - 2m_0E_1 = -16(m_0c)^2$$

or

$$E_1 = 7m_0c^2.$$

E_1 includes the rest energy of the projectile, so that the kinetic energy of the projectile at threshold is

$$\begin{aligned} K_1 &= E_1 - m_0c^2 \\ &= 6m_0c^2. \end{aligned}$$

The argument here can be applied to the production of other particles, for instance, to the production of a negative proton in the reaction

$$p^+ + p^+ \to p^+ + p^+ + (p^+ + p^-).$$

Since the proton rest mass is 0.94 GeV, the threshold kinetic energy for the production of negative protons is 6(0.94) GeV = 5.64 GeV. The Bevatron at the Lawrence Radiation Laboratory, Berkeley, California, was designed to accelerate protons to 6 GeV to allow this process to be observed. Owen Chamberlain and Emilio Segré received the Nobel Prize in 1959 for producing negative protons, or *antiprotons.*

14.5 Concluding Remarks

The special theory of relativity, far from representing a complete break with classical physics, has a heavy flavor of newtonian mechanics in its insistence on the equivalence of inertial frames. Essentially, Einstein generalized the work of Newton by bringing classical mechanics into accord with the requirements of electromagnetic theory.

Fundamentally, however, the emphases of special relativity are not the same as those of newtonian physics. Einstein's rejection of unobservable concepts like absolute space and time and his insistence on operational definitions related to observation were much more far-reaching than were Newton's efforts in this direction. Einstein laid the groundwork for the analysis of observables which was essential in the development of modern quantum mechanics. In addition, he made significant contributions to our philosophical understanding of how man obtains knowledge of the world.

As we have seen in this chapter, one of Einstein's great contributions was recognition of the power of transformation theory as an organizing principle in physics. Transformation theory unifies and simplifies the concepts of special relativity and has served as a knowledgeable guide in the search for new laws.

However, in spite of its power and harmony, special relativity is not a complete dynamical theory since it is inadequate to deal with accelerating reference frames. To Einstein this was a fundamental defect. According to Mach's principle of equivalence it is impossible to distinguish locally between an inertial system in a gravitational field and an accelerating coordinate system in free space. Therefore, by the principle of relativity, the frames must be equally valid for the description of physical phenomena. Since special relativity is incapable of dealing with accelerating reference frames, it is inherently incapable of dealing with gravitational fields.

Einstein went far toward removing these difficulties with his general theory of relativity, published in 1916. The general theory deals with transformations between all coordinate systems, not just inertial systems. It is essentially a theory of gravitation, since it is possible to "produce" a gravitational field merely by changing coordinate systems. From this point of view the effect of gravity is regarded as a local distortion in the geometry of space. Even in the gravitational field of the sun, however, effects attributable to general relativity are small and difficult to detect. For example, the deflection of starlight by the sun, one of the most dramatic effects predicted, amounts to only 1.7 seconds of arc.

General relativity's greatest impact has been on cosmology, since gravity is the only important force in the universe at large. Its role in terrestrial physics has been minor, however, partly because the effects are small and partly because so far it has not been extended to include electromagnetism. In contrast, special relativity has a multitude of applications and is part of the working knowledge of every physicist.

Einstein's impact on the twentieth century is difficult to assess in its entirety. He altered and enlarged our perceptions of the natural world, and in this respect he ranks among the great figures of Western thought.

Problems

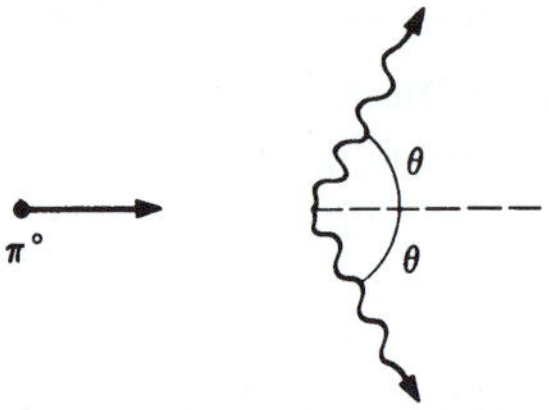

14.1 A neutral pi meson, rest mass 135 MeV, decays symmetrically into two photons while moving at high speed. The energy of each photon in the laboratory system is 100 MeV.

a. Find the meson's speed V. Express your answer as a ratio V/c.

b. Find the angle θ in the laboratory system between the momentum of each photon and the initial line of motion.

Ans. $\theta \approx 51°$

14.2 A high energy photon (γ ray) collides with a proton at rest. A neutral pi meson (π°) is produced according to the reaction

$$\gamma + p \rightarrow p + \pi^\circ.$$

What is the minimum energy the γ ray must have for this reaction to occur? The rest mass of a proton is 938 MeV, and the rest mass of a π° is 135 MeV.

Ans. Approximately 154 MeV

14.3 A high energy photon (γ ray) hits an electron and produces an electron-positron pair according to the reaction

$$\gamma + e^- \rightarrow e^- + (e^- + e^+).$$

What is the minimum energy the γ ray must have for the reaction to occur?

14.4 A particle of rest mass M spontaneously decays from rest into two particles with rest masses m_1 and m_2. Show that the energies of the particles are

$$E_1 = (M^2 + m_1{}^2 - m_2{}^2)c^2/2M \qquad E_2 = (M^2 - m_1{}^2 + m_2{}^2)c^2/2M.$$

14.5 A nucleus of rest mass M_1 moving at high speed with kinetic energy K_1 collides with a nucleus of rest mass M_2 at rest. A nuclear reaction occurs according to the scheme

$$\text{nucleus } 1 + \text{nucleus } 2 \rightarrow \text{nucleus } 3 + \text{nucleus } 4.$$

The rest masses of nuclei 3 and 4 are M_3 and M_4.

The rest masses are related by

$$(M_3 + M_4)c^2 = (M_1 + M_2)c^2 + Q,$$

where $Q > 0$. Find the minimum value of K_1 required to make the reaction occur, in terms of M_1, M_2, and Q.

Ans. clue. If $M_1 = M_2 = Q/c^2$, then $K_1 = 5Q/2$

14.6 A rocket of initial mass M_0 starts from rest and propels itself forward along the x axis by emitting photons backward.

a. Show that the four-momentum of the rocket's exhaust in the initial rest system can be written

$$\underset{\rightarrow}{\mathbf{p}} = \gamma M_f v(-1,0,0,i),$$

where M_f is the final mass of the rocket. (Note that this result is valid for the exhaust as a whole even though the photons are Doppler-shifted.)

b. Show that the final velocity of the rocket relative to the initial frame is

$$v = \frac{x^2 - 1}{x^2 + 1} c,$$

where x is the ratio of the rocket's initial mass to final mass, M_0/M_f.

14.7 Construct a four-vector representing acceleration. For simplicity, consider only straight line motion along the x axis. Let the instantaneous four-velocity be $\underset{\rightarrow}{\mathbf{u}}$.

Ans. clue. norm $= a^2/(1 - u^2/c^2)^3$, where $a = du/dt$

14.8 The function $f(x,t) = A \sin 2\pi[(x/\lambda) - \nu t]$ represents a sine wave of frequency ν and wavelength λ. The wave propagates along the x axis with velocity = wavelength $\times$ frequency $= \lambda\nu$. $f(x,t)$ can represent a light wave; A then corresponds to some component of the electromagnetic field which constitutes the light signal, and the wavelength and frequency satisfy $\lambda\nu = c$.

Consider the same wave in the coordinate system x', y', z', t' moving along the x axis at velocity v. In this reference frame the wave has the form

$$f'(x',t') = A' \sin 2\pi \left(\frac{x'}{\lambda'} - \nu' t'\right).$$

a. Show that the velocity of light is correctly given provided that $1/\lambda'$ and ν' are components of a four-vector k given in the x, y, z, t system by

$$\underset{\rightarrow}{\mathbf{k}} = 2\pi \left(\frac{1}{\lambda}, 0, 0, \frac{i\nu}{c}\right).$$

b. Using the result of part *a*, derive the result for the longitudinal Doppler shift by evaluating the frequency in a moving system.

c. Extend the analysis of part *b* to find the expression for the transverse Doppler shift by considering a wave propagating along the y axis.

INDEX

Abraham, M., 492
Acceleration, 13
  centripetal, 36, 359
  Coriolis, 36, 359
  invariance of, in Newtonian mechanics, 454
  nonuniform, 22
  in polar coordinates, 36
  radial, 36
  relativistic transformation of, 486, 533
  in rotating coordinates, 358
  tangential, 37
Air suspension gyroscope, 332
Air track, 53
Amplitude of simple harmonic motion, 411
Angular frequency, 411
Angular momentum, 233ff.
  conservation of, 305
  definition of, 233
  and effective potential, 385
  and fixed axis rotation, 248
  and kinetic energy, 314, 370
  orbital, 262
  quantization of, 272
  spin, 262
  vector nature of, 288
Angular velocity, 289
Antiparticles, 506, 533
Antiprotons, 533
Apogee, 396
Approximations, numerical, 39
Atomic clock, 470
Atwood's machine, 104, 254
Average value, 413

Balmer, J., 271
Base vectors, 10
Binary star system, 212
Binomial series, 41
Bohr, N., 270
Bohr atom, 270
Bounded orbits, 386
Bucherer, A. H., 492

c, role in relativity, 455
Capstan, 107
Capture cross section of a planet, 241
Cavendish, H., 81
Center of mass, 116, 145
  angular momentum of, 261
  kinetic energy of, 264
Center of mass coordinates, 127
  collisions and, 190
  relativistic, 531
Center of percussion, 260
Centimeter, 67
Central force motion, 378
  constants of motion, 380
  law of equal areas, 382
  reduction to one-body problem, 378
  and two-body problem, 382
Centrifugal force, 359
Centrifugal potential energy, 381
Centripetal acceleration, 36
cgs system of units, 67
Chamberlain, O., 534
Chasle's theorem, 274
Circular motion, 17, 25
Cockroft, J. D., 498
Coefficient of friction, 93
Collisions, 187
  and center of mass coordinates, 190
  and conservation laws, 188
  elastic, 188
  inelastic, 188
  relativistic, 490, 496, 533
Commutativity:
  of infinitesimal rotations, 322
  of vector operations, 4
Complex variables, 433
Compton, A. H., 503
Compton effect, 503
Conical pendulum, 77, 161, 237, 245
Conservation:
  of angular momentum, 305

Conservation:
of energy, 169, 184
of four-momentum, 529
of momentum, 122
Conservative force, 163, 215
Constant energy surface, 211
Constants of motion, 380
Constraints, 70, 74
Contact forces, 87
Contour lines, 211
Coordinates:
cartesian, 12
polar, 27
Coriolis acceleration, 36, 359
Coriolis force, 359
and deflection of a falling mass, 362
and weather systems, 364
Cosines, law of, 5
Cosmic ray, 512
Coulomb's law, 86
Crab nebula, 510
Curl, 218

Damped oscillator, 414, 435
Damping time, 418
Del (operator), 207
De Moivre's theorem, 434
Derivative:
partial, 202
of a vector, 15, 23
Diatomic molecule, 179
Dicke, R. H., 354
Differentials, 45, 204
Dike, 143
Dimension of a physical quantity, 18
Displacement, 11
Doorstop, 259
Doppler effect, 475, 507, 530
Doppler navigation, 479
Dot product, 5
Dredl, 290
Dyne, 67

Earth, as a rotating reference system, 362, 366
Earth-moon-sun system, 125
and precession of the equinoxes, 296
and tide, 348
Eccentricity, 392
Effective potential, 381
Einstein, A., 272, 451, 534
Electric field, 87
Electron:
mass of, 492
motion due to radio wave, 22
Electrostatic force, 86
Ellipse, 392, 403
Elliptic integral, 278
Elliptic orbit, 398
Energy, 152ff.
conservation of, 184, 495
diagrams, 176, 383
kinetic, 156
mathematical aspects of, 202
potential, 168
relativistic, 493
and stability, 174
surface, 211
total, 169
and work, 160
English system of units, 67
Eötvös, R., 354
Equal areas, law of, 240, 382
Equilibrium, 175, 322
Equivalence principle, 346, 369
Erg, 185
Escape velocity, 157, 162
Ether, 446
Euler's equations, 320
Euler's theorem, 232
Event, 462
Exponential function, 44

Fictitious force, 62, 344
Field:
electric, 87
gravitational, 85

FitzGerald, G. F., 450, 459
Fizeau, H. L., 460, 475
Foot (unit), 67
Force, 58
  conservative, 163
    criteria for, 215
  contact, 87
  diagram, 68
  electric, 86
  fictitious, 344
  of friction, 92
  gravitational, 80
  inertial (*see* Force, fictitious)
  relativistic, 528
  and transport of momentum, 139
  units of, 67
  vector nature of, 59, 524
  viscous, 95
Foucault pendulum, 366
Four-momentum, 527
Four-vector, 525
Four-velocity, 525
Frequency, 411
Fresnal drag coefficient, 460
Friction, 92
  coefficient of, 93
  fluid, 95
Frisch, D. H., 469

g (acceleration of gravity), 83
  variation with altitude, 83
  variation with latitude (problem), 374
G (gravitational constant), 80
Galilean transformations, 340, 453
Gas, pressure of, 144
General theory of relativity, 535
Gradient operator, 207, 210
Gram, 67
Grandfather's clock, 256
Gravitational mass, 81, 352
Gravitational red shift, 369
Graviton, 501
Gravity, 80
Gravity:
  gravitational field, 85
    of spherical shell, 101
  and tides, 352
  and weight, 84
Gyrocompass, 301
Gyroscope, 295, 328
  nutation of, 331

Hall, D. B., 469
Halley's comet, 407
Harmonic oscillator, 410
  damped, 414, 435
  forced, 421, 436
Hertz, H., 502
Hertz (unit), 411
Hohmann transfer orbit, 408
Hooke, R., 97
Hooke's law, 97
Horsepower, 186
Hyperbola, 392

Impulse, 130
Inelastic collisions, 188
Inertia, 372
Inertial force (*see* Force, fictitious)
Inertial mass, 81, 356
Inertial system, 55, 340, 455
Infinitesimal rotations, 322
Initial conditions, 99
International system of units, 67
Invariants, 520
Inverse square law:
  electric, 86
  gravitational, 80
  motion under, 389

Joule, J. P., 185
Joule (unit), 156

Kater's pendulum, 258
Kaufmann, W., 23

Kennedy-Thorndike experiment, 459
Kepler's laws, 240, 400
Kilogram, definition, 66
Kinematical equations, formal solution, 19
Kinetic energy, 156
  and center of mass motion, 264
  in collisions, 188
  of rotating rigid body, 313
  of two-body system, 383

Lariat trick, 336
Laws of motion, 53
Length:
  contraction of, 466
  unit of, 66
Light:
  electromagnetic theory of, 445
  particle model of, 501
  speed of, 445, 451
    constancy of, 508
    in moving medium, 451
Line integral, 159, 166
Linear air track, 53
Linear restoring force, 97
Lorentz, H. A., 450, 457
Lorentz contraction, 466
Lorentz invariant, 487
Lorentz transformations, 455
  and four-vectors, 523

Mach, E., 369, 443
Mach's principle, 369
Many-particle system:
  angular momentum of, 305
  momentum of, 113
Mascons, 390
Mass, 56
  gravitational, 353
  inertial, 353
  relativistic, 490
  standard of, 66
  unit of, 67
Maxwell, J. C., 445
Meson decay, 468
Meter (unit), 66
Metric system, 67
Michelson, A. A., 445, 448
Michelson-Morley experiment, 445
Millikan, R. A., 502
Minkowski force, 528
Minkowski space, 521
mks system of units, 67
Molecule, diatomic, 179
Moment of inertia, 249, 309
  and parallel axis theorem, 252
  and principal axes, 313
Momentum, 112
  angular (*see* Angular momentum)
  conservation of, 122, 490, 529
  and the flow of mass, 133
  relativistic, 490
  transport, 139
Momentum-energy four vector, 527
Morley, F. W., 450
Motion, 19
  in accelerating coordinate system, 347
  circular, 17, 25, 34
  in conservative systems, 168
  laws of, 53
  in nonconservative systems, 182
  in plane polar coordinates, 27
  relation to acceleration, 20
  on rotating earth, 368
  along a straight line, 34
  in uniform gravitational field, 21
Muon decay, 468

Neutino, 501
Newton, I., 52, 353, 368, 442
Newton (unit), 67
Newton's law of gravitation, 80

Newton's laws of motion, 52ff., 442
first law, 55
second law, 56
third law, 59
Nonconservative force, 182
Normal force, 92
Nutation, 332

Operational definition, 57
Orbital angular momentum, 262
Orbits, 382
bounded, 386
elliptic, 394
hyperbolic, 393
under inverse square force, 385
perturbed, 388

Pair production, 505, 533
Parabola, 392
Parallel axis theorem, 252
Partial derivatives, 202
Pendulum:
inverted, 164
Kater's, 258
physical, 257
simple, 255
period versus amplitude, 256
Perigee, 396
Period of motion, 411
Perturbed orbit, 388
Phase, 411
Photoelectric effect, 502
Photon, 501
rest mass of, 512
Physical pendulum, 257
Planck, M., 272
Planets:
motion of, 390
orbits of, table, 395
perturbation of, 391
Polar coordinates, 27
acceleration in, 36
velocity in, 30
Pole vaulter paradox, 486
Potential energy, 168
effective, 385
gradient of, 211
relation to force, 173, 206, 214
surface, 211
Pound, R. V., 370
Pound, 67
Power, 186
Precession:
of equinoxes, 300
of gyroscope, 296, 331
torque-free, 317, 331
Pressure of a gas, 144
Principal axes, 313
Principia, 440, 452*n*.
Principle of equivalence, 346, 369
Principle of relativity, 451
Principle of superposition, 58
Products of inertia, 309
Proper time, 468, 524
Pulleys, 90
Pulsar, 510

Q (quality factor), 418

Radiation pressure, 502
Radius of gyration, 257
Reduced mass, 179, 191, 379
Relative velocity, 48
Relativity:
general theory, 535
special theory, 450
Resonance, 423
curve, 427
Rest energy, 491
Rest mass, 491
Rigid body motion, 288, 308
Rocket, 136
relativistic, 536
Rossi, B., 465
Rotating bucket experiment, 368
Rotating coordinate system, 355

Rotating coordinate transformation, 371
Rotating vectors, 25, 294, 297
Rotations, noncommutativity of, 285, 322
Rutherford, E., 271

Satellite orbit, 396
Scalar, 308, 520
  invariants, 521
Scalar product, 5
Second, definition of, 66
Segré, E., 534
Series:
  binomial, 41
  Taylor's, 42
SI (international system of units), 67
Simple harmonic motion, 97, 154, 410
Simple pendulum, 255
Simultaneity, 463
Skew rod, 292–294, 312
Slug (unit), 67
Small oscillations, 178
Smith, J. H., 469
Spacelike interval, 466
Special theory of relativity, 451
Speed of light:
  in empty space, 445
  in a moving medium, 474
Spin angular momentum, 262
Stability, 174
  of rotating objects, 304, 322
Standards and units, 64
Stokes' theorem, 225
Superposition of forces, 58, 82
Synchronous satellite, 104
System of units, 67

Tangential acceleration, 36
Taylor's series, 42
Teeter toy, 175, 181
Tension, 87
  and atomic forces, 91
Tensor, 520
Tensor of inertia, 311
Thomson, J. J., 271
Tide, 348
Time, 44
  dilation, 468, 524
  unit of, 66
Time constant, 418
Timelike interval, 466
Torque, 238
Torque-free precession, 317, 324
Total mechanical energy, 169
Trajectory, 21
Transformation properties:
  of a four-vector, 523
  of physical laws, 520
  of a vector, 516
Transformations:
  Galilean, 340, 453
  Lorentz, 455, 523
Twin paradox, 480

Uniform precession, 296
Unit vectors, 3, 10
Units, 18, 67
Universal gravitation, 80
  constant of, 81

Vector operator, 207
Vectors, 2
  addition, 4
  and area, 7
  base vectors, 10, 28
  components, 8
  derivative of, 15, 23
  displacement vector, 11
  four-dimensional, 523
  multiplication: of cross product, 6
    of scalar (dot) product, 5
  orthogonal, 10
  position vector, 11
  rotating, 25

Vectors:
- subtraction, 4
- transformation properties of, 516
- unit, 3

Velocity, 13
- angular, 289
- average, 13
- four-, 525
- in polar coordinates, 30
- radial, 33
- relative, 48
- relativistic transformation of, 472, 526
- tangential, 33

Vibration eliminator, 427
Viscosity, 95

Walton, E. T. S., 498
Watt (unit), 186
Weather systems, 364
Weight, 68, 84
Work, 156, 160
Work-energy theorem, 160
- in one dimension, 156
- for rotation, 267

Work function, 502
World line, 524
World point, 524